Springer Series in Statistics

Advisors:
P. Bickel, P. Diggle, S. Fienberg, U. Gather,
I. Olkin, S. Zeger

For further volumes:
http://www.springer.com/series/692

P.P.B. Eggermont · V.N. LaRiccia

Maximum Penalized Likelihood Estimation

Volume II: Regression

Springer

P.P.B. Eggermont
Department of Food and Resource Economics
University of Delaware
Newark, DE 19716
USA
eggermon@udel.edu

V.N. LaRiccia
Department of Food and Resource Economics
University of Delaware
Newark, DE 19716
USA
lariccia@udel.edu

ISBN 978-1-4614-1712-5 e-ISBN 978-0-387-68902-9
DOI 10.1007/b12285
Springer Dordrecht Heidelberg London New York

Softcover reprint of the hardcover 1st edition 2009

Printed on acid-free paper

Springer is part of Springer Science+Business Media (www.springer.com)

To Jeanne and Tyler

To Cindy

Preface

This is the second volume of a text on the theory and practice of maximum penalized likelihood estimation. It is intended for graduate students in statistics, operations research, and applied mathematics, as well as researchers and practitioners in the field. The present volume was supposed to have a short chapter on nonparametric regression but was intended to deal mainly with inverse problems. However, the chapter on nonparametric regression kept growing to the point where it is now the only topic covered. Perhaps there will be a Volume III. It might even deal with inverse problems. But for now we are happy to have finished Volume II.

The emphasis in this volume is on smoothing splines of arbitrary order, but other estimators (kernels, local and global polynomials) pass review as well. We study smoothing splines and local polynomials in the context of reproducing kernel Hilbert spaces. The connection between smoothing splines and reproducing kernels is of course well-known. The new twist is that letting the inner product depend on the smoothing parameter opens up new possibilities: It leads to asymptotically equivalent reproducing kernel estimators (without qualifications) and thence, via uniform error bounds for kernel estimators, to uniform error bounds for smoothing splines and, via strong approximations, to confidence bands for the unknown regression function. It came as somewhat of a surprise that reproducing kernel Hilbert space ideas also proved useful in the study of local polynomial estimators. Throughout the text, the reproducing kernel Hilbert space approach is used as an "elementary" alternative to methods of metric entropy. It reaches its limits with least-absolute-deviations splines, where it still works, and total-variation penalization of nonparametric least-squares problems, where we miss the optimal convergence rate by a power of $\log n$ (for sample size n).

The reason for studying smoothing splines of arbitrary order is that one wants to use them for data analysis. The first question then is whether one can actually compute them. In practice, the usual scheme based on spline interpolation is useful for cubic smoothing splines only. For splines of arbitrary order, the *Kalman filter* is the bee's knees. This, in fact, is the traditional meeting ground between smoothing splines and reproducing kernel Hilbert spaces, by way of the identification of the standard smoothing problem for Gaussian processes having continuous sample paths with "generalized" smoothing spline estimation in nonparametric regression problems. We give a detailed account, culminating in the Kalman filter algorithm for spline smoothing.

The second question is how well smoothing splines of arbitrary order work. We discuss simulation results for smoothing splines and local and global polynomials for a variety of test problems. (We avoided the usual pathological examples but did include some nonsmooth examples based on the Cantor function.) We also show some results on confidence bands for the unknown regression function based on undersmoothed quintic smoothing splines with remarkably good coverage probabilities.

Acknowledgments

When we wrote the preface for Volume I, we had barely moved to our new department, Food and Resource Economics, in the College of Agriculture and Natural Resources. Having spent the last nine years here, the following assessment is time-tested: Even without the fringe benefits of easy parking, seeing the U.S. Olympic skating team practice, and enjoying (the first author, anyway) the smell of cows in the morning, we would be fortunate to be in our new surroundings. We thank the chair of the department, Tom Ilvento, for his inspiring support of the Statistics Program, and the faculty and staff of FREC for their hospitality.

As with any intellectual endeavor, we were influenced by many people and we thank them all. However, six individuals must be explicitly mentioned: first of all, Zuhair Nashed, who keeps our interest in inverse problems alive; Paul Deheuvels, for his continuing interest and encouragement in our project; Luc Devroye, despite the fact that this volume hardly mentions density estimation; David Mason, whose influence on critical parts of the manuscript speaks for itself; Randy Eubank, for his enthusiastic support of our project and subtly getting us to study the extremely effective Kalman filter; and, finally, John Kimmel, editor extraordinaire, for his continued reminders that we repeatedly promised him we would be done by next Christmas. This time around, we almost made that deadline.

Newark, Delaware — *Paul Eggermont* and *Vince LaRiccia*
January 14, 2009

Contents

Contents of Volume I

Notations, Acronyms and Conventions

The numbering and referencing conventions are as follows. Items of importance, such as formulas, theorems, and exercises, are labeled as $(Y.X)$, with Y the current section number, and X the current (consecutive) item number. The exceptions are tables and figures, which are independently labeled following the same system. A reference to Item $(Y.X)$ is to the item with number X in section Y of the current chapter. References to items outside the current chapter take the form $(Z.Y.X)$, with Z the chapter number, and X and Y as before. References to (other) sections within the same chapter resp. other chapters take the form $\S\, Y$ resp. $\S\, Z.Y$.

References to the literature take the standard form AUTHOR (YEAR) or AUTHOR#1 and AUTHOR#2 (YEAR), and so on. The references are arranged in alphabetical order by the first author and by year.

We tried to limit our use of acronyms to some very standard ones, as in the following list.

iid	independent, identically distributed
rv	random variable
m(p)le	maximum (penalized) likelihood estimation (or estimator)
pdf(s)	probability density function(s)
a.s.	almost surely, or, equivalently, with probability 1

Some of the standard notations throughout the text are as follows.

$\mathbb{1}(x \in A)$	The indicator function of the set A.
$\mathbb{1}_A(x)$	Also the indicator function of the set A.
$\mathbb{1}(x \leqslant X)$	The indicator function of the event $\{\, x \leqslant X \,\}$.
$(\, x\,)_+$	The maximum of 0 and x (x a real-valued expression).
$\asymp$, $\asymp_{\text{as}}$	Asymptotic equivalence, and the almost sure version; see Definition (1.3.6).
$\approx\!\approx$, $\approx\!\approx_d$	Asymptotic equivalence with a very mild rate, and the in-distribution version; see Notation (22.5.10).
$=_{\text{as}}$	Almost sure equality.
$\leqslant_{\text{as}}$	Almost surely less than or equal; likewise for $\geqslant_{\text{as}}$.

$\det(A)$	The determinant of the square matrix A. Equivalently, the product of the eigenvalues of A, counting *algebraic* multiplicities.
$\det_+(A)$	The product of the positive eigenvalues of the square matrix A, counting *algebraic* multiplicities. Presumably, useful only if A is semi-positive-definite, in which case *algebraic* and *geometric* multiplicities coincide.
$\text{trace}(A)$	The trace of the square matrix A, defined as the sum of the diagonal elements of A. If A is symmetric, then equivalently, the sum of the eigenvalues of A.
f_o	The "true" regression function to be estimated.
f_h	This is typically the large-sample asymptotic estimator under consideration. For random designs, it is the expectation of the (smoothing spline) estimator, conditioned on the design.
$f^{nh,m}$	The estimator based on the data, for sample size n and smoothing/regularization parameter h, with m explicitly denoting the "order" of the estimator; see Chapter 23.
f^{nh}	The estimator based on the data, for sample size n and smoothing/regularization parameter h, the "order" being known from the context.
h, H	The smoothing parameter (h: deterministic; H : random).
$\mathcal{F}_n(\alpha)$	An interval of smoothing parameters; see (22.2.14).
$\mathcal{G}_n(\alpha)$	Another interval of smoothing parameters, see Theorem (14.6.13).
$\mathcal{H}_n(\alpha)$	Yet another interval of smoothing parameters, see (14.6.2).
$C^m(a, b)$	The vector space of functions that are m times continuously differentiable on the interval $[a, b]$. For $m = 0$, one writes $C(a, b)$. With the norm $$\| f \|_\infty + \| f^{(m)} \|_\infty ,$$ it becomes a Banach space. See Appendix 3 in Volume I.
$\| \cdot \|_p$	The $L^p(D)$ norm, $1 \leqslant p \leqslant \infty$, but also the ℓ^p vector norm on finite-dimensional spaces $\mathbb{R}^m$, as well as the induced matrix norm: If $A \in \mathbb{R}^{n \times m}$ then $$\| A \|_p = \max\{ \| Ax \|_p : \| x \|_p = 1 \} .$$
$L^p(D)$	Space of (equivalence classes of measurable) functions f on $D \subset \mathbb{R}$ with $\| f \|_p < \infty$.
$\| \cdot \|$	(Without subscript) The $L^2(D)$ norm.
$\langle \cdot, \cdot \rangle$	(Without subscript) The standard inner product on $L^2(D)$.
$W^{m,p}(D)$	Sobolev space of all functions in $L^p(D)$ with m-th derivative in $L^p(D)$ also.

$\langle \cdot , \cdot \rangle_{m,h}$ The inner product on $W^{m,2}(0,1)$ defined by

$$\langle f , \varphi \rangle_{m,h} \stackrel{\text{def}}{=} \langle f , \varphi \rangle + h^{2m} \langle f^{(m)} , \varphi^{(m)} \rangle .$$

$\| f \|_{m,h}$ The norm on $W^{m,2}(0,1)$ induced by the inner product above,

$$\| f \|_{m,h}^2 = \langle f , f \rangle_{m,h} .$$

Some or most of the time, we write $\| \cdot \|_{mh}$.

$\| f \|_{W^{m,2}(0,1)}$ The standard norm on $W^{m,2}(0,1)$; viz. the norm above for $h = 1$,

$$\| f \|_{W^{m,2}(0,1)}^2 = \| f \|_{m,1}^2 .$$

$\| f \|_{h,W^{m,p}(0,1)}$ $\| f \|_{h,W^{m,p}(0,1)} = \| f \|_{L^p(0,1)} + h^m \| f^{(m)} \|_{L^p(0,1)}$,

$\mathcal{R}_{mh}$ The reproducing kernel of the reproducing kernel Hilbert space $W^{m,2}(0,1)$ with the inner product $\langle \cdot , \cdot \rangle_{mh}$ for $m \geqslant 1$.

$\langle \cdot , \cdot \rangle_{\omega mh}$ The inner product on $W^{m,2}(0,1)$ defined by

$$\langle f , \varphi \rangle_{m,h} \stackrel{\text{def}}{=} \langle f , \varphi \rangle_{L^2((0,1),\omega)} + h^{2m} \langle f^{(m)} , \varphi^{(m)} \rangle .$$

Here, ω is a nonnegative weight function, typically the design density in a regression problem.

$\| \cdot \|_{\omega mh}$ The norm on $W^{m,2}(0,1)$ corresponding to the inner product above.

$\| \cdot \|_{\omega mh,p}$ Norms on $W^{m,p}(0,1)$ defined in (22.2.18) as

$$\| f \|_{\omega mh,p} = \| f \|_{L^p((0,1),\omega)} + h^m \| f^{(m)} \|_{L^p(0,1)} .$$

$\mathfrak{R}_{\omega mh}$ The reproducing kernel of the reproducing kernel Hilbert space $W^{m,2}(0,1)$ with the inner product $\langle \cdot , \cdot \rangle_{\omega mh}$ for $m \geqslant 1$ and quasi-uniform design density ω.

ω, Ω_o, Ω_n The design density, the cumulative distribution function, and the empirical distribution function of the random design $X_1, X_2, \cdots, X_n$.

$\mathfrak{S}_{nh}$ The reproducing kernel random sum function (deterministic design)

$$\mathfrak{S}_{nh}(t) \stackrel{\text{def}}{=} \tfrac{1}{n} \sum_{i=1}^{n} d_{in} \mathfrak{R}_{mh}(x_{in}, t) ,$$

or (random design)

$$\mathfrak{S}_{nh}(t) \stackrel{\text{def}}{=} \tfrac{1}{n} \sum_{i=1}^{n} D_i \mathfrak{R}_{mh}(X_i, t) .$$

$\mathfrak{S}_{\omega nh}$ — The reproducing kernel random sum function for random designs, with quasi-uniform design density ω,

$$\mathfrak{S}_{nh}(t) \stackrel{\text{def}}{=} \frac{1}{n} \sum_{i=1}^{n} D_i \, \mathfrak{R}_{\omega mh}(X_i, t) .$$

$\mathcal{W}_m$ — Weighted Sobolev space of all functions in $L^2(0,1)$ with m-th derivative satisfying $| f |_{\mathcal{W}_m} < \infty$.

$\langle \cdot , \cdot \rangle_{\mathcal{W}_m}$ — A semi inner product on $\mathcal{W}_m$, defined by

$$\langle f , g \rangle_{\mathcal{W}_m} = \int_0^1 [\, x(1-x) \,]^m \, f^{(m)}(x) \, g^{(m)}(x) \, dx .$$

$| \cdot |_{\mathcal{W}_m}$ — The semi-norm on $\mathcal{W}_m$ defined by

$$| f |_{\mathcal{W}_m}^2 = \langle f , f \rangle_{\mathcal{W}_m} .$$

$\langle\langle \cdot , \cdot \rangle\rangle_{h,\mathcal{W}_m}$ — Inner products on $\mathcal{W}_m$ defined by

$$\langle\langle f , g \rangle\rangle_{h,\mathcal{W}_m} = \langle f , g \rangle + h^{2m} \langle f , g \rangle_{\mathcal{W}_m} .$$

$\| \cdot \|_{h,\mathcal{W}_m}$ — Norms on $\mathcal{W}_m$ defined by

$$\| f \|_{h,\mathcal{W}_m}^2 = \langle\langle f , f \rangle\rangle_{h,\mathcal{W}_m} .$$

R_{mh} — The reproducing kernel of the reproducing kernel Hilbert space $\mathcal{W}_m(0,1)$ under the inner product

$$\langle\langle \cdot , \cdot \rangle\rangle_{h,\mathcal{W}_m}$$

for $m \geqslant 2$.

$T_m(\,\cdot\,, x)$ — The m-th order Taylor polynomial of f_o around x; see (16.2.13).

$R_{hm}(\,\cdot\,, x)$ — The leading term of the local bias in local polynomial regression; see (16.6.6).

$a \vee b$ — The function with values $[\, a \vee b \,](x) = \max\big(a(x) , b(x) \big)$.

$a \wedge b$ — The function with values $[\, a \wedge b \,](x) = \min\big(a(x) , b(x) \big)$.

$\delta f(x; h)$ — The Gateaux variation of f at x in the direction h, see § 10.1.

$*$ — Convolution, possibly restricted, as in

$$[\, f * g \,](x) = \int_D f(x - y) \, g(y) \, dy , \quad x \in D ,$$

for $f, g \in L^p(D)$ for suitable p, with $D \subset \mathbb{R}$.

$\circledast$ — "Reverse" convolution, possibly restricted, as in

$$[\, f \circledast g \,](x) = \int_D f(y - x) \, g(y) \, dy , \quad x \in D .$$

12

Nonparametric Regression

1. What and why?

In this volume, we study univariate nonparametric regression problems. The prototypical example is where one observes the data

$$(Y_i, X_i) , \quad i = 1, 2, \cdots, n , \tag{1.1}$$

following the model

$$Y_i = f_o(X_i) + \varepsilon_i , \quad i = 1, 2, \cdots, n , \tag{1.2}$$

where $\varepsilon_1, \varepsilon_2, \cdots, \varepsilon_n$ are independent normal random variables with mean 0 and unknown variance σ^2. The object is to estimate the (smooth) function f_o and construct inferential procedures regarding the model (1.2). However, the real purpose of this volume is to outline a down-to-earth approach to nonparametric regression problems that may be emulated in other settings. Here, in the introductory chapter, we loosely survey what we will be doing and which standard topics will be omitted. However, before doing that, it is worthwhile to describe various problems in which the need for nonparametric regression arises.

We start with the standard multiple linear regression model. Let $d \geqslant 1$ be a fixed integer, and consider the data

$$\bigl(Y_i, X_{i1}, X_{i2}, \cdots, X_{id} \bigr) , \quad i = 1, 2, \cdots, n , \tag{1.3}$$

where $X_{i1}, X_{i2}, \cdots, X_{id}$ are the *predictors* and Y_i is the response, following the model

$$Y_i = \sum_{j=0}^{d} \beta_j X_{ij} + \varepsilon_i , \quad i = 1, 2, \cdots, n , \tag{1.4}$$

where $X_{io} = 1$ for all i, and $\varepsilon_1, \varepsilon_2, \cdots, \varepsilon_n$ are independent Normal$(0, \sigma^2)$ random variables, with $\beta_0, \beta_1, \cdots, \beta_d$ and σ^2 unknown.

The model (1.4) is denoted succinctly and to the point as

$$Y = X\beta + \varepsilon , \tag{1.5}$$

P.P.B. Eggermont, V.N. LaRiccia, *Maximum Penalized Likelihood Estimation*,
Springer Series in Statistics, DOI 10.1007/b12285_1,

with $Y = (Y_1, Y_2, \cdots, Y_n)^T$, $\beta = (\beta_0, \beta_1, \cdots, \beta_d)^T$, and $X \in \mathbb{R}^{n\times(d+1)}$ defined as

$$(1.6) \qquad X = \begin{bmatrix} 1 & X_{1,1} & X_{1,2} & \cdots & X_{1,d} \\ 1 & X_{2,1} & X_{2,2} & \cdots & X_{2,d} \\ \vdots & \vdots & \vdots & \ddots & \vdots \\ 1 & X_{n,1} & X_{n,2} & \cdots & X_{n,d} \end{bmatrix} .$$

Moreover, with $\varepsilon = (\varepsilon_1, \varepsilon_2, \cdots, \varepsilon_n)^T$, we have

$$(1.7) \qquad \varepsilon \sim \text{Normal}(0, \sigma^2 I) .$$

If the columns of the design matrix X are linearly independent, then the estimator of β is the unique solution $b = \widehat{\beta}$ of the least-squares problem

$$(1.8) \qquad \text{minimize} \quad \| Xb - Y \|^2 \quad \text{subject to} \quad b \in \mathbb{R}^{d+1} .$$

Here, $\| \cdot \|$ denotes the Euclidean norm.

(1.9) REMARK. Had it been up to statisticians, the mysterious, threatening, unidentified stranger in the B-movies of eras gone by would not have been called Mr. X but rather Mr. β. See(!), e.g., VORHAUS (1948).

Continuing, we have that $\widehat{\beta}$ is an unbiased estimator of β,

$$(1.10) \qquad \widehat{\beta} - \beta \sim \text{Normal}(0, \sigma^2 (X^T X)^{-1}) ,$$

and $S^2 = \| Y - X\widehat{\beta} \|^2/(n-d-1)$ is an unbiased estimator of σ^2, with

$$(1.11) \qquad \frac{(n-d-1)\, S^2}{\sigma^2} \sim \chi^2(n-d-1) .$$

Moreover, $\widehat{\beta}$ and S^2 are independent, and the usual inferential procedures apply. See, e.g., SEBER (1977).

Note that in (1.4) the response is modeled as growing linearly with the predictors. Of course, not all multiple linear regression problems are like this. In Figure 1.1(b), we show the Wastewater Data Set of HETRICK and CHIRNSIDE (2000), referring to the treatment of wastewater with the white rot fungus, *Phanerochaete chrysosporium*. In this case, we have one predictor (elapsed time). The response is the concentration of magnesium as an indicator of the progress of the treatment (and of the growth of the fungus). It is clear that the response does not grow linearly with time. Moreover, the variance of the noise seems to change as well, but that is another story. The traditional approach would be to see whether suitable transformations of the data lead to "simple" models. Indeed, in Figure 1.1(a), we show the scatter plot of $\log_{10}\log_{10}(1/Y)$ versus $\log_{10} X$, omitting the first two data points (with *concentration* = 0). Apart from the first five data points, corresponding to very small concentrations, the

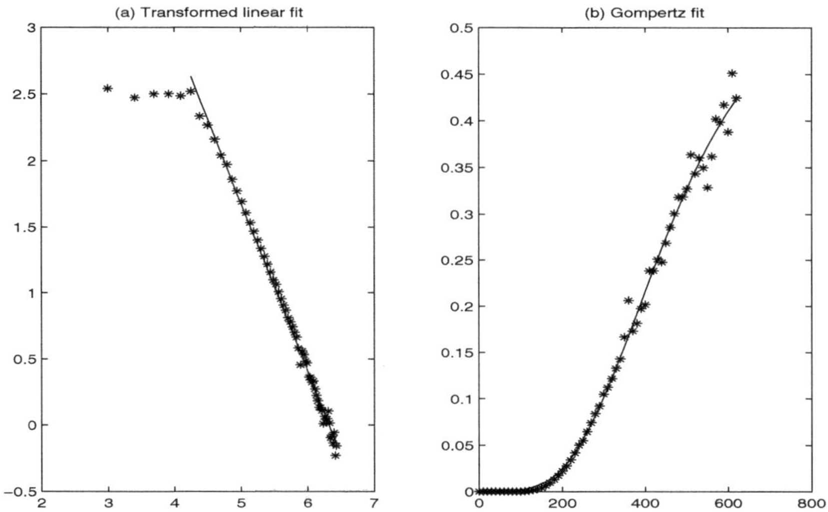

Figure 1.1. The Wastewater Data Set of HETRICK and CHIRNSIDE (2000): Magnesium concentration vs. time. (a) The linear least-squares fit of $\log_{10}\log_{10}(1/\text{concentration})$ vs. $\log_{10}(\text{time})$ with the first seven observations removed and the scatter plot of the transformed data with the first two data points omitted. (b) The original data and the Gompertz fit.

straight-line model suggests itself:

$$\log_{10}\log_{10}(1/Y_i) = \beta_0 + \beta_1 \log_{10} X_i + \varepsilon_i \ , \quad i = 1, 2, \cdots, n \ . \tag{1.12}$$

The least-squares fit is shown as well. Within the context of simple linear regression, it seems clear that the model (1.12) must be rejected because of systematic deviations (data point lying on one side of the fitted line over long stretches). Even so, this model seems to give tremendously accurate predictions! Of course, having to omit five data points is awkward. As an alternative, parametric models suggest themselves, and here we use one of the oldest, from GOMPERTZ (1825),

$$G(x \mid \beta) = \beta_1 \exp\left(-\beta_3^{-1} \exp\bigl(-\beta_3(x - \beta_2)\bigr)\right) \ , \quad x > 0 \ . \tag{1.13}$$

(See § 23.5 for others.) In Figure 1.1(b), we show the unadulterated Wastewater Data Set, together with the least-squares fit with the Gompertz model, obtained by solving

$$\begin{aligned} &\text{minimize} \quad \sum_{i=1}^{n} \bigl| Y_i - G(X_i \mid \beta) \bigr|^2 \\ &\text{subject to} \quad \beta \geqslant 0 \text{ (componentwise)} \ . \end{aligned} \tag{1.14}$$

This works admirably. The only trouble, if we may call it such, is between 150 and 200 time units, where the Gompertz curve lies above the data points, but not by much. This hardly has practical significance, so para-

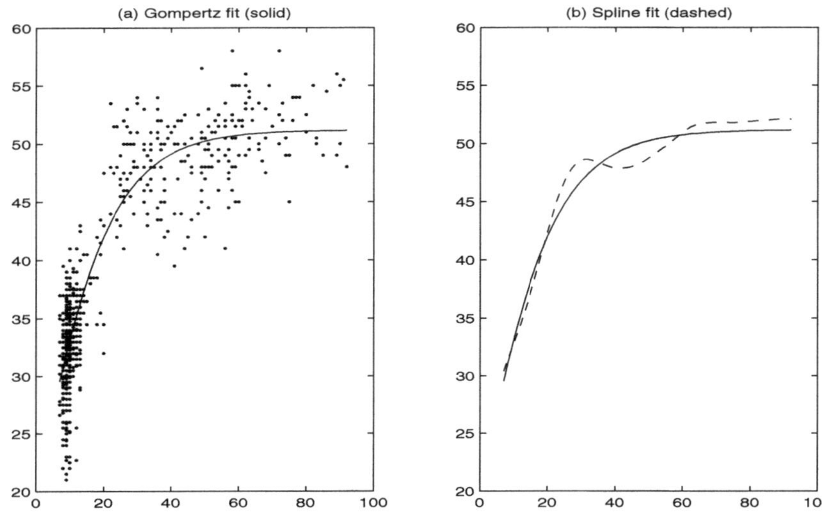

Figure 1.2. The Wood Thrush Data Set of BROWN and ROTH (2004): Weight vs. age. (a) The scatter plot of the data and the Gompertz fit. (b) The Gompertz fit (solid) and the smoothing spline fit (dashed). Is the weight loss between ages 30 and 45 days real ?

metric modeling works here, or so it seems. In § 6, we will spot a fly in the ointment when considering *confidence bands* for parametric regression functions. A complete (?) analysis of the Wastewater Data Set is shown in § 23.7.

At this point, we should note that at least one aspect of the Wastewater Data has been ignored, viz. that it represents longitudinal data: With some poetic license, one may view the data as the growth curve of one individual fungus colony. However, the population growth curve is the one of interest, but we have only one sample curve.

We now consider average (or population) growth data where the errors are uncorrelated. In Figure 1.2(a), the scatter plot of weight versus age of the wood thrush, *Hylocichla mustelina*, as obtained by BROWN and ROTH (2004), is shown. For nestlings, age can be accurately estimated; for older birds, age is determined from ring data, with all but the "oldest" data discarded. Thus, the data may be modeled *parametrically* or *nonparametrically* as

$$Y_i = f_o(X_i) + \varepsilon_i \ , \quad i = 1, 2, \cdots, n \ , \tag{1.15}$$

with the weight Y_i and age X_i. The noise ε_i is the result of measurement error and random individual variation, and $y = f_o(x)$ is the population growth curve. In avian biology, it is customary to use *parametric* growth curves. In Figure 1.2(a), we also show the least-squares fit for the Gompertz model $f_o = G(\cdot \mid \beta)$. A cursory inspection of the graph does not reveal any problems, although BROWN and ROTH (2004) noted a large discrepancy between observed and predicted average adult weights. Questions do arise when we consider *nonparametric* estimates of f_o in (1.15). In Figure 1.2(b), we show the Gompertz fit again, as well as a nonparametric estimator (a smoothing spline). The nonparametric estimator suggests a decrease in weight between ages 30 and 45 days. The biological explanation for this weight loss is that the parents stop feeding the chick and that the chick is spending more energy catching food than it gains from it until it learns to be more efficient. This explanation is convincing but moot if the apparent dip is not "real". While one could envision parametric models that incorporate the abandonment by the parents and the learning behavior of the chicks, it seems more straightforward to use nonparametric models and only assume that the growth curve is nice and smooth. We should note here that a more subtle effect occurs when considering wingspan versus age, comparable to the growth spurt data analysis of GASSER, MÜLLER, KOHLER, MOLINARI, and PRADER (1984). Also, an important, and often detrimental, feature of parametric models is that the associated *confidence bands* are typically quite narrow, whether the model is correct or not. We come back to this for the Wood Thrush Data Set in § 23.6.

While on the subject of parametric versus nonparametric regression, we should note that GOMPERTZ (1825) deals with mortality data. For a comparison of various parametric and nonparametric models in this field, see DEBÓN, MONTES, and SALA (2005, 2006). In § 2, we briefly discuss (nonparametric) spline smoothing of mortality data as introduced by WHITTAKER (1923).

We return to the multiple regression model (1.5) and discuss three more "nonparametric" twists. Suppose one observes an additional covariate, t_i,

$$(1.16) \qquad \bigl(Y_i, X_{i1}, X_{i2}, \cdots, X_{id}, t_i \bigr) , \quad i = 1, 2, \cdots, n .$$

One possibility is that the data are being collected over time, say with

$$(1.17) \qquad 0 \leqslant t_1 \leqslant t_2 \leqslant \cdots \leqslant t_n \leqslant T ,$$

for some finite final time T.

In the *partially linear model,* the extra covariate enters into the model in some arbitrary but smooth, additive way but is otherwise uninteresting. Thus, the model is

$$(1.18) \qquad Y_i = \sum_{j=0}^{d} \beta_j X_{ij} + f_o(t_i) + \varepsilon_i , \quad i = 1, 2, \cdots, n ,$$

for a nice, smooth function f_o. Interest is still in the distribution of the estimators of the β_j, but now only asymptotic results are available; see, e.g., HECKMAN (1988) and ENGLE, GRANGER, RICE, and WEISS (1986) and § 13.8. In DEMICCO, LIN, LIU, REJTŐ, BELDONA, and BANCROFT (2006), interest is in explaining the effect of holidays on daily hotel revenue. They first used a linear model incorporating the (categorical) effects of day of the week and week of the year. However, this left a large noise component, so that holiday effects could not be ascertained. Adding the time covariate (consecutive day of the study) provided a satisfactory model for the revenue throughout the year, except for the holidays, which have their own discernible special effects. (In (1.18), the time covariate is taken care of by f_o.)

We now move on to *varying coefficients* models or, more precisely, *time-varying* linear models. If we may think of the data as being collected over time, then the question arises as to whether the model (1.5) is a good fit regardless of "time" or whether the coefficients change with time. In the event of an abrupt change at some point in time t_m, one can test whether the two data sets

$$\left(Y_i, X_{i,1}, X_{i,2}, \cdots, X_{i,d}\right), \quad i = 1, 2, \cdots, m-1,$$

and

$$\left(Y_i, X_{i,1}, X_{i,2}, \cdots, X_{i,d}\right), \quad i = m, m+1, \cdots, n,$$

are the same or not. This is then commonly referred to as a *change point* problem. However, if the change is not abrupt but only gradual, then it makes sense to view each β_j as a smooth function of time, so a reasonable model would be

$$(1.19) \qquad Y_i = \sum_{j=0}^{d} \beta_j(t_i)\, X_{ij} + \varepsilon_i , \quad i = 1, 2, \cdots, n ,$$

and we must estimate the functions $\beta_j(t)$ (and perhaps test whether they are constant). See, e.g., JANDHYALA and MACNEIL (1992).

Note that, for each i, we have a standard linear model (good) but only one observation per model (bad). This is where the *smoothness* assumption on the coefficients $\beta_j(t)$ comes in: For nearby t_i, the models are almost the same and we may pretend that the nearby data all pertain to the same original model at time t_i. So, one *should* be able to estimate the coefficient functions with reasonable accuracy. See CHIANG, RICE, and WU (2001) and EGGERMONT, EUBANK, and LARICCIA (2005).

The final nonparametric twist to multiple regression is by way of the generalized linear model

$$(1.20) \qquad Y_i = f_o\big([X\beta]_i \big) + \varepsilon_i , \quad i = 1, 2, \cdots, n ,$$

where now f_o (or rather its inverse) is called the *link* function. In the standard approach, f_o is a fixed known function, depending on the application. The model (1.12) almost fits into this framework. In the nonparametric approach, the function f_o is an unknown, smooth function. See MCCULLAGH and NELDER (1989) and GREEN and SILVERMAN (1990). Unfortunately, of these three nonparametric twists of the multiple regression model, the partially linear model is the only one to be discussed in this volume.

All of the nonparametric models above are attempts at avoiding the "full" nonparametric model for the data (1.3),

$$Y_i = f_o(X_{i,1}, X_{i,2}, \cdots, X_{i,d}) + \varepsilon_i \;, \quad i = 1, 2, \cdots, n \;, \tag{1.21}$$

which allows for arbitrary interactions between the predictors. Here, f_o is a smooth multivariate function. The case $d = 2$ finds application in spatial statistics; e.g., in precision farming. The case $d = 3$ finds application in seismology. It is a good idea to avoid the full model if one of the models (1.18), (1.19), or (1.22) below applies because of the *curse of dimensionality*: The full version requires orders of magnitude more data to be able to draw sensible conclusions.

Another attempt at avoiding the full model (1.21) is the "separable" model

$$Y_i = \sum_{j=1}^{d} f_{o,j}(X_{ij}) + \varepsilon_i \;, \quad i = 1, 2 \cdots, n \;, \tag{1.22}$$

with "smooth" functions $f_{o,1}, f_{o,2}, \cdots, f_{o,d}$. This is somewhat analogous to the varying coefficients model. For more on additive linear models, see BUJA, HASTIE, and TIBSHIRANI (1989).

As said, we only consider the univariate case: There are still plenty of interesting and half-solved problems both in theory and practice.

2. Maximum penalized likelihood estimation

We now survey various estimators for the univariate nonparametric regression problem and make precise the notion of smoothness. The general approach taken here is to view *everything* as regularized maximum likelihood estimation: GRENANDER (1981) calls it a first principle. Indeed, since all of statistics is fitting models to data, one must be able to judge the quality of a fit, and the likelihood is the first rational criterion that comes to mind. (In hypothesis testing, this leads to likelihood ratio tests.)

It is perhaps useful to briefly state the maximum likelihood principle in the parametric case and then "generalize" it to the nonparametric setting. To keep things simple, consider the univariate density estimation problem. Suppose one observes independent, identically distributed (iid) data, with shared probability density function (pdf) g_o,

$$X_1, X_2, \cdots, X_n, \text{ iid with pdf } g_o \;, \tag{2.1}$$

where g_o belongs to a parametric family of probability density functions,

$$\mathcal{F} = \{\, f(\,\cdot\,|\,\theta) \;|\; \theta \in \Theta \,\} \;. \tag{2.2}$$

Here, Θ is the parameter space, a subset of $\mathbb{R}^d$ for "small" d ($d \leqslant 3$, say). Thus, $g_o = f(\,\cdot\,|\,\theta_o)$ for some unknown $\theta_o \in \Theta$. The objective is to estimate θ_o or g_o.

The maximum likelihood estimation problem is

$$\text{(2.3)} \qquad \text{minimize} \quad -\tfrac{1}{n}\sum_{i=1}^{n} \log f(X_i \,|\, \theta) \quad \text{subject to} \quad \theta \in \Theta \; .$$

This formulation puts the emphasis on the parameter. Alternatively, one may wish to emphasize the associated probability density function and reformulate (2.3) as

$$\text{(2.4)} \qquad \text{minimize} \quad -\tfrac{1}{n}\sum_{i=1}^{n} \log g(X_i) \quad \text{subject to} \quad g \in \mathcal{F} \; .$$

In this formulation, there are two ingredients: the likelihood function and the (parametric) family of pdfs. Maximum likelihood estimation "works" if the family $\mathcal{F}$ is nice enough.

In the nonparametric setting, one observes the data as in (2.1), but now with the pdf g_o belonging to some (nonparametric) family $\mathcal{F}$. One may view $\mathcal{F}$ as a parametric family as in (2.2) with Θ a subset of an infinite-dimensional space. However, some care must be exercised. By way of example, the naive maximum likelihood estimation problem,

$$\text{(2.5)} \qquad \begin{array}{ll} \text{minimize} & -\tfrac{1}{n}\sum_{i=1}^{n} \log g(X_i) \\ \text{subject to} & g \text{ is a continuous pdf} \end{array}$$

has no solution: The ideal unconstrained "solution" is $g = \gamma$, with

$$\gamma = \tfrac{1}{n}\sum_{i=1}^{n} \delta(\,\cdot\, - X_i) \; ,$$

which puts probability $\frac{1}{n}$ at every observation, but this is not a pdf. However, one may approximate γ by pdfs in such a way that the negative log likelihood tends to $-\infty$. So, (2.5) has no solution. The moral of this story is that, even in nonparametric maximum likelihood estimation, the minimization must be constrained to suitable subsets of pdfs. See GRENANDER (1981) and GEMAN and HWANG (1982). The nonparametrics kick in by letting the constraint set get bigger with increasing sample size. The authors' favorite example is the GOOD and GASKINS (1971) estimator, obtained by restricting the minimization in (2.5) to pdfs g satisfying

$$\text{(2.6)} \qquad \int_{\mathbb{R}} \frac{|\, g'(x) \,|^2}{g(x)} \, dx \leqslant C_n$$

for some constant C_n, which must be estimated based on the data; see § 5.2 in Volume I. Of course, suitable variations of (2.6) suggest themselves. A different type of constraint arises from the requirement that the pdf be piecewise convex/concave with a finite number of inflection points; see MAMMEN (1991a). Again, nonparametric maximum likelihood works if the nonparametric family is nice enough. In an abstract setting, GRENANDER

(1981) carefully defines what "nice enough" means and refers to nice enough families as "sieves".

After these introductory remarks, we consider maximum penalized likelihood estimation for various nonparametric regression problems. We start with the simplest case of a deterministic design (and change the notation from § 1 somewhat). Here, we observe the data $y_{1,n}, y_{2,n}, \cdots, y_{n,n}$, following the model

$$y_{in} = f_o(x_{in}) + d_{in} \ , \quad i = 1, 2, \cdots, n \ , \tag{2.7}$$

where the x_{in} are design points and $d_n = (d_{1,n}, d_{2,n}, \cdots, d_{n,n})^T$ is the random noise. Typical assumptions are that the d_{in}, $i = 1, 2, \cdots, n$, are uncorrelated random variables with mean 0 and common variance σ^2,

$$\mathbb{E}[\, d_n \,] = 0 \ , \quad \mathbb{E}[\, d_n d_n^T \,] = \sigma^2 I \ , \tag{2.8}$$

where σ^2 is unknown. We refer to this as the Gauss-Markov model in view of the Gauss-Markov theorem for linear regression models. At times, we need the added condition that the d_{in} are independent, identically distributed random variables, with a finite moment of order $\kappa > 2$,

$$\mathbb{E}[\, | \, d_{1,n} \, |^{\kappa} \,] < \infty \ . \tag{2.9}$$

Another, more restrictive but generally made, assumption is that the d_{in} are iid Normal$(0, \sigma^2)$ random variables, again with the variance σ^2 usually unknown,

$$d_n \sim \text{Normal}(\, 0 \, , \sigma^2 \, I \,) \ . \tag{2.10}$$

This is referred to as the Gaussian model.

Regarding the design, we typically consider more-or-less uniform designs, such as

$$x_{in} = \frac{i-1}{n-1} \ , \quad i = 1, 2, \cdots, n \ . \tag{2.11}$$

In general, for random designs, we need the design density to be bounded and bounded away from 0.

We now consider maximum likelihood estimation in the Gaussian model (2.7)–(2.11). Assuming both f_o and σ are unknown, the maximum likelihood problem for estimating them is

$$\begin{array}{ll} \text{minimize} & (2\,\sigma^2)^{-1} \sum_{i=1}^{n} \, | \, f(x_{in}) - y_{in} \, |^2 + n \log \sigma \\ \text{subject to} & f \text{ is continuous} \ , \ \sigma > 0 \ . \end{array} \tag{2.12}$$

As expected, this leads to unacceptable estimators. In particular, for any continuous function f that interpolates the data,

$$f(x_{in}) = y_{in} \ , \quad i = 1, 2, \cdots, n \ , \tag{2.13}$$

letting $\sigma \to 0$ would make the negative log-likelihood in (2.12) tend to $-\infty$. Of course, in (2.13), we then have a definite case of *overfitting* the data. The only way around this is to restrict the minimization in (2.12) to suitably "small" sets of functions. This is the aforementioned method of sieves of GRENANDER (1981). In the regression context, the simplest examples of sieves are sequences of nested finite-dimensional subspaces or nested compact subsets of $L^2(0,1)$.

The classic example of nested, finite-dimensional subspaces of $L^2(0,1)$ is the polynomial sieve: Letting $\mathcal{P}_r$ be the set of all polynomials of order r (degree $\leqslant r-1$), the polynomial sieve is

$$\mathcal{P}_1 \subset \mathcal{P}_2 \subset \mathcal{P}_3 \subset \cdots \subset L^2(0,1) \,. \tag{2.14}$$

The "sieved" maximum likelihood estimation problem is then defined as

$$\begin{aligned} &\text{minimize} \quad (2\sigma^2)^{-1} \sum_{i=1}^{n} |\, f(x_{in}) - y_{in} \,|^2 + n \log \sigma \\ &\text{subject to} \quad f \in \mathcal{P}_r \,, \ \sigma > 0 \,, \end{aligned} \tag{2.15}$$

where r has to be suitably chosen. Note that this minimization problem works out nicely: For given f, the optimal σ^2 is given by

$$\sigma_n^2(f) = \tfrac{1}{n} \sum_{i=1}^{n} |\, f(x_{in}) - y_{in} \,|^2 \,, \tag{2.16}$$

and then the problem is to minimize $\log \sigma_n^2(f)$, which despite appearances is a plain least-squares problem. Theoretically, from the usual mean-squared error point of view, the polynomial sieve works remarkably well. See § 15.2, where some other sieves of finite-dimensional spaces are discussed as well. In practice, for small sample sizes, things are not so great; see § 23.4.

The typical example of a "compact" sieve arises by imposing a bound on the size of the m-th derivative of f_o for some integer $m \geqslant 1$,

$$\| f_o^{(m)} \| \leqslant C \,, \tag{2.17}$$

where $\| \cdot \|$ denotes the usual $L^2(0,1)$ norm. The implicit assumption in (2.17) is that f_o belongs to the Sobolev space $W^{m,2}(0,1)$, where, for $1 \leqslant p \leqslant \infty$ ($p = 1, 2, \infty$ are the cases of interest),

$$W^{m,p}(0,1) = \left\{ f \in C^{m-1}(0,1) \;\middle|\; \begin{array}{l} f^{(m-1)} \text{ is absolutely} \\ \text{continuous, } \| f^{(m)} \|_p < \infty \end{array} \right\} , \tag{2.18}$$

with $\| \cdot \|_p$ denoting the $L^p(0,1)$ norm. In (2.18), the set $C^k(0,1)$ is the vector space of all k times continuously differentiable functions on $[\,0\,,\,1\,]$. The requirement that $f^{(m-1)}$ be absolutely continuous implies that $f^{(m)}$ exists almost everywhere and is integrable. However, (for $p > 1$) the additional requirement that $f^{(m)} \in L^p(0,1)$ is imposed. Thus, the sieve is the continuous scale of nested subsets

$$\mathfrak{F}_C = \{\, f \in W^{m,2}(0,1) \;:\; \| f^{(m)} \| \leqslant C \,\} \,, \quad 0 < C < \infty \,. \tag{2.19}$$

(The sets $\mathfrak{F}_C$ are not compact subsets of $L^2(0,1)$, but their intersections with bounded, closed subsets of $L^2(0,1)$ are.) Then, the sieved maximum likelihood estimation problem is

$$\text{(2.20)} \qquad \begin{array}{ll} \text{minimize} & (2\,\sigma^2)^{-1} \sum_{i=1}^{n} |\, f(x_{in}) - y_{in} \,|^2 + n \log \sigma \\ \text{subject to} & f \in \mathfrak{F}_C \;,\; \sigma > 0 \;. \end{array}$$

Now for known C the set $\mathfrak{F}_C$ is a closed and convex subset of $W^{m,2}(0,1)$, so the use of Lagrange multipliers leads to the equivalent maximum penalized likelihood problem

$$\text{(2.21)} \qquad \begin{array}{ll} \text{minimize} & (2\,\sigma^2)^{-1} \sum_{i=1}^{n} |\, f(x_{in}) - y_{in} \,|^2 + n \log \sigma + \lambda \,\|\, f^{(m)} \,\|^2 \\ \text{subject to} & f \in W^{m,2}(0,1) \;,\; \sigma > 0 \;, \end{array}$$

where $\lambda > 0$ is the smoothing parameter, chosen such that the solution satisfies $\|\, f^{(m)} \|^2 = C$. (Very rarely will one have $< C$.) However, if C is unknown and must be estimated, then we may as well take (2.21) as the starting point and consider choosing λ instead of C.

(2.22) Exercise. Show that the problems (2.20) and (2.21) are equivalent.

The problem (2.21) may be treated similarly to (2.15). Start out by taking $\lambda = n\,h^{2m}/(2\,\sigma^2)$. Then, given f, explicitly minimizing over σ^2 leads to the problem

$$\text{(2.23)} \qquad \begin{array}{ll} \text{minimize} & \frac{1}{n} \sum_{i=1}^{n} |\, f(x_{in}) - y_{in} \,|^2 + h^{2m} \,\|\, f^{(m)} \,\|^2 \\ \text{subject to} & f \in W^{m,2}(0,1) \;. \end{array}$$

Ignoring questions of existence and uniqueness for now, the solution is denoted by f^{nh}. The parameter h in (2.23) is the smoothing parameter: It determines how smooth the solution is. For $h \to 0$, we are back to (2.12), and for $h \to \infty$ we get $\|\, (f^{nh})^{(m)} \,\| \longrightarrow 0$; i.e., f^{nh} tends to a polynomial of degree $\leqslant m - 1$ (and so f^{nh} is very smooth). We discuss smoothing parameter selection in § 7 and Chapter 18.

The solution of (2.23) is a spline of order $2m$, or degree $2m-1$. They are commonly referred to as smoothing splines. In practice, the case $m = 2$ is predominant, and the corresponding splines are called cubic smoothing splines. But in Chapter 23 we present evidence for the desirability of using higher-order smoothing splines. The traditional definition of splines is discussed in Chapter 19 together with the traditional computational details. The nontraditional computations by way of the Kalman filter are explored in Chapter 20.

Returning to (2.21), given f^{nh}, the estimator for σ is given by

$$\sigma_{nh}^2 = \tfrac{1}{n} \sum_{i=1}^{n} | f^{nh}(x_{in}) - y_{in} |^2 + h^{2m} \| (f^{nh})^{(m)} \|^2 , \tag{2.24}$$

although the extra term $h^{2m} \| (f^{nh})^{(m)} \|^2$ raises eyebrows, of course.

The constraint (2.17) and the maximum penalized likelihood problem (2.23) are the most common, but alternatives abound. The choice

$$\| f^{(m)} \|_1 \leqslant C \tag{2.25}$$

is made frequently as well, mostly for the case $m = 1$. For technical reasons, for the case $m = 1$ and continuous functions f, this is recast as

$$| f |_{BV} \stackrel{\text{def}}{=} \sum_{i=1}^{n} | f(x_{in}) - f(x_{i-1,n}) | \leqslant C , \tag{2.26}$$

which leads to total-variation penalization. See §§ 17.2 and 17.3 for the fine print. At this stage, we should mention the "graduated" curves of Whittaker (1923), who considers equally spaced (mortality) data. Here, the constraint (2.17) for $m = 3$ is enforced in the form (2.23), except that the L^2-norm of the third derivative is replaced by the sum of squared third-order (divided) differences. Thus, the constraint is that

$$\sum_{i=2}^{n-2} \big| f(x_{i+2,n}) - 3 f(x_{i+1,n}) + 3 f(x_{in}) - f(x_{i-1,n}) \big|^2 \leqslant C . \tag{2.27}$$

The corresponding maximum penalized likelihood estimation problem is then phrased in terms of the function values $f(x_{1,n}), f(x_{2,n}), \cdots, f(x_{n,n})$ only, so one may well argue that one is not estimating a function. Perhaps for this reason, these "graduated" curves are not used (much) anymore.

A combinatorial method (for lack of a better term) of constructing a sieve is obtained by considering (continuous) functions, the m-th derivative of which has a finite number of sign changes. Thus, for $s = 0$ or $s = 1$, let

$$\mathcal{S}_{mq}^{s} = \left\{ f \in W^{m,2}(0,1) : \begin{array}{c} \exists\, 0 = t_0 < t_1 < t_2 < \cdots < t_{q-1} < t_q = 1 \\ s\,(-1)^j f^{(m)} \geqslant 0 \text{ on } [\,t_j, t_{j+1}\,] \text{ for all } j \end{array} \right\} ,$$

and consider the constrained maximum likelihood problem

$$\begin{array}{ll} \text{minimize} & (2\sigma^2)^{-1} \sum_{i=1}^{n} | f(x_{in}) - y_{in} |^2 + n \log \sigma \\ \text{subject to} & f \in \mathcal{S}_{mq}^{+} \cup \mathcal{S}_{mq}^{-} . \end{array} \tag{2.28}$$

Here, the "order" q must be estimated from the data; e.g., by way of Schwarz's Bayesian Information Criterion; see Schwarz (1978). There are theoretical and practical issues since the sets $\mathcal{S}_{mq}^{\pm}$ are not convex. (They are if the "inflection" points are fixed once and for all.) Regardless, this procedure works quite well; see Mammen (1991a, 1991b) and Diewert and WALES (1998). Unfortunately, we will not consider this further. It should be noted that sometimes qualitative information is available, which should be used in the problem formulation. Some typical instances arise when f_o is nonnegative and/or increasing, which are special cases of (2.28) for known m, q, and s.

What happens to the interpretation of smoothing splines as maximum penalized likelihood estimators when the noise is not iid normal? If the noise is iid (but not necessarily normal) and satisfies (2.8)–(2.9), then it may be approximated by iid normal noise, e.g., by way of the construction of KOMLÓS, MAJOR, and TUSNÁDY (1976); see §§ 22.2 and 22.4. So, perhaps, we may then refer to (2.20) and (2.21) as *asymptotic* maximum penalized likelihood problems. However, there is a perplexing twist to this. Consider the regression problem (2.7), (2.11) with iid two-sided exponential noise; i.e., $d_{1,n}, d_{2,n}, \cdots, d_{n,n}$ are iid with common pdf

$$f_d(t) = (2\lambda)^{-1} \exp(-|t|/\lambda), \quad -\infty < t < \infty, \tag{2.29}$$

for some unknown constant $\lambda > 0$. Then, with the constraint (2.17) on f_o, the penalized maximum likelihood problem leads to the roughness-penalized least-absolute-deviations problem

$$\begin{aligned} &\text{minimize} && \tfrac{1}{n} \sum_{i=1}^{n} | f(x_{in}) - y_{in} | + h^{2m} \| f^{(m)} \|^2 \\ &\text{subject to} && f \in W^{m,2}(0,1) . \end{aligned} \tag{2.30}$$

Surely, under the noise model (2.29), this should give better results than the smoothing spline problem (2.22), but the normal approximation scheme, which applies here with even more force because of the finite exponential moments, suggests that asymptotically it should *not* make a difference. It gets even more perplexing when one realizes that (2.30) is extremely useful when the noise does not even have an expectation, e.g., for Cauchy noise. Then, it is customary to assume that $d_{1,n}, d_{2,n}, \cdots, d_{n,n}$ are iid with

$$\text{median}(d_{1,n}) = 0 . \tag{2.31}$$

So here we go from maximum likelihood estimation for the two-sided exponential noise model to the momentless noise model with zero median. This is a b i g step! See §§ 17.4 and 17.5. There are more general versions than (2.30), where the absolute values are replaced by a general function,

$$\begin{aligned} &\text{minimize} && \tfrac{1}{n} \sum_{i=1}^{n} \Psi(f(x_{in}) - y_{in}) + h^{2m} \| f^{(m)} \|^2 \\ &\text{subject to} && f \in W^{m,2}(0,1) . \end{aligned} \tag{2.32}$$

with, e.g., Ψ the HUBER (1981) function, $\Psi(t) = \min\{ t^2, \varepsilon \}$, where ε has to be chosen appropriately. See, e.g., COX (1981).

Now consider the case where the noise is independent and normal but not identically distributed, as in the regression model

$$(2.33) \qquad y_{in} = f_o(x_{in}) + \sigma_o(x_{in})\, d_{in} \ , \qquad i = 1, 2, \cdots, n \ ,$$

under the assumptions (2.10)–(2.11), for a nice variance function $\sigma_o^2(t)$. Now, (2.23) is of course not a maximum likelihood problem, and should be replaced by

$$(2.34) \qquad \begin{array}{ll} \text{minimize} & \displaystyle\sum_{i=1}^{n} \Big\{ \frac{|\, f(x_{in}) - y_{in}\,|^2}{2\,\sigma^2(x_{in})} + \log \sigma^2(x_{in}) \Big\} + \lambda \,\|\, f^{(m)} \,\|^2 \\ \text{subject to} & f \in W^{m,2}(0,1) \ , \qquad \sigma > 0 \ , \end{array}$$

but it is clear that the condition $\sigma > 0$ is not "effective". Take any f with $f(x_{1,n}) = y_{1,n}$, and let $\sigma(x_{1,n}) \to 0$. It is unclear whether a condition like $\sigma(t) \geqslant \delta$, $t \in [0, 1]$, would save the day (and δ would have to be estimated). We pursue this further in § 22.4, where for the model (2.33) we consider a two-stage process for mean and variance estimation.

We next discuss *local* maximum likelihood estimation for the regression problem with iid normal noise, (2.7)–(2.11). Here, the objective of estimating the regression function f_o is replaced by that of estimating f_o at a fixed point $t \in [0, 1]$ (but perhaps at many points). So, fix $t \in [0, 1]$, and consider the graph of f_o near t for "small" h,

$$\mathcal{G}(f_o, t, h) = \{\, (x, f_o(x)) \ : \ |\, x - t\,| \leqslant h \,\}$$

(Yes, h is going to be the smoothing parameter.) Surely, $\mathcal{G}(f_o, t, h)$ gives us no information about the graph of f_o near some point s "far away" from t and conversely. *Mutatis mutandis*, the data near the point s, $\{\, (x_{in}, y_{in}) \ : \ |\, x_{in} - s\,| \leqslant h \,\}$, contains no information about $\mathcal{G}(f_o, t, h)$. So, when estimating the graph $\mathcal{G}(f_o, t, h)$, one should really only consider the data $\{\, (x_{in}, y_{in}) \ : \ |\, x_{in} - t\,| \leqslant h \,\}$. Since the joint pdf of these data is

$$\prod_{|x_{in} - t| \leqslant h} \frac{1}{\sqrt{2\pi}\,\sigma} \exp\Big(- \frac{|\, f_o(x_{in}) - y_{in}\,|^2}{2\,\sigma_o^2} \Big) \ ,$$

the local maximum likelihood problem is

$$(2.35) \qquad \text{minimize} \quad (2\sigma^2)^{-1} \sum_{|x_{in} - t| \leqslant h} |\, f(x_{in}) - y_{in}\,|^2 + n_{t,h} \log \sigma \ ,$$

where $n_{t,h}$ is the number of design points satisfying $|\, x_{in} - t\,| \leqslant h$. Apparently, there is still a need for sieves. It is customary to use the polynomial sieve and to (re)write (2.35) as

$$(2.36) \qquad \begin{array}{ll} \text{minimize} & (2\sigma^2)^{-1} \displaystyle\sum_{i=1}^{n} A_h(x_{in} - t)\, \big|\, f(x_{in}) - y_{in}\,\big|^2 + n_{t,h} \log \sigma \\ \text{subject to} & f \in \mathcal{P}_r \ , \end{array}$$

where $A_h(t) = h^{-1}A(h^{-1}t)$ and $A(t) = 1\!\!1(|t| \leqslant \frac{1}{2})$ is the indicator function of the interval $[-\frac{1}{2}, \frac{1}{2}]$. In practice, the kernel A may be any pdf, including ones with unbounded support, but then $n_{t,h}$ may need to be redefined. Let $f(x) = p^{n,r,h}(x; t)$, $x \in [0, 1]$, denote the solution of (2.36). Then, the estimator of $f_o(t)$ is taken to be $f^{n,A,r,h}(t) = p^{n,r,h}(t; t)$. One refers to $f^{n,A,r,h}$ as the local polynomial estimator. This originated with STONE (1977) for first degree polynomials and CLEVELAND (1979) and CLEVELAND and DEVLIN (1988) in the general case. The case of local constant polynomials may be solved explicitly as the Nadaraya-Watson estimator,

$$(2.37) \qquad f^{n,A,h}(t) = \frac{\frac{1}{n}\sum_{i=1}^{n} y_{in}\, A_h(x - x_{in})}{\frac{1}{n}\sum_{i=1}^{n} A_h(x - x_{in})}, \quad x \in [0, 1],$$

due to NADARAYA (1964) and WATSON (1964) (but their setting was different, see (2.39) below). In (2.37), one may even take kernels of arbitrary order, as long as one keeps an eye on the denominator. The Nadaraya-Watson estimator is a popular subject of investigation amongst probabilists; see, e.g, DEVROYE and WAGNER (1980), EINMAHL and MASON (2000), and DEHEUVELS and MASON (2004, 2007). One drawback of local maximum likelihood estimation is that it is not easy to enforce global constraints. Of course, nonnegativity of the regression function is a local property and does fit right in, but nonnegativity of the first derivative, say (monotonicity), already seems problematic. A detailed study of local polynomial estimators is made in Chapter 16 (but we do not consider constraints).

The least-absolute-deviations version of (2.36) suggests itself when the noise is momentless, satisfying (2.31); see, e.g., TSYBAKOV (1996).

The distinction between smoothing splines and local polynomial estimators is not as dramatic as one might think; e.g., the smoothing spline and Nadaraya-Watson estimators are equivalent in a strong sense (if we ignore the boundary problems of the Nadaraya-Watson estimator) for a certain exponentially decaying kernel A. See § 21.9 for the fine print.

So far, we have only discussed deterministic designs. In the classical nonparametric regression problem, the design points x_{in} themselves are random. Then (2.7) becomes

$$(2.38) \qquad y_i = f_o(X_i) + D_i\,, \quad i = 1, 2, \cdots, n\,,$$

where $X_1, X_2, \cdots, X_n$ are iid random variables and (2.8)–(2.9) hold *conditional* on the design. So now there are two sources of randomness (the design and the noise in the responses), but in Chapters 16 and 21 we show that the randomness due to the design is mostly harmless, at least if the design density is bounded and bounded away from 0. (This implies that we must consider regression on a closed, bounded interval.) The classical interpretation of estimating $f_o(x)$ at a fixed x is that one must estimate the conditional expectation

$$f_o(x) = \mathbb{E}[\, Y \mid X = x \,] . \tag{2.39}$$

The maximum penalized likelihood approach leads to the same problems (2.12). The interpretation of conditional expectation (2.39) does not work in the bad case of (2.31) and must be replaced by the conditional median,

$$f_o(x) = \text{median}[\, Y \mid X = x \,] . \tag{2.40}$$

However, we still refer to this as a regression problem.

EXERCISE: (2.22).

3. Measuring the accuracy and convergence rates

Having agreed that the object of nonparametric regression is the estimation of the regression function, there remains the question of deciding what we mean by a good estimator. In loose terms, a good estimator is one that means what it says and says what it means: It should show those features of the regression function we are interested in, if and only if they are present. Of course, this must be qualified by adding "with high probability" or some such thing. In this interpretation, the estimation problem is slowly being transformed into one of testing and inference. As in testing, to paraphrase NEYMAN (1937), p. 192, one needs to have a good idea of the kind of features to which the estimator should be sensitive, but this may beg the question. By way of illustration, for the Wood Thrush Data Set of BROWN and ROTH (2004) discussed in § 1, one needs a good estimator before one even suspects the presence of some features (the "dip"). This exploratory step seems to require that the maximal error of the estimator, generically denoted by $\widehat{f}_n$, is "small" or perhaps "as small as possible". Thus, one measure of interest is the so-called *uniform error*,

$$\| \widehat{f}_n - f_o \|_\infty \stackrel{\text{def}}{=} \sup_x \, | \widehat{f}_n(x) - f_o(x) \,| , \tag{3.1}$$

where the supremum is over the relevant range of x. The discrete version

$$| \widehat{f}_n - f_o |_\infty \stackrel{\text{def}}{=} \max_{1 \leqslant i \leqslant n} | \widehat{f}_n(x_{in}) - f_o(x_{in}) | \tag{3.2}$$

may be of interest as well.

The uniform error, discrete or not, is studied in §§ 14.5–14.7, § 16.8, and § 21.5. Unfortunately, the uniform error is not a convenient object of study, unlike the mean squared error and the discrete version,

$$\| \widehat{f}_n - f_o \|^2 \quad \text{and} \quad | \widehat{f}_n - f_o |^2 \stackrel{\text{def}}{=} \tfrac{1}{n} \sum_{i=1}^{n} | \widehat{f}_n(x_{in}) - f_o(x_{in}) |^2 . \tag{3.3}$$

This is especially true in connection with linear least-squares problems or problems that act like them, including estimators that are linear in the data $y_{1,n}, y_{2,n}, \cdots, y_{n,n}$.

There are other ways of measuring the error, such as the L^1 error, see DEVROYE, GYÖRFI, KRZYŻAK, and LUGOSI (1994) and HENGARTNER and WEGKAMP (2001), or the Prohorov metric studied by MARRON and TSYBAKOV (1995), but in this volume, we concentrate on the uniform and mean squared error.

Once it has been decided how to measure the error of the estimator(s) $\widehat{f}_n$, the optimal estimator is then defined as the one that realizes the smallest possible error,

$$\min_{\widehat{f}_n} \| \widehat{f}_n - f_o \|_p \tag{3.4}$$

(for the chosen p; $p = 2$ or $p = \infty$, but perhaps for the discrete versions). Here, the minimization is over all possible estimators $\widehat{f}_n$; i.e., all continuous functions of the data, be they linear or not. This is in fact a tall order since it is not so easy to account for all estimators. Added to that, one has to deal with the fact that realizing the minimum is a random event (with probability 0 at that). So, perhaps, one should maximize the probability of coming close to the minimum,

$$\text{maximize} \quad \mathbb{P}\Big[\, \| \widehat{f}_n - f_o \|_p \leqslant (1 + \varepsilon_n) \min_{\widehat{f}_n} \| \widehat{f}_n - f_o \|_p \Big] , \tag{3.5}$$

for a reasonable choice of ε_n. Ideally, one would want $\varepsilon_n \longrightarrow 0$ as $n \longrightarrow \infty$ or would like to find the smallest sequence of ε_n for which the probability tends to 1,

$$\mathbb{P}\Big[\, \| \widehat{f}_n - f_o \|_p \leqslant (1 + \varepsilon_n) \min_{\widehat{f}_n} \| \widehat{f}_n - f_o \|_p \Big] \longrightarrow 1 . \tag{3.6}$$

Unfortunately, that does not say much about the small-sample case. In fact, it may be more realistic to replace ε_n by a small constant. Another way to account for the randomness is to consider expectations. So, one may wish to achieve the expected minimum or the minimum expectation, say in the form

$$\mathbb{E}\Big[\, \| \widehat{f}_n - f_o \|_p \Big] \leqslant (1 + \varepsilon_n) \mathbb{E}\Big[\min_{\widehat{f}_n} \| \widehat{f}_n - f_o \|_p \Big] , \tag{3.7}$$

again with $\varepsilon_n \longrightarrow 0$.

All of the above depend on the unknown regression function f_o. As already discussed in § 2, it is not unreasonable to make some nonparametric assumptions about f_o; e.g., that it belongs to some class $\mathfrak{F}$ of nice functions. A standard example is

$$\mathfrak{F}_C = \{\, f \in W^{m,2}(0,1) \,:\, \| f^{(m)} \| \leqslant C \,\} \tag{3.8}$$

for some known m and constant C. This certainly ought to facilitate the study of optimality. Another possibility now presents itself in the form of *minimax* estimation: Can one come close to realizing the minimax loss,

$$\inf_{\widehat{f}_n} \sup_{f_o \in \mathfrak{F}_C} \| \widehat{f}_n - f_o \|_p , \tag{3.9}$$

or perhaps the minimax risk, by first taking expectations. See, e.g., BARRON, BIRGÉ, and MASSART (1999), EFROMOVICH (1999), and references therein. Unfortunately, we shall not be concerned with this.

This is the right time to mention the Bayesian point of view. Rather than assuming that the constant C in (3.8) is known, it seems more reasonable to consider the regression function to be a random element of $W^{m,2}(0,1)$ and prescribe a distribution for f_o. Of course, one hopes that the particular distribution does not affect the estimator too much. The classic distribution for f_o is the diffuse prior of KIMELDORF and WAHBA (1970a, 1970b). So then it makes sense to replace the supremum over $\mathfrak{F}$ by the expectation over $W^{m,p}(0,1)$ according to this (improper) distribution,

$$\min_{\widehat{f}_n} \mathbb{E}_{f_o} \big[\| \widehat{f}_n - f_o \|_p \big] \tag{3.10}$$

(with $p = 2$, presumably). Unfortunately, for the diffuse prior, this explanation is somewhat suspect. See § 6 and Chapter 20. For the modern implementation of the Bayesian point of view, see, e.g., GRIFFIN and STEEL (2008).

It is clear that it is next to impossible to establish optimality in the small-sample case, except perhaps by way of simulation studies. It is instructive to contrast the present nonparametric situation with that of linear regression or parametric density estimation. There, the central concept is that of unbiased minimum variance estimators of the parameters in the context of exponential families. It is very unfortunate that this has no analogue in nonparametric regression: *In nonparametric problems, there are no unbiased estimators with finite variance.* Fortunately, by allowing estimators with a small bias, one can get finite or even small variances, to the overall effect of having a small mean squared error, say. (Unfortunately, this has implications for the practical use of asymptotic distributions for questions of inference such as confidence bands, which assume that the bias is negligible. Fortunately, it can be overcome by undersmoothing. See § 7 and § 23.5.)

Typically, the construction of such biased estimators involves a smoothing parameter, such as h in the smoothing spline estimator (2.23) or in the local polynomial estimator (2.36). This smoothing parameter provides for a trade-off between bias and variance of the estimator(s). With f^{nh} denoting such an estimator, one typically has for the global variance, regardless of the regression function, that

$$\mathbb{E}[\, \| f^{nh} - \mathbb{E}[f^{nh}] \|^2 \,] = \mathcal{O}(\,(nh)^{-1}\,) \,, \tag{3.11}$$

for $h \to 0$, $nh \to \infty$, and for the global bias if $f_o \in W^{m,2}(0,1)$,

$$\| f_o - \mathbb{E}[f^{nh}] \|^2 = \mathcal{O}(\, h^{2m}\,) \,. \tag{3.12}$$

Combining the two results gives

$$\mathbb{E}[\, \| f^{nh} - f_o \|^2 \,] = \mathcal{O}(\, h^{2m} + (nh)^{-1}\,) \,, \tag{3.13}$$

and one gets the smallest order by choosing $h \asymp n^{-1/(2m+1)}$. Then,

$$\min_{h>0} \mathbb{E}[\, \| f^{nh} - f_o \|^2 \,] = \mathcal{O}(\, n^{-2m/(2m+1)}\,) \,. \tag{3.14}$$

In the next section and the remainder of this text, this gets treated in great detail.

The question now arises as to whether (3.14) is indeed the best possible rate. Note that things would get considerably more complicated if instead of

$$\min_{h>0} \mathbb{E}\Big[\, \| f^{nh} - f_o \|^2 \,\Big]$$

one had to consider

$$\mathbb{E}\Big[\, \min_{h>0} \| f^{nh} - f_o \|^2 \,\Big] \; !$$

However, it does not matter. For the measures of the accuracy that interest us in this volume STONE (1980, 1982) has shown that if $f_o \in W^{m,2}(0,1)$, then the best possible rate is

$$\min_{\widehat{f}_n} \mathbb{E}\Big[\, \| \widehat{f}_n - f_o \|^2 \,\Big] = \mathcal{O}(\, n^{-2m/(2m+1)}\,) \,, \tag{3.15}$$

and if $f_o \in W^{m,\infty}(0,1)$, the best possible rate for the uniform error is

$$\| \widehat{f}_n - f_o \|_\infty = \mathcal{O}(\,(\, n^{-1} \log n\,)^{m/(2m+1)}\,) \quad \text{almost surely} \,. \tag{3.16}$$

In fact, under very weak conditions on the design and the density of the noise, STONE (1982) proves it in the following form. First, when specialized to our setting, he shows for $f_o \in W^{m,p}(0,1)$, $1 \leqslant p < \infty$, and a suitably small constant $c > 0$ that

$$\lim_{n\to\infty} \inf_{\widehat{f}_n} \sup_{f_o \in \mathfrak{F}_C} \mathbb{P}\Big[\, \| \widehat{f}_n - f_o \|_{L^p(0,1)} \geqslant c\, n^{-m/(2m+1)} \,\Big] = 1 \,, \tag{3.17}$$

and then he exhibits an estimator such that for another constant $c > 0$,

$$\lim_{n\to\infty} \sup_{f_o \in \mathfrak{F}_C} \mathbb{P}\Big[\, \| \widehat{f}_n - f_o \|_{L^p(0,1)} \geqslant c\, n^{-m/(2m+1)} \,\Big] = 0 \,. \tag{3.18}$$

Thus, (3.17) says that the rate $n^{-m/(2m=1)}$ cannot be improved; (3.18) says that it is an achievable rate. For $p = \infty$, the rates are to be replaced by $(n^{-1} \log n)^{-m/(2m+1)}$. On the one hand, (3.17) and (3.18) are stronger than (3.15)–(3.16), since it is uniformly over classes $\mathfrak{F}_C$, but on the other hand, one cannot get results on the expectations merely from (3.17)–(3.18). However, we shall let that pass. For other optimality results, see SPECKMAN (1985) and YATRACOS (1988). For the smoothing spline estimator, NUSSBAUM (1985) even calculates the constant, implicit in the $\mathcal{O}$ term in (3.12), and shows that this constant (the PINSKER (1980) constant) is best possible.

It is noteworthy that in the modern approach to optimality results, as, e.g., in IBRAGIMOV and HAS'MINSKII (1982), one studies the white noise analogue of the regression model (2.7),

$$(3.19) \qquad Y(x)\,dx = f_o(x)\,dx + \lambda(x)\,dW(x)\ , \quad x \in [\,0\,,\,1\,]\ ,$$

where $W(x)$ is the standard Wiener process, and $\lambda(x)$ is a nice function (incorporating the variance of the noise and the design density) and one uses the asymptotic equivalence of the two models. See, e.g., NUSSBAUM (1985), BROWN, CAI, and LOW (1996), and BROWN, CAI, LOW, and ZHANG (2002) for a precise statement. The equation (3.19) must be interpreted in the weak L^2 sense; i.e., (3.19) is equivalent to

$$(3.20) \qquad \langle\, Y\,,\,g\,\rangle = \langle\, f_o\,,\,g\,\rangle + U(g) \quad \text{for all} \quad g \in L^2(0,1)\ ,$$

where $U(g)$, $g \in L^2(0,1)$, are jointly normal with mean 0 and covariance

$$(3.21) \qquad \mathbb{E}\big[\,U(f)\,U(g)\,\big] = \int_0^1 f(x)\,g(x)\,\lambda^2(x)\,dx\ , \quad f,\,g \in L^2(0,1)\ .$$

The study of white noise regression (3.19) originated with PINSKER (1980). The equivalence between the two regression models was shown only later. Needless to say, we shall not study this further. However, in § 8 and Chapter 22, we do discuss the equivalence for smoothing spline estimators and connect it with the asymptotic distribution theory of $\|\, f^{nh} - f_o \,\|_\infty$ and $\max_{1\leqslant i\leqslant n} |\, f^{nh}(x_{in}) - f_o(x_{in})\,|$.

As a final remark on the white noise model (3.19), we note that it has its drawbacks; e.g., notions such as residual sums of squares do not have an analogue.

In the remainder of this text, we typically assume that $f_o \in W^{m,p}(0,1)$ for suitable m and p, and prove the optimal rate of convergence, but taking the optimality for granted. In a few cases, the techniques used do not yield the optimal rate; e.g., for the total-variation penalized least-squares estimator in Chapter 17.

4. Smoothing splines and reproducing kernels

The connection between smoothing splines and reproducing kernel Hilbert spaces is well-known and originated in the treatment of state-space models for time series data. It can be traced back at least as far as LOÈVE (1948) via PARZEN (1961). For the present context, see WAHBA (1978). In this approach, one puts a "diffuse" prior distribution on the regression function in the form

$$f^{(m)}(t) = \lambda \frac{dW(t)}{dt} \,, \quad t \in [0, 1] \,, \tag{4.1}$$

where $W(t)$ is the standard Wiener process. We are not interested in this, *except* that it leads to efficient ways to compute smoothing splines by way of the Kalman filter; see Chapter 20.

Here, we take an applied mathematics viewpoint: We push the probability problems ahead of us until they become "simple" and their solutions are "well-known". The quotation marks are necessary since the probability theory we rely on is of recent vintage: DEHEUVELS and MASON (2004)and EINMAHL and MASON (2005). So, consider the smoothing spline estimation problem (2.23). The first problem is whether the objective function is continuous. This amounts to the expression $f(x_{in})$ making sense. Surely, if $f \in W^{m,2}(0,1)$, then f is a continuous function, so we know what $f(x_{in})$ means. Moreover, if x_{in} changes a little, then so does $f(x_{in})$. However, since we want to minimize over f, this is the wrong kind of continuity: We need $f(x_{in})$ to change a little when f changes a little. The Sobolev embedding theorem (see, e.g., ADAMS (1975) or MAZ'JA (1985)) tells us that, for each integer $m \geqslant 1$, there exists a constant c_m such that, for all $x \in [0, 1]$ and for all $f \in W^{m,2}(0,1)$,

$$| f(x) | \leqslant c_m \| f \|_{W^{m,2}(0,1)} \,, \tag{4.2}$$

where the standard norm $\| \cdot \|_{W^{m,2}(0,1)}$ on $W^{m,2}(0,1)$ is defined as

$$\| f \|_{W^{m,2}(0,1)} = \left\{ \| f \|^2 + \| f^{(m)} \|^2 \right\}^{1/2} . \tag{4.3}$$

Inspection of the smoothing spline problems suggests that the "proper" norms on $W^{m,2}(0,1)$ should be

$$\| f \|_{m,h} = \left\{ \| f \|^2 + h^{2m} \| f^{(m)} \|^2 \right\}^{1/2} , \tag{4.4}$$

with the associated inner products $\langle \cdot , \cdot \rangle_{m,h}$ defined by

$$\langle f , g \rangle_{m,h} = \langle f , g \rangle + h^{2m} \langle f^{(m)} , g^{(m)} \rangle \,. \tag{4.5}$$

Here, $\langle \cdot , \cdot \rangle$ is the usual $L^2(0,1)$ inner product on $L^2(0,1)$,

$$\langle f , g \rangle = \int_0^1 f(t)\, g(t)\, dt \,. \tag{4.6}$$

A simple scaling argument shows that (4.2) implies that, for all $0 < h \leqslant 1$, all $x \in [0, 1]$, and all $f \in W^{m,2}(0,1)$,

$$| f(x) | \leqslant c_m \, h^{-1/2} \| f \|_{m,h} \,, \tag{4.7}$$

with the same constant c_m as in (4.2). In § 13.2, we prove this from scratch.

(4.8) EXERCISE. Prove (4.7) starting from (4.2). [Hint: Let $0 < h \leqslant 1$, and let $f \in W^{m,2}(0,1)$. Define $g(x) = f(\,h\,x\,)$, $x \in [\,0\,,\,1\,]$. Then, obviously, $g \in W^{m,2}(0,1)$ and one checks that (4.2) applied to g reads in terms of f as

$$| \, f(\,h\,x\,) \,| \leqslant c_m \left\{ h^{-1} \int_0^h | \, f(\,t\,) \,|^2 \, dt + h^{2m-1} \int_0^h | \, f^{(m)}(\,t\,) \,|^2 \, dt \right\}^{1/2} .$$

Now, we surely may extend the integration in the integrals on the right to all of $[\,0\,,\,1\,]$, so that this shows that (4.7) holds for all $x \in [\,0\,,\,h\,]$. Then, it should be a small step to get it to work for all $x \in [\,0\,,\,1\,]$.]

Now, (4.7) says that the linear functionals $\ell_x : W^{m,2}(0,1) \to \mathbb{R}$,

$$\ell_x(\,f\,) = f(x)\;, \quad f \in W^{m,2}(0,1)\;, \tag{4.9}$$

are continuous, and so may be represented as inner products. Thus, there exists a function (the reproducing kernel) $\mathcal{R}_{mh}(x,y)$, $x, y \in [\,0\,,\,1\,]$, such that $\mathcal{R}_{mh}(x,\,\cdot\,) \in W^{m,2}(0,1)$ for every $x \in [\,0\,,\,1\,]$, and the reproducing kernel Hilbert space trick works:

$$f(x) = \langle\, f\,,\, \mathcal{R}_{mh}(x,\,\cdot\,)\,\rangle_{m,h} \tag{4.10}$$

for all $x \in [\,0\,,\,1\,]$ and all $f \in W^{m,2}(0,1)$. Moreover,

$$\|\, \mathcal{R}_{mh}(x,\,\cdot\,)\,\|_{m,h} \leqslant c_m \, h^{-1/2}\;, \tag{4.11}$$

with the same constant c_m as in (4.7).

The reproducing kernel Hilbert space trick turns out to be tremendously significant for our analysis of smoothing splines. To begin with, it implies the existence of smoothing splines (uniqueness is no big deal). For now, we take this for granted. Denote the solution of (2.23) by f^{nh}, and let

$$\varepsilon^{nh} = f^{nh} - f_o\;. \tag{4.12}$$

It is a fairly elementary development to show that

$$\frac{1}{n} \sum_{i=1}^{n} |\, \varepsilon^{nh}(x_{in})\,|^2 + h^{2m} \,\|\, (\varepsilon^{nh})^{(m)} \,\|^2 = S_n(\,\varepsilon^{nh}\,) - h^{2m} \,\langle\, f_o^{(m)}\,,\, (\varepsilon^{nh})^{(m)} \,\rangle\;, \tag{4.13}$$

where, for $f \in W^{m,2}(0,1)$,

$$S_n(\,f\,) \stackrel{\text{def}}{=} \frac{1}{n} \sum_{i=1}^{n} d_{in} \, f(x_{in})\;. \tag{4.14}$$

See §§ 13.3 and 13.4. In nonparametric estimation, this goes back at least as far as VAN DE GEER (1987) and in applied mathematics, at least as far as RIBIÈRE (1967). Now, the left-hand side of (4.13) is just about the same as $\| \varepsilon^{nh} \|_{m,h}^2$ (but that requires proof) and

$$- \langle f_o^{(m)}, (\varepsilon^{nh})^{(m)} \rangle \leqslant \| f_o^{(m)} \| \, \| (\varepsilon^{nh})^{(m)} \| \leqslant h^{-m} \| f_o^{(m)} \| \, \| \varepsilon^{nh} \|_{m,h} ,$$

so, if we can establish a bound of the form

$$(4.15) \qquad S_n(\varepsilon^{nh}) \leqslant \eta_{nh} \| \varepsilon^{nh} \|_{m,h} ,$$

with (to relieve the suspense) $\eta_{nh} \asymp (nh)^{-1/2}$ in a suitable sense, then (4.13) would show that

$$(4.16) \qquad \| \varepsilon^{nh} \|_{m,h} \leqslant \eta_{nh} + h^m \| f_o^{(m)} \| ,$$

and we are in business.

The difficulty in establishing (4.15) is that ε^{nh} is random; i.e., ε^{nh} is a function of the d_{in}. For a fixed, deterministic f, one would have that

$$(4.17) \qquad | S_n(f) | \leqslant \zeta_n \left\{ \tfrac{1}{n} \sum_{i=1}^{n} | f(x_{in}) |^2 \right\}^{1/2} ,$$

with $\zeta_n \asymp n^{-1/2}$ in a suitable sense.

To show (4.15), we use the reproducing kernel Hilbert space trick (4.10). Since $\varepsilon^{nh} \in W^{m,2}(0,1)$, then, for all $t \in [\,0\,,\,1\,]$,

$$\varepsilon^{nh}(t) = \langle \varepsilon^{nh} , \mathcal{R}_{mh}(t, \cdot) \rangle_{m,h} ,$$

and so

$$(4.18) \quad \tfrac{1}{n} \sum_{i=1}^{n} d_{in} \, \varepsilon^{nh}(x_{in}) = \langle \varepsilon^{nh} , \mathfrak{S}_{nh} \rangle_{m,h} \leqslant \| \varepsilon^{nh} \|_{m,h} \| \mathfrak{S}_{nh} \|_{m,h} ,$$

with

$$(4.19) \qquad \mathfrak{S}_{nh}(t) = \tfrac{1}{n} \sum_{i=1}^{n} d_{in} \, \mathcal{R}_{mh}(x_{in}, t) , \quad t \in [\,0\,,\,1\,] .$$

Now, it is easy to show that, under the assumptions (2.8) on the noise,

$$(4.20) \qquad \mathbb{E}[\, \| \mathfrak{S}_{nh} \|^2 \,] \leqslant c_m^2 \, \sigma^2 \, (nh)^{-1} ,$$

so that with (4.16)

$$(4.21) \qquad \mathbb{E}[\, \| f^{nh} - f_o \|^2 \,] = \mathcal{O}((nh)^{-1} + h^{2m}) .$$

For $h \asymp n^{-1/(2m+1)}$, this implies the optimal rate (see § 3)

$$(4.22) \qquad \mathbb{E}[\, \| f^{nh} - f_o \|^2 \,] = \mathcal{O}(n^{-2m/(2m+1)}) .$$

So, these expected convergence rates were relatively easy to obtain. More sophisticated results require other methods, but probabilists typically consider random sums like (4.19) with convolution kernels only, and with compact support at that. DEHEUVELS (2000) is one exception. It should be noted that the kernels $\mathcal{R}_{mh}$ have very useful convolution-like properties, which we prove, and make heavy use of, in Chapters 14 and 21.

The significance of the reproducing kernels does not end here. It turns out that the inequality (4.18) is just about sharp, to the effect that the smoothing spline estimator is "equivalent" to φ^{nh} defined by

$$(4.23)\qquad \varphi^{nh}(t) = \int_o^1 \mathcal{R}_{mh}(t,x)\, f_o(x)\, dx + \mathfrak{S}_{nh}(t)\ ,\quad t\in[0,1]\ ,$$

in the sense that, e.g.,

$$(4.24)\qquad \mathbb{E}[\,\|\, f^{nh} - \varphi^{nh}\,\|_{m,h}^2\,] = \mathcal{O}(\,h^{4m} + (nh)^{-2}\,)\ .$$

Note that this is the *square* of the bound on $\mathbb{E}[\,\|\, f^{nh} - f_o\,\|_{m,h}^2\,]$! Following SILVERMAN (1984), we should call $\mathcal{R}_{mh}$ the equivalent *reproducing* kernels, but we shall usually just call them equivalent kernels, although they are not kernels in the sense of (convolution) kernel estimators. See Remark (8.30).

The equivalence $f^{nh} \approx \varphi^{nh}$ comes about because φ^{nh} is the solution of the C-spline problem

$$(4.25)\qquad \begin{array}{ll} \text{minimize} & \|\, f - f_o\,\|^2 \\ & \displaystyle -\ \tfrac{2}{n}\sum_{i=1}^{n} d_{in}\,(\,f(x_{in}) - f_o(x_{in})\,) + h^{2m}\,\|\, f^{(m)}\,\|^2 \\ \text{subject to} & f\in W^{m,2}(0,1)\ , \end{array}$$

and this is very close to the original smoothing spline problem (2.23). (The integral $\|\, f - f_o\,\|^2$ behaves like the sum $\frac{1}{n}\sum_{i=1}^{n}|\, f(x_{in}) - f_o(x_{in})\,|^2$.) Moreover, with obvious adjustments, this works for nice random designs as well. See § 8, § 14.7, and Chapter 21 for the fine print. As a rather weak consequence, this would give that

$$(4.26)\qquad \mathbb{E}[\,\|\, f^{nh} - \varphi^{nh}\,\|_\infty^2\,] = \mathcal{O}(\,h^{4m-1} + h^{-1}(nh)^{-2}\,)\ .$$

So, this leads to the representation $f^{nh} \approx \mathbb{E}[\,f^{nh}\,] + \mathfrak{S}_{nh}$ in the uniform norm and may be used to derive uniform error bounds on the spline estimator. In particular, this gives

$$(4.27)\qquad \|\, f^{nh} - f_o\,\|_\infty = \mathcal{O}(\,h^m + \sqrt{(nh)^{-1}\log n}\,)\quad \text{in probability}\ ,$$

essentially uniformly in h, but consult § 14.7 for the precise details. The bound (4.27) gives rise to the rate $(n^{-1}\log n)^{m/(2m+1)}$, which is again optimal, as discussed in § 3.

(4.28) REMARK. The probabilistic content of the smoothing spline problem (4.23) is the behavior of the random sums $S_n(\,f^{nh} - f_o\,)$, which we reduced to the study of the sums $\mathfrak{S}_{nh}$ by way of the (in)equalities

$$S_n(\,f^{nh} - f_o\,) = \langle\,\mathfrak{S}_{nh}\,,\, f^{nh} - f_o\,\rangle_{m,h} \leqslant \|\,\mathfrak{S}_{nh}\,\|_{m,h}\,\|\, f^{nh} - f_o\,\|_{m,h}\ .$$

This seems a useful way of isolating the probability considerations. The "modern" treatment of the random sums involves the notion of Kolmogorov or metric entropy. Its use in probability was pioneered by DUDLEY (1978).

In statistics, it goes back at least as far as YATRACOS (1985) and VAN DE GEER (1987). See also VAN DE GEER and WEGKAMP (1996). The idea is to consider the supremum of $|S_n(f)|$ over suitable classes of functions $\mathfrak{F}$. In the present setting, this seems to take the form

$$\mathfrak{F} = \left\{ f \in W^{m,2}(0,1) \;\middle|\; \| f \|_\infty \leqslant K_1 \;,\; \| f^{(m)} \| \leqslant K_2 \right\}$$

for suitable K_1 and K_2. The goal is to give probability bounds on

$$\sup_{f \in \mathfrak{F}} |S_n(f)| / (K_1^{\alpha} K_2^{1-\alpha})$$

for suitable α (dependent on m). Obviously, the "size" of the sets $\mathfrak{F}$ comes into play, and the Kolmogorov entropy seems to capture it. The methods of Kolmogorov entropy come into their own in nonlinear or non-Hilbert space settings, as well as for *minimax* estimation. We leave it at this. See, e.g., VAN DE GEER (2000), DEVROYE, GYÖRFI, and LUGOSI (1996), DEVROYE and LUGOSI (2000b), and GYÖRFI, KOHLER, KRZYŻAK, and WALK (2002). We note that the precise probabilistic treatment of the sums $\mathfrak{S}_{nh}$ also seems to require notions of Kolmogorov entropy or the related notion of Vapnik-Chervonenkis classes; see EINMAHL and MASON (2005). An imprecise treatment (but fortunately accurate enough for our purpose) does not require it; see § 14.6.

(4.29) REMARK. In outline, this section is quite similar to the discussion of linear smoothing splines in COX (1981). However, the details are completely different, so as to make the overall result almost unrecognizable. It should be pointed out that the traditional approach to smoothing splines is by way of a detailed and precise study of the eigenvalues and eigenfunctions of the hat matrix; see (6.8). The determination of the asymptotic behavior of the eigenvalues of these matrices is by no means trivial; see COX (1988a, 1988b), UTRERAS (1981), COX and O'SULLIVAN (1990), NUSSBAUM (1985), and GOLUBEV and NUSSBAUM (1990). In this respect, the modern approach by way of the white noise model (3.19) is "easier" because then one must study the eigenvalues of the two-point boundary value problem

$$\begin{aligned} &(-h^2)^m \, u^{(2m)} + u = f \quad \text{on} \quad (0,1) \;, \\ &u^{(k)}(0) = u^{(k)}(1) = 0 \quad \text{for} \quad k = m, \cdots, 2m-1 \;. \end{aligned}$$

OUDSHOORN (1998) quotes DUISTERMAAT (1995), who gives the final word on the spectral analysis of this problem. Note that the Green's function for this boundary value problem is precisely the reproducing kernel $\mathcal{R}_{mh}$! We also note that OUDSHOORN (1998) studies piecewise smooth regression, with a finite (small) number of jumps in the regression function, and after carefully estimating the jump locations obtains the usual convergence rates with the optimal PINSKER (1980) constant; i.e., as if there were no jumps!

EXERCISE: (4.8).

5. The local error in local polynomial estimation

Does the approach to smoothing splines outlined in § 3 extend to the study of other estimators? Not surprisingly, it works for smoothing spline estimators in the time-varying coefficients problem (1.19); see EGGERMONT, EUBANK, and LARICCIA (2005). In Chapter 17, we show that the answer is yes for the least-absolute-deviations smoothing spline of (2.30) and almost yes for the total-variation penalized least-squares estimators induced by (2.26). In the latter case, it is "almost yes" because we miss the optimal rate by a power of $\log n$. In this section, the implications of this approach to local polynomial estimation are briefly discussed. We consider the "uniform" deterministic design, but the same ideas apply to random designs.

Let $r \geqslant 0$ be a fixed integer, and let $\mathcal{P}_r$ denote the vector space of all polynomials of order $\leqslant r$ (degree $\leqslant r-1$). For fixed $t \in [0, 1]$, consider the problem (2.36), repeated in the form

$$\text{(5.1)} \qquad \begin{aligned} &\text{minimize} \quad \tfrac{1}{n} \sum_{i=1}^{n} A_h(x_{in} - t) \, | \, p(x_{in}) - y_{in} \, |^2 \\ &\text{subject to} \quad p \in \mathcal{P}_r \, . \end{aligned}$$

Here, A is a nice pdf on the line, and $A_h(t) = h^{-1} A(h^{-1} t)$ for all t. One needs to assume that

$$\text{(5.2)} \qquad \begin{aligned} &\sup_{t \in \mathbb{R}} \, (1 + t^{2r}) A(t) < \infty \quad \text{and} \\ &A > 0 \text{ on some open interval } (-\delta, \delta) \text{ with } \delta > 0 \, . \end{aligned}$$

Proceeding in the programmatic (dogmatic ?) approach of § 3, one should ask whether the objective function in (5.1) is a continuous function of p on $\mathcal{P}_r$. Ignoring the obvious answer (yes!), this raises the question of which norm to select for $\mathcal{P}_r$. In view of the objective function and the principle that sums turn into integrals, the choice

$$\text{(5.3)} \qquad \| \, p \, \|_{A,h,\,t} = \Big\{ \int_0^1 A_h(x - t) \, | \, p(x) \, |^2 \, dx \Big\}^{1/2} \, ,$$

together with the associated inner product, denoted by $\langle \cdot, \cdot \rangle_{A,h,t}$, suggests itself. Then, the continuity of the objective function is established if there exists an inequality of the form

$$| \, p(x_{in}) \, | \leqslant c \, \| \, p \, \|_{A,h,\,t} \, ,$$

for a constant "c" which does not depend on x_{in}, t, and h. Of course, the inequality does not make much sense for $| \, x_{in} - t \, | \gg h$, so surely the proposed inequality must be modified to something like

$$(5.4)\qquad A_h(x_{in} - t)\,|\,p(x_{in})\,|^2 \leqslant c\,h^{-1}\,\|\,p\,\|_{A,h,t}^2 \ .$$

At this stage, the factor h^{-1} on the right is a wild but reasonable guess. Now, using the common representation of polynomials,

$$p(x) = \sum_{q=0}^{r-1} p^{(q)}(\,t\,)\,(\,x - t\,)^q/q!\ ,$$

the inequality (5.4) would hold if for another suitable constant

$$\sum_{q=0}^{r-1} A_h(x_{in} - t)\,|\,p^{(q)}(\,t\,)\,|^2 |\,x_{in} - t\,|^2 \leqslant c\,h^{-1}\,\|\,p\,\|_{A,h,t}^2 \ .$$

With the first part of the assumption (5.2), this would hold if: *There exists a constant* c_r *such that, for all* $0 < h \leqslant \frac{1}{2}$, *all* $t \in [\,0\,,\,1\,]$, *and all* $p \in \mathcal{P}_r$,

$$(5.5)\qquad h^q\,|\,p^{(q)}(\,t\,)\,| \leqslant c_r\,\|\,p\,\|_{A,h,t}\ ,\qquad q = 0, 1, \cdots, r-1\ .$$

This is indeed the case; see § 16.2.

The programmatic approach now says that we should conclude from (5.5) (for $q = 0$) that $\mathcal{P}_r$ with the inner product $\langle\,\cdot\,,\,\cdot\,\rangle_{A,h,t}$ is a reproducing kernel Hilbert space, but that seems a bit of overkill.

Thus, leaving the programmatic part for a moment, and ignoring questions of uniqueness (solutions always exist), let $p^{n,r,h}(\,\cdot\,|\,t\,)$ denote the solution of (5.1). Now, ultimately, the error is defined as $p^{n,r,h}(\,t\,|\,t\,) - f_o(\,t\,)$, but initially it is better to define it as

$$(5.6)\qquad \varepsilon(x) = p^{n,r,h}(x\,|\,t\,) - T_r(x\,|\,t\,)\ ,\quad x \in [\,0\,,\,1\,]\ ,$$

where

$$(5.7)\qquad T_r(x\,|\,t\,) = \sum_{q=0}^{r-1} f_o^{(q)}(\,t\,)\,(\,x - t\,)^q/q!$$

is the Taylor polynomial of order r of f_o around the point t. So, in particular, $T_r(\,t\,|\,t\,) = f_o(\,t\,)$. Then, as in (4.13), one derives the equality

$$(5.8)\qquad \frac{1}{n}\sum_{i=1}^{n} A_h(x_{in} - t)\,|\,\varepsilon(x_{in})\,|^2 = \frac{1}{n}\sum_{i=1}^{n} A_h(x_{in} - t)\,d_{in}\,\varepsilon(x_{in}) + \frac{1}{n}\sum_{i=1}^{n} A_h(x_{in} - t)\,\big(\,f_o(x_{in}) - T_r(x_{in}\,|\,t\,)\big)\,\varepsilon(x_{in})\ .$$

The expression on the left is the discrete *local error*; the continuous version suggests itself. Now, one has to bound the two sums on the right of (5.8). The last sum corresponds to the bias of the estimator. For the first sum, one indeed uses that $\mathcal{P}_r$ with the advertised inner product is a reproducing kernel Hilbert space without having to acknowledge it. All of this works and leads (presumably) to the bound

$$(5.9)\qquad \mathbb{E}[\,\|\,\varepsilon\,\|_{A,h,t}^2\,] = \mathcal{O}\big(\,(nh)^{-1} + h^{2r}\,\big)\ .$$

Unfortunately, it appears that we have solved the wrong problem! While it is true that the above development leading to (5.9) works quite nicely, is the real interest not in the pointwise error

$$p^{n,r,h}(t\,|\,t) - f_o(t) = p^{n,r,h}(t\,|\,t) - T_r(t\,|\,t)\ ?$$

The answer is yes, and our bacon is saved by the reproducing kernel inducing inequality (5.5), which implies that

$$|\,\varepsilon(t)\,| \leqslant c\,\|\,\varepsilon\,\|_{A,h,t}\ , \tag{5.10}$$

and we get pointwise error bounds (and bounds on global measures of the pointwise error).

At this stage, the point is clear: The programmatic approach here leads to the notion of the local error. We should add that the local error is a much nicer object than the pointwise error, especially when selecting the smoothing parameter. In § 18.9, we show that selecting the optimal h for the (global) pointwise error is the "same" as selecting it for the (global measures) of the local error, in the style of FAN and GIJBELS (1995a).

6. Computation and the Bayesian view of splines

The computation of sieved and local polynomial estimators is relatively straightforward, although the straightforward methods are relatively hazardous. See Chapter 19 for the details. Here we highlight the computation of smoothing splines. We discuss two methods, the standard one based on spline interpolation, and the other based on Kalman filtering.

Spline smoothing and interpolation. It is perhaps surprising that the solution of the smoothing spline problem

$$\begin{array}{ll} \text{minimize} & \frac{1}{n}\sum\limits_{i=1}^{n} |\,f(x_{in}) - y_{in}\,|^2 + h^{2m}\,\|\,f^{(m)}\,\|^2 \\ \text{subject to} & f \in W^{m,2}(0,1) \end{array} \tag{6.1}$$

can be computed exactly. One explanation is that the solution of (6.1) is a *natural polynomial spline* of order $2m$, which is uniquely determined by its values at the design points. For now, we assume that the design points are distinct. Thus, if we let

$$\widehat{y}^{nh} = \bigl(\,f^{nh}(x_{1,n}),\,f^{nh}(x_{2,n}),\,\cdots,f^{nh}(x_{n,n})\,\bigr)^T\ , \tag{6.2}$$

then $f^{nh}(x)$, $x \in [\,0\,,\,1\,]$, is the solution to the spline interpolation problem

$$\begin{array}{ll} \text{minimize} & \|\,f^{(m)}\,\|^2 \\ \text{subject to} & f \in W^{m,2}(0,1)\ , \\ & f(x_{in}) = [\,\widehat{y}^{nh}\,]_i,\ i = 1,2,\cdots,n\ . \end{array} \tag{6.3}$$

The solution of (6.3) is a piecewise-polynomial function of order $2m$ (degree $\leqslant 2m-1$) with so-called knots at the design points $x_{1,n}, x_{2,n}, \cdots, x_{n,n}$, i.e., in the common representation of polynomials,

$$f(x) = \sum_{j=0}^{2m-1} a_{ij} (x - x_{in})^j , \quad x \in [x_{in}, x_{i+1,n}] , \tag{6.4}$$

for $i = 1, 2, \cdots, n-1$, and the pieces must connect smoothly in the sense that f should be $2m-2$ times continuously differentiable. (Note that the solution is smoother than the condition $f \in W^{m,2}(0,1)$ suggests, the more so as m gets larger.) This imposes conditions at the interior design points,

$$\lim_{x \to x_{in}-0} (f^{nh})^{(k)}(x) = \lim_{x \to x_{in}+0} (f^{nh})^{(k)}(x) , \quad i = 2, 3 \cdots, n-1 , \tag{6.5}$$

for $k = 0, 1, \cdots, 2m-2$, which translate into linear equations in the unknown coefficients a_{ij}. There are also the natural boundary conditions to be satisfied, but for now we shall leave it at that. Obviously, also $a_{i,0} = [\widehat{y}^{nh}]_i$. The net result is that the a_{ij} can be expressed linearly in terms of $\widehat{y}^{nh}$.

The solution of (6.1) being a polynomial spline, it follows that the minimization may be restricted to polynomial splines, and then $\| f^{(m)} \|^2$ is a quadratic form in $\widehat{y} = \big(f(x_{1,n}), f(x_{2,n}), \cdots, f(x_{n,n}) \big)^T$,

$$\| f^{(m)} \|^2 = \langle \widehat{y}, M \widehat{y} \rangle , \tag{6.6}$$

for a semi-positive-definite matrix M. In (6.6), $\langle \cdot, \cdot \rangle$ denotes the usual inner product on $\mathbb{R}^n$. The Euclidean norm on $\mathbb{R}^n$ is denoted by $\| \cdot \|$. So, the minimization problem (6.1) takes the form

$$\text{minimize} \quad \| \widehat{y} - y_n \|^2 + nh^{2m} \langle \widehat{y}, M\widehat{y} \rangle \quad \text{subject to} \quad \widehat{y} \in \mathbb{R}^n , \tag{6.7}$$

the solution of which is readily seen to be

$$\widehat{y}^{nh} = R_{nh}\, y_n , \tag{6.8}$$

in which R_{nh} is the so-called hat matrix

$$R_{nh} = (I + nh^{2m} M)^{-1} . \tag{6.9}$$

For $m \geqslant 2$, the matrix M is dense ("all" of its elements are nonzero) but can be expressed as $S T^{-1} S^T$, where S and T are banded matrices with roughly $2m+1$ nonzero bands. For $m = 1$, this is very simple (see below); for $m = 2$, some nontrivial amount of work is required; for $m = 3$ and beyond, it becomes painful. See §§ 19.3 and 19.5.

The case $m = 1$ is instructive for illustrating a second weakness of the approach above. In this case, the polynomial spline is of order 2 and is piecewise linear, so that for $x \in [x_{in}, x_{i+1,n}]$

$$f(x) = \frac{x_{i+1,n} - x}{\Delta_{in}} [\widehat{y}]_i + \frac{x - x_{in}}{\Delta_{in}} [\widehat{y}]_{i+1} , \tag{6.10}$$

for $i = 1, 2, \cdots, n-1$, and $f(x) = \widehat{y}_1$ for $x \in [\,0\,, x_{1,n}\,]$, and likewise on $[\,x_{n,n}\,, 1\,]$. Here, $\Delta_{in} = x_{i+1,n} - x_{in}$ are the spacings. Then, it is a simple exercise to show that

$$\| f' \|^2 = \langle\, \widehat{y}\,, M_1\, \widehat{y}\,\rangle \tag{6.11}$$

with a tridiagonal matrix M_1 : For $i, j = 1, 2, \cdots, n$

$$[\,M_1\,]_{i,j} = \begin{cases} -\,\Delta_{i-1,n}^{-1} & , \quad j = i-1\,, \\ \Delta_{i-1,n}^{-1} + \Delta_{in}^{-1} & , \quad j = i\,, \\ -\,\Delta_{in}^{-1} & , \quad j = i+1\,, \end{cases} \tag{6.12}$$

in which formally $\Delta_{0,n} = \Delta_{n,n} = \infty$. The unmentioned components of M_1 are equal to 0.

One purpose of the above is to show that the matrix M_1 is badly scaled (its components may be of different orders of magnitude) if some of the Δ_{in} are much smaller than the average, so that the matrix M_1, and consequently the matrix $I + nh^{2m} M_1$, becomes ill-conditioned. This implies that the computation of $\widehat{y}^{nh}$ in (6.8)–(6.9) may become numerically unstable. Note that this is a feature of the computational scheme since the smoothing spline problem itself is perfectly stable. Moreover, the numerical instability gets worse with increasing m. It should also be noted that, in the scheme above, independent replications in the observations need special though obvious handling. See, e.g., EUBANK (1999), Chapter 5, Exercise 4. See also Exercises (19.3.16) and (19.5.53). The way around all this comes from unsuspected quarters.

The Bayesian interpretation. A long time ago, KIMELDORF and WAHBA (1970a, 1970b) noticed that the smoothing spline problem (6.1) has the same solution as a standard estimation problem for a certain Gaussian process (estimate a sample path, having observed some points on the sample path subject to noise). This is based on the identification of linear estimation problems for Gaussian processes with penalized least-squares problems in the reproducing kernel Hilbert space naturally associated with the Gaussian process, due to LOÈVE (1948) and PARZEN (1961). See § 20.3. The pertinent Gaussian process, denoted by

$$X(x)\,, \quad x \geqslant 0\,, \tag{6.13}$$

has white noise as its m-th derivative,

$$X^{(m)}(x) = \kappa\, \frac{dW(x)}{dx}\,, \quad x \in [\,0\,, 1\,]\,. \tag{6.14}$$

Here, $W(x)$ is the standard Wiener process, and $\kappa > 0$ is a parameter, depending on the sample size. The Taylor expansion with exact remainder,

$$X(x) = \sum_{j=0}^{m-1} X^{(j)}(0)\, \frac{x^j}{j!} + \int_0^x \frac{(x-t)^{m-1}}{(m-1)!}\, X^{(m)}(\,t\,)\, dt\,, \tag{6.15}$$

shows that the distribution of $X(0), X'(0), X''(0), \cdots, X^{(m-1)}(0)$ still remains to be specified. Although it is possible to treat them as parameters to be estimated, usually one puts a so-called *diffuse prior* on them. This takes the form

$$(6.16) \qquad \bigl(X(0), X'(0), \cdots, X^{(m-1)}(0) \bigr)^T \sim \text{Normal}(0, \xi^2 D)$$

for some positive-definite matrix D, and one lets $\xi \to \infty$. (In this introductory section, we just keep ξ fixed.) The choice of D turns out to be immaterial. Moreover, $\bigl(X(0), X'(0), \cdots, X^{(m-1)}(0) \bigr)^T$ is assumed to be independent of the Wiener process in (6.14). Then, $X(x)$ is a Gaussian process with mean 0 and covariance $\mathbb{E}[X(x)X(y)] = R_{m,\xi,\kappa}(x,y)$,

$$(6.17) \qquad R_{m,\xi,\kappa}(x,y) = \xi^2 \vec{x}^T D \vec{y} + \kappa^2 V(x,y),$$

where $\vec{x} = (1, x, x^2/2!, \cdots, x^{m-1}/(m-1)!)^T$, and likewise for $\vec{y}$, and

$$(6.18) \qquad V(x,y) = \int_0^{x\wedge y} \frac{(x-t)^{m-1}}{(m-1)!} \frac{(y-t)^{m-1}}{(m-1)!} \, dt .$$

Reproducing kernels again. It is well-known that the covariance function of a stochastic process is the reproducing kernel of a Hilbert space. In the present case, $R_{m,\xi,\kappa}(x,y)$ is the reproducing kernel of the Sobolev space $W^{m,2}(0,1)$ with the inner product, for all $f, g \in W^{m,2}(0,1)$,

$$(6.19) \qquad \langle\langle f, g \rangle\rangle_{m,\xi,\kappa} \stackrel{\text{def}}{=} \xi^{-2} \sum_{j=0}^{m-1} f^{(j)}(0)\, g^{(j)}(0) + \kappa^{-2} \langle f^{(m)}, g^{(m)} \rangle ,$$

where now $\langle \cdot, \cdot \rangle$ denotes the $L^2(0,1)$ inner product. Denote the norm induced by (6.19) by $\|\cdot\|_{m,\xi,\kappa}$. One verifies that indeed

$$f(x) = \langle\langle f, R_{m,\xi,\kappa}(x, \cdot) \rangle\rangle_{m,\xi,\kappa}$$

for all $x \in [0,1]$ and all $f \in W^{m,2}(0,1)$, see § 20.3.

Estimation for stochastic processes. There are many estimation problems for stochastic processes. Here, consider the data smoothing problem

$$(6.20) \qquad \begin{array}{ll} \text{estimate} & X(x_{in}), \ i = 1, 2, \cdots, n, \\ \text{given} & y_{in} = X(x_{in}) + d_{in}, \ i = 1, 2, \cdots, n, \end{array}$$

with $d_{1,n}, d_{2,n}, \cdots, d_{n,n}$ iid normal noise, as in (2.11), independent of X. For the prediction problem, see (6.35) below.

To solve the data smoothing problem (6.20), observe that the prior joint distribution of the points on the sample path, i.e., of the vector

$$(6.21) \qquad X_n = \bigl(X(x_{1,n}), X(x_{2,n}), \cdots, X(x_{1,n}) \bigr)^T ,$$

is given by

$$(6.22) \qquad X_n \sim \text{Normal}(0, R),$$

with $R \in \mathbb{R}^{n\times n}$, and $R_{ij} = R_{m,\xi,\kappa}(x_{in}, x_{jn})$ for $i,j = 1,2,\cdots,n$. The posterior distribution given the data then follows, and the estimation problem (6.20) is equivalent to (has the same solution as)

$$\text{(6.23)} \qquad \text{minimize} \quad \sum_{i=1}^{n} |\, X(x_{in}) - y_{in}\,|^2 + \sigma^2 \langle\, X_n \,,\, R^{-1} X_n \,\rangle \,.$$

The minimizer satisfies $X_n - y_n + \sigma^2 R^{-1} X_n = 0$, or

$$X_n = \sigma^{-2} R\,(\, X_n - y_n \,)\,.$$

Thus, we may take $X_n = R\,Y_n$, where $Y_n = (\, Y_1, Y_2, \cdots, Y_n\,)^T$ is to be determined. We now take a big conceptual leap, and say that we search for solutions of the form

$$\text{(6.24)} \qquad f(x) = \sum_{i=1}^{n} Y_i \, R_{m,\xi,\kappa}(x_{in}, x)$$

for $x = x_{jn}$, $j = 1,2,\cdots,n$, but in (6.24) we consider the whole path. Note that $f \in W^{m,2}(0,1)$ is random. Now, an easy calculation shows that

$$\text{(6.25)} \qquad \begin{aligned} \langle\, X_n \,,\, R^{-1} X_n \,\rangle = \langle\, Y_n \,,\, R\,Y_n \,\rangle &= \sum_{i,j=1}^{n} Y_i\, Y_j\, R_{m,\xi,\kappa}(x_{in}, x_{jn}) \\ &= \Big\|\, \sum_{i=1}^{n} Y_i\, R_{m,\xi,\kappa}(x_{in}, \cdot\,) \,\Big\|_{m,\xi,\kappa}^2 = \|\, f \,\|_{m,\xi,\kappa}^2 \,. \end{aligned}$$

Thus, the estimation problem (6.20) is equivalent to

$$\text{(6.26)} \qquad \begin{aligned} &\text{minimize} \quad \sum_{i=1}^{n} |\, f(x_{in}) - y_{in}\,|^2 + \sigma^2 \,\|\, f \,\|_{m,\xi,\kappa}^2 \\ &\text{subject to} \quad f \in W^{m,2}(0,1) \end{aligned}$$

in the sense that the solution $f = \widehat{f}$ of (6.26) is also the solution of the estimation problem (6.20); i.e., $\widehat{X}(x) = \widehat{f}(x)$ for all $x \geqslant 0$.

It is now clear that, formally at least, letting $\xi \to \infty$ in (6.26) gives that

$$\|\, f \,\|_{m,\xi,\kappa}^2 \longrightarrow \kappa^{-2}\, \|\, f^{(m)} \,\|^2 \,,$$

and then (6.26) turns into the standard smoothing spline problem.

State-space models. So far, all of this has little bearing on computations, but now a new possibility emerges by exploiting the Markov structure of the process (6.14)–(6.16). First, introduce the state vector at "time" x for the Gaussian process,

$$\text{(6.27)} \qquad \mathbb{S}(x) = \big(\, X(x),\, X'(x) \,\cdots, X^{(m-1)}(x) \,\big)^T \,.$$

Now, let $x < y$. Substituting (6.14) into the Taylor expansion (6.15) gives

$$\text{(6.28)} \qquad X(y) = \sum_{j=1}^{m-1} X^{(j)}(x)\, \frac{(y-x)^j}{j!} + \kappa \int_x^y \frac{(\,y - t\,)^{m-1}}{(m-1)!}\, dW(\,t\,) \,,$$

and similarly for the derivatives $X^{(j)}(y)$. Then, for $i = 1,2,\cdots,n-1$,

$$\text{(6.29)} \qquad \mathbb{S}(x_{i+1,n}) = Q(x_{i+1,n}\,|\,x_{in})\, \mathbb{S}(x_{in}) + \mathbb{U}(x_{i+1,n}\,) \,,$$

with suitable deterministic *transition* matrices $Q(x_{i+1,n} \,|\, x_{in})$ and where the $\mathbb{U}(x_{in})$ are independent multivariate normals with covariance matrices $\Sigma_i = \mathbb{E}[\,\mathbb{U}(x_{in})\,\mathbb{U}(x_{in})^T\,]$ given by

$$(6.30)\qquad [\,\Sigma_i\,]_{k,\ell} = \int_{x_{in}}^{x_{i+1,n}} \frac{(x_{i+1,n} - t\,)^{m-k-1}}{(m-k-1)\,!}\,\frac{(x_{i+1,n} - t\,)^{m-\ell-1}}{(m-\ell-1)\,!}\,dt\;.$$

Now, in view of the equivalence between (6.20) and (6.26), the spline smoothing problem is equivalent to

$$(6.31)\qquad \begin{array}{ll} \text{estimate} & \mathbb{S}(x_{in})\,,\ i = 1, 2, \cdots, n \\ \text{given the data} & y_{in} = \mathrm{e}_1^T\,\mathbb{S}(x_{in}) + d_{in}\,,\ i = 1, 2, \cdots, n\;. \end{array}$$

Here, $\mathrm{e}_1 = (\,1,\,0,\,\cdots,0)^T \in \mathbb{R}^m$. Note that $\mathbb{S}(0)$ has the prior (6.16). Now, analogous to (6.23), one can formally derive the maximum *a posteriori* likelihood problem, yielding

$$(6.32)\qquad \begin{array}{ll} \text{minimize} & \displaystyle\sum_{i=1}^{n} \frac{|\,\mathrm{e}_1^T\,\mathbb{S}(x_{in}) - y_{in}\,|^2}{\sigma^2} + \big\langle\,\mathbb{S}\,,\,\big(\,\xi^2\,\mathbb{D}_n + \kappa^2\,\mathbb{V}_n\,\big)^{-1}\,\mathbb{S}\,\big\rangle \\ \text{subject to} & \mathbb{S} \in \mathbb{R}^{mn} \end{array}$$

for the appropriate matrices $\mathbb{D}_n$ and $\mathbb{V}_n$, and where

$$\mathbb{S} = \big(\,\mathbb{S}(x_{1,n})^T, \mathbb{S}(x_{2,n})^T, \cdots, \mathbb{S}(x_{n,n})^T\,\big)^T\,.$$

Note that the solution is given by

$$(6.33)\qquad \widehat{\mathbb{S}} = \big(\,E + (\,\widetilde{\xi}^2\,\mathbb{D}_n + \widetilde{\kappa}^2\,\mathbb{V}_n\,)^{-1}\,\big)^{-1}\,\mathbb{Y}_n\;,$$

where $\mathbb{Y}_n = \big(\,y_{1,n}\,\mathrm{e}_1^T, y_{2,n}\,\mathrm{e}_1^T, \cdots, y_{n,n}\,\mathrm{e}_1^T\,\big)^T$, and E is block-diagonal with diagonal blocks $\mathrm{e}_1\mathrm{e}_1^T$.

(6.34) REMARK. Note that in (6.32) all the components of the $\mathbb{S}(x_{in})$ are required for the computation of the spline function $f(x)$: On $[\,x_{in}\,,\,x_{i+1,n}\,]$, find the polynomial of order $2m$ such that

$$p^{(j)}(x_{in}) = [\,\mathbb{S}(x_{in})\,]_j \quad , \quad p^{(j)}(x_{i+1,n}) = [\,\mathbb{S}(x_{i+1,n})\,]_j\;,$$

for $j = 0, 1, \cdots, m-1$. This is a simple Hermite-Birkhoff interpolation problem; see, e.g., KINCAID and CHENEY (1991).

The problems (6.31) and (6.32) are computationally just as hard as the original problem, so nothing has been gained. However, let us consider the prediction problem

$$(6.35)\qquad \begin{array}{ll} \text{estimate} & \mathbb{S}(x_{in}) \\ \text{given the data} & y_{jn} = \mathrm{e}_1^T\,\mathbb{S}(x_{jn}) + d_{jn}\,,\ i = 1, 2, \cdots, i-1\;. \end{array}$$

In other words, one wishes to predict $\mathbb{S}(x_{in})$ based on data up to $x_{i-1,n}$ (the past). Here, the state-space model (6.29) comes to the fore. The crux of the matter is that the $\mathbb{S}(x_{in})$ conditioned on the past are normally distributed, and one can easily compute the conditional covariance matrices in terms of the past ones. The central idea is the construction of the "innovations",

$$\widetilde{y}_i = y_{in} + \sum_{j=1}^{i-1} a_{ij}\, y_{jn} \,, \tag{6.36}$$

for suitable (recursively computed) coefficients a_{ij}, such that $\widetilde{y}_1, \widetilde{y}_2, \cdots, \widetilde{y}_n$ are independent, mean 0, normal random variables. So, $\mathbb{E}[\,\widetilde{y}\,\widetilde{y}^T\,]$ is a diagonal matrix.

The prediction problem (6.35) was formulated and solved (for general transition matrices and covariance matrices, and for general linear observations) by KALMAN (1960). The resulting procedure (now) goes by the name of the *Kalman filter*.

Cholesky factorization. The only objection to the above is that we are not really interested in the prediction problem. However, one verifies that, as a by-product of (6.36), the Kalman filter produces the Cholesky factorization of the system matrix in (6.33),

$$\mathbb{L}\,\mathbb{L}^T = I + \sigma^2\,(\,\xi^2\,\mathbb{D}_n + \kappa^2\,\mathbb{V}_n\,)^{-1} \,, \tag{6.37}$$

where $\mathbb{L}$ is a lower block-bidiagonal matrix. Then $\widehat{f}_n$ can be computed in $\mathcal{O}(\,m^2 n\,)$ operations. For the precise details, see Chapter 20.

Two final observations: First, all of this is easily implemented, regardless of the value of m; one only needs to compute the transition and covariance matrices. Second, small spacings do not cause any problem. In this case, the transition matrices are close to the identity, and the covariance matrices are small. Also, replications in the design are handled seamlessly. Although it is no doubt more efficient to handle replications explicitly, especially when there are many such as in the Wood Thrush Data Set of BROWN and ROTH (2004), no harm is done if one fails to notice.

Bayesian smoothing parameter selection. The Bayesian model allows for the estimation of the parameters σ^2, ξ^2, and κ^2. Since $y_n = X_n + d_n$, and since X_n and d_n are independent, it follows that

$$y_n \sim \text{Normal}\Big(\,0\,,\,\sigma^2\,(\,I + \widetilde{\xi}^2\,D_n + \widetilde{\kappa}^2\,V_n\,)\,\Big) \,, \tag{6.38}$$

where $\widetilde{\xi}^2 = \xi^2/\sigma^2$, and likewise for $\widetilde{\kappa}^2$. It is an exercise to show that

$$(\,I + \widetilde{\xi}^2\,D_n + \widetilde{\kappa}^2\,V_n\,)^{-1} = I - R_{n\widetilde{\kappa}} \,, \tag{6.39}$$

with $R_{n\widetilde{\kappa}}$ the hat matrix for the problem (6.26). The negative log-likelihood of σ^2, $\widetilde{s}$, and $\widetilde{\kappa}$ is then

$$(\,2\,\sigma^2\,)^{-1}\,\langle\, y_n\,,\,(\,I - R_{n\widetilde{\kappa}}\,)\,y_n\,\rangle + n\,\log\sigma + \log\det(\,I - R_{n\widetilde{\kappa}}\,)^{-1} \,.$$

This must be minimized over σ, $\widetilde{s}$ and $\widetilde{\kappa}$. Minimizing over σ^2 gives $\widehat{\sigma}^2 = \langle\, y_n\,,\,(\,I - R_{n\widetilde{\kappa}}\,)\,y_n\,\rangle$. Then, apart from constants not depending on

$\widetilde{s}$ and $\widetilde{\kappa}$, the negative log-likelihood is $\frac{1}{2}\,n \log L(\widetilde{s}\,,\widetilde{\kappa}\,)$, where

$$L(\widetilde{s}\,,\widetilde{\kappa}\,) = \frac{\langle\, y_n\,,\,(\,I - R_{n\widetilde{\kappa}}\,)\,y_n\,\rangle}{\big(\,\det(\,I - R_{n\widetilde{\kappa}}\,)\,\big)^{1/n}}\;. \tag{6.40}$$

In the diffuse limit $\widetilde{\xi} \to \infty$, some extra work must be done, resulting in the GML procedure of BARRY (1983, 1986) and WAHBA (1985),

$$\mathrm{GML}(h) = \frac{\langle\, y_n\,,\,(\,I - R_{nh}\,)\,y_n\,\rangle}{\big(\,\det_+(\,I - R_{nh}\,)\,\big)^{1/(n-m)}}\;, \tag{6.41}$$

with R_{nh} the usual hat matrix (6.9), and for positive-definite matrices A, $\det_+(\,A\,)$ denotes the product of the positive eigenvalues. See Chapter 20 for the details. This procedure appears to work quite well; see Chapter 23.

Bayesian confidence bands. The last topic is that of confidence intervals for the $X(x_{in})$. One verifies that the distribution of X_n of (6.21), conditioned on the data y_n, is

$$X_n \,|\, y_n \;\sim\; \mathrm{Normal}(\,\widehat{X}_n\,,\,(\,\sigma^{-2}\,I + R^{-1}\,)^{-1}\,)\;. \tag{6.42}$$

The *individual* $100(1-\alpha)\%$ confidence intervals for the $X(x_{in})$ are then

$$\widehat{X}(x_{in}) \pm z_{\alpha/2}\,\sqrt{[\,(\,\sigma^{-2}\,I + R^{-1}\,)^{-1}\,]_{i,i}}\;, \quad i = 1, 2, \cdots, n\;, \tag{6.43}$$

with z_α the $1-\alpha$ quantile of the standard normal distribution. The Bayesian interpretation now suggests that these confidence intervals should work for the regression problem (2.7) under the normal noise assumption (2.11) and the smoothness condition (2.17), reformulated as $f_o \in W^{m,2}(0,1)$.

There are several subtle points in the Bayesian interpretation of smoothing splines. The most important one is that the sample paths of the process (6.14)–(6.15) do not lie in $W^{m,2}(0,1)$. (Probabilists force us to add "with probability 1".) Consequently, the exact meaning of Bayesian, as opposed to frequentist, confidence bands for the regression function is not so clear. We shall not come back to Bayesian confidence intervals, but for more see, e.g., WAHBA (1983) or NYCHKA (1988). For more on frequentist confidence bands, see § 8 and Chapter 22.

7. Smoothing parameter selection

The standard nonparametric regression estimators (smoothing splines, local polynomials, kernels, sieves) all depend on a smoothing parameter which more or less explicitly controls the trade-off between the bias and the variance of the estimator. In Figure 7.1, we show the quintic smoothing spline of the regression function in the Wood Thrush Data Set of BROWN and ROTH (2004). In case (a), we have $h \to \infty$, so we have the least-squares quadratic-polynomial estimator. In case (d), for $h \to 0$, we get the quintic

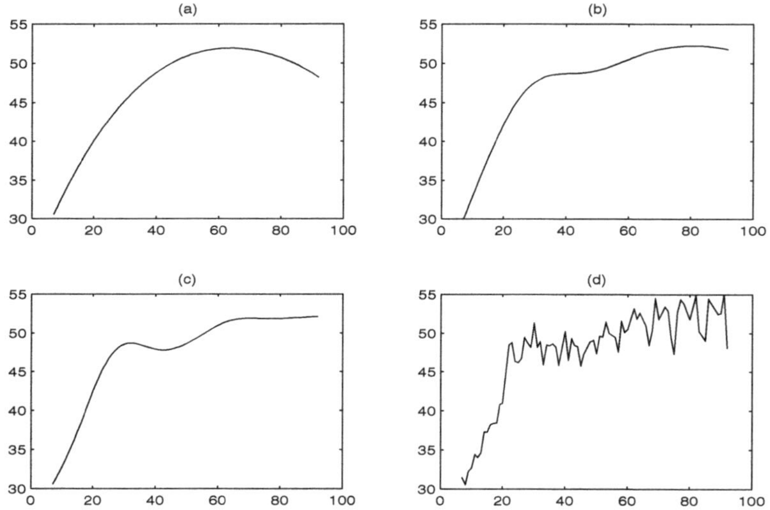

Figure 7.1. Quintic smoothing spline estimators for the Wood Thrush Data Set of Brown and Roth (2004) for four smoothing parameters. In case (a), we have oversmoothing; in (d), undersmoothing. Which one of (b) or (c) is right ?

spline interpolator of the data. (See § 19.2 for details.) Both of these can be ruled out as being "right". The two remaining estimators are more reasonable, but which one is "right"? Obviously, there is a need for *rational* or *data-driven* methods for selecting the smoothing parameter. This is done here and in Chapter 18.

The smoothing parameter can take many forms. For smoothing splines, it is the weight of the penalization as well as the order of the derivative in the penalization; in local polynomial estimation, it is the size of the local interval of design points included in the local estimation (or, not quite equivalently, the number of design points included in the local estimation) and also the order of the local polynomial; in sieves of nested compact subsets (e.g., $\mathfrak{F}_C$, $C > 0$, as in (2.19)), it is the constant C; in sieves of nested finite-dimensional subspaces, it is the dimension of the subspace that matters. In this section, we discuss the optimal selection of the smoothing parameter. The "optimal" part requires a criterion by which to judge the selected smoothing parameter which in essence refers back to what one means by an "optimal" estimator of *the regression function*. This should be contrasted with so-called plug-in methods, where the optimality usually refers to unbiased minimum variance estimation of the asymptotic smoothing parameter. It should also be contrasted with the GML procedure discussed in § 6, which is not obviously related to the optimality of the regression estimator. Finally, it should be contrasted with methods of complexity penalization for sieves of nested finite-dimensional subspaces (in terms of the dimension of the subspace), see, e.g, BARRON, BIRGÉ, and MASSART (1999), and the minimum description length methods of RISSANEN (1982), see BARRON, RISSANEN, and YU (1998), RISSANEN (2000), HANSEN and YU (2001), and GRÜNWALD, MYUNG, and PITT (2005). Unfortunately, we shall not consider these.

We settle on the optimality criterion for nonparametric regression estimators suggested by § 3 and, for notational convenience, consider a single smoothing parameter h belonging to some parameter space $\mathbb{H}$. Thus, for smoothing splines or local polynomials the parameter h could denote both the order m and the "usual" smoothing parameter.

Consider a family of estimators f^{nh}, $h \in \mathbb{H}$. The goal is to select the smoothing parameter $h = H$, such that f^{nH} is optimal in the sense that

$$\| f^{nH} - f_o \| = \min_{h \in \mathbb{H}} \| f^{nh} - f_o \| \tag{7.1}$$

for a suitable (semi-)norm on $W^{m,2}(0,1)$. Of course, the statement (7.1) should probably be given a probabilistic interpretation, but for now we leave it at this. We shall almost exclusively deal with the discrete L^2 norm; i.e., we wish to select $h = H$ such that

$$\| \widehat{y}^{nH} - y_o \| = \min_{h \in \mathbb{H}} \| y^{nh} - y_o \| \tag{7.2}$$

(Euclidean norms on $\mathbb{R}^n$), with

$$\begin{aligned} \widehat{y}^{nh} &= \left(f^{nh}(x_{1,n}), f^{nh}(x_{2,n}), \cdots, f^{nh}(x_{1,n}) \right)^T \quad \text{and} \\ y_o &= \left(f_o(x_{1,n}), f_o(x_{2,n}), \cdots, f_o(x_{1,n}) \right)^T . \end{aligned} \tag{7.3}$$

It would be interesting to also consider the (discrete) maximum norm, but the authors do not know how to do this.

(7.4) REMARK. Note that $\| f^{nh} - f_o \|_{L^2(0,1)}$ cannot be expressed in terms of $\| y^{nh} - y_o \|$, but it is possible to do this for

$$\| f^{nh} - \mathcal{N}_n f_o \|_{L^2(0,1)} ,$$

where $\mathcal{N}_n f_o$ is the natural spline interpolator (6.3) for the noiseless data $(x_{in}, f_o(x_{in}))$, $i = 1, 2, \cdots, n$.

Here, we discuss two methods for selecting h in the smoothing spline estimator f^{nh} of (2.23): the C_L estimator of MALLOWS (1972) and the GCV estimator of GOLUB, HEATH, and WAHBA (1979) and CRAVEN and WAHBA (1979) using the nil-trace estimator of LI (1986).

MALLOWS (1972) proceeds as follows. Since the goal is to minimize the unknown functional $\| \widehat{y}^{nh} - y_o \|^2$, one must first estimate it for all h. Recalling that $\widehat{y}^{nh} = R_{nh}\, y_n$ and that R_{nh} is symmetric, one calculates

$$\mathbb{E}[\, \| \widehat{y}^{nh} - y_o \|^2 \,] = \| (I - R_{nh})\, y_o \|^2 + \sigma^2 \operatorname{trace}(R_{nh}^2) . \tag{7.5}$$

Since it seems clear that any estimator of $\| \widehat{y}^{nh} - y_o \|^2$ must be based on $\| \widehat{y}^{nh} - y_n \|^2 = \| (I - R_{nh}) y_n \|^2$, let us compute

$$(7.6) \qquad \mathbb{E}[\, \| \widehat{y}^{nh} - y_n \|^2 \,] = \| (I - R_{nh}) y_o \|^2 + \sigma^2 \operatorname{trace}(\, (I - R_{nh})^2 \,) \; .$$

Then, a pretty good (one observes the mathematical precision with which this idea is expressed) estimator of $\| \widehat{y}^{nh} - y_o \|^2$ is obtained in the form

$$(7.7) \qquad \mathfrak{M}(h) = \| \widehat{y}^{nh} - y_n \|^2 + 2\,\sigma^2 \operatorname{trace}(\, R_{nh} \,) - n\,\sigma^2 \; .$$

This is the C_L functional of MALLOWS (1972). So, MALLOWS (1972) proposes to select h by minimizing $\mathfrak{M}(h)$ over $h > 0$. An obvious drawback of this procedure is that σ^2 is usually not known and must be estimated. However, for any reasonable estimator of the variance σ^2, this procedure works quite well.

We can avoid estimating the variance by using the zero-trace (nil-trace) idea of LI (1986). This in fact leads to the celebrated GCV procedure of CRAVEN and WAHBA (1979) and GOLUB, HEATH, and WAHBA (1979) and hints at the optimality of GCV in the sense of (7.1). Note that we wish to estimate y_o and that $\widehat{y}^{nh}$ is a biased estimator of y_o with (small) variance proportional to $(nh)^{-1}$. Of course, we also have an unbiased estimator of y_o in the guise of y_n with largish variance σ^2. So, why not combine the two? So, the new and improved estimator of y_o is $\widehat{y}^{nh}_{\alpha} = \alpha\, \widehat{y}^{nh} + (1-\alpha)\, y_n$ for some as yet unspecified α, and we wish to estimate h so as to

$$(7.8) \qquad \text{minimize} \quad \| y_o - \widehat{y}^{nh}_{\alpha} \|^2 \quad \text{over} \quad h > 0 \; .$$

Repeating the considerations that went into the estimator $\mathfrak{M}(h)$ of (7.7), one finds that an unbiased estimator of the new error is

$$(7.9) \quad \mathfrak{M}(\alpha, h) = \alpha^2 \, \| (I - R_{nh}) y_n \|^2 - 2\,\alpha\,\sigma^2 \operatorname{trace}(\, I - R_{nh} \,) + 2\,n\sigma^2 - n\,\sigma^2 \; .$$

The rather bizarre way of writing $2\,n\sigma^2 - n\,\sigma^2$ instead of plain $n\sigma^2$ will become clear shortly. Now, rather than choosing α so as to reduce the bias or overall error, LI (1986) now chooses $\alpha = \alpha_o$ such that

$$(7.10) \qquad -2\,\alpha_o\,\sigma^2 \operatorname{trace}(\, I - R_{nh} \,) + 2\,n\,\sigma^2 = 0 \; .$$

Thus, $\alpha_o = \{\, \frac{1}{n} \operatorname{trace}(\, I - R_{nh} \,) \,\}^{-1}$, and then $\mathfrak{M}(\alpha_o, h) = \mathit{GCV}(h)$ with

$$(7.11) \qquad \mathit{GCV}(h) = \frac{\| (I - R_{nh}) y_n \|^2}{\{\, \frac{1}{n} \operatorname{trace}(\, I - R_{nh} \,) \,\}^2} - n\,\sigma^2 \; .$$

(7.12) REMARK. It turns out that α_o is close to 1, which is exactly what one wants. If instead of (7.10) one takes $\alpha = \alpha_1$, where

$$-2\,\alpha_1 \operatorname{trace}(\, I - R_{nh} \,) + n\,\sigma^2 = 0 \; ,$$

then one would get $\alpha_1 = \frac{1}{2}\,\alpha_o$, so that $\alpha_1 \approx \frac{1}{2}$. The resulting estimator of y_o would not be good. At any rate, they cannot both be good.

Thus, the LI (1986) estimator of h is obtained by minimizing $GCV(h)$ over $h > 0$. However, LI (1986) suggests the use of $\widehat{y}_{\alpha}^{nh}$, with α chosen to minimize $\mathfrak{M}(\alpha, h)$ instead of the spline estimator; see Exercise (18.2.16). This is in effect a so-called Stein estimator; see LI (1985). We denoted $\mathfrak{M}(\alpha_o, h)$ by $GCV(h)$ because it is the GCV score of CRAVEN and WAHBA (1979) and GOLUB, HEATH, and WAHBA (1979), who derive the GCV score as ordinary cross validation in a coordinate-free manner. In ordinary cross validation, one considers the model

$$y_{in} = [\, y_o \,]_i + d_{in} \,, \quad i = 1, 2, \cdots, n \,, \tag{7.13}$$

and one would like to judge any estimator by how well it predicts future observations. Of course, there are no future observations, so instead one may leave out one observation and see how well we can "predict" this observation based on the remaining data. Thus, for some j, consider the smoothing spline estimator $\widehat{y}_{\text{-}j}^{nh}$ based on the data (7.13) *with the j-th datum removed.* Thus $\widehat{y} = \widehat{y}_{\text{-}j}^{nh}$ is the solution of

$$\text{minimize} \quad \sum_{\substack{i=1 \\ i \neq j}}^{n} |\, \widehat{y}_i - y_{in} \,|^2 + nh^{2m} \langle\, \widehat{y} \,,\, M \,\widehat{y} \,\rangle \tag{7.14}$$

over $\widehat{y} \in \mathbb{R}^n$; cf. (6.7). Then, $[\, \widehat{y}_{\text{-}j}^{nh} \,]_j$ is the "predictor" of y_{jn}. Of course, one does this for all indices j. Then the score

$$OCV(h) = \sum_{j=1}^{n} |\, [\, \widehat{y}_{\text{-}j}^{nh} \,]_j - y_{jn} \,|^2 \tag{7.15}$$

is minimized over $h > 0$ to get an estimator with the best predictive power. This also goes under the name of the *leave-one-out* procedure. This idea under the name of *predictive (residual) sums of squares* (PRESS) goes back to ALLEN (1974). The general idea goes back at least as far as MOSTELLER and WALLACE (1963).

To get to the coordinate-free version of ordinary cross validation, one selects an orthogonal matrix Q and considers the model

$$[\, Q \, y_n \,]_i = [\, Q \, y_o \,]_i + [\, Q \, d_n \,]_i \,, \quad i = 1.2. \cdots, n \,. \tag{7.16}$$

Now, repeat ordinary cross validation. In particular, let $\widehat{y} = \widehat{y}_{\text{-}j}^{nhQ}$ be the solution of

$$\text{minimize} \quad \sum_{\substack{i=1 \\ i \neq j}}^{n} |\, [\, Q \, \widehat{y} \,]_i - [\, Q \, y_n \,]_i \,|^2 + nh^{2m} \langle\, \widehat{y} \,,\, M \,\widehat{y} \,\rangle \tag{7.17}$$

and consider the score

$$OCV(h, Q) = \sum_{j=1}^{n} |\, [\, Q \, \widehat{y}_{\text{-}j}^{nhQ} \,]_j - [\, Q \, y_n \,]_j \,|^2 \,. \tag{7.18}$$

Improbable as it sounds, with the proper choice of the orthogonal matrix Q, one obtains that $OCV(h, Q) = GCV(h)$. (Note that the GCV functional is coordinate-free because the Euclidean norm and the traces are so.) A drawback of this derivation of the GCV functional is that it is difficult to connect it to our sense of optimality of the smoothing parameter. The same "objection" applies to the GML estimator of the previous section.

Two final remarks: First, one more time, we emphasze that the GCV and C_L procedures may also be used to estimate the order m by also minimizing over m in a reasonable range, say $1 \leqslant m \leqslant \sqrt{n}$. Second, both the C_L and the zero-trace approaches may be applied to any estimator which is linear in the data.

Data splitting. The methods discussed above may be extended to the discrete L^2 error of any linear estimator, but they break down for other measures of the error. So, how can one choose $h = H$ such that

$$\| \widehat{y}^{nH} - y_o \|_p = \min_{h>0} \| \widehat{y}^{nh} - y_o \|_p \tag{7.19}$$

for $1 \leqslant p \leqslant \infty$, $p \neq 2$? Here, as a general solution to the problem of what to do when one does not know what to do, we discuss data-splitting methods Instead of leaving one out, now one leaves out half the data.

Start with splitting the model (2.7) in two,

$$y_{in} = f_o(x_{in}) + d_{in} \,, \quad i = 1, 3, 5, \cdots, 2k-1 \,, \tag{7.20}$$

and

$$y_{in} = f_o(x_{in}) + d_{in} \,, \quad i = 2, 4, \cdots, 2k \,, \tag{7.21}$$

where $k = \lceil n/2 \rceil$ is the smallest integer which is not smaller than $n/2$.

It is useful to introduce the following notation. Let $y_{[1]n}$ and $r_{[1]n}$ be the data and sampling operator for the model (7.20),

$$\begin{aligned} y_{[1]n} &= \bigl(y_{1,n} \,, \; y_{3,n} \,, \; \cdots \; y_{2k-1,n} \bigr) \,, \\ r_{[1]n} f &= \bigl(f(x_{1,n}) \,, \; f(x_{3,n}) \,, \; \cdots \; f(x_{2k-1,n}) \bigr) \,, \end{aligned} \tag{7.22}$$

and likewise for $y_{[2]n}$ and $r_{[2]n}$ applied to the model (7.21). Also, let $y_{[1]o}$ and $y_{[2]o}$ denote the noiseless data for the two models.

Now, let $f^{[1]nh}$ be a not necessarily linear estimator of f_o in the model (7.20), and select h by solving

$$\min_{h>0} \| y_{[2]n} - r_{[2]n} f^{[1]nh} \|_p \,. \tag{7.23}$$

For symmetry reasons, the objective function should perhaps be changed to $\| y_{[2]n} - r_{[2]n} f^{[1]nh} \|_p + \| y_{[1]n} - r_{[1]n} f^{[2]nh} \|_p$.

It seems clear that such a procedure would give good results, but the task at hand is to show that this procedure is optimal in the sense that

$$\frac{\| y_{[2]n} - r_{[2]n} f^{[1]nH} \|_p}{\displaystyle\min_{h>0} \| y_{[2]o} - r_{[2]n} f^{[1]nh} \|_p} \longrightarrow_{\text{as}} 1 \,, \tag{7.24}$$

where H is selected via (7.23) under mild conditions on the nonlinear behavior of $f^{[1]nh}$. How one would do that the authors do not know, but GYÖRFI, KOHLER, KRZYŻAK, and WALK (2002) do.

How good are the GCV and GML estimators ? There are two types of optimality results regarding smoothing parameter selection: our favorite, the optimality of the resulting regression estimator along the lines of § 3 and Theorem (7.31) below, and the optimality of the smoothing parameter itself in the sense of asymptotically unbiased minimum variance.

The prototype of the optimality of any method for selecting the smoothing parameter $h = H$ is the following theorem of LI (1985) for a fixed order m. Recall the definition of $\widehat{y}^{nh}$ in (6.2) and the hat matrix R_{nh} in (6.8). The assumptions on the regression problem

$$y_{in} = f_o(x_{in}) + d_{in} \ , \quad i = 1, 2, \cdots, n \ , \tag{7.25}$$

are that

(7.26) $f_o \in W^{k,2}(0,1)$ for some $k \geqslant 1$ but

(7.27) f_o is not a polynomial of order $\leqslant m$;

(7.28) $d_n = (d_{1,n}, d_{2,n}, \cdots, d_{n,n})^T \sim \text{Normal}(0, \sigma^2 I)$;

(7.29) $\inf\limits_{h>0} \frac{1}{n} \| \widehat{y}^{nh} - y_o \|^2 \longrightarrow 0$ in probability ,

and finally, with $h = H$ the random smoothing parameter,

(7.30) $\big(\frac{1}{n} \operatorname{trace}(R_{nH}) \big)^2 / \big(\frac{1}{n} \operatorname{trace}(R_{nH}^2) \big) \longrightarrow 0$ in probability .

(7.31) THEOREM (LI, 1985). *Let $m \geqslant 1$ be fixed. Under the assumptions (7.26)–(7.30),*

$$\frac{\| \widehat{y}^{nH} - y_o \|^2}{\min\limits_{h>0} \| \widehat{y}^{nh} - y_o \|^2} \longrightarrow 1 \quad \textit{in probability} .$$

Several things stand out. First, only iid normal noise is covered. Second, there are no restrictions on the smoothing parameter h, so there is no *a priori* restriction of h to some reasonable deterministic range, such as $\frac{1}{n} \leqslant h \leqslant 1$. Third, the condition (7.27) is needed since, for polynomials of order m, the optimal h is $h = \infty$, so that $\inf_{h>0} \| \widehat{y}^{nh} - y_o \|^2 = \mathcal{O}(n^{-1})$. The authors do not know whether the GCV procedure would match this, let alone optimally in the sense of the theorem, but see the counterexample of LI (1986). Fourth, the condition (7.29) merely requires the existence of a *deterministic* smoothing parameter h for which $\frac{1}{n} \| \widehat{y}^{nh} - y_o \|^2 \to 0$, and that we have shown. Actually, the condition (7.29) is needed only to show that (7.30) holds. This is an unpleasant condition, to put it mildly. For the *equidistant* design (2.10), LI (1985) shows it using (7.29) and some precise knowledge of the eigenvalues,

$$0 = \lambda_{1,n} = \cdots = \lambda_{m,n} < \lambda_{m+1,n} \leqslant \cdots \lambda_{n,n} ,$$

of the matrix M in (6.9), provided by CRAVEN and WAHBA (1979). However, even for general, asymptotically uniform designs, such knowledge is not available.

So, where do things go wrong? Note that a crucial step in the proof is the following trick. With $\lambda_{in}(h) = (1 + nh^{2m} \lambda_{in})^{-1}$, the eigenvalues of the hat matrix R_{nh}, we have

$$\| R_{nh} d_n \|^2 - \mathbb{E}[\, \| R_{nh} d_n \|^2 \,] = S_n(h) \tag{7.32}$$

where, for $j = 1, 2, \cdots, n$,

$$S_j(h) = \sum_{i=1}^{j} | \lambda_{in}(h) |^2 \{ | \delta_{in} |^2 - \sigma^2 \} \tag{7.33}$$

and

$$\delta_n = (\delta_{1,n}, \delta_{2,n}, \cdots, \delta_{n,n})^T \sim \text{Normal}(0 , \sigma^2 I) . \tag{7.34}$$

In other words, $\delta_n = d_n$ in distribution, but only if $d_n \sim \text{Normal}(0 , \sigma^2 I)$. (Then, Abel's summation-by-parts lemma yields the bound

$$| S_n(h) | \leqslant \lambda_{1,n}(h) \max_{1 \leqslant j \leqslant n} | S_j(h) | = \mathcal{O}(\sqrt{n^{-1} \log \log n}) \tag{7.35}$$

almost surely, uniformly in $h > 0$; see SPECKMAN (1982, 1985).)

So, (7.33) breaks down for nonnormal noise, and even for normal noise, if there is not enough information on the eigenvalues. This is the case for quasi-uniform designs: A design is quasi-uniform if there exists a bounded measurable function ω which is bounded away from 0 and a constant c such that, for all $f \in W^{1,1}(0,1)$,

$$\Bigl| \tfrac{1}{n} \sum_{i=1}^{n} f(x_{in}) - \int_0^1 f(t)\, \omega(t)\, dt \Bigr| \leqslant c\, n^{-1} \| f' \|_{L^1(0,1)} . \tag{7.36}$$

(Asymptotically uniform designs would have $\omega(t) \equiv 1$.)

Various ways around these problems suggest themselves. The use of the eigenvalues might be avoided by using the equivalent kernel representation

$$[R_{nh} d_n]_j \approx \tfrac{1}{n} \sum_{i=1}^{n} d_{in} \mathfrak{R}_{\omega m h}(x_{in}, x_{jn}) , \tag{7.37}$$

and information about the traces of $\mathfrak{R}_{\omega m h}$ is readily available (see § 8 and Chapter 21). Also, in (7.37) one may employ strong approximation ideas,

$$\tfrac{1}{n} \sum_{i=1}^{n} d_{in} \mathfrak{R}_{\omega m h}(x_{in}, x_{jn}) \approx \tfrac{1}{n} \sum_{i=1}^{n} Z_{in} \mathfrak{R}_{\omega m h}(x_{in}, x_{jn}) , \tag{7.38}$$

where $Z_{1,n}, Z_{2,n}, \cdots, Z_{n,n}$ are iid Normal$(0 , \sigma^2)$, provided the d_{in} are iid and have a high enough moment,

$$\mathbb{E}[d_{1,n}] = 0 \quad , \quad \mathbb{E}[| d_{1,n} |^4] < \infty \tag{7.39}$$

(see §§ 22.2 and 22.4). However, these approximations require that the smoothing parameter belong to the aforementioned reasonable interval.

To summarize, the authors would have liked to prove the following theorem, but do not know how to do it.

(7.40) THE MISSING THEOREM. *Let* $m \geqslant 1$ *be fixed. Assume that the design is quasi-uniform in the sense of* (7.36), *that the noise satisfies* (7.39), *and that* f_o *satisfies* (7.26)–(7.27). *Then, the GCV estimator* H *of the smoothing parameter* h *satisfies*

$$\frac{\| \widehat{y}^{nH} - y_o \|^2}{\min_{h>0} \| \widehat{y}^{nh} - y_o \|^2} \longrightarrow 1 \quad \textit{in probability} .$$

We repeat, the authors do not know how to prove this (or whether it is true, for that matter). Restricting h beforehand might be necessary.

With The Missing Theorem in hand, it is easy to show that for h_n the smallest minimizer of $\| \widehat{y}^{nh} - y_o \|$, and $f_o \in W^{m,2}(0,1)$,

$$(7.41) \qquad \Bigl| \frac{H}{h_n} - 1 \Bigr| = \mathcal{O}\bigl(n^{-1/(2m+1)} \bigr) \quad \text{in probability} .$$

While this is not very impressive, it is strong enough for determining the asymptotic distribution of the (weighted) uniform error and determining confidence bands. See § 8 and Chapter 22.

And what about the optimality of the GML procedure? Here, not much is known one way or the other. For the Bayesian model, STEIN (1990) shows that the GCV and GML estimators of h have the same asymptotic mean, but the GML estimated h has a much smaller variance. Unfortunately, it is hard to tell what that means for the frequentist optimality in the sense of Theorem (7.31) or The Missing Theorem (7.40).

8. Strong approximation and confidence bands

In the final two chapters, Chapters 21 and 22, we take the idea of the equivalent reproducing kernel representation of smoothing splines, briefly discussed in § 4, and drill it into the ground. We already noted that the equivalent reproducing kernel representation leads to uniform error bounds. Here, we discuss the logical next step of obtaining confidence bands for the unknown regression function. Unfortunately, we stop just short of nonparametric hypothesis testing, such as testing for monotonicity in the Wood Thrush Data Set of BROWN and ROTH (2004) or testing for a parametric fit as in HART (1997). However, these two chapters lay the foundation for such results.

The general regression problem. The development is centered on the following general regression problem with quasi-uniform designs: One observes the random sample

$$(8.1) \qquad (X_1, Y_1), (X_2, Y_2), \cdots, (X_n, Y_n)$$

of the random variable (X, Y), with $X \in [0, 1]$ almost surely, and one wishes to estimate

$$(8.2) \qquad f_o(x) = \mathbb{E}[Y \mid X = x], \quad x \in [0, 1].$$

Of course, some assumptions are needed.

(8.3) Assumptions. (a) $f_o \in W^{m,2}(0,1)$ for some integer $m \geqslant 1$;

(b) $\mathbb{E}[\,|Y|^4\,] < \infty$;

(c) $\sigma^2(x) = \mathbb{V}\mathrm{ar}[Y \mid X = x]$ is Lipschitz continuous and positive on $[0, 1]$;

(d) The random design $\mathbb{X}_n = (X_1, X_2, \cdots, X_n)$ is *quasi-uniform*; i.e., the marginal pdf ω of the random variable X is bounded and bounded away from 0: There exist positive constants $\omega_{\min}$ and $\omega_{\max}$ such that

$$0 < \omega_{\min} \leqslant \omega(x) \leqslant \omega_{\max} < \infty, \quad x \in [0, 1].$$

The signal-plus-noise model. The model (8.1) and the goal (8.2) are not very convenient. We prefer to work instead with the usual model

$$(8.4) \qquad Y_i = f_o(X_i) + D_i, \quad i = 1, 2, \cdots, n,$$

where, roughly speaking, $\mathbb{D}_n = (D_1, D_2, \cdots, D_n)$ is a random sample of the random variable $D = Y - f_o(X)$. Now, the D_i are independent, conditioned on the design $\mathbb{X}_n$, since the conditional joint pdf satisfies

$$(8.5) \qquad f_{\mathbb{D}_n | \mathbb{X}_n}(d_1, d_2, \cdots, d_n \mid x_1, x_2, \cdots, x_n) = \prod_{i=1}^{n} f_{D|X}(d_i \mid x_i).$$

Later on, there is a need to order the design. Let

$$X_{1,n} < X_{2,n} < \cdots < X_{n,n}$$

be the order statistics of the design and let the conformally rearranged Y_i and D_i be denoted by $Y_{i,n}$ and $D_{i,n}$. Thus, the model (8.4) becomes

$$(8.6) \qquad Y_{i,n} = f_o(X_{i,n}) + D_{i,n}, \quad i = 1, 2, \cdots, n.$$

Now, the $D_{i,n}$ are no longer independent, but in view of (8.5), conditioned on the design they still are. (The conditional distribution of, say, $D_{1,n}$ only depends on the value of $X_{1,n}$, not on $X_{1,n}$ being the smallest design point.) So, in what follows, we will happily condition on the design. For the unconditional treatment, see, e.g., Györfi, Kohler, Krzyżak, and Walk (2002).

Equivalent reproducing kernels. Of course, we estimate f_o by the smoothing spline estimator $f = f^{nh}$, the solution of

$$(8.7)\qquad \begin{aligned} &\text{minimize} \quad \tfrac{1}{n} \sum_{i=1}^{n} |\, f(X_i) - Y_i \,|^2 + h^{2m} \,\|\, f^{(m)} \,\|^2 \\ &\text{subject to} \quad f \in W^{m,2}(0,1) \ . \end{aligned}$$

Replacing sums by integrals, this suggests that the "proper" inner products on $W^{m,2}(0,1)$ should be

$$(8.8)\qquad \langle\, f \,,\, g \,\rangle_{\omega mh} = \langle\, f \,,\, g \,\rangle_{L^2((0,1),\omega)} + h^{2m} \,\langle\, f^{(m)} \,,\, g^{(m)} \,\rangle \ ,$$

where $\langle\, \cdot \,,\, \cdot \,\rangle_{L^2((0,1),\omega)}$ is a weighted $L^2(0,1)$ inner product,

$$\langle\, f \,,\, g \,\rangle_{L^2((0,1),\omega)} = \int_0^1 f(x)\, g(x)\, \omega(x)\, dx \ .$$

The associated norm is denoted by $\|\cdot\|_{\omega mh}$. Since ω is quasi-uniform, the norms $\|\cdot\|_{\omega mh}$ and $\|\cdot\|_{m,h}$ are equivalent, uniformly in $h > 0$. In particular, for all $f \in W^{m,2}(0,1)$ and all $h > 0$,

$$(8.9)\qquad \omega_{\min} \,\|\, f \,\|_{m,h}^2 \leqslant \|\, f \,\|_{\omega mh}^2 \leqslant \omega_{\max} \,\|\, f \,\|_{m,h}^2 \ .$$

So then, for all $f \in W^{m,2}(0,1)$, all $0 < h \leqslant 1$, and all $x \in [\,0\,,\,1\,]$,

$$(8.10)\qquad |\, f(x) \,| \leqslant c_m \, h^{-1/2} \,\|\, f \,\|_{m,h} \leqslant c_m \, \omega_{\min}^{-1/2} \, h^{-1/2} \,\|\, f \,\|_{\omega mh} \ ,$$

whence $W^{m,2}(0,1)$ with the inner product $\langle\, \cdot \,,\, \cdot \,\rangle_{\omega mh}$ is a reproducing kernel Hilbert space. Denoting the reproducing kernel by $\mathfrak{R}_{\omega mh}$, we get that

$$(8.11)\qquad f(x) = \langle\, f \,,\, \mathfrak{R}_{\omega mh}(x, \,\cdot\,) \,\rangle_{\omega mh}$$

for all relevant f, h, and x.

As in § 4, one proves under the Assumptions (8.3) that, conditional on the design $\mathbb{X}_n$,

$$(8.12)\qquad f^{nh}(x) \approx \int_0^1 \mathfrak{R}_{\omega mh}(x, t)\, f(t)\, \omega(t)\, dt + \mathfrak{S}_{nh}(x)$$

with

$$(8.13)\qquad \mathfrak{S}_{nh}(x) = \tfrac{1}{n} \sum_{i=1}^{n} D_i \, \mathfrak{R}_{\omega mh}(X_i, x) \ .$$

This is again the basis for uniform error bounds as in (4.27). For the precise statement, see § 21.1.

Confidence bands. As stated before, the goal is to provide confidence bands, interpreted in this text as simultaneous confidence intervals for the components of the vector y_o in terms of the vector $\widehat{y}^{nh}$, defined in (7.3).

(For now, we assume that the smoothing parameter h is chosen deterministically.) Thus, we want to determine intervals

$$[\,\widehat{y}^{nh}\,]_i \pm c_{\alpha nh}\,\Omega_{inh}\;, \quad i = 1, 2, \cdots, n\;, \tag{8.14}$$

such that, for $0 < \alpha < 1$,

$$\mathbb{P}\left[\; \frac{|\,[\,\widehat{y}^{nh}\,]_i - [\,y_o\,]_i\,|}{c_{\alpha nh}\,\Omega_{inh}} > 1 \quad \text{for all } i \;\right] = \alpha\;. \tag{8.15}$$

Of course, this must be taken with a grain of salt; certainly asymptotic considerations will come into play. Note that the constants $c_{\alpha nh}$ are the quantiles of the random variables

$$\max_{1 \leqslant i \leqslant n} \frac{|\,[\,\widehat{y}^{nh}\,]_i - [\,y_o\,]_i\,|}{\Omega_{inh}}\;, \tag{8.16}$$

a weighted uniform norm of the estimation error. The Ω_{inh} determine the shape of the confidence band. The actual widths of the individual confidence intervals are equal to $2\,c_{\alpha nh}\,\Omega_{inh}$. The widths Ω_{inh} can be chosen arbitrarily (e.g., just simply $\Omega_{inh} = 1$ for all i, n, and h,) but there are good reasons to be more circumspect. Ideally, as in Beran (1988), one would like to choose the Ω_{inh} such that the individual confidence intervals have equal coverage,

$$\mathbb{P}\left[\; \frac{|\,[\,\widehat{y}^{nh}\,]_i - [\,y_o\,]_i\,|}{c_{\alpha nh}\,\Omega_{inh}} > 1 \;\right] = \gamma\;, \tag{8.17}$$

for all $i = 1, 2, \cdots, n$, and of course (8.15) holds. (Note that $\gamma < \alpha$.)

When $D_1, D_2, \cdots, D_n$ are iid Normal$(0, \sigma^2)$ (conditional on the design), then the choice

$$\Omega_{inh} = \sqrt{\mathrm{Var}[\,[\,\widehat{y}^{nh}\,]_i \,|\, \mathbb{X}_n\,]} = \sigma^2 \sqrt{[\,R_{nh}^2\,]_{i,i}} \tag{8.18}$$

would work since, for all i,

$$[\,\widehat{y}^{nh}\,]_i \sim \text{Normal}\Big(\, R_{nh}\,y_o\,,\; \sigma^2\,[\,R_{nh}^2\,]_{i,i}\Big)\;. \tag{8.19}$$

The caveat is that the bias $R_{nh}\,y_o - y_o$ must be negligible compared with the noise in the estimator,

$$\Big|\,[\,R_{nh}\,y_o\,]_{i,i} - [\,y_o\,]_i\,\Big| \ll \sqrt{\mathrm{Var}[\,[\,\widehat{y}^{nh}\,]_i \,|\, \mathbb{X}_n\,]}\;, \tag{8.20}$$

for all i. In fact, this requires *undersmoothing* of the estimator (which decreases the bias and increases the variance). It seems reasonable to make the choice (8.18) even if the D_i are not iid normals.

How do we establish confidence bands? Under the assumption (8.20), our only worry is the noise part of the solution of the smoothing spline problem (8.7). The equivalent reproducing kernel representation then gives

$$\widehat{y}^{nh} - \mathbb{E}[\,\widehat{y}^{nh} \,|\, \mathbb{X}_n\,] \approx \mathfrak{S}_{nh}\;, \tag{8.21}$$

where, in a slight notational change, $\mathfrak{S}_{nh} \in \mathbb{R}^n$, with $[\,\mathfrak{S}_{nh}\,]_i = \mathfrak{S}_{nh}(X_i)$ for all i. Thus, we must deal with the distribution of the random variable $U_{nh}(\mathbb{D}_n \,|\, \mathbb{X}_n\,)$, where

$$(8.22)\quad U_{nh}(\mathbb{D}_n|\mathbb{X}_n\,) = \max_{1\leqslant j\leqslant n} \frac{\frac{1}{n}\sum\limits_{i=1}^{n} D_i\,\mathfrak{R}_{\omega mh}(X_i, X_j\,)}{\sqrt{\frac{1}{n^2}\sum\limits_{i=1}^{n} \mathbb{V}\mathrm{ar}[\,D_i\,|\,X_i\,]\,|\,\mathfrak{R}_{\omega mh}(X_i, X_j)\,|^2}} .$$

The fact that the D_i are independent but not identically distributed complicates matters, but resorting to the strong approximation theorem of Sakhanenko (1985) (see the more easily accessible Shao (1995) or Zaitsev (2002)), under the moment condition of Assumption (8.3), there exist iid Normal(0, 1) random variables $Z_{1,n}, Z_{2,n}, \cdots, Z_{n,n}$ (conditional on the design) such that

$$(8.23)\quad \max_{1\leqslant i\leqslant n} \Bigl|\, \sum_{j=1}^{i} \bigl(\, D_{j,n} - \sigma(X_{jn})\, Z_{jn}\,\bigr) \Bigr| = \mathcal{O}(\,n^{1/4}\,) \quad \text{in probability} .$$

Using summation by parts, one then shows that uniformly in h (in a reasonable range) and uniformly in $y \in [\,0\,,\,1\,]$,

$$(8.24)\quad \frac{1}{n}\sum_{i=1}^{n} D_{in}\,\mathfrak{R}_{\omega mh}(X_{in}, y) \approx \frac{1}{n}\sum_{i=1}^{n} \sigma(X_{in})\, Z_{in}\,\mathfrak{R}_{\omega mh}(X_{in}, y) .$$

For a precise statement, see § 22.4. It is a little more work to show that

$$(8.25)\quad U_{nh}(\mathbb{D}_n|\mathbb{X}_n\,) \approx \max_{1\leqslant j\leqslant n} \frac{\frac{1}{n}\sum\limits_{i=1}^{n} \sigma(X_i)\, Z_i\,\mathfrak{R}_{\omega mh}(X_i, X_j\,)}{\sqrt{\frac{1}{n^2}\sum\limits_{i=1}^{n} \sigma^2(X_i)\,|\,\mathfrak{R}_{\omega mh}(X_i, X_j)\,|^2}} .$$

Finally, assuming that $\sigma(x)$ is positive and Lipschitz continuous, one shows further that one may get rid of $\sigma(X_i)$:

$$(8.26)\quad U_{nh}(\mathbb{D}_n|\mathbb{X}_n\,) \approx \max_{1\leqslant j\leqslant n} \frac{\frac{1}{n}\sum\limits_{i=1}^{n} Z_i\,\mathfrak{R}_{\omega mh}(X_i, X_j\,)}{\sqrt{\frac{1}{n^2}\sum\limits_{i=1}^{n} |\,\mathfrak{R}_{\omega mh}(X_i, X_j)\,|^2}} .$$

This may be rephrased as

$$(8.27)\quad U_{nh}(\mathbb{D}_n|\mathbb{X}_n\,) \approx U_{nh}(\mathbb{Z}_n|\mathbb{X}_n\,) ;$$

in other words, the confidence bands for the non-iid noise model (8.4) are just about the same as for the iid Normal(0, 1) model

$$(8.28)\quad Y_i' = f_o(X_i) + Z_i\,, \quad i = 1, 2, \cdots, n .$$

This forms the basis for *simulating* the critical values $c_{\alpha nh}$ in (8.15). To some extent, this is necessary, because the (asymptotic) distribution of $U_{nh}(\mathbb{Z}_n|\mathbb{X}_n)$ for nonuniform designs does not appear to be known; see § 22.6. For uniform designs, the distribution is known asymptotically because one shows that

$$(8.29) \qquad U_{nh}(\mathbb{Z}_n \,|\, \mathbb{X}_n) \approx \sup_{0 \leqslant x \leqslant h^{-1}} \int_{\mathbb{R}} \mathfrak{B}_m(x-t)\, dW(t) \,,$$

where $W(t)$ is the standard Wiener process and the kernel $\mathfrak{B}_m$ is given by Lemma (14.7.11). Except for the conditioning on the design, the development (8.23)–(8.29) closely follows KONAKOV and PITERBARG (1984). See §§ 22.4 and 22.6.

(8.30) REMARK. SILVERMAN (1984) introduced the notion of equivalent kernels for smoothing splines. In effect, he showed that, away from the boundary,

$$\mathfrak{R}_{\omega mh}(x,y) \approx \mathfrak{B}_{m,\lambda(x)}(x-y) \,,$$

where $\lambda(x) = h \,/\, \omega(x)^{1/2m}$, with $\omega(t)$ the design density and $\mathfrak{B}_{m,h}$ given by Lemma (14.7.11). See § 21.9 for the precise statement. It is interesting to note that SILVERMAN (1984) advanced the idea of equivalent kernels in order to increase the mindshare of smoothing splines against the onslaught of kernel estimators, but we all know what happened.

9. Additional notes and comments

Ad § 1: A variation of the partially linear model (1.18) is the partially parametric model; see, e.g., ANDREWS (1995) and references therein.

Ad § 2: The authors are honor-bound to mention kernel density estimation as the standard way of estimating densities nonparametrically; see DEVROYE and GYÖRFI (1985) and of course Volume I.

The first author was apparently the only one who thought that denoting the noise by *din* was funny. After having it explained to him, the second author now also thinks it is funny.

Ad § 3: The statement "there are no unbiased estimators with finite variance" is actually a definition of ill-posed problems in a stochastic setting, but a reference could not be found. For nonparametric density estimation, DEVROYE and LUGOSI (2000a) actually *prove* it.

Ad § 4: BIANCONCINI (2007) applies the equivalent reproducing kernel approach to spline smoothing for noisy time series analysis.

13

Smoothing Splines

1. Introduction

In this section, we begin the study of nonparametric regression by way of smoothing splines. We wish to estimate the regression function f_o on a bounded interval, which we take to be $[\,0\,,\,1\,]$, from the data $y_{1,n}, \cdots, y_{n,n}$, following the model

$$y_{in} = f_o(x_{in}) + d_{in} \;, \quad i = 1, 2, \cdots, n \;. \tag{1.1}$$

Here, the x_{in} are design points (in this chapter, the design is deterministic) and $d_n = (d_{1,n}, d_{2,n}, \cdots, d_{n,n})^T$ is the random noise. Typical assumptions are that $d_{1,n}, d_{2,n}, \cdots, d_{n,n}$ are uncorrelated random variables, with mean 0 and common variance, i.e.,

$$\mathbb{E}[\,d_n\,] = 0 \quad , \quad \mathbb{E}[\,d_n d_n^T\,] = \sigma^2 I \;, \tag{1.2}$$

where σ is typically unknown. We refer to this as the Gauss-Markov model, in view of the Gauss-Markov theorem for linear regression models. At times, we need the added condition that

$$\begin{aligned} &d_{1,n}, d_{2,n}, \cdots, d_{n,n} \quad \text{are iid and} \\ &\mathbb{E}[\,d_{1,n}\,] = 0 \quad , \quad \mathbb{E}[\,|\,d_{1,n}\,|^\kappa\,] < \infty \;, \end{aligned} \tag{1.3}$$

for some $\kappa > 2$. A typical choice is $\kappa = 4$.

A more restrictive but generally made assumption is that the d_{in} are iid normal random variables with mean 0 and again with the variance σ^2 usually unknown, described succinctly as

$$d_n \sim \text{Normal}(\,0\,, \sigma^2 I\,) \;. \tag{1.4}$$

This is referred to as the Gaussian model.

Regarding the regression function, the typical *nonparametric* assumption is that f_o is smooth. In this volume, this usually takes the form

$$f_o \in W^{m,2}(0,1) \tag{1.5}$$

for some integer m, $m \geqslant 1$. Recall the definition of the Sobolev spaces $W^{m,p}(a,b)$ in (12.2.18). Assumptions of this kind are helpful when the

P.P.B. Eggermont, V.N. LaRiccia, *Maximum Penalized Likelihood Estimation*,
Springer Series in Statistics, DOI 10.1007/b12285_2,

data points are spaced close together, so that the changes in the function values $f_o(x_{in})$ for neighboring design points are small compared with the noise. The following exercise shows that in this situation one can do better than merely "connecting the dots". Of course, it does not say how much better.

(1.6) EXERCISE. Let $x_{in} = i/n$, $i = 1, 2, \cdots, n$, and let y_{in} satisfy (1.1), with the errors satisfying (1.2). Assume that f_o is twice continuously differentiable. For $i = 2, 3, \cdots, n-1$,
(a) show that

$$\tfrac{1}{4}\{f_o(x_{i-1,n}) + 2\,f_o(x_{in}) + f_o(x_{i+1,n})\} = f_o(x_{in}) + (1/n)^2\,f_o{}''(\theta_{in})$$

for some $\theta_{in} \in (\,x_{i-1,n}\,,\,x_{i+1,n}\,)$;
(b) compute the mean and the variance of

$$z_{in} = \tfrac{1}{4}\{\,y_{i-1,n} + 2\,y_{in} + y_{i+1,n}\,\}\ ;$$

(c) compare the mean squared errors

$$\mathbb{E}[\,|\,z_{in} - f_o(x_{in})\,|^2\,] \quad\text{and}\quad \mathbb{E}[\,|\,y_{in} - f_o(x_{in})\,|^2\,]$$

with each other (the case $n \to \infty$ is interesting).

In this chapter, we study the smoothing spline estimator of differential order m, the solution to

$$\text{(1.7)}\qquad \begin{array}{ll} \text{minimize} & \frac{1}{n}\sum\limits_{i=1}^{n} |\,f(x_{in}) - y_{in}\,|^2 + h^{2m}\,\|\,f^{(m)}\,\|^2 \\ \text{subject to} & f \in W^{m,2}(0,1)\ . \end{array}$$

(The factor $\frac{1}{n}$ appears for convenience; this way, the objective function is well-behaved as $n \to \infty$. The funny choice h^{2m} vs. h^2 or h is more convenient later on, although this is a matter of taste.) The solution is denoted by f^{nh}. The parameter h in (1.7) is the smoothing parameter. In this chapter, we only consider deterministic choices of h. Random (data-driven) choices are discussed in Chapter 18, and their effects on the smoothing spline estimator are discussed in Chapter 22.

The solution of (1.7) is a spline of polynomial order $2m$. In the literature, the case $m = 2$ is predominant, and the corresponding splines are called cubic splines. The traditional definition of splines is discussed in Chapter 19 together with the traditional computational details. The modern way to compute splines of arbitrary order is discussed in Chapter 20.

The following questions now pose themselves: Does the solution of (1.7) exist and is it unique (see § 3), and how accurate is the estimator (see § 4 and § 14.7)? To settle these questions, the reproducing kernel Hilbert space setting of the smoothing spline problem (1.7) is relevant, in which $W^{m,2}(0,1)$ is equipped with the inner products

$$\text{(1.8)}\qquad \langle\, f\,,\,g\,\rangle_{m,h} = \langle\, f\,,\,g\,\rangle + h^{2m}\,\langle\, f^{(m)}\,,\,g^{(m)}\,\rangle\ ,$$

where $\langle \cdot , \cdot \rangle$ is the usual $L^2(0,1)$ inner product. Then, $W^{m,2}(0,1)$ with the $\langle \cdot , \cdot \rangle_{m,h}$ inner product is a reproducing kernel Hilbert space with the reproducing kernel indexed by the smoothing parameter h. Denoting the reproducing kernel by $\mathcal{R}_{mh}(s, t)$, this then gives the reproducing kernel property

$$(1.9) \qquad f(x) = \langle f , \mathcal{R}_{mh}(x , \cdot) \rangle_{m,h} , \quad x \in [0,1] ,$$

for all $f \in W^{m,2}(0,1)$ and all h, $0 < h \leqslant 1$.

The reproducing kernel shows up in various guises. For uniform designs and pure-noise data, the smoothing spline estimator is approximately the same as the solution ψ^{nh} of the semi-continuous version of the smoothing spline problem (1.7), viz.

$$(1.10) \qquad \begin{array}{ll} \text{minimize} & \| f \|^2 - \frac{2}{n} \sum\limits_{i=1}^{n} y_{in} f(x_{in}) + h^{2m} \| f^{(m)} \|^2 \\ \text{subject to} & f \in W^{m,2}(0,1) . \end{array}$$

The authors are tempted to call ψ^{nh} the C-spline estimator, C being short for "continuous", even though ψ^{nh} is not a polynomial spline. The reproducing kernel now pops up in the form

$$(1.11) \qquad \psi^{nh}(t) = \frac{1}{n} \sum_{i=1}^{n} y_{in} \mathcal{R}_{mh}(t , x_{in}) , \quad x \in (0,1) ,$$

because $\mathcal{R}_{mh}(t, x)$ is the Green's function for the Sturm-Liouville boundary value problem

$$(1.12) \qquad \begin{array}{lll} (-h^2)^m u^{(2m)} + u = w & , & t \in (0,1) , \\ u^{(\ell)}(0) = u^{(\ell)}(1) = 0 & , & m \leqslant \ell \leqslant 2m-1 . \end{array}$$

That is, the solution of (1.12) is given by

$$(1.13) \qquad u(t) = \int_0^1 \mathcal{R}_{mh}(t, x) w(x) \, dx , \quad t \in [0,1] .$$

With suitable modifications, this covers the case of point masses (1.11). In § 14.7, we show that the smoothing spline estimator is extremely well-approximated by

$$(1.14) \qquad \varphi^{nh}(t) = \int_0^1 \mathcal{R}_{mh}(t, x) f_o(x) \, dx + \frac{1}{n} \sum_{i=1}^{n} d_{in} \mathcal{R}_{mh}(t , x_{in})$$

for all $t \in [0,1]$. In effect, this is the equivalent *reproducing* kernel approximation of smoothing splines, to be contrasted with the equivalent kernels of SILVERMAN (1984). See also § 21.8. The reproducing kernel setup is discussed in § 2.

In § 5, we discuss the need for boundary corrections and their construction by way of the Bias Reduction Principle of EUBANK and SPECKMAN (1990b). In §§ 6–7, we discuss the boundary splines of OEHLERT (1992), which avoids rather than corrects the problem. Finally, in § 9, we briefly discuss the estimation of derivatives of the regression function.

EXERCISE: (1.6).

2. Reproducing kernel Hilbert spaces

Here we begin the study of the smoothing spline estimator for the problem (1.1)–(1.2). Recall that the estimator is defined as the solution to

$$\text{(2.1)} \qquad \begin{aligned} &\text{minimize} \quad L_{nh}(f) \stackrel{\text{def}}{=} \tfrac{1}{n} \sum_{i=1}^{n} |\, f(x_{in}) - y_{in} \,|^2 + h^{2m} \,\|\, f^{(m)} \,\|^2 \\ &\text{subject to} \quad f \in W^{m,2}(0,1) \; . \end{aligned}$$

Here, h is the smoothing parameter and m is the differential order of the smoothing spline. The solution of (2.1) is a spline of polynomial order $2m$ (or polynomial degree $2m-1$). At times, we just speak of the *order* of the spline, but the context should make clear which one is meant.

The design points are supposed to be (asymptotically) uniformly distributed in a sense to be defined precisely in Definition (2.22). For now, think of the equally spaced design $x_{in} = t_{in}$ with

$$\text{(2.2)} \qquad t_{in} = \frac{i-1}{n-1} \; , \quad i = 1, 2, \cdots, n \; .$$

The first question is of course whether the point evaluation functionals $f \longmapsto f(x_{in})$, $i = 1, 2, \cdots, n$, are well-defined. This has obvious implications for the existence and uniqueness of the solution of (2.1). Of course, if these point evaluation functionals are well-defined, then we are dealing with reproducing kernel Hilbert spaces. In Volume I, we avoided them more or less (more!) successfully, but see the KLONIAS (1982) treatment of the maximum penalized likelihood density estimator of GOOD and GASKINS (1971) in Exercise (5.2.64) in Volume I. For spline smoothing, the use of reproducing kernel Hilbert spaces will have far-reaching consequences.

The setting for the problem (2.1) is the space $W^{m,2}(0,1)$, which is a Hilbert space under the inner product

$$\text{(2.3)} \qquad \langle\, f \,,\, \varphi \,\rangle_{W^{m,2}(0,1)} = \langle\, f \,,\, \varphi \,\rangle + \langle\, f^{(m)} \,,\, \varphi^{(m)} \,\rangle$$

and associated norm

$$\text{(2.4)} \qquad \|\, f \,\|_{W^{m,2}(0,1)} = \{\, \|\, f \,\|^2 + \|\, f^{(m)} \,\|^2 \,\}^{1/2} \; .$$

Here, $\langle \cdot \,,\, \cdot \rangle$ denotes the usual $L^2(0,1)$ inner product. However, the norms

$$\text{(2.5)} \qquad \|\, f \,\|_{m,h} = \{\, \|\, f \,\|^2 + h^{2m} \,\|\, f^{(m)} \,\|^2 \,\}^{1/2}$$

and corresponding inner products

$$\langle f , \varphi \rangle_{m,h} = \langle f , \varphi \rangle + h^{2m} \langle f^{(m)} , \varphi^{(m)} \rangle \tag{2.6}$$

are useful as well. Note that, for each $h > 0$ the norms (2.4) and (2.5) are equivalent, but not uniformly in h. (The "equivalence" constants depend on h.) We remind the reader of the following definition.

(2.7) DEFINITION. Two norms $\| \cdot \|_U$ and $\| \cdot \|_W$ on a vector space V are equivalent if there exists a constant $c > 0$ such that

$$c \, \| v \|_U \leqslant \| v \|_W \leqslant c^{-1} \| v \|_U \quad \text{for all } v \in V \, .$$

(2.8) EXERCISE. Show that the norms (2.4) and (2.5) are equivalent.

We are now in a position to answer the question of whether the $f(x_{in})$ are well-defined for $f \in W^{m,2}(0,1)$ in the sense that $| f(x_{in}) | \leqslant c \| f \|_{m,h}$ for a suitable constant. This amounts to showing that $W^{m,2}(0,1)$ is a reproducing kernel Hilbert space; see Appendix 3, § 7, in Volume I.

In what follows, it is useful to introduce an abbreviation of the L^2 norm of a function $f \in L^2(0,1)$ restricted to an interval $(a,b) \subset (0,1)$,

$$\| f \|_{(a,b)} \stackrel{\text{def}}{=} \left\{ \int_a^b | f(x) |^2 \, dx \right\}^{1/2} , \tag{2.9}$$

but please do not confuse $\| \cdot \|_{(a,b)}$ (with parentheses) with $\| \cdot \|_{m,h}$ (without them).

(2.10) LEMMA. *There exists a constant* c_1 *such that, for all* $f \in W^{1,2}(0,1)$, *all* $0 < h \leqslant 1$, *and all* $x \in [\,0\,,\,1\,]$,

$$| f(x) | \leqslant c_1 \, h^{-1/2} \, \| f \|_{1,h} \, .$$

PROOF. The inequality

$$| f(x) - f(y) | = \left| \int_y^x f'(t) \, dt \, \right| \leqslant | x - y |^{1/2} \, \| f' \|$$

implies that every $f \in W^{m,2}(0,1)$ is (uniformly) continuous on $(0,1)$.

Consider an interval $[\,a, a+h\,] \subset [\,0\,,\,1\,]$. An appeal to the Intermediate Value Theorem shows the existence of a $y \in (a, a+h)$ with

$$| f(y) | = h^{-1/2} \, \| f \|_{(a,a+h)} \, .$$

From the inequalities above, we get, for all $x \in (a, a+h)$,

$$\begin{aligned} | f(x) | &\leqslant | f(y) | + | f(x) - f(y) | \\ &\leqslant h^{-1/2} \, \| f \|_{(a,a+h)} + h^{1/2} \, \| f' \|_{(a,a+h)} \\ &\leqslant h^{-1/2} \, \| f \| + h^{1/2} \, \| f' \| \, , \end{aligned}$$

and thus, after some elementary manipulations

$$| f(x) | \leqslant c\, h^{-1/2} \{ \| f \|^2 + h^2 \| f' \|^2 \}^{1/2}$$

with $c = \sqrt{2}$. Q.e.d.

(2.11) LEMMA [CONTINUITY OF POINT EVALUATIONS]. *Let $m \geqslant 1$ be an integer. There exists a constant c_m such that, for all $f \in W^{m,2}(0,1)$, all $0 < h \leqslant 1$, and all $x \in (0,1)$,*

$$| f(x) | \leqslant c_m\, h^{-1/2} \| f \|_{m,h} \, .$$

The proof goes by induction on m, as per the next two lemmas.

(2.12) INTERPOLATION LEMMA. *Let $m \geqslant 1$ be an integer. There exists a constant $c_m \geqslant 1$ such that, for all $f \in W^{m+1,2}(0,1)$ and all $0 < h \leqslant 1$,*

(a) $$\| f^{(m)} \| \leqslant c_m\, h^{-m} \| f \|_{m+1,h} \, ,$$

and, with $\theta = 1/(m+1)$,

(b) $$\| f^{(m)} \| \leqslant c_m \| f \|^{\theta} \{ \| f \|_{W^{m+1,2}(0,1)} \}^{1-\theta} \, .$$

Note that the inequality (b) of the lemma implies that

$$\| f \|_{W^{m,2}(0,1)} \leqslant \widetilde{c}_m \| f \|^{\theta} \| f \|^{1-\theta}_{W^{m+1,2}(0,1)}$$

for another constant $\widetilde{c}_m$. So ignoring this constant, after taking logarithms, the upper bound on $\log \| f \|_{W^{m,2}(0,1)}$ is obtained by linear interpolation on $\log \| f \|_{W^{x,2}(0,1)}$ between $x = 0$ and $x = m+1$, hence the name.

PROOF. From (a) one obtains that

$$\| f^{(m)} \| \leqslant c_m\, h^{-m} \| f \| + c_m\, h \| f \|_{W^{m+1}(0,1)} \, .$$

Now, take h such that $h^{m+1} = \| f \| / \| f \|_{W^{m+1,2}(0,1)}$ and (b) follows, for a possible larger constant c_m. (Note that indeed $h \leqslant 1$.)

The case $m = 1$ of the lemma is covered by the main inequality in the proof of Lemma (5.4.16) in Volume I. The proof now proceeds by induction.

Let $m \geqslant 1$. Suppose that the lemma holds for all integers up to and including m. Let $f \in W^{m+2,2}(0,1)$. Applying the inequality (a) with $m = 1$ to the function $f^{(m)}$ gives

$$\| f^{(m+1)} \| \leqslant c_1\, h^{-1} \left(\| f^{(m)} \| + h^2 \| f^{(m+2)} \| \right) . \tag{2.13}$$

Now, apply the inequality (b) of the lemma for m, so

$$\begin{aligned} \| f^{(m)} \| &\leqslant c_m \| f \|^{\theta} \{ \| f \|_{W^{m+1,2}(0,1)} \}^{1-\theta} \\ &\leqslant c_m \| f \| + c_m \| f \|^{\theta} \| f^{(m+1)} \|^{1-\theta} \, , \end{aligned} \tag{2.14}$$

since $(x+y)^\alpha \leqslant x^\alpha + y^\alpha$ for all positive x and y and $0 < \alpha \leqslant 1$. Substituting this into (2.13) gives, for suitable constants $\widetilde{c}_m$ and $\widetilde{c}_1$,

$$(2.15) \quad \| f^{(m+1)} \| \leqslant \widetilde{c}_m \, h^{-1} \| f \|^\theta \| f^{(m+1)} \|^{1-\theta} + \widetilde{c}_1 \, h^{-1} \left(\| f \| + h^2 \| f^{(m+2)} \| \right) .$$

Since $h \leqslant 1$, then $h^{-1} \leqslant h^{-m-1}$, so that

$$h^{-1} \left(\| f \| + h^2 \| f^{(m+2)} \| \right) \leqslant h^{-m-1} \| f \| + h \| f^{(m+2)} \| \leqslant h^{-m-1} \| f \|_{m+2,h} .$$

Substituting this into (2.15) gives

$$(2.16) \quad \| f^{(m+1)} \| \leqslant \widetilde{c}_m \, h^{-1} \| f \|^\theta \| f^{(m+1)} \|^{1-\theta} + \widetilde{c}_1 \, h^{-m-1} \| f \|_{m+2,h} .$$

This is an inequality of the form $x^p \leqslant a\,x + b$ with $p > 1$, which implies that $x^p \leqslant a^q + q\,b$, where $1/q = 1 - (1/p)$. See Exercise (4.10). This gives

$$\| f^{(m+1)} \| \leqslant (\widetilde{c}_m \, h^{-1})^{m+1} \| f \| + (m+1) \, \widetilde{c}_1 \, h^{-m-1} \| f \|_{m+1,h} .$$

This implies the inequality (a) for $m+1$. Q.e.d.

(2.17) LEMMA. *Let $m \geqslant 1$ be an integer. There exists a constant k_m such that, for all $f \in W^{m+1,2}(0,1)$ and all $0 < h < 1$,*

$$\| f \|_{m,h} \leqslant k_m \| f \|_{m+1,h} .$$

PROOF. Lemma (2.12) says that $h^m \| f^{(m)} \| \leqslant c_m \| f \|_{m+1,h}$. Now, squaring both sides and then adding $\| f \|^2$ gives the lemma, with $k_m^2 = 1 + c_m^2$. Q.e.d.

We now put all of the above together to show that the smoothing spline problem is "well-behaved" from various points of view.

Reproducing kernel Hilbert spaces. Lemma (2.11) shows that, for fixed $x \in [\,0\,,\,1\,]$, the linear functional $\ell(f) = f(x)$ is bounded on $W^{m,2}(0,1)$. Thus, the vector space $W^{m,2}(0,1)$ with the inner product (2.6) is a reproducing kernel Hilbert space and, for each $x \in (0,1)$, there exists an $\mathcal{R}_{m,h,x} \in W^{m,2}(0,1)$ such that, for all $f \in W^{m,2}(0,1)$,

$$f(x) = \langle \, \mathcal{R}_{m,h,x} \, , \, f \, \rangle_{m,h} .$$

It is customary to denote $\mathcal{R}_{m,h,x}(y)$ by $\mathcal{R}_{mh}(x,y)$. Applying the above to the function $f = \mathcal{R}_{mh}(y, \,\cdot\,)$, where $y \in [\,0\,,\,1\,]$ is fixed, then gives

$$\mathcal{R}_{mh}(y,x) = \langle \, \mathcal{R}_{mh}(x, \,\cdot\,) \, , \, \mathcal{R}_{mh}(y, \,\cdot\,) \, \rangle_{m,h} \quad \text{for all} \quad x \in [\,0\,,\,1\,] ,$$

whence $\mathcal{R}_{mh}(x,y) = \mathcal{R}_{mh}(y,x)$. Moreover, Lemma (2.11) implies that

$$\| \mathcal{R}_{mh}(x, \,\cdot\,) \|_{m,h}^2 = \mathcal{R}_{mh}(x,x) \leqslant c_m \, h^{-1/2} \| \mathcal{R}_{mh}(x, \,\cdot\,) \|_{m,h} ,$$

and the obvious conclusion may be drawn. We summarize this in a lemma.

(2.18) Reproducing Kernel Lemma. *Let $m \geqslant 1$ be an integer, and let $0 < h \leqslant 1$. Then $W^{m,2}(0,1)$ with the inner product $\langle \cdot , \cdot \rangle_{m,h}$ is a reproducing kernel Hilbert space, with kernel $\mathcal{R}_{mh}(x,y)$, such that, for all $f \in W^{m,2}(0,1)$ and all x,*

$$f(x) = \langle \mathcal{R}_{mh}(x, \cdot) , f \rangle_{m,h} \quad \text{for all} \quad x \in [0,1] .$$

Moreover, there exists a c_m such that, for all $0 < h \leqslant 1$, and all x,

$$\| \mathcal{R}_{mh}(x, \cdot) \|_{m,h} \leqslant c_m \, h^{-1/2} .$$

Random sums. The reproducing kernel Hilbert space framework bears fruit in the consideration of the random sums

$$\tfrac{1}{n} \sum_{i=1}^{n} d_{in} \, f(x_{in}) ,$$

where $f \in W^{m,2}(0,1)$ is random, i.e., is allowed to depend on the noise vector $d_n = (d_{1,n}, d_{2,n}, \cdots , d_{n,n})$. In contrast, define the "simple" random sums

$$\mathfrak{S}^{nh}(x) = \tfrac{1}{n} \sum_{i=1}^{n} d_{in} \, \mathcal{R}_{mh}(x_{in}, \cdot) , \tag{2.19}$$

where the randomness of the functions f is traded for the dependence on a smoothing parameter.

(2.20) Random Sum Lemma. *Let $m \geqslant 1$. For all $f \in W^{m,2}(0,1)$ and all noise vectors $d_n = (d_{1,n}, d_{2,n}, \cdots , d_{n,n})$,*

$$\Bigl| \tfrac{1}{n} \sum_{i=1}^{n} d_{in} \, f(x_{in}) \Bigr| \leqslant \| f \|_{m,h} \, \| \mathfrak{S}^{nh} \|_{m,h} .$$

Moreover, if d_n satisfies (1.2), then there exists a constant c such that

$$\mathbb{E}[\, \| \mathfrak{S}^{nh} \|_{m,h}^{2} \,] \leqslant c \, (nh)^{-1}$$

for all h, $0 < h \leqslant 1$, and all designs.

Proof. Since $f \in W^{m,2}(0,1)$, the reproducing kernel Hilbert space trick of Lemma (2.18) gives

$$f(x_{in}) = \langle \mathcal{R}_{mh}(x_{in}, \cdot) , f \rangle_{m.h} ,$$

and consequently

$$\tfrac{1}{n} \sum_{i=1}^{n} d_{in} \, f(x_{in}) = \langle \mathfrak{S}^{nh} , f \rangle_{m,h} ,$$

which gives the upper bound

$$\tfrac{1}{n} \sum_{i=1}^{n} d_{in} \, f(x_{in}) \leqslant \| f \|_{m,h} \, \| \mathfrak{S}^{nh} \|_{m,h} .$$

Note that all of this holds whether f is random or deterministic.

Now, one verifies that

$$\| \mathfrak{S}^{nh} \|^2 = n^{-2} \sum_{i,j=1}^{n} d_{in}\, d_{jn} \left\langle \mathcal{R}_{mh}(x_{in}, \cdot\,) , \mathcal{R}_{mh}(x_{jn}, \cdot\,) \right\rangle_{m,h} ,$$

and so, under the assumption (1.2),

$$\mathbb{E}\Big[\| \mathfrak{S}^{nh} \|^2 \Big] = \sigma^2 n^{-2} \sum_{i=1}^{n} \| \mathcal{R}_{mh}(x_{in}, \cdot\,) \|_{m,h}^2 .$$

The bound from the previous lemma on $\| \mathcal{R}_{mh}(x_{in}, \cdot\,) \|_{m,h}^2$ now completes the proof. Q.e.d.

(2.21) EXERCISE. Show that, under the assumptions of Lemma (2.20),

$$\sup \left\{ \frac{\left| \frac{1}{n} \sum_{i=1}^{n} d_{in}\, f(x_{in}) \right|}{\| f \|_{m,h}} \;\middle|\; \begin{array}{l} f \in W^{m,2}(0,1) \\ f \not\equiv 0 \end{array} \right\} = \| \mathfrak{S}^{nh} \|_{m,h} .$$

In other words, the supremum is attained by the solution of the pure-noise version of (1.10); i.e., with $y_{in} = d_{in}$ for all i.

Quadrature. The reproducing kernel Hilbert spaces setup of Lemma (2.18) shows that the linear functionals $\ell_{i,n}(f) = f(x_{in})$, $i = 1, 2, \cdots, n$, are continuous on $W^{m,2}(0,1)$ for $m \geqslant 1$. So the problem (2.1) starts to make sense. Along the same lines, we need to be able to compare

$$\tfrac{1}{n} \sum_{i=1}^{n} | f(x_{in}) |^2 \quad \text{and} \quad \| f \|^2$$

with each other, at least for $f \in W^{m,2}(0,1)$. In effect, this is a requirement on the design, and is a quadrature result for specific designs.

(2.22) DEFINITION. We say that the design x_{in}, $i = 1, 2, \cdots, n$, is asymptotically uniform if there exists a constant c such that, for all $n \geqslant 2$ and all $f \in W^{1,1}(0,1)$,

$$\left| \tfrac{1}{n} \sum_{i=1}^{n} f(x_{in}) - \int_0^1 f(t)\, dt \right| \leqslant c\, n^{-1} \| f' \|_{L^1(0,1)} .$$

(2.23) REMARK. The rate n^{-1} could be lowered to $\big(n^{-1} \log n \big)^{1/2}$ but seems to cover most cases of interest. Random designs require their own treatment; see Chapter 21.

(2.24) LEMMA. *The design (2.2) is asymptotically uniform. In fact, for every $f \in W^{1,1}(0,1)$,*

$$\Bigl| \tfrac{1}{n} \sum_{i=1}^{n} f(t_{in}) - \int_0^1 f(t)\,dt \Bigr| \leqslant \tfrac{1}{n-1} \| f' \|_1 \ .$$

PROOF. The first step is the identity,

$$\tfrac{1}{n} \sum_{i=1}^{n} c_{in} = \tfrac{1}{n-1} \sum_{i=1}^{n-1} \{ a_{in} c_{in} + b_{in} c_{i+1,n} \} \ , \tag{2.25}$$

for all c_{in}, $i = 1, 2, \cdots, n$, where $a_{in} = (n-i)/n$, $b_{in} = i/n$. Of course, we take $c_{in} = f(t_{in})$. Then, with the intervals $\omega_{in} = (t_{in}, t_{i+1,n})$,

$$\tfrac{1}{n-1} \{ a_{in} f(t_{in}) + b_{in} f(t_{i+1,n}) \} - \int_{\omega_{in}} f(t)\,dt =$$

$$a_{in} \int_{\omega_{in}} \{ f(t_{in}) - f(t) \}\,dt + b_{in} \int_{\omega_{in}} \{ f(t_{i+1,n}) - f(t) \}\,dt \ .$$

Now, for $t \in \omega_{in}$,

$$| f(t_{in}) - f(t) | = \Bigl| \int_t^{t_{in}} f'(s)\,ds \Bigr| \leqslant \int_{\omega_{in}} | f'(s) |\,ds \ ,$$

so, after integration over ω_{in}, an interval of length $1/(n-1)$,

$$\int_{\omega_{in}} | f(t_{in}) - f(t) |\,dt \leqslant \tfrac{1}{n-1} \int_{\omega_{in}} | f'(t) |\,dt \ .$$

The same bound applies to $\int_{\omega_{in}} | f(t_{i+1,n}) - f(t) |\,dt$. Then, adding these bounds gives

$$\Bigl| \tfrac{1}{n-1} \{ a_{in} f(t_{in}) + b_{in} f(t_{i+1,n}) \} - \int_{\omega_{in}} f(t)\,dt \Bigr| \leqslant \tfrac{1}{n-1} \int_{\omega_{in}} | f'(t) |\,dt \ ,$$

and then adding *these* over $i = 1, 2, \cdots, n-1$, together with the triangle inequality, gives the required result. Q.e.d.

(2.26) EXERCISE. Show that the design $t_{in} = i/(n+1)$, $i = 1, 2, \cdots, n$, is asymptotically uniform and likewise for $t_{in} = (i - \frac{1}{2})/n$.

(2.27) QUADRATURE LEMMA. *Let $m \geqslant 1$. Assuming the design is asymptotically uniform, there exists a constant c_m such that, for all $f \in W^{m,2}(0,1)$, all $n \geqslant 2$, and all h, $0 < h \leqslant \frac{1}{2}$,*

$$\Bigl| \tfrac{1}{n} \sum_{i=1}^{n} | f(x_{in}) |^2 - \| f \|^2 \Bigr| \leqslant c_m (nh)^{-1} \| f \|_{m,h}^2 \ .$$

PROOF. Let $m \geqslant 1$. As a preliminary remark, note that, for $f \in W^{m,2}(0,1)$, we have of course that $\| f^2 \|_1 = \| f \|^2$ and that

$$\| (f^2)' \|_1 = 2 \| f f' \|_1 \leqslant 2 \| f \| \| f' \| = 2 h^{-1} \| f \| \{ h \| f' \| \} \\ \leqslant h^{-1} \{ \| f \|^2 + h^{-1} \| f' \|^2 \} = h^{-1} \| f \|_{1,h}^2 ,$$

where we used Cauchy-Schwarz and the inequality $2ab \leqslant a^2 + b^2$.

Then, for $n \geqslant 2$, by the asymptotic uniformity of the design,

$$\left| \tfrac{1}{n} \sum_{i=1}^{n} | f(x_{in}) |^2 - \| f \|^2 \right| \leqslant c n^{-1} \| (f^2)' \|_1 , \tag{2.28}$$

which by the above, may be further bounded by

$$c (nh)^{-1} \| f \|_{1,h}^2 \leqslant \widetilde{c} (nh)^{-1} \| f \|_{m,h}^2$$

for an appropriate constant $\widetilde{c}$, the last inequality by Lemma (2.17). This is the lemma. Q.e.d.

The following is an interesting and useful exercise on the multiplication of functions in $W^{m,2}(0,1)$.

(2.29) EXERCISE. (a) Show that there exists a constant c_m such that, for all h, $0 < h \leqslant \frac{1}{2}$,

$$\| f g \|_{m,h} \leqslant c_m h^{-1/2} \| f \|_{m,h} \| g \|_{m,h} \quad \text{for all} \quad f, \ g \in W^{m,2}(0,1) .$$

(b) Show that the factor $h^{-1/2}$ is sharp for $h \to 0$.

EXERCISES: (2.8), (2.21), (2.26), (2.29).

3. Existence and uniqueness of the smoothing spline

In this section, we discuss the existence and uniqueness of the solution of the smoothing spline problem. Of course, the quadratic nature of the problem makes life very easy, and it is useful to consider that first. Note that in Lemma (3.1) below there are no constraints on the design. We emphasize that, throughout this section, the sample size n and the smoothing parameter h remain fixed.

(3.1) QUADRATIC BEHAVIOR LEMMA. *Let $m \geqslant 1$, and let φ be a solution of (2.1). Then, for all $f \in W^{m,2}(0,1)$,*

$$\tfrac{1}{n} \sum_{i=1}^{n} | f(x_{in}) - \varphi(x_{in}) |^2 + h^{2m} \| (f - \varphi)^{(m)} \|^2 =$$

$$\tfrac{1}{n} \sum_{i=1}^{n} (f(x_{in}) - y_{in}) (f(x_{in}) - \varphi(x_{in})) + h^{2m} \langle f^{(m)} , (f - \varphi)^{(m)} \rangle .$$

PROOF. Since $L_{nh}(f)$ is quadratic, it is convex, and thus, see, e.g., Chapter 10 in Volume I or Chapter 3 in TROUTMAN (1983), it has a Gateaux variation (directional derivative) at each $\varphi \in W^{m,2}(0,1)$. One verifies that it is given by

$$\text{(3.2)} \quad \delta L_{nh}(\varphi\,,\, f-\varphi) = 2\,h^{2m}\,\langle\, \varphi^{(m)}\,,\, (f-\varphi)^{(m)}\,\rangle \; + \\ \tfrac{2}{n}\sum_{i=1}^{n}\big(\,\varphi(x_{in}) - y_{in}\,\big)\big(\,f(x_{in}) - \varphi(x_{in})\,\big)\;,$$

so that

$$\text{(3.3)} \quad L_{nh}(f) - L_{nh}(\varphi) - \delta L_{nh}(\varphi\,,\, f-\varphi) = \\ h^{2m}\,\|\,(f-\varphi)^{(m)}\,\|^2 + \tfrac{1}{n}\sum_{i=1}^{n}|\,f(x_{in}) - \varphi(x_{in})\,|^2\;.$$

In fact, this last result is just an identity for quadratic functionals. Now, by the necessary and sufficient conditions for a minimum, see, e.g., Theorem (10.2.2) in Volume I or Proposition (3.3) in TROUTMAN (1983), the function φ solves the problem (2.1) if and only if

$$\text{(3.4)} \qquad \delta L_{nh}(\varphi\,,\, f-\varphi) = 0 \quad \text{for all} \quad f \in W^{m,2}(0,1)\;.$$

Then, the identity (3.3) simplifies to

$$\text{(3.5)} \quad \tfrac{1}{n}\sum_{i=1}^{n}|\,f(x_{in}) - \varphi(x_{in})\,|^2 + h^{2m}\,\|\,(f-\varphi)^{(m)}\,\|^2 = L_{nh}(f) - L_{nh}(\varphi)\;.$$

Now, in (3.3), interchange f and φ to obtain

$$L_{nh}(f) - L_{nh}(\varphi) = -\tfrac{1}{n}\sum_{i=1}^{n}|\,f(x_{in}) - \varphi(x_{in})\,|^2 - h^{2m}\,\|\,(\,f-\varphi\,)^{(m)}\,\|^2 + \\ \tfrac{2}{n}\sum_{i=1}^{n}\big(\,f(x_{in}) - y_{in}\,\big)\big(\,f(x_{in}) - \varphi(x_{in})\big) + \\ 2\,h^{2m}\,\langle\, f^{(m)}\,,\,(\,f-\varphi\,)^{(m)}\,\rangle\;.$$

Finally, substitute this into (3.5), move the negative quadratics to the left of the equality, and divide by 2. This gives the lemma. Q.e.d.

(3.6) UNIQUENESS LEMMA. *Let $m \geqslant 1$, and suppose that the design contains at least m distinct points. Then the solution of (2.1) is unique.*

PROOF. Suppose φ and ψ are solutions of (2.1). Since $L_{nh}(\varphi) = L_{nh}(\psi)$, then, by (3.5),

$$\tfrac{1}{n}\sum_{i=1}^{n}|\,\varphi(x_{in}) - \psi(x_{in})\,|^2 + h^{2m}\|\,(\,\varphi-\psi\,)^{(m)}\,\|^2 = 0\;.$$

It follows that $(\,\varphi-\psi\,)^{(m)} = 0$ almost everywhere, and so $\varphi - \psi$ is a polynomial of degree $\leqslant m-1$. And of course

$$\varphi(x_{in}) - \psi(x_{in}) = 0\;, \quad i = 1, 2, \cdots, n.$$

Now, if there are (at least) m distinct design points, then this says that the polynomial $\varphi - \psi$ has at least m distinct zeros. Since it has degree $\leqslant m-1$, the polynomial vanishes everywhere. In other words, $\varphi = \psi$ everywhere. Q.e.d.

(3.7) EXISTENCE LEMMA. *Let* $m \geqslant 1$. *For any design, the smoothing spline problem* (2.1) *has a solution.*

PROOF. Note that the functional L_{nh} is bounded from below (by 0), and so its infimum over $W^{m,2}(0,1)$ is finite. Let $\{ f_k \}_k$ be a minimizing sequence. Then, using Taylor's theorem with exact remainder, write

$$f_k(x) = p_k(x) + [\, T f_k^{(m)} \,](x) \; , \tag{3.8}$$

where p_k is a polynomial of order m, and for $g \in L^2(0,1)$,

$$Tg(x) = \int_0^x \frac{(x-t)^{m-1}}{(m-1)!} \, g(t) \, dt \; . \tag{3.9}$$

Note that the Arzelà-Ascoli theorem implies the compactness of the operator $T : L^2(0,1) \longrightarrow C[\,0\,,1\,]$.

Now, since without loss of generality $L_{nh}(f_k) \leqslant L_{nh}(f_1)$, it follows that

$$\| f_k^{(m)} \|^2 \leqslant h^{-2m} L_{nh}(f_1)$$

and so $\{ f_k^{(m)} \}_k$ is a bounded sequence in $L^2(0,1)$. Thus, it has a weakly convergent subsequence, which we denote again by $\{ f_k^{(m)} \}_k$, with weak limit denoted by φ_o. Then, by the weak lower semi-continuity of the norm,

$$\lim_{k \to \infty} \| f_k^{(m)} \|^2 \geqslant \| \varphi_o \|^2 \; . \tag{3.10}$$

Moreover, since T is compact, it maps weakly convergent sequences into strongly convergent ones. In other words,

$$\lim_{k \to \infty} \| T f_k^{(m)} - T\varphi_o \|_\infty = 0 \; . \tag{3.11}$$

Now, consider the restrictions of the f_k to the design points,

$$r_n f_k \stackrel{\text{def}}{=} \big(f_k(x_{1,n}), f_k(x_{2,n}), \cdots, f_k(x_{n,n}) \big) \; , \quad k = 1, 2, \cdots \; .$$

We may extract a subsequence from $\{ f_k \}_k$ for which the corresponding sequence $\{ r_n f_k \}_k$ converges in $\mathbb{R}^n$ to some vector v_o. Then it is easy to see that, for the corresponding polynomials,

$$\lim_{k \to \infty} p_k(x_{in}) = [\, v_o \,]_i - T\varphi_o(x_{in}) \; , \quad i = 1, 2, \cdots, n \; .$$

All that there is left to do is claim that there exists a polynomial p_o of order m such that

$$p_o(x_{in}) = [\, v_o \,]_i - T\varphi_o(x_{in}) \; , \quad i = 1, 2, \cdots, n \; ,$$

the reason being that the vector space $\{ r_n p : p \in \mathcal{P}_m \}$, where $\mathcal{P}_m$ is the vector space of all polynomials of order m, is finite-dimensional, and hence closed; see, e.g., HOLMES (1975).

So now we are in business: Define $\psi_o = p_o + T\varphi_o$, and it is easy to see that, for the (subsub) sequence in question,

$$\lim_{k\to\infty} L_{nh}(f_k) \geqslant L_{nh}(\psi_o) ,$$

so that ψ_o minimizes $L_{nh}(f)$ over $f \in W^{m,2}(0,1)$. Q.e.d.

(3.12) EXERCISE. The large-sample asymptotic problem corresponding to the finite-sample problem (2.1) is defined by

$$\begin{aligned} &\text{minimize} \quad L_{\infty h}(f) \stackrel{\text{def}}{=} \| f - f_o \|^2 + h^{2m} \| f^{(m)} \|^2 \\ &\text{subject to} \quad f \in W^{m,2}(0,1) . \end{aligned}$$

(a) Compute the Gateaux variation of $L_{\infty h}$, and show that

$$L_{\infty h}(f) - L_{\infty h}(\varphi) - \delta L_{\infty h}(\varphi , f - \varphi) = \| f - \varphi \|_{m,h}^2 .$$

(b) Show that $L_{\infty h}$ is strongly convex and weakly lower semi-continuous on $W^{m,2}(0,1)$.

(c) Conclude that the solution of the minimization problem above exists and is unique.

(3.13) EXERCISE. Consider the C-spline estimation problem (1.10), repeated here for convenience:

$$\begin{aligned} &\text{minimize} \quad \| f \|^2 - \tfrac{2}{n} \sum_{i=1}^{n} y_{in} f(x_{in}) + h^{2m} \| f^{(m)} \|^2 \\ &\text{subject to} \quad f \in W^{m,2}(0,1) . \end{aligned}$$

Show the existence and uniqueness of the solution of this problem. You should not need the asymptotic uniformity of the design.

As mentioned before, the convexity approach to showing existence and uniqueness is a heavy tool, but it makes for an easy treatment of convergence rates of the spline estimators, see § 4. It has the additional advantage that we can handle constrained problems without difficulty. Let $\mathcal{C}$ be a closed, convex subset of $W^{m,2}(0,1)$, and consider the problems

$$\begin{aligned} &\text{minimize} \quad \tfrac{1}{n} \sum_{i=1}^{n} | f(x_{in}) - y_{in} |^2 + h^{2m} \| f^{(m)} \|^2 \\ &\text{subject to} \quad f \in \mathcal{C} , \end{aligned} \tag{3.14}$$

and

$$\text{(3.15)}\qquad \begin{aligned} &\text{minimize} \quad \| f \|^2 - \tfrac{2}{n} \sum_{i=1}^{n} f(x_{in})\, y_{in} + h^{2m} \| f^{(m)} \|^2 \\ &\text{subject to} \quad f \in \mathcal{C} \ . \end{aligned}$$

(3.16) THEOREM. *The solution of the constrained smoothing spline problem (3.13) exists, and if there are at least m distinct design points, then it is unique. For the constrained problem (3.14), the solution always exists and is unique.*

(3.17) EXERCISE. Prove it!

Finally, we consider the Euler equations for the problem (2.1). One verifies that they are given by

$$\text{(3.18)}\qquad \begin{aligned} &(-h^2)^m\, u^{(2m)} + \tfrac{1}{n} \sum_{i=1}^{n} \big(u(x_{in}) - y_{in} \big)\, \delta(\,\cdot - x_{in}) = 0 \quad \text{in } (0,1) \ , \\ &u^{(k)}(0) = u^{(k)}(1) = 0 \ , \quad k = m, m+1, \cdots, 2m-1 \ . \end{aligned}$$

Here $\delta(\,\cdot - x_{in})$ is the unit point mass at $x = x_{in}$. (For the two endpoints, this requires the proper interpretation: Assume that they are moved into the interior of $[\,0\,,\,1\,]$ and take limits.) The boundary conditions in (3.18) go by the name of "natural" boundary conditions in that they are automagically associated with the problem (2.1). As an alternative, one could prescribe boundary values; e.g., if one knew $f_o^{(k)}(x)$, $k = 0, 1, \cdots, m-1$, at the endpoints $x = 0$, $x = 1$. In this case, the minimization in (2.1) could be further restricted to those functions f with the same boundary values, and the boundary conditions in (3.18) would be replaced by

$$\text{(3.19)}\qquad u^{(k)}(0) = f_o^{(k)}(0) \ , \ u^{(k)}(1) = f_o^{(k)}(1) \ , \quad 0 \leqslant k \leqslant m-1 \ .$$

(3.20) EXERCISE. (a) Verify that (3.18) are indeed the Euler equations for the smoothing spline problem (2.1) and that
(b) the unique solution of the Euler equations solves (2.1) and vice versa. [Hint: See § 10.5 in Volume I.]

(3.21) EXERCISE. (a) Show that the Euler equations for the C-spline problem discussed in (3.15) are given by

$$\begin{aligned} &(-h^2)^m\, u^{(2m)} + u = \tfrac{1}{n} \sum_{i=1}^{n} y_{in}\, \delta(\,\cdot - x_{in}) \quad \text{in } (0,1) \ , \\ &u^{(k)}(0) = u^{(k)}(1) = 0 \ , \quad k = m, m+1, \cdots, 2m-1 \ . \end{aligned}$$

(b) Verify that the solution is given by

$$\psi^{nh}(\,t\,) = \tfrac{1}{n} \sum_{i=1}^{n} y_{in}\, \mathcal{R}_{mh}(x_{in}, t\,) \ , \quad t \in [\,0\,,\,1\,] \ .$$

(c) Show that the unique solution of the Euler equations solves (2.1) and vice versa. [Hint: § 10.5.]

Exercises: (3.12), (3.13), (3.17), (3.20), (3.21).

4. Mean integrated squared error

We are now ready to investigate the asymptotic error bounds for the smoothing spline estimator. We recall the model

$$(4.1) \qquad y_{in} = f_o(x_{in}) + d_{in} \ , \quad i = 1, 2, \cdots, n \ ,$$

in which the noise vector $d_n = (d_{1,n}, d_{2,n}, \cdots, d_{n,n})^T$ satisfies the Gauss-Markov conditions

$$(4.2) \qquad \mathbb{E}[\, d_n \,] = 0 \ , \quad \mathbb{E}[\, d_n \, d_n^T \,] = \sigma^2 I \ ,$$

and f_o is the function to be estimated. The design is supposed to be asymptotically uniform; see Definition 2.22. Regarding the unknown function f_o, we had the assumption

$$(4.3) \qquad f_o \in W^{m,2}(0,1) \ .$$

The smoothing spline estimator, denoted by f^{nh}, is the solution to

$$(4.4) \qquad \begin{array}{ll} \text{minimize} & L_{nh}(f) \stackrel{\text{def}}{=} \frac{1}{n} \sum\limits_{i=1}^{n} |\, f(x_{in}) - y_{in} \,|^2 + h^{2m} \|\, f^{(m)} \,\|^2 \\ \text{subject to} & f \in W^{m,2}(0,1) \ . \end{array}$$

It is useful to introduce the abbreviation ε^{nh} for the error function,

$$(4.5) \qquad \varepsilon^{nh} \equiv f^{nh} - f_o \ .$$

(4.6) Theorem. *Let $m \geqslant 1$. Suppose the Markov conditions (4.1) and (4.2) hold and that $f_o \in W^{m,2}(0,1)$. If x_{in}, $i = 1, 2, \cdots, n$, is asymptotically uniform, then for all $n \geqslant 2$ and all h, $0 < h \leqslant \frac{1}{2}$, with $nh \to \infty$,*

$$\zeta^{nh} \,\|\, f^{nh} - f_o \,\|_{m,h}^2 \leqslant \Big\{ \,\|\, \mathfrak{S}^{nh} \,\|_{m,h} + h^m \,\|\, f_o^{(m)} \,\| \,\Big\}^2 \ ,$$

where $\zeta^{nh} \to 1$. Here, $\mathfrak{S}^{nh}$ is given by (2.19).

(4.7) Corollary. *Under the same conditions as in the previous theorem, for $h \asymp n^{-1/(2m+1)}$ (deterministically),*

$$\mathbb{E}[\,\|\, f^{nh} - f_o \,\|_{m,h}^2 \,] = \mathcal{O}\big(n^{-2m/(2m+1)} \big) \ .$$

Proof of Theorem (4.6). The approach to obtaining error bounds is via the Quadratic Behavior Lemma (3.1) for $L_{nh}(f)$. This gives the equality

$$(4.8) \quad \tfrac{1}{n}\sum_{i=1}^{n} |\,\varepsilon^{nh}(x_{in})\,|^2 + h^{2m}\,\|\,(\varepsilon^{nh})^{(m)}\,\|^2 = \tfrac{1}{n}\sum_{i=1}^{n} d_{in}\,\varepsilon^{nh}(x_{in}) - h^{2m}\,\langle\, f_o^{(m)}\,,\,(\varepsilon^{nh})^{(m)}\,\rangle\;.$$

Of course, first we immediately use Cauchy-Schwarz,

$$-\,\langle\, f_o^{(m)}\,,\,(\varepsilon^{nh})^{(m)}\,\rangle \leqslant \|\,f_o^{(m)}\,\|\,\|\,(\varepsilon^{nh})^{(m)}\,\|\;.$$

Second, by the Random Sum Lemma (2.20), the random sum in (4.8) may be bounded by $\|\,\varepsilon^{nh}\,\|\,\|\,\mathfrak{S}^{nh}\,\|_{m,h}$.

Third, by the Quadrature Lemma (2.27), the sum on the left of (4.8) may be bounded from below by

$$\zeta^{nh}\,\|\,\varepsilon^{nh}\,\|_{m,h}^2 \leqslant \tfrac{1}{n}\sum_{i=1}^{n} |\,\varepsilon^{nh}(x_{in})\,|^2 + h^{2m}\|\,(\varepsilon^{nh})^{(m)}\,\|^2\;,$$

with $\zeta^{nh} = 1 - c_m(nh)^{-1}$. So, under the stated conditions, then $\zeta^{nh} \to 1$.

It follows from (2.10) that then

$$(4.9) \qquad \zeta^{nh}\,\|\,\varepsilon^{nh}\,\|_{m,h}^2 \leqslant \|\,\varepsilon^{nh}\,\|_{m,h}\Bigl\{\,\|\,\mathfrak{S}^{nh}\,\|_{m,h} + h^m\,\|\,f_o^{(m)}\,\|\,\Bigr\}\;,$$

where we used that $h^{2m}\,\|\,(\varepsilon^{nh})^{(m)}\,\| \leqslant h^m\,\|\,\varepsilon^{nh}\,\|_{m,h}$. The theorem follows by an appeal to the following exercise. Q.e.d.

(4.10) Exercise. Let a and b be positive real numbers, and let $p > 1$. If the nonnegative real number x satisfies $x^p \leqslant a\,x + b$, then

$$x^p \leqslant a^q + q\,b\;,$$

in which q is the dual exponent of p; i.e., $(1/p) + (1/q) = 1$.

(4.11) Exercise. Show that the bounds of Theorem (4.6) and Corollary (4.7) apply also to $\frac{1}{n}\sum_{i=1}^{n} |\,f^{nh}(x_{in}) - f_o(x_{in})\,|^2$.

The above is a concise treatment of the smoothing spline problem. The reader should become very comfortable with it since variations of it will be used throughout the text.

Can the treatment above be improved? The only chance we have is to avoid Cauchy-Schwarz in

$$-\,\langle\, f_o^{(m)}\,,\,(\varepsilon^{nh})^{(m)}\,\rangle \leqslant \|\,f_o^{(m)}\,\|\,\|\,(\varepsilon^{nh})^{(m)}\,\|\;,$$

following (4.8). Under the special smoothness condition and natural boundary conditions,

$$(4.12) \quad f_o \in W^{2m,2}(0,1)\;,\quad f^{(\ell)}(0) = f^{(\ell)}(1) = 0\;,\quad m \leqslant \ell \leqslant 2m-1\;,$$

this works. Results like this go by the name of superconvergence, since the accuracy is much better than guaranteed by the estimation method.

(4.13) SUPER CONVERGENCE THEOREM. *Assume the conditions of Theorem* (4.6). *If the regression function* $f_o \in W^{2m,2}(0,1)$ *satisfies the natural boundary conditions* (4.12), *then*

$$\| f^{nh} - f_o \|_{m,h}^2 \leqslant \Big\{ \| \mathfrak{S}^{nh} \|_{m,h}^2 + h^{2m} \| f_o \|_{W^{2m,2}(0,1)} \Big\}^2 ,$$

and for $h \asymp n^{-1/(4m+1)}$ *(deterministically),*

$$\mathbb{E}[\, \| f^{nh} - f_o \|_{m,h}^2 \,] = \mathcal{O}\big(n^{-4m/(4m+1)} \big) .$$

PROOF. The natural boundary conditions (4.11) allow us to integrate by parts m times, without being burdened with boundary terms. This gives

$$- \langle f_o^{(m)} , (\varepsilon^{nh})^{(m)} \rangle = (-1)^{m+1} \langle f_o^{(2m)} , \varepsilon^{nh} \rangle \leqslant \| f_o^{(2m)} \| \, \| \varepsilon^{nh} \| ,$$

and, of course, $\| \varepsilon^{nh} \| \leqslant \| \varepsilon^{nh} \|_{m,h}$. Thus, in the inequality (4.9), we may replace $h^m \| f_o^{(m)} \|$ by $h^{2m} \| f_o^{(2m)} \|$, and the rest follows. Q.e.d.

A brief comment on the condition (4.12) in the theorem above is in order. The smoothness assumption $f_o \in W^{2m,2}(0,1)$ is quite reasonable, but the boundary conditions on f_o are inconvenient, to put it mildly. In the next two sections, we discuss ways around the boundary conditions. In the meantime, the following exercise is useful in showing that the boundary conditions of Theorem (4.13) may be circumvented *at a price* (viz. of periodic boundary conditions).

(4.14) EXERCISE. Let $m \geqslant 1$ and $f_o \in W^{2m,2}(0,1)$. Prove the bounds of Theorem (4.13) for the solution of

$$\begin{aligned} &\text{minimize} && L_{nh}(f) \stackrel{\text{def}}{=} \tfrac{1}{n} \sum_{i=1}^{n} | f(x_{in}) - y_{in} |^2 + h^{2m} \| f^{(m)} \|^2 \\ &\text{subject to} && f \in W^{m,2}(0,1) , \\ & && \text{and for} \quad k = 0, 1, \cdots, m-1 , \\ & && f^{(k)}(0) = f_o^{(k)}(0) , \; f^{(k)}(1) = f_o^{(k)}(1) . \end{aligned}$$

The following exercise discusses what happens when the boundary conditions in Theorem (4.13) are only partially fulfilled. This finds a surprising application to boundary corrections; i.e., for obtaining estimators for which the conclusions of Theorem (4.13) remain valid. See § 5.

(4.15) EXERCISE. Let $1 \leqslant k \leqslant m$. Suppose that $f_o \in W^{m+k}(0,1)$ satisfies

$$f_o^{(\ell)}(0) = f_o^{(\ell)}(1) = 0 , \quad \ell = m, \cdots, m+k-1 .$$

Show that $\| f^{nh} - f_o \|^2 \leqslant \{ \| \mathfrak{S}^{nh} \|_{m,h} + h^{m+k} \| f_o^{(m+k)} \| \}^2$.

We finish with some exercises regarding constrained estimation and the C-spline problem (1.10).

(4.16) EXERCISE. (a) Derive the error bounds of Theorem (4.7) and Corollary (4.8) for the constrained smoothing spline problem

$$\begin{aligned} &\text{minimize} \quad \tfrac{1}{n} \sum_{i=1}^{n} | f(x_{in}) - y_{in} |^2 + h^{2m} \| f^{(m)} \|^2 \\ &\text{subject to} \quad f \in \mathcal{C} , \end{aligned}$$

where $\mathcal{C}$ is a closed and convex subset of $W^{m,2}(0,1)$. Assume that $f_o \in \mathcal{C}$.
(b) Do likewise for the constrained version of (1.10).

(4.17) EXERCISE. Show that the error bounds of Theorems (4.6) and (4.13) also apply to the solution of the C-spline estimation problem (1.10).

An alternative approach. We now consider an alternative development based on the observation that there are three sources of "error" in the smoothing spline problem (4.4). The obvious one is the noise in the data. Less obvious is that the roughness penalization is the source of bias, and finally there is the finiteness of the data. Even if the data were noiseless, we still could not estimate f_o perfectly due to the finiteness of the design. We need to introduce the finite noiseless data problem,

$$(4.18) \qquad \begin{aligned} &\text{minimize} \quad \tfrac{1}{n} \sum_{i=1}^{n} | f(x_{in}) - f_o(x_{in}) |^2 + h^{2m} \| f^{(m)} \|^2 \\ &\text{subject to} \quad f \in W^{m,2}(0,1) , \end{aligned}$$

as well as the large-sample asymptotic noiseless problem,

$$(4.19) \qquad \begin{aligned} &\text{minimize} \quad \| f - f_o \|^2 + h^{2m} \| f^{(m)} \|^2 \\ &\text{subject to} \quad f \in W^{m,2}(0,1) . \end{aligned}$$

In the exercise below, we (i.e., you) will analyze these problems.
The following simple exercise is quite useful.

(4.20) EXERCISE. Show that, for all real numbers A, B, a, b

$$| A - b |^2 - | A - a |^2 + | B - a |^2 - | B - b |^2 = 2\,(a - b)(A - B) .$$

(4.21) EXERCISE. Let $f_o \in W^{m,2}(0,1)$. Let f^{nh} be the solution of (4.4) and f_{hn} the solution of (4.18). Show that, for $nh \to \infty$ and h bounded,

$$\| f^{nh} - f_{hn} \|_{m,h} \leqslant \| \mathfrak{S}^{nh} \|_{m,h} ,$$

with $\mathfrak{S}^{nh}$ as in (2.19).

The bias due to the finiteness of the data is considered next.

(4.22) EXERCISE. Let $f_o \in W^{m,2}(0,1)$. Let f_h be the solution of (4.19) and f_{hn} the solution of (4.18). Show that, for a suitable constant c, as $h \to 0$ and nh large enough,

$$\| f_{hn} - f_h \|_{m,h} \leqslant c\,(nh)^{-1} \, \| f_h - f_o \|_{m,h} \;.$$

(4.23) EXERCISE. Show that the solution f_h of (4.19) satisfies

$$\| f_h - f_o \|_{m,h}^2 \leqslant h^{2m} \, \| f_o^{(m)} \|^2 \;.$$

We may now put these exercises together.

(4.24) EXERCISE. Prove Theorem (4.6) using Exercises (4.21)–(4.23).

(4.25) EXERCISE. Prove the analogue of Theorem (4.6) for the C-spline estimator of (1.10) straightaway (or via the detour).

EXERCISES: (4.10), (4.11), (4.14), (4.15), (4.16), (4.17), (4.20), (4.21), (4.22), (4.23), (4.24), (4.25).

5. Boundary corrections

In this section and the next, we take a closer look at the smoothness and boundary conditions (4.12), repeated here for convenience:

$$(5.1) \qquad f_o \in W^{2m,2}(0,1) \;,$$

$$(5.2) \qquad f_o^{(\ell)}(0) = f_o^{(\ell)}(1) \;, \quad m \leqslant \ell \leqslant 2m-1 \;.$$

In Theorem (4.13), we showed that, under these circumstances, the smoothing spline estimator f^{nh} of order $2m$ (degree $2m-1$) has expected error

$$(5.3) \qquad \mathbb{E}[\, \| f^{nh} - f_o \|^2 \,] = \mathcal{O}\big(n^{-4m/(4m+1)} \big) \;,$$

at least for $h \asymp n^{1/(4m+1)}$ (deterministic), and that the improvement over the bounds of Corollary (4.7) is due to *bias reduction*. The variance part remains unchanged. It follows from STONE (1982) (see the discussion in § 12.3) that (5.3) is also the asymptotic lower bound. See also RICE and ROSENBLATT (1983). However, away from the boundary, (5.3) holds regardless of whether (5.2) holds. Thus, the question is whether one can compute *boundary corrections* to achieve the global error bound (5.3).

Returning to the conditions (5.1)–(5.2), in view of STONE (1982), one cannot really complain about the smoothness condition, but the boundary condition (5.2) makes (5.3) quite problematic. By way of example, if $f^{(m)}(0) \neq 0$, then one does not get any decrease in the global error, and so the bound (5.3) is achievable only for smoothing splines of polynomial order $4m$. It would be nice if the smoothing spline estimator of order $2m$ could be suitably modified such that (5.3) would apply under the sole condition (5.1). This would provide a measure of *adaptation*: One may underestimate (guess) the smoothness of f_o by a factor 2 if we may characterize the distinction $f_o \in W^{m,2}(0,1)$ vs. $f_o \in W^{2m,2}(0,1)$ in this way.

There is essentially only one boundary correction method, viz. the application of the Bias Correction Principle of EUBANK and SPECKMAN (1990b) by HUANG (2001) as discussed in this section. The relaxed boundary splines of OEHLERT (1992) avoid the problem rather than correcting it; see § 6.

(5.4) THE BIAS REDUCTION PRINCIPLE (EUBANK and SPECKMAN, 1990b). Suppose one wishes to estimate a parameter $\theta_o \in \mathbb{R}^n$ and has available two estimators, each one flawed in its own way. One estimator, $\widetilde{\theta}$, is unbiased,

$$\mathbb{E}[\,\widetilde{\theta}\,] = \theta_o \ , \tag{5.5}$$

and each component has finite variance but otherwise has no known good properties. The other estimator, $\widehat{\theta}$, is biased but nice,

$$\mathbb{E}[\,\widehat{\theta}\,] = \theta_o + Ga + b \ , \tag{5.6}$$

for some $G \in \mathbb{R}^{n \times m}$, and $a \in \mathbb{R}^m$, $b \in \mathbb{R}^n$. It is assumed that G is known but that a and b are not. Let Π_G be the orthogonal projector onto the range of G. (If G has full column rank, then $\Pi_G = G(G^T G)^{-1} G^T$.)

Then, the estimator

$$\theta^{\#} = \widehat{\theta} + \Pi_G \left(\widetilde{\theta} - \widehat{\theta} \right) \tag{5.7}$$

satisfies

$$\mathbb{E}[\,\theta^{\#}\,] = \theta_o + \gamma \tag{5.8}$$

with

$$\|\,\gamma\,\| \leqslant \min\{\,\|\,Ga+b\,\|\,,\,\|\,b\,\|\,\}$$

and

$$\mathbb{E}[\,\|\,\theta^{\#} - \mathbb{E}[\,\theta^{\#}\,]\,\|^2\,] \leqslant \mathbb{E}[\,\|\,\widehat{\theta} - \mathbb{E}[\,\widehat{\theta}\,]\,\|^2\,] + \lambda\, m \ .$$

Here, $\lambda = \lambda_{\max}(\mathbb{V}\mathrm{ar}[\,\widetilde{\theta}\,])$ is the largest eigenvalue of $\mathbb{V}\mathrm{ar}[\,\widetilde{\theta}\,]$.

PROOF OF THE BIAS REDUCTION PRINCIPLE. One verifies that

$$c^{\#} \stackrel{\text{def}}{=} \mathbb{E}[\,\theta^{\#} - \theta_o\,] = (\,I - \Pi_G\,)\,(\,Ga + b\,) \ .$$

Now, since Π_G is an orthogonal projector, so is $I - \Pi_G$, and therefore $\|\,c^{\#}\,\| \leqslant \|\,Ga{+}b\,\|$. On the other hand, $(\,I{-}\Pi_G)G = O$, so $c^{\#} = (\,I{-}\Pi_G\,)\,b$, and $\|\,c^{\#}\,\| \leqslant \|\,b\,\|$.

For the variance part, it is useful to rewrite $\theta^{\#}$ as

$$\theta^{\#} = (\,I - \Pi_G)\,\widehat{\theta} + \Pi_G \widetilde{\theta} \ ,$$

so that by Pythagoras' theorem

$$\| \theta^\# - \mathbb{E}[\theta^\#] \|^2 = \| (I - \Pi_G)(\widehat{\theta} - \mathbb{E}[\widehat{\theta}]) \|^2 + \| \Pi_G(\widetilde{\theta} - \mathbb{E}[\widetilde{\theta}]) \|^2 .$$

The first term on the right is bounded by $\| \widetilde{\theta} - \mathbb{E}[\widehat{\theta}] \|^2$. For the second term, we have

$$\mathbb{E}\Big[\| \Pi_G(\widetilde{\theta} - \mathbb{E}[\widetilde{\theta}]) \|^2 \Big] = \operatorname{trace}\Big(\Pi_G \operatorname{\mathbb{V}ar}[\widetilde{\theta}] \Pi_G^T \Big) .$$

Let $\Lambda = \lambda I$. Then, $\Lambda - \operatorname{\mathbb{V}ar}[\widetilde{\theta}]$ is semi-positive-definite, so that

$$\operatorname{trace}(\Lambda - \operatorname{\mathbb{V}ar}[\widetilde{\theta}]) \geqslant 0 .$$

It follows that

$$\begin{aligned} &\mathbb{E}[\| \Pi_G(\widetilde{\theta} - \mathbb{E}[\widetilde{\theta}]) \|^2] \\ &\qquad = \operatorname{trace}(\Pi_G \Lambda \Pi_G^T) - \operatorname{trace}(\Pi_G (\Lambda - \operatorname{\mathbb{V}ar}[\widetilde{\theta}]) \Pi_G^T) \\ &\qquad \leqslant \operatorname{trace}(\Pi_G \Lambda \Pi_G^T) = \lambda \operatorname{trace}(\Pi_G \Pi_G^T) = \lambda m . \end{aligned}$$

The bound on the variance of $\theta^\#$ follows. Q.e.d.

The Bias Reduction Principle is useful when Ga is much larger than b and m is small. Under these circumstances, the bias is reduced dramatically, whereas the variance is increased by only a little. Note that $\Pi_G(\widetilde{\theta} - \widehat{\theta})$ is a "correction" to the estimator $\widehat{\theta}$.

We now wish to apply the Bias Reduction Principle to compute boundary corrections to the spline estimator of § 3. Actually, corrections to the values $f^{nh}(x_{in})$, $i = 1, 2, \cdots, n$, will be computed. For corrections to the spline function, see Exercise (5.21).

For the implementation of this scheme, the boundary behavior of the smoothing spline estimator must be described in the form (5.6). Thus, the boundary behavior must be "low-dimensional".

(5.9) The asymptotic behavior of the bias of the smoothing spline estimator near the boundary. Let $f_o \in W^{2m,2}(0,1)$. Then, $f_o^{(k)}$ is continuous for $k = 0, 1, \cdots, 2m-1$. Now, for $k = m, \cdots, 2m-1$, let L_k and R_k be polynomials (yet to be specified), and consider

$$p_o(x) = \sum_{\ell=m}^{2m-1} f_o^{(\ell)}(0)\, L_k(x) + f_o^{(\ell)}(1)\, R_k(x) . \tag{5.10}$$

We now wish to choose the L_k and R_k such that $g_o \stackrel{\text{def}}{=} f_o - p_o \in W^{2m,2}(0,1)$ satisfies

$$g_o^{(k)}(0) = g_o^{(k)}(1) = 0 , \quad k = m, \cdots, 2m-1 . \tag{5.11}$$

One verifies that it is sufficient that, for all k,

$$(5.12)\qquad \begin{aligned} &L_k^{(\ell)}(0) = L_k^{(\ell)}(1) = 0 \quad \text{for } \ell = m, \cdots, 2m-1 \;, \\ &\text{except that} \quad L_k^{(k)}(0) = 1 \;, \\ &\text{and} \quad R_k(x) = (-1)^k \, L_k(1-x) \;. \end{aligned}$$

The construction of the L_k is an exercise in Hermite-Birkhoff interpolation; see KINCAID and CHENEY (1991). For the case $m = 2$, see Exercise (5.20).

Now, let g_h be the solution to

$$\begin{aligned} &\text{minimize} \quad \| f - g_o \|^2 + h^{2m} \, \| f^{(m)} \|^2 \\ &\text{subject to} \quad f \in W^{m,2}(0,1) \;, \end{aligned}$$

and construct the functions $L_{k,h}$ similarly, based on the $L_{k,o} = L_k$. Let

$$(5.13)\qquad \eta_{k,h} \stackrel{\text{def}}{=} h^{-k} \{ L_{k,h} - L_k \} \;, \quad k = m, \cdots, 2m-1 \;.$$

Then, g_h satisfies $\| g_h - g_o \|_{m,h}^2 = \mathcal{O}(h^{4m})$, and by Exercise (4.15) applied to noiseless data,

$$(5.14)\qquad \| \eta_{k,h} \|_{m,h}^2 = \mathcal{O}(1) \;, \quad k = m, \cdots, 2m-1 \;.$$

By linearity, it follows that

$$(5.15)\qquad f_h = f_o + \sum_{k=m}^{2m-1} h^k \{ f_o^{(k)}(0) \, \eta_{k,h} + f_o^{(k)}(1) \, \zeta_{k,h} \} + \varepsilon_h \;,$$

with $\| \varepsilon_h \|_{m,h} = \mathcal{O}(h^{2m})$, and $\zeta_{k,h} = (-1)^k \, \eta_{k,h}$ for all k. Of course, by the Quadrature Lemma (2.27), the corresponding bounds hold for the sums:

$$(5.16)\qquad \tfrac{1}{n} \sum_{i=1}^{n} | \varepsilon_h(x_{in}) |^2 = \mathcal{O}(h^{4m}),$$

$$(5.17)\qquad \tfrac{1}{n} \sum_{i=1}^{n} | \eta_{k,h}(x_{in}) |^2 = \mathcal{O}(1) \;.$$

(5.18) COMPUTING BOUNDARY CORRECTIONS (HUANG, 2001). The Bias Reduction Principle may now be applied to compute boundary corrections. In the notation of the Bias Reduction Principle (5.4), take

$$\theta_o = \big(f_o(x_{1,n}), f_o(x_{2,n}), \cdots, f_o(x_{n,n}) \big)^T$$

and consider the estimators

$$\widehat{\theta} = \big(f^{nh}(x_{1,n}), f^{nh}(x_{2,n}), \cdots, f^{nh}(x_{n,n}) \big)^T$$

and

$$\widetilde{\theta} = \big(y_{1,n}, y_{2,n}, \cdots, y_{n,n} \big)^T \;.$$

Then, $\widetilde{\theta}$ is an unbiased estimator of θ_o. The asymptotic behavior of $\widehat{\theta}$ is described by

$$\mathbb{E}[\, \widehat{\theta} \,] = \theta_o + F \, a_0 + G \, a_1 + \varepsilon_h \;,$$

with ε_h as in (5.15) and $F, G \in \mathbb{R}^{n\times m}$, given by

$$F_{i,k} = \eta_{k,h}(x_{in}) \ , \quad G_{i,k} = \zeta_{k,h}(x_{in})$$

for $i = 1, 2, \cdots, n$ and $k = m, \cdots, 2m-1$. The vectors a_0 and a_1 contain the (unknown) derivatives of f_o at the endpoints. The estimator $\theta^{\#}$ may now be computed as per (5.7) and satisfies

$$\begin{aligned} \mathbb{E}\Big[\, \tfrac{1}{n} \sum_{i=}^{n} |\, \theta_i^{\#} - f_o(x_{in}) \,|^2 \Big] &\leqslant \mathbb{E}\Big[\, \tfrac{1}{n} \sum_{i=}^{n} |\, (\varepsilon_{k,h})_i \,|^2 \Big] + 2\, m\, n^{-1}\, \sigma^2 \\ &= \mathcal{O}\big(h^{4m} + (nh)^{-1} \big) \ . \end{aligned} \tag{5.19}$$

The boundary behavior (5.15) is due to Rice and Rosenblatt (1983). We consider an analogous result for trigonometric sieves.

(5.20) Exercise. (a) Let $m = 2$. Verify that

$$L_2(x) = \tfrac{1}{4}\,(\,1 - x\,)^2 - \tfrac{1}{10}\,(\,1 - x\,)^5 \ , \ L_3(x) = -\tfrac{1}{12}\,(\,1 - x)^4 + \tfrac{1}{20}\,(\,1 - x\,)^5$$

satisfy (5.12), and verify (5.11).
(b) Verify (5.16).
(c) Prove that the bounds (5.17) are sharp.
(d) Prove (5.19).

(5.21) Exercise. Suppose we are not interested in $f^{nh}(x_{in})$, $i = 1, \cdots, n$, but in the actual spline $f^{nh}(x)$, $x \in [\,0\,,1\,]$. Formulate an algorithm to compute the boundary correction to the spline function. [Hint: One may think of the spline estimator as being given by its coefficients; in other words, it is still a finite-dimensional object. Unbiased estimators of f_o do not exist, but we do have an unbiased estimator of the spline interpolant of f_o using the data $f_o(x_{in})$, $i = 1, 2, \cdots, n$, which is a very accurate approximation to f_o. See Chapter 19 for the details on spline interpolation.]

Exercises: (5.20), (5.21).

6. Relaxed boundary splines

In this section, we discuss the solution of Oehlert (1992) to the boundary correction problem for smoothing splines. His approach is to avoid the problem altogether by modifying the roughness penalization in the smoothing spline problem (4.4). The choice of penalization by Oehlert (1992) is actually quite fortuitous: It is easy to analyze the resulting estimator, much in the style of §§ 2 and 3, but the choice itself is *magic*.

We operate again under the Gauss-Markov model (1.1)–(1.2) with asymptotically uniform designs; see Definition (2.22). For now, suppose that

$$f_o \in W^{2m,2}(0,1) \ . \tag{6.1}$$

In general, under these circumstances, the smoothing spline estimator of polynomial order $2m$, defined as the solution to (2.1), has mean integrated squared error $\mathcal{O}\big(n^{-2m/(2m+1)} \big)$, whereas the smoothness assumption (6.1) should allow for an error $\mathcal{O}\big(n^{-4m/(4m+1)} \big)$.

It is worthwhile to repeat the motivation of Oehlert (1992) for his suggested modification of (4.4). He observes that the global variance of the estimator is $\mathcal{O}\big((nh)^{-1} \big)$ and that the squared bias is $\mathcal{O}\big(h^{4m} \big)$ away from the boundary points but, in general, is only $\mathcal{O}\big(h^{2m} \big)$ near the boundary. Thus, it would be a good idea to reduce the bias near the boundary if this could be done without dramatically increasing the variance. His way of doing this is to downweight the standard roughness penalization near the endpoints. There would appear to be many ways of doing this, until one has to do it. Indeed, the analysis of Oehlert (1992) and the analysis below show that quite a few "things" need to happen.

The particular suggestion of Oehlert (1992) is as follows. Let $m \geqslant 1$ be an integer, and consider the vector space of functions defined on $(0,1)$

$$\mathcal{W}_m = \left\{ f \;\middle|\; \begin{array}{l} \forall\, \delta : 0 < \delta < \tfrac{1}{2} \implies f \in W^{m,2}(\,\delta, 1-\delta\,) \\ |\, f \,|_{\mathcal{W}_m} < \infty \end{array} \right\} , \tag{6.2}$$

where the semi-norm $|\cdot|_{\mathcal{W}_m}$ is defined by way of

$$|\, f \,|^2_{\mathcal{W}_m} \stackrel{\text{def}}{=} \int_0^1 \{\, x(1-x)\,\}^m \,|\, f^{(m)}(x)\,|^2 \, dx \; . \tag{6.3}$$

The relaxed boundary spline estimator of the regression function is then defined as the solution $f = \psi^{nh}$ of the problem

$$\begin{array}{ll} \text{minimize} & RLS(\,f\,) \stackrel{\text{def}}{=} \frac{1}{n} \sum\limits_{i=1}^{n} |\, f(x_{in}) - y_{in}\,|^2 + h^{2m} \,|\, f \,|^2_{\mathcal{W}_m} \\ \text{subject to} & |\, f \,|_{\mathcal{W}_m} < \infty \; . \end{array} \tag{6.4}$$

Of course, the existence and uniqueness of the solution must be established, and the objective function $RLS(\,f\,)$ must be well-defined on $\mathcal{W}_m$. There are some difficulties in the case $m = 1$ that require some extra conditions on the design (asymptotic uniformity does not suffice, it appears). So, at the crucial moment, we assume that $m \geqslant 2$. Also, the assumption (6.1) may be replaced by the condition

$$f_o \in \mathcal{W}_{2m} \; . \tag{6.5}$$

The difficulties for $m = 1$ are illustrated in the following exercise.

(6.6) Exercise. Show that the function

$$f(x) = \big|\, \log\{\, x(1-x)\,\}\,\big|^{\alpha} \; , \quad x \in (0,1) \, ,$$

belongs to $\mathcal{W}_1$ for $\alpha < \frac{1}{2}$ but not for $\alpha = \frac{1}{2}$.

The two "final" results are as follows. Note that there are almost no conditions in the existence and uniqueness theorem.

(6.7) THEOREM. *Let* $m \geqslant 1$. *The solution of* (6.4) *exists. If the design contains at least* m *distinct points, then the solution is unique.*

(6.8) THEOREM. *Let* $m \geqslant 2$. *Assume the Gauss-Markov model* (1.1)–(1.2), *and that the design is asymptotically uniform. Assuming* $f_o \in \mathcal{W}_{2m}$, *the solution* ψ^{nh} *of* (6.4) *satisfies*

$$\mathbb{E}[\, \| \psi^{nh} - f_o \|^2 \,] = \mathcal{O}\big(n^{-4m/(4m+1)} \big) ,$$

provided $h \asymp n^{-1/(4m+1)}$ *(deterministically).*

We now set out to prove Theorems (6.7) and (6.8). The proof essentially follows the development in §§ 2, 3, and 4: The relevant lemmas all have useful analogues, but some of the proofs are simple computations in terms of an orthogonal basis for the Hilbert spaces in question. This orthogonal basis (the Legendre polynomials, suitably scaled) was already featured prominently in OEHLERT (1992), and we are not above using it. In fact, this constitutes the *magic* of the particular choice of penalization. This section is devoted to preliminaries, analogous to § 2. In the next section, the existence, uniqueness, and convergence rates are established. We note that, for $m = 1$, Theorem (6.7) holds for a modified design, not including the endpoints of the interval; see Exercise (6.41).

Reproducing kernel Hilbert spaces. For $h > 0$, define the inner products on $\mathcal{W}_m$,

(6.9) $$\langle\!\langle\, f \,,\, g \,\rangle\!\rangle_{h,\mathcal{W}_m} = \langle\, f \,,\, g \,\rangle + h^{2m} \langle\, f \,,\, g \,\rangle_{\mathcal{W}_m} ,$$

where

(6.10) $$\langle\, f \,,\, g \,\rangle_{\mathcal{W}_m} = \int_0^1 [\, x(1-x) \,]^m \, f^{(m)}(x) \, g^{(m)}(x) \, dx ,$$

and the associated norms $\| \cdot \|_{h,\mathcal{W}_m}$ by way of

(6.11) $$\| f \|_{h,\mathcal{W}_m}^2 = \langle\!\langle\, f \,,\, f \,\rangle\!\rangle_{h,\mathcal{W}_m} .$$

It is obvious that, with all these norms, $\mathcal{W}_m$ is a Hilbert space. Moreover, these norms are equivalent, but not uniformly in h; see Definition (2.7) and Exercise (2.8).

At this point, we introduce the shifted Legendre polynomials, which behave very nicely in all of the $\mathcal{W}_m$. As mentioned before, OEHLERT (1992) already made extensive use of this.

First, we summarize the relevant properties of the Legendre polynomials. One way to define the standard Legendre polynomials is through the recurrence relations

$$
(6.12)\qquad \begin{aligned} &P_{-1}(x) = 0 \,, \quad P_0(x) = 1 \,, \\ &(k+1)\, P_{k+1}(x) = (2k+1)\, x\, P_k(x) - k\, P_{k-1}(x) \,, \quad k \geqslant 0 \,. \end{aligned}
$$

The shifted, normalized Legendre polynomials are here defined as

$$
(6.13)\qquad Q_k(x) = (2k+1)^{1/2}\, P_k(2x-1) \,, \quad k \geqslant 0 \,.
$$

They satisfy the following orthogonality relations:

$$
(6.14)\qquad \langle\, Q_k \,,\, Q_\ell \,\rangle_{L^2(0,1)} = \begin{cases} 1 \,, & \text{if} \quad k = \ell \,, \\ 0 \,, & \text{otherwise} \,, \end{cases}
$$

$$
(6.15)\qquad \langle\, Q_k \,,\, Q_\ell \,\rangle_{\mathcal{W}_m} = \begin{cases} 0 \,, & \text{if} \quad k \neq \ell \,, \\ \dfrac{4^m\,(k+m)!}{(k-m)!} \,, & \text{if} \quad k = \ell \,. \end{cases}
$$

Note that the last inner product vanishes (also) for $k = \ell < m$. We also have the pointwise bounds

$$
(6.16)\qquad \begin{aligned} &|\, Q_k(x) \,| \leqslant (2k+1)^{1/2} && \text{for all } 0 \leqslant x \leqslant 1 \text{ and } k \geqslant 0 \,, \\ &|\, Q_k(x) \,| \leqslant c\, \{\, x(1-x) \,\}^{-1/4} && \text{for all } 0 < x < 1 \text{ and } k \geqslant 1 \end{aligned}
$$

for a suitable constant c independent of k and x. A handy reference for all of this is SANSONE (1959). Note that (6.14)–(6.15) prove the following lemma.

(6.17) LEMMA. *For all $h > 0$ and all $m \geqslant 1$,*

$$
\langle\!\langle\, Q_k \,,\, Q_\ell \,\rangle\!\rangle_{h,\mathcal{W}_m} = \begin{cases} 1 + (2h)^{2m}\, \lambda_{k,m} \,, & \textit{if} \quad k = \ell \,, \\ 0 \,, & \textit{otherwise} \,, \end{cases}
$$

where $\lambda_{k,m} = (k+m)!/(k-m)!$. Moreover, there exist constants $c_m > 1$ such that, for all k, $k \geqslant m$, we have $(c_m)^{-1} \leqslant k^{-2m}\, \lambda_{k,m} \leqslant c_m$.

It follows that Q_k, $k \geqslant 0$, is an orthonormal basis for $L^2(0,1)$ and an orthogonal basis for $\mathcal{W}_m$. Also, it gives us a handy expression for the norms on $\mathcal{W}_m$, but we shall make them handier yet. For $f \in L^2(0,1)$, define

$$
(6.18)\qquad \widehat{f}_k = \langle\, f \,,\, Q_k \,\rangle \,, \quad k \geqslant 0 \,.
$$

The following lemma is immediate.

(6.19) LEMMA. *Let $m \geqslant 1$. For all $h > 0$ and $f \in \mathcal{W}_m$,*

$$
f = \sum_{k \geqslant 0} \widehat{f}_k\, Q_k \,,
$$

with convergence in the $\mathcal{W}_m$*-topology, and for all* $f\,,\ g \in \mathcal{W}_m\,$,

$$\langle\!\langle\, f\,,\, g\,\rangle\!\rangle_{h,\mathcal{W}_m} = \sum_{k \geqslant 0} \big(\, 1 + (2h)^{2m}\, \lambda_{k,m}\,\big)\ \widehat{f}_k\ \widehat{g}_k\ .$$

Finally, for all $f \in \mathcal{W}_m\,$,

$$\|\, f\,\|^2_{h,\mathcal{W}_m} = \sum_{k \geqslant 0} \big(\, 1 + (2h)^{2m}\, \lambda_{k,m}\,\big)\, |\, \widehat{f}_k\,|^2 < \infty\ .$$

The representation above for the norms is nice, but the behavior of the $\lambda_{k,m}$ is a bit of a bummer. So, let us define the equivalent norms

$$(6.20)\qquad |\!|\!|\, f\,|\!|\!|_{h,\mathcal{W}_m} = \Big\{\, \sum_{k \geqslant 0} \big(\, 1 + (2hk)^{2m}\,\big)\, |\, \widehat{f}_k\,|^2\,\Big\}^{1/2}\ .$$

(6.21) Lemma. *Let* $m \geqslant 1$. *The norms* $|\!|\!| \cdot |\!|\!|_{h,\mathcal{W}_m}$ *and* $\|\cdot\|_{h,\mathcal{W}_m}$ *are equivalent, uniformly in* $h\,$, $0 < h \leqslant 1$*; i.e., there exists a constant* γ_m *such that, for all* $h\,$, $0 < h \leqslant 1$, *and all* $f \in \mathcal{W}_m$,

$$(\gamma_m)^{-1}\, \|\, f\,\|_{h,\mathcal{W}_m} \leqslant |\!|\!|\, f\,|\!|\!|_{h,\mathcal{W}_m} \leqslant \gamma_m\, \|\, f\,\|_{h,\mathcal{W}_m}\ .$$

Proof. By Lemma (6.17), we obviously have

$$1 + (2h)^{2m}\, \lambda_{k,m} \leqslant c_m\, \big(\, 1 + (2hk)^{2m}\,\big)\ ,$$

with the same c_m as in Lemma (6.17). Thus, $c_m^{-1}\, \|\, f\,\|_{h,\mathcal{W}_m} \leqslant |\!|\!|\, f\,|\!|\!|_{h,\mathcal{W}_m}$.
Also, for $k \geqslant m$, the lower bound of Lemma (6.17) on $\lambda_{k,m}$ is useful. For $0 < h \leqslant 1$ and $0 \leqslant k < m$, we have

$$\frac{1 + (2hk)^{2m}}{1 + (2m)^{2m}} \leqslant 1 = 1 + (2h)^{2m}\, \lambda_{k,m}\ ,$$

so that with $\gamma_m = \max\{\, c_m\,,\ 1 + (2m)^{2m}\,\}$,

$$|\!|\!|\, f\,|\!|\!|_{h,\mathcal{W}_m} \leqslant \gamma_m\, \|\, f\,\|_{h,\mathcal{W}_m}\ .$$

The lemma follows. Q.e.d.

We are now ready to show that the point evaluations $x \mapsto f(x)$ are bounded linear functionals on $\mathcal{W}_m$; in other words, that the $\mathcal{W}_m$ are reproducing kernel Hilbert spaces. First, define the functions

$$(6.22)\qquad \Phi_h(x) = \min\big(\, h^{-1/2}\,,\ \{\, x(1-x)\,\}^{-1/4}\,\big)\ .$$

(6.23) Lemma (The case $m = 2$). *There exists a constant* c *such that, for all* h, $0 < h \leqslant 1$, *and all* $f \in \mathcal{W}_2\,$,

$$|\, f(x)\,| \leqslant c_m\, \Phi_h(x)\, h^{-1/2}\, \|\, f\,\|_{h,\mathcal{W}_2} \quad \textit{for all}\ \ x \in (0,1)\ .$$

PROOF. Using the representation of Lemma (6.19) for $f \in \mathcal{W}_m$, we get

$$| f(x) | \leqslant \sum_{k \geqslant 0} | \widehat{f}_k | \, | Q_k(x) | \, .$$

Now, with the second inequality of (6.16),

$$| f(x) | \leqslant c \{ x(1-x) \}^{-1/4} \sum_{k \geqslant 0} | \widehat{f}_k | \, ,$$

and with Cauchy-Schwarz, the last series is bounded by

$$||| f |||_{h,\mathcal{W}_2} \Big\{ \sum_{k \geqslant 0} \big(1 + (2hk)^4 \big)^{-1} \Big\}^{1/2} .$$

Now, the infinite series is dominated by

$$\int_0^\infty \big(1 + (2hx)^4 \big)^{-1} dx = h^{-1} \int_0^\infty \big(1 + (2t)^4 \big)^{-1} dt = c h^{-1} \, ,$$

for a suitable constant. Thus,

$$| f(x) | \leqslant c h^{-1/2} \{ x(1-x) \}^{-1/4} ||| f |||_{h,\mathcal{W}_2} \, . \tag{6.24}$$

For all x, we use the first bound of (6.16). With Cauchy-Schwarz, this gives the bound

$$| f(x) | \leqslant ||| f |||_{h,\mathcal{W}_2} \Big\{ \sum_{k \geqslant 0} \frac{2k+1}{1 + (2hk)^4} \Big\}^{1/2} .$$

Now, we may drop the +1 in the numerator, and then the infinite series behaves like

$$\int_0^\infty \frac{2x}{1 + (2hx)^4} \, dx = h^{-2} \int_0^\infty \frac{2t}{1 + (2t)^4} \, dt = c h^{-2}$$

for (another) constant c. Thus,

$$| f(x) | \leqslant c h^{-1} ||| f |||_{h,\mathcal{W}_2} \, . \tag{6.25}$$

By the equivalence of the norms, uniformly in h, $0 < h \leqslant 1$, the lemma follows from (6.24) and (6.25). Q.e.d.

(6.26) LEMMA. *For all $m \geqslant 1$, there exists a constant c_m such that, for all $f \in \mathcal{W}_{m+1}$ and all h, $0 < h \leqslant 1$,*

$$\| f \|_{h,\mathcal{W}_m} \leqslant c_m \| f \|_{h,\mathcal{W}_{m+1}} \, .$$

PROOF. With the representation of Lemma (6.19) and (6.20),

$$||| f |||^2_{h,\mathcal{W}_m} \leqslant c \, ||| f |||^2_{h,\mathcal{W}_{m+1}} \, ,$$

where

$$c = \sup \Big\{ \frac{1 + (2hk)^{2m}}{1 + (2hk)^{2m+2}} \;\Big|\; k \geqslant 0 , \; 0 < h \leqslant 1 \Big\} \leqslant \sup_{t > 0} \frac{1 + t^{2m}}{1 + t^{2m+2}} < \infty \, .$$

Together with the equivalence of the norms, that is all that there is to it. Q.e.d.

The final result involving the Legendre polynomials or, more to the point, the equivalent norms, is an integration-by-parts formula.

(6.27) LEMMA. *Let* $m \geqslant 1$. *For all* $f \in \mathcal{W}_m$ *and all* $g \in \mathcal{W}_{2m}$,

$$\langle f , g \rangle_{\mathcal{W}_m} \leqslant \| f \| \, \| g \|_{1,\mathcal{W}_{2m}} .$$

PROOF. Using the representation of Lemma (6.19), and Lemma (6.17), the inner product may be written as, and then bounded by,

$$\sum_{k \geqslant m} (1 + 2^{2m} \lambda_{k,m}) \, \widehat{f}_k \, \widehat{g}_k \leqslant c_m \sum_{k \geqslant m} (1 + (2k)^{2m}) \, | \widehat{f}_k | \, | \widehat{g}_k | .$$

Now, with Cauchy-Schwarz, the right-hand side may be bounded by

$$\| f \| \Bigl\{ \sum_{k \geqslant 0} (1 + (2k)^{2m})^2 \, | \widehat{g}_k |^2 \Bigr\}^{1/2} ,$$

and, in turn, the infinite series may be bounded by

$$2 \sum_{k \geqslant 0} (1 + (2k)^{4m}) \, | \widehat{g}_k |^2 = 2 \, ||| \, g \, |||_{1,\mathcal{W}_{2m}}^2 .$$ Q.e.d.

(6.28) REMARK. The reason we called Lemma (6.27) an integration-by-parts formula is because it is. Recall that

$$\langle f , g \rangle_{\mathcal{W}_m} = \int_0^1 \{ x(1-x) \}^m f^{(m)}(x) \, g^{(m)}(x) \, dx ,$$

so that integrating by parts m times gives

$$\int_0^1 g^{(2m)}(x) \, (-D)^m \Bigl\{ \{ x(1-x) \}^m f^{(m)}(x) \Bigr\} \, dx ,$$

where D denotes differentiation with respect to x, *provided* the boundary terms vanish. Showing that they do is harder than it looks (e.g., are the boundary values actually defined?), but the expansion in Legendre polynomials avoids the issue.

Quadrature. The last technical result deals with quadrature. The only hard part is an embedding result where apparently, the Legendre polynomials are of no use. We must slug it out; cf. the proof of Lemma (2.10).

(6.29) EMBEDDING LEMMA. *There exists a constant c such that*

$$\| f \|_{W^{1,2}(0,1)} \leqslant c \| f \|_{1,\mathcal{W}_2} \quad \textit{for all } f \in \mathcal{W}_2 .$$

PROOF. For $x, y \in (0,1)$, with $y < x$,

$$| f'(x) - f'(y) | \leqslant \int_y^x | f''(t) | \, dt \leqslant c(x,y) \, | f |_{\mathcal{W}_2}$$

with
$$c^2(x,y) = \int_y^x [\, t(1-t)\,]^{-2} \, dt \; .$$

It follows that, for any closed subinterval $[\, a\, , b\,] \subset (0,1)$, we have

$$c(x,y) \leqslant C(a,b) \, | x - y |^{1/2} \quad \text{for } x,\, y \in [\, a\, , b\,] \; ,$$

for a suitable constant $C(a,b)$. Thus, f' is continuous in $(0,1)$.

Now, let $M = \left(\frac{1}{4}, \frac{3}{4}\right)$, and choose $y \in M$ such that

$$| f'(y) |^2 = 2 \, \| f' \|_M^2$$

in the notation of (2.10). This is possible by the Mean Value Theorem. Then,

$$\int_0^y | f'(t) |^2 \, dt \leqslant 2 \, | f'(y) |^2 + 2 \int_0^y | f'(t) - f'(y) |^2 \, dt \; .$$

Now, by Hardy's inequality (see Lemma (6.31) and Exercise (6.32) below),

$$\begin{aligned} \int_0^y | f'(t) - f'(y) |^2 \, dt &\leqslant 4 \int_0^y t^2 \, | f''(t) |^2 \, dt \\ &\leqslant 64 \int_0^y [\, t(1-t)\,]^2 \, | f''(t) |^2 \, dt . \end{aligned}$$

Also, $| f'(y) |^2 = 2 \, \| f' \|_M^2$. By the Interpolation Lemma (2.12) with $h = 1$, we have the bound

$$\| f' \|_M^2 \leqslant c \, \| f \|_M^2 + c_1 \, \| f'' \|_M^2 \tag{6.30}$$

for constants c and c_1 independent of y (since y is bounded away from 0). Since, on the interval M, the weight function $\{\, x(1-x)\,\}^2$ is bounded from below by $\frac{9}{256}$, then

$$\int_{\frac{1}{4}}^{\frac{3}{4}} | f''(t) |^2 \, dt \leqslant c \int_{\frac{1}{4}}^{\frac{3}{4}} \{\, x(1-x)\,\}^2 \, | f''(t) |^2 \, dt$$

with $c = 256/9$, so that $\| f'' \|_M^2 \leqslant c \, | f |_{\mathcal{W}_2}^2$ and we obtain

$$\| f' \|_{(0,y)}^2 \leqslant c \, \| f \|_{1,\mathcal{W}_2}^2 \; .$$

The same bound applies to $\| f' \|_{(y,1)}^2$. The lemma follows. Q.e.d.

To prove the version of Hardy's inequality alluded to above, we quote the following result from HARDY, LITTLEWOOD, and POLYA (1951).

(6.31) LEMMA. *Let* $K : \mathbb{R}^+ \times \mathbb{R}^+ \to \mathbb{R}^+$ *be homogeneous of degree* -1*; i.e.,* $K(tx, ty) = t^{-1} K(x,y)$ *for all nonnegative* t, x, *and* y. *Then, for all* $f \in L^2(\mathbb{R}^+)$,

$$\int_{\mathbb{R}^+} \Big| \int_{\mathbb{R}^+} K(x,y)\, f(y)\, dy \Big|^2 dx \leqslant k \, \| f \|^2_{L^2(\mathbb{R}^+)} \,,$$

where $k = \displaystyle\int_{\mathbb{R}^+} y^{-1/2}\, K(1,y)\, dy$.

PROOF. For nonnegative $f, g \in L^2(\mathbb{R}^*)$,

$$\begin{aligned} &\int_{\mathbb{R}^+} f(x) \int_{\mathbb{R}^+} K(x,y)\, g(y)\, dy\, dx \\ &\qquad = \int_{\mathbb{R}^+} f(x) \int_{\mathbb{R}^+} x\, K(x,xy)\, g(xy)\, dy\, dx \quad \text{(change of variable)} \\ &\qquad = \int_{\mathbb{R}^+} f(x) \int_{\mathbb{R}^+} K(1,y)\, g(xy)\, dy\, dx \quad \text{(homogeneity)} \\ &\qquad = \int_{\mathbb{R}^+} K(1,y) \int_{\mathbb{R}^+} f(x)\, g(xy)\, dx\, dy \quad \text{(Fatou)} \,. \end{aligned}$$

Now, with Cauchy-Schwarz,

$$\begin{aligned} \int_{\mathbb{R}^+} f(x)\, g(xy)\, dx\, dy &\leqslant \| f \|_{L^2(\mathbb{R}^+)} \Big\{ \int_{\mathbb{R}^+} |\, g(xy)\,|^2\, dx \Big\}^{1/2} \\ &\leqslant y^{-1/2}\, \| f \|_{L^2(\mathbb{R}^+)}\, \| g \|_{L^2(\mathbb{R}^+)} \,, \end{aligned}$$

the last equality by a change of variable. Thus, for all nonnegative f, $g \in L^2(\mathbb{R}^+)$,

$$\int_{\mathbb{R}^+} f(x) \int_{\mathbb{R}^+} K(x,y)\, g(y)\, dy\, dx \leqslant k\, \| f \|_{L^2(\mathbb{R}^+)}\, \| g \|_{L^2(\mathbb{R}^+)} \,,$$

with the constant k as advertised. Obviously, then this holds also for *all* $f, g \in L^2(\mathbb{R}^+)$. Finally, take

$$f(x) = \int_{\mathbb{R}^+} K(x,y)\, g(y)\, dy \,, \quad x \in \mathbb{R}^+ \,,$$

and we are in business. Q.e.d.

(6.32) EXERCISE. (a) Show that the function

$$K(x,y) = y^{-1}\, 1\!\!1(\, x < y\,) \,, \quad x, y > 0 \,,$$

is homogeneous of degree -1.

(b) Show the following consequence of Lemma (6.31): For all integrable functions f on $\mathbb{R}^+$,

$$\int_{\mathbb{R}^+} \Big| \int_x^\infty y^{-1}\, f(y)\, dy \Big|^2 dx \leqslant 4 \int_{\mathbb{R}^+} x^2\, |\, f(x)\,|^2\, dx \,.$$

(c) Use (b) to show that, for all functions f with a measurable derivative,

$$\int_0^\infty |f(t)|^2\,dt \leqslant 4 \int_0^\infty t^2\,|f'(t)|^2\,dt \ .$$

(d) Use (c) to show that, for all functions f with a measurable derivative,

$$\int_0^T |f(t) - f(T)|^2\,dt \leqslant 4 \int_0^T t^2\,|f'(t)|^2\,dt \ .$$

(6.33) LEMMA. *Let $m \geqslant 2$. For asymptotically uniform designs, there exists a constant c_m such that, for all $f \in \mathcal{W}_m$ and all h, $0 < h \leqslant 1$,*

$$\Bigl|\, \tfrac{1}{n} \sum_{i=1}^{n} |f(x_{in})|^2 - \|f\|^2 \,\Bigr| \leqslant c_m\,(nh^2)^{-1}\,\|f\|_{h,\mathcal{W}_m} \ .$$

PROOF. By Definition (2.22), the left-hand side is bounded by

$$c n^{-1}\,\|(f^2)'\|_1 \ .$$

Now,

$$\|(f^2)'\|_1 \leqslant 2\,\|f\|\,\|f'\| \leqslant 2\,\|f\|\,\|f\|_{W^{1,2}(0,1)} \leqslant c\,\|f\|\,\|f\|_{1,\mathcal{W}_2} \ ,$$

the last inequality by Lemma (6.29). Finally for $0 < h \leqslant 1$,

$$\|f\|_{1,\mathcal{W}_2} \leqslant h^{-2}\,\|f\|_{h,\mathcal{W}_2} \ ,$$

and of course $\|f\| \leqslant \|f\|_{h,\mathcal{W}_2}$. Thus,

$$\|(f^2)'\|_1 \leqslant c\,h^{-2}\,\|f\|^2_{h,\mathcal{W}_2} \ . \tag{6.34}$$

The lemma then follows from Lemma (6.26). Q.e.d.

(6.35) REMARK. The inequality (6.34) does not appear to be sharp as far as the rate h^{-2} is concerned. The example $f(x) = (x - \lambda)_+^2$ (for appropriate λ) shows that the rate $h^{-3/2}$ may apply. Verify this. What is the best possible rate? (The authors do not know.)

Reproducing kernels. We finally come to the existence of the reproducing kernels, implied by Lemmas (6.23) and (6.26), and its consequences for random sums.

(6.36) THEOREM [REPRODUCING KERNEL HILBERT SPACES]. *Let $m \geqslant 2$. Then, $\mathcal{W}_m$ is a reproducing kernel Hilbert space with reproducing kernels $R_{m,h}(x,y)$, $x, y \in [0,1]$, so that, for all $f \in \mathcal{W}_m$,*

$$f(x) = \langle\!\langle\, R_{m,h}(x,\cdot\,)\,,\, f \,\rangle\!\rangle_{h,\mathcal{W}_m} \ , \qquad \textit{for all } x \in [\,0\,,\,1\,] \ ,$$

and, for a suitable constant c_m not depending on h,

$$\|\, R_{m,h}(x,\cdot\,)\,\|_{h,\mathcal{W}_m} \leqslant c_m\,\Phi_h(x)\,h^{-1/2} \ .$$

The reproducing kernel Hilbert space setting has consequences for random sums. For the noise vector d_n and design x_{in}, $i = 1, 2, \cdots, n$, let

$$(6.37) \qquad S^{nh}(t) = \tfrac{1}{n} \sum_{i=1}^{n} d_{in}\, R_{mh}(x_{in}, t) , \quad t \in [0, 1] .$$

(6.38) LEMMA. *Let $m \geqslant 2$. Then, for all $f \in \mathcal{W}_m$ and h, $0 < h \leqslant 1$, and for all designs,*

$$\Bigl| \tfrac{1}{n} \sum_{i=1}^{n} d_{in}\, f(x_{in}) \Bigr| \leqslant \| f \|_{h,\mathcal{W}_m} \, \| S^{nh} \|_{h,\mathcal{W}_m} .$$

If, moreover, the noise vector d_n satisfies (1.2) and the design is asymptotically uniform, then there exists a constant c such that, for all n and all h, $0 < h \leqslant 1$, with $nh^2 \to \infty$,

$$\mathbb{E}[\, \| S^{nh} \|_{h,\mathcal{W}_m}^2 \,] \leqslant c\,(nh)^{-1} .$$

PROOF. For the first inequality, use the representation

$$f(x_{in}) = \langle\!\langle\, f \,,\, R_{mh}(x_{in}, \cdot\,) \,\rangle\!\rangle_{h.\mathcal{W}_m}$$

to see that the sum equals $\langle\!\langle\, f \,,\, S^{nh} \,\rangle\!\rangle_{h.\mathcal{W}_m}$. Then Cauchy-Schwarz implies the inequality.

For the second inequality, note that the expectation equals

$$n^{-2} \sum_{i,j=1}^{n} \mathbb{E}[\, d_{in}\, d_{jn} \,] \, \langle\!\langle\, R_{mh}(x_{in}, \cdot\,) \,,\, R_{mh}(x_{jn}, \cdot\,) \,\rangle\!\rangle_{h.\mathcal{W}_m} .$$

By the assumption (1.2) and Theorem (6.36), this equals and may be bounded as

$$(6.39) \quad \sigma^2 n^{-2} \sum_{i=1}^{n} \| R_{mh}(x_{in}, \cdot\,) \|_{h,\mathcal{W}_m}^2 \leqslant c\,(nh)^{-1} \cdot \tfrac{1}{n} \sum_{i=1}^{n} |\, \Phi_h(x_{in}) \,|^2$$

for a suitable constant c. By the asymptotic uniformity of the design, see Definition (2.22), we have

$$(6.40) \qquad \tfrac{1}{n} \sum_{i=1}^{n} |\, \Phi_h(x_{in}) \,|^2 \leqslant \| \Phi_h \|^2 + c\, n^{-1/2}\, \| \Phi_h^2 \|_{W^{1,1}(0,1)} .$$

Now, $\| \Phi_h^2 \|_1 = \| \Phi_h \|^2$ and

$$\| \Phi_h \|^2 \leqslant \int_0^1 \{\, x(1-x) \,\}^{-1/2} \, dx = \pi .$$

Also,

$$\| \{ \Phi_h^2 \}' \|_1 = \int_a^{1-a} \Bigl| \frac{d}{dx} \{\, x(1-x) \,\}^{-1/2} \Bigr| \, dx ,$$

where a is the smallest solution of

$$h^{-1/2} = \{\, x(1-x)\,\}^{-1/4} \,.$$

So, $a \asymp h^2$. Now, $\{\, x(1-x)\,\}^{-1/4}$ is decreasing on $(\, a \,,\, \frac{1}{2}\,)$, and so

$$\int_a^{\frac{1}{2}} \Bigl|\, \frac{d}{dx} \{\, x(1-x)\,\}^{-1/2} \Bigr|\, dx = \{\, a(1-a)\,\}^{-1/2} - 2 \asymp h^{-1} \,,$$

and the same bound applies to the integral over $(\, \frac{1}{2} \,,\, 1-a\,)$. To summarize, all of this shows that, for a suitable constant c,

$$\|\, \Phi_h \,\|^2 \leqslant \pi \quad , \quad \|\, \{\, \Phi_h^2 \,\}' \,\|_1 \leqslant c\, h^{-1} \,, \quad \text{and so} \quad \|\, \Phi_h^2 \,\|_{W^{1,1}(0,1)} \leqslant c\, h^{-1} \,,$$

and then (6.40) shows that

$$\tfrac{1}{n} \sum_{i=1}^{n} |\, \Phi_h(x_{in}) \,|^2 \leqslant \pi + c\,(nh^2)^{-1/2} \,.$$

This implies the advertised bound on $\mathbb{E}[\,\|\, S^{nh} \,\|_{h,\mathcal{W}_m}^2\,]$. Q.e.d.

(6.41) EXERCISE. Some of the results in this section also hold for $m = 1$. (a) Show that, (6.24) holds for $m = 1$ and that instead of the uniform bound we have

$$|\, f(x) \,| \leqslant c\, h^{-1} \,|\, \log\{\, x(1-x)\,\} \,|^{1/2} \,|\!|\!|\, f \,|\!|\!|_{h,\mathcal{W}_1} \,.$$

(b) Show that, for $a = 1/n$,

$$\int_a^{1-a} |\, \log\{\, x(1-x)\,\} \,|^{1/2} \,|\, f\,'(x) \,|\, dx \leqslant c\,(\, \log n \,)^2 \,|\, f \,|_{\mathcal{W}_1} \,.$$

(c) Prove the case $m = 1$ of Lemma (6.38) for the designs

$$x_{in} = i/(n-1) \quad \text{and} \quad x_{in} = (i - \tfrac{1}{2})/(n-1) \,, \quad i = 1, 2, \cdots, n \,.$$

Indeed, for $m = 1$, the requirement on the designs is the asymptotic uniformity of Definition (2.22) together with the assumption that

(d) $$\sup_{n \geqslant 1} \tfrac{1}{n} \sum_{i=1}^{n} \{\, x_{in}\,(\, 1 - x_{in} \,)\,\}^{-1/2} < \infty \,.$$

[Hint: For (a), proceed analogously to the proof of Lemma (2.10). For (b), Cauchy-Schwarz does it. For (c), use (6.39), but with Φ_h replaced by $\Psi_h(x) = \{\, x\,(\, 1 - x \,)\,\}^{-1/4}$.]

EXERCISES: (6.6), (6.32), (6.40).

7. Existence, uniqueness, and rates

In this section, we actually prove Theorems (6.7) and (6.8). This pretty much goes along the lines of §§ 3 and 4. We start out with the quadratic behavior.

(7.1) LEMMA. *Let $m \geqslant 1$, and let f^{nh} be a solution of* (6.4). *Then, for all $f \in \mathcal{W}_m$ and all $h > 0$,*

$$\frac{1}{n} \sum_{i=1}^{n} | f(x_{in}) - f^{nh}(x_{in}) |^2 + h^{2m} | f - f^{nh} |_{\mathcal{W}_m}^2 =$$
$$\frac{1}{n} \sum_{i=1}^{n} \bigl(f(x_{in}) - y_{in} \bigr) \bigl(f(x_{in}) - f^{nh}(x_{in}) \bigr) + h^{2m} \langle f , f - f^{nh} \rangle_{\mathcal{W}_m} .$$

(7.2) UNIQUENESS LEMMA. *Let $m \geqslant 1$, and suppose that the design contains at least m distinct points. Then the solution of* (6.4) *is unique.*

(7.3) EXERCISE. Prove it. [Hint: Copy the proofs of Lemmas (3.1) and (3.6) with some cosmetic changes.]

We go on to prove the convergence rates of Theorem (6.8).

PROOF OF THEOREM (6.8). The starting point is the quadratic behavior of Lemma (7.1). After the usual manipulations with

$$\varepsilon^{nh} = f^{nh} - f_o , \tag{7.4}$$

this gives the equality

$$\frac{1}{n} \sum_{i=1}^{n} | \varepsilon^{nh}(x_{in}) |^2 + h^{2m} | \varepsilon^{nh} |_{\mathcal{W}_m}^2 = \frac{1}{n} \sum_{i=1}^{n} d_{in}\, \varepsilon^{nh}(x_{in}) - h^{2m} \langle f_o , \varepsilon^{nh} \rangle_{\mathcal{W}_m} . \tag{7.5}$$

Now, we just need to apply the appropriate results.

For the bias part, note that by Lemma (6.27)

$$- h^{2m} \langle f_o , \varepsilon^{nh} \rangle_{\mathcal{W}_m} \leqslant h^{2m} \| \varepsilon^{nh} \| \, \| f_o \|_{1,\mathcal{W}_{2m}} . \tag{7.6}$$

For the random sum on the right of (7.5), we use Lemma (6.38), so

$$\frac{1}{n} \sum_{i=1}^{n} d_{in}\, \varepsilon^{nh}(x_{in}) \leqslant \| \varepsilon^{nh} \|_{h,\mathcal{W}_m} \| S^{nh} \|_{h,\mathcal{W}_m} . \tag{7.7}$$

For the sum on the left-hand side of (7.5), Lemma (6.33) provides the lower bound

$$\| \varepsilon^{nh} \|^2 - c\,(nh^2)^{-1} \| \varepsilon^{nh} \|_{h,\mathcal{W}_m} ,$$

so that

$$\zeta^{nh} \| \varepsilon^{nh} \|_{h,\mathcal{W}_m}^2 \leqslant \frac{1}{n} \sum_{i=1}^{n} | \varepsilon^{nh}(x_{in}) |^2 + h^{2m} | \varepsilon^{nh} |_{\mathcal{W}_m}^2 , \tag{7.8}$$

with $\zeta^{nh} \to 1$, provided $nh^2 \to \infty$.

Substituting (7.6), (7.7), and (7.6) into (7.5) results in

$$\zeta^{nh} \| \varepsilon^{nh} \|^2_{h,\mathcal{W}_m} \leqslant \| S^{nh} \|_{h,\mathcal{W}_m} \| \varepsilon^{nh} \|_{h,\mathcal{W}_m} + c\,h^{2m} \| \varepsilon^{nh} \| \, \| f_o \|_{1,\mathcal{W}_{2m}} ,$$

and the right-hand side may be bounded by

$$\| \varepsilon^{nh} \|_{h,\mathcal{W}_m} \{ \| S^{nh} \|_{h,\mathcal{W}_m} + c\,h^{2m} \| f_o \|_{1,\mathcal{W}_{2m}} \} ,$$

so that, for a different c,

$$\| \varepsilon^{nh} \|^2_{h,\mathcal{W}_m} \leqslant c \| S^{nh} \|^2_{h,\mathcal{W}_m} + c\,h^{4m} \| f_o \|^2_{1,\mathcal{W}_{2m}} .$$

Finally, Lemma (6.38) gives that

$$\mathbb{E}[\| \varepsilon^{nh} \|^2_{h,\mathcal{W}_m}] = \mathcal{O}((nh)^{-1} + h^{4m}) ,$$

provided, again, that $nh^2 \to \infty$. For the optimal choice $h \asymp n^{-1/(4m+1)}$, this is indeed the case, and then

$$\mathbb{E}[\| \varepsilon^{nh} \|^2_{h,\mathcal{W}_m}] = \mathcal{O}(n^{-4m/(4m+1)}) .$$

This completes the proof. Q.e.d.

Finally, we prove the existence of the solution of (6.4). The following (compactness) result is useful. Define the mapping $T : \mathcal{W}_m \to C[0,1]$ by

$$Tf(x) = \int_{\frac{1}{2}}^{x} (x-t)^{m-1} f^{(m)}(t)\,dt , \quad x \in [0,1] . \tag{7.9}$$

(7.10) LEMMA. *Let $m \geqslant 2$. There exists a constant c such that, for all $f \in \mathcal{W}_m$ and all $x, y \in [0,1]$,*

$$| Tf(x) - Tf(y) | \leqslant c \, | y - x |^{1/2} \, | f |_{\mathcal{W}_m} .$$

PROOF. First we show that T is bounded. Let $0 < x \leqslant \frac{1}{2}$. Note that

$$| Tf(x) |^2 \leqslant c(x) \left| \int_{\frac{1}{2}}^{x} \{ t(1-t) \}^m \, | f^{(m)}(t) |^2 \, dt \right|$$

with

$$c(x) = \left| \int_{\frac{1}{2}}^{x} \frac{(x-t)^{2m-2}}{\{ t(1-t) \}^m} \, dt \right| .$$

Now, for $0 < x < t \leqslant \frac{1}{2}$, we have

$$0 \leqslant \frac{(t-x)^{2m-2}}{\{ t(1-t) \}^m} \leqslant 2^m (t-x)^{m-2} \leqslant 4$$

since $m \geqslant 2$. It follows that $c(x) \leqslant 2$ on the interval $0 \leqslant x \leqslant \frac{1}{2}$. The same argument applies to the case $x \geqslant \frac{1}{2}$, so that

$$\text{(7.11)} \qquad | \, Tf(x) \, | \leqslant 2 \, | \, f \, |_{\mathcal{W}_m} \, .$$

Thus, T is a bounded linear mapping from $\mathcal{W}_m$ into $L^\infty(0,1)$ in the $| \cdot |_{\mathcal{W}_m}$ topology on $\mathcal{W}_m$.

Let $f \in \mathcal{W}_m$, and set $g = Tf$. Then, from (7.11), the function g is bounded, so surely $\| \, g \, \| \leqslant 2 \, | \, f \, |_{\mathcal{W}_m}$. Of course, $g^{(m)} = f^{(m)}$ (almost everywhere), so that $| \, g \, |_{\mathcal{W}_m} = | \, f \, |_{\mathcal{W}_m}$. It follows that

$$\| \, g \, \|_{1,\mathcal{W}_m} \leqslant 3 \, | \, f \, |_{\mathcal{W}_m} \, .$$

Now, by the Embedding Lemma (6.29), for a suitable constant c,

$$\| \, g' \| \leqslant c \, \| \, g \, \|_{1,\mathcal{W}_2} \leqslant c_m \, \| \, g \, \|_{1,\mathcal{W}_m} \leqslant \widetilde{c} \, | \, f \, |_{\mathcal{W}_m} \, .$$

It follows that, for all x, $y \in [\, 0 \, , \, 1 \,]$,

$$| \, g(x) - g(y) \, | = \left| \int_y^x g'(\, t \,) \, dt \right| \leqslant | \, x - y \, |^{1/2} \, \| \, g' \, \| \leqslant \widetilde{c} \, | \, x - y \, |^{1/2} \, | \, f \, |_{\mathcal{W}_m} \, ,$$

as was to be shown. Q.e.d.

(7.12) COROLLARY. *Let $m \geqslant 2$. Then the mapping $T : \mathcal{W}_m \to C[\, 0 \, , \, 1 \,]$ is compact in the $| \cdot |_{\mathcal{W}_m}$ topology on $\mathcal{W}_m$.*

PROOF. This follows from the Arzelà-Ascoli theorem. Q.e.d.

(7.13) LEMMA. *Let $m \geqslant 2$. Then the relaxed boundary smoothing problem* (6.4) *has a solution.*

PROOF. Obviously, the objective function $RLS(f)$ of (6.4) is bounded from below (by 0), so there exists a minimizing sequence, denoted by $\{ \, f_k \, \}_k$. Then, obviously,

$$h^{2m} \, | \, f_k \, |^2_{\mathcal{W}_m} \leqslant RLS(f_k) \leqslant RLS(f_1) \, ,$$

the last inequality without loss of generality. Thus, there exists a subsequence, again denoted by $\{ \, f_k \, \}_k$, for which $\{ \, f_k^{(m)} \, \}_k$ converges, in the weak topology on $\mathcal{W}_m$ induced by the $| \cdot |_{\mathcal{W}_m}$ semi-norm, to some element φ_o. Then,

$$| \, \varphi_o \, |^2_{\mathcal{W}_m} \leqslant \liminf_{k \to \infty} \, | \, f_k \, |^2_{\mathcal{W}_m} \, ,$$

and by the compactness of T in this setting, then

$$\lim_{k \to \infty} \, \| \, Tf_k - T\varphi_o \, \|_\infty = 0 \, .$$

Finally, use Taylor's theorem with exact remainder to write

$$f_k(x) = p_k(x) + Tf_k(x)$$

for suitable polynomials p_k. Now, proceed as in the proof of the Existence Lemma (3.7) for smoothing splines. Consider the restrictions of the f_k to the design points,

$$r_n\, f_k \stackrel{\text{def}}{=} \big(\, f_k(x_{1,n}),\, f_k(x_{2,n}),\, \cdots\,,\, f_k(x_{n,n})\big)\ , \quad k = 1, 2, \cdots\ .$$

We may extract a subsequence from $\{\, f_k \,\}_k$ for which $\{\, r_n\, f_k \,\}_k$ converges in $\mathbb{R}^n$ to some vector v_o. Then, for the corresponding polynomials,

$$\lim_{k\to\infty} p_k(x_{in}) = [\, v_o\,]_i - T\varphi_o(x_{in})\ , \quad i = 1, 2, \cdots, n\ ,$$

and there exists a polynomial p_o of order m such that

$$(7.14) \qquad p_o(x_{in}) = [\, v_o\,]_i - T\varphi_o(x_{in})\ , \quad i = 1, 2, \cdots, n\ .$$

Finally, define $\psi_o = p_o + T\varphi_o$, and then, for the (subsub) sequence in question,

$$\lim_{k\to\infty} \mathit{RLS}(f_k) \geqslant \mathit{RLS}(\psi_o)\ ,$$

so that ψ_o minimizes $\mathit{RLS}(f)$ over $f \in \mathcal{W}_m$. Q.e.d.

(7.15) EXERCISE. Prove Theorem (6.8) for the case $m = 1$ when the design is asymptotically uniform in the sense of Definition (2.22) and satisfies condition (d) of Exercise (6.41).

EXERCISES: (7.3), (7.15).

8. Partially linear models

The gravy train of the statistical profession is undoubtedly data analysis by means of the linear model

$$(8.1) \qquad y_{in} = x_{in}^T \beta_o + d_{in}\ , \quad i = 1, 2, \cdots, n\ ,$$

in which the vectors $x_{in} \in \mathbb{R}^d$ embody the design of the experiment (d is some fixed integer $\geqslant 1$), $\beta_o \in \mathbb{R}^d$ are the unknown parameters to be estimated, $y_n = (y_{1,n}, y_{2,n}, \cdots, y_{n,n})^T$ are the observed response variables, and the collective noise is $d_n = (d_{1,n}, d_{2,n}, \cdots, d_{n,n})^T$, with independent components, assumed to be normally distributed,

$$(8.2) \qquad d_n \sim \text{Normal}(0, \sigma^2 I)\ ,$$

with σ^2 unknown. The model (8.1) may be succinctly described as

$$(8.3) \qquad y_n = X_n\, \beta_o + d_n\ ,$$

with the design matrix $X_n = (x_{1,n}\,|\, x_{2,n}\,|\, \cdots \,|\, x_{n,n})^T \in \mathbb{R}^{n\times d}$.

If X_n has full column rank, then the maximum likelihood estimator of β_o is given by

$$(8.4) \qquad \widehat{\beta}_n = (\,X_n^T X_n)^{-1} X_n^T\, y_n$$

and is normally distributed,

$$(8.5) \qquad \sqrt{n}\,(\,\widehat{\beta}_n - \beta_o\,) \sim N\big(0, \sigma^2\,(\,X_n^T X_n\,)^{-1}\,\big) \ ,$$

and the train is rolling.

In this section, we consider the *partially* linear model

$$(8.6) \qquad y_{in} = z_{in}^T\,\beta_o + f_o(x_{in}) + d_{in} \ , \quad i = 1, 2, \cdots, n \ ,$$

where $z_{in} \in \mathbb{R}^d$, $x_{in} \in [\,0\,,\,1\,]$ (as always), the function f_o belongs to $W^{m,2}(0,1)$ for some integer $m \geqslant 1$, and

(8.7) the d_{in} are iid, zero-mean random variables with a finite fourth moment.

In analogy with (8.3), this model may be described as

$$(8.8) \qquad y_n = Z_n\,\beta_o + r_n f_o + d_n \ ,$$

with $r_n f_o = \big(\,f_o(x_{1,n}), f_o(x_{2,n}), \cdots, f_o(x_{n,n})\,\big)^T$. Thus, r_n is the *restriction operator* from $[\,0\,,\,1\,]$ to the x_{in}.

Such models arise in the standard regression context, where interest is really in the model $y_n = Z_n\,\beta_o + d_n$ but the additional covariates x_{in} cannot be ignored. However, one does not wish to assume that these covariates contribute linearly or even parametrically to the response variable. See, e.g., ENGLE, GRANGER, RICE, and WEISS (1986), GREEN, JENNISON, and SEHEULT (1985), or the introductory example in HECKMAN (1988). Amazingly, under reasonable (?) conditions, one still gets best asymptotically normal estimators of β_o; that is, asymptotically, the contribution of the nuisance parameter f_o vanishes.

In this section, we exhibit asymptotically normal estimators of β_o and also pay attention to the challenge of estimating f_o at the optimal rate of convergence.

The assumptions needed are as follows. We assume that the x_{in} are deterministic and form a uniformly asymptotic design; e.g., equally spaced as in

$$(8.9) \qquad x_{in} = \frac{i-1}{n-1} \ , \quad i = 1, 2, \cdots, n \ .$$

The z_{in} are assumed to be random, according to the model

$$(8.10) \qquad z_{in} = g_o(x_{in}) + \varepsilon_{in} \ , \quad i = 1, 2, \cdots, n \ ,$$

in which

(8.11) the ε_{in} are mutually independent, zero-mean random variables, with finite fourth moment, and

$$\mathbb{E}[\,\varepsilon_{in}\,\varepsilon_{in}^T\,] = V \in \mathbb{R}^{d \times d} \ ,$$

with V positive-definite. Moreover, the ε_{in} are independent of d_n in the model (8.6).

In (8.10), $g_o(x) = \mathbb{E}[\, z \mid x \,]$ is the conditional expectation of z and is assumed to be a smooth function of x; in particular,

$$g_o \in W^{1,2}(0,1) \; . \tag{8.12}$$

(Precisely, *each component* of g_o belongs to $W^{1,2}(0,1)$.) Regarding f_o, we assume that, for some integer $m \geqslant 1$,

$$f_o \in W^{m,2}(0,1) \; . \tag{8.13}$$

Below, we study two estimators of β_o, both related to smoothing spline estimation. However, since the model (8.3) and the normality result (8.5) constitute the guiding light, the methods and notations used appear somewhat different from those in the previous sections.

The simplest case. To get our feet wet, we begin with the case in which $g_o(x) = 0$ for all x, so that

$$\begin{aligned}&\text{the } z_{in} \text{ are mutually independent, zero-mean}\\ &\text{random variables, with finite fourth moment,}\\ &\text{independent of the } d_{in}\text{, and satisfy}\\ &\qquad\qquad \mathbb{E}[\, z_{in}\, z_{in}^T \,] = V \; ,\\ &\text{with } V \text{ positive-definite}.\end{aligned} \tag{8.14}$$

Under these circumstances, by the strong law of large numbers,

$$\tfrac{1}{n}\, Z_n^T Z_n \longrightarrow_{\text{as}} V \; . \tag{8.15}$$

The estimator under consideration goes by the name of the partial spline estimator, the solution to

$$\begin{aligned}&\text{minimize} && \tfrac{1}{n}\, \| \, Z_n\,\beta + r_n f - y_n \, \|^2 + h^{2m} \, \| \, f^{(m)} \, \|^2\\ &\text{subject to} && \beta \in \mathbb{R}^d \; , \; f \in W^{m,2}(0,1) \; .\end{aligned} \tag{8.16}$$

One verifies that the solution (β^{nh}, f^{nh}) exists and is unique *almost surely*, and that f^{nh} is an ordinary ("natural") spline function of polynomial order $2m$ with the x_{in} as knots. With (8.4) in mind, we wish to express the objective function in (8.16) in linear algebra terms. For fixed β, the Euler equations (3.18) applied to (8.15) imply that the natural spline function f is completely determined in terms of its function values at the knots, encoded in the vector $r_n f$. Then, there exists a symmetric, semi-positive-definite matrix $M \in \mathbb{R}^{n \times n}$, depending on the knots x_{in} only, such that

$$\| \, f^{(m)} \, \|^2 = (r_n f)^T M \, r_n f \quad \text{for all natural splines } f \; . \tag{8.17}$$

So, the problem (8.16) may be written as

$$\begin{aligned}&\text{minimize} && \tfrac{1}{n}\, \| \, Z_n\,\beta + r_n f - y_n \, \|^2 + h^{2m} \, (r_n f)^T M \, r_n f\\ &\text{subject to} && \beta \in \mathbb{R}^d \; , \; r_n f \in \mathbb{R}^n \; .\end{aligned} \tag{8.18}$$

Here, the notation $r_n f$ is suggestive but otherwise denotes an arbitrary vector in $\mathbb{R}^n$.

The solution (β, f) to (8.18) is uniquely determined by the normal equations

$$\begin{aligned} Z_n^T (Z_n \beta + r_n f - y_n) &= 0 \ , \\ (I + n\, h^{2m} M) r_n f + Z_n \beta - y_n &= 0 \ . \end{aligned} \tag{8.19}$$

Eliminating $r_n f$, we get the explicit form of the partial spline estimator

$$\beta^{nh} = \bigl(Z_n^T (I - S_h) Z_n \bigr)^{-1} Z_n^T (I - S_h) y_n \ , \tag{8.20}$$

in which

$$S_h = (I + n\, h^{2m} M)^{-1} \tag{8.21}$$

is the natural smoothing spline operator. Note that S_h is symmetric and positive-definite. The following exercise is useful.

(8.22) Exercise. Let $\delta_n = (\delta_{1,n}, \delta_{2,n}, \cdots, \delta_{n,n})^T \in \mathbb{R}^n$ and let $f = \varphi$ be the solution to

$$\begin{aligned} &\text{minimize} \quad \tfrac{1}{n} \sum_{i=1}^{n} | f(x_{in}) - \delta_{in} |^2 + h^{2m} \| f^{(m)} \|^2 \\ &\text{subject to} \quad f \in W^{m,2}(0,1) \ . \end{aligned}$$

Show that $r_n \varphi = S_h \delta_n$.

In view of the model (8.8), we then get that

$$\beta^{nh} - \beta_o = \text{variation} + \text{bias} \ , \tag{8.23}$$

with

$$\begin{aligned} \text{variation} &= \bigl(Z_n^T (I - S_h) Z_n \bigr)^{-1} Z_n^T (I - S_h) d_n \ , \\ \text{bias} &= \bigl(Z_n^T (I - S_h) Z_n \bigr)^{-1} Z_n^T (I - S_h) r_n f_o \ . \end{aligned} \tag{8.24}$$

In the above, we tacitly assumed that $Z_n^T (I - S_h) Z_n$ is nonsingular. Asymptotically, this holds by (8.15) and the fact that

$$\tfrac{1}{n} Z_n^T S_h Z_n = \mathcal{O}_P\bigl((nh)^{-1} \bigr) \ , \tag{8.25}$$

as we show below. The same type of argument shows that

$$\tfrac{1}{n} Z_n^T S_h d_n = \mathcal{O}_P\bigl((nh)^{-1} \bigr) \ , \tag{8.26}$$

$$\tfrac{1}{n} Z_n^T (I - S_h) r_n f_o = \mathcal{O}_P\bigl((nh)^{-1/2} h^m \bigr) \ . \tag{8.27}$$

This gives

$$\beta^{nh} - \beta_o = (Z_n^T Z_n)^{-1} Z_n^T d_n + \mathcal{O}_P\bigl((nh)^{-1} + n^{-1/2} h^{m-1/2} \bigr) \ ,$$

and the asymptotic normality of $\beta^{nh} - \beta_o$ follows for the appropriate h (but (8.25)–(8.27) need proof).

(8.28) THEOREM. *Under the assumptions* (8.2), (8.7), *and* (8.9)–(8.14),

$$\sqrt{n}\,(\,\beta^{nh} - \beta_o) \longrightarrow_{\rm d} \Upsilon \sim \text{Normal}(\,0\,,\,\sigma^2\,V\,)\ ,$$

provided $h \to 0$ *and* $nh^2 \to \infty$.

Note that Theorem (8.28) says that β^{nh} is asymptotically a minimum variance unbiased estimator.

(8.29) EXERCISE. Complete the proof of the theorem by showing that

$$(\,Z_n^T Z_n)^{-1} Z_n^T d_n \longrightarrow_{\rm d} U \sim \text{Normal}(\,0\,,\,\sigma^2\,V^{-1}\,)\ .$$

PROOF OF (8.26). Note that $d \geqslant 1$ and $Z_n^T \in \mathbb{R}^{d \times n}$. We actually pretend that $d = 1$.

In accordance with Exercise (8.22), let $\mathfrak{z}^{nh}$ be the natural spline of order $2m$ with the x_{in} as knots satisfying $r_n \mathfrak{z}^{nh} = S_h Z_n$. Then,

$$\| \, \mathfrak{z}^{nh} \, \|_{m,h} = \mathcal{O}_P\big(\,(nh)^{-1/2}\,\big)\ ; \tag{8.30}$$

see the Random Sum Lemma (2.20). Now,

$$\tfrac{1}{n}\, Z_n^T S_h\, d_n = \tfrac{1}{n}\, d_n^T S_h\, Z_n = \tfrac{1}{n} \sum_{i=1}^{n} d_{in}\, \mathfrak{z}^{nh}(x_{in})\ ,$$

so that in the style of the Random Sum Lemma (2.20),

$$\tfrac{1}{n}\, Z_n^T S_h\, d_n = \Big\langle\, \tfrac{1}{n} \sum_{i=1}^{n} d_{in}\, \mathcal{R}_{mh}(\,\cdot\,, x_{in})\,,\, \mathfrak{z}^{nh} \,\Big\rangle_{m,h}\ ,$$

whence

$$\big|\, \tfrac{1}{n}\, Z_n^T S_h\, d_n \,\big| \leqslant \|\, \tfrac{1}{n} \sum_{i=1}^{n} d_{in}\, \mathcal{R}_{mh}(\,\cdot\,, x_{in})\, \|_{m,h}\, \|\, \mathfrak{z}^{nh}\, \|_{m,h}\ . \tag{8.31}$$

Finally, observe that

$$\mathbb{E}\Big[\, \|\, \tfrac{1}{n} \sum_{i=1}^{n} d_{in}\, \mathcal{R}_{mh}(\,\cdot\,, x_{in})\, \|_{m,h}^2 \,\Big] = \mathcal{O}\big(\,(nh)^{-1}\,\big) \tag{8.32}$$

by assumption (8.14). Thus, (8.26) follows for $d = 1$. Q.e.d.

(8.33) EXERCISE. Clean up the proof for the case $d \geqslant 2$. Note that $Z_n^T S_h\, d_n \in \mathbb{R}^d$, so we need not worry about the choice of norms.

(8.34) EXERCISE. Prove (8.25) for $d = 1$ by showing that

$$\tfrac{1}{n}\, Z_n^T S_h\, Z_n = \Big\langle\, \tfrac{1}{n} \sum_{i=1}^{n} z_{in}\, \mathcal{R}_{mh}(\,\cdot\,, x_{in})\,,\, \mathfrak{z}^{nh} \,\Big\rangle_{m,h}\ ,$$

with $\mathfrak{z}^{nh}$ as in the proof of (8.26) and properly bounding the expression on the right. Then, do the general case $d \geqslant 2$.

PROOF OF (8.27). Note that $S_h r_n f_o = r_n f_{hn}$, with f_{hn} the solution to the noise-free problem (4.18). Then, the results of Exercises (4.22) and (4.23) imply that $\| f_o - f_{hn} \|_{m,h} = \mathcal{O}(h^m)$. Thus,

$$\tfrac{1}{n} Z_n^T (I - S_h) r_n f_o = \tfrac{1}{n} \sum_{i=1}^{n} z_{in} (f_o(x_{in}) - f_{hn}(x_{in}))$$

$$= \Big\langle \tfrac{1}{n} \sum_{i=1}^{n} z_{in} \mathcal{R}_{mh}(\cdot , x_{in}) , f_o - f_{hn} \Big\rangle_{m,h} ,$$

so that

$$\big| \tfrac{1}{n} Z_n^T (I - S_h) r_n f_o \big| \leqslant \big\| \tfrac{1}{n} \sum_{i=1}^{n} z_{in} \mathcal{R}_{mh}(\cdot , x_{in}) \big\|_{m,h} \| f_o - f_{hn} \|_{m,h} ,$$

and the rest is old hat. Q.e.d.

(8.35) EXERCISE. Show that, under the conditions of Theorem (8.28),

$$\| f^{nh} - f_o \|_{m,h}^2 = \mathcal{O}_P\big((nh)^{-1} + h^{2m} \big) .$$

Thus, for $h \asymp n^{-1/(2m+1)}$, we get the optimal convergence rate for f^{nh} as well as the asymptotic normality of β^{nh}.

Arbitrary designs. We now wish to see what happens when the z_{in} do not satisfy (8.14) but only (8.11). It will transpire that one can get asymptotic normality of β^{nh} but not the optimal rate of convergence for f^{nh}, at least not at the same time. Thus, the lucky circumstances of Exercise (8.35) fail to hold any longer. However, a fix is presented later.

Again, as estimators we take the solution (β^{nh}, f^{nh}) of (8.18), and we need to see in what form (8.25)–(8.27) hold. When all is said and done, it turns out that (8.27) is causing trouble, as we now illustrate.

It is useful to introduce the matrix G_n,

$$(8.36) \qquad G_n = r_n g_o = \big[g_o(x_{1,n}) \,|\, g_o(x_{2,n}) \,|\, \cdots \,|\, g_o(x_{n,n}) \big]^T \in \mathbb{R}^{n \times d} ,$$

and define

$$(8.37) \qquad \widetilde{Z}_n = Z_n - G_n .$$

Thus, $\widetilde{Z}_n$ shares the properties of Z_n for the simplest case.

TRYING TO PROVE (8.27) FOR ARBITRARY DESIGNS. Write

$$\tfrac{1}{n} Z_n^T (I - S_h) r_n f_o = \tfrac{1}{n} \widetilde{Z}_n^T (I - S_h) r_n f_o + \tfrac{1}{n} G_n^T (I - S_h) r_n f_o .$$

For the first term on the right, we do indeed have

$$\tfrac{1}{n} \widetilde{Z}_n^T (I - S_h) r_n f_o = \mathcal{O}_P\big((nh)^{-1/2} h^m \big) ,$$

see the proof of (8.27) for the simplest case.

For the second term, recall that $(I - S_h)\, r_n f_o = r_n(f_o - f_{hn})$, and this is $\mathcal{O}(h^m)$. Thus, we only get

$$\tag{8.38} \tfrac{1}{n}\, G_n^T\, (I - S_h)\, r_n f_o = \mathcal{O}(h^m) \ .$$

Moreover, it is easy to see that in general this is also an asymptotic lower bound. Thus (8.27) must be suitably rephrased. Q.e.d.

So, if all other bounds stay the same, asymptotic normality is achieved only if $h \ll n^{-1/(2m)}$, but then we do not get the optimal rate of convergence for the estimator of f_o since the required $h \asymp n^{-1/(2m+1)}$ is excluded.

(8.39) EXERCISE. Prove (suitable modifications of) (8.25) and (8.26) for the arbitrary designs under consideration. Also, verify the asymptotic normality of β^{nh} for $nh^2 \to \infty$ and $nh^{2m} \to 0$.

So, what is one to do? From a formal mathematical standpoint, it is clear that a slight modification of (8.38), and hence a slight modification of (8.27), would hold, viz.

$$\tag{8.40} \tfrac{1}{n}\, G_n^T (I - S_h)^2\, r_n f_o = \mathcal{O}(h^{m+1}) \ ,$$

but then everything else must be modified as well. All of this leads to two-stage estimators, in which the conditional expectation $g_o(x) = \mathbb{E}[\, z \mid x \,]$ is first estimated nonparametrically and then the estimation of β_o and f_o is considered.

(8.41) EXERCISE. Prove (8.40) taking

$$\tfrac{1}{n}\, G_n^T (I - S_h)^2\, r_n f_o = \tfrac{1}{n} \sum_{i=1}^{n} \{\, g_o(x_{in}) - g_{hn}(x_{in}) \,\} \{\, f_o(x_{in}) - f_{hn}(x_{in}) \,\}$$

as the starting point. (g_{hn} is defined analogously to f_{hn}.)

Two-stage estimators for arbitrary designs. Suppose we estimate the conditional expectation $g_o(x)$ by a smoothing spline estimator g^{nh} (componentwise). In our present finite-dimensional context, then

$$\tag{8.42} r_n\, g^{nh} = S_h\, Z_n \ .$$

With this smoothing spline estimator of g_o, let

$$\tag{8.43} G^{nh} = r_n g^{nh} = [\, g^{nh}(x_{1,n}) \,|\, g^{nh}(x_{2,n}) \,|\, \cdots , g^{nh}(x_{n,n}) \,]^T \in \mathbb{R}^{n \times d} \ ,$$

and define

$$\tag{8.44} \mathcal{Z}^{nh} = Z_n - G^{nh} \ .$$

Now, following CHEN and SHIAU (1991), consider the estimation problem

$$\tag{8.45} \begin{array}{ll} \text{minimize} & \tfrac{1}{n} \|\, \mathcal{Z}^{nh}\, \beta + r_n\, \varphi - y_n \,\|^2 + h^{2m} \,\|\, \varphi^{(m)} \,\|^2 \\ \text{subject to} & \beta \in \mathbb{R}^d \ , \ \varphi \in W^{m,2}(0,1) \ . \end{array}$$

The solution is denoted by $(\beta^{nh,1}, f^{nh,1})$. Note that, with $\varphi = f + (g^{nh})^T\beta$, the objective function may also be written as

$$\tfrac{1}{n}\| \, Z_n \, \beta + r_n f - y_n \, \| + h^{2m} \, \| \, \{ \, f + (g^{nh})^T \, \beta \, \}^{(m)} \, \|^2 \ ;$$

in other words, the (estimated) conditional expectation is part of the roughness penalization, with the same smoothing parameter h.

It is a straightforward exercise to show that this two-stage estimator of β_o is given by

$$\beta^{nh,1} = \big(\mathcal{Z}_n^T (I - S_h)^3 \, \mathcal{Z}_n \big)^{-1} \, \mathcal{Z}_n^T (I - S_h)^2 \, y_n \ , \tag{8.46}$$

so that

$$\beta^{nh,1} - \beta_o = \text{variation} + \text{bias} \ , \tag{8.47}$$

with

$$\begin{aligned} \text{variation} &= \big(\mathcal{Z}_n^T (I - S_h)^3 \, \mathcal{Z}_n \big)^{-1} \, \mathcal{Z}_n^T (I - S_h)^2 \, d_n \ , \\ \text{bias} &= \big(\mathcal{Z}_n^T (I - S_h)^3 \, \mathcal{Z}_n \big)^{-1} \, \mathcal{Z}_n^T (I - S_h)^2 \, r_n f_o \ . \end{aligned} \tag{8.48}$$

The crucial results to be shown are

$$\tfrac{1}{n} \, \mathcal{Z}_n^T \, \mathcal{Z}_n \longrightarrow_{\text{as}} V \ , \tag{8.49}$$

$$\tfrac{1}{n} \, \mathcal{Z}_n^T (-3 \, S_h + 3 \, {S_h}^2 - {S_h}^3) \, \mathcal{Z}_n = \mathcal{O}_P\big((nh)^{-1} \big) \ , \tag{8.50}$$

$$\tfrac{1}{n} \, Z_n^T \, (-2 \, S_h + {S_h}^2) \, d_n = \mathcal{O}_P\big((nh)^{-1} \big) \ , \tag{8.51}$$

$$\tfrac{1}{n} \, Z_n^T \, (I - S_h)^2 \, r_n f_o = \mathcal{O}_P\big(n^{-1/2} \, h^{m-1/2} + h^{m+1} \big) \ , \tag{8.52}$$

with V as in (8.11). They are easy to prove by the previously used methods. All of this then leads to the following theorem.

(8.53) THEOREM. *Under the assumptions* (8.2), (8.7), *and* (8.9)–(8.13),

$$\sqrt{n} \, (\beta^{nh,1} - \beta_o) \longrightarrow_{\text{d}} \Upsilon \sim \text{Normal}(0 \, , \sigma^2 \, V) \ ,$$

provided $nh^2 \to \infty$ *and* $nh^{2m+2} \to 0$.

(8.54) EXERCISE. (a) Prove (8.49) through (8.52).
(b) Assume that $g_o \in W^{m,2}(0,1)$. Prove that, for $h \asymp n^{-1/(2m+1)}$, we get the asymptotic normality of $\beta^{nh,1}$ as advertised in Theorem (8.53) as well as the optimal rate of convergence for the estimator of f_o, viz. $\| \, f^{nh,1} - f_o \, \| = \mathcal{O}_P\big(n^{-m/(2m+1)} \big)$.

We finish this section by mentioning the estimator

$$\beta^{nh,2} = \big(\mathcal{Z}_n^T (I - S_h)^2 \, \mathcal{Z}_n \big)^{-1} \, \mathcal{Z}_n^T (I - S_h)^2 \, y_n \tag{8.55}$$

of SPECKMAN (1988), who gives a piecewise regression interpretation. (To be precise, SPECKMAN (1988) considers kernel estimators, not just smoothing splines.) The estimator for f_o is then given by

$$r_n f^{nh,2} = S_h (y_n - Z_n \beta^{nh,2}) . \tag{8.56}$$

The asymptotic normality of $\beta^{nh,2}$ may be shown similarly to that of $\beta^{nh,1}$.

(8.57) EXERCISE. State and prove the analogue of Theorem (8.53) for the estimator $\beta^{nh,2}$.

This completes our discussion of spline estimation in partially linear models. It is clear that it could be expanded considerably. By way of example, the smoothing spline $r_n g^{nh} = S_h Z_n$, see (8.42), applies to each component of Z_n separately, so it makes sense to have different smoothing parameters for each component so

$$r_n (g^{nh})_j = S(h_j) (Z_n)_j , \quad j = 1, 2, \cdots , d , \tag{8.58}$$

with the notation $S(h) \equiv S_h$. Here $(Z_n)_j$ denotes the j-th column of Z_n.

EXERCISES: (8.22), (8.29), (8.33), (8.34), (8.35), (8.39), (8.41), (8.54), (8.57).

9. Estimating derivatives

Estimating derivatives is an interesting problem with many applications. See, e.g., D'AMICO and FERRIGNO (1992) and WALKER (1998), where cubic and quintic splines are considered. In this section, we briefly discuss how smoothing splines may be used for this purpose and how error bounds may be obtained.

The problem is to estimate $f_o'(x)$, $x \in [\,0\,,\,1\,]$, in the model

$$y_{in} = f_o(x_{in}) + d_{in} , \quad i = 1, 2, \cdots , n , \tag{9.1}$$

under the usual conditions (4.1)–(4.4). We emphasize the last condition,

$$f_o \in W^{m,2}(0,1) . \tag{9.2}$$

As the estimator of f_o', we take $(f^{nh})'$, the derivative of the spline estimator. We recall that, under the stated conditions,

$$\mathbb{E}[\, \| f^{nh} - f_o \|_{m,h}^2 \,] = \mathcal{O}\big(n^{-2m/(2m+1)} \big) , \tag{9.3}$$

provided $h \asymp n^{-1/(2m+1)}$ (deterministically); see Corollary (4.7). Now, recall Lemma (2.17),

$$\| \varphi \|_{k,h} \leqslant c_m \| \varphi \|_{m,h} ,$$

valid for all $\varphi \in W^{m,2}(0,1)$ and for all $k = 0, 1, \cdots , m$, with a constant c_m depending on m only. Applying this to the problem at hand with

$k = 1$ yields

$$\mathbb{E}[\, h^2 \, \| \, (f^{nh} - f_o)' \, \|^2 \,] = \mathcal{O}\big(n^{-2m/(2m+1)} \big) \; ,$$

so that

$$\mathbb{E}[\, \| \, (f^{nh} - f_o)' \, \|^2 \,] = \mathcal{O}\big(n^{-2(m-1)/(2m+1)} \big) \; , \tag{9.4}$$

provided $h \asymp n^{-1/(2m+1)}$. This argument applies to all derivatives of order $< m$. We state it as a theorem.

(9.5) THEOREM. *Assume the conditions* (4.1) *through* (4.4) *and that the design is asymptotically uniform. Then, for* $d = 1, 2, \cdots, m-1$,

$$\mathbb{E}[\, \| \, (f^{nh})^{(d)} - f_o^{(d)} \, \|^2 \,] = \mathcal{O}\big(n^{-2(m-d)/(2m+1)} \big) \; ,$$

provided $h \asymp n^{-1/(2m+1)}$.

(9.6) EXERCISE. Prove the remaining cases of the theorem.

Some final comments are in order. It is not surprising that we lose accuracy in differentiation compared with plain function estimation. However, it is surprising that the asymptotically optimal value of h does not change (other than through the constant multiplying $n^{-1/(2m+1)}$). We also mention that RICE and ROSENBLATT (1983) determine the optimal convergence rates as well as the constants. Inasmuch as we get the optimal rates, the proof above is impeccable. Of course, our proof does not give any indication *why* these are the correct rates. The connection with kernel estimators through the "equivalent" kernels might provide some insight; see Chapter 14.

EXERCISE: (9.6).

10. Additional notes and comments

Ad § 1: Nonparametric regression is a huge field of study, more or less (less!) evenly divided into smoothing splines, kernel estimators, and local polynomials, although wavelet estimators are currently in fashion. It is hard to do justice to the extant literature. We mention WAHBA (1990), EUBANK (1999), HÄRDLE (1990), GREEN and SILVERMAN (1990), FAN and GIJBELS (1996), ANTONIADIS (2007), and GYÖRFI, KOHLER, KRZYŻAK, and WALK (2002) as general references.

Ad § 2: For everything you might ever need to know about the Sobolev spaces $W^{m,2}(0,1)$, see ADAMS (1975), MAZ'JA (1985), and ZIEMER (1989). The statement and proof of the Interpolation Lemma (2.12) comes essentially from ADAMS (1975).

The reference on reproducing kernel Hilbert spaces is ARONSZAJN (1950). MESCHKOWSKI (1962) and HILLE (1972) are also very informative. For a survey of the use of reproducing kernels in statistics and probability, see BERLINET and THOMAS-AGNAN (2004). For more on Green's functions, see, e.g., STAKGOLD (1967).

The definition (2.22) of asymptotically uniform designs is only the tip of a sizable iceberg; see AMSTLER and ZINTERHOF (2001) and references therein.

Ad § 3: RICE and ROSENBLATT (1983) refer to the natural boundary conditions (3.18) as unnatural (boundary) conditions, which is wrong in the technical sense but accurate nevertheless.

Ad § 6: The Embedding Lemma (6.13) is the one-dimensional version of a result in KUFNER (1980). Of course, the one-dimensional version is much easier than the multidimensional case.

Ad § 8: The "simplest" case of the partially linear model was analyzed by HECKMAN (1988). RICE (1986a) treated arbitrary designs and showed that asymptotic normality of β^{nh} requires undersmoothing of the spline estimator of f_o. The two-stage estimator $\beta^{nh,1}$ and the corresponding asymptotic normality and convergence rates are due to CHEN and SHIAU (1991). The estimator $\beta^{nh,2}$ was introduced and studied by SPECKMAN (1988) for "arbitrary" kernels S_h. CHEN (1988) considered piecewise polynomial estimators for f_o. Both of these authors showed the asymptotic normality of their estimators and the optimal rate of convergence of the estimator for f_o. Bayesian interpretations may be found in EUBANK (1988) and HECKMAN (1988). EUBANK, HART, and SPECKMAN (1990) discuss the use of the trigonometric sieve combined with boundary correction using the Bias Reduction Principle; see § 15.4. Finally, BUNEA and WEGKAMP (2004) study model selection in the partially linear model.

14

Kernel Estimators

1. Introduction

We continue the study of the nonparametric regression problem, in which one wishes to estimate a function f_o on the interval $[\,0\,,\,1\,]$ from the data $y_{1,n}, y_{2,n}, \cdots, y_{n,n}$, following the model

$$(1.1) \qquad y_{in} = f_o(x_{in}) + d_{in} \;, \quad i = 1, 2, \cdots, n \;.$$

Here, $d_n = (d_{1,n}, d_{2,n}, \cdots, d_{n,n})^T$ is the random noise. One recalls that the d_{in}, $i = 1, 2, \cdots, n$, are assumed to be uncorrelated random variables with mean 0 and common variance, σ^2

$$(1.2) \qquad \mathbb{E}[\,d_n\,] = 0 \quad , \quad \mathbb{E}[\,d_n d_n^T\,] = \sigma^2 I \;,$$

where σ is typically unknown, and that we refer to (1.1)–(1.2) as the Gauss-Markov model. When deriving uniform error bounds, we need the added condition that the d_{in} are independent, identically distributed (iid) random variables and that

$$(1.3) \qquad \mathbb{E}[\,|\,d_{in}\,|^{\kappa}\,] < \infty$$

for some $\kappa > 2$.

The design points (we are mostly concerned with deterministic designs) are assumed to be asymptotically uniform in the sense of Definition (13.2.22). As an example, think of

$$(1.4) \qquad x_{in} = \frac{i-1}{n-1} \;, \quad i = 1, 2, \cdots, n \;.$$

In this chapter, we consider one of the early competitors of smoothing splines, the Nadaraya-Watson kernel estimator,

$$(1.5) \qquad f^{nh}(x) = \frac{\frac{1}{n}\sum\limits_{i=1}^{n} y_{in}\, A_h(x - x_{in})}{\frac{1}{n}\sum\limits_{i=1}^{n} A_h(x - x_{in})} \;, \quad x \in (0,1) \;,$$

P.P.B. Eggermont, V.N. LaRiccia, *Maximum Penalized Likelihood Estimation*,
Springer Series in Statistics, DOI 10.1007/b12285_3,

in which $A_h(x) = h^{-1} A(h^{-1}x)$ for some nice pdf A. We point out now that in (1.5) we have a *convolution* kernel. Later, we comment on the important case where the x_{in} are random; see (1.8)–(1.12). In theory and practice, the estimator (1.5) is not satisfactory near the boundary of the interval, and either *boundary corrections* must be made (see § 4) or one must resort to *boundary kernels* (see § 3). When this is done, the usual convergence rates obtain: For appropriate boundary kernels, we show in § 2 that, for $m \geqslant 1$ and $f_o \in W^{m,2}(0,1)$, under the model (1.1)–(1.2),

$$\mathbb{E}[\, \| f^{nh} - f_o \|_2^2 \,] = \mathcal{O}\big(n^{-2m/(2m+1)} \big) \ , \tag{1.6}$$

provided $h \asymp n^{-1/(2m+1)}$ (deterministically). In § 4, we show the same results for regular kernels after boundary corrections have been made by means of the Bias Reduction Principle of Eubank and Speckman (1990b).

We also consider uniform error bounds on the kernel estimator: Under the added assumption of iid noise and (1.3), then almost surely

$$\| f^{nh} - f_o \|_\infty = \mathcal{O}\big((n^{-1} \log n)^{m/(2m+1)} \big) \ , \tag{1.7}$$

provided $h \asymp (n^{-1} \log n)^{1/(2m+1)}$, again deterministically, but in fact the bounds hold uniformly in h over a wide range. This is done in §§ 5 and 6. Uniform error bounds for smoothing splines are also considered by way of the equivalent reproducing kernel approach outlined in (13.1.10)–(13.1.14). The hard work is to show that the reproducing kernels are sufficiently convolution-like in the sense of Definitions (2.5) and (2.6), so that the results of § 6 apply. In § 7, we work out the details of this program.

So far, we have only discussed deterministic designs. In the classical nonparametric regression problem, the design points x_{in} themselves are random; that is, the data consist of

$$(X_1, Y_1), (X_2, Y_2), \cdots, (X_n, Y_n) \ , \tag{1.8}$$

an iid sample of the random variable (X, Y), with

$$\mathbb{E}[\, D \mid X = x \,] = 0 \quad , \quad \sup_{x \in [0,1]} \mathbb{E}[\, | D |^\kappa \mid X = x \,] < \infty \ , \tag{1.9}$$

for some $\kappa > 2$. The classical interpretation is now that estimating $f_o(x)$ at a fixed x amounts to estimating the conditional expectation

$$f_o(x) = \mathbb{E}[\, Y \mid X = x \,] \ . \tag{1.10}$$

Formally, with Bayes' rule, the conditional expectation may be written as

$$f_o(x) = \frac{1}{f_X(x)} \int_{\mathbb{R}} y \, f_{X,Y}(x,y) \, dy \ , \tag{1.11}$$

where $f_{X,Y}$ is the joint density of (X, Y) and f_X is the marginal density of X. Now, both of these densities may be estimated by kernel density estimators, $\widehat{f}_X(x) = \frac{1}{n} \sum_{i=1}^{n} A_h(x - X_i)$, and

$$\widehat{f}_{X,Y}(x,y) = \frac{1}{n} \sum_{i=1}^{n} A_h(y - Y_i) A_h(x - X_i) ,$$

for a suitable A and smoothing parameter h. Then, if $\int_{\mathbb{R}} y\, A_h(y)\, dy = 0$, one sees that

$$\int_{\mathbb{R}} y\, \widehat{f}_{X,Y}(x,y)\, dy = \frac{1}{n} \sum_{i=1}^{n} Y_i\, A_h(x - X_i) .$$

The result is the Nadaraya-Watson estimator

$$f^{nh}(x) = \frac{1}{n} \sum_{i=1}^{n} Y_i\, A_h(x, X_i) , \tag{1.12}$$

where the Nadaraya-Watson kernel $A_h(x,y)$ is given by

$$A_h(x,y) \equiv A_h(x,y \,|\, X_1, X_1, \cdots, X_n) = \frac{A_h(x-y)}{\frac{1}{n} \sum_{i=1}^{n} A_h(x - X_i)} . \tag{1.13}$$

This goes back to NADARAYA (1964, 1965) and WATSON (1964). In Chapter 16, we come back to the Nadaraya-Watson estimator for *random* designs as a special case of local polynomial estimators. See also § 21.9.

In the remainder of this chapter, the estimators above are studied in detail. In § 2, we study the mean integrated squared error for boundary kernel estimators. The boundary kernels themselves are discussed in § 3. Boundary corrections, following EUBANK and SPECKMAN (1990b), are discussed in § 4. Uniform error bounds for kernel estimators are considered in §§ 5 and 6. The same result holds for smoothing spline estimators; see § 7. In effect, there we construct a semi-explicit representation of the kernel in the equivalent kernel formulation of the spline estimator.

2. Mean integrated squared error

In the next few sections, we study kernel estimators for the standard regression problem (1.1)–(1.2). The basic form of a kernel estimator is

$$f^{nh}(x) = \frac{1}{n} \sum_{i=1}^{n} y_{in}\, A_h(x - x_{in}) , \quad x \in [\,0\,,\,1\,] , \tag{2.1}$$

in which $0 < h \leqslant 1$ and

$$A_h(x) = h^{-1} A(\,h^{-1}x\,) , \quad x \in [\,0\,,\,1\,] , \tag{2.2}$$

for some kernel A. This presumes that the design is asymptotically uniform. For "arbitrary" design points (deterministic or random), the full Nadaraya-

Watson estimator is

$$(2.3) \qquad f^{nh}(x) = \tfrac{1}{n} \sum_{i=1}^{n} y_{in} \, A_h(x, x_{in}) \, ,$$

where $A_h(x, t) \equiv A_h(x, t \mid x_{1,n}, x_{2,n}, \cdots, x_{nn})$ depends on all the design points; see (1.13). It should be noted that, even for uniform designs, the Nadaraya-Watson estimator is better near the endpoints than (2.1) but still not good enough.

Returning to (2.1) for uniform designs, it is well-known that *boundary corrections* must be made if we are to adequately estimate f_o near the endpoints of the interval $[0, 1]$. Thus, the general, boundary-corrected kernel estimator is

$$(2.4) \qquad f^{nh}(x) = \tfrac{1}{n} \sum_{i=1}^{n} y_{in} \, A_h(x, x_{in}) \, , \quad x \in [0, 1] \, ,$$

with $0 < h \leqslant 1$. In (2.4), the family of kernels $A_h(x, y)$, $0 < h \leqslant 1$, is assumed to be convolution-like in the following sense.

(2.5) Definition. *We say a family of kernels A_h, $0 < h \leqslant 1$, defined on $[0, 1] \times [0, 1]$, is convolution-like if there exists a constant c such that, for all $x \in [0, 1]$ and all h, $0 < h \leqslant 1$,*

$$\| A_h(x, \cdot) \|_{L^1(0,1)} \leqslant c \, ; \; \| A_h(\cdot, x) \|_\infty \leqslant c h^{-1} \, ; \; | A_h(x, \cdot) |_{BV} \leqslant c h^{-1} \, .$$

Here, $| f |_{BV}$ denotes the total variation of the function f over $[0, 1]$. See § 17.2 for more details.

(2.6) Definition. *Let $m \geqslant 1$ be an integer. We say a family of kernels A_h, $0 < h \leqslant 1$, defined on $[0, 1] \times [0, 1]$, is convolution-like of order m if it is convolution-like and if for some constant c, for all $x \in [0, 1]$,*

$$\text{(a)} \qquad \int_0^1 A_h(x, t) \, dt = 1 \, ,$$

$$\text{(b)} \qquad \int_0^1 (x - t)^k \, A_h(x, t) \, dt = 0 \, , \quad k = 1, 2, \cdots, m-1 \, ,$$

$$\text{(c)} \qquad \int_0^1 | x - t |^m \, | A_h(x, t) | \, dt \leqslant c h^m \, .$$

We refer to these extra conditions as moment conditions.

(2.7) Exercise. Let $K \in L^1(\mathbb{R}) \cap BV(\mathbb{R})$, and for $0 < h \leqslant 1$ define

$$A_h(x, y) = h^{-1} K\big(h^{-1}(x - y)\big) \, , \quad x, y \in [0, 1] \, .$$

Show that the family A_h, $0 < h \leqslant 1$, is convolution-like in the sense of Definition (2.5). (This justifies the name convolution-like.)

In the next section, we survey some families of boundary kernels $A_h(x, y)$, $0 < h \leqslant 1$, satisfying the conditions above. In § 7, we show that the reproducing kernels $\mathcal{R}_{mh}$ for $W^{m,2}(0,1)$ are convolution-like as well.

We now provide bounds on $\mathbb{E}[\, \| f^{nh} - f_o \|_2^2 \,]$, the mean integrated squared error. This may be written as

$$\text{(2.8)} \qquad \mathbb{E}[\, \| f^{nh} - f_o \|_2^2 \,] = \| f_{hn} - f_o \|_2^2 + \mathbb{E}[\, \| f^{nh} - f_{hn} \|_2^2 \,] ,$$

where $f_{hn}(x) = \mathbb{E}[\, f^{nh}(x) \,]$, or

$$\text{(2.9)} \qquad f_{hn}(x) = \tfrac{1}{n} \sum_{i=1}^{n} A_h(x, x_{in})\, f_o(x_{in}) \;, \quad x \in [\, 0 \,,\, 1 \,] \;.$$

Thus f_{hn} is a discretely smoothed version of f_o. It is also useful to introduce the continuously smoothed version

$$\text{(2.10)} \qquad f_h(x) = \int_0^1 A_h(x, t)\, f_o(t)\, dt \;, \quad x \in [\, 0 \,,\, 1 \,] \;.$$

As usual, in our study of the error, we begin with the bias term. This is quite similar to the treatment of the bias in kernel density estimation in Volume I and is standard approximation theory in the style of, e.g., SHAPIRO (1969), although there is one twist. For present and later use, we formulate some general results on the bias.

(2.11) LEMMA. *Let $1 \leqslant p \leqslant \infty$, and let $m \in \mathbb{N}$. If $f \in W^{m,p}(0,1)$, then*

$$\| f_{hn} - f_h \|_p \leqslant (nh)^{-1} \, \| f_o \|_{W^{m,p}(0,1)} \;.$$

PROOF. Note that $f_{hn}(x) - f_h(x)$ for fixed x deals with a quadrature error. In fact, we have for its absolute value, uniformly in x,

$$\begin{aligned} &\left| \tfrac{1}{n} \sum_{i=1}^{n} A_h(x, x_{in})\, f_o(x_{in}) - \int_0^1 A_h(x, t)\, f_o(t)\, dt \right| \\ &\qquad \leqslant c\, n^{-1} \, \| \{ A_h(x, \cdot\,)\, f_o(\,\cdot\,) \}' \|_{L^1(0,1)} \leqslant c_1 \, (nh)^{-1} \, \| f_o \|_{W^{1,1}(0,1)} \\ &\qquad \leqslant c_2 \, (nh)^{-1} \, \| f_o \|_{W^{m,p}(0,1)} \;. \end{aligned}$$

The first inequality follows from the asymptotic uniformity of the design; see Definition (13.2.22). The lemma now follows upon integration. Q.e.d.

(2.12) LEMMA. *Under the conditions of Lemma* (2.11),

$$\| f_h - f_o \|_p \leqslant c\, h^m \, \| f_o \|_{W^{m,p}} \;.$$

(2.13) EXERCISE. Prove Lemma (2.12). [Hint: See the proof of Theorem (4.2.9) in Volume I.]

Turning to the variance of the kernel estimator, we have

$$(2.14) \qquad f^{nh}(x) - f_{hn}(x) = \tfrac{1}{n} \sum_{i=1}^{n} d_{in} \, A_h(x, x_{in}) \;, \quad x \in [\,0\,,1\,] \;,$$

and the following lemma.

(2.15) LEMMA. *If the noise vector d_n satisfies the Gauss-Markov conditions (1.2), then, for $nh \to \infty$,*

$$\mathbb{E}\Big[\, \| \tfrac{1}{n} \sum_{i=1}^{n} d_{in} \, A_h(\,\cdot\,, x_{in}) \, \|_2^2 \, \Big] = \mathcal{O}\big((nh)^{-1} \big) \; .$$

PROOF. Apart from the factor σ^2, the expected value equals

$$\tfrac{1}{n} \int_0^1 \Big\{ \tfrac{1}{n} \sum_{i=1}^{n} | \, A_h(x, x_{in}) \, |^2 \Big\} \, dx \; .$$

By the convolution-like properties of the kernels A_h, see Definition (2.5), this may be bounded by

$$c \, n^{-1} \sup_{x \in [\,0\,,1\,]} \| \, A_h(x, \,\cdot\,) \, \|_\infty \int_0^1 \Big\{ \tfrac{1}{n} \sum_{i=1}^{n} | \, A_h(x, x_{in}) \, | \Big\} \, dx \leqslant$$

$$c_1 \, (nh)^{-1} \int_0^1 \Big\{ \tfrac{1}{n} \sum_{i=1}^{n} | \, A_h(x, x_{in}) \, | \Big\} \, dx \; .$$

Now, as in the proof of Lemma (2.11), we have, for each x,

$$\Big| \, \tfrac{1}{n} \sum_{i=1}^{n} | \, A_h(x, x_{in}) \, | - \int_0^1 | \, A_h(x, t) \, | \, dt \, \Big| \leqslant \frac{c_2}{n} \, | \, A_h(x, \,\cdot\,) \, |_{BV} \leqslant \frac{c_3}{nh} \; ,$$

and the lemma follows. (See § 17.2 for a clarification of some of the details.) Q.e.d.

Together, the results for the bias and variance give the bound on the mean integrated square error.

(2.16) THEOREM. *Let $m \geqslant 1$ and $f_o \in W^{m,2}(0,1)$. If the family of kernels A_h, $0 < h \leqslant 1$, is convolution-like of order m, then for the Gauss-Markov model (1.1)–(1.2),*

$$\mathbb{E}[\, \| \, f^{nh} - f_o \, \|_2^2 \,] = \mathcal{O}\big(n^{-2m/(2m+1)} \big) \; ,$$

provided $h \asymp n^{-1/(2m+1)}$ (deterministically).

(2.17) EXERCISE. Prove the theorem.

(2.18) EXERCISE. Here, we consider bounds on the discrete sums of squares error. In particular, show that, under the conditions of Theorem (2.16) for

$h \asymp n^{-1/(2m+1)}$ (deterministic),

$$\text{(2.19)} \qquad \mathbb{E}\Big[\, \tfrac{1}{n} \sum_{i=1}^{n} |\, f^{nh}(x_{in}) - f_o(x_{in}) \,|^2 \Big] = \mathcal{O}\big(\, n^{-2m/(2m+1)} \,\big) \ ,$$

as follows. Let $\varepsilon \equiv f^{nh} - f_{hn}$, $\delta \equiv f_{hn} - f_h$, $\eta \equiv f_h - f_o$. Show that

(a) $\mathbb{E}\Big[\, \frac{1}{n} \sum_{i=1}^{n} |\, \varepsilon(x_{in}) \,|^2 - \|\, \varepsilon \,\|_2^2 \Big] = \mathcal{O}\big(\, (nh)^{-2} \,\big)$.

(b) $\frac{1}{n} \sum_{i=1}^{n} |\, \delta(x_{in}) \,|^2 - \|\, \delta \,\|_2^2 = \mathcal{O}\big(\, (nh)^{-2} \,\big)$.

(c) $\frac{1}{n} \sum_{i=1}^{n} |\, \eta(x_{in}) \,|^2 - \|\, \eta \,\|_2^2 = \mathcal{O}\big(\, (nh)^{-1}\, h^m \,\big)$.

(d) Finally, conclude that (2.19) holds. [Hint: (a) through (c) are quadrature results. See Definition (13.2.22) and use Lemma (17.2.20)]

In the next section, we discuss some convolution-like boundary kernels. Boundary *corrections* are considered in § 4.

EXERCISES: (2.13), (2.17), (2.18).

3. Boundary kernels

In this section, we discuss some boundary kernels to be used in nonparametric regression. The necessity of boundary kernels is best illustrated by considering the classical Nadaraya-Watson estimator in the form (2.3)

$$\text{(3.1)} \qquad f^{nh}(x) = \tfrac{1}{n} \sum_{i=1}^{n} y_{in}\, A_h(x, x_{in}) \ , \quad x \in [\, 0 \, , \, 1 \,] \ .$$

Here, we might take A to be the Epanechnikov kernel, $A(x) = \frac{3}{4}\,(1-x)_+^2$. Even for noiseless data; i.e., $y_{in} = f_o(x_{in})$, there is trouble at the endpoints. If f_o has two continuous derivatives, one verifies that

$$\text{(3.2)} \qquad f^{nh}(0) = f_o(0) + \tfrac{1}{2}\, h\, f\,'(0) + \mathcal{O}\big(\, (nh)^{-1} + h^2 \,\big)$$

for $nh \to \infty$, $h \to 0$. In other words, the bias of the estimator is only $\mathcal{O}\big(\, h \,\big)$ instead of $\mathcal{O}\big(\, h^2 \,\big)$. This adversely affects the global error as well, no matter how that error is measured. Thus, convolution kernels as in (3.1) will not do. Of course, it is easy to see that convolution kernels are not necessarily (convolution-like) of order m, at least not for x near the endpoints of the interval $[\, 0 \, , \, 1 \,]$.

The families of kernels described here are more or less explicitly constructed to satisfy the moment conditions of Definition (2.6). The remaining conditions do not usually cause problems. An important class of boundary kernels is constructed by considering variational problems, i.e., we are considering "optimal" kernels. On the practical side, this may have severe drawbacks since typically each point near the boundary *requires its own kernel*, and so ease of construction is a consideration.

Christoffel-Darboux-Legendre kernels. The simplest kernels to be described are based on reproducing kernel ideas. Note that the moment conditions of Definition (2.6) imply that

$$\text{(3.3)} \qquad \int_0^1 P(y)\, A_h(x,y)\, dy = P(x)\ , \quad x \in [\,0\,,\,1\,]\ ,$$

for all polynomials P of degree $\leqslant m-1$.

(3.4) EXERCISE. Prove (3.3)!

Now, in a loose sense, (3.3) says that $A_h(x,y)$ is a reproducing kernel for the (m-dimensional) Hilbert space $\mathcal{P}_m$ consisting of all polynomials of degree $\leqslant m-1$. Let us first consider the case $h=1$. Choosing the normalized, shifted Legendre polynomials Q_k as the basis, see (13.6.13) shows that *the* reproducing kernel for $\mathcal{P}_m$ with the $L^2(0,1)$ inner product is

$$\text{(3.5)} \qquad B(x,y) = \sum_{k=0}^{m-1} Q_k(\,x\,)\, Q_k(\,y\,)\ , \quad x,y \in [\,0\,,\,1\,]\ .$$

Using the Christoffel-Darboux formula, see, e.g., SANSONE (1959), this may be rewritten as

$$\text{(3.6)} \quad B(x,y) = \frac{m+1}{2\sqrt{4\,m^2-1}} \; \frac{Q_{m-1}(\,x\,)\,Q_m(\,y\,) - Q_m(\,x\,)\,Q_{m-1}(\,y\,)}{y-x}\ .$$

It is worth noting that

$$\text{(3.7)} \qquad B(x,y) = B(y,x) = B(1-x, 1-y)\ , \quad x,y \in [\,0\,,\,1\,]\ .$$

Since the cases $m=2$ and $m=4$ are of most practical interest, we state them here. In the extended notation $B_m(x,y)$ to reflect the order of the kernel, with $X = 2x-1,\ Y = 2y-1$,

$$\text{(3.8)} \qquad B_2(x,y) = \tfrac{3}{4}\,XY + \tfrac{1}{4}$$

and

$$\text{(3.9)} \qquad \begin{aligned} 12\,B_4(x,y) = {} & +525\,X^3\,Y^3 \\ & -315\,X^3\,Y + 135\,X^2\,Y^2 - 315\,X\,Y^3 \\ & -45\,X^2 + 225\,Y\,X - 45\,Y^2 \\ & +27\ . \end{aligned}$$

Now, we construct the family of kernels $A_h(x,y)$, $0 < h \leqslant 1$. Away from the boundary, we want a convolution kernel. For $\frac{1}{2}h \leqslant x \leqslant 1-\frac{1}{2}h$, this is accomplished by centering a y-interval of length h at x. (Note that the center of $B(x,y)$ is $(x,y) = (\frac{1}{2},\frac{1}{2})$.) So, for $|\,y-x\,| \leqslant \frac{1}{2}h$, we take

$h^{-1}B(\frac{1}{2}, \frac{1}{2} + h^{-1}(x-y))$. For the piece $0 \leqslant x < \frac{1}{2}h$, we make sure that the kernel transitions continuously in x, but we center the y-interval at $x = \frac{1}{2}h$. That is, we take $h^{-1}B(h^{-1}x, 1 - h^{-1}y)$. For $1 - \frac{1}{2}h < x \leqslant 1$, we use reflection.

To summarize, we define the boundary kernel as

$$(3.10) \quad \mathbb{CDL}_h(x,y) = \begin{cases} \dfrac{1}{h} B\Big(\dfrac{x}{h}, 1 - \dfrac{y}{h}\Big) & , \quad 0 \leqslant x < \frac{1}{2}h \,, \\ \dfrac{1}{h} B\Big(\frac{1}{2}, \frac{1}{2} + \dfrac{(x-y)}{h}\Big) & , \quad \frac{1}{2}h \leqslant x \leqslant 1 - \frac{1}{2}h \,, \\ \dfrac{1}{h} B\Big(\dfrac{1-x}{h}, 1 - \dfrac{1-y}{h}\Big) , & \quad 1 - \frac{1}{2}h < x \leqslant 1 \,, \end{cases}$$

and all $y \in [0, 1]$, with $B(x,y) = 0$, for $(x,y) \notin [0,1] \times [0,1]$. We refer to the kernels above as *Christoffel-Darboux-Legendre* kernels.

(3.11) EXERCISE. Show that the family of kernels $\mathbb{CDL}_h(x,y)$, $0 < h \leqslant 1$, is indeed convolution-like of order m.

Weighted convolution kernels. A second approach to boundary kernels starts with a convolution kernel A of order m with support in $[-1, 1]$. If $0 < h \leqslant \frac{1}{2}$ is the smoothing parameter, then for $x \in [h, 1-h]$, the kernel estimator (3.1) may be expected to work satisfactorily but not so on the two remaining boundary intervals, the reason being that the kernel is convolution-like but not of order m. One way of fixing this in a smooth way was suggested by GASSER and MÜLLER (1979) (see also HART and WEHRLY, 1992): Let $0 < h \leqslant \frac{1}{2}$ be the smoothing parameter. For $x \in [0, h]$, replace the kernel $A_h(x-y)$ by $p(h^{-1}(x-y))\, A_h(x-y)$ for some polynomial p. This polynomial should be chosen such that the moment conditions of Definition (2.6) do hold for this value of x. This amounts to

$$\int_0^{x+h} p(\,(x-y)/h\,)\, A_h(x-y)\,(x-y)^k\, dy = \begin{cases} 1\,, & k = 0\,, \\ 0\,, & 1 \leqslant k \leqslant m-1\,. \end{cases}$$

After the change of variable $t = h^{-1}(x-y)$, this gives

$$(3.12) \qquad \int_{-1}^{q} p(t)\, A(t)\, t^k\, dt = \begin{cases} 1\,, & k = 0\,, \\ 0\,, & 1 \leqslant k \leqslant m-1\,, \end{cases}$$

with $q = h^{-1}x$. Note that this requires a different polynomial p for each value of $q \in [0, 1]$, and that for $q = 1$ we may take $p(t) \equiv 1$. Moreover, the polynomial p may be expected to vary smoothly with q.

The question is, of course, whether such a polynomial p exists for each value of q. A possible construction runs as follows. Take p to be a polynomial of degree $\leqslant m-1$, and write it as

$$(3.13) \qquad p(t) = \sum_{\ell=0}^{k-1} p_\ell \, t^\ell \,.$$

Then, (3.12) is equivalent to

$$(3.14) \qquad \sum_{\ell=0}^{m-1} p_\ell \int_{-1}^{q} A(t)\, t^{\ell+k} \, dt = \begin{cases} 1 \,, & k=0 \,, \\ 0 \,, & 1 \leqslant k \leqslant m-1 \,. \end{cases}$$

This is a system of m linear equations in m unknowns. Does it have a solution for all $q \in [\,0\,,\,1\,]$? One instance is easy.

(3.15) EXERCISE. Show that (3.14) has a unique solution for all $q \in [\,0\,,\,1\,]$ if A is a nonnegative kernel. [Hint: If

$$\sum_{\ell=0}^{m-1} b_\ell \int_{-1}^{q} A(t)\, t^{\ell+k} \, dt = 0 \,, \quad 0 \leqslant k \leqslant m-1 \,,$$

show that then

$$\int_{-1}^{q} A(t) \Bigl| \sum_{\ell=0}^{m-1} b_\ell \, t^\ell \Bigr|^2 dt = 0 \,,$$

and draw your conclusions.]

In general, we are in trouble. It seems wise to modify the scheme (3.12) by weighting only the positive part of A. Thus, the convolution kernel A is modified to

$$\widetilde{A}(x) = \{\, 1\!\!1(\, A(x) < 0\,) + p(\,x)\, 1\!\!1(\, A(x) \geqslant 0\,)\,\}\, A(x)$$

for some polynomial p of degree $\leqslant m-1$. Note that then

$$\widetilde{A}_h(x) = \{\, 1\!\!1(\, A_h(x) < 0\,) + p(\,x/h)\, 1\!\!1(\, A_h(x) \geqslant 0\,)\,\}\, A_h(x) \,,$$

and the equations (3.12) reduce to

$$(3.16) \qquad \int_{-1}^{q} \widetilde{A}(t)\, t^k \, dt = \begin{cases} 1 \,, & k=0 \,, \\ 0 \,, & 1 \leqslant k \leqslant m-1 \,. \end{cases}$$

It is easy to show along the lines of Exercise (3.15) that there is a unique polynomial p of degree $\leqslant m-1$ such that (3.16) is satisfied.

(3.17) EXERCISE. Prove the existence and uniqueness of the polynomial p of degree $\leqslant m-1$ that satisfies (3.16).

Optimal kernels. The last class of boundary kernels to be considered are the kernels that are optimal in the sense of minimizing the (asymptotic) mean squared error. However, there are two such types of kernels.

For fixed $x \in [0, 1]$, consider the estimator of $f_o(x)$,

$$f^{nh}(x) = \tfrac{1}{n} \sum_{i=1}^{n} y_{in} \, A_h(x, x_{in}) \ , \tag{3.18}$$

based on the family of kernels $A_h(x,y)$, $0 < h \leqslant 1$. The mean squared error may be written as

$$\mathbb{E}[\, | \, f^{nh}(x) - f_o(x) \, |^2 \,] = \text{bias} + \text{variance} \tag{3.19}$$

with

$$\text{bias} = \Bigl| \, \tfrac{1}{n} \sum_{i=1}^{n} f_o(x_{in}) \, A_h(x, x_{in}) - f_o(x) \, \Bigr|^2 \ , \tag{3.20}$$

$$\text{variance} = \sigma^2 \, n^{-2} \sum_{i=1}^{n} | \, A_h(x, x_{in}) \, |^2 \ . \tag{3.21}$$

With the quadrature result of § 17.2,

$$\sigma^{-2} \, \text{variance} = n^{-1} \, \| \, A_h(x, \,\cdot\,) \, \|^2 + \mathcal{O}(\, (nh)^{-2} \,) \ . \tag{3.22}$$

For the bias, we have from Lemma (2.11)

$$\text{bias} = \Bigl| \int_0^1 A_h(x,y) \, (\, f_o(y) - f_o(x) \,) \, dy + \mathcal{O}(\, (nh)^{-1} \,) \Bigr|^2 \ . \tag{3.23}$$

The last Big O term actually depends on $| \, f_o \, |_{BV}$, but this may be ignored (it is fixed).

If $A_h(x,y)$, $0 < h \leqslant 1$, is a family kernels of order m, and if $f_o^{(m)}(x)$ is continuous at x, then

$$\text{bias} = \Bigl| \, \frac{f_o^{(m)}(x)}{m!} \int_0^1 (x-y)^m \, A_h(x,y) \, dy + o(\, h^m \,) + \mathcal{O}(\, (nh)^{-1} \,) \, \Bigr| \ , \tag{3.24}$$

and the integral is $\mathcal{O}(\, h^m \,)$; see property (c) of Definition (2.6).

Now, choose $0 \leqslant x \leqslant h$ such that $q = h^{-1}x$ is fixed. For this x, suppose that $A_h(x,y)$ is of the form

$$A_h(x,y) = h^{-1} B(\, q, h^{-1}(x-y) \,)$$

with $B(q, t) = 0$ for $| \, t \, | > 1$. Then,

$$\int_0^1 (x-y)^m \, A_h(x,y) \, dy = h^m \int_{-1}^{q} t^m \, B(q, t) \, dt \ .$$

Replacing the "bias + variance" by their leading terms gives

$$\mathbb{E}[\, | \, f^{nh}(x) - f_o(x) \, |^2 \,] = c_1 \, h^m \int_{-1}^{q} t^m \, B(q, t) \, dt + c_2 \, (nh)^{-1} \int_{-1}^{q} | \, B(q, t) \, |^2 \, dt + \cdots \ . \tag{3.25}$$

Here c_1 and c_2 depend only on f_o and do not interest us.

After minimization over h, the "optimal" kernel $B(q, \cdot\,)$ is then given by the kernel B, for which

$$(3.26)\qquad \Bigl\{ \int_{-1}^{q} t^m\, B(t)\, dt \Bigr\} \Bigl\{ \int_{-1}^{q} |\, B(t)\,|^2\, dt \Bigr\}^m$$

is as small as possible, subject to the constraints

$$(3.27)\qquad \int_{-1}^{q} t^k\, B(t)\, dt = \begin{cases} 1\,, & k = 0\,, \\ 0\,, & k = 1, 2, \cdots, m-1\,. \end{cases}$$

It is customary to add the constraint that B is a polynomial of order m; see MÜLLER (1993b).

An alternative set of asymptotically optimal kernels is obtained if we bound the bias by

$$(3.28)\qquad \text{bias} \leqslant \Bigl|\, \frac{f_o^{(m)}(\theta)}{m\,!} \int_0^1 |\, x - y\,|^m\, |\, A_h(x,y)\,|\, dy\, \Bigr|^2 .$$

This leads to the Zhang-Fan kernels, for which

$$(3.29)\qquad \Bigl\{ \int_{-1}^{q} |\, t\,|^m\, |\, B(t)\,|\, dt \Bigr\} \Bigl\{ \int_{-1}^{q} |\, B(t)\,|^2\, dt \Bigr\}^m$$

is as small as possible, subject to the constraints (3.27). For $q = 1$, they were discussed at length in § 11.2 in Volume I and, of course, also by ZHANG and FAN (2000).

(3.30) EXERCISE. Another class of kernels is obtained by minimizing the variance of the kernel estimator, see MÜLLER (1993b). Asymptotically, cf. the proof of Lemma (2.15), this amounts to solving the problem

$$\begin{aligned} &\text{minimize} \quad \int_0^1 \int_0^1 |\, A_h(x,y)\,|^2\, dx\, dy \\ &\text{subject to} \quad \text{the conditions (2.6)(a), (b) hold}\,. \end{aligned}$$

It seems that the Christoffel-Legendre-Darboux kernels ought to be the solution. Is this in fact true ?

EXERCISES: (3.11), (3.15), (3.17), (3.30).

4. Asymptotic boundary behavior

In this section, we compute boundary corrections for the Nadaraya-Watson estimator

$$(4.1)\qquad f^{nh}(x) = \tfrac{1}{n} \sum_{i=1}^{n} y_{in}\, A_h(x, x_{in})\,,$$

where the kernel $A_h(x,y) = A_h(x,y \mid x_{1,n}, x_{2,n}, x_{nn})$ is given by (1.13). Here, the Bias Reduction Principle of EUBANK and SPECKMAN (1990b), discussed in § 13.5, gets another outing. Actually, the only worry is the asymptotic behavior of the boundary errors of the estimators and the verification that they lie in a low-dimensional space of functions. Here, to a remarkable degree, the treatment parallels that for smoothing splines.

The assumptions on the convolution kernel are that

$$\text{(4.2)} \qquad A \text{ is bounded, integrable, has bounded variation and satisfies} \\ \int_{\mathbb{R}} x^k\, A(x)\, dx = \begin{cases} 1\,, & k=0\,, \\ 0\,, & k=1,\cdots,m-1\,, \end{cases}$$

Thus, A is a kernel of order (at least) m. For convenience, we also assume that A has bounded support,

$$\text{(4.3)} \qquad A(x) = 0 \quad \text{for} \quad |\,x\,| > 1\,.$$

(4.4) THEOREM. *Let $m \geqslant 1$ and $f_o \in W^{m,2}(0,1)$. Under the assumptions (4.2)–(4.3) on the kernel, there exist continuous functions $\varepsilon_{h,k}$ and $\delta_{h,k}$, $k = 0, 1, \cdots, m-1$, such that, for all $x \in [\,0\,,\,1\,]$,*

$$\mathbb{E}[\, f^{nh}(x)\,] = f_o(x) + b(x) + \eta_{nh}(x)\,,$$

where
$$b(x) = \sum_{k=0}^{m-1} \{\, f_o^{(k)}(0)\, \varepsilon_{h,k}(x) + f_o^{(k)}(1)\, \delta_{h,k}(x)\, \}\,,$$

and
$$\tfrac{1}{n} \sum_{i=1}^{n} |\, \eta_{nh}(x_{in})\,|^2 = \mathcal{O}\big(\, h^{2m} + (nh)^{-2}\,\big)\,,$$

$$\|\, \eta_{nh}\,\|^2 = \mathcal{O}\big(\, h^{2m} + (nh)^{-2}\,\big)\,.$$

In the proof of the theorem, the functions $\varepsilon_{h,k}$ and $\delta_{h,k}$ are explicitly determined. Note that the theorem deals with the bias of the Nadaraya-Watson estimators. The variance part needs no further consideration.

It is useful to introduce some notation. Let $f_{hn} = \mathbb{E}[\, f^{nh}\,]$, so

$$\text{(4.5)} \qquad f_{hn}(x) = \tfrac{1}{n} \sum_{i=1}^{n} f_o(x_{in})\, A_h(x, x_{in})\,, \quad x \in [\,0\,,\,1\,]\,.$$

Let f_h be the formal limit of f_{hn} as $n \to \infty$; i.e.,

$$\text{(4.6)} \qquad f_h(x) = \mathcal{A}_h f(x)\,, \quad x \in [\,0\,,\,1\,]\,,$$

where, for any integrable function f,

$$\text{(4.7)} \qquad \mathcal{A}_h f(x) = \frac{\displaystyle\int_0^1 f_o(\,t\,)\, A_h(x-t)\, dt}{\displaystyle\int_0^1 A_h(x-t)\, dt}\,, \quad x \in [\,0\,,\,1\,]\,.$$

The hard work in proving the theorem is to show that the "usual" error bounds apply when $f_o \in W^{m,2}(0,1)$ satisfies the boundary conditions

$$f_o^{(k)}(0) = f_o^{(k)}(1) = 0 \,, \quad k = 0, 1, \cdots, m-1 \,. \tag{4.8}$$

Of course, only the bias in the estimator needs consideration.

(4.9) LEMMA. *Let $f_o \in W^{m,2}(0,1)$ satisfy the boundary conditions (4.8). Under the assumptions (4.2)–(4.3) on the kernel,*

$$\| f_h - f_o \|^2 = \mathcal{O}(h^{2m}) \,.$$

PROOF. Since A has support in $[-1\,,\,1]$ and is a kernel of order m, then for the L^2 norm over the interval $(\,h\,,\,1-h\,)$ we have

$$\| f_h - f_o \|_{(h,1-h)} = \mathcal{O}(h^m) \,.$$

For the remaining, pieces we have by Taylor's theorem with exact remainder

$$f_o(x) = \int_0^x \frac{(x-t)^{m-1}}{(m-1)!} \, f_o^{(m)}(t)\,dt \,,$$

so that by Cauchy-Schwarz

$$| f_o(x) |^2 \leqslant c\, h^{2m-1} \, \| f_o^{(m)} \|^2 \quad \text{uniformly in } x,\ 0 \leqslant x \leqslant 2h \,. \tag{4.10}$$

It follows that $\| f_o \|_{(0,h)} = \mathcal{O}(h^m)$.

Likewise, for $0 \leqslant x \leqslant h$, the elementary inequality

$$\left| \int_0^1 A_h(x-t)\, f_o(t)\,dt \right| \leqslant \| A \|_1 \sup_{0 \leqslant t \leqslant 2h} | f_o(t) |$$

combined with (4.10) shows that $\| \mathcal{A}_h f_o \|_{(0,h)} = \mathcal{O}(h^m)$ as well. Thus,

$$\| \mathcal{A}_h f_o - f_o \|_{(0,h)}^2 = \mathcal{O}(h^{2m}) \,.$$

The same bound applies to the integral over $(\,1-h\,,\,1\,)$. Q.e.d.

(4.11) EXERCISE. Under the conditions of Lemma 4.10, show that

$$\mathbb{E}[\, \| f^{nh} - f_o \|^2 \,] = \mathcal{O}(n^{-2m/(2m+1)}) \,,$$

provided $h \asymp n^{-1/(2m+1)}$.

We now investigate what happens when the boundary conditions (4.8) do not hold. For functions L_k, R_k, $k = 0, 1, \cdots, m-1$, to be determined, define

$$p_o(x) = \sum_{k=0}^{m-1} \{\, f_o^{(k)}(0)\, L_k(x) + f_o^{(k)}(1)\, R_k(x) \,\} \,. \tag{4.12}$$

We wish to determine the L_k and R_k such that $\varphi_o \stackrel{\text{def}}{=} f_o - p_o \in W^{m,2}(0,1)$ satisfies the boundary conditions (4.8); i.e.,

$$\varphi_o^{(k)}(0) = \varphi_o^{(k)}(1) = 0 \,, \quad k = 0, 1, \cdots, m-1 \,. \tag{4.13}$$

One verifies that for this purpose it is sufficient that, for each k,

$$
\begin{aligned}
& L_k^{(\ell)}(0) = L_k^{(\ell)}(1) = 0 \quad \text{for} \quad \ell = 0, 1, \cdots, m-1 \\
(4.14) \qquad & \text{except that} \quad L_k^{(k)}(0) = 1 \quad \text{and} \\
& R_k(x) = (-1)^k \, L_k(1 - x) \; .
\end{aligned}
$$

(4.15) EXERCISE. Take $m = 4$. Show that the choices

$$L_k(x) = (x-1)^4 \, P_k(x)/k! \; ,$$

with

$$
\begin{aligned}
P_0(x) &= 1 + 4x + 10x^2 + 5x^3 \; , \\
P_1(x) &= x + 4x^2 + 10x^3 \; , \\
P_2(x) &= x^2 + 4x^3 \; , \quad \text{and} \\
P_3(x) &= x^3 \; ,
\end{aligned}
$$

work. (There obviously is a pattern here, but ... .)

With the choices for the L_k and R_k above, we do indeed have that $\varphi_o \in W^{m,2}(0,1)$ satisfies (4.13). By Lemma (4.9), then

$$\| \mathcal{A}_h \varphi_o - \varphi_o \|^2 = \mathcal{O}(h^{2m}).$$

It follows that $\mathcal{A}_h f_o(x) - f_o(x) = \mathcal{A}_h p_o - p_o + \eta$, where $\eta = \mathcal{A}_h \varphi_o - \varphi_o$. All of this essentially proves Theorem (4.4), with

$$(4.16) \qquad \varepsilon_{h,k} = \mathcal{A}_h L_k - L_k \quad , \quad \delta_{h,k} = \mathcal{A}_h R_k - R_k \; .$$

(4.17) EXERCISE. Write a complete proof of Theorem (4.4), in particular, show the bound

$$\tfrac{1}{n} \sum_{i=1}^{n} | \, \eta(x_{in}) \, |^2 = \mathcal{O}(h^{2m} + (nh)^{-1}) \; .$$

[Hint: Exercise (2.18) should come in handy.]

At this point, everything is set up for the application of the Bias Reduction Principle (13.5.4) of EUBANK and SPECKMAN (1990b) to computing boundary corrections. We leave the details as an exercise.

(4.18) EXERCISE. Work out the details of computing boundary corrections so that the resulting estimator has an expected mean squared error of order $n^{-2m/(2m+1)}$.

(4.19) EXERCISE. EUBANK and SPECKMAN (1990b) prove Theorem (4.4) using the much simpler functions $L_k(x) = x^k / k!$. Verify that this works!

EXERCISES: (4.11), (4.15), (4.17), (4.18), (4.19).

5. Uniform error bounds for kernel estimators

We now consider bounds on the uniform error of kernel estimators. To start out, we shall be rather modest in our goals. In the next section, we stretch the result to its limit.

We make the following assumptions.

(5.1) The noise d_{in}, $i = 1, 2, \cdots, n$, is iid and satisfies the Gauss-Markov model (1.1)–(1.2). Moreover,

$$\mathbb{E}[\,|\,d_{in}\,|^{\kappa}\,] < \infty \qquad \text{for some } \kappa > 2\,. \tag{5.2}$$

(5.3) The family of kernels $A_h(x, y)$, $0 \leqslant h \leqslant 1$, are convolution-like of order m in the sense of Definition (2.5).

(5.4) The design $x_{1,n}, x_{2,n}, \cdots, x_{n,n}$ is asymptotically uniform in the sense of Definition (13.2.22).

(5.5) $f_o \in W^{m,\infty}(0, 1)$ for some integer $m \geqslant 1$.

(5.6) THEOREM. *Assuming* (5.1) *through* (5.5), *the kernel estimator* f^{nh} *given by* (2.4) *satisfies*

$$\|\, f^{nh} - f_o \,\|_\infty = \mathcal{O}\big(\, (n^{-1} \log n)^{\,m/(2m+1)} \,\big) \quad \textit{almost surely},$$

provided $h \asymp (n^{-1} \log n)^{\,1/(2m+1)}$ *(deterministically).*

The first step in the analysis of kernel estimators is of course the bias-variance decomposition, although the decomposition via the triangle inequality only provides an upper bound. In the notations of (2.4) and (2.9),

$$\|\, f^{nh} - f_o \,\|_\infty \leqslant \|\, f^{nh} - f_{hn} \,\|_\infty + \|\, f_{hn} - f_o \,\|_\infty \;. \tag{5.7}$$

The bias term $\|\, f_{hn} - f_o \,\|_\infty$ is covered by Lemmas (2.11)–(2.12). All that is left here is to study the variance term $\|\, f^{nh} - f_{hn} \,\|_\infty$. Of course,

$$f^{nh}(x) - f_{hn}(x) = \tfrac{1}{n} \sum_{i=1}^{n} d_{in}\, A_h(x, x_{in}) \;. \tag{5.8}$$

(5.9) THEOREM. *Under the assumptions* (5.1)–(5.3), *for deterministic* h, *satisfying* $h \geqslant c\,(\,n^{-1} \log n\,)^{1-2/\kappa}$ *for some positive constant* c,

$$\Big\|\, \tfrac{1}{n} \sum_{i=1}^{n} d_{in}\, A_h(\,\cdot\,, x_{in})\,\Big\|_\infty = \mathcal{O}\Big(\big(\,(nh)^{-1} \log n\,\big)^{1/2}\Big) \quad \textit{almost surely}.$$

(5.10) NOTE. Comparison with HÄRDLE, JANSSEN, and SERFLING (1988) and EINMAHL and MASON (2000, 2005) reveals that, in the theorem, the factor $\log n$ should in fact be $\log(1/h)$. However, for the values of h we are considering, this makes no difference.

(5.11) EXERCISE. Prove Theorem (5.6) based on Theorem (5.9) and Lemmas (2.11)–(2.12).

The proof of Theorem (5.9) closely follows that of Lemma 2.2 in HÄRDLE, JANSSEN, and SERFLING (1988). First we reduce the supremum over the interval $[0, 1]$ to the maximum over a finite number of points, then truncate the noise, and finally apply Bernstein's inequality. However, the first step is implemented quite differently. In fact, there is a zeroth step, in which the "arbitrary" family of kernels $A_h(x, y)$, $0 \leqslant h \leqslant 1$, is replaced by the one-sided exponential family $g_h(x - y)$, $0 \leqslant h \leqslant 1$, defined by

$$(5.12) \qquad \begin{aligned} g(x) &= \exp(-x)\, 1\!\!1(x \geqslant 0)\,, \\ g_h(x) &= h^{-1} g(h^{-1} x)\,, \quad x \in \mathbb{R}\,. \end{aligned}$$

It is a nice exercise to show that $h\, g_h$ is the fundamental solution of the initial value problem

$$(5.13) \qquad \begin{aligned} h\, u' + u &= v \quad \text{on } (0, 1)\,, \\ u(0) &= a\,; \end{aligned}$$

i.e., for $1 \leqslant p \leqslant \infty$ and $v \in L^p(0,1)$, the solution u of the initial value problem (5.13) satisfies $u \in L^p(0,1)$ and is given by

$$(5.14) \qquad u(x) = h\, g_h(x)\, u(0) + \int_0^1 g_h(x - z)\, v(z)\, dz\,, \quad x \in [0, 1]\,.$$

Note that this amounts to an *integral representation* of $u \in W^{1,p}(0,1)$,

$$(5.15) \qquad u(x) = h\, g_h(x)\, u(0) + \int_0^1 g_h(x - z) \{ h\, u'(z) + u(z) \}\, dz\,,$$

for $x \in [0, 1]$. (In fact, this is a reproducing kernel trick.)

(5.16) EXERCISE. (a) Show that all the solutions of the *homogeneous* differential equation

$$h\, u' + u = 0 \quad \text{on} \quad \mathbb{R}$$

are given by $u(x) = c \exp(-x/h)$, $c \in \mathbb{R}$.
(b) For the inhomogeneous differential equation

$$h\, u' + u = v \quad \text{on} \quad (0, 1)\,,$$

try the ansatz $u(x) = c(x) \exp(-x/h)$, where $c(x)$ is differentiable. Show that if u satisfies the differential equation, then

$$c'(x) \exp(-x/h) = v(x)\,,$$

and so a solution is

$$c(x) = \int_0^x v(t) \exp(t/h)\, dt \ , \quad x \in [0, 1] \,.$$

Then,

$$u_p(x) = \int_0^x v(t) \exp(-(x-t)/h)\, dt \ , \quad x \in [0, 1] \,,$$

is a *particular* solution of the inhomogeneous differential equation.
(c) Now try the ansatz

$$u(x) = c \exp(-x/h) + u_p(x)$$

to solve (5.13). This should give $c = h\,a = h\,u(0)$. [The technique under (b) is called "variation of constants"; see, e.g., SÁNCHEZ (1968).]

(5.17) LEMMA. *Assume that the family A_h, $0 < h \leqslant 1$, is convolution-like in the sense of Definition (2.5). Then, there exists a constant c such that, for all h, $0 < h \leqslant 1$, all $D_1, D_2, \cdots, D_n \in \mathbb{R}$, and for arbitrary designs $x_{1,n}, x_{2,n}, \cdots, x_{n,n} \in [0, 1]$*

$$\Big\| \tfrac{1}{n} \sum_{i=1}^{n} D_i\, A_h(x_{in}, \cdot) \Big\|_\infty \leqslant c \Big\| \tfrac{1}{n} \sum_{i=1}^{n} D_i\, g_h(x_{in} - \cdot) \Big\|_\infty \,.$$

PROOF. For $t \in [0, 1]$, let

$$S_{nh}(t) \stackrel{\text{def}}{=} \tfrac{1}{n} \sum_{i=1}^{n} D_i\, A_h(t, x_{in}) \ , \quad s^{nh}(t) \stackrel{\text{def}}{=} \tfrac{1}{n} \sum_{i=1}^{n} D_i\, g_h(x_{in} - t) \,.$$

Assume that A_h is differentiable with respect to its first argument. Then,

$$|\, A_h(\cdot, t)\,|_{BV} = \|\, A_h'(\cdot, t)\,\|_{L^1(0,1)} \,,$$

where the prime $'$ denotes differentiation with respect to the first argument. Now, applying the integral representation formula (5.15) to the function $u = A_h(t, \cdot)$ (for fixed t), we obtain, for all x

$$A_h(t, x) = h\, g_h(x)\, A_h(t, 0) + \int_0^1 g_h(x - z) \left\{ h\, A_h'(t, z) + A_h(t, z) \right\} dz \,.$$

Next, take $x = x_{in}$ and substitute the formula above into the expression for $S^{nh}(t)$. This gives

$$S_{nh}(t) = h\, A_h(0, t)\, s^{nh}(0) + \int_0^1 \left\{ h\, A_h'(t, z) + A_h(t, z) \right\} s^{nh}(z)\, dz \,.$$

Now, straightforward bounding gives

$$\|\, S_{nh}\,\|_\infty \leqslant C\, |\, s^{nh}(0)\,| + C_1 \,\|\, s^{nh}\|_\infty \,,$$

where $C = h\, \|\, A_h(0, \cdot)\,\|_\infty$ and

$$C_1 = \sup_{t \in [0,1]} \|\, A_h(t, \cdot)\,\|_{L^1(0,1)} + h\, |\, A_h(t, \cdot)\,|_{BV} \,.$$

So, since the family of kernels A_h, $0 < h \leqslant 1$, is convolution-like, the constants C and C_1 are bounded uniformly in h, and then for $C_2 = C + C_1$, we have $\| S_{nh} \|_\infty \leqslant C_2 \| s^{nh} \|_\infty$. (Note that here we took $\| \cdot \|_\infty$ to be the "everywhere" supremum, not just the essential or "almost everywhere" supremum.)

The extension to the case where A_h is not necessarily differentiable with respect to its first argument follows readily. Q.e.d.

(5.18) EXERCISE. Prove Lemma (5.17) for the case where the A_h are not differentiable. [Hint: Let $\lambda > 0$ and let $\mathcal{R}_{1,\lambda}$ be the reproducing kernel of the space $W^{1,2}(0,1)$ as in Lemma (13.2.18) with $m = 1$ and $h = \lambda$. Replace A_h by

$$A_{h\lambda}(x,y) = \int_0^1 \int_0^1 A_h(s,t)\, \mathcal{R}_{1,\lambda}(s,x)\, \mathcal{R}_{1,\lambda}(t,y)\, ds\, dt\,, \quad x, y \in [0,1]\,.$$

Show that $| A_{h\lambda}(\cdot, t) |_{BV} \leqslant | A_h(\cdot, t) |_{BV}$ for all $\lambda > 0$ and that

$$\lim_{\lambda \to 0} \| \tfrac{1}{n} \sum_{i=1}^{n} d_{in} A_{h\lambda}(x_{in}, t) \|_\infty = \| S_{nh} \|_\infty\,.$$

Take it from there.]

(5.19) EXERCISE. Let Ω_n be the empirical distribution function of the design $X_1, X_2, \cdots, X_n$, and let Ω_o be the design distribution function. Let

$$[A_h(d\Omega_n - d\Omega_o)](t) = \int_0^1 A_h(x,t) \left(d\Omega_n(x) - d\Omega_o(x) \right)\,.$$

For the "reverse" convolution, we write

$$[g_h \circledast (d\Omega_n - d\Omega_o)](t) = \int_0^1 g_h(x - t) \left(d\Omega_n(x) - d\Omega_o(x) \right)\,.$$

Show that $\| A_h(d\Omega_n - d\Omega_o) \|_\infty \leqslant c \| g_h \circledast (d\Omega_n - d\Omega_o) \|_\infty$ for a suitable constant.

The next step is to replace the suprema of $s^{nh}(x)$ over $x \in [0,1]$ by the maximum over a finite number of points. It is easier to consider the supremum of $s^{nh}(x)$ over $x \in \mathbb{R}$. Obviously, $s^{nh}(x) \to 0$ for $|x| \to \infty$. Now, elementary calculus is sufficient. From (5.12), one verifies that

$$\text{(5.20)} \qquad \frac{d}{dz} g_h(x_{in} - z) = h^{-1} g_h(x_{in} - z)\,, \quad z \neq x_{in}\,,$$

and so, for $z \neq x_{in}$, $i = 1, 2, \cdots, n$,

$$\frac{d}{dz} s^{nh}(z) = \tfrac{1}{n} \sum_{i=1}^{n} d_{in} g_h(x_{in} - z) = \frac{1}{nh} \sum_{i=1}^{n} d_{in} g_h(x_{in} - z) = \frac{1}{h} s^{nh}(z)\,.$$

It follows that the derivative vanishes if and only if the function itself vanishes. So this way we obtain very uninteresting local extrema, if any.

Consequently, the absolute extrema occur at points where the function is not differentiable; i.e., at the x_{in}. This proves the following lemma.

(5.21) LEMMA. *For all designs* $x_{1,n}, x_{2,n}, \cdots, x_{n,n} \in [0, 1]$ *and all noise components* $d_{1,n}, d_{2,n}, \cdots, d_{n,n}$,

$$\Big\| \tfrac{1}{n} \sum_{i=1}^{n} d_{in}\, g_h(x_{in} - \cdot\,) \Big\|_\infty = \max_{1 \leqslant j \leqslant n} \Big| \tfrac{1}{n} \sum_{i=1}^{n} d_{in}\, g_h(x_{in} - x_{jn}) \Big| .$$

We now wish to bound the maximum of the sums by way of exponential inequalities. One way or another, exponential inequalities require exponential moments, but the assumption (5.2) is obviously not sufficient for the purpose. We must first *truncate* the d_{in}. Since we do not know *yet* at what level, we just call it M.

(5.22) TRUNCATION LEMMA. *Let* $M > 0$, *and let*

$$\gamma_{in} = d_{in}\, 1\!\!1(\, |\, d_{in}\,| \geqslant M\,) , \quad i = 1, 2, \cdots, n .$$

Then, almost surely,

$$\max_{1 \leqslant j \leqslant n} \Big| \tfrac{1}{n} \sum_{i=1}^{n} \gamma_{in}\, g_h(x_{in} - x_{jn}) \Big| = \mathcal{O}(\, h^{-1} M^{1-\kappa}\,) .$$

PROOF. For each j, we have the inequality

$$\Big| \tfrac{1}{n} \sum_{i=1}^{n} \gamma_{in}\, g_h(x_{in} - x_{jn}) \Big| \leqslant h^{-1} \tfrac{1}{n} \sum_{i=1}^{n} |\, \gamma_{in}\,| ,$$

and, since $|\,\gamma_{in}\,| \geqslant M$ and $\kappa > 2$, this may be bounded by

$$h^{-1} M^{1-\kappa} \cdot \tfrac{1}{n} \sum_{i=1}^{n} |\, \gamma_{in}\,|^\kappa \leqslant h^{-1} M^{1-\kappa} \cdot \tfrac{1}{n} \sum_{i=1}^{n} |\, d_{in}\,|^\kappa .$$

By the Kolmogorov Strong Law of Large Numbers, the last sum tends to $\mathbb{E}[\,|\, d_{1,n}\,|^\kappa\,]$ almost surely. This expectation is finite by assumption. Q.e.d.

(5.23) IMPROVED TRUNCATION LEMMA. *Under the conditions* (5.1) *and* (5.2), *there exists a continuous function* $\psi : [0, \infty\,) \to [0, \infty\,)$ *that tends to* 0 *at* ∞ *such that*

$$\max_{1 \leqslant j \leqslant n} \Big| \tfrac{1}{n} \sum_{i=1}^{n} \gamma_{in}\, g_h(x_{in} - x_{jn}) \Big| \leqslant C_n\, h^{-1} M^{1-\kappa}\, \psi(M) ,$$

where C_n *does not depend on* h *or* M, *and* $C_n \longrightarrow (\mathbb{E}[\,|\, d_{1,n}\,|^\kappa\,])^{1/2}$ *almost surely.*

PROOF. Let X be a positive random variable with distribution function F. If $\mathbb{E}[\,X^\kappa\,] < \infty$, then

$$\varphi(x) \stackrel{\text{def}}{=} \int_x^\infty z^\kappa\, dF(z) , \quad z \geqslant 0 ,$$

is decreasing and tends to 0 as $z \to \infty$. But then

$$\int_0^\infty \frac{z^\kappa}{\sqrt{\varphi(z)}}\, dF(z) = \int_0^\infty \frac{-\,d\varphi(z)}{\sqrt{\varphi(z)}} = 2\sqrt{\varphi(0)}\ .$$

With $\psi(x) = 2\sqrt{\varphi(x)}$, this may be more suggestively written as

$$\mathbb{E}[\,X^\kappa/\psi(X)\,] = \big(\mathbb{E}[\,X^\kappa\,]\big)^{1/2} < \infty\ .$$

Now, applying the equation above to the case $X = |\,d_{in}\,|$, one obtains that

$$\Big|\, \tfrac{1}{n}\sum_{i=1}^n d_{in}\, \mathbb{1}(|\,d_{in}\,| > M\,) \,\Big| \leqslant M^{1-\kappa}\,\psi(M)\cdot \tfrac{1}{n}\sum_{i=1}^n \frac{|\,d_{in}\,|^\kappa}{\psi(|\,d_{in}\,|)}\ ,$$

and this tends almost surely to $M^{1-\kappa}\,\psi(M)\,\big(\mathbb{E}[\,|\,d_{1,n}\,|^\kappa\,]\big)^{1/2}$. The rest of the proof of the Truncation Lemma (5.22) now applies. Q.e.d.

We now continue with the truncated random variables

(5.24) $$\theta_{in} = d_{in}\,\mathbb{1}(\,|\,d_{in}\,| \leqslant M\,)\ ,\quad i = 1, 2, \cdots, n\ ,$$

and consider the sums

(5.25) $$G_j^{nh} = \tfrac{1}{n}\sum_{i=1}^n \theta_{in}\, g_h(x_{in} - x_{jn})\ ,\quad j = 1, 2, \cdots, n\ .$$

The last step consists of applying Bernstein's inequality; see Appendix 4.

(5.26) Lemma. *Under the conditions* (5.1), (5.2), *and* (5.4), *for all* $t > 0$ *and* $j = 1, 2, \cdots, n$,

$$\mathbb{P}[\,|\,G_j^{nh}\,| > t\,] \leqslant 2\exp\Big(\frac{-\,nh\,t^2}{c\,\sigma^2 + \frac{2}{3}\,t\,M}\Big)\ ,$$

where $c \equiv c_{nh} = 1 + \mathcal{O}\big(\,(nh)^{-1}\,\big)$.

Proof. This is an application of Bernstein's inequality. Note that we are dealing with sums of the form

$$G_j^{nh} = \sum_{i=1}^n \Theta_{ij} \quad \text{in which} \quad \Theta_{ij} = \tfrac{1}{n}\,\theta_{in}\, g_h(x_{in} - x_{jn})\ .$$

Since the θ_{in} are bounded by M, then

(5.27) $$|\,\Theta_{ij}\,| \leqslant (nh)^{-1}\, M\ .$$

For the variances $V_j = \sum_{i=1}^n \mathbb{V}\mathrm{ar}[\,\Theta_{ij}\,]$, we obtain by independence that

$$V_j \leqslant \sigma^2 \sum_{i=1}^n n^{-2}\,|\,g_h(x_{in} - x_{jn})\,|^2 \leqslant \sigma^2\,(nh)^{-1}\cdot \tfrac{1}{n}\sum_{i=1}^n (g^2)_h(x_{in} - x_{jn})\ ,$$

so that, by an appeal to Lemma (13.2.24),

$$(5.28)\qquad V_j \leqslant \sigma^2\,(nh)^{-1}\int_0^1 (g^2)_h(x-x_{jn})\,dx + \mathcal{O}\big((nh)^{-2}\big)\ .$$

Consequently, the upper bound of V_j hardly depends on j: Note that

$$\int_0^1 (g^2)_h(x-x_{jn})\,dx \leqslant \int_0^\infty \mathrm{e}^{-2x}\,dx = \tfrac{1}{2}\ ,$$

so that $V_j \leqslant (nh)^{-1}\,V$, where

$$V = \sigma^2\,\{\,\tfrac{1}{2} + \mathcal{O}\big((nh)^{-1}\big)\,\}\ .$$

From Bernstein's inequality, we then get, for each j

$$\mathbb{P}[\,|\,G_j^{nh}\,| > t\,] \leqslant 2\,\exp\Big(\frac{-\,nh\,t^2}{2\,V + \frac{2}{3}\,t\,M}\Big)\ ,$$

which completes the proof. Q.e.d.

(5.29) COROLLARY. *For all $t > 0$,*

$$\mathbb{P}\Big[\max_{1\leqslant j\leqslant n}\,|\,G_j^{nh}\,| > t\,\Big] \leqslant 2\,n\,\exp\Big(\frac{-\,nh\,t^2}{c\,\sigma^2 + \frac{2}{3}\,t\,M}\Big)\ ,$$

where $c \equiv c_{nh} = 1 + \mathcal{O}\big((nh)^{-1}\big)$.

PROOF. Obviously,

$$\mathbb{P}\Big[\max_{1\leqslant j\leqslant n}\,|\,G_j^{nh}\,| > t\,\Big] \leqslant \sum_{j=1}^{n}\,\mathbb{P}\Big[\,|\,G_j^{nh}\,| > t\,\Big]\ ,$$

and that does it. Q.e.d.

(5.30) COROLLARY TO THE COROLLARY. *If $\big((nh)^{-1}\log n\big)^{1/2} M$ remains bounded as $n\to\infty$, $h\to 0$, then almost surely*

$$\max_{1\leqslant j\leqslant n}\Big|\,\tfrac{1}{n}\sum_{i=1}^{n}\theta_{in}\,g_h(x_{in}-x_{jn})\Big| = \mathcal{O}\big(\,((nh)^{-1}\log n\,)^{1/2}\,\big)\ .$$

PROOF. Suppose that $\big((nh)^{-1}\log n\big)^{1/2} M \leqslant K$ for some constant K. Let C be the positive solution of the quadratic equation

$$\frac{C^2}{c\,\sigma^2 + \frac{2}{3}\,C\,K} = 3$$

with the same constant c as in Corollary (5.29). Choose t as

$$t = t_n = C\,\big((nh)^{-1}\log n\big)^{1/2}\ .$$

Then, $t\, M \leqslant C\, K$ and so

$$\frac{nh\, t_n^2}{c\, \sigma^2 + \frac{2}{3}\, t\, M} \geqslant \frac{C^2 \log n}{c\, \sigma^2 + \frac{2}{3}\, C\, K} = 3 \log n \ .$$

Then, Corollary (5.29) implies

$$\mathbb{P}\Big[\max_{1 \leqslant j \leqslant n} |\, G_j^{nh} \,| > t_n \Big] = \mathcal{O}\big(n^{-2} \big) \ ,$$

and the result follows by Borel-Cantelli. Q.e.d.

We must now reconcile Corollary (5.30) with the Improved Truncation Lemma (5.23). Thus, Corollary (5.30) imposes an upper bound on M,

$$M \,\big((nh)^{-1} \log n \big)^{1/2} = \mathcal{O}(\,1\,)$$

but the (improved) truncation error should not exceed the almost sure bound of the corollary, so

$$h^{-1}\, M^{1-\kappa}\, \psi(M) = \mathcal{O}\big(\,(\,(nh)^{-1} \log n\,)^{1/2}\,\big)$$

for some positive function ψ with $\psi(x) \to 0$ as $x \to \infty$. Note that $1 - \kappa < 0$. So, combining these two bounds and ignoring constants as well as the function ψ, we must choose M such that

$$\big((nh)^{-1} \log n \big)^{(\kappa-1)/2} \leqslant M^{1-\kappa} \leqslant h \,\big((nh)^{-1} \log n \big)^{1/2} \ .$$

This is possible if the leftmost side is smaller than the rightmost side, which holds if $h \geqslant \big(n^{-1} \log n \big)^{1-2/\kappa}$. Then, one may pick M to be anywhere in the prescribed range. The choice

$$M = (\, nh/ \log n\,)^{1/2} \tag{5.31}$$

suggests itself, and one verifies that it "works".

(5.32) EXERCISE. Prove Theorem (5.9). [Hint : Check the Improved Truncation Lemma (5.23) and the exponential inequality of Lemma (5.26) with the choice (5.31) for M, and that both give the correct orders of magnitude.]

Kernel density estimation. For later use and as an introduction to random designs, we consider uniform error bounds for kernel density estimation with convolution-like kernels on $[\,0\,,\,1\,]$. Here, we are not interested in the bias. So, assume that

(5.33) $X_1, X_2, \cdots, X_n$ are independent and identically distributed, with common probability density function ω with respect to Lebesgue measure on $(0,1)$. Moreover, there exists a positive constant ω_2 such that $\omega(\,t\,) \leqslant \omega_2$ a.e. $t \in [\,0\,,\,1\,]$.

Let Ω_o be the distribution function of X, and let Ω_n be the empirical distribution function of $X_1, X_2, \cdots, X_n$,

$$(5.34) \qquad \Omega_n(x) = \tfrac{1}{n} \sum_{i=1}^{n} 1\!\!1(X_i \leqslant x) .$$

We use the shorthand notation

$$(5.35) \qquad [\,\mathcal{A}_h dF\,](x) = \int_0^1 A_h(x,z)\, dF(z) \;, \quad x \in [\,0\,,\,1\,] \;,$$

where F is any function of bounded variation.

(5.36) THEOREM (UNIFORM BOUNDS FOR KERNEL DENSITY ESTIMATION). *Let X_1, X_2, $\cdots$, X_n satisfy* (5.33), *and assume that the family of kernels A_h, $0 < h < 1$, is convolution-like. Then, for any positive constant α and h satisfying $\alpha(\,n \log n\,)^{-1} \leqslant h \leqslant \frac{1}{2}$,*

$$\limsup_{n\to\infty} \frac{\|\, \mathcal{A}_h(d\Omega_n - d\Omega_o)\,\|_\infty}{\sqrt{(nh)^{-1}\log n}} < \infty \quad \textit{almost surely} .$$

The first step is to see whether Lemma (5.17) is applicable to random designs. It is (and we may as well consider regression sums).

(5.37) LEMMA. *Assume that the family A_h, $0 < h \leqslant 1$, is convolution-like in the sense of Definition* (2.5). *Then, there exists a constant c such that, for all h, $0 < h \leqslant 1$, all $D_1, D_2, \cdots, D_n \in \mathbb{R}$, and for every strictly positive design $X_1, X_2, \cdots, X_n \in (\,0\,,1\,]$,*

$$\|\, \tfrac{1}{n} \sum_{i=1}^{n} D_i\, A_h(X_i, \cdot\,)\,\|_\infty \leqslant c \,\|\, \tfrac{1}{n} \sum_{i=1}^{n} D_i\, g_h(X_i - \cdot\,)\,\|_\infty \,.$$

PROOF. One may repeat the proof of Lemma (5.17), with one slight emendation. Note that since all X_i are strictly positive, then $s^{nh}(\,t\,)$ is continuous at $t = 0$; i.e.,

$$\lim_{z\to 0} s^{nh}(z) = s^{nh}(0) \;,$$

and so $|\, s^{nh}(0)\,| \leqslant \|\, s^{nh}\,\|_\infty$, where now $\|\;\cdot\;\|_\infty$ denotes the usual almost everywhere supremum norm. Q.e.d.

Next, we continue simply with the pointwise result. Recall the definition of "reverse" convolutions in Exercise (5.19).

(5.38) LEMMA. *Under the assumption* (5.33), *for all $x \in [\,0\,,\,1\,]$, $t > 0$,*

$$\mathbb{P}\Big[\,|\,[\,g_h \circledast (d\Omega_n - d\Omega_o)\,](x)\,| > t\,\Big] \leqslant 2 \exp\Big(\frac{-\,nh\,t^2}{\omega_2 + \frac{2}{3}\,t} \Big) \,.$$

PROOF. Consider, for fixed x,

$$[\mathcal{A}_h d\Omega_n](x) = \tfrac{1}{n} \sum_{i=1}^{n} g_h(X_i - x) = \sum_{i=1}^{n} \theta_i$$

with $\theta_i = \frac{1}{n} g_h(X_i - x)$. Then, $\theta_1, \theta_2, \cdots, \theta_n$ are iid and

$$(5.39) \qquad |\theta_i| \leqslant (nh)^{-1} .$$

For the variances $\mathbb{V}\mathrm{ar}[\theta_i]$, we obtain

$$\begin{aligned}\mathbb{V}\mathrm{ar}[\theta_i] &= n^{-2} \{ [(g_h)^2 \circledast d\Omega_o](x) - ([g_h \circledast d\Omega_o](x))^2 \} \\ &\leqslant n^{-2} [(g_h)^2 \circledast d\Omega_o](x) \leqslant \tfrac{1}{2} \omega_2 n^{-2} h^{-1} ,\end{aligned}$$

with ω_2 an upper bound on the density w. So,

$$(5.40) \qquad V \stackrel{\text{def}}{=} \sum_{i=1}^{n} \mathbb{V}\mathrm{ar}[\theta_i] \leqslant \tfrac{1}{2} \omega_2 (nh)^{-1} .$$

Now, Bernstein's inequality gives the bound of the lemma. Q.e.d.

We would now like to use the inequalities of Lemmas (5.37) and (5.21), but for Lemma (5.21) the term $g_h \circledast d\Omega_o$ spoils the fun. We will fix that when the time comes. Then, however, a new difficulty appears in that we have to consider

$$\max_{1 \leqslant j \leqslant n} \Big| \tfrac{1}{n} \sum_{i=1}^{n} g_h(X_i - X_j) \Big| .$$

Now, for $j = n$ say, the random variables $X_i - X_n$, $i = 2, 3, \cdots, n$, are not independent. So, to get the analogue of Lemma (5.38) for $t = X_n$, we first condition on X_n and then average over X_n.

(5.41) COROLLARY. *Under the conditions of Lemma (5.38), we have, for all $j = 1, 2, \cdots, n$,*

$$\mathbb{P}\Big[\big| [g_h \circledast (d\Omega_n - d\Omega_o)](X_j) \big| > t \Big] \leqslant 2 \exp\Big(\frac{-\frac{1}{4} nh\, t^2}{\omega_2 + \frac{2}{3} t} \Big) ,$$

provided $t \geqslant 2(1+\omega_2)(nh)^{-1}$, where ω_2 is an upper bound on the density.

PROOF. For notational convenience, consider the case $j = n$. Note that

$$\begin{aligned}[g_h \circledast d\Omega_n](X_n) &= (nh)^{-1} + \tfrac{1}{n} \sum_{i=1}^{n-1} g_h(X_i - X_n) \\ &= (nh)^{-1} + \tfrac{n-1}{n} [g_h \circledast d\Omega_{n-1}](X_n) ,\end{aligned}$$

so that its expectation, conditioned on X_n, equals

$$\mathbb{E}\Big[[g_h \circledast d\Omega_n](X_n) \ \Big| \ X_n \Big] = (nh)^{-1} + \tfrac{n-1}{n} [g_h \circledast d\Omega_o](X_n) .$$

From Lemma (5.38), we obtain that almost surely

$$\mathbb{P}\Big[\,\big|\,[\,g_h \circledast (d\Omega_{n-1} - d\Omega_o)\,](X_n)\,\big| > t \;\Big|\; X_n\Big] \leqslant 2\,\exp\Big(\,\frac{-\,(n-1)h\,t^2}{\omega_2 + \frac{2}{3}\,t}\,\Big)\;.$$

Note that this is a nonasymptotic upper bound that, moreover, does not involve X_n. It follows that

$$\mathbb{P}\Big[\,\big|\,[\,g_h \circledast (d\Omega_{n-1} - d\Omega_o)\,](X_n)\,\big| > t\,\Big] = \mathbb{E}\Big[\,\mathbb{P}\Big[\,\big|\,[\,g_h \circledast (d\Omega_{n-1} - d\Omega_o)\,](X_n)\,\big| > t \;\Big|\; X_n\Big]\Big]$$

has the same bound. Finally, note that

$$[\,g_h \circledast (d\Omega_n - d\Omega_o)\,](X_n) = \varepsilon_{nh} + \tfrac{n-1}{n}\,[\,g_h \circledast (d\Omega_{n-1} - d\Omega_o)\,](X_n)\;,$$

where

$$\varepsilon_{nh} = (nh)^{-1} - \tfrac{1}{n}\,[\,g_h \circledast d\Omega_o\,](X_n)\;,$$

which has the nonasymptotic upper bound

$$|\,\varepsilon_{nh}\,| \leqslant (nh)^{-1} + (nh)^{-1}\,\omega_2 \leqslant c_2\,(nh)^{-1}$$

with $c_2 = 1 + \omega_2$. It follows that

$$\begin{aligned}
&\mathbb{P}\Big[\,\big|\,[\,g_h \circledast (d\Omega_n - d\Omega_o)\,](X_n)\,\big| > t\,\Big] \\
&\qquad \leqslant \mathbb{P}\Big[\,\big|\,[\,g_h \circledast (d\Omega_{n-1} - d\Omega_o)\,](X_n)\,\big| > \tfrac{n}{n-1}\,(\,t - c_2\,(nh)^{-1}\,)\Big] \\
&\qquad \leqslant 2\,\exp\Big(\,\frac{-\,nh\,(\,t - c_2\,(nh)^{-1}\,)^2}{\omega_2 + \frac{2}{3}(\,t - c_2\,(nh)^{-1})}\,\Big)\;.
\end{aligned}$$

For $t \geqslant 2\,c_2\,(nh)^{-1}$, this is bounded by the expression in the lemma. Q.e.d.

We now turn to the issue of how to apply Lemma (5.21). Let t_{in} be defined as

$$\text{(5.42)} \qquad t_{in} = \frac{i-1}{n-1}\;, \quad i = 1, 2, \cdots, n\;,$$

and let U_n denote the transformed empirical distribution function,

$$\text{(5.43)} \qquad U_n(x) = \tfrac{1}{n}\sum_{i=1}^{n} \mathbb{1}\big(\,t_{in} \leqslant \Omega_o(x)\,\big)\;, \quad x \in [\,0\,,\,1\,]\;.$$

In other words, if Ω_o^{inv} denotes the (left-continuous) inverse of Ω_o, then U_n is the empirical distribution function of $\Omega_o^{\text{inv}}(\,t_{in})$, $i = 1, 2, \cdots, n$.

(5.44) LEMMA. *For convolution-like kernels A_h, $0 < h < 1$,*

$$\|\,\mathcal{A}_h(dU_n - d\Omega_o)\,\|_\infty \leqslant c\,(nh)^{-1}\;.$$

PROOF. In view of the identity

$$(5.45)\qquad \mathcal{A}_h d\Omega_o(x) = \int_{\mathbb{R}} A_h(x,y)\, d\Omega_o(y) = \int_0^1 A_h\big(x, \Omega_o^{\text{inv}}(z)\big)\, dz\ ,$$

the quadrature error $QE(x) = [\,\mathcal{A}_h(dU_n - d\Omega_o)\,](x)$ may be written as

$$QE(x) = \tfrac{1}{n} \sum_{i=1}^{n} A_h\big(x, \Omega_o^{\text{inv}}(t_{in})\big) - \int_0^1 A_h\big(x, \Omega_o^{\text{inv}}(z)\big)\, dz\ .$$

Since $t_{in} = (i-1)/(n-1)$, $i = 1, 2, \cdots, n$, then Lemma (13.2.24) implies that, for all $x \in [\,0\,,\,1\,]$, all n, and all $h > 0$ with $nh \to \infty$,

$$(5.46)\qquad |\, QE(x)\, | \leqslant (n-1)^{-1}\, \big|\, A_h\big(x, \Omega_o^{\text{inv}}(\,\cdot\,)\big)\, \big|_{BV}\ .$$

Now, since Ω_o^{inv} is an increasing function with range $[\,0\,,\,1\,]$, the total variation is equal to $\big|\, A_h(x,\,\cdot\,)\, \big|_{BV}$, which is $\mathcal{O}\big(\,h^{-1}\big)$, uniformly in x. Q.e.d.

PROOF OF THEOREM (5.36). Lemma (5.44) gives the deterministic inequality (even though a random function is involved)

$$(5.47)\qquad \|\, \mathcal{A}_h(d\Omega_o - d\Omega_n)\, \|_\infty \leqslant \|\, \mathcal{A}_h(\,dU_n - d\Omega_n)\, \|_\infty + c\,(nh)^{-1}\ .$$

Applying Lemma (5.37), we obtain, for a suitable constant

$$(5.48)\qquad \|\, \mathcal{A}_h(dU_n - d\Omega_n)\, \|_\infty \leqslant c\, \|\, g_h \circledast (dU_n - d\Omega_n)\, \|_\infty\ .$$

Now, Lemma (5.21) gives that $\|\, g_h \circledast (dU_n - d\Omega_n)\, \|_\infty = \mu_n$ with

$$\mu_n = \max_{1 \leqslant i \leqslant n} \big\{\, |\, [g_h \circledast (d\Omega_o - d\Omega_n)](X_i)\, |\, ,\, |\, [g_h \circledast (d\Omega_o - d\Omega_n)](\,t_{in})\, |\, \big\}\ .$$

Then, using Lemma (5.44) once more, we have the deterministic inequality

$$(5.49)\qquad \|\, \mathcal{A}_h(d\Omega_o - d\Omega_n)\, \|_\infty \leqslant 2\, c\,(nh)^{-1} + \mu_n\ .$$

Finally, since μ_n is the maximum over $2n$ terms, Corollary (5.41) yields for $t \geqslant 2\,(1+\omega_2)\,(nh)^{-1}$ that

$$\mathbb{P}\Big[\, \mu_n > t\, \Big] \leqslant 4\, n\, \exp\Big(\frac{-\,\frac{1}{4}\, nh\, t^2}{\omega_2 + \frac{2}{3}\, t} \Big)\ .$$

Now, the choice $t = t_n = C\,\big(\,(nh)^{-1} \log n\,\big)^{1/2}$ with C large enough would give that $\mathbb{P}[\, \mu_n > t\,] \leqslant c\, n^{-2}$ for another suitable constant c, and with Borel-Cantelli then

$$\mu_n = \mathcal{O}\big(\,\big(\,(nh)^{-1} \log n\,\big)^{1/2}\big) \quad \text{almost surely}\ .$$

The caveat is that Corollary (5.41) requires that

$$t_n = C\,\big(\,(nh)^{-1} \log n\,\big)^{1/2} \geqslant 2\,(1+\omega_2\,)\,(nh)^{-1}\ ,$$

which is the case for $h \geqslant \alpha\,(\,n \log n\,)^{-1}$ if $\alpha > 0$. If necessary, just increase C to $C = 2(1+\omega_2)\,\alpha^{-1/2}$. Q.e.d.

EXERCISES: (5.11), (5.16), (5.18), (5.19), (5.32).

6. Random designs and smoothing parameters

In this section, we consider uniform error bounds for kernel regression estimators with random smoothing parameters. There are two approaches to this. On the one hand, one might assume that the smoothing parameter has a nice asymptotic behavior. Denoting the random smoothing parameter by $H \equiv H_n$, one might assume that there exists a deterministic sequence $\{ h_n \}_n$ such that $H_n / h_n - 1 = \mathcal{O}_P\big(n^{-\beta} \big)$ for a suitable $\beta > 0$, and proceed from there, as in DEHEUVELS and MASON (2004), where the exact constant is obtained. The alternative is to prove results uniformly in the smoothing parameter over a wide range, as in EINMAHL and MASON (2005), but the price one pays is that the exact constant is out of reach. In this section, we obtain a somewhat weaker version of the EINMAHL and MASON (2005) results by tweaking the results of the previous section. All of this applies to random designs as well, and we concentrate on this case.

To simplify matters, we shall ignore bias issues. It turns out that, under the same conditions, the result of the previous section,

$$\text{(6.1)} \qquad \| \tfrac{1}{n} \sum_{i=1}^{n} D_i \, A_h(\,\cdot\, , X_i) \|_\infty = \mathcal{O}\big(\sqrt{(nh)^{-1} \log n } \,\big) \ , \ ,$$

in fact holds almost surely, uniformly in $h \in \mathcal{H}_n(\alpha)$ for $\alpha > 0$, where

$$\text{(6.2)} \qquad \mathcal{H}_n(\alpha) = \Big[\alpha \left(n^{-1} \log n \right)^{1-2/\kappa} , \tfrac{1}{2} \Big] \ .$$

It even applies to regression problems with random designs. This result is weaker than the results of EINMAHL and MASON (2005), who show that

$$\| f^{nh} - \mathbb{E}[f^{nh}] \|_\infty = \mathcal{O}\big(\sqrt{ (nh)^{-1} \{ \log(1/h) \vee \log\log n \} } \,\big)$$

uniformly in $h \in \mathcal{H}_n(\alpha)$. Since our interest is in the case $h \asymp n^{-1/(2m+1)}$ or $h \asymp (n^{-1} \log n)^{1/(2m+1)}$ (random or deterministic), the improved rate of EINMAHL and MASON (2005) is inconsequential. Again, the uniformity in h implies that (6.1) holds for *random (data-driven)* h in this range! Clearly, the range is so big that it hardly constitutes a constraint. For kernel density estimation, there is a similar result.

Since we are not interested in the bias, the random design model is as follows. Let $(X_1, D_1), (X_2, D_2), \cdots (X_n, D_n)$ be a random sample of the bivariate random variable (X, D) with $X \in [\,0\,,\,1\,]$. Assume that

$$\text{(6.3)} \qquad \mathbb{E}[\, D \,|\, X = x \,] = 0 \ , \quad x \in [\,0\,,\,1\,] \ ,$$

and that, for some $\kappa > 2$,

$$\text{(6.4)} \qquad \sup_{x \in [\,0\,,\,1\,]} \mathbb{E}[\, |\, D \,|^\kappa \,|\, X = x \,] < \infty \ .$$

Of course, then

$$\sigma^2 \stackrel{\text{def}}{=} \sup_{x\in[0,1]} \mathbb{E}[\,D^2 \mid X = x\,] \,. \tag{6.5}$$

We assume that

(6.6) the design $X_1, X_2, \cdots, X_n$ is random, with a bounded design density as in (5.33) .

Below, we need a version of Lemma (5.26) for random designs. Analogous to (5.25), (re)define G_j^{nh} as

$$G_j^{nh} = \tfrac{1}{n} \sum_{i=1}^{n} \theta_i\, g_h(X_i - X_j) \,, \quad j = 1, 2, \cdots, n \,, \tag{6.7}$$

with $\theta_i = D_i\, \mathbb{1}(|\,D_i\,| < M)$.

(6.8) LEMMA. *Assume that* (6.3) *through* (6.6) *hold. Then, for* $1 \leqslant j \leqslant n$ *and* $t \geqslant 2\,(nh)^{-1} M$,

$$\mathbb{P}[\,|\,G_j^{nh}\,| > t\,] \leqslant 2 \exp\Bigl(\frac{-\tfrac{1}{4}\, n h\, t^2}{\omega_2\, \sigma^2 + \tfrac{2}{3}\, t\, M} \Bigr) \,.$$

PROOF. Consider the case $j = n$. Then,

$$G_n^{nh} = \varepsilon_{nh} + \tfrac{n-1}{n}\, S_{n-1}(X_n) \,,$$

where $\varepsilon_{nh} = (nh)^{-1}\, \theta_n$ satisfies the bound $|\,\varepsilon_{nh}\,| \leqslant (nh)^{-1}\, M$, and

$$S_{n-1}(\,t\,) = \tfrac{1}{n-1} \sum_{i=1}^{n-1} \theta_i\, g_h(X_i - t\,) \,.$$

We now wish to apply Bernstein's inequality to bound $S_{n-1}(X_n)$. We have, obviously,

$$\bigl|\, \tfrac{1}{n-1}\, \theta_i\, g_h(X_i - X_n) \,\bigr| \leqslant \bigl(\, (n-1)\, h \,\bigr)^{-1} M \,,$$

and for the conditional variances,

$$\mathbb{V}\mathrm{ar}\bigl[\, \tfrac{1}{n-1}\, \theta_i\, g_h(X_i - X_n\,) \;\bigm|\; X_n \,\bigr] \leqslant \tfrac{1}{2}\, \sigma^2\, \omega_2\, (n-1)^{-2}\, h^{-1} \,, \tag{6.9}$$

and so

$$V = \sum_{i=1}^{n-1} \mathbb{V}\mathrm{ar}\bigl[\, \tfrac{1}{n-1}\, \theta_i\, g_h(X_i - X_n\,) \;\bigm|\; X_n \,\bigr] \leqslant \tfrac{1}{2}\, \sigma^2\, \omega_2\, \bigl(\, (n-1)\, h \,\bigr)^{-1} \,.$$

From Bernstein's inequality, then almost surely

$$\mathbb{P}\Bigl[\, |\, S_{n-1}(X_n) \,| > t \;\Bigm|\; X_n \,\Bigr] \leqslant 2 \exp\Bigl(\frac{-\,(n-1)\, h\, t^2}{\omega_2\, \sigma^2 + \tfrac{2}{3}\, t\, M} \Bigr)$$

and, as before, the same bound holds for the unconditional probability $\mathbb{P}\big[\,|\,S_{n-1}(X_n)\,|>t\,\big]$. Finally,

$$\mathbb{P}\Big[\,|\,G_n^{nh}\,|>t\,\Big]\leqslant\mathbb{P}\Big[\,|\,S_{n-1}(X_n)\,|>\frac{n}{n-1}\,\{\,t-(nh)^{-1}M\,\}\Big]$$

$$\leqslant 2\,\exp\Big(\frac{-nh\,\{\,t-(nh)^{-1}M\,\}^2}{\omega_2\,\sigma^2+\frac{2}{3}\,\{\,t-(nh)^{-1}M\,\}\,M}\Big)\;.$$

For $t\geqslant 2\,(nh)^{-1}M$, the bound of the lemma follows. Q.e.d.

(6.10) EXERCISE. Verify (6.9).

(6.11) REMARK. The following two theorems are weak versions of two theorems of EINMAHL and MASON (2005). They prove that, under the same conditions, the factors $\log n$ may be replaced by $\log(1/h)\vee\log\log n$. This requires a heavy dose of modern empirical process theory (metric entropy ideas and such). Again, since we are interested in the "optimal" smoothing parameter $h\asymp n^{-1/(2m+1)}$, we do not lose much. EINMAHL and MASON (2005) (only) consider convolution kernels with compact support but allow the design to have unbounded support. These two assumptions can be interchanged: The convolution kernel may have unbounded support and the design compact support; see MASON (2006). Using the Green's function trick of Lemma (5.17), it then also works for convolution-like families, as stated in the theorems above, with the improved $\log(1/h)\vee\log\log n$.

(6.12) THEOREM. *Under the assumptions* (6.3)-(6.6) *for any convolution-like family of kernels* A_h, $0<h\leqslant 1$, *we have almost surely*

$$\limsup_{n\to\infty}\;\sup_{h\in\mathcal{H}_n(\alpha)}\;\frac{\big\|\,\frac{1}{n}\sum_{i=1}^{n}D_i\,A_h(\,\cdot\,,X_i)\,\big\|_\infty}{\sqrt{(nh)^{-1}\,\log n}}<\infty\;.$$

(6.13) THEOREM. *Assume* (6.3) *through* (6.6) *and that the family of kernels* A_h, $0<h\leqslant 1$, *is convolution-like. Let*

$$w^{nh}(\,t\,)=\frac{1}{n}\sum_{i=1}^{n}A_h(\,t\,,X_i)\;,\qquad t\in[\,0\,,1\,]\;.$$

Then, for $\alpha>0$, *we have almost surely*

$$\limsup_{n\to\infty}\;\sup_{h\in\mathcal{G}_n(\alpha)}\;\frac{\|\,w^{nh}-\mathbb{E}[\,w^{nh}\,]\,\|_\infty}{\sqrt{(nh)^{-1}\,\log n}}<\infty\;.$$

Here, $\mathcal{G}_n(\alpha)=\big[\,\alpha\,(\,n\log n\,)^{-1}\,,\,\frac{1}{2}\,\big]$.

PROOF OF THEOREM (6.12). We may assume that $A_h(x,\,t\,)=g_h(x-t)$ for all x, t, courtesy of Lemma (5.37). The first real observation is that the truncation causes no problems. Let

$$(6.14)\qquad T_n \stackrel{\text{def}}{=} \sup_{h\in\mathcal{H}_n(\alpha)} \frac{\max\limits_{1\leqslant j\leqslant n} \bigl|\, \frac{1}{n} \sum\limits_{i=1}^{n} \gamma_i\, g_h(X_i - X_j)\,\bigr|}{\sqrt{(nh)^{-1}\log n}}$$

with $\gamma_i = D_i\, 1\!\!1(|\, D_i\,| > M)$. One verifies that the Improved Truncation Lemma (5.22) applies to random designs as well. Consequently, we get

$$T_n \leqslant C_n \sup_{h\in\mathcal{H}_n(\alpha)} h^{-1}\,\bigl(\, nh/\log n\,\bigr)^{1/2}\, M^{1-\kappa}\, \psi(M)$$

with $M = \bigl(\, nh/\log n\,\bigr)^{1/2}$. Of course, since $h^{-1}\,\bigl(\, nh/\log n\,\bigr)^{1/2}\, M^{1-\kappa}$ remains bounded for $h \in \mathcal{H}_n(\alpha)$ and $\psi(M) \to 0$, the supremum over $h \in \mathcal{H}_n(\alpha)$ tends to 0. Also, C_n does not depend on the design, h, or M and tends to a constant almost surely. It follows that

$$(6.15)\qquad \limsup_{n\to\infty}\; T_n = 0 \quad \text{almost surely}\,.$$

Next, we deal with the truncated sums. Let

$$(6.16)\qquad \mathfrak{G}(h) \stackrel{\text{def}}{=} \frac{\max\limits_{1\leqslant j\leqslant n} |\, G_j^{nh}\,|}{\sqrt{(nh)^{-1}\log n}}\;,$$

with G_j^{nh} as in (6.7). (In this notation, the dependence on n is suppressed.) Then, Lemma (6.8) implies, for $h \in \mathcal{H}_n(\alpha)$ and $t \geqslant 2\,(nh)^{-1}\, M$, that

$$(6.17)\qquad \mathbb{P}\Bigl[\, \mathfrak{G}(h) > t\,\Bigr] \leqslant 2\,n\,\exp\Bigl(\, \frac{-\,\frac{1}{4}\,t^2\,\log n}{\omega_2\,\sigma^2 + \frac{2}{3}\,t}\,\Bigr)\;,$$

when M is chosen as in (5.31).

We now must come up with a bound on $\sup_{h\in\mathcal{H}_n(\alpha)}\, \mathfrak{G}_n(h)$. Here, metric entropy ideas would seem to be indispensable; i.e., somehow, the supremum must be approximated by the maximum over a finite subset of $\mathcal{H}_n(\alpha)$. We shall fake our way around this problem, but either way some knowledge about $\mathfrak{G}(h)$ as a function of h is required. From Exercise (6.24) below, we infer that $\max\limits_{1\leqslant j\leqslant n} |\, G_j^{nh}\,|$ is a decreasing function of h, so

$$(6.18)\qquad \max_{1\leqslant j\leqslant n} |\, G_j^{n\lambda}\,| \leqslant \max_{1\leqslant j\leqslant n} |\, G_j^{nh}\,| \qquad \text{for} \quad 0 < h < \lambda\,.$$

It follows that

$$(6.19)\qquad \mathfrak{G}(\lambda) \leqslant \sqrt{\frac{h}{\lambda}}\; \mathfrak{G}(h)\;, \quad 0 < h < \lambda\,.$$

Now, we cover the "big" interval $\mathcal{H}_n(\alpha)$ by a bunch of small intervals. Let

$$(6.20)\qquad \lambda_\ell = \alpha\,\bigl(\, n^{-1}\log n\,\bigr)^{1-2/\kappa} \cdot 4^\ell\;, \quad \ell = 0, 1, \cdots, L\;,$$

with L chosen such that $\frac{1}{2} \leqslant \lambda_L < 1$. Then,

$$\mathcal{H}_n(\alpha) \subset \bigcup_{\ell=0}^{L-1} [\lambda_\ell, \lambda_{\ell+1}] . \tag{6.21}$$

As for the choice of L, taking

$$L = \left\lfloor \frac{\log(1/\alpha) + (1 - 2/\kappa)(\log n - \log\log n)}{\log 4} \right\rfloor ,$$

where $\lfloor x \rfloor$ denotes the largest integer not exceeding x, will do. One verifies that $L \asymp \log n$ for $n \to \infty$ and, since $\lambda_{\ell+1}/\lambda_\ell = 4$, (6.19) implies

$$\sup_{h \in [\lambda_\ell, \lambda_{\ell+1}]} \mathfrak{G}(h) \leqslant 2\,\mathfrak{G}(\lambda_\ell) . \tag{6.22}$$

Now, we are in business. By (6.21) and (6.22), we have

$$\begin{aligned} \mathbb{P}\Bigl[\sup_{h \in \mathcal{H}_n(\alpha)} \mathfrak{G}(h) > t \Bigr] &\leqslant \sum_{\ell=0}^{L-1} \mathbb{P}\Bigl[\sup_{h \in [\lambda_\ell, \lambda_{\ell+1}]} \mathfrak{G}(h) > t \Bigr] \\ &\leqslant \sum_{\ell=0}^{L-1} \mathbb{P}\Bigl[\mathfrak{G}(\lambda_\ell) > \tfrac{1}{2} t \Bigr] \\ &\leqslant 2\,n\,L \exp\Bigl(\frac{-\frac{1}{16} t^2 \log n}{\omega_2\,\sigma^2 + \frac{2}{3} t} \Bigr) , \end{aligned}$$

the last step by (6.17). Then, for t a large enough constant (not depending on n), we get

$$\mathbb{P}\Bigl[\sup_{h \in \mathcal{H}_n(\alpha)} \mathfrak{G}(h) > t \Bigr] \leqslant 2\,n\,L\,n^{-3} \leqslant c\,n^{-2} ,$$

and then, by Borel-Cantelli,

$$\limsup_{n\to\infty} \sup_{h \in \mathcal{H}_n(\alpha)} \mathfrak{G}(h) < \infty \quad \text{almost surely} . \tag{6.23}$$

Combined with (6.15), this completes the proof. Q.e.d.

(6.24) EXERCISE. Here, we consider the monotonicity properties of the one-sided exponential family g_h, $h > 0$, of (5.12).
(a) Show that, for λ, $h > 0$,

$$g_\lambda = (h/\lambda)\, g_h + (1 - h/\lambda)\, g_\lambda * g_h .$$

Here, $*$ denotes convolution on the line,

$$[f * g](t) = \int_{\mathbb{R}} f(t - \tau)\, g(\tau)\, d\tau , \quad t \in \mathbb{R} .$$

(b) Let F have bounded variation on $\mathbb{R}$. Show that, for $0 < h < \lambda$,

$$\| g_\lambda \circledast dF \|_{L^\infty(0,1)} \leqslant \| g_h \circledast dF \|_{L^\infty(0,1)} .$$

(c) Prove (6.18). [Hint: For (b), for $0 < h < \lambda$, observe that (a) says that g_λ is a convex combination of g_h and $g_\lambda * g_h$, with the latter also being a convex combination.]

(6.25) EXERCISE. Show that Theorem (6.12) also holds for deterministic, asymptotically uniform designs.

PROOF OF THEOREM (6.13). The proof is analogous to the random design regression case. The hard work has been done in Corollary (5.41). Again, we may assume that $A_h(x, t) = g_h(x - t)$ for all x, t. Let

$$G_j^{nh} = [\, g_h \circledast (d\Omega_n - d\Omega_o)\,](X_j) \;, \quad j = 1, 2, \cdots, n \;,$$

and

$$\mathfrak{G}(h) = \frac{\max\limits_{1 \leqslant j \leqslant n} |\, G_j^{nh} \,|}{\sqrt{(nh)^{-1} \log n}} \;. \tag{6.26}$$

Recall that the numerator equals $\|\, g_h \circledast (\, d\Omega_n - d\Omega_o\,) \,\|_\infty$. Consider the cover of the interval $\mathcal{G}_n(\alpha)$,

$$\mathcal{G}_n(\alpha) \subset \bigcup_{\ell=0}^{L-1} [\, \lambda_\ell, \lambda_{\ell+1}\,] \;,$$

where $\lambda_\ell = \alpha\,(\, n \log n\,)^{-1} \cdot 4^\ell$, $\ell = 0, 1, \cdots, L$, with L chosen such that $\frac{1}{2} \leqslant \lambda_L < 1$; i.e.,

$$L = \Big\lfloor \frac{\log(1/\alpha) + \log n + \log\log n}{\log 4} \Big\rfloor \;.$$

Then, from the monotonicity result of Exercise (6.24)(b) and the remark following (6.26),

$$\sup_{h \in [\, \lambda_\ell, \lambda_{\ell+1}\,]} \mathfrak{G}(h) \leqslant 2\, \mathfrak{G}(\lambda_\ell) \;,$$

and so, with Corollary (5.41), as in the derivation of (6.17),

$$\begin{aligned} \mathbb{P}\Big[\sup_{h \in \mathcal{G}_n(\alpha)} \mathfrak{G}(h) > t \,\Big] &\leqslant \sum_{\ell=0}^{L-1} \mathbb{P}\Big[\sup_{h \in [\, \lambda_\ell, \lambda_{\ell+1}\,]} \mathfrak{G}(h) > t \,\Big] \\ &\leqslant \sum_{\ell=0}^{L-1} \mathbb{P}\Big[\, \mathfrak{G}(\lambda_\ell) > \tfrac{1}{2} t \,\Big] \\ &\leqslant 2\, n\, L \, \exp\Big(\frac{-\frac{1}{16} t^2 \log n}{\omega_2\, \sigma^2 + \frac{2}{3} t \sqrt{(nh)^{-1} \log n}} \Big) \;, \end{aligned}$$

provided

$$t \geqslant 2(1 + \omega_2)\,(\, nh \, \log n\,)^{-1/2} \;. \tag{6.27}$$

Again, for t a large enough constant C, we get

$$\mathbb{P}\Big[\sup_{h\in\mathcal{G}_n(\alpha)} \mathfrak{G}(h) > C\Big] \leqslant c\,n^{-2}$$

and Borel-Cantelli carries the day. The only caveat is that (6.27) must be satisfied; i.e.,

$$2(1+\omega_2)\,(\,nh\,\log n\,)^{-1/2} \leqslant C$$

for the previously chosen constant C. This holds for $h \geqslant \alpha(\,n\log n\,)^{-1}$ for any fixed $\alpha > 0$ (just take C bigger if necessary). Q.e.d.

(6.28) EXERCISE. Consider the following variation on Theorem (6.13). Let $f \in W^{1,\infty}(\mathbb{R})$ be a deterministic function and let $A \in W^{1,1}(\mathbb{R})$ be a kernel (no further properties provided). Consider

$$S^{nh}(x) = \tfrac{1}{n}\sum_{i=1}^{n} A_h(x - X_i)\,f(X_i)\ ,$$

and set $S_h(x) = \mathbb{E}[\,S^{nh}(x)\,]$. Show that, for any positive constant α and all h satisfying $\alpha(\,n\log n\,)^{-1} \leqslant h \leqslant \frac{1}{2}$,

$$\limsup_{n\to\infty} \frac{\|\,S^{nh} - S_h\,\|_\infty}{\sqrt{(nh)^{-1}\log n}} \leqslant \eta \sup_{x\in[0,1]} \|\,A_h(x - \,\cdot\,)f(\,\cdot\,)\,\|_{h,W^{1,1}(0,1)}\ ,$$

where $\eta < \infty$ almost surely. Here,

$$\|\,\varphi\,\|_{h,W^{1,1}(0,1)} = \|\,\varphi\,\|_{L^1(0,1)} + h\,\|\,\varphi'\,\|_{L^1(0,1)}\ .$$

[Hint: Show that $A_h(x - z)\,f(z)$, $0 < h \leqslant 1$, is convolution-like in the sense of Definition (2.5), and proceed as in the proof of the theorem.]

EXERCISES: (6.9), (6.24), (6.25).

7. Uniform error bounds for smoothing splines

In this section, we consider bounds on the uniform error of smoothing splines with deterministic, asymptotically uniform designs. The approach is to approximate the spline estimator by the C-spline estimator of (13.1.10) and show that the kernel in the representation (13.1.11) of the C-spline is such that the results of the previous section apply. The case of random designs is discussed in Chapter 21. When all is said and done, the following theorem applies.

(7.1) THEOREM. *Let* $m \geqslant 1$ *and* $f_o \in W^{m,\infty}(0,1)$. *Under the Gauss-Markov model* (1.1)–(1.2), *with iid noise satisfying the moment condition*

$$\mathbb{E}[\,d_{1,n}\,] = 0\quad,\quad \mathbb{E}[\,|\,d_{1,n}\,|^{\kappa}\,] < \infty \quad \textit{for some } \kappa > 3\ ,$$

and with an asymptotically uniform design, the smoothing spline estimator satisfies almost surely

$$\limsup_{n\to\infty} \sup_{h\in\mathcal{H}_n(\alpha)} \frac{\| f^{nh} - f_o \|_\infty}{\sqrt{h^{2m} + (nh)^{-1} \log n}} < \infty \ .$$

Moreover,

$$\| f^{nh} - f_o \|_\infty = \mathcal{O}\big((n^{-1} \log n)^{m/(2m+1)} \big) \quad \textit{in probability}$$

whenever $h \asymp (n^{-1} \log n)^{1/(2m+1)}$ *in probability.*

Here, $\mathcal{H}_n(\alpha)$ was defined in (6.2). Note that the theorem allows a random choice of h. As outlined above, the proof consists of two parts. First, we show that the smoothing spline estimator is very close to the C-spline estimator of (13.1.10) in the precise sense of the following lemma.

(7.2) LEMMA. *Let* $m \geqslant 1$ *and* $f_o \in W^{m,2}(0,1)$. *Let* f^{nh} *be the spline estimator of* (13.1.7) *and* ψ^{nh} *the C-spline estimator of* (13.1.10). *Under the conditions of Theorem* (7.1), *we have almost surely*

$$\limsup_{n\to\infty} \sup_{h\in\mathcal{H}_n(\alpha)} \frac{\| f^{nh} - \psi^{nh} - \mathbb{E}[\, f^{nh} - \psi^{nh} \,] \|_{m,h}}{(nh)^{-1} \log n} < \infty \ .$$

We postpone the proof until the end of this section but do note that already here we will have to appeal to Theorem (6.13) on *density* estimation. Of course, the bound,

$$\limsup_{n\to\infty} \sup_{h\in\mathcal{H}_n(\alpha)} \frac{\| f^{nh} - \psi^{nh} - \mathbb{E}[\, f^{nh} - \psi^{nh} \,] \|_\infty}{h^{-1/2} \, (nh)^{-1} \log n} < \infty \ ,$$

follows from the lemma by the reproducing kernel Hilbert space property of the space $W^{m,2}(0,1)$. Then, the proof of Theorem (7.1) proceeds by observing that ψ^{nh} is the kernel estimator,

$$\psi^{nh}(x) = \tfrac{1}{n} \sum_{i=1}^{n} y_{in} \, \Gamma_{mh}(x, x_{in}) \ , \quad x \in [\,0\,,\,1\,] \ , \tag{7.3}$$

in which $\Gamma_{mh}(x,y)$ is the Green's function for the boundary value problem (13.1.12), repeated here for convenience,

$$\begin{aligned} &(-h^2)^m \, u^{(2m)} + u = w \quad \text{on } (0,1) \ , \\ &u^{(k)}(0) = u^{(k)}(1) = 0 \ , \quad k = m, m+1, \cdots, 2m-1 \ . \end{aligned} \tag{7.4}$$

This boundary value problem constitutes the Euler equations for the C-spline problem (13.1.10) for the choice

$$w = \tfrac{1}{n} \sum_{i=1}^{n} y_{in} \, \delta(\,\cdot\, - y) \ ; \tag{7.5}$$

see Exercise (13.3.21). Below, we show that the Green's function is in fact the reproducing kernel of the Hilbert space $W^{m,2}(0,1)$ with the inner product $\langle \,\cdot\, , \,\cdot\, \rangle_{m,h}$; see § 13.2. Thus, $\Gamma_{mh}(x, t) = \mathcal{R}_{mh}(x, t)$ for all x , t .

So, the actual second part of the proof of Theorem (7.1) consists of showing that the family of kernels $\Gamma_{mh}(x,y)$, $h>0$, is convolution-like of order m, see Definition (2.6), after which we may apply Theorem (6.12) regarding the uniform error of kernel regression estimators.

(7.6) THEOREM. *For $\ell = 0, 1, \cdots, m$, the families of kernels*

$$h^{\ell}\, \Gamma_{mh}^{(\ell)} \;, \quad 0 < h < 1 \;,$$

are convolution-like. For $\ell = 0$, they are convolution-like of order m. Moreover, there exist constants c_m and γ_m, depending only on the order m, such that, for all $0 < h < 1$, all s, $t \in [\,0\,,\,1\,]$, and $\ell = 0, 1, \cdots, m$,

$$|\, h^{\ell}\, \Gamma_{mh}^{(\ell)}(s,\, t)\,| \leqslant c_m\, h^{-1} \exp(\, - \gamma_m\, h^{-1}\, |\, s - t\,|\,) \;.$$

Some of these properties are easy to verify, and we do that now.

(7.7) LEMMA.

$$\int_0^1 (x-y)^k\, \Gamma_{mh}(x,y)\, dy = \begin{cases} 1\,, & k = 0\,, \\ 0\,, & k = 1, 2, \cdots, m-1\,. \end{cases}$$

PROOF. By Exercise (3.4), it suffices to show that

$$\int_0^1 y^k\, \Gamma_{mh}(x,y)\, dy = x^k \;, \quad k = 0, 1, \cdots, m-1 \;. \tag{7.8}$$

Let $p_k(x) = x^k$. Note that the problem

$$\text{minimize} \quad \|\, \varphi - p_k\,\|^2 + h^{2m}\, \|\, \varphi^{(m)}\,\|^2 \tag{7.9}$$

has the solution

$$\varphi(x) = \int_0^1 p_k(y)\, \Gamma_{mh}(x,y)\, dy \;, \quad x \in [\,0\,,\,1\,] \;. \tag{7.10}$$

On the other hand, also note that $p_k^{(m)}(x) \equiv 0$ for $0 \leqslant k \leqslant m-1$, so that $\varphi = p_k$ is also a solution of (7.9). By the uniqueness of the solution, (7.8) follows. Q.e.d.

The remaining properties of $\Gamma_{mh}(x,y)$ involve absolute values and so are not easily related to problems like (7.9). It seems that the only recourse we have is to find an appropriate more or less explicit representation for $\Gamma_{mh}(x,y)$. The following representation and bounds on $\Gamma_{mh}(x,y)$ turn out to be sufficient for the purpose. The construction is essentially due to

MESSER and GOLDSTEIN (1993), the only difference being the treatment of the boundary conditions.

(7.11) LEMMA (MESSER and GOLDSTEIN, 1993). *Define the function* $\mathfrak{B}_{m,h}(x)$, $x \in \mathbb{R}$, *by its Fourier transform*

$$\widehat{\mathfrak{B}_{m,h}}(\omega) = \left(1 + (2\pi h\omega)^{2m}\right)^{-1}, \quad \omega \in \mathbb{R},$$

and let

$$\varphi_{\ell,h}(x) = \begin{cases} \exp(h^{-1}\varpi_\ell\, x) & , \quad \ell = 0, 1, \cdots, m-1, \\ \exp(h^{-1}\varpi_\ell\,(x-1)) , & \ell = m, m+1, \cdots, 2m-1, \end{cases}$$

where

$$\varpi_\ell = \exp\Bigl(\frac{2\ell + m + 1}{2m}\,\pi i\Bigr), \quad \ell = 0, 1, \cdots, 2m-1.$$

Then, for a suitable $h_o > 0$ *and for all* $h < h_o$, *there exist functions* $a_{\ell,h}$ *and positive constants* c_m, κ_m *such that, for all* $x, y \in [0, 1]$,

$$\Gamma_{mh}(x,y) = \mathfrak{B}_{m,h}(x-y) + \sum_{\ell=0}^{2m-1} h^{-1}\,\varphi_{\ell,h}(x)\, a_{\ell,h}(y)$$

and

$$\sup_{0\leqslant\ell\leqslant m-1} |\, a_{\ell,h}(y)\,| \leqslant c_m \exp(-h^{-1}\kappa_m y),$$

$$\sup_{m\leqslant\ell\leqslant 2m-1} |\, a_{\ell,h}(y)\,| \leqslant c_m \exp(-h^{-1}\kappa_m (1-y)).$$

Moreover, $\Gamma_{mh}(x,y) = \Gamma_{mh}(y,x)$ *for all* x, $y \in [0, 1]$.

Before proving this, we show that this actually provides the required convolution-kernel-like properties of the Green's function. Some useful, easily proved information regarding the functions $\mathfrak{B}_{m,h}$ and $\varphi_{\ell,h}$ is stated in the next lemma.

(7.12) LEMMA. *Let* $m \geqslant 1$. *There exist positive constants* κ_m *and* c_m *such that, for* $x \in [0, 1]$ *and* $t \in \mathbb{R}$,

(a) $$|\,\varphi_{\ell,h}(x)\,| \leqslant \exp(-h^{-1}\kappa_m x), \quad 0 \leqslant \ell \leqslant m-1,$$

(b) $$|\,\varphi_{\ell,h}(x)\,| \leqslant \exp(h^{-1}\kappa_m (x-1)), \quad m \leqslant \ell \leqslant 2m-1,$$

(c) $$|\,\mathfrak{B}^{(k)}_{m,h}(t)\,| \leqslant c_m\, h^{-k-1} \exp(-h^{-1}\kappa_m\,|\,t\,|), \quad 0 \leqslant k \leqslant 2m-1.$$

(d) $$\sup_{h>0}\,\sup_{\ell} \int_0^1 h^{-1}\,|\,\varphi_{\ell,h}(y)\,|\, dy < \infty,$$

(e) $$\sup_{h>0}\,\sup_{x\in[0,1]} \int_0^1 |\,\mathfrak{B}_{m,h}(x-y)\,|\, dy < \infty.$$

A comparison between these bounds and the symmetry of the Green's function suggests that $a_{\ell,h}$ and $\varphi_{\ell,h}$ are just about the same, but we shall restrain ourselves.

(7.13) EXERCISE. Prove Lemma (7.12).

The remaining properties of the kernels $\Gamma_{mh}(x,y)$, $h>0$, now follow.

(7.14) LEMMA. *There exists a constant c such that, for all h, $0<h<1$,*

(a)
$$\sup_{x\in[0,1]} \int_0^1 |\,\Gamma_{mh}(x,y)\,|\,dy \leqslant c\,,$$

(b)
$$\sup_{x\in[0,1]} \int_0^1 |\,x-y\,|^m\,|\,\Gamma_{mh}(x,y)\,|\,dy \leqslant c\,h^m\,,$$

(c)
$$\sup_{x\in[0,1]} |\,\Gamma_{mh}(x,\,\cdot\,)\,|_{BV} \leqslant c\,h^{-1}\,.$$

PROOF. Part (a) follows from the triangle inequality and Lemma (7.12), parts (d)-(e). Part (c) follows similarly, upon noting that

$$|\,\Gamma_{mh}(x,\,\cdot\,)\,|_{BV} = \int_0^1 \left|\,\frac{\partial}{\partial y}\Gamma_{mh}(x,y)\,\right| dy = \int_0^1 \left|\,\frac{\partial}{\partial y}\Gamma_{mh}(y,x)\,\right| dy\,,$$

since $\Gamma_{mh}(x,y)$ is symmetric.

Part (b) is a bit more involved. The first step is again the triangle inequality. The straightforward part is that with $z\equiv h^{-1}x$,

$$\begin{aligned}\int_0^1 |\,x-y\,|^m\,|\,\mathfrak{B}_{m,h}(x-y)\,|\,dy &= h^m \int_0^{h^{-1}} |\,z-y\,|^m\,|\,\mathfrak{B}_m(z-y)\,|\,dy\\ &\leqslant h^m \int_{\mathbb{R}} |\,y\,|^m\,|\,\mathfrak{B}_m(y)\,|\,dy \leqslant c\,h^m\,.\end{aligned}$$

We next consider

$$I_{\ell,h}(x) \overset{\text{def}}{=} \int_0^1 |\,x-y\,|^m\,h^{-1}\,|\,\varphi_{\ell,h}(x)\,|\,|\,a_{\ell,h}(y)\,|\,dy.$$

For $0\leqslant\ell\leqslant m-1$, we use $|\,x-y\,|^m \leqslant x^m+y^m$. This gives the upper bound for $I_{\ell,h}$

$$\begin{aligned}I_{\ell,h}(x) \leqslant x^m\,\varphi_{\ell,h}(x) \int_0^1 h^{-1}\,|\,a_{\ell,h}(y)\,|\,dy\,+&\\ \varphi_{\ell,h}(x)\int_0^1 h^{-1}\,y^m\,|\,a_{\ell,h}(y)\,|\,dy\,.&\end{aligned}$$

By Lemma (7.11), the first integral is bounded by

$$c_m \int_0^1 h^{-1} \exp(-h^{-1}\kappa_m y)\, dy \leqslant c_m \int_0^\infty \exp(-\kappa_m y)\, dy = c_m$$

and the second one by

$$c_m \int_0^1 h^{-1}\, y^m \exp(-h^{-1}\kappa_m y)\, dy \leqslant$$
$$c_m h^m \int_0^\infty y^m \exp(-\kappa_m y)\, dy = (m-1)!\, c_m\, h^m \ .$$

Also, by Lemma (7.12)(a) and the substitution $z = h^{-1}x$,

$$\sup_{x\in[\,0\,,\,1\,]} x^m\, \varphi_{\ell,h}(x) \leqslant \sup_{x\in[\,0\,,\,1\,]} x^m \exp(-\kappa_m h^{-1} x\,)$$
$$\leqslant h^m \sup_{z\geqslant 0} z^m \exp(-\kappa_m z\,) = \kappa_m^{-m}\, h^m \ .$$

And, of course, $|\,\varphi_{\ell,h}(x)\,| \leqslant 1$ for the relevant x. The result is that

$$\sup_{x\in[\,0\,,\,1\,]}\ \sup_{0<h<h_o}\ \sum_{\ell=0}^{m-1} I_{\ell,h}(x) = \mathcal{O}\big(\,h^m\,\big) \ .$$

For $m \leqslant \ell \leqslant 2m-1$, the starting point is that, for all $x\,,\ y \in [\,0\,,\,1\,]$,

$$|\,x-y\,|^m \leqslant |\,1-x\,|^m + |\,1-y\,|^m \ ,$$

with the final result that

$$\sup_{x\in[\,0\,,\,1\,]}\ \sup_{0<h<h_o}\ \sum_{\ell=m}^{2m-1} I_{\ell,h}(x) = \mathcal{O}\big(\,h^m\,\big)\,.$$

We leave the details as an exercise. The lemma follows. Q.e.d.

(7.15) Exercise. Show that $\displaystyle\sup_{x\in[\,0\,,\,1\,]}\ \sup_{0<h<h_o}\ \sum_{\ell=m}^{2m-1} I_{\ell,h}(x) = \mathcal{O}\big(\,h^m\,\big)\,.$

Proof of Lemma (7.11). This proof seems to be a standard exercise, but the details are quite involved. Messer and Goldstein (1993) just state the result, although undoubtedly they had the following proof in mind.

The proof consists of four parts. First, we prove the existence of the Green's function and show that it is symmetric. Second, we construct a fundamental solution of the boundary value problem (7.4). Then, all homogeneous solutions of the differential equation are determined. In the last step, we construct the Green's function in the form stated and show the bound on the coefficients.

The Green's function is the reproducing kernel. Consider the boundary value problem (7.4) with $w \in L^2(0,1)$. We already know that the solution exists and is unique. Moreover, the solution satisfies

$$\| u \|_{m,h} \leqslant \| w \| ,$$

so that $u \in W^{m,2}(0,1)$. By the Reproducing Kernel Lemma (13.2.18), we may represent u as

$$u(x) = \langle \mathcal{R}_{mh}(x, \cdot\,), u \rangle_{m,h} ,$$

where $\mathcal{R}_{mh}$ is a reproducing kernel of the space $W^{m,2}(0,1)$. After integration by parts, making good use of the boundary conditions of (7.4), the representation above is seen to equal

$$\begin{aligned} u(x) &= \int_0^1 \mathcal{R}_{mh}(x,y) \{ (-h^2)^m u^{(2m)}(y) + u(y) \} \, dy \\ &= \int_0^1 \mathcal{R}_{mh}(x,y) \, w(y) \, dy . \end{aligned}$$

Thus, $\Gamma_{mh}(x,y) = \mathcal{R}_{mh}(x,y)$. The symmetry of $\Gamma_{mh}(x,y)$ follows.

The fundamental solution. Consider the boundary value problem on the line

$$\text{(7.16)} \qquad \begin{aligned} (-h^2)^m \, u^{(2m)} + u = w \quad &\text{on } (-\infty, \infty) , \\ u^{(k)}(x) \longrightarrow 0 \text{ as } |x| \to \infty , \quad &k = m, m+1, \cdots, 2m-1 , \end{aligned}$$

with $w \in L^2(\mathbb{R})$. The easiest way to solve this problem is by means of Fourier transforms; see Volume I, Appendix 2. Letting

$$\widehat{u}(\omega) = \int_{\mathbb{R}} u(x) \, \mathrm{e}^{-2\pi i \omega x} \, dx ,$$

one obtains

$$\widehat{u}(\omega) = \frac{\widehat{w}(\omega)}{1 + (2\pi h \omega)^{2m}} , \qquad \omega \in \mathbb{R} ,$$

and consequently u is given as a convolution, $u = \mathfrak{B}_{m,h} * w$, with

$$\widehat{\mathfrak{B}}_{m,h}(\omega) = \left(1 + (2\pi h \omega)^{2m} \right)^{-1} .$$

It follows that $\mathfrak{B}_{m,h}(x-y)$ is the Green's function for the boundary value problem (7.16) and a fundamental solution for (7.4). Exercise (7.13) supplies the required properties of $\mathfrak{B}_{m,h}$.

All homogeneous solutions. Consider the differential equation

$$\text{(7.17)} \qquad (-h^2)^m \, u^{(2m)} + u = 0 \quad \text{on } (0,1) .$$

The homogeneous solutions are of the form

$$u(x) = \exp(\, i \lambda x)$$

for suitable constants λ. Substituting this into the differential equation shows that λ must satisfy $(h\lambda)^{2m} + 1 = 0$, and one verifies that the solutions are given by $\lambda = -i\,h^{-1}\varpi_\ell$, $0 \leqslant \ell \leqslant 2m-1$. This gives the homogeneous solutions

$$u_\ell(x) = \exp(\,h^{-1}\varpi_\ell\, x\,)\ , \quad \ell = 0, 1, \cdots, 2m-1\ .$$

It is useful to scale the u_ℓ such that

$$\max_{x \in [\,0\,,\,1\,]} |\,u_\ell(x)\,| = 1\ ,$$

with the maximum occurring at either $x = 0$ or $x = 1$. This leads to the $2\,m$ homogeneous solutions $\varphi_{\ell,h}$, $\ell = 0, 1, \cdots, 2m-1$.

Since these solutions are obviously linearly independent, they are a basis for the set of all homogeneous solutions of the differential equation (7.17).

Taking care of the boundary conditions. We now construct the Green's function as a linear combination of the fundamental solution and the basic homogeneous solutions in the form

$$\Gamma_{mh}(x, y) = \mathfrak{B}_{m,h}(x-y) + \sum_{\ell=0}^{2m-1} h^{-1}\varphi_{\ell,h}(x)\, a_{\ell,h}(y)\ . \tag{7.18}$$

The coefficients $a_{\ell,h}(y)$ are to be determined such that the boundary conditions of (7.4) are satisfied. This leads to the system of linear equations

$$\begin{aligned} &\mathfrak{B}_m^{(k)}\big(h^{-1}(x-y)\big) + \sum_{\ell=0}^{2m-1} \varpi_\ell^k\, \varphi_{\ell,h}(x)\, a_{\ell,h}(y) = 0 \\ &\text{for } x = 0, 1, \text{ and } k = m, m+1, \cdots, 2m-1\ . \end{aligned} \tag{7.19}$$

We must show that the $a_{\ell,h}$ exist, so that $\Gamma_{mh}(x, y)$ may indeed be represented by (7.18) and so that the bounds of Lemma (7.11) apply.

The bounds of the $a_{\ell,h}$. We must study the system (7.19). Note that it is reasonable to partition it into two blocks of equations, corresponding to the boundary conditions at $x = 0$ and $x = 1$. It turns out that, for $h \to 0$, this partitioning amounts to an asymptotic decoupling, and two $m \times m$ systems of equations result with coefficient matrices independent of h. The existence of the solution, as well as the bounds on them, may then be read off.

We now write (7.19) in matrix vector notation and implement the partitioning. Thus, we partition the unknown $a_{\ell,h}$ into two blocks,

$$b_0 = \begin{bmatrix} a_{0,h}(y) \\ a_{1,h}(y) \\ \vdots \\ a_{m-1,h}(y) \end{bmatrix}\ , \quad b_1 = \begin{bmatrix} a_{m,h}(y) \\ a_{m+1,h}(y) \\ \vdots \\ a_{2m-1,h}(y) \end{bmatrix}\ , \tag{7.20}$$

and do likewise for the right-hand sides,

$$\text{rhs}_0 = \begin{bmatrix} \mathfrak{B}_m^{(m)}(-h^{-1}y) \\ \mathfrak{B}_m^{(m+1)}(-h^{-1}y) \\ \vdots \\ \mathfrak{B}_m^{(2m-1)}(-h^{-1}y) \end{bmatrix} , \quad \text{rhs}_1 = \begin{bmatrix} \mathfrak{B}_m^{(m)}(h^{-1}(1-y)) \\ \mathfrak{B}_m^{(m+1)}(h^{-1}(1-y)) \\ \vdots \\ \mathfrak{B}_m^{(2m-1)}(h^{-1}(1-y)) \end{bmatrix} .$$

The coefficient matrix is partitioned as

$$A = \begin{bmatrix} P & R \\ S & Q \end{bmatrix} \tag{7.21}$$

with

$$P = \begin{bmatrix} P_{m,0} & P_{m,1} & \cdots & P_{m,m-1} \\ P_{m+1,0} & P_{m+1,1} & \cdots & P_{m+1,m-1} \\ \vdots & \vdots & \vdots & \vdots \\ P_{2m-1,0} & P_{2m-1,1} & \cdots & P_{2m-1,m-1} \end{bmatrix} , \tag{7.22}$$

and similarly for the other matrices, and

$$\begin{aligned} P_{k,\ell} &= \varpi_\ell^k &&, \quad \ell = 0, 1, \cdots, m-1, \\ Q_{k,\ell} &= \varpi_\ell^k &&, \quad \ell = m, m+1, \cdots, 2m-1, \\ S_{k,\ell} &= \varpi_\ell^k \exp(-h^{-1}\varpi_\ell) &&, \quad \ell = m, m+1, \cdots, 2m-1, \\ R_{k,\ell} &= \varpi_\ell^k \exp(-h^{-1}\varpi_\ell) &&, \quad \ell = 0, 1, \cdots, m-1, \end{aligned} \tag{7.23}$$

and $k = m, m+1, \cdots, 2m-1$. The system (7.19) then takes the form

$$\begin{bmatrix} P & R \\ S & Q \end{bmatrix} \begin{bmatrix} b_0 \\ b_1 \end{bmatrix} = - \begin{bmatrix} \text{rhs}_0 \\ \text{rhs}_1 \end{bmatrix} . \tag{7.24}$$

We now study this system more carefully. To summarize what follows, we show that, for $h \to 0$ we have essentially

$$P\,b_0 = \text{rhs}_0 , \quad Q\,b_1 = \text{rhs}_1 .$$

Now, Lemma (7.12) gives nice bounds on the right-hand sides of the equations above, and the matrices P and Q are independent of h. This just about completes the proof, but, of course, the "essentially" in the above must be precisely quantified, as we now do.

Note that by Lemma (7.12)

$$\begin{aligned} \| \text{rhs}_0 \|_\infty &\leqslant c \exp\bigl(-h^{-1}\kappa_m y\bigr) , \\ \| \text{rhs}_1 \|_\infty &\leqslant c \exp\bigl(-h^{-1}\kappa_m (1-y)\bigr) , \end{aligned} \tag{7.25}$$

as well as

$$\begin{aligned} \| R \|_\infty &\leqslant m\, c_m \exp(-h^{-1}\kappa_m) , \\ \| S \|_\infty &\leqslant m\, c_m \exp(-h^{-1}\kappa_m) , \end{aligned} \tag{7.26}$$

uniformly in h. Here, $\| \cdot \|_\infty$ denotes the max-norm on $\mathbb{R}^m$, as well as the induced matrix norm on $\mathbb{R}^{m \times m}$.

Now, the matrices P and Q, being Vandermonde matrices, see KINCAID and CHENEY (1991), are nonsingular (and they do not depend on h). It follows that, for some $h_o > 0$ and all $h < h_o$, the matrix

$$B \stackrel{\text{def}}{=} \begin{bmatrix} P^{-1} & 0 \\ 0 & Q^{-1} \end{bmatrix} A = \begin{bmatrix} I & P^{-1}R \\ Q^{-1}S & I \end{bmatrix} \tag{7.27}$$

is diagonally dominant, see Exercise (7.32) below, and so is invertible, with a bounded inverse.

The new system of equations then reads as

$$\begin{bmatrix} I & P^{-1}R \\ Q^{-1}S & I \end{bmatrix} \begin{bmatrix} b_0 \\ b_1 \end{bmatrix} = - \begin{bmatrix} P^{-1}\text{rhs}_0 \\ Q^{-1}\text{rhs}_1 \end{bmatrix} . \tag{7.28}$$

and the new right-hand sides satisfy the same bounds as before. It follows that

$$\sup_{h<h_o} \| b_0 \|_\infty < \infty \quad , \quad \sup_{h<h_o} \| b_1 \|_\infty < \infty . \tag{7.29}$$

Moreover, from (7.28),

$$b_0 = - P^{-1}\text{rhs}_0 - P^{-1} R \, b_1 . \tag{7.30}$$

Now, the bound (7.29) on b_1 and the bound (7.26) on R imply that, for all $y \in [\,0\,,1\,]$,

$$\begin{aligned} \| b_0 \|_\infty &= \mathcal{O}\big(\exp(h^{-1}\kappa_m y) \big) + \mathcal{O}\big(\exp(h^{-1}\kappa_m) \big) \\ &= \mathcal{O}\big(\exp(h^{-1}\kappa_m y) \big) . \end{aligned} \tag{7.31}$$

A similar derivation applies to b_1. Q.e.d.

(7.32) EXERCISE. (a) Show that P and Q in (7.21) are nonsingular.
(b) Show that the matrix B in (7.27) is diagonally dominant, i.e., there exists a constant $r < 1$ and such that all $0 < h < h_o$,

$$\| P^{-1} R \|_\infty \leqslant r \;, \quad \| Q^{-1} S \|_\infty \leqslant r \;,$$

and that this implies

$$\sup_{h<h_o} \| B \|_\infty \leqslant (\,1 - r\,)^{-1} \,,$$

as well as (7.28). [Hint: For (a), show that $P^T z = 0$ implies $z = 0$ by interpreting the statement $P^T z = 0$ as saying that a certain polynomial of degree $\leqslant m-1$ has m distinct zeros. Part (b) is an example of the Banach *contraction principle.*]

Finally, we give the proof of Lemma (7.2).

PROOF OF LEMMA (7.2). This proof is entirely along the lines of the material in §§ 13.3 and 13.4 regarding smoothing spline problems. However, as a simplification, note that by linearity we may assume that $f_o = 0$, so that $\mathbb{E}[\, f^{nh}\,] = 0$ and $\mathbb{E}[\,\psi^{nh}\,] = 0$.

Let $\varepsilon = f^{nh} - \psi^{nh}$. From the (in)equalities of the Quadratic Behavior Lemma (13.3.1) for the smoothing spline problem (13.1.7) and the C-spline problem (13.1.10), we obtain that

$$(7.33) \qquad \| \,\varepsilon\, \|^2 + 2\, h^{2m} \, \| \,\varepsilon^{(m)}\, \|^2 + \tfrac{1}{n} \sum_{i=1}^{n} |\, \varepsilon(x_{in})\,|^2 \leqslant \text{rhs} \; ,$$

with

$$\text{rhs} = \tfrac{1}{n} \sum_{i=1}^{n} \bigl\{ \,|\, \psi^{nh}(x_{in})\,|^2 - |\, f^{nh}(x_{in})\,|^2 \,\bigr\} - \bigl\{ \,\|\, \psi^{nh}\,\|^2 - \|\, f^{nh}\,\|^2 \,\bigr\} \; .$$

Using the asymptotic uniformity of the design, see Definition (13.2.22), we obtain, for a suitable constant c

$$\text{rhs} \leqslant c\, n^{-1} \, \| \, \{ (f^{nh})^2 - (\psi^{nh})^2 \,\}' \, \|_{L^1(0,1)}$$

and so, with Cauchy-Schwarz, and with the same constant c

$$\text{rhs} \leqslant c\, n^{-1} \bigl\{ \,\| \,\psi^{nh} + f^{nh}\, \| \, \| \,\varepsilon'\, \| + \| \,(\psi^{nh} + f^{nh})'\, \| \, \| \,\varepsilon\, \| \,\bigr\} \; .$$

This implies $\text{rhs} \leqslant c\,(nh)^{-1} \, \| \,\psi^{nh} + f^{nh}\, \|_{1,h} \, \| \,\varepsilon\, \|_{1,h}$, and so

$$(7.34) \qquad \text{rhs} \leqslant c_m \,(nh)^{-1} \, \| \,\psi^{nh} + f^{nh}\, \|_{m,h} \, \| \,\varepsilon\, \|_{m,h} \; .$$

Together with (7.33), this yields

$$(7.35) \qquad \| \,\varepsilon\, \|_{mh} \leqslant c\,(nh)^{-1} \bigl\{ \,\| \, f^{nh}\, \|_{mh} + \| \,\psi^{nh}\, \|_{mh} \,\bigr\} \; .$$

Now, from Theorem (13.4.6) applied with $f_o = 0$,

$$(7.36) \qquad \begin{aligned} \| \, f^{nh}\, \|_{m,h} &\leqslant c \, \Bigl\| \, \tfrac{1}{n} \sum_{i=1}^{n} d_{in} \, \Gamma_{mh}(x_{in}, \,\cdot\,) \, \Bigr\|_{m,h} \\ &\leqslant c \, \Bigl\| \, \tfrac{1}{n} \sum_{i=1}^{n} d_{in} \, \Gamma_{mh}(x_{in}, \,\cdot\,) \, \Bigr\|_{\infty} + \\ &\qquad c\, h^m \, \Bigl\| \, \tfrac{1}{n} \sum_{i=1}^{n} d_{in} \, \Gamma_{mh}^{(m)}(x_{in}, \,\cdot\,) \, \Bigr\|_{\infty} \; , \end{aligned}$$

where $\Gamma_{mh}^{(m)}(x_{in}, \,\cdot\,)$ denotes the derivative with respect to the second argument. So, by Theorem (6.12), almost surely,

$$(7.37) \qquad \limsup_{n\to\infty} \; \sup_{h \in \mathcal{H}_n(\alpha)} \; \frac{\| \, f^{nh}\, \|_{m,h}}{\sqrt{(nh)^{-1} \log n}} < \infty \; .$$

The analogous bounds hold for ψ^{nh}. With (7.35), the lemma follows. Q.e.d.

(7.38) EXERCISE. (a) Fill in the details of (7.36).
(b) Prove the analogue of (7.37) for ψ^{nh}, the C-spline estimator.

(7.39) EXERCISE. Make sure of all the details for a proof of Theorem (7.1).

EXERCISES: (7.13), (7.15), (7.32), (7.38), (7.39).

8. Additional notes and comments

Ad § 2: For the use of the bias reduction principle to construct boundary corrections for the Nadaraya-Watson estimator, see EUBANK and SPECKMAN (1990b).

Ad § 3: For more extensive work on boundary kernels, see MÜLLER (1993a, 1993b), MÜLLER and STADTMÜLLER (1999), and references therein.

Ad § 4: Regarding boundary corrections and boundary kernels, RICE (1984a) must be mentioned also.

Ad § 5: HÄRDLE, JANSSEN, and SERFLING (1988) prove the asymptotic order of the decay on the uniform error for the Nadaraya-Watson kernel estimator, including random designs. EINMAHL and MASON (2000, 2005) provide the almost sure limiting behavior using modern methods surrounding the notion of metric entropy, that are outside the scope of this work. DEHEUVELS and MASON (2004) provide the precise (in probability) limiting behavior for a weighted uniform error, including the constant, for random sequences of the smoothing parameter that lie in the "correct" range (in probability). CSÖRGŐ and RÉVÉSZ (1981, Section 6.3) determine the asymptotic distribution of the uniform error for nearest neighbor regression estimators (not covered in this text).

Ad § 6: The big surprise here is that the "elementary" Bernstein inequality is sufficiently powerful to get the stated results.

Ad § 7: The comparison of the smoothing spline estimator with the C-spline estimator typically goes by the name of the "equivalent kernel" interpretation of the spline. Since we are interested in uniform errors, we need a few more properties of the equivalent kernel than are usually discussed in the literature. See COX (1981), SILVERMAN (1984), MESSER (1991), COX (1984), NYCHKA (1995), and CHIANG, RICE, and WU (2001). The case of quasi-uniform random designs is treated in Chapter 21.

Regarding Green's functions and fundamental solutions, see, e.g., STAKGOLD (1967), and for the connection with reproducing kernels, see, e.g., MESCHKOWSKI (1962).

15

Sieves

1. Introduction

In this chapter, we study nonparametric regression estimators based on sieves. Here, a sieve is taken to be a nested sequence of finite-dimensional subspaces of the ambient L^2 space. This is somewhat different from the alternative interpretation of a sieve as a nested sequence of compact subsets of the L^2 space; see § 12.2. Either way, a sieved estimator is defined as the solution to a minimization problem, e.g., least-squares or maximum likelihood, with the solution constrained to be a member of the sieve.

To avoid more generalities, we again consider the Gauss-Markov model

$$y_{in} = f_o(x_{in}) + d_{in} \; , \quad i = 1, 2, \cdots , n \; , \tag{1.1}$$

with the noise $d_n = (\, d_{1,n}, d_{2,n}, \cdots , d_{n,n})^T$ satisfying

$$\mathbb{E}[\, d_n \,] = 0 \; , \quad \mathbb{E}[\, d_n d_n^T \,] = \sigma^2 \, I \; , \tag{1.2}$$

and with an asymptotically uniform design, e.g., $x_{in} = z_{in}$, with

$$z_{in} = \frac{i-1}{n-1} \; , \quad i = 1, 2, \cdots , n \; . \tag{1.3}$$

Regarding the unknown function f_o, the customary smoothness condition is imposed; that is, for some integer $m \geqslant 1$,

$$f_o \in W^{m,2}(0,1) \; . \tag{1.4}$$

On to sieves. A sieve is a nested sequence of finite-dimensional subspaces of continuous functions,

$$V_1 \subset V_2 \subset \cdots \subset V_r \subset \cdots \subset L^2(0,1) \; , \tag{1.5}$$

that is dense in $L^2(0,1)$; i.e., for all $f \in L^2(0,1)$,

$$\lim_{r \to \infty} \inf \{ \, \| \, v - f_o \, \| : v \in V_r \, \} = 0 \; . \tag{1.6}$$

Thus, any $f \in L^2(0,1)$ can be approximated to arbitrary accuracy by some element from some subspace V_r. Typical examples of sieves are the

P.P.B. Eggermont, V.N. LaRiccia, *Maximum Penalized Likelihood Estimation*,
Springer Series in Statistics, DOI 10.1007/b12285_4,

polynomial sieve and the sieve of step functions, discussed below. The *sieved* least-squares estimator of f_o is now defined as the solution v^{nr} of

$$\text{(1.7)} \qquad \begin{aligned} &\text{minimize} \quad \tfrac{1}{n} \sum_{i=1}^{n} |\, v(x_{in}) - y_{in} \,|^2 \\ &\text{subject to} \quad v \in V_r \ . \end{aligned}$$

The notation v^{nr} is analogous to the notations f^{nh} and φ^{nh} in the context of spline or kernel estimation. In particular, the parameter r, indicative of the dimension of V_r, plays the same role as the smoothing parameter h in the roughness-penalized least-squares spline estimator.

Of course, the usual questions must be answered. Existence and uniqueness usually not being a problem, we turn our attention to the error in the estimators. From the theory of linear regression, we know that the variance depends only on the dimension of the subspaces,

$$\text{(1.8)} \qquad \mathbb{E}\Big[\, \tfrac{1}{n} \sum_{i=1}^{n} \big|\, v^{nr}(x_{in}) - \mathbb{E}[\, v^{nr}(x_{in}) \,] \,\big|^2 \Big] = \frac{\sigma^2 \dim(V_r)}{n} \ ,$$

regardless of the particular choice of the subspaces V_r. Thus, the variance tends to ∞ as $r \to \infty$. The bias typically behaves as

$$\text{(1.9)} \qquad \tfrac{1}{n} \sum_{i=1}^{n} \big|\, f_o(x_{in}) - \mathbb{E}[\, v^{nr}(x_{in}) \,] \,\big|^2 \approx \inf\{\, \|\, v - f_o \,\|^2 : v \in V_r \,\} \ ,$$

and this tends to 0 as $r \to \infty$ by the defining property (1.6) of a sieve. Consequently, a proper balance between the conflicting goals of a small bias and a small variance must be struck. In particular, data-driven methods for choosing r are essential, but in this chapter only deterministic choices are considered.

The prototypical example of a sieve is the polynomial sieve,

$$\text{(1.10)} \qquad \mathcal{P}_1 \subset \mathcal{P}_2 \subset \cdots \mathcal{P}_r \subset \cdots \subset L^2(0,1) \ ,$$

where $\mathcal{P}_r$ is the vector space of all polynomials of degree $\leqslant r-1$ (or order $\leqslant r$). Thus, any $p \in \mathcal{P}_r$ may be written as

$$\text{(1.11)} \qquad p(x) = \sum_{k=0}^{r-1} p_k \, x^k \ , \quad x \in [\,0\,,1\,] \ ,$$

where $p_k \in \mathbb{R}$ for all k. Note that $\dim(\mathcal{P}_r) = r$. The denseness property (1.6) holds by virtue of the Weierstrass approximation theorem; see, e.g., Feinerman and Newman (1974) or Shapiro (1969). The sieved least-squares estimation problem is defined analogously to (1.7) or, more precisely, as a special case of it. We denote the polynomially sieved estimator by p^{nr}. The existence and, for $n > r$, the uniqueness of p^{nr} are easy to establish: It is not even posed as an exercise! As to convergence rates in § 2 we show that

$$\text{(1.12)} \qquad \mathbb{E}\Big[\,\tfrac{1}{n}\sum_{i=1}^{n} |\,p^{nr}(x_{in}) - f_o(x_{in})\,|^2\,\Big] = \mathcal{O}\big(\,n^{-2m/(2m+1)}\,\big)\ ,$$

provided $r \asymp n^{1/(2m+1)}$ (deterministically). This is the same optimal rate as for spline and kernel estimators. Indeed, asymptotically, there is nothing that distinguishes these types of estimators. However, in the small-sample case, the polynomial sieve seems to have definite disadvantages. Apparently, this may be traced to the fact that a discrete smoothing parameter, such as the degree of a polynomial or the dimension of a subspace, does not offer enough flexibility in fitting the data.

Of course, the natural way to increase the flexibility of polynomially sieved estimators is to consider subspaces of piecewise polynomial functions. This may be achieved by dividing the interval $[\,0\,,\,1\,]$ into two halves,

$$[\,0\,,\,1\,] = \omega_1 \cup \omega_2\ ,$$

and by replacing the subspaces $\mathcal{P}_r$ by

$$\text{(1.13)} \qquad \mathcal{P}_{r_1,r_2} = \big\{\, f \in C^{s-1}[\,0\,,\,1\,] : f\,\big|_{\omega_j} \in \mathcal{P}_{r_j}\ ,\ j = 1,2\,\big\}\ ;$$

i.e., the subspaces of piecewise polynomial functions that fit smoothly (up to order s) together. One verifies that

$$\text{(1.14)} \qquad \dim(\,\mathcal{P}_{r_1,r_2}\,) = r_1 + r_2 - s\ .$$

This process may be continued recursively until we get to the terminal point of many short subintervals with piecewise constant polynomials or piecewise linear, continuous functions. Alternatively, this process may be implemented in a data-driven way to achieve adaptation to the local smoothness of f_o ; see, e.g., DONOHO and JOHNSTONE (1998) and references therein. This should be especially effective when the function to be estimated exhibits several scales of change. Then, when possible, it is advantageous to go one step further by introducing adaptive designs, in which the data are gathered in stages: At each stage, the current estimator of f_o may be used in deciding where to put the new design points. In § 5, we study this idea in a simple setting with deterministic designs to illustrate what can theoretically be achieved this way.

We discuss two more examples of sieves. The sieve of trigonometric polynomials is studied as an example of a "natural" sieve that does not work: Regardless of smoothness, the expected squared L^2 error can be as large as $n^{-1/2}$ (rather than close to n^{-1} as one might hope). Surprisingly, this is due to bad boundary behavior, and we spend considerable effort in constructing boundary corrections following EUBANK and SPECKMAN (1990a). As an example of a good sieve, we study the sieve of natural splines of a fixed order with a variable number of (variable) knots.

The remainder of this chapter is put together as follows. In § 2, the polynomial sieve is studied in detail. In particular, a proof of the error bound (1.12) is presented. Also, the difficulties of obtaining optimal uniform error bounds are briefly addressed. The trigonometric-polynomial sieve and spline sieve are discussed in §§ 4 and 5. In § 6, the piecewise polynomial setup is discussed. Derivatives are discussed in § 3.

2. Polynomials

In this section, we study least-squares estimation with the polynomial sieve

$$\mathcal{P}_1 \subset \mathcal{P}_2 \subset \cdots \subset \mathcal{P}_r \subset \cdots \subset L^2(0,1) , \tag{2.1}$$

where $\mathcal{P}_r$ is the vector space of all polynomials of degree $\leqslant r-1$. So, in the common representation, any $p \in \mathcal{P}_r$ may be written as

$$p(x) = \sum_{k=0}^{r-1} p_k x^k , \quad x \in [\,0\,,1\,] , \tag{2.2}$$

with real coefficients $p_0, p_1, \cdots, p_{r-1}$. We also refer to r as the *order* of the polynomials. The polynomially sieved estimator p^{nr} is defined as the solution to the problem,

$$\text{minimize} \quad \tfrac{1}{n} \sum_{i=1}^{n} |\, p(x_{in}) - y_{in}\,|^2 \quad \text{subject to} \quad p \in \mathcal{P}_r \, . \tag{2.3}$$

Obviously, solutions always exist, and for $n \geqslant r$ the solution is unique.

It turns out that the polynomially sieved estimator admits the usual convergence rates under the usual nonparametric assumptions.

(2.4) THEOREM. *Let $m \geqslant 1$ and $f_o \in W^{m,2}(0,1)$. Under the Gauss-Markov model* (1.1)–(1.2), *the polynomially sieved estimator p^{nr} satisfies*

$$\mathbb{E}[\,\|\, p^{nr} - f_o \,\|^2\,] = \mathcal{O}\big(\, n^{-2m/(2m+1)} \,\big) ,$$

$$\mathbb{E}\Big[\, \tfrac{1}{n} \sum_{i=1}^{n} |\, p^{nr}(x_{in}) - f_o(x_{in})\,|^2 \,\Big] = \mathcal{O}\big(\, n^{-2m/(2m+1)} \,\big) ,$$

provided $r \asymp n^{1/(2m+1)}$ deterministically.

Note that the order of the polynomial acts as a smoothing parameter and that the "optimal" one depends on m, the degree of smoothness of f_o. This should be contrasted with spline estimators, where the smoothness also enters as the order of the spline. Similar observations apply to kernel and local polynomial estimators; see the next chapter.

How would one go about proving the theorem? No doubt, under the Gauss-Markov model (1.1)–(1.2), the starting point is the nice formula for the variance of the estimator.

(2.5) LEMMA. *Let* $p_{rn} = \mathbb{E}[\, p^{nr} \,]$. *Then*

$$\mathbb{E}\Big[\, \tfrac{1}{n} \sum_{i=1}^{n} |\, p^{nr}(x_{in}) - p_{rn}(x_{in}) \,|^2 \,\Big] = \frac{r\,\sigma^2}{n} \; .$$

Moreover, for $r^2/n \to 0$ *and asymptotically uniform designs,*

$$\mathbb{E}[\, \|\, p^{nr} - p_{rn} \,\|^2 \,] = \frac{r\,\sigma^2}{n} \, \big(1 + o(1) \big) \; .$$

PROOF. The first part of the lemma is obvious. To prove the second part, let $p \in \mathcal{P}_r$. Observe that, for asymptotically uniform designs, Definition (13.2.22) gives

$$\Big|\, \tfrac{1}{n} \sum_{i=1}^{n} |\, p(x_{in}) \,|^2 - \|\, p \,\|^2 \,\Big| \leqslant c\, n^{-1} \,\|\, (p^2)' \,\|_1 \; ,$$

and, of course, $\|\, (p^2)' \,\|_1 \leqslant 2 \,\|\, p \,\| \,\|\, p' \,\|$. Here, we need a bound on $\|\, p' \,\|$ in terms of $\|\, p \,\|$. In Lemma (2.6) below, we prove that $\|\, p' \,\| \leqslant c\, r^2 \,\|\, p \,\|$ for a suitable constant c independent of r and p. Thus, for another constant c,

$$\Big|\, \tfrac{1}{n} \sum_{i=1}^{n} |\, p(x_{in}) \,|^2 - \|\, p \,\|^2 \,\Big| \leqslant c\, n^{-1} r^2 \,\|\, p \,\|^2 \; .$$

Now, apply this with $p = \varepsilon = p^{nr} - p_{rn}$. Then,

$$\|\, \varepsilon \,\|^2 \leqslant \tfrac{1}{n} \sum_{i=1}^{n} |\, \varepsilon(x_{in}) \,|^2 + c\, n^{-1}\, r^2 \,\|\, \varepsilon \,\|^2 \; .$$

For $r^2/n \to 0$, the second part of the lemma follows. Q.e.d.

(2.6) LEMMA. *There exists a constant* c *such that, for all* $r \geqslant 1$ *and all polynomials* $p \in \mathcal{P}_r$, *we have* $\|\, p' \,\| \leqslant c\, r^2 \,\|\, p \,\|$.

PROOF. Let $r \geqslant 1$, and take $p \in \mathcal{P}_r$. From (13.6.29), we obtain

$$\|\, p' \,\|^2 \leqslant c \,\{\, \|\, p \,\|^2 + |\, p \,|_{\mathcal{W}_2}^2 \,\} \; .$$

Now, write p as $p = \sum_{k<r} \widehat{p}_k \, Q_k$, for suitable $\widehat{p}_k$, with Q_k the scaled Legendre polynomials; see (13.6.13). Then, $\|\, p \,\|^2 = \sum_{k<r} |\, \widehat{p}_k \,|^2$ and

$$|\, p \,|_{\mathcal{W}_2}^2 = \sum_{k<r} |\, \widehat{p}_k \,|^2 \,|\, Q_k \,|_{\mathcal{W}_2}^2 \leqslant c \sum_{k<r} |\, \widehat{p}_k \,|^2 \, k^4 \leqslant c\, r^4 \,\|\, p \,\|^2 \; . \qquad \text{Q.e.d.}$$

We now discuss the bias $\|\, p_{rn} - f_o \,\|^2$ as well as its discrete analogue. The final result is as follows.

(2.7) LEMMA. *Let $m \geqslant 1$. For asymptotically uniform designs, there exists a constant c such that, for all $f_o \in W^{m,2}(0,1)$,*

$$\frac{1}{n} \sum_{i=1}^{n} | p_{rn}(x_{in}) - f_o(x_{in}) |^2 \leqslant c\, r^{-2m} \| f_o \|^2_{W^{m,2}(0,1)}$$

and
$$\| p_{rn} - f_o \|^2 \leqslant c\, r^{-2m} \| f_o \|^2_{W^{m,2}(0,1)} ,$$

provided r/n is small enough.

The remainder of this section deals with its proof. It is obvious that $p_{rn} = \mathbb{E}[\, p^{nr}\,]$ is the solution to the problem

$$\text{(2.8)} \qquad \begin{array}{ll} \text{minimize} & \frac{1}{n} \sum_{i=1}^{n} | p(x_{in}) - f_o(x_{in}) |^2 \\ \text{subject to} & p \in \mathcal{P}_r , \end{array}$$

and equally obvious that we should compare this with the continuous version (minimize $\| p - f_o \|^2$ subject to $p \in \mathcal{P}_r$), but this actually leads to the following problem, with the solution denoted by $p = \pi_r$:

$$\text{(2.9)} \qquad \text{minimize} \quad \| p - f_o \|^2_{1,h} \quad \text{subject to} \quad p \in \mathcal{P}_r ,$$

where $h \equiv r^{-1}$. For the Sobolev space norms $\| f \|_{m,h}$, see (13.2.5). This comes about by the usual quadrature result.

(2.10) LEMMA. *Let $m \geqslant 1$. For asymptotically uniform designs, there exists a constant c such that, for all $r \in \mathbb{N}$ and $f_o \in W^{m,2}(0,1)$ (and $h \equiv r^{-1}$),*

$$\frac{1}{n} \sum_{i=1}^{n} | p^{nr}(x_{in}) - f_o(x_{in}) |^2 \leqslant \Bigl\{ 1 + \frac{c\, r}{n} \Bigr\} \| \pi_r - f_o \|^2_{1,h} .$$

PROOF. By the Quadrature Lemma (13.2.27), for all $p \in \mathcal{P}_r$,

$$\begin{aligned} \frac{1}{n} \sum_{i=1}^{n} | p(x_{in}) - f_o(x_{in}) |^2 &\leqslant \| p - f_o \|^2 + c\,(nh)^{-1} \| p - f_o \|^2_{1,h} \\ &\leqslant \bigl\{ 1 + c\,(nh)^{-1} \bigr\} \| p - f_o \|^2_{1,h} . \end{aligned}$$

Now, the left-hand side is minimized over all $p \in \mathcal{P}_r$ by $p = p_{rn}$. Then, the left-hand side is a lower bound for the right-hand side for all $p \in \mathcal{P}_r$. Thus, we may take p to be *its* minimizer. Q.e.d.

So, if we can find appropriate upper bounds on $\| \pi_r - f_o \|_{1,h}$, then we have proven Lemma (2.7). The problem (2.9) is one of (best) polynomial approximation in the Sobolev space $W^{m,2}(0,1)$. Now, in dealing with polynomials in $W^{m,2}(0,1)$, we realize that we would be much happier if the setting was the space $\mathcal{W}_m$ because then we could just compute things.

Recall the scaled Legendre polynomials Q_k, $k = 0, 1, \cdots$, of (13.6.13) and their orthogonality properties (13.6.14)–(13.6.15) in the space $\mathcal{W}_m$. The one difficulty is the weighting of the derivatives in the $\mathcal{W}_m$ norm. Here is a simple trick to get around this.

First, "squeeze" the function $f_o \in W^{m,2}(0,1)$ to live on $[\,\frac{1}{4}, \frac{3}{4}\,]$,

$$(2.11) \qquad g_o(x) = f\big(\,2(x - \tfrac{2}{4})\,\big)\ , \quad \tfrac{1}{4} \leqslant x \leqslant \tfrac{3}{4}\ .$$

Then, extend g_o by its Taylor polynomial at $x = \frac{1}{4}$ and $x = \frac{3}{4}$,

$$(2.12) \qquad g_o(x) = \sum_{k<m} \frac{f_o^{(k)}(0)}{k!}\,\big(\,2(x - \tfrac{1}{4})\,\big)^k\ , \quad 0 \leqslant x \leqslant \tfrac{1}{4}\ ,$$

and likewise on $\frac{3}{4} \leqslant x \leqslant 1$. This transformation of f_o is denoted by Sf_o,

$$(2.13) \qquad Sf_o = g_o \quad \text{on} \quad [\,0\,,\,1\,]\ .$$

Note that $f_o^{(k)}(0)$, $k = 0, 1, \cdots, m-1$, are well-defined for $f \in W^{m,2}(0,1)$ and that Sf_o is $m-1$ times continuously differentiable on $[\,0\,,\,1\,]$. In fact, $Sf_o \in W^{m,2}(0,1)$. Moreover, $(Sf_o)^{(m)} = 0$ outside $[\,\frac{1}{4}\,,\,\frac{3}{4}\,]$.

(2.14) LEMMA. *There exists a constant c_m such that, for all $f \in W^{m,2}(0,1)$,*

$$\|\, Sf\,\|_{W^{m,2}(0,1)} \leqslant c_m \,\|\, f\,\|_{W^{m,2}(0,1)}\ ,$$

and for all h, $0 < h \leqslant 1$,

$$\|\, f\,\|_{m,h} \leqslant c_m \,\|\, Sf\,\|_{h,\mathcal{W}_m}\ .$$

PROOF. Note that, by a change of integration variable,

$$(2.15) \qquad \|\, f\,\|^2_{(0,1)} = 2\,\|\, Sf\,\|^2_{(\frac{1}{4},\frac{3}{4})} \leqslant 2\,\|\, Sf\,\|^2_{(0,1)}\ ,$$

where we revived the notation (13.2.9) for $L^2(a,b)$ integrals. Likewise,

$$(2.16) \quad \|\, f^{(m)}\,\|^2_{(0,1)} = 2^{1-2m}\,\|\,(\,Sf\,)^{(m)}\,\|^2_{(\frac{1}{4},\frac{3}{4})}$$

$$\leqslant c_m \int_{\frac{1}{4}}^{\frac{3}{4}} \{\, x(1-x)\,\}^m \,|\,(Sf)^{(m)}(x)\,|^2\,dx = c_m\,|\,Sf\,|^2_{\mathcal{W}_m}\ ,$$

where $c_m = 2^{1-2m}\,(16/3)^m$. So, for all $f \in W^{m,2}(0,1)$, it follows that

$$\|\, f\,\|^2 + h^{2m}\,\|f^{(m)}\,\|^2 \leqslant 2\,\|\, Sf\,\|^2 + c_m\,h^{2m}\,|\,Sf\,|^2_{\mathcal{W}_m}\ .$$

The second inequality of the lemma follows.

For the first inequality, recall that $(Sf)^{(m)} = 0$ outside $[\,\frac{1}{4}, \frac{3}{4}\,]$, and so

$$\|\,(Sf)^{(m)}\,\|^2 = \|\,(Sf)^{(m)}\,\|^2_{(\frac{1}{4},\frac{3}{4})} = 2^{2m-1}\,\|\, f^{(m)}\,\|^2$$

by (2.16). By (2.15), one gets

$$\|\, Sf\,\|^2_{(\frac{1}{4},\frac{3}{4})} = \tfrac{1}{2}\,\|\, f\,\|^2\ .$$

Finally, by virtue of the Point Evaluation Lemma (13.2.11),

$$\| Sf \|_{(0,\frac{1}{4})} \leqslant \sum_{k<m} | f^{(k)}(0) | \int_0^{\frac{1}{4}} \big(2(x-1/4) \big)^{2k} \, dx \leqslant c_m \| f \|_{W^{m,2}(0,1)} .$$

The same bound holds for $\| Sf \|^2_{(\frac{3}{4},1)}$. The first inequality of the lemma follows. Q.e.d.

For polynomials, there is a more convenient form of the squeeze operation. Define the operator T on the set of all polynomials by

$$(2.17) \qquad [Tp](x) = p\big(2(x - \tfrac{1}{4}) \big) , \quad 0 \leqslant x \leqslant 1 .$$

(2.18) LEMMA. *There exists a constant* c_m *such that, for all* $f \in W^{m,2}(0,1)$,

$$\min_{p \in \mathcal{P}_r} \| p - f \|^2_{m,h} \leqslant c_m \min_{p \in \mathcal{P}_r} \| p - Sf \|^2_{h,\mathcal{W}_m} .$$

PROOF. By a slight modification of the proof of Lemma (2.14), we get that

$$\| p - f \|^2 \leqslant 2 \| Tp - Sf \|^2 \quad \text{and} \quad \| (p-f)^{(m)} \|^2 \leqslant c_m \, | Tp - Sf |^2_{\mathcal{W}_m} ,$$

so that, for all $p \in \mathcal{P}_r$,

$$\| p - f \|^2_{1,h} \leqslant \tfrac{8}{3} \| Tp - Sf \|^2_{h,\mathcal{W}_1} .$$

Now, minimizing first the left-hand side and only then the right-hand side does the trick. Note that $T(\mathcal{P}_r) = \mathcal{P}_r$. Q.e.d.

Since Lemma (2.14) implies that $Sf_o \in \mathcal{W}_m$, the final task is to bound

$$\min_{p \in \mathcal{P}_r} \| p - g \|_{h,\mathcal{W}_1} \leqslant \min_{p \in \mathcal{P}_r} \| p - g \|^2_{h,\mathcal{W}_m}$$

for arbitrary $g \in \mathcal{W}_m$, where the inequality is from Lemma (13.6.26).

(2.19) LEMMA. *Let* $m \geqslant 1$. *There exists a constant* c_m *such that, for all* $r \in \mathbb{N}$, *and* $g \in \mathcal{W}_m$ *(with* $h \equiv r^{-1}$ *),*

$$\min_{p \in \mathcal{P}_r} \| p - g \|^2_{h,\mathcal{W}_m} \leqslant c_m \, r^{-2m} \, | g |^2_{\mathcal{W}_m} .$$

PROOF. The proof is computational. We prove the bounds above for the norm $||| \cdot |||_{h,\mathcal{W}_m}$ of (13.6.20), which are equivalent to the norms $\| \cdot \|_{h,\mathcal{W}_m}$, uniformly in h, $0 < h \leqslant 1$. Now, let $g \in \mathcal{W}_m$. Then, see Lemma (13.6.19),

$$g = \sum_{k=1}^{\infty} \widehat{g}_k \, Q_k , \quad ||| g |||^2_{h,\mathcal{W}_m} = \sum_{k=0}^{\infty} \big(1 + (2hk)^{2m} \big) \, | \widehat{g}_k |^2 .$$

The best polynomial approximant to g is $p = \sum_{k<r} \widehat{g}_k \, Q_k$, and then

$$\begin{aligned} ||| \, p - g \, |||^{\,2}_{h,\mathcal{W}_m} &= \sum_{k \geqslant r} \big(1 + (2hk)^{2m} \big) \, | \, \widehat{g}_k \, |^2 \\ &\leqslant c_{m,r} \sum_{k \geqslant r} (2k)^{2m} \, | \, \widehat{g}_k \, |^2 \leqslant c_{m,r} \, | \, g \, |^{\,2}_{\mathcal{W}_m} \, , \end{aligned}$$

where

$$c_{m,r} = \sup_{k \geqslant r} \, \frac{1 + (2k/r)^{2m}}{(2k)^{2m}} = \sup_{k \geqslant r} \, \frac{1 + (2k/r)^{2m}}{r^{2m} \, (2k/r)^{2m}} \leqslant \sup_{t \geqslant 2} \, \frac{1 + t^{2m}}{r^{2m} \, t^{2m}} \, .$$

The supremum occurs at $t = 2$. So, $c_{m,r} \leqslant r^{-2m} \, (1 + 2^{-4m})$. Q.e.d.

(2.20) EXERCISE. Assemble all the pieces for a proof of Lemma (2.7).

(2.21) EXERCISE. Prove the following elaboration of Lemma (2.19):

$$\min_{p \in \mathcal{P}_r} \, \| \, p - g \, \|^{\,2}_{r^{-1},\mathcal{W}_d} \leqslant c \, r^{-2m} \, | \, g \, |^{\,2}_{\mathcal{W}_m} \, , \qquad d = 1, 2, \cdots , m \, .$$

EXERCISES: (2.20), (2.21).

3. Estimating derivatives

It seems that we are in a bind when it comes to estimating derivatives in the context of the polynomial sieve. Of course, it is clear how to estimate f_o', viz. by the derivative of the estimator of f_o, but how should one go about obtaining error bounds? Compared with the case of smoothing splines, there is an obvious need for a family of norms involving derivatives. As might be expected, the "natural" candidates are the $| \cdot |_{\mathcal{W}_m}$ norms, already put to good use in § 2. This realization is in fact due to COX (1988b). Indeed, in this setting, the usual results under the usual nonparametric assumptions apply.

(3.1) THEOREM. *Let $m \geqslant 2$, and suppose that $f_o \in W^{m,2}(0,1)$. Then, for $1 \leqslant d \leqslant m - 1$,*

$$\mathbb{E}\big[\, | \, p^{nr} - f_o \, |^{\,2}_{\mathcal{W}_d} \, \big] = \mathcal{O}\big(\, n^{-2(m-d)/(2m+1)} \, \big) \, ,$$

provided $r \asymp n^{1/(2m+1)}$.

We note that the theorem gives a bound on a weighted norm; in particular, there is downweighting near the endpoints. Thus, we may expect the derivatives to behave badly near the boundary. The proof of the theorem depends on the following "natural" lemma. (A comparison with Lemma (2.6) is informative.)

(3.2) LEMMA. *Let* $m \geqslant 1$. *There exists a constant* c_m *such that, for all* $r \geqslant 1$ *and all polynomials* $p \in \mathcal{P}_r$,

$$| p |_{\mathcal{W}_m} \leqslant c_m \, r^m \, \| p \| \, .$$

First, we show how the lemma above helps in proving Theorem (3.1).

PROOF OF THEOREM (3.1). Observe that

$$\mathbb{E}[\, | p^{nr} - f_o |_{\mathcal{W}_m}^2 \,] = \mathbb{E}[\, | p^{nr} - p_{rn} |_{\mathcal{W}_m}^2 \,] + \mathbb{E}[\, | p_{rn} - f_o |_{\mathcal{W}_m}^2 \,] \, ,$$

with $p_{rn} = \mathbb{E}[\, p^{nr} \,]$. Now, by Lemma (3.2),

$$\mathbb{E}[\, | p^{nr} - p_{rn} |_{\mathcal{W}_m}^2 \,] \leqslant c \, r^{2m} \, \mathbb{E}[\, \| p^{nr} - p_{rn} \|^2 \,] = c \, \sigma^2 \, r^{2m} \, n^{-1} \, r \, .$$

Next, with $h \equiv r^{-1}$,

$$\begin{aligned} | p_{rn} - f_o |_{\mathcal{W}_m}^2 &\leqslant 2 \, | p_{rn} - \pi_r |_{\mathcal{W}_m}^2 + 2 \, | \pi_r - f_o |_{\mathcal{W}_m}^2 \\ &\leqslant c \, r^{2m} \| \, p_{rn} - \pi_r \, \|^2 + 2 \, | \pi_r - f_o |_{\mathcal{W}_m}^2 \\ &\leqslant 2 \, c \, r^{2m} \, \| \, p_{rn} - f_o \, \|^2 + 2 \, c \, r^{2m} \, \| \, \pi_r - f_o \, \|^2 + 2 \, | \pi_r - f_o |_{\mathcal{W}_m}^2 \\ &\leqslant 2 \, c \, r^{2m} \, \| \, p_{rn} - f_o \, \|^2 + 2 \, c \, r^{2m} \, \| \, \pi_r - f_o \, \|_{h,\mathcal{W}_m}^2 \, , \end{aligned}$$

where we used Lemma (3.2). Now, by Lemma (2.7), the first term on the right is $\mathcal{O}(\, 1 \,)$. For the second term, Lemma (2.19) gives a $\mathcal{O}(\, 1 \,)$ bound as well. Then Lemmas (2.5) and (2.7) imply that

$$\mathbb{E}[\, \| \, p^{nr} - f_o \, \|_{h,\mathcal{W}_m} \,] = \mathcal{O}(\, n^{-1} \, r + r^{-2m} \,) \, ,$$

which is the "same" asymptotic bound as for $\mathbb{E}[\, \| \, p^{nr} - f_o \, \|^2 \,]$. Finally, for $d = 1, 2, \cdots, m-1$, Lemma (13.6.26) gives

$$\mathbb{E}[\, \| \, p^{nr} - f_o \, \|_{h,\mathcal{W}_d}^2 \,] \leqslant c \, \mathbb{E}[\, \| \, p^{nr} - f_o \, \|_{h,\mathcal{W}_m}^2 \,] \, ,$$

so that

$$\mathbb{E}[\, | p^{nr} - f_o |_{\mathcal{W}_m}^2 \,] = \mathcal{O}(\, r^{2d} \, (\, n^{-1} \, r + r^{-2m} \,) \,) \, ,$$

and the theorem follows. Q.e.d.

So, all we need to do is prove Lemma (3.2). Note that the proof is computational.

PROOF OF LEMMA (3.2). Write

$$p(x) = \sum_{k<r} \widehat{p}_k \, Q_k(x) \, , \quad x \in [\, 0 \, , \, 1 \,] \, ,$$

with Q_k as in § 2 and § 13.6. Then, $\| p \|^2 = \sum_{k<r} | \widehat{p}_k |^2$, as well as

$$| p |_{\mathcal{W}_m}^2 = \sum_{k<r} | \widehat{p}_k |^2 \, | Q_k |_{\mathcal{W}_m}^2 \leqslant c \sum_{k<r} | \widehat{p}_k |^2 \, k^{2m} \leqslant c \, r^{2m} \, \| p \|^2 \; ,$$

where we used (13.6.15). Q.e.d.

4. Trigonometric polynomials

We now turn our attention to least-squares estimation using the trigonometric sieve. At first glance, this seems but a trivial modification of the polynomial sieve, especially if one thinks of a sieve as being generated by an orthonormal basis for $L^2(0,1)$. This turns out to be wrong: The trigonometric-sieve least-squares estimator has a disappointing convergence rate, independent of the smoothness of the function being estimated. The reason is that the trigonometric basis for $L^2(0,1)$ is *not* a basis for $W^{m,2}(0,1)$, $m \geqslant 1$. This defect may be completely fixed by changing the estimation problem and/or the sieve or by using the Bias Reduction Principle, the details of which are left as exercises.

The trigonometric sieve is defined as

$$\mathcal{T}_1 \subset \mathcal{T}_2 \subset \cdots \subset \mathcal{T}_r \subset \cdots \subset L^2(0,1) \; , \tag{4.1}$$

with $\mathcal{T}_r$ the vector space of trigonometric polynomials of degree $\leqslant r$. Thus, in the common and useful representation, any $t \in \mathcal{T}_r$ may be written as

$$t\,(x) = \sum_{k=-r}^{r} c_k \, e^{2\pi i k x} \; , \quad x \in [\,0\,,\,1\,] \; , \tag{4.2}$$

with complex c_k and $c_{-k} = \overline{c_k}$ for all k. Equivalently, one may write

$$t\,(x) = \sum_{k=0}^{r} \{\, a_k \, \cos(2\pi k x) + b_k \, \sin(2\pi k x) \,\} \; , \quad x \in [\,0\,,\,1\,] \; , \tag{4.3}$$

with real coefficients a_k, b_k, but notationally (4.2) is more convenient. Note that the coefficient b_0 is not actually present in (4.3), so there are only $2r+1$ real coefficients. In other words, $\mathcal{T}_r$ as a vector space over the reals has dimension equal to $2r+1$.

As in the previous section, we study least-squares estimation for the Gauss-Markov model (1.1)–(1.2) for smooth f_o and asymptotically uniform designs. The trigonometric-sieve least-squares estimator of f_o is the solution, denoted by t^{nr}, to the problem

$$\text{minimize} \quad \tfrac{1}{n} \sum_{i=1}^{n} | \, t(x_{in}) - y_{in} \, |^2 \quad \text{subject to} \quad t \in \mathcal{T}_r \; . \tag{4.4}$$

We would like to prove the bound $\mathcal{O}\big(n^{-2m/(2m+1)} \big)$ for the mean squared error of the estimator, but this bound turns out to be wrong. This is due to the bias not being small enough. Of course, the variance is exactly as expected. To state this precisely, it is useful to introduce the noise-free version of (4.4),

$$\text{(4.5)} \qquad \text{minimize} \quad \tfrac{1}{n} \sum_{i=1}^{n} | t(x_{in}) - f_o(x_{in}) |^2 \quad \text{subject to} \quad t \in \mathcal{T}_r \ .$$

The solution is denoted by t_{rn}. Of course, $t_{rn} = \mathbb{E}[\, t^{nr} \,]$.

(4.6) THEOREM. *Let* $m \geqslant 1$ *and* $f_o \in W^{m,2}(0,1)$. (a) *The variance of* t^{nr} *satisfies*

$$\mathbb{E}\Bigl[\tfrac{1}{n} \sum_{i=1}^{n} | t^{nr}(x_{in}) - t_{rn}(x_{in}) |^2 \Bigr] = \frac{(2r+1)\,\sigma^2}{n} \ .$$

(b) *For asymptotically uniform designs, the bias satisfies*

$$\tfrac{1}{n} \sum_{i=1}^{n} | t_{rn}(x_{in}) - f_o(x_{in}) |^2 = \mathcal{O}\bigl(r^{-1} + n^{-1} r^{-1/2} \bigr) \ .$$

The variance part needs no further comment. Regarding the bias, the extra term $n^{-1} r^{-1/2}$ is somewhat of a blemish, but since it is much smaller than the variance, it has no effect on the final convergence rate. More seriously, it turns out that the stated upper bound is sharp in the sense that, in general, the term r^{-1} cannot be replaced by r^{-k} for any $k > 1$. That being the case, the upper bound on the mean squared error

$$\text{(4.7)} \qquad \mathbb{E}\Bigl[\tfrac{1}{n} \sum_{i=1}^{n} | t^{nr}(x_{in}) - f_o(x_{in}) |^2 \Bigr] = \mathcal{O}\bigl(n^{-1/2} \bigr) \ ,$$

achieved with $r \asymp n^{1/2}$, is sharp irrespective of the value of m governing the smoothness of f_o as long as $m \geqslant 1$. The conclusion is that the trigonometric sieve does not provide good estimators.

We now set out to prove Theorem (4.6)(b) and in the process pave the way for the application of the Bias Reduction Principle. The proof of the bounds on the bias follows the same plan as that for the polynomial sieve. The first part of the proof is to replace the sum by an integral but to that end we need bounds on derivatives of trigonometric polynomials.

(4.8) LEMMA. *Let* $r \geqslant 1$. *For all* $t \in \mathcal{T}_r$, *we have* $\| t' \| \leqslant 2\pi r \, \| t \|$.

PROOF. Write $t \in \mathcal{T}_r$ as $t(x) = \sum_{|k| \leqslant r} \widehat{t}_k \, e^{2\pi i k x}$. Then, $\| t \|^2 = \sum_{|k| \leqslant r} | \widehat{t}_k |^2$ and

$$\| t' \|^2 = \sum_{|k| \leqslant r} | 2\pi i k \, \widehat{t}_k |^2 \leqslant (2\pi r)^2 \, \| t \|^2 \ . \qquad \text{Q.e.d.}$$

The next step in bounding the bias is to observe the following. Let τ_r be the solution to the problem

$$\text{(4.9)} \qquad \text{minimize} \quad \| t - f_o \|_{1,h}^2 \quad \text{subject to} \quad t \in \mathcal{T}_r \ ,$$

where $h \equiv r^{-1}$. Note that τ_r is the same for $h = 0$. The comparison of the solutions of the problems (4.5) and (4.9) is now effected as follows.

(4.10) LEMMA. *Let $m \geqslant 1$. There exists a constant c such that, for all $f_o \in W^{m,2}(0,1)$, and all $r \geqslant 1$ and $t \in \mathcal{T}_r$ (with $h \equiv r^{-1}$),*

$$\tfrac{1}{n} \sum_{i=1}^{n} | \, t_{rn}(x_{in}) - f_o(x_{in}) \, |^2 \leqslant \Bigl(1 + \frac{c}{nh} \Bigr) \, \| \, \tau_r - f_o \, \|_{1,h}^2 \; .$$

(4.11) EXERCISE. Prove the lemma. [Hint: See the proof of Lemma (2.10).]

The final task is now to exhibit bounds on $\| \, \tau_r - f_o \, \|$ and $\| \, (\tau_r - f_o)' \, \|$. It is useful to observe that τ_r has an explicit representation. If

$$f_o(x) = \sum_{k \in \mathbb{Z}} (\widehat{f}_o)_k \, \mathrm{e}^{2\pi i k x} \; , \quad \text{a.e. } x \in (0,1) \; , \tag{4.12}$$

then τ_r is given by

$$\tau_r(x) = \sum_{|k| \leqslant r} (\widehat{f}_o)_k \, \mathrm{e}^{2\pi i k x} \; , \quad \text{every } x \in (0,1) \; . \tag{4.13}$$

(4.14) LEMMA. *Let $m \geqslant 1$ and assume $f_o \in W^{m,2}(0,1)$. Then*

$$\| \, \tau_r - f_o \, \|^2 \leqslant c \, r^{-1} \, \| \, f_o \, \|_{W^{1,2}(0,1)}^2 \qquad \text{and} \qquad \| \, (\, \tau_r - f_o \,)' \, \|^2 \leqslant \| \, f_o' \, \|^2 \; ,$$

and, in general, these bounds cannot be improved.

PROOF. The proof is entirely computational. Since $f_o \in W^{m,2}(0,1)$, then $f_o \in L^2(0,1)$, and so

$$f_o(x) = \sum_{k \in \mathbb{Z}} (\widehat{f}_o)_k \, \varepsilon^{2\pi i k x} \; , \qquad \text{a.e. } x \in [\, 0 \, , \, 1 \,] \; , \tag{4.15}$$

and

$$\| \, f_o \, \|^2 = \sum_{k \in \mathbb{Z}} | \, (\widehat{f}_o)_k \, |^2 < \infty \; . \tag{4.16}$$

It follows that

$$\| \, \tau_r - f_o \, \|^2 = \sum_{|\, k \,| > r} | \, (\widehat{f}_o)_k \, |^2 \; . \tag{4.17}$$

Thus, we must study the rate of decay of the Fourier coefficients $(\widehat{f}_o)_k$. Integration by parts on

$$(\widehat{f}_o)_k = \int_0^1 f_o(x) \, \mathrm{e}^{-2\pi i k x} \, dx \tag{4.18}$$

gives, for $k \neq 0$,

$$2\pi i k (\widehat{f}_o)_k = f_o(0) - f_o(1) + \varphi_k \; , \tag{4.19}$$

with

$$\varphi_k = \int_0^1 f_o'(x) \, \mathrm{e}^{-2\pi i k x} \, dx \; . \tag{4.20}$$

Since $f_o' \in L^2(0,1)$, then

$$(4.21)\qquad \sum_{|k|>r} k^{-2} |\, \varphi_k \,|^2 \leqslant r^{-2} \sum_{|k|>r} |\, \varphi_k \,|^2 \leqslant r^{-2} \, \| \, f_o' \, \|^2 \; ,$$

so that, using Exercise (4.23) below and Cauchy-Schwarz,

$$(4.22)\qquad \sum_{|k|>r} \bigl|\, (\widehat{f_o})_k \,\bigr|^2 = 2 \,|\, (2\pi)^{-1} \{\, f_o(0) - f_o(1) \,\} \,|^2 \, r^{-1} + \mathcal{O}(\, r^{-3/2} \,)$$

for $r \to \infty$. The first bound of the lemma now follows from (4.17).

Finally, for the second inequality,

$$\| \, (\, \tau_r - f_o \,)' \, \|^2 = \sum_{|k|>r} (2\pi k)^2 \,|\, (\widehat{f_o})_k \,|^2 \leqslant \| \, f_o{}' \, \|^2 \; ,$$

and this too cannot be improved upon uniformly in $f_o \in W^{m,2}(0,1)$. Q.e.d.

(4.23) Exercise. Show that $\sum_{k=r+1}^{\infty} k^{-2} = r^{-1} + \mathcal{O}(\, r^{-2} \,)$, and use this to show (4.22).

(4.24) Exercise. Complete the proof of Theorem (4.6)(b).

Now, the question is if and how we can do better than Theorem (4.6).

The trigonometric representation of $f_o \in W^{m,2}(0,1)$. Let us return to (4.15)–(4.19). Integration by parts $m-1$ more times gives

$$(4.25)\qquad (\widehat{f_o})_k = \sum_{\ell=0}^{m-1} \frac{f_o^{(\ell)}(0) - f_o^{(\ell)}(1)}{(2\pi i k\,)^{\ell+1}} + \int_0^1 \frac{f_o^{(m)} \, \mathrm{e}^{-2\pi i k x}}{(2\pi i k\,)^m} \, dx \; .$$

Substitution of (4.25) into (4.15) reveals the importance of the special Fourier series

$$(4.26)\qquad A_\ell(x) = \sum_{k \neq 0} (2\pi i k)^{-\ell} \, \mathrm{e}^{2\pi i k x} \; , \qquad \ell = 1, 2, \cdots, m-1 \; .$$

Note that, for $\ell \geqslant 2$, these Fourier series converge absolutely. For $\ell = 1$, initially one has only convergence in $L^2(0,1)$. As the following exercise makes clear, two applications of Abel's summation-by-parts lemma show that the series for $\ell = 1$ converges for all $x \neq 0, 1$.

(4.27) Exercise. Use Abel's summation-by-parts lemma to show that, for $\exp(2\pi i x) \neq 1$,

$$\sum_{k=1}^{\infty} (2\pi i k)^{-1} \, \mathrm{e}^{2\pi i k x} = -(2\pi i)^{-1} + \sum_{k=1}^{\infty} (2\pi i (k+1)\, k\,)^{-1} \, s_k(x)$$

with $s_k(x) = \{\, 1 - \mathrm{e}^{2\pi i (k+1) x} \,\} / \{\, 1 - \mathrm{e}^{2\pi i x} \,\}$. Conclude that the Fourier series (4.26) with $\ell = 1$ converges for all $x \neq 0, 1$.

The functions A_ℓ are related to the well-known Bernoulli polynomials; see, e.g., ZYGMUND (1968), Volume I, Chapter II. They are defined recursively by $B_0(x) = 1$ and

$$(4.28)\qquad \left.\begin{aligned} &B_\ell'(x) = \ell\, B_{\ell-1}(x) \\ &\int_0^1 B_\ell(x)\, dx = 0 \end{aligned}\right\} \quad \text{for } \ell = 1, 2, \cdots .$$

Explicitly, the first few polynomials are

$$(4.29)\qquad \begin{aligned} &B_1(x) = x - \tfrac{1}{2} \quad , \quad B_2(x) = x^2 - x + \tfrac{1}{6} \, , \\ &B_3(x) = x^3 - \tfrac{3}{2}\, x^2 + \tfrac{1}{2}\, x \, . \end{aligned}$$

(4.30) LEMMA. *The Fourier series of the Bernoulli polynomials are*

$$B_\ell(x) = \sum_{k \neq 0} \ell\,!\, (2\pi i k)^{-\ell}\, e^{2\pi i k x} \quad , \quad x \in [\,0\,,\,1\,] \,, \quad \ell = 1, 2, \cdots .$$

(4.31) EXERCISE. Prove the lemma. [Hint: *Compute* the Fourier coefficients of B_ℓ using integration by parts and (4.28).]

Returning to (4.25), we have proven the following lemma.

(4.32) LEMMA. *Let* $m \geqslant 1$ *and* $f_o \in W^{m,2}(0,1)$. *Then,*

$$f_o(x) = (\widehat{f}_o)_o + \sum_{\ell=0}^{m-1} \frac{f_o^{(\ell)}(0) - f_o^{(\ell)}(1)}{\ell\,!}\, B_\ell(x) + \sum_{k \neq 0} \frac{c_k\, e^{2\pi i k x}}{(2\pi i k)^m} \;,$$

where $c_k = \big(\widehat{f_o^{(m)}}\big)_k$, *the* k*-th Fourier coefficient of* $f_o^{(m)}$.

We are finally ready to determine the asymptotic behavior of $\tau_r - f_o$. First, an exercise on how well the B_ℓ can be approximated by trigonometric polynomials.

(4.33) EXERCISE. Show that $\displaystyle\min_{t \in \mathcal{T}_r} \|\, t - B_\ell \,\|^2 \asymp r^{-2\ell-1} \;, \quad r \to \infty \,.$
[Hint: Compare this with (4.13).]

(4.34) THEOREM. *Let* $m \geqslant 1$ *and* $f_o \in W^{m,2}(0,1)$. *Then, for* $r \geqslant 1$,

$$f_o(x) - \tau_r(x) = \sum_{\ell=0}^{m-1} \frac{f_o^{(\ell)}(0) - f_o^{(\ell)}(1)}{\ell\,!}\, BE_\ell(x) + \sum_{|k|>r} \frac{c_k\, e^{2\pi i k x}}{(2\pi i k)^m} \;,$$

where $c_k = \big(\widehat{f_o^{(m)}}\big)_k$ *and* $BE_\ell(x) = \displaystyle\sum_{|k|>r} \ell\,!\, (2\pi i k)^{-\ell}\, e^{2\pi i k x} \;, \quad \ell = 1, 2, \cdots .$

The necessary and sufficient conditions for $\| \tau_r - f_o \|^2$ to attain the usual rates now follow.

(4.35) COROLLARY. *Let* $m \geqslant 1$. *For asymptotically uniform designs and for all* $f_o \in W^{m,2}(0,1)$ *and* $r \geqslant 1$,

$$\| \tau_r - f_o \|^2 = \mathcal{O}(r^{-2m})$$

if and only if

$$f_o^{(\ell)}(0) = f_o^{(\ell)}(1) \ , \quad \ell = 0, 1, \cdots, m-1 \ ;$$

i.e., if and only if f_o *is periodic of order* m. *Then, for* $m \geqslant 2$, *also*

$$\| (\tau_r - f_o)^{(d)} \|^2 = \mathcal{O}(r^{-2(m-d)}) \ , \quad d = 1, 2, \cdots, m-1 \ .$$

The treatment above shows the necessity of boundary corrections of one sort or another for the trigonometric-sieve estimator. There are several ways to achieve this: The estimation problem itself may be modified by expanding the sieve or by applying the Bias Reduction Principle of EUBANK and SPECKMAN (1990a). We phrase each one as an exercise.

(4.36) EXERCISE. Consider estimating f_o in the model (1.1) through (1.4) by $p^{nm} + t^{nr}$, the (combined) solution of

$$\text{minimize} \quad \tfrac{1}{n} \sum_{i=1}^{n} | p(x_{in}) + t(x_{in}) - y_{in} |^2 \quad \text{subject to} \quad p \in \mathcal{P}_m \ , \ t \in \mathcal{T}_r \ ,$$

with $\mathcal{P}_m$ the set of all polynomials of degree $\leqslant m-1$. In effect, the subspace $\mathcal{T}_r$ is replaced by $\mathcal{P}_m + \mathcal{T}_r$.

(a) Compute the variance ($\equiv$ the expected residual sum of squares) of the estimator.

(b) Show that the bias is $\mathcal{O}(r^{-2m})$.

(c) Show the usual bound on the expected mean squared error.

(4.37) EXERCISE. For the model (1.1) through (1.4), use the Bias Reduction Principle of EUBANK and SPECKMAN (1990a) to compute boundary corrections for the estimator t^{nr} of (4.4), so that the resulting estimator has mean squared error $\mathcal{O}(n^{-2m/2m+1})$.

(4.38) EXERCISE. Investigate the sieve of cosines,

$$\mathfrak{C}_1 \subset \mathfrak{C}_2 \subset \cdots \subset \mathfrak{C}_r \subset \cdots \subset L^2(0,1) \ ,$$

where $\mathfrak{C}_r$ is the vector space of all polynomials of degree $\leqslant r-1$ in $\cos(\pi x)$. Thus, an arbitrary element of $\mathfrak{C}_r$ may be written as

$$\gamma(x) = \sum_{k=0}^{r-1} \gamma_k \cos(\pi k x) \ , \quad x \in [\,0\,,\,1\,] \ .$$

For the model (1.1)–(1.4), show that $\mathbb{E}[\, \| \gamma^{nr} - f_o \|^2 \,] = \mathcal{O}(\, n^{-2/3} \,)$, provided $r \asymp n^{1/3}$, and that this rate cannot be improved in general. (Figure out what γ^{nr} is.) This does not contradict Lemma (4.14) since $\mathfrak{C}_r \not\subset \mathcal{T}_d$ for $r \geqslant 2$ and any $d \geqslant 1$.

EXERCISES: (4.11), (4.23), (4.24), (4.27), (4.31), (4.33), (4.36), (4.37), (4.38).

5. Natural splines

To motivate the spline sieve, it is useful to review a few aspects of the smoothing spline estimator of Chapter 13. For ease of presentation, consider the case $m = 2$ (cubic splines). In § 19.2, we give the standard computational scheme for natural cubic spline interpolation, which shows that the cubic smoothing spline estimator is uniquely determined by the values of the spline at the design points,

$$(5.1) \qquad a_i = f^{nh}(x_{in}) \ , \quad i = 1, 2, \cdots, n \ .$$

So, the cubic smoothing spline estimator is an object with dimension n. Now, one way of viewing the role of the roughness penalization is that it reduces the effective dimensionality of the estimator to something very much less than n. With hindsight, in view of the bound on the variance, one might say that the effective dimension is $\asymp (nh)^{-1}$.

With the above in mind, if we (meaning you; the authors certainly won't) are not allowing roughness penalization, then we must find other ways to reduce the dimensionality of the estimator. One way of doing this is to disentangle the dual role of the x_{in} as design points and as knots for the spline estimator. Thus, it seems reasonable to select as knots for the spline only a few of the design points. Thus, for $\mathcal{J} \subset \{\, x_{2,n}, \cdots, x_{n-1,n} \,\}$ but never including the endpoints $x_{1,n}$ and $x_{n,n}$, let $\mathfrak{S}(\mathcal{J})$ be the space of natural spline functions of order $2m$ with knots at the $x_{j,n} \notin \mathcal{J}$. So, $\mathfrak{J}$ is the set of knots that were removed.

For a given design, this leads to a lattice of spaces of spline functions,

$$(5.2) \qquad \mathfrak{S}(\emptyset) \supset \mathfrak{S}(\{\, j \,\}) \supset \mathfrak{S}(\{\, j, k \,\}) \supset \cdots \ ,$$

for $2 \leqslant j \leqslant n-1, 2 \leqslant k \leqslant j-1, \cdots$, as opposed to the linear ordering of a sieve. The task is to select the "optimal" set of knots and, moreover, do it in a data-driven way. This is a rather daunting task. Here, we limit ourselves to showing what is possible as far as asymptotic convergence rates are concerned.

In the context of nice functions $f \in W^{m,2}(0,1)$, it is reasonable to take the knots to be more or less equally spaced. For notational convenience, we consider r equally spaced knots z_{sr} , $s = 1, 2, \cdots, r$, which include the endpoints, and let

$$(5.3)\qquad \mathcal{S}_r = \begin{cases} \text{the vector space of natural spline functions of order} \\ 2m \text{ on the knot sequence } z_{1,r} < z_{2,r} < \cdots < z_{r,r} \ . \end{cases}$$

From the discussion above, we have that

$$(5.4)\qquad \dim(\mathcal{S}_r) = r \ .$$

The nonparametric regression estimator is the solution f^{nr} to

$$(5.5)\qquad \begin{aligned} &\text{minimize} \quad \tfrac{1}{n} \sum_{i=1}^{n} |\, f(x_{in}) - y_{in} \,|^2 \\ &\text{subject to} \quad f \in \mathcal{S}_r \ . \end{aligned}$$

As usual, existence and uniqueness of the solution are no problem. Regarding convergence rates, we have the familiar result under the familiar conditions.

(5.6) THEOREM. *Let* $m \geqslant 1$ *and* $f_o \in W^{m,2}(0,1)$. *Then,*

$$\mathbb{E}\Big[\, \tfrac{1}{n} \sum_{i=1}^{n} |\, f^{n,r}(x_{in}) - f_o(x_{in}) \,|^2 \,\Big] = \mathcal{O}\big(\, n^{-2m/(2m+1)} \,\big) \ ,$$

$$\mathbb{E}\Big[\, \|\, f^{n,r} - f_o \,\|^2 \,\Big] = \mathcal{O}\big(\, n^{-2m/(2m+1)} \,\big) \ ,$$

provided that $r \asymp n^{1/(2m+1)}$.

PROOF. Let $f_{r,n} = \mathbb{E}[\, f^{n,r} \,]$. As usual, we decompose the squared error into bias and variance and obtain from (5.4) that

$$(5.7)\qquad \tfrac{1}{n} \sum_{i=1}^{n} |\, f^{n,r}(x_{in}) - f_{r,n}(x_{in}) \,|^2 = \frac{r\,\sigma^2}{n} \ .$$

For the bias, note that $f_{r,n}$ solves the problem (5.5), with the y_{in} replaced by $f_o(x_{in})$. Now, let $\varphi = \varphi_{r,h}$ be the solution to

$$(5.8)\qquad \begin{aligned} &\text{minimize} \quad \tfrac{1}{r} \sum_{s=1}^{r} \big|\, \varphi(z_{sr}) - f_o(z_{sr}) \,\big|^2 + h^{2m} \,\|\, \varphi^{(m)} \,\|^2 \\ &\text{subject to} \quad f \in W^{m,2}(0,1) \ , \end{aligned}$$

with $h > 0$ unspecified as of yet. Note that in (5.8) only the knots of $\mathcal{S}_r$ are used as the design points, so that $\varphi_{r,h} \in \mathcal{S}_r$.

Now, by Exercises (13.2.22) and (13.2.23), with f_h the solution of the noiseless C-spline problem (13.4.19),

$$\begin{aligned} \|\, \varphi_{r,h} - f_o \,\|_{m,h} &\leqslant \|\, \varphi_{r,h} - f_h \,\|_{m,h} + \|\, f_h - f_o \,\|_{m,h} \\ &\leqslant c\,\{\, 1 + (r\,h)^{-1} \,\}\, h^m \,\|\, f_o^{(m)} \,\| \ , \end{aligned}$$

provided $r\,h$ is large enough. Thus, $h \asymp r^{-1}$ would work. Then, by the Quadrature Lemma (13.2.27),

$$\tfrac{1}{n}\sum_{i=1}^{n} \bigl|\,\varphi_{r,h}(x_{in}) - f_o(x_{in})\,\bigr|^2 \leqslant \{\,1 + c\,(nh)^{-1}\,\}\,\|\,\varphi_{r,h} - f_o\,\|_{m,h}^2$$

$$\leqslant c\,h^{2m}\,\|\,f_o^{(m)}\,\|^2\;,$$

provided $h \asymp r^{-1}$, see Exercises (13.4.22)-(13.4.23). Since $f_{r,n}$ is the minimizer over $\mathcal{S}_r$, then obviously, for *all* h,

$$\tfrac{1}{n}\sum_{i=1}^{n} \bigl|\,f_{r,n}(x_{in}) - f_o(x_{in})\,\bigr|^2 \leqslant \tfrac{1}{n}\sum_{i=1}^{n} \bigl|\,\varphi_{r,h}(x_{in}) - f_o(x_{in})\,\bigr|^2\;,$$

and so

$$\tfrac{1}{n}\sum_{i=1}^{n} \bigl|\,f^{n,r}(x_{in}) - f_o(x_{in})\,\bigr|^2 \leqslant \frac{r\,\sigma^2}{n} + c\,r^{-2m}\,\|\,f_o^{(m)}\,\|^2\;.$$

With $r \asymp n^{-1/(2m+1)}$, the result follows. Q.e.d.

(5.9) EXERCISE. Prove the remaining part of Theorem (5.6).

Of course, the hard part is to choose the number of knots and the knots themselves in a data-driven way. This requires two things: some criterion to judge how good a specific selection of the knots is and a way to improve a given knot selection. There are all kinds of reasonable schemes to go about this; see, e.g., ZHOU and SHEN (2001), HE, SHEN, and SHEN (2001), and MIYATA and SHEN (2005), and references therein.

6. Piecewise polynomials and locally adaptive designs

In this section, we give an exploratory discussion of the estimation of regression functions exhibiting several (here two) scales of change. Of course, it is well-known that this requires *local* smoothing parameters, but to be really effective it also requires *locally adaptive* designs. (This may not be an option for random designs.)

The estimators discussed up to this point (smoothing splines, kernels and the various sieved estimators) were studied as global estimators, depending on a single smoothing parameter, and were shown to work well for smooth functions having one scale of change. Here, we investigate what happens for functions exhibiting more than one scale. Figure 6.1 illustrates what we have in mind. There, graphs of the functions

(6.1) $$f(x\,;\lambda) = \varphi(x) + \lambda^{1/2}\,\varphi(\,\lambda\,x\,)\;,\quad x \in [\,0\,,\,1\,]\;,$$

with

(6.2) $$\varphi(x) = 4\,\mathrm{e}^{-x} - 16\,\mathrm{e}^{-2\,x} + 12\,\mathrm{e}^{-3\,x}\;,\quad x \in \mathbb{R}$$

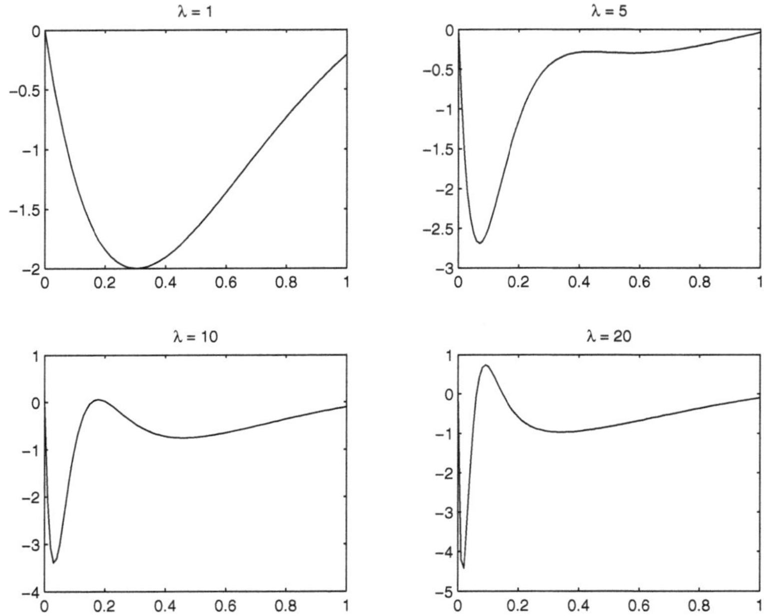

Figure 6.1. Some graphs of functions with two scales, the more so as λ increases. The example is based on Wahba (1990).

(Wahba, 1990), are shown for $\lambda = 1, 5, 10,$ and 20. Clearly, for $\lambda = 10$ and $\lambda = 20$, the function $f(\,\cdot\,;\lambda)$ exhibits two scales of change, whereas for $\lambda = 1$ there is only one. Thus, the estimators discussed so far are well-suited for estimating $f_o = f(\,\cdot\,; 1\,)$ but definitely not so for $f_o = f(\,\cdot\,; 20\,)$.

In considering the estimation of the function $f_o = f(\,\cdot\,; 20\,)$, the problem is that a single smoothing parameter cannot handle the two scales and accommodations must be made for *local* smoothing parameters. Taking a few liberties, it is useful to model $f(\,\cdot\,;\lambda)$ as follows. Partition the interval $[\,0\,,\,1\,]$ as

$$(6.3) \qquad [\,0\,,\,1\,] = \omega_1 \cup \omega_2 \quad \text{with} \quad \omega_1 = [\,0\,,\,\lambda^{-1}\,] \quad \text{and} \quad \omega_2 = [\,\lambda^{-1}\,,\,1\,]\,,$$

and set

$$(6.4) \qquad f_o(x;\lambda) = \begin{cases} \lambda^{1/2}\,\psi(\lambda x)\,, & x \in [\,0\,,\,\lambda^{-1}\,]\,, \\ \vartheta(x) \qquad , & x \in [\,\lambda^{-1}\,,\,1\,]\,, \end{cases}$$

for nice functions ψ and ϑ. Now, in the context of the polynomial sieve, it is clear that the solution lies in the *piecewise*-polynomial sieve:

$$(6.5) \qquad \mathcal{PP}_{r,s} = \left\{ f \;:\; \begin{array}{ll} f = p_1 \in \mathcal{P}_r & \text{on} \quad \omega_1 \\ f = p_2 \in \mathcal{P}_s & \text{on} \quad \omega_2 \end{array} \right\}.$$

Note that now, there are two smoothing parameters, to wit r and s.

Now, the piecewise polynomially sieved estimator is defined as

$$\text{(6.6)} \qquad \text{minimize} \quad \tfrac{1}{n} \sum_{i=1}^{n} |\, f(x_{in}) - y_{in} \,|^2 \quad \text{subject to} \quad f \in \mathcal{PP}_{r,s} \;.$$

Regardless of the design, one may expect to see some improvement over the polynomial sieve, but the clincher is to allow for *adaptive* designs. It is clear that, circumstances permitting, each of the intervals ω_1 and ω_2 should contain (roughly) equal numbers of design points. For n even, the optimal(?) adaptive design consists of $n/2$ points in the interval ω_1 and the remaining $n/2$ points in ω_1 with $k = n/2$ and $\delta = 1 - \lambda^{-1}$.

$$\text{(6.7)} \qquad z_{in} = \begin{cases} \dfrac{i-1}{(k-1)\,\lambda} \;, & i = 1, 2, \cdots, k \;, \\[2ex] \dfrac{1}{\lambda} + \dfrac{i-k}{n-k}\,\delta \;, & i = k+1, \cdots, n \;. \end{cases}$$

We finish by deriving error bounds for the piecewise-polynomial sieve estimator p^{nrs} of f_o in the model (1.1)–(1.2) with the design (6.7). We take $f_o = f_o(\,\cdot\,;\lambda)$ and prove error bounds uniformly in $\lambda \to \infty$. For simplicity, the degrees of the polynomials on the two pieces are taken to be equal, and we do not enforce continuity of the estimator.

The piecewise-polynomial estimator p^{nrr} is defined as the solution to the problem (6.6). However, since we did not enforce continuity, it decomposes into two separate problems, to wit

$$\text{(6.8)} \qquad \text{minimize} \quad \tfrac{2}{n} \sum_{i=1}^{n/2} |\, p(z_{in}) - y_{in} \,|^2 \quad \text{subject to} \quad p \in \mathcal{P}_r$$

on the interval ω_1 and

$$\text{(6.9)} \qquad \text{minimize} \quad \tfrac{2}{n} \sum_{i=n/2+1}^{n} |\, p(z_{in}) - y_{in} \,|^2 \quad \text{subject to} \quad p \in \mathcal{P}_r$$

on ω_2. Of course, the key is that the first problem may be transformed into a nice problem on $[\,0\,,\,1\,]$, as can the second one (in a much less dramatic fashion).

(6.10) THEOREM. *Let $m \geqslant 1$. In (6.4), assume that $\psi, \vartheta \in W^{m,2}(0,1)$. Then, for the model (1.1)–(1.2) with the design (6.7), the piecewise-polynomial estimator p^{nrr} of $f_o(\,\cdot\,,\lambda)$ satisfies*

$$\mathbb{E}[\,\|\, p^{nrr} - f_o(\,\cdot\,;\lambda)\,\|^2\,] = \mathcal{O}\big(\, n^{-2m/(2m+1)}\,\big) \quad \textit{uniformly in } \lambda > 2 \;,$$

provided $r \asymp n^{1/(2m+1)}$ (deterministically).

PROOF. Let $p = q^{nr}$ denote the solution of (6.8). Define the polynomial π^{nr} by

$$\text{(6.11)} \qquad \pi^{nr}(x) = q^{nr}(x/\lambda) \;, \quad x \in [\,0\,,\,1\,] \;.$$

Now, rewrite the problem (6.8) as

$$\text{(6.12)} \qquad \text{minimize} \quad \tfrac{2}{n} \sum_{i=1}^{n/2} |\, p(w_{in}) - y_{in} \,|^2 \quad \text{subject to} \quad p \in \mathcal{P}_r$$

with

$$\text{(6.13)} \qquad w_{in} = \frac{i-1}{n/2} \;, \quad i = 1, 2, \cdots, n/2 \; .$$

(Recall that n is even.) Then, one verifies that $p = \pi^{nr}$ solves (6.12) and that

$$\text{(6.14)} \qquad y_{in} = \lambda^{1/2}\, \psi(w_{in}) + d_{in} \;, \quad x \in [\,0\,,\,1\,] \; .$$

So, apart from the factor $\lambda^{1/2}$, this is our standard regression problem for the polynomial sieve. It follows that the variance is

$$\text{(6.15)} \qquad \mathbb{E}[\, \| \, \pi^{nr} - \pi_{rn} \, \|^2 \,] = \frac{\sigma^2\, r}{n/2} \;,$$

where $\pi_{rn} = \mathbb{E}[\, \pi^{nr} \,]$, but the bias has a factor λ,

$$\text{(6.16)} \qquad \mathbb{E}[\, \| \, \pi_{rn} - \lambda^{1/2}\, \psi \, \|^2 \,] = \mathcal{O}(\, \lambda\, r^{-2m} \,) \;,$$

uniformly in $\lambda > 2$. Equivalently,

$$\text{(6.17)} \qquad \mathbb{E}\Bigl[\, \| \, q^{nr} - f_o \, \|^2_{(0,\lambda^{-1})} \,\Bigr] = \mathcal{O}(\, (\lambda n)^{-1} r + r^{-2m} \,) \;,$$

again uniformly in $\lambda > 2$. Here, we dug up the notation (13.2.9).

The error on $[\, \lambda^{-1} \,,\, 1\,]$ may be treated similarly, leading to

$$\text{(6.18)} \qquad \mathbb{E}[\, \| \, p^{nrr} - f_o \, \|^2_{(\lambda^{-1},1)} \,] = \mathcal{O}(\, n^{-1} r + r^{-2m} \,) \;,$$

so that the overall error is

$$\text{(6.19)} \qquad \mathbb{E}[\, \| \, p^{nrr} - f_o \, \|^2 \,] = \mathcal{O}(\, n^{-1} r + r^{-2m} \,)$$

uniformly in $\lambda > 2$. The theorem follows. Q.e.d.

The discussion above may be adapted to smoothing splines or kernels. For piecewise smoothing spline estimation problems, it is tempting to assemble everything in the form

$$\text{(6.20)} \qquad \begin{aligned} &\text{minimize} \quad \tfrac{1}{n} \sum_{i=1}^{n} |\, f(x_{in}) - y_{in} \,|^2 + \int_0^1 w(\,t\,)\, |\, f^{(m)}(\,t\,) \,|^2 \, dt \\ &\text{subject to} \quad f \in W^{m,2}(0,1) \;, \end{aligned}$$

where

$$\text{(6.21)} \qquad w(\,t\,) = h_1\, 1\!\!1(x \in \omega_1\,) + h_2\, 1\!\!1(x \in \omega_2\,) \; .$$

Here, h_ℓ is shorthand for h^{2m} with whatever value of h is appropriate on the interval ω_ℓ. The problem (6.20) was suggested by ABRAMOVICH and STEINBERG (1996) with the choice

$$w(t) = h^{2m} \, | \, f_o^{(m)}(t) \, |^{-2} \, , \quad t \in [\, 0 \, , 1 \,] \tag{6.22}$$

(which must be estimated). See also PINTORE, SPECKMAN, and HOLMES (2000). Although ABRAMOVICH and STEINBERG (1996) do not suggest adaptive designs, this would be an obvious enhancement.

(6.23) EXERCISE. For the model (1.1)–(1.2), (6.1)–(6.7), formulate the piecewise smoothing spline problem, and prove error bounds uniformly in λ. A method for smoothing the piecewise smooth estimators, such as $p^{n,r,s}$, is given in BROWN, EGGERMONT, LARICCIA, and ROTH (2008). For splines, the method (6.20) suggests itself also.

7. Additional notes and comments

Ad § 2: The analysis of sieved estimators begins and ends with COX (1988b). He considers designs where the design points have asymptotically a Beta density and shows that essentially the same treatment of the bias applies. However, he prefers to do everything in terms of orthogonal bases, even *defining* the sieves that way, whereas in this text bases are only a convenience. On the other hand, COX (1988b) considers quite general sieves, often defined with the aid of orthogonal bases for $L^2(0,1)$. The polynomial sieve is just an instance of his theory. For the modern treatment of sieves, see, e.g., BARRON, BIRGÉ, and MASSART (1999) and EFROMOVICH (1999).

Results such as Lemma (2.6) go under the name of Markov-Bernstein inequalities for polynomials; see, e.g., RAHMAN and SCHMEISSER (2002). The order r^2 of the bound of Lemma (2.6) is sharp. For the best (implicitly defined) constant, see HILLE, SZEGŐ, and TAMARKIN (1937) and KROÓ (2008).

We picked up the shrinking trick (2.14) from FEINERMAN and NEWMAN (1974), p. 36, in a somewhat different context, but boundaries always cause problems. (Their only concern is (best) polynomial approximation in the uniform norm.)

Ad § 3: The estimation of derivatives material originated in COX (1988b), in a somewhat different form.

Ad § 4: Any serious study of Fourier series has to start with the classical approach of, e.g., ZYGMUND (1968). For a more abstract approach, see, e.g., DYM and MCKEAN (1972).

16

Local Polynomial Estimators

1. Introduction

Having studied three widely differing nonparametric regression estimators, it is perhaps time for a comparative critique. In the authors' view, the strength of the smoothing spline and sieved estimators derive from the maximum likelihood and/or minimum principles. A weakness is that the estimators are constructed in a global manner, even though the estimators are essentially local (as they should be). Contrast this with kernel estimators, which are nothing if not local. It is thus natural to attempt a synthesis of these two principles in the form of maximum local likelihood estimation. In theory at least, this combines the best of both worlds.

This chapter provides an additional twist to nonparametric regression problems by concentrating on random designs. This causes some extra problems in that there are now two sources of variation: the randomness of the design and the noise in the responses. Of course, the approach is to separate the two as much as possible, toward which conditioning on the design goes a long way. (For smoothing splines with random designs, see Chapters 21 and 22.)

We briefly describe the nonparametric regression problem with random designs. As before, the problem is to estimate a function f_o on the interval $[\,0\,,\,1\,]$, but now the data

$$(1.1) \qquad (X_1, Y_1), (X_2, Y_2), \cdots, (X_n, Y_n) \; ,$$

are a random sample of the bivariate random variable (X, Y) with

$$(1.2) \qquad f_o(x) = \mathbb{E}[\,Y \mid X = x\,] \; , \quad x \in [\,0\,,\,1\,] \; .$$

We assume that the random variable Y has bounded conditional variances,

$$(1.3) \qquad \sigma^2(x) = \mathbb{V}\mathrm{ar}[\,Y \mid X = x\,] \; , \quad \text{and} \quad \sigma^2 = \sup_{x \in [0,1]} \sigma^2(x) < \infty \; ,$$

with $\sigma(x)$ and σ typically unknown. The marginal density ω of the generic design point X is assumed to be bounded and bounded away from

P.P.B. Eggermont, V.N. LaRiccia, *Maximum Penalized Likelihood Estimation*,
Springer Series in Statistics, DOI 10.1007/b12285_5,

zero; i.e., there exist positive constants ω_1 and ω_2 such that

$$\omega_1 \leqslant \omega(x) \leqslant \omega_2 \ , \quad x \in [\,0\,,\,1\,] \ . \tag{1.4}$$

We then call the random design *quasi-uniform*. It is convenient to rewrite the model (1.1)–(1.3) as

$$Y_i = f_o(X_i) + D_i \ , \quad i = 1, 2, \cdots, n \ , \tag{1.5}$$

where $(X_1, D_1), (X_2, D_2), \cdots, (X_n, D_n)$ form a random sample of the random variable (X, D), and

$$\mathbb{E}[\,D \,|\, X\,] = 0 \quad , \quad \mathbb{E}[\,D^2 \,|\, X\,] = \sigma^2(X) \ . \tag{1.6}$$

The usual nonparametric assumptions on f_o will be discussed later.

We now describe the local polynomial estimators. First, choose the order m of the polynomials and, as before, let $\mathcal{P}_m$ be the vector space of all polynomials of order $\leqslant m$ (degree $\leqslant m - 1$). Let A be a symmetric nonnegative kernel (pdf) and, as usual, for $h > 0$, define the scaled kernel A_h by $A_h(x) = h^{-1} A(h^{-1}x)$. We assume that A has bounded variation, is not completely silly near $t = 0$, and decays fast enough at infinity,

$$\begin{gathered} |\,A\,|_{BV} < \infty \ , \\ \int_0^1 A(\,t\,)\,dt > 0 \quad , \quad \int_{-1}^0 A(\,t\,)\,dt > 0 \ , \\ \sup_{t \in \mathbb{R}} \,(\,1 + |\,t\,|^{3m}\,)\,A(\,t\,) < \infty \quad , \quad \int_{\mathbb{R}} (\,1 + |\,t\,|^{3m}\,)\,A(\,t\,)\,dt < \infty \ . \end{gathered} \tag{1.7}$$

(For a discussion of bounded variation; see § 17.2.)

Now, let $x \in [\,0\,,\,1\,]$. The *local polynomial estimation* problem is to

$$\begin{aligned} &\text{minimize} \quad \tfrac{1}{n} \sum_{i=1}^{n} A_h(x - X_i)\,|\,p(X_i) - Y_i\,|^2 \\ &\text{subject to} \quad p \in \mathcal{P}_m \ . \end{aligned} \tag{1.8}$$

We denote the solution of (1.8) by $p = p^{nhm}(\,\cdot\,;\,x\,)$. Observe that the solution depends on x since it appears as a parameter in (1.8). The estimator of f_o is then taken to be

$$f^{nhm}(x) = p^{nhm}(x\,;x) \ , \quad x \in [\,0\,,\,1\,] \ . \tag{1.9}$$

This is the local polynomial regression estimator, but it is clear that "local polynomial" refers to the *construction* of these estimators only. (*Locally* polynomial estimators are a different kettle of fish altogether; see § 15.5.)

Before proceeding, we mention some difficulties. First is the practical issue of the existence and uniqueness of $p^{nhm}(\,\cdot\,;\,x\,)$. From what we know about (weighted) polynomial least-squares regression, the solution of (1.8) always exists. It is unique, *provided* there are enough (distinct) design points; i.e.,

$$\sum_{i=1}^{n} 1\!\!1\bigl(A_h(x - X_i) > 0 \bigr) \geqslant m \; . \tag{1.10}$$

Thus, if A has compact support, this leads to a minor annoyance that may be resolved as follows. Since the existence of solutions of (1.8) is not a problem, let us enforce uniqueness by (quite arbitrarily) choosing the minimum norm solution of (1.8); i.e., the solution to

$$\begin{aligned} &\text{minimize} \quad \int_0^1 A_h(x-y)\,|\,p(y)\,|^2\,dy \\ &\text{subject to} \quad p \text{ solves (1.8)} \; . \end{aligned} \tag{1.11}$$

Another equally arbitrary choice might be to take the solution of (1.8) with the lowest order.

A theoretical issue concerns the function-theoretic properties of the local polynomial estimator: Is it measurable, continuous, and so on? We shall not unduly worry about this, but the following computations for the cases $m = 1$ and $m = 2$ indicate that in general there are no problems.

The case $m = 1$ gives the *local constant* estimator,

$$f^{nh,1}(x) = \frac{\frac{1}{n}\sum\limits_{i=1}^{n} Y_i\, A_h(x - X_i)}{\frac{1}{n}\sum\limits_{i=1}^{n} A_h(x - X_i)} \; , \quad x \in [\,0\,,1\,] \; . \tag{1.12}$$

This is the Nadaraya-Watson kernel estimator, already encountered in Chapter 14, except that, in general, the kernel A in (1.12) need not be nonnegative. The development for local polynomial estimators with nonnegative kernels goes through for general Nadaraya-Watson estimators, as we show in § 10.

For the case $m = 2$, easy but cumbersome calculations show that the *local linear* estimator is given by

$$f^{nh,2}(x) = \frac{[\mathbb{Y}_n A_h](x)[P_n B_h](x) - [P_n A_h](x)[\mathbb{Y}_n B_h](x)}{[P_n A_h](x)\,[P_n C_h](x) - |\,[P_n B_h](x)\,|^2} \; , \tag{1.13}$$

where the operators P_n and $\mathbb{Y}_n$ are defined by

$$\begin{aligned} P_n\, g(x) &= \tfrac{1}{n}\sum_{i=1}^{n} g(x - X_i) \; , \\ \mathbb{Y}_n\, g(x) &= \tfrac{1}{n}\sum_{i=1}^{n} Y_i\, g(x - X_i) \; , \quad x \in [\,0\,,1\,] \; , \end{aligned} \tag{1.14}$$

and the kernels B_h and C_h are defined as usual, with $B(x) = xA(x)$ and $C(x) = x^2 A(x)$ for all x.

It seems clear that the explicit formula for the local linear estimator is a mixed blessing. Moreover, it gets worse with increasing m. However, the explicit computations for the local constant and linear estimators serve another purpose: They indicate that the local polynomial estimators as defined by (1.11) are at least measurable functions, so that the "error" $\| f^{nhm} - f_o \|^2$ (squared L^2 integral) in fact makes sense. The conclusion remains valid if the smoothing parameter is a measurable function of x and the data (X_i, Y_i), $i = 1, 2, \cdots, n$, as would be the case for data-driven choices of h.

Because of the random design, for local polynomial (and other) estimators the usual expected error bounds under the usual conditions need to be replaced by conditional expected error bounds. To indicate the required conditional expectations, we denote the design by $\mathbb{X}_n = (X_1, X_2, \cdots, X_n)$ and write

$$\mathbb{E}[\cdots \mid \mathbb{X}_n] \equiv \mathbb{E}[\cdots \mid X_1, X_2, \cdots, X_n] \,. \tag{1.15}$$

We may now state the following theorem.

(1.16) THEOREM. *Let $m \geqslant 1$ and $f_o \in W^{m,2}(0,1)$. For the quasi-uniform random-design model* (1.1)–(1.3), *the solution f^{nhm} of* (1.8) *satisfies*

$$\mathbb{E}\Bigl[\, \| f^{nhm} - f_o \|^2 \, \Big| \, \mathbb{X}_n \Bigr] =_{\text{as}} \mathcal{O}\bigl(n^{-2m/(2m+1)} \bigr) \,,$$

$$\mathbb{E}\Bigl[\, \tfrac{1}{n} \sum_{i=1}^{n} | f^{nhm}(X_i) - f_o(X_i) |^2 \, \Big| \, \mathbb{X}_n \Bigr] =_{\text{as}} \mathcal{O}\bigl(n^{-2m/(2m+1)} \bigr) \,,$$

almost surely, provided $h \asymp n^{-1/(2m+1)}$ (deterministically).

The theorem above permits a refinement, sort of, under the sole extra condition that $f_o \in W^{m+1,2}(0,1)$.

(1.17) THEOREM. *In addition to the conditions of Theorem* (1.16), *assume that $f_o \in W^{m+1,2}(0,1)$. Then, there exists a deterministic function r_{hm}, depending on f_o, such that*

$$\mathbb{E}\Bigl[\, \| f^{nhm} - f_o + r_{hm} \|^2 \, \Big| \, \mathbb{X}_n \Bigr] =_{\text{as}} \mathcal{O}\bigl(n^{(2m+2)/(2m+3)} \bigr) \,,$$

provided $h \asymp n^{-1/(2m+3)}$.

Of course, the theorem does not say much about the function r_{hm}. If the design density is continuous, then we may take

$$r_{hm}(x) = h^m \, f_o^{(m)}(x) \, \varrho_m\bigl(\lambda(x,h)\bigr) \,, \quad x \in [\,0\,,\,1\,] \,, \tag{1.18}$$

where $\lambda(x,h) = 0 \vee \min\bigl(x/h \,,\, 1 \,,\, (1-x)/h \bigr)$, and ϱ_m is a nice continuous function, not depending on h. The function λ is the hip-roof function: $\lambda(x,h) = 1$ for $h < x < 1-h$, and linearly drops off to 0 at $x=0$ and $x=1$. The precise details are explicated in § 7. For odd m, it gets better: Then,

$\varrho_m(1) = 0$ and, away from the boundary, the rate $\mathcal{O}\big(n^{-(2m+2)/(2m+3)} \big)$ for the expected squared error applies. As for smoothing splines, this result may be classified as *superconvergence*: The estimator is more accurate than is warranted by the approximation power of local polynomials of order m; See § 7.

We also consider uniform error bounds on the local polynomial estimators: Under the added assumptions that $f_o \in W^{m,\infty}(0,1)$ and that

$$(1.19) \qquad \sup_{x\in[0,1]} \mathbb{E}[\,|\,D_1\,|^\kappa \,|\, X = x\,] < \infty \;,$$

for some $\kappa > 2$, then uniformly over a wide range of h values,

$$(1.20) \quad \|\, f^{nhm} - f_o \,\|_\infty = \mathcal{O}\big(h^{2m} + \big((nh)^{-1} \log n\big)^{1/2} \big) \quad \text{almost surely} \;.$$

(Compare this with smoothing splines and kernels.)

In the remainder of this chapter, a detailed study of local polynomial estimators is presented. In § 2, the connection between the pointwise squared error $|\, f^{nhm}(x) - f_o(x)\,|^2$ and the *local* squared error

$$\tfrac{1}{n} \sum_{i=1}^{n} A_h(x - X_i)\,\big|\, p^{nhm}(X_i;x) - f_o(X_i)\,\big|^2$$

is pointed out. In § 3, we exhibit a decoupling of the randomness of the design from the randomness of the responses, even if this sounds like a contradiction in terms. In §§ 4 and 5, Theorem (1.16) is proved. The asymptotic behavior of the bias and variance is studied in §§ 6 and 7. Uniform error bounds and derivative estimation are considered in §§ 8 and 9.

2. Pointwise versus local error

In this section, we make some preliminary observations on the analysis of local polynomial estimators. We keep m and the kernel A fixed and denote the resulting estimators by $p^{nh}(\,\cdot\,,x)$ and f^{nh}. As a programmatic point, we would like to analyze the least-squares problem (1.8) as is and would like to ignore the availability of explicit formulas for the estimators.

Of course, the standard treatment applies. Since the problem (1.8) is quadratic, we get the nice behavior around its minimum, in the form of

$$(2.1) \quad \tfrac{1}{n} \sum_{i=1}^{n} A_h(x - X_i)\,|\, p(X_i) - p^{nh}(X_i)\,|^2 = \\ \tfrac{1}{n} \sum_{i=1}^{n} A_h(x - X_i)\,\{\,|\, p(X_i) - Y_i\,|^2 - |\, p^{nh}(X_i) - Y_i\,|^2\,\}\;,$$

valid for all $p \in \mathcal{P}_m$.

(2.2) EXERCISE. Prove (2.1).

Thus, if this approach works at all, it will lead to bounds on the *local* squared error,

$$(2.3) \qquad \tfrac{1}{n} \sum_{i=1}^{n} A_h(x - X_i) \, | \, p^{nh}(X_i \,; x) - t_m(X_i \,; x) \, |^2 \,,$$

where t_m is a polynomial of order m approximating f_o in some suitable sense. Of course, interest is not in (2.3) but in the pointwise error,

$$(2.4) \qquad | \, p^{nh}(x \,; x) - t_m(x \,; x) \, |^2 \,,$$

provided we insist on $t_m(x\,;x) = f_o(x)$. Thus, the two must be connected. There are two aspects to this, expressed in Lemmas (2.5) and (2.7).

(2.5) POINT EVALUATION LEMMA. *Let $m \geqslant 1$. If the design density ω is quasi-uniform in the sense of assumption* (1.4), *then there exists a constant c, depending on m, ω, and A only, such that, for all $p \in \mathcal{P}_m$, all $x \in [\,0\,,\,1\,]$, all $0 < h \leqslant \frac{1}{2}$, all $1 \leqslant r < \infty$, and all $k = 0, 1, \cdots, m-1$,*

$$| \, h^k \, p^{(k)}(x) \, |^r \leqslant c \int_0^1 A_h(x - y) \, \omega(y) \, | \, p(y) \, |^r \, dy \,.$$

For the following lemma, recall the definition in Theorem (14.6.13),

$$(2.6) \qquad \mathcal{G}_n(\alpha) = \bigl[\, \alpha \, (n \log n)^{-1} \,,\, \tfrac{1}{2} \,\bigr] \,.$$

(2.7) QUADRATURE LEMMA. *Let $m \geqslant 1$ and $\alpha > 0$. Let*

$$p(\,t\,) \equiv p(\,t \,|\, X_1, \cdots, X_n, Y_1 \cdots, Y_n) \,, \qquad t \in \mathbb{R} \,,$$

be a random polynomial of order $\leqslant m$. Let

$$\Delta = \tfrac{1}{n} \sum_{i=1}^{n} A_h(x - X_i) \, p(X_i) - \int_0^1 A_h(x - t\,) \, \omega(\,t\,) \, p(\,t\,) \, dt \,.$$

Then,

$$| \, \Delta \, | \leqslant \eta^{nh} \int_0^1 A_h(x - t\,) \, \omega(\,t\,) \, | \, p(\,t\,) \, | \, dt$$

(note the absolute value on the right), where $\eta^{nh} = \eta^{nh}(\mathbb{X}_n)$ depends on the design only but not on the noise, and

$$(2.8) \qquad \lim_{n \to \infty} \; \sup_{h \in \mathcal{G}_n(\alpha)} \; \frac{\eta^{nh}}{\sqrt{(nh)^{-1} \log n}} < \infty \quad \textit{almost surely} \,.$$

PROOF OF LEMMA (2.5). In view of the quasi-uniformity of ω, see (1.4), we may as well assume that $\omega(\,t\,) = 1$ for all t.

The proof to follow should actually be read from bottom to top. Obviously, for $1 \leqslant r < \infty$,

$$\int_0^1 A(\,t\,) \, | \, q(\,t\,) \, |^r \, dt \quad \text{and} \quad \int_{-1}^0 A(\,t\,) \, | \, q(\,t\,) \, |^r \, dt$$

are r-th powers of norms of $q \in \mathcal{P}_m$. Since $q^{(k)}(0)$, $k = 0, 1, \cdots, m-1$, are bounded linear functionals of $q \in \mathcal{P}_m$, there exists a positive constant c such that, for all $q \in \mathcal{P}_m$, and all k,

$$
(2.9) \qquad | q^{(k)}(0) |^r \leqslant c \int_0^1 A(t) \, | q(t) |^r \, dt \; .
$$

Now, for $p \in \mathcal{P}_m$, define $q \in \mathcal{P}_m$ by $q(t) = p(x - ht)$ for all t. Then, $q^{(k)}(0) = h^k p^{(k)}(x)$ for all k, and

$$
\int_0^1 A(t) \, | p(x - ht) |^r \, dt = \int_{x-h}^{x} A_h(x - t) \, | p(t) |^r \, dt \; ,
$$

so that (2.9) implies

$$
(2.10) \qquad | h^k p^{(k)}(x) |^r \leqslant c \int_{x-h}^{x} A_h(x - t) \, | p(t) |^r \, dt \; .
$$

Similarly, one derives the inequality

$$
(2.11) \qquad | h^k p^{(k)}(x) |^r \leqslant c \int_{x}^{x+h} A_h(x - t) \, | p(t) |^r \, dt \; .
$$

Now, for $x \in [0, 1]$ and $0 < h \leqslant \frac{1}{2}$, at least one of the intervals $[x-h, h]$ and $[x, x+h]$ lies completely inside the interval $[0, 1]$, so that from either (2.10) or (2.11) one obtains the desired inequalities. Q.e.d.

(2.12) EXERCISE. (a) An attempt at a more precise version of the Point Evaluation Lemma (2.5) reads

$$
| p(x) |^2 \int_0^1 A_h(x - t) \, \omega(t) \, d(t) =
$$
$$
\int_0^1 A_h(x - y) \, | p(y) |^2 \, \omega(y) \, dy + \varepsilon_m(h) \int_0^1 B_h(x - y) \, | p'(y) |^2 \, dy \; ,
$$

where $B(x) = | x |^2 A(x)$, and $\varepsilon_m(h) = \mathcal{O}\big(h^2 \big)$ as $h \to 0$. Prove this!
(b) Pointwise, part (a) is hardly an improvement over Lemma (2.5) but the integrated version is. To see this, show that

$$
\int_0^1 \int_0^1 B_h(x - y) \, | p'(y) |^2 \, dy \, dx \leqslant c \int_0^1 \int_0^1 A_h(x - y) \, | p(y) |^2 \, \omega(y) \, dy \, dx \; .
$$

[Hint for (a): Use the Taylor expansion of $p(y)$ around $y = x$ and compute the integral in the Point Evaluation Lemma. For (a) as well as (b), one still needs to play tricks with equivalent norms on $\mathcal{P}_m$, as in the proof of Lemma (2.5).]

PROOF OF THE QUADRATURE LEMMA (2.7). Obviously, we wish to factor out the randomness of the polynomial $p \in \mathcal{P}_m$. Let $p_k = h^k p^{(k)}(x)/k!$.

Then, $p(t) = \sum_{k=0}^{m-1} p_k \{h^{-1}(t-x)\}^k$ and $|\Delta| \leqslant \sum_{k=0}^{m-1} |p_k|\,|\Delta_{kh}|$, where

$$\Delta_{kh} = \tfrac{1}{n} \sum_{i=1}^{n} A_{k,h}(x - X_i) - \int_0^1 A_{k,h}(x-t)\,\omega(t)\,dt ,$$

in which $A_{k,h}(z) = \{h^{-1}z\}^k A_h(z)$. It follows that

$$|\Delta| \leqslant \Bigl(\max_{1\leqslant k\leqslant m-1} |\Delta_{kh}| \Bigr) \sum_{k=0}^{m-1} |p_k| .$$

Now, by Theorem (14.6.13), one obtains that

$$\max_{1\leqslant k\leqslant m-1} |\Delta_{kh}| \leqslant \eta^{nh} ,$$

with η^{nh} satisfying the uniform bound of (2.8). Finally, by the Point Evaluation Lemma (2.5), for a suitable constant c,

$$\sum_{k=0}^{m-1} |p_k| \leqslant c \int_0^1 A_h(x-t)\,|p(t)|\,dt ,$$

and the lemma follows. Q.e.d.

The analysis of the local polynomial estimators now proceeds by taking p in (2.1) as $p = T_m(\,\cdot\,; x)$, the m-th-order Taylor polynomial of f_o centered at the point x,

$$T_m(t;x) = \sum_{k<m} f_o^{(k)}(x)\,(t-x)^k / k! . \tag{2.13}$$

Then, we may proceed as in the chapter on spline estimation (§ 13.4). However, the authors prefer the more leisurely route of the following sections.

EXERCISES: (2.2), (2.12).

3. Decoupling the two sources of randomness

Here we begin the actual study of the local polynomial estimation problem (1.8), which we aim to do in a number of simple steps. In particular, we begin by decoupling the randomness of the design from the noise in the responses. This may be done by means of an appropriately chosen asymptotic version of the local polynomial minimization problem (1.8).

In this and the following sections, the kernel A and the order m of the local polynomials are kept fixed and $x \in [0, 1]$ denotes a generic point. To keep formulas tidy, we write $p^{nh}(t;x)$ simply as $p^{nh}(t)$ and introduce the following functionals, for nice functions φ, ψ,

$$
\begin{aligned}
L^{nh}(\varphi) &= \tfrac{1}{n} \sum_{i=1}^{n} D_i \, A_h(x - X_i)\, \varphi(X_i) \,, \\
B^{nh}(\varphi, \psi) &= \tfrac{1}{n} \sum_{i=1}^{n} A_h(x - X_i)\, \varphi(X_i)\, \psi(X_i) \,, \\
B_h(\varphi, \psi) &= \int_0^1 A_h(x - t)\, \varphi(t)\, \psi(t)\, \omega(t)\, dt \,, \\
Q^{nh}(\varphi) &= B^{nh}(\varphi, \varphi) \quad , \quad Q_h(\varphi) = B_h(\varphi, \varphi) \,.
\end{aligned} \tag{3.1}
$$

Here, L, B and Q stand for "Linear", "Bilinear", and "Quadratic", respectively. We draw attention to the fact that the D_i are unknown, so that $L^{nh}(\varphi)$ is not an empirical functional. Nevertheless, it is useful to observe that the problem (1.8) is equivalent to

$$
\begin{aligned}
&\text{minimize} \quad Q^{nh}(p - f_o) - 2\, L^{nh}(p - f_o) \\
&\text{subject to} \quad p \in \mathcal{P}_m \,.
\end{aligned} \tag{3.2}
$$

The first step in the analysis is to replace the problem (3.2) with a much simpler one, viz. by replacing Q^{nh} with Q_h,

$$
\begin{aligned}
&\text{minimize} \quad Q_h(p - f_o) - 2\, L^{nh}(p - f_o) \\
&\text{subject to} \quad p \in \mathcal{P}_m \,.
\end{aligned} \tag{3.3}
$$

This may be called a semi-asymptotic problem. Note that (3.3) is not an empirical problem despite similarities with (3.2).

(3.4) EXERCISE. (a) Show that (3.2) and (1.8) have the same solutions.
(b) Show that the solution of (3.3) exists and is unique.

We denote the solution of (3.3) as $p = \pi^{nh}$, dropping the dependence on x in the notation. Note that $\pi^{nh}(x)$ is not an estimator of $f_o(x)$ in the statistically accepted sense. Much more importantly, observe that π^{nh} is linear in the noise $D_1, D_2, \cdots, D_n$, so that π_h, defined as the conditional expectation,

$$
\pi_h(t) = \mathbb{E}\big[\, \pi^{nh}(t) \,\big|\, \mathbb{X}_n \,\big] \qquad t \in [0, 1] \,, \tag{3.5}
$$

solves the problem (3.3) with $D_i = 0$ for all i. In other words, π_h solves the problem

$$
\text{minimize} \quad Q_h(p - f_o) \quad \text{subject to} \quad p \in \mathcal{P}_m \,. \tag{3.6}
$$

The implication is that $\mathbb{E}\big[\, \pi^{nh}(t) \,\big|\, \mathbb{X}_n \,\big]$ does not depend on $\mathbb{X}_n$ anymore and so equals $\mathbb{E}[\, \pi^{nh}(t) \,]$. In effect, the problem (3.3) implements a decoupling of the two sources of randomness alluded to in § 1. The randomness of the design shows up only in the difference $p^{nh} - \pi^{nh}$, which turns out to be asymptotically negligible compared with the error of π^{nh} or the error in

p^{nh}. This is the content of Lemma (3.16) below. But first we give an exercise and an annoying but necessary lemma on the effect of approximating f_o by its Taylor polynomial.

(3.7) EXERCISE. Convince yourself that the existence and uniqueness of the solution of (3.6) follows from Exercise (3.4).

Assume that the kernel A satisfies (1.7). For convenience, we assume that the kernel is differentiable, so

$$A \in W^{1,1}(\mathbb{R}) . \tag{3.8}$$

Define the kernels A_k, $k = 0, 1, 2, \cdots$, by

$$A_k(t) = t^k A(t) , \quad -\infty < t < \infty , \tag{3.9}$$

and $A_{k,h}(t) = h^{-1} A_k(h^{-1}t)$. Define the kernels $C_k(x) = [\mathcal{T}_m A_k](x)$ by

$$[\mathcal{T}_m A_k](x) = \int_{|x|}^{\infty} |A_{k+2m-1}(t)| \, dt , \quad -\infty < x < \infty , \tag{3.10}$$

and as usual set $C_{k,h} = (\mathcal{T}_m A_k)_h$ or $C_{k,h}(x) = h^{-1} C_k(h^{-1}x)$. One checks that the C_k are bounded and integrable,

$$C_k \in L^\infty(\mathbb{R}) \cap L^1(\mathbb{R}) , \quad k = 0, 1, \cdots, m-1 , \tag{3.11}$$

since by (1.7)

$$\begin{aligned} \int_1^\infty C_k(x)\, dx &= \int_1^\infty \int_x^\infty t^{k+2m-1} A(t)\, dt\, dx \\ &\leqslant \int_1^\infty x^{-m+k-1} \int_0^\infty t^{3m} A(t)\, dt\, dx < \infty \end{aligned}$$

for $k = 0, 1, \cdots, m-1$.

We have use for the transforms $\mathfrak{C}_{k,h}$, defined for all $f \in W^{m,2}(0,1)$ by

$$[\mathfrak{C}_{k,h} f](x) = \int_0^1 C_{k,h}(x-y) \, | f^{(m)}(y) |^2 \, dy , \quad x \in [0, 1] . \tag{3.12}$$

(3.13) LEMMA. *For $f_o \in W^{m,2}(0,1)$, let T_m be the m-th-order Taylor polynomial of f_o centered at x; see (2.13). For $f \in W^{m,2}(0,1)$, define*

$$Q_{k,h}(f) = \int_0^1 | A_{k,h}(x-t) | \, | f(t) |^2 \, \omega(t)\, dt . \tag{3.14}$$

Then, there exists a constant c, not depending on f_o, such that, for all h, $0 < h \leqslant \frac{1}{2}$, and all $k = 0, 1, \cdots, m-1$,

$$Q_{k,h}(T_m - f_o) \leqslant c\, h^{2m} \, [\mathfrak{C}_{k,h} f_o](x) . \tag{3.15}$$

PROOF. This is approximation theory. Using Taylor's formula with exact remainder, see (13.3.8)–(13.3.9), we get with Cauchy-Schwarz that

$$| T_m(t) - f_o(t) |^2 \leqslant c\, | t - x |^{2m-1} \left| \int_x^t | f_o^{(m)}(y) |^2 \, dy \right| .$$

The absolute values around the last integral are necessary for the case $t < x$. It follows that

$$Q_{k,h}(T_m - f_o) \leqslant c\, h^{2m-1} \int_0^1 \left| A_{k+2m-1,h}(x - t) \right| \left| \int_x^t | f_o^{(m)}(y) |^2 \, dy \right| dt ,$$

where we used that the design density ω is bounded.

Now, split the interval of integration into the ranges $0 < t < x$ and $x < t < 1$. Then, the integral over $0 < t < x$ equals

$$\int_0^x | A_{k+2m-1,h}(x - t) | \int_t^x | f_o^{(m)}(y) |^2 \, dy \, dt = \int_0^x | f_o^{(m)}(y) |^2 \int_0^y | A_{k+2m-1}(x - t) | \, dt \, dy$$

and the integral over $x < t < 1$ equals

$$\int_x^1 | A_{k+2m-1,h}(x - t) | \int_x^t | f_o^{(m)}(y) |^2 \, dy \, dt = \int_x^1 | f_o^{(m)}(y) |^2 \int_y^1 | A_{k+2m-1,h}(x - t) | \, dt \, dy .$$

Note that for $x < y$ a simple change of variables (and extending the integration region) gives

$$\int_y^1 | A_{k+2m-1,h}(x - t) | \, dt \leqslant \int_{-\infty}^{(x-y)/h} | A_{k+2m-1}(\tau) | \, d\tau = h\, C_{k,h}(x - y) ,$$

and likewise for the other case. Adding these two integrals leads to the inequality (3.15). Q.e.d.

(3.16) LEMMA. *Let $m \geqslant 1$ and $\alpha > 0$. Under the quasi-uniformity assumption (1.4) on the design, there exists a constant c such that for $h \to 0$ and $nh \to \infty$, and for all $f_o \in W^{m,2}(0,1)$, with T_m given by (2.13),*

$$Q_h(p^{nh} - \pi^{nh}) \leqslant \zeta^{nh} \left\{ Q_h(\pi^{nh} - T_m) + h^{2m} \left[\mathfrak{C}_h\, f_o \right](x) \right\} ,$$

with $\mathfrak{C}_h$ having a kernel $C_h(x) = h^{-1} C(h^{-1}x)$, with $C \in L^\infty(\mathbb{R}) \cap L^1(\mathbb{R})$, and where $\zeta^{nh} \equiv \zeta^{nh}(\mathbb{X}_n)$ depends on the design but not on the noise, and

$$\lim_{n\to\infty} \sup_{h \in \mathcal{G}_n(\alpha)} \frac{| \zeta^{nh} |}{(nh)^{-1} \log n} < \infty \quad \textit{almost surely}.$$

PROOF. Let $\varepsilon = \pi^{nh} - p^{nh}$. Then, in the usual fashion, one derives that

$$(3.17) \qquad Q^{nh}(\varepsilon) = B^{nh}(\pi^{nh} - f_o , \varepsilon) - B_h(\pi^{nh} - f_o , \varepsilon) .$$

This must be further rewritten as

$$(3.18)\quad Q^{nh}(\varepsilon) \leqslant B^{nh}(\pi^{nh} - T_m, \varepsilon) - B_h(\pi^{nh} - T_m, \varepsilon) + B^{nh}(T_m - f_o, \varepsilon) - B_h(T_m - f_o, \varepsilon) .$$

Now, since $(\pi^{nh} - T_m)\,\varepsilon \in \mathcal{P}_{2m-1}$, Lemma (2.7) gives

$$(3.19)\quad \begin{aligned} B^{nh}(\pi^{nh} - T_m, \varepsilon) - B_h(\pi^{nh} - T_m, \varepsilon) &\leqslant \eta_1^{nh}\, B_h(\,|\,\pi^{nh} - T_m\,|, |\,\varepsilon\,|\,) \\ &\leqslant \eta_1^{nh} \{\, Q_h(\varepsilon)\,\}^{1/2} \{\, Q_h(\pi^{nh} - T_m)\,\}^{1/2} , \end{aligned}$$

with the required bound on η_1^{nh}; i.e.,

$$(3.20)\quad \lim_{n\to\infty} \sup_{h\in\mathcal{G}_n(\alpha)} \frac{|\,\eta_1^{nh}\,|}{\sqrt{(nh)^{-1}\log n}} < \infty \quad \text{almost surely} .$$

For the other term, we write $\varepsilon(t) = \sum_{k<m} \varepsilon_k \{\, h^{-1}(t-x)\,\}^k / k!$, with $\varepsilon_k = h^k\, \varepsilon^{(k)}(x)$ for all k. Then,

$$(3.21)\quad \Delta \stackrel{\text{def}}{=} B^{nh}(T_m - f_o, \varepsilon) - B_h(T_m - f_o, \varepsilon) = \sum_{k<m} \varepsilon_k\, \Delta_{k,h} ,$$

with

$$\Delta_{k,h} = \int_0^1 A_{k,h}(x-t)\,\bigl(T_m(t) - f_o(t)\bigr)\,\bigl(d\Omega_n(t) - d\Omega_o(t)\bigr) .$$

Here, Ω_n is the empirical distribution and Ω_o the actual distribution of the design. By Exercise (14.6.28),

$$(3.22)\quad |\,\Delta_{k,h}\,| \leqslant \eta_k^{nh}\, \|\, A_{k,h}(x - \cdot\,)\,\bigl(T_m(\cdot\,) - f_o(\cdot\,)\bigr)\,\|_{h,W^{1,1}(0,1)} ,$$

with the required bound (3.20) on η_k^{nh}. Now, using Cauchy-Schwarz and Lemma (3.13),

$$\begin{aligned} &\|\, A_{k,h}(x - \cdot\,)\,\bigl(T_m(\cdot\,) - f_o(\cdot\,)\bigr)\,\|_{L^1(0,1)} \\ &\qquad\leqslant c\,\{\, Q_{k,h}(T_m - f_o)\,\}^{1/2} \leqslant c_1\, h^m\, \{\, [\,\mathfrak{C}_{k,h} f_o\,](x)\,\}^{1/2} . \end{aligned}$$

Likewise, one obtains that, for suitable constants,

$$\begin{aligned} &\|\, A_{k,h}(x - \cdot\,)\,\{T_m(\cdot\,) - f_o(\cdot\,)\}'\,\|_{L^1(0,1)} \\ &\qquad\leqslant c\,\{\, Q_{k,h}(T_m' - f_o')\,\}^{1/2} \leqslant c_1\, h^{m-1}\, [\,\mathfrak{C}_{k-1,h} f_o\,](x) . \end{aligned}$$

Finally, observe that

$$\|\, \{\, A_{k,h}(x - \cdot\,)\,\}'\,\bigl(T_m(\cdot\,) - f_o(\cdot\,)\bigr)\,\|_{L^1(0,1)} \leqslant c\,h^{-1}\,\{\, Q_{k,h}'(T_m - f_o)\,\}^{1/2} ,$$

where

$$Q_{k,h}'(f) = h^{-1} \int_0^1 B_{k,h}(x-t)\,|\,f(t)\,|^2\, dt ,$$

in which $B_k(t) = A_k'(t)$ and $B_{k,h}(t) = h^{-1} B_k(h^{-1} t)$. Thus,

$$Q_{k,h}'(T_m - f_o) \leqslant h^{m-1} [\mathfrak{B}_{k,h} f_o](x) ,$$

where $\mathfrak{B}_{k,h}$ is the operator with kernels $B_{k,h} = (\mathcal{T}_m B_k)_h$. It follows that

$$\| A_{k,h}(x - \cdot\,)\,(T_m(\,\cdot\,) - f_o(\,\cdot\,))\,\|_{h,W^{1,1}(0,1)} \leqslant c\,h^m \{ [\mathfrak{D}_{k,h} f_o](x) \}^{1/2} ,$$

where $\mathfrak{D}_{k,h}$ has kernel

(3.23) $$D_{k,h} = \mathcal{T}_m(A_k) + \mathcal{T}_{m-1}(A_k) + \mathcal{T}_m(A_k') .$$

From (3.22) and Lemma (2.5), we then get that

(3.24) $$|\,\Delta\,| \leqslant c\,\eta_m^{nh} \{ Q_h(\varepsilon) \}^{1/2} \{ [\mathfrak{C}_h f_o](x) \}^{1/2} ,$$

where η_m^{nh} satisfies (3.20), and with $\mathfrak{C}_h$ having the kernel C_h,

(3.25) $$C_h = \sum_{k<m} D_{k,h} .$$

From (3.17), (3.19), and (3.24), it then follows that

(3.26) $$Q^{nh}(\varepsilon) \leqslant \eta_m^{nh} \{ Q_h(\varepsilon) \}^{1/2} \times \\ \{ \{ Q_h(\pi^{nh} - T_m) \}^{1/2} + h^m \{ [\mathfrak{C}_h f_o](x) \}^{1/2} \} .$$

Finally, with the Quadrature Lemma (2.7),

$$Q_h(\varepsilon) - \zeta^{nh} \{ Q_h(\varepsilon) \} \leqslant Q^{nh}(\varepsilon) ,$$

with ζ^{nh} satisfying the bound (2.10). Then, (3.26) takes on the form

$$Q_h(\varepsilon) \leqslant c\,\eta_{m+1}^{nh} \{ Q_h(\varepsilon) \}^{1/2} \times \\ \{ \{ Q_h(\pi^{nh} - T_m) \}^{1/2} + h^m \{ [\mathfrak{C}_h f_o](x) \}^{1/2} \} ,$$

and the lemma follows. Q.e.d.

(3.27) EXERCISE. Verify the (in)equalities (3.17), (3.18), and (3.19).

EXERCISES: (3.4), (3.7), (3.27).

4. The local bias and variance after decoupling

In this section, we discuss the local bias and variance of the "decoupled" estimator π^{nh}, defined as the solution of (3.3). Without further ado, we give the following lemma.

(4.1) LEMMA (LOCAL VARIANCE). *Assuming* (1.5)–(1.6), *there exists a constant* c *such that, for all* $x \in [\,0\,,1\,]$,

$$\mathbb{E}[\, Q_h(\pi^{nh} - \pi_h) \,|\, \mathbb{X}_n \,] \leqslant c\,(nh)^{-1} , \quad \textit{almost surely}.$$

PROOF. Let $\varepsilon = \pi^{nh} - \pi_h$. The starting point is the quadratic equality describing the behavior of $Q_h(p - f_o) - 2\,L^{nh}(p - f_o)$ about its minimizer $p = \pi^{nh}$. This yields

$$(4.2) \qquad Q_h(\varepsilon) = Q_h(\pi_h - f_o) - Q_h(\pi^{nh} - f_o) + 2\,L^{nh}(\varepsilon) \; .$$

Likewise, for the functional $Q_h(p - f_o)$ with its minimizer π_h,

$$(4.3) \qquad Q_h(\varepsilon) = Q_h(\pi^{nh} - f_o) - Q_h(\pi_h - f_o) \; ,$$

so that, adding these inequalities, we obtain

$$(4.4) \qquad Q_h(\varepsilon) = L^{nh}(\varepsilon) \; .$$

Writing $\varepsilon(t) = \sum\limits_{k<m} \varepsilon_k \left\{ h^{-1}(t-x) \right\}^k$, with $\varepsilon_k = h^k\, \varepsilon^{(k)}(x) \,/\, k\,!$, results in

$$(4.5) \qquad L^{nh}(\varepsilon) = \sum_{k<m} \varepsilon_k \cdot \tfrac{1}{n} \sum_{i=1}^{n} D_i\, A_{k,h}(x - X_i) \; ,$$

with $A_{k,h}$ as in (3.9). It then follows that

$$(4.6) \quad \mathbb{E}\Bigl[\, \bigl| L^{nh}(\varepsilon) \bigr|^2 \,\Bigm|\, \mathbb{X}_n \Bigr] \leqslant \mathbb{E}\Bigl[\sum_{k<m} |\,\varepsilon_k\,|^2 \,\Bigm|\, \mathbb{X}_n \Bigr] \cdot \mathbb{E}\Bigl[\sum_{k<m} S_k \,\Bigm|\, \mathbb{X}_n \Bigr] \; ,$$

where

$$(4.7) \qquad S_k = \Bigl| \tfrac{1}{n} \sum_{i=1}^{n} D_i\, A_{k,h}(x - X_i) \Bigr|^2 \; , \qquad k = 0, 1, \cdots, m-1 \; .$$

Now, let $B_{k,h}(z) = h^{-1}\,(z/h)^k\,(\,A(h^{-1}z)\,)^2$. Then,

$$\mathbb{E}[\, S_k \mid \mathbb{X}_n \,] \leqslant \sigma^2\,(nh)^{-1} \cdot \tfrac{1}{n} \sum_{i=1}^{n} B_{k,h}(x - X_i) \; ;$$

see (1.3). By the Quadrature Lemma (2.7),

$$\tfrac{1}{n} \sum_{i=1}^{n} B_{k,h}(x - X_i) \leqslant \int_0^1 B_{k,h}(x - t)\,\omega(t)\,dt + \zeta^{nh} \; ,$$

with the advertised bound on ζ^{nh}. It follows that

$$\mathbb{E}\bigl[\, L^{nh}(\varepsilon) \bigm| \mathbb{X}_n \bigr] \leqslant c\,(nh)^{-1/2}\,(\,1 + \eta^{nh}\,) \; ,$$

and then the same bound applies to $\mathbb{E}[\, Q_h(\varepsilon) \mid \mathbb{X}_n \,]$. Q.e.d.

(4.8) EXERCISE. Verify the inequalities (4.2) and (4.3).

(4.9) LEMMA (LOCAL BIAS). *Suppose that $f_o \in W^{m,2}(0,1)$, and let T_m be the m-th-order Taylor polynomial of f_o centered at x; see (2.13). Then, there exists a constant c, not depending on f_o, such that, for $0 < h \leqslant \frac{1}{2}$,*

$$Q_h(\,\pi_h - T_m\,) \leqslant h^{2m}\,[\,\mathfrak{C}_h f_o\,](x) \; ,$$

with $\mathfrak{C}_h$ *as in Lemma* (3.16).

PROOF. Since π_h minimizes $Q_h(p - f_o)$ over $p \in \mathcal{P}_m$, then the usual convexity-type (in)equality gives

$$(4.10)\quad Q_h(\pi_h - T_m) = Q_h(T_m - f_o) - Q_h(\pi_h - f_o) \leqslant Q_h(T_m - f_o)\,.$$

Lemma (3.13) clinches the deal. Q.e.d.

EXERCISE: (4.8).

5. Expected pointwise and global error bounds

In this section, we collect the results from the previous sections in a formal proof of Theorem (1.16). Recall that the kernel A and the order m are fixed and that $T_m(\,\cdot\,;x)$ is the m-th-order Taylor polynomial of f_o around the point x; see (2.13). For convenience, f^{nhm}, p^{nhm}, etc., are abbreviated as f^{nh}, p^{nh}, and so on. Moreover, $p^{nh}(\,\cdot\,;x)$ is denoted simply as p^{nh} and similarly for $\pi^{nh}(\,\cdot\,;x)$, $\pi_h(\,\cdot\,;x)$, and $T_m(\,\cdot\,;x)$. Also recall that π^{nh} is the solution of the semi-asymptotic problem (3.3). Let

$$(5.1)\qquad \varphi^{nh}(x) = \pi^{nh}(x\,;x) \quad\text{and}\quad \varphi_h(x) = \mathbb{E}[\,\varphi^{nh}(x) \mid \mathbb{X}_n\,]$$

for $x \in [\,0\,,1\,]$. It is useful to define the function ψ_h by

$$(5.2)\qquad \psi_h(x) = [\,\mathfrak{C}_h f_o\,](x)\,, \quad x \in [\,0\,,1\,]\,,$$

with $\mathfrak{C}_h$ as in Lemma (3.16).

Theorem (1.16) is a simple consequence of the following lemma.

(5.3) LEMMA. *Let* $m \geqslant 1$ *and* $\alpha > 0$. *For the quasi-uniform random-design model* (1.1)–(1.3), *there exists a constant* c *such that, for all functions* $f_o \in W^{m,2}(0,1)$, *and for all* $h \to 0$, $nh \to \infty$, *and* $x \in [\,0\,,1\,]$,

$$(a)\quad \mathbb{E}[\,|\,f^{nh}(x) - \varphi^{nh}(x)\,|^2 \mid \mathbb{X}_n\,] \leqslant c\,\zeta^{nh}\,\{\,(nh)^{-1} + h^{2m}\,\psi_h(x)\,\}\,,$$

$$(b)\quad \mathbb{E}[\,|\,\varphi^{nh}(x) - \varphi_h(x)\,|^2 \mid \mathbb{X}_n\,] \leqslant c\,(nh)^{-1}\,,$$

$$(c)\quad |\,\varphi_h(x) - f_o(x)\,|^2 \leqslant c\,h^{2m}\,\psi_h(x)\,,$$

with $\zeta^{nh} = \mathcal{O}\big(\,(nh)^{-1}\log n\,\big)$ *almost surely, uniformly in* $h \in \mathcal{G}_n(\alpha)$.

PROOF. To start, Lemma (2.5) implies that

$$\begin{aligned}|\,\varphi_h(x) - f_o(x)\,|^2 &= |\,\pi_h(x\,;x) - T_m(x\,;x)\,|^2\\ &\leqslant c\,Q_h(\pi_h - T_m) \leqslant c\,h^{2m}\,\psi_h(x)\end{aligned}$$

uniformly in x by Lemma (4.9). This is (c).

Similarly, for part (b),

$$| \varphi^{nh}(x) - \varphi_h(x) |^2 = | \pi^{nh}(x\,;x) - \pi_h(x\,;x) |^2 \leqslant c\, Q_h(\, \pi^{nh} - \pi_h\,) \;,$$

and so by Lemma (4.1), almost surely, uniformly in x,

$$\mathbb{E}[\, | \varphi^{nh}(x) - \varphi_h(x) |^2 \mid \mathbb{X}_n \,] \leqslant c\,(nh)^{-1} \;.$$

For part (a), we get likewise

$$\begin{aligned}\mathbb{E}[\, | f^{nh}(x) - \varphi^{nh}(x) |^2 \mid \mathbb{X}_n \,] &= \mathbb{E}[\, | p^{nh}(x;x) - \pi^{nh}(x;x) |^2 \mid \mathbb{X}_n \,] \\ &\leqslant c\,\mathbb{E}[\, Q_h(p^{nh} - \pi^{nh}\,) \mid \mathbb{X}_n \,] \\ &\leqslant \zeta^{nh} \{ \mathbb{E}[\, Q_h(\,\pi^{nh} - T_m\,) \mid \mathbb{X}_n \,] + h^{2m}\, [\mathfrak{C}_h\, f_o\,](x) \} \;,\end{aligned}$$

where we used in sequence Lemma (2.5) and Lemma (3.8). Now, by the bias/variance decomposition and Lemmas (4.1) and (4.9), we get

$$\begin{aligned}\mathbb{E}[\, Q_h(\,\pi^{nh} - T_m\,) \mid \mathbb{X}_n \,] &= \mathbb{E}[\, Q_h(\,\pi^{nh} - \pi_h\,) \mid \mathbb{X}_n \,] + Q_h(\,\pi_h - T_m\,) \\ &\leqslant c\,(nh)^{-1} + h^{2m}\, \psi_h(x) \;.\end{aligned}$$

The lemma follows. Q.e.d.

(5.4) Exercise. Prove Theorem (1.16) by bias/variance decomposition and integration. Note that $\| \mathfrak{C}_h f_o \|_{L^1(0,1)} \leqslant c \| f_o^{(m)} \|_2^2$.

(5.5) Exercise. Prove the bound of Theorem (1.16) on

$$\mathbb{E}\left[\, \tfrac{1}{n} \sum_{i=1}^{n} | f^{nh}(X_i) - f_o(X_i) |^2 \,\middle|\, \mathbb{X}_n \right] \;.$$

The detour $f^{nh}(X_i) \longrightarrow \varphi^{nh}(X_i) \longrightarrow f_o(X_i)$ suggests itself.

Exercises: (5.4), (5.5).

6. The asymptotic behavior of the error

In this section, we discuss the asymptotic behavior of the bias of the local polynomial estimator of order m as advertised in Theorem (1.17), as well as the variance. The bias may be treated using the methods of the previous sections. The variance is treated with elementary methods based on an explicit representation of the local polynomials, even though the variance is the same in every representation. To make all of this work, the study of the bias merely requires that $f_o^{(m)}$ be continuous. For the variance, the design density and the conditional variance of the noise need to be continuous. However, to make life a little easier, in addition to (1.4) and

(1.7), we assume in the appropriate places that

(6.1) the design density ω is continuous on $[\,0\,,\,1\,]$,

(6.2) $\sigma^2(\,t\,) = \mathbb{V}\text{ar}[\,Y \mid X = t\,]$ is continuous on $[\,0\,,\,1\,]$,

(6.3) $$f_o \in W^{m+1,2}(0,1) \ .$$

We start with the asymptotic bias of $f^{nh}(x)$ for fixed $x \in [\,0\,,\,1\,]$. Of course, Lemma (5.3)(a) says that we may ignore the difference between $f^{nh}(x)$ and $\varphi^{nh}(x)$. In part (c) of that lemma, we proved that, for the semi-asymptotic local polynomial estimator of order m,

(6.4) $$|\,\varphi_h(x) - f_o(x)\,|^2 \leqslant c\,h^{2m}\,\psi_h(x) \ ,$$

with ψ_h given by (5.2), under suitable conditions. The hardest part is *guessing* what the asymptotic behavior of $f^{nh}(x)$ might be. Recall that (6.4) arose from the inequalities (4.10) in the proof of Lemma (4.9),

$$|\,\varphi_h(x) - f_o(x)\,|^2 \leqslant c\,Q_h(\,\pi_h - T_m\,) \leqslant c\,Q_h(\,T_m - f_o\,) \ ,$$

which suggests that the leading term of the bias is determined by the missing next term of the Taylor polynomial of f_o, viz.

(6.5) $$N_m(\,t\,;\,x) = f_o^{(m)}(x)\,(\,t\,-x\,)^m/m!\ ,$$

Let $p = P_{hm}(\,\cdot\,;\,x)$ be the solution to

(6.6) minimize $Q_h(\,p - N_m\,)$ subject to $p \in \mathcal{P}_m$,

and define r_{hm} by

(6.7) $$r_{hm}(x) = P_{hm}(x\,;\,x)\ , \quad x \in [\,0\,,\,1\,]\ .$$

Now, by linearity, $p = \pi_h - P_{hm}(\,\cdot\,;\,x)$ is the solution to

(6.8) minimize $Q_h(\,p - \{\,f_o - N_m(\,\cdot\,;\,x)\,\}\,)$
subject to $p \in \mathcal{P}_m$,

and it follows from Lemma (5.3)(c) that

(6.9) $$|\,\varphi_h(x) + r_{hm}(x) - \{\,f_o(x) - N_m(x\,;x\,)\,\}\,|^2 \leqslant c\,h^{2m}\int_0^1 C_h(x-t\,)\,|\,f_o^{(m)}(\,t\,) - f_o^{(m)}(x)\,|^2\,dt \ .$$

Manipulations similar to those in the proof of Lemma (3.13) show that, for a kernel D_h sharing the properties of C_h as stated in Lemma (3.16),

(6.10) $$\int_0^1 C_h(x-t\,)\,|\,f_o^{(m)}(\,t\,) - f_o^{(m)}(x)\,|^2\,dt \leqslant h^2\int_0^1 D_h(x-t\,)\,|\,f_o^{(m+1)}(\,t\,)\,|^2\,dt \ ;$$

see Exercise (6.13) below. Combining (6.9) and (6.10) proves that

$$(6.11) \qquad | \varphi_h(x) + r_{hm}(x) - f_o(x) |^2 \leqslant c\, h^{2m+2}\, \psi_{h,m+1}(x) \ ,$$

with

$$(6.12) \qquad \int_0^1 \psi_{h,m+1}(t)\, dt \leqslant c \| f_o^{(m+1)} \|^2 \ .$$

Thus, $r_{hm}(x)$ is the leading term in the bias of $\varphi^{nh}(x)$. Compared with Lemma (5.3), the only extra requirement is that $f_o \in W^{m+1,2}(0,1)$, which is reasonable in view of the resulting bound of h^{2m+2} in (6.11). Of course, the precise behavior of $r_{hm}(x)$ is still lacking. The generic behavior is

$$r_{hm}(x) \approx c_o\, h^m\, f_o^{(m)}(x)$$

for a suitable constant c_o, depending on the kernel A, the order m, and the design density, but under special circumstances it may in fact be

$$r_{hm}(x) = \mathcal{O}(\, h^{m+1}\,) \ .$$

In the next section, we get to the bottom of this. A "computational" formula for r_{hm} is explored below in Exercise (6.41).

(6.13) EXERCISE. Verify that $p = \pi_h - P_{hm}$ indeed solves (6.7).

(6.14) EXERCISE. Show (6.10), with the advertised properties of D_h.

The above shows the following lemma.

(6.15) LEMMA (ASYMPTOTIC BIAS). *There exists a constant c such that, under the conditions of Theorem (1.16) supplemented with the assumption that all $f_o \in W^{m+1,2}(0,1)$ and $x \in [\,0\,,\,1\,]$,*

$$\text{(a)} \qquad | f_h(x) - f_o(x) + r_{hm}(x) |^2 \leqslant c\, h^{2m+2}\, \psi_{h,m+1}(x) \ ,$$

$$\text{(b)} \qquad \| f_h - f_o + r_{hm} \|^2 \leqslant c\, h^{2m+2} \| f_o^{(m+1)} \|^2 \ ,$$

where r_{hm} is given by (6.6)–(6.7).

It is useful to also consider the "local" bias, i.e., $Q_h(\,\pi_h - f_o\,)$. First, write f_o as

$$(6.16) \qquad f_o(t) = T_m(t\,;x) + N_m(t\,;x) + R_{m+1}(t\,;x) \ ,$$

with N_m as in (6.5) and

$$(6.17) \qquad R_{m+1}(t\,;x) = (\,m!\,)^{-1} \int_x^t (\,\tau - x\,)^m\, f_o^{(m+1)}(\tau)\, d\tau \ .$$

Then, we may write π_h as

$$(6.18) \qquad \pi_h = T_m + P_{hm} + S_{h,m+1}$$

with $p = P_{hm}$ the solution to (6.6) and $p = S_{h,m+1}$ the solution of

$$\text{(6.19)} \qquad \text{minimize} \quad Q_h(p - R_{m+1}) \quad \text{subject to} \quad p \in \mathcal{P}_m \ .$$

Then,

$$\text{(6.20)} \quad Q_h(\pi_h - f_o) = Q_h(P_{hm} - N_m) + Q_h(S_{h,m+1} - R_{m+1}) + \text{rem} \ ,$$

with

$$\text{(6.21)} \qquad |\,\text{rem}\,| \leqslant \{\, Q_h(P_{hm} - N_m)\,\}^{1/2} \{\, Q_h(S_{h,m+1} - R_{m+1})\,\}^{1/2} \ ,$$

where we used Cauchy-Schwarz.

Of course, for a suitable constant c not depending on f_o,

$$\text{(6.22)} \qquad Q_h(S_{h,m+1} - R_{m+1}) \leqslant c\, h^{2m+2} \, [\, \mathfrak{C}_h\,(f_o')\,](x) \ .$$

For $Q_h(P_{hm} - N_m)$, we may factor out $h^m f_o^{(m)}(x)$, ultimately leading to

$$h^{-2m} Q_h(P_{hm} - N_m) \longrightarrow \gamma_m \,|\, f_o^{(m)}(x)\,|^2 \ ,$$

where

$$\text{(6.23)} \qquad \gamma_m = \min_{q \in \mathcal{P}_m} \int_{-1}^{1} A(t)\, \frac{|\, q(t) - t^m\,|^2}{(m!)^2}\, \omega(t)\, dt \ ,$$

provided A has compact support in $[-1, 1]$ and $x \in [h, 1-h]$. If, in addition, the design density is continuous, then

$$\text{(6.24)} \qquad h^{-2m} Q_h(P_{hm} - N_m) \longrightarrow \gamma_m\, \omega(x)\,|\, f_o^{(m)}(x)\,|^2 \ .$$

All of this proves the following lemma.

(6.25) LEMMA (LOCAL BIAS). *Under the assumptions of Theorem* (1.16), *supplemented by* (6.1)–(6.3),

$$Q_h(\pi_h - f_o) = \gamma_m\, \omega(x)\,|\, f_o^{(m)}(x)\,|^2\, h^{2m} + \text{const} \cdot h^{2m+1} \ ,$$

with γ_m *given by* (6.23) *and where, for a suitable constant* k,

$$|\,\text{const}\,| \leqslant k\,(\,\|\, f_o^{(m)}\,\|^2 + \|\, f_o^{(m+1)}\,\|^2\,) \ .$$

We now consider the variance of $f^{nh}(x)$ for fixed $x \in [0, 1]$. As before, it suffices to consider the variance of $\varphi^{nh}(x) \equiv \pi^{nh}(x\,;\,x)$. By linearity, $p = \pi^{nh} - \pi_h$ is the solution to the problem

$$\text{(6.26)} \qquad \begin{array}{ll} \text{minimize} & Q_h(p) - 2\, L^{nh}(p) \\ \text{subject to} & p \in \mathcal{P}_m \ . \end{array}$$

To proceed, we apparently need an explicit representation of $p \in \mathcal{P}_m$. Here, we settle on a computational approach with

$$\text{(6.27)} \qquad p(t) = \sum_{k<m} p_k \,\{\, h^{-1}(t - x)\,\}^k \ , \quad t \in [0, 1] \ .$$

(In the next section, we use *orthogonal* polynomials.) Then, (6.26) is equivalent to the quadratic minimization problem

$$\text{(6.28)}\qquad \begin{array}{ll} \text{minimize} & \langle\, p\,,\, \mathbb{A}p\,\rangle - 2\,\langle\, p\,,\, \mathbb{C}\,\mathbb{D}_n\,\rangle \\ \text{subject to} & p \in \mathbb{R}^m\;, \end{array}$$

where $p = (\,p_0, p_1, \cdots, p_{m-1}\,)^T$, and $\mathbb{D}_n = (\,D_0, D_1, \cdots, D_{m-1}\,)^T$. The matrix $\mathbb{C}$ is defined by

$$\text{(6.29)}\qquad [\,\mathbb{C}\,\mathbb{D}_n\,]_k = \tfrac{1}{n}\sum_{i=1}^{n} D_i\, A_{h,k}(\,x - X_i\,)\;,$$

in which $A_{h,k}(z) = \{\,z/h\,\}^k\, A_h(z)$ as in (3.9), and $\mathbb{A} \in \mathbb{R}^{m\times m}$ is given by

$$\text{(6.30)}\qquad [\,\mathbb{A}\,]_{k,\ell} = \int_0^1 A_{h,k+\ell}(\,x - t\,)\,\omega(t)\,dt \quad \text{for } k,\,\ell = 0, 1, \cdots, m-1\;.$$

Then, writing $\varepsilon = \pi^{nh} - \pi_h$ and $\varepsilon(\,t\,) = \sum_{k<m} \varepsilon_k\,\{\,h^{-1}(\,t - x\,)\,\}^k$, we get

$$\text{(6.31)}\qquad \varepsilon = \langle\, e_1\,,\, \mathbb{A}^{-1}\,\mathbb{C}\,\mathbb{D}_n\,\rangle\;,$$

where e_1 is the first unit vector in the standard basis for $\mathbb{R}^m$. Of course, $\mathbb{E}[\,\varepsilon \mid \mathbb{X}_n\,] = 0$, and

$$\text{(6.32)}\qquad \mathbb{E}\big[\,\varepsilon^2 \,\big|\, \mathbb{X}_n\,\big] = (nh)^{-1}\,\langle\, e_1\,,\, \mathbb{A}^{-1}\mathbb{B}\mathbb{A}^{-1}\,e_1\,\rangle\;,$$

where $\mathbb{B} \in \mathbb{R}^{m\times m}$ is defined as $\mathbb{B} = \mathbb{C}\,\Sigma^2\,\mathbb{C}^T$, with Σ^2 the diagonal matrix with diagonal entries $\sigma^2(X_i)$. Specifically, then

$$\text{(6.33)}\qquad [\,\mathbb{B}\,]_{k,\ell} = n^{-1}\,h\,\sum_{i=1}^{n} \sigma^2(X_i)\,A_{h,k}(x - X_i)\,A_{h,\ell}(x - X_i)$$

for $k\,,\,\ell = 0, 1, \cdots, m-1$. (Note that the dependence on h is just right.)

The dependence on h as well as the design density somewhat obscure the result (6.32). If the design density ω is continuous and quasi-uniform and the conditional variance $\sigma^2(x)$ is continuous, then for $n \to \infty$

$$\begin{aligned} \mathbb{A} &\longrightarrow \int_0^1 A_{h,k+\ell}(x - \,t\,)\,dt\;, \\ h^{-1}\,\mathbb{B} &\longrightarrow \int_0^1 A_{h,k}(x - \,t\,)\,A_{h,\ell}(x - \,t\,)\,\sigma^2(\,t\,)\,\omega(\,t\,)\,dt\;, \end{aligned}$$

almost surely, so that

$$\text{(6.34)}\qquad \mathbb{A}^{-1}\mathbb{B}\mathbb{A}^{-1} \longrightarrow \sigma^2(x)\,\{\omega(x)\}^{-1}\,\mathbb{A}_u^{-1}\mathbb{B}_u\mathbb{A}_u^{-1}\;,$$

where $\mathbb{A}_u$ and $\mathbb{B}_u$ are equal to the matrices $\mathbb{A}$ and $\mathbb{B}$ for the uniform design density and variance; i.e., for the case $\omega(\,t\,) = 1,\ \sigma^2(\,t\,) = 1,\ t \in [\,0\,,\,1\,]$.

Regarding the dependence on h, note that, for $h < x < 1-h$ and the kernel A having support in $[-1, 1]$,

$$(6.35)\qquad \begin{aligned} [\mathbb{A}_u]_{k,\ell} &= \int_{-1}^{1} t^{k+\ell} A(t)\,dt , \\ [\mathbb{B}_u]_{k,\ell} &= \int_{-1}^{1} t^{k+\ell} A^2(t)\,dt , \quad k,\ell = 0,1,\cdots,m-1 . \end{aligned}$$

It then follows that

$$(6.36)\qquad c_o = \langle e_1 , \mathbb{A}_u^{-1}\mathbb{B}_u\mathbb{A}_u^{-1} e_1 \rangle$$

is a positive constant, and so $\mathbb{V}\mathrm{ar}[\pi^{nh}(x) - \pi_h(x) \,|\, \mathbb{X}_n] \asymp (nh)^{-1}$.

For $0<x<h$, one gets "incomplete" integrals in (6.35), with the integration bounds depending on x/h. For $1-h < x < 1$, a similar observation applies.

(6.37) Exercise. Show that (a) $\mathbb{A}_u$ and $\mathbb{B}_u$ are positive-definite, so that (b) $\mathbb{A}_u^{-1}\mathbb{B}_u\mathbb{A}_u^{-1}$ is positive-definite, which implies that (c) for $h<x<1-h$, the expression $\langle e_1 , \mathbb{A}_u^{-1}\mathbb{B}_u\mathbb{A}_u^{-1} e_1 \rangle$ denotes a positive constant not depending on x or h.

We summarize the above in the following lemma.

(6.38) Lemma (Asymptotic Variance). *For the quasi-uniform random-design model* (1.1)–(1.3), *supplemented with the conditions* (6.1)–(6.2), *there exists a constant* c_o, *depending on the order* m *and the kernel* A *only, such that for* $h \to 0$, $nh \to \infty$, *almost surely,*

$$\text{(a)}\quad \mathbb{E}[\,|\varphi^{nh}(x) - \varphi_h(x)|^2 \,|\, \mathbb{X}_n\,] = c_o\,\frac{\sigma^2(x)}{\omega(x)}\,(nh)^{-1}\,(1+o(1)) ,$$

$$\text{(b)}\quad \mathbb{E}[\,\|\varphi^{nh} - \varphi_h\|^2 \,|\, \mathbb{X}_n\,] = c_o\,(nh)^{-1}\,(1+o(1)) \int_0^1 \frac{\sigma^2(x)}{\omega(x)}\,dx .$$

Along the same lines, one may obtain the asymptotic behavior of the local variance $\mathbb{E}[\,Q_h(\pi^{nh} - \pi_h) \,|\, \mathbb{X}_n\,]$. With $\varepsilon = (\varepsilon_0, \varepsilon_1, \cdots .\varepsilon_{m-1})^T$, we may write $Q_h(\pi^{nh} - \pi_h) = \langle \mathbb{D}_n , \mathbb{C}\,\mathbb{A}^{-1}\mathbb{C}^T\mathbb{D}_n \rangle$, and it follows that

$$(6.39)\qquad \mathbb{E}[\,Q_h(\pi^{nh} - \pi_h) \,|\, \mathbb{X}_n\,] = (nh)^{-1}\,\mathrm{trace}(\mathbb{A}^{-1}\mathbb{B})$$

with $\mathbb{B} = \mathbb{C}\,\Sigma^2\,\mathbb{C}^T$ as before; see (6.32)–(6.33). So, if the kernel A has compact support in $[-1, 1]$ and $x \in [h, 1-h]$, then almost surely

$$nh\,\mathbb{E}[\,Q_h(\pi^{nh} - \pi_h) \,|\, \mathbb{X}_n\,] \longrightarrow \sigma^2(x)\,\mathrm{trace}(\mathbb{A}_u^{-1}\mathbb{B}_u) .$$

(6.40) Lemma (Local Variance). *Under the conditions of Lemma* (6.38),

$$\mathbb{E}[\,Q_h(\pi^{nh} - \pi_h) \,|\, \mathbb{X}_n\,] = \frac{\sigma^2(x)}{nh}\,\mathrm{trace}(\mathbb{A}_u^{-1}\mathbb{B}_u)\,(1+o(1)) .$$

We finish with an "explicit" formula for the leading term of the bias, r_{hm}.

(6.41) EXERCISE. Show that P_{hm} may be written explicitly as follows. Let $\beta = (\beta_0, \beta_1, \cdots, \beta_{m-1})^T$ be defined as

$$\beta_k = \int_0^1 A_{h,k+m}(t-x)\,\omega(t)\,dt\ , \quad k = 0, 1, \cdots, m-1\ ,$$

and define $\gamma = (\gamma_0, \gamma_1, \cdots, \gamma_{m-1})^T$ by $\gamma = \mathbb{B}^{-1}\beta$, with $\mathbb{B}$ as in (6.33). (Note that γ is *deterministic*.) Then,

$$P_{hm}(t; x) = \frac{f_o^{(m)}(x)}{m!} \sum_{k<m} \gamma_k \{h^{-1}(t-x)\}^k\ , \quad t \in [0, 1]\ .$$

EXERCISES: (6.13), (6.14), (6.37), (6.41).

7. Refined asymptotic behavior of the bias

In this section, we determine the precise behavior of the leading term of the bias of the local polynomial estimator of order m using "our" methods. Since we assumed already that f_o had one extra degree of smoothness, we shall do the same for the design density ω and assume that it is Lipschitz continuous; i.e., there exists a constant, denoted by $|\omega|_{\text{Lip}}$, such that

(7.1) $$|\omega(t) - \omega(x)| \leqslant |\omega|_{\text{Lip}}\, |t-x| \quad \text{for all } t, x \in [0, 1]\ .$$

For convenience, we also assume that the kernel A has compact support,

(7.2) $$A(t) = 0 \quad \text{for} \quad |t| > 1\ .$$

In the previous section, the function r_{hm}, defined in (6.7)–(6.8), was identified as the leading term of the bias in f^{nh}. Of course, r_{hm} depends on everything, but its dependence on f_o and h ought to be relatively simple. Thus, it is easy to see that

(7.3) $$r_{hm}(x) = (1/m!)\, h^m\, f_o^{(m)}(x)\, \varrho_{hm}(x)\ ,$$

where ϱ_{hm} is defined as follows. Let

(7.4) $$M_{hm}(t; x) = \{h^{-1}(t-x)\}^m\ , \quad t \in [0, 1]\ ,$$

and let $p = \mu_{hm}(\,\cdot\,; x)$ be the solution to

(7.5) $$\text{minimize} \quad Q_h(p - M_{hm}) \quad \text{subject to} \quad p \in \mathcal{P}_m\ .$$

Then, set

(7.6) $$\varrho_{hm}(x) = \mu_{hm}(x; x)\ .$$

Obviously, the dependence on f_o is neatly factored out, but the factor h^m is somewhat tendentious since ϱ_{hm} is still allowed to depend on h. However, to show that we are on the right track, note that for the uniform design density, A satisfying (7.2) and $h < x < 1-h$, we have

$$Q_h(p - M_{hm}) = \int_{-1}^{1} A(t)\,|\,q(t) - t^m\,|^2\,dt \ ,$$

where $q(t) = p(x - ht)$. Thus, the minimizer over q does not depend on h, and then neither does its value at 0. Of course, it must be shown that the design density may be replaced locally by the uniform density, and this we set out to do. The net result is the following theorem.

(7.7) THEOREM. (a) *Assuming that the design density is quasi-uniform and satisfies* (7.1), *and the kernel satisfies* (1.7) *and* (7.2), *then* r_{hm}, *defined in* (6.7)–(6.8), *satisfies, for* $h \to 0$,

$$r_{hm}(x) = \frac{h^m\, f_o^{(m)}(x)}{m\,!}\,\varrho(\,\lambda(x,h)\,) + \mathcal{O}(\,h^{m+1}\,) \ ,$$

where $\lambda(x,h) = 0 \vee \min(\,x/h\,,\,1\,,\,(1-x)/h\,)$.
(b) *If, moreover, the order* m *is odd, then* $r_{hm}(x) = \mathcal{O}(\,h^{m+1}\,)$ *uniformly in* $x \in [\,h\,,\,1-h\,]$.

(7.8) EXERCISE. Under the assumptions of Theorem (1.17), prove that, for odd m,

$$\mathbb{E}[\,\|\,f^{nh} - f_o\,\|^2_{(h,1-h)}\,] = \mathcal{O}(\,n^{-(2m+2)/(2m+3)}\,) \ ,$$

provided $h \asymp n^{-1/(2m+3)}$. Here we resurrected the notation (13.2.9).

We go on to the proof of Theorem (7.7). The first step is to get rid of the design density. To that end, define the quadratic functional $\mathfrak{Q}_h$ by

$$\mathfrak{Q}_h(\,g\,) = \int_0^1 A_h(x-t)\,|\,g(t)\,|^2\,dt \ . \tag{7.9}$$

Thus, $\mathfrak{Q}_h(\,g\,) = Q_h(\,g\,)$ in the case where the design density is uniform. Let $p = \nu_{hm}(\,\cdot\,;\,x\,)$ be the solution to

$$\text{minimize} \quad \mathfrak{Q}_h(\,p - M_{hm}\,) \quad \text{subject to} \quad p \in \mathcal{P}_m \ . \tag{7.10}$$

(7.11) LEMMA. *If the design density* ω *is quasi-uniform and satisfies* (7.1), *then there exists a constant* c *such that, for all* $h > 0$,

$$|\,\mu_{hm}(x\,;\,x) - \nu_{hm}(x\,;\,x)\,| \leqslant c\,h \qquad \textit{for all } x \in [\,0\,,\,1\,] \ .$$

PROOF. By Lemma (2.5), it suffices to show that

$$Q_h(\,\mu_{hm}(\,\cdot\,;\,x) - \nu_{hm}(\,\cdot\,;\,x)\,) \leqslant c\,h\,|\,\omega\,|_{\mathrm{Lip}} \ .$$

The by now familiar arguments regarding local polynomial minimization problems are applicable here. Let $\nu_{hm}(\,\cdot\,;x)$ and $\mu_{hm}(\,\cdot\,;x)$ be denoted by ν_{hm} and μ_{hm}, and set $\varepsilon = \nu_{hm} - \mu_{hm}$. The convexity (in)equalities for the problems (7.5) and (7.10) give

$$\mathfrak{Q}_h(\varepsilon) = \mathfrak{Q}_h(\mu_{hm} - M_{hm}) - \mathfrak{Q}_h(\nu_{hm} - M_{hm}) \,,$$

$$Q_h(\varepsilon) = Q_h(\nu_{hm} - M_{hm}) - Q_h(\mu_{hm} - M_{hm}) \,.$$

Multiply the first inequality by $\omega(x)$ and add the two inequalities. The Lipschitz continuity of ω then yields

$$Q_h(\varepsilon) + \omega(x)\,\mathfrak{Q}_h(\varepsilon) \leqslant h\,|\,\omega\,|_{\mathrm{Lip}} \int_0^1 A_h(x-t)\,|\,b(t)\,|\,dt \,,$$

where

$$b(t) = |\,\mu_{hm}(t) - M_{hm}(t)\,|^2 - |\,\nu_{hm}(t) - M_{hm}(t)\,|^2 \,.$$

Now, dropping the arguments t for now, write $b(t)$ as

$$b = -\,|\,\nu_{hm} - \mu_{hm}\,|^2 - 2\,(\mu_{hm} - M_{hm})\,(\nu_{hm} - \mu_{hm}) \,,$$

and apply Cauchy-Schwarz, to conclude that

$$Q_h(\varepsilon) + \omega(x)\,\mathfrak{Q}_h(\varepsilon) \leqslant$$
$$h\,|\,\omega\,|_{\mathrm{Lip}} \Bigl(\mathfrak{Q}_h(\varepsilon) + 2\,\{\, \mathfrak{Q}_h(\varepsilon) \cdot \mathfrak{Q}_h(\mu_{hm} - M_{hm})\,\}^{1/2} \Bigr) \,.$$

The inequality above implies that

$$\mathfrak{Q}_h(\varepsilon) \leqslant \frac{h\,|\,\omega\,|_{\mathrm{Lip}}}{\omega(x)} \Bigl(\mathfrak{Q}_h(\varepsilon) + 2\,\{\, \mathfrak{Q}_h(\varepsilon)\,\}^{1/2}\,\{\, \mathfrak{Q}_h(\mu_{hm} - M_{hm})\,\}^{1/2} \Bigr) \,,$$

which implies, for $h \longrightarrow 0$, that

$$\mathfrak{Q}_h(\varepsilon) \leqslant \gamma^2(h)\,\mathfrak{Q}_h(\mu_{hm} - M_{hm}) \,,$$

where

$$\gamma(h) = \frac{2\,h\,|\,\omega\,|_{\mathrm{Lip}}}{\omega(x) - h\,|\,\omega\,|_{\mathrm{Lip}}} \longrightarrow \frac{2\,h\,|\,\omega\,|_{\mathrm{Lip}}}{\omega(x)} \quad \text{for } h \to 0 \,.$$

Finally, apply Lemma 4.12 with the uniform density to obtain the bound

$$\mathfrak{Q}_h(\mu_{hm} - M_{hm}) \leqslant c\,h^{2m}\,g_h(x) \,,$$

where
$$g_h(x) = \int_0^1 C_h(x,t)\,|\,M_{hm}^{(m)}(t)\,|^2\,dt \,.$$

Since $M_{hm}^{(m)}(t) = m!\,h^{-m}$, it follows that $Q_h(\mu_{hm} - M_{hm})$ is bounded uniformly in x and h. The lemma follows. Q.e.d.

(7.12) Exercise. Prove Lemma (7.11) under the assumption

$$\int_0^1 \int_0^1 A_{h,k}(x-t)\,|\,\omega(t) - \omega(x)\,|\,dt\,dx \leqslant c\,h \,, \quad 0 \leqslant k < m \,,$$

with $A_{k,h}$ as in (3.9). (This is a weaker than Lipschitz continuity, (7.1).)

The next step is to show that ν_{hm} is nice.

(7.13) LEMMA. *Assume that the kernel A satisfies (7.2). Then, there exists a continuous function ϱ on $[0, 1]$ such that*

$$\nu_{hm}(x\,;x) = \varrho(\,\lambda(x,h)\,)\;, \quad x \in [\,0\,,1\,]\;.$$

Moreover, if m is odd, then $\varrho(1) = 0$.

PROOF. Keep in mind that $A(t) = 0$ for $|\,t\,| > 1$. Then, for h satisfying $h < x < 1 - h$, we have $[\,x - h\,,\, x + h\,] \subset [\,0\,,1\,]$, so that

$$\mathfrak{Q}_h(\,g\,) = \int_{x-h}^{x+h} A_h(\,x - t\,)\,|\,g(\,t\,)\,|^2\,dt = \int_{-1}^{1} A(\,t\,)\,|\,g(\,x + h\,t\,)\,|^2\,dt\;.$$

It follows that

$$\mathfrak{Q}_h(\,p - M_{hm}\,) = \int_0^1 A(\,t\,)\,|\,q(\,t\,) - t^m\,|^2\,dt\;, \tag{7.14}$$

where $q(\,t\,) = p(\,x - h\,t\,)$. Thus, for $h < x < 1 - h$, the solution q of

$$\begin{aligned} &\text{minimize} \quad \int_0^1 A(\,t\,)\,|\,q(\,t\,) - t^m\,|^2\,dt \\ &\text{subject to} \quad q \in \mathcal{P}_m \end{aligned} \tag{7.15}$$

does not depend on h, but then neither does $\nu_{hm}(x\,;x) = q(0)$. Thus, we may define $\varrho(1)$ as $\varrho(1) = q(0)$, and this does not depend on h.

For $0 \leqslant x < h$, things change because then the interval $[\,x - h\,,\, x + h\,]$ is not contained in $[\,0\,,1\,]$. Let $\theta = x/h$. Then,

$$\begin{aligned} \mathfrak{Q}_h(\,g\,) &= \int_0^{x+h} A_h(\,x - t\,)\,|\,g(\,t\,)\,|^2\,dt \\ &= \int_{-h}^{x} A_h(\,t\,)\,|\,g(\,x - t\,)\,|^2\,dt = \int_{-1}^{\theta} A(\,t\,)\,|\,g(\,x - h\,t\,)\,|^2\,dt\;. \end{aligned}$$

Thus, in this case, the problem (7.10) is equivalent to

$$\begin{aligned} &\text{minimize} \quad \int_{-1}^{\theta} A(\,t\,)\,|\,q(\,t\,) - t^m\,|^2\,dt \\ &\text{subject to} \quad q \in \mathcal{P}_m\;, \end{aligned} \tag{7.16}$$

and it follows that the solution depends on θ. Moreover, it is easy to show that the solution depends continuously on θ. Then, $\nu_{hm}(x\,;x) = q(0)$ and so define $\varrho(\theta) = q(0)$.

The case $1 - h < x < 1$ may be treated similarly. This proves part (a). Part (b) follows from the symmetry of A. In that case, the solution q of (7.10) is odd also, in which case $q(0) = 0$. Q.e.d.

(7.17) EXERCISE. Show that the solution $q(\,\cdot\,;\theta)$ of (7.16) depends continuously on θ. In particular, show that, for θ, $\eta \in [\,0\,,\,1\,]$,

$$| \, q(0\,;\theta) - q(0\,;\eta) \, | \leqslant c \, | \, \theta - \eta \, | \ .$$

[Hint: The easiest way to see it is by choosing a computational model. The alternative is to use "our" methods, in which case the following result is useful: There exists a constant K such that, for all $p \in \mathcal{P}_m$ and for all θ, $\eta \in [\,0\,,\,1\,]$ with $\theta > \eta$,

$$\int_\eta^\theta A(\,t\,)\,|\,p(\,t\,)\,|^2\,dt \leqslant K\,a(\theta,\eta) \int_{-1}^\eta A(\,t\,)\,|\,p(\,t\,)\,|^2\,dt \ ,$$

where $a(\theta,\eta) = \displaystyle\int_\eta^\theta A(\,t\,)\,dt$.

Prove this (the equivalent norms in the proof of Lemma (2.5) should be a helpful idea) and use it.]

(7.18) EXERCISE. Show that the solution of (7.10) is odd when m is odd.

(7.19) REMARK. It is interesting to investigate the oddity of the odd solution of (7.15) a bit further. In particular, the study of the *orthogonal polynomials* associated with the least-squares problem (7.10) suggests itself. The relevant inner product is

$$\langle\, f\,,\,g\,\rangle_A = \int_{-1}^1 A(\,t\,)\,f(\,t\,)\,g(\,t\,)\,dt \tag{7.20}$$

for f, $g \in L^2(-1,1)$. The orthogonal polynomials $P_k(\,t\,)$, $k = 0, 1, 2, \cdots$, are uniquely defined by the requirement that $P_k(\,t\,)$ is a polynomial in t of exact degree k and

$$\langle\, P_k\,,\,P_\ell\,\rangle_A = \begin{cases} 1 & \text{for } k = \ell\ , \\ 0 & \text{for } k \neq \ell \text{ and } k,\ \ell = 0,1,2,\ \cdots \ . \end{cases} \tag{7.21}$$

This implies that P_k has k distinct zeros inside $(0,1)$. Moreover, the orthogonal polynomials satisfy a three-point recurrence relation: There exist constants a_k, b_k, c_k, $k = 1, 2, \cdots$, such that

$$P_{k+1}(\,t\,) = (\,a_k\,t + b_k\,)\,P_k(\,t\,) - c_k\,P_{k-1}(\,t\,)\ ; \tag{7.22}$$

see, e.g., SANSONE (1959). It is a standard exercise to show that the symmetry of the weight function $A(\,t\,)$ implies that $b_k = 0$ for all k. Consequently, the polynomial $P_k(\,t\,)$ is even for even k and odd for odd k,

$$P_k(-\,t\,) = (-1)^k\,P_k(\,t\,) \quad \text{for all } t\ , \tag{7.23}$$

and it follows that $P_k(0) = 0$ for odd k.

With the aid of the orthogonal polynomials, the solution of (7.15) may be written as

$$q(t) = \sum_{k=0}^{m-1} \langle t^m , P_k \rangle_A P_k(t) , \tag{7.24}$$

where we abused notation somewhat: $\langle t^m , P_k \rangle_A$ should be interpreted as

$$\langle t^m , P_k \rangle_A = \int_{-1}^{1} A(t)\, t^m \, P_k(t)\, dt \ .$$

Of course,

$$t^m = \sum_{k=0}^{m} \langle t^m , P_k \rangle_A P_k(t) \ . \tag{7.25}$$

It follows that

$$q(t) - t^m = \langle g , P_m \rangle_A P_m(t) , \tag{7.26}$$

and this vanishes at $t = 0$ when m is odd, so then $q(0) = 0$.

What happens for even m? Of course, (7.26) implies that $q(t)$ and t^m coincide at any of the m zeros of P_m inside the interval $(-1, 1)$. So, even for even m, one can achieve super convergence by choosing as the estimator

$$f^{nh}(x) = p^{nhm}(x \,; x + h\, z_m) , \tag{7.27}$$

in the notation analogous to (1.9), for any zero z_m of P_m. (This is rarely considered in the literature.)

(7.28) EXERCISE. Prove the statements regarding the orthogonal polynomials, to wit (7.22) through (7.26), and the details of (7.27).

EXERCISES: (7.8), (7.12), (7.17), (7.18), (7.28).

8. Uniform error bounds for local polynomials

In previous chapters, we have successfully studied (optimal) uniform error bounds for kernel estimators and, by studying the "equivalent" kernel estimator, for smoothing spline estimators. What is the situation for local polynomial estimators? If one could approximate local polynomial estimators with kernel estimators with nice kernels, there would be no problem, but the explicit representation (1.13) does not inspire confidence, and this is only the local *linear* estimator. Nevertheless, DONY, EINMAHL, and MASON (2006) bite the bullet and, assuming (1.1)–(1.4) and (1.19), show that

$$\lim_{n\to\infty} \sup_{h\in\mathcal{H}_n(\alpha)} \frac{\| f^{nhm} - \mathbb{E}[\, f^{nhm}\,] \|_\infty}{\sqrt{(nh)^{-1}\{\log(1/h) \vee \log\log n\}}} < \infty \tag{8.1}$$

almost surely. Here, see (14.6.2), the range of the smoothing parameter is

$$(8.2) \qquad \mathcal{H}_n(\alpha) = \Bigl[\, \alpha\, \bigl(n^{-1} \log n\,\bigr)^{1-2/\kappa} \,,\, \tfrac{1}{2} \,\Bigr] \;.$$

The result (8.1) is our goal, apart from the $\log(1/h) \vee \log\log n$ business. Also, we will replace $\mathbb{E}[\, f^{nhm} \,]$ by $\mathbb{E}\bigl[\, f^{nhm} \bigm| \mathbb{X}_n \,\bigr]$. However, we do not want to follow the approach of DONY, EINMAHL, and MASON (2006). The alternative is to abandon any connection with kernel estimators and to use Lemma (2.5) as the starting point. Thus, it would suffice to get bounds on

$$(8.3) \qquad \sup_{x\in[\,0\,,\,1\,]} \int_0^1 A_h(x-t)\,|\,p^{nhm}(\,t\,;x) - T_m(\,t\,;x)\,|^2\,\omega(\,t\,)\,dt \;.$$

This has mostly been done already in §§ 2 through 4.

To set the stage, we assume the regression model (1.4)-(1.6), with the kernel A satisfying (1.7). Regarding f_o, we assume

$$(8.4) \qquad f_o \in W^{m,\infty}(0,1) \;.$$

(8.5) THEOREM. *Let* $m \geqslant 1$ *and* $\alpha > 0$. *Assuming* (1.4)–(1.7) *and* (8.4),

$$\lim_{n\to\infty} \sup_{h\in\mathcal{H}_n(\alpha)} \frac{\|\, f^{nh} - f_o \,\|_\infty}{\sqrt{h^{2m} + (nh)^{-1}\log n}} < \infty \quad \textit{almost surely}\;.$$

(8.6) COROLLARY. *Under the conditions of Theorem* (8.5), *if*

$$H \asymp (n^{-1}\log n)^{-1/(2m+1)} \quad \textit{in probability},$$

then $\|\, f^{nH} - f_o \,\|_\infty = \mathcal{O}\bigl(\, (n^{-1}\log n)^{-m/(2m+1)} \,\bigr)$ *in probability*.

PROOF OF THEOREM (8.5). Lemma (2.5) implies that it suffices to provide the appropriate bounds on $\|\, R \,\|_\infty$, where $R(x) = \{\, Q_h(\,p^{nh} - T_m\,)\}^{1/2}$. (Recall the dependence of Q_h as well as p^{nh} and T_m on x.) By the triangle inequality,

$$R(x) \leqslant \{\, Q_h(\,p^{nh} - \pi^{nh}\,)\,\}^{1/2} + \{\, Q_h(\,\pi^{nh} - T_m\,)\,\}^{1/2} \;.$$

By Lemma (3.16),

$$\{\, Q_h(\,p^{nh} - \pi^{nh}\,)\,\}^{1/2} \leqslant \eta^{nh} \Bigl(\, \{\, Q_h(\,\pi^{nh} - T_m\,)\,\}^{1/2} + h^m \,\|\, f_o^{(m)} \,\|_\infty \Bigr) \,,$$

where $\eta^{nh} = \mathcal{O}\bigl(\, ((nh)^{-1}\log n)^{1/2} \,\bigr)$ almost surely, uniformly in $h \in \mathcal{H}_n(\alpha)$, and where we used the bound $[\,\mathfrak{C}_h f_o\,](x) \leqslant c\,\|\, f_o \,\|_\infty^2$. Thus,

$$R(x) \leqslant (\,1 + \eta^{nh}\,)\,\{\, Q_h(\,\pi^{nh} - T_m\,)\,\}^{1/2} + \eta^{nh}\, h^m \,\|\, f_o^{(m)} \,\|_\infty \;.$$

Now, again by the triangle inequality,

$$\{\, Q_h(\,\pi^{nh} - T_m\,)\,\}^{1/2} \leqslant \{\, Q_h(\,\pi^{nh} - \pi_h\,)\,\}^{1/2} + \{\, Q_h(\,\pi_h - T_m\,)\,\}^{1/2} \;.$$

Lemma (4.9) tells us that $\{\, Q_h(\,\pi_h - T_m\,)\,\}^{1/2} \leqslant h^m \,\|\, f_o^{(m)} \,\|_\infty^2$, and so

$$R(x) \leqslant (\,1 + \eta^{nh}\,) \Bigl(\,\{\, Q_h(\,\pi^{nh} - \pi_h\,)\,\}^{1/2} + h^m \,\|\, f_o^{(m)} \,\|_\infty \Bigr) \; .$$

The theorem now follows from Lemma (8.7) below. Q.e.d.

(8.7) LEMMA (VARIANCE). *Under the assumptions* (1.4)–(1.6) *and* (1.7), *provided* $h = \mathcal{O}\bigl(\,(n^{-1}\log n)^{1/3}\,\bigr)$,

$$\sup_{x\in(0,1)} \; Q_h(\,\pi^{nh} - \pi_h\,) = \mathcal{O}_P\bigl(\,(nh)^{-1}\,\log n\,\bigr) \; .$$

(Note that only an upper bound on h *is specified.)*

PROOF. The proof follows that of Lemma (4.1) almost to the letter. Let $\varepsilon = \pi^{nh}(\,\cdot\,;x) - \pi_h(\,\cdot\,;x)$, and again write

$$\varepsilon(\,t\,) = \sum_{k<m} \eta_k \,\{\, h^{-1}(t\,-x)\,\}^k$$

with $\varepsilon_k = h^k\,\varepsilon^{(k)}(x)/\,k\,!$, $k = 0, 1, \cdots, m-1$. Now, recall (4.4)–(4.5) and use Lemma (2.5) to conclude that

$$Q_h(\,\varepsilon\,) = L^{nh}(\,\varepsilon\,) \leqslant \sum_{k<m} |\,\varepsilon_k\,|\,|\,S_{knh}(x)\,| \leqslant c\,\{\,Q_h(\,\varepsilon\,)\,\}^{1/2} \max_{k<m} |\,S_{knh}(x)\,| \; ,$$

where

$$S_{knh}(x) = \Bigl\|\, \tfrac{1}{n} \sum_{i=1}^{n} D_i\, A_{h,k}(\,\cdot\, - X_i)\,\Bigr\|_\infty$$

with $A_{h,k}$ as in (3.9). So, for another constant c,

$$\sup_{x\in(0,1)} \; Q_h(\,\varepsilon\,) \leqslant c \max_{k<m} \,\|\, \tfrac{1}{n} \sum_{i=1}^{n} D_i\, A_h(\,\cdot\, - X_i)\,\|_\infty^2 \; .$$

Finally, Theorem (14.6.12) provides the closing argument. Q.e.d.

9. Estimating derivatives

Here, we give a brief discussion of how to estimate derivatives in the local polynomial estimation context. The obvious estimator of the k-th-order derivative of the conditional expectation function $f_o(x) = \mathbb{E}[\,Y\,|\,X = x\,]$ is

(9.1) $$\varphi^{nh(k)}(x) = (p^{nh})^{(k)}(x;x) \; , \quad x \in [\,0\,,\,1\,] \; ,$$

where, for $k = 1, 2, \cdots, m-1$,

(9.2) $$(p^{nh})^{(k)}(\,t\,;x) = \frac{d^{\,k}}{d\,t^{\,k}}\, p^{nh}(\,t\,;x) \; .$$

Moreover, Lemma (2.5) provides for ready-made pointwise error bounds,

(9.3) $$|\,\varphi^{nh(k)}(x) - f_o^{(k)}(x)\,|^2 \leqslant c\,h^{-2k}\, Q_h(\,p^{nh}(\,\cdot\,;x) - T_m(\,\cdot\,;x)\,) \; ,$$

for $k < m$, so that, with Lemma (5.3), under the usual conditions

$$\text{(9.4)} \qquad \mathbb{E}\Big[\, \| \, \varphi^{nh(k)} - f_o^{(k)} \, \|^2 \, \Big| \, \mathbb{X}_n \Big] \leqslant c\, h^{-2k} \, \{ \, h^{2m} + (nh)^{-1} \, \}$$

(plus negligible terms), which leads to the usual bounds.

(9.5) THEOREM. *Assume the conditions* (1.4) *through* (1.6). *Let* $m \geqslant 1$, *and suppose that* $f_o \in W^{m,2}(0,1)$. *Then, for* $k = 1, 2, \cdots, m-1$,

$$\mathbb{E}\Big[\, \| \, \varphi^{nh(k)} - f_o^{(k)} \, \|^2 \, \Big| \, \mathbb{X}_n \Big] = \mathcal{O}\big(\, n^{-2(m-k)/(2m+1)} \, \big)$$

almost surely, provided $h \asymp n^{-1/(2m+1)}$ *(deterministically).*

(9.6) EXERCISE. Verify the statements (9.3) and (9.4).

Note that there are no differentiability conditions on the kernel!

EXERCISE: (9.6).

10. Nadaraya-Watson estimators

Finally, we discuss the Nadaraya-Watson estimator in the case of convolution or convolution-like kernels of order m for $m \geqslant 1$. In particular, we prove the "usual" convergence rates. This appears to be new for $m \geqslant 3$, where the kernels have to take on negative values. For such kernels, it is not obvious that the Nadaraya-Watson estimator can be interpreted as a local polynomial estimator of order 1, but asymptotically everything is under control. This characterization of the Nadaraya-Watson estimator as the solution of a quadratic minimization problem is key to its study.

The main emphasis here is on the accurate approximation of the bias, conditioned on the design. (Theorem (14.6.12) covers the noise part.) In particular, asymptotically, the approximation error should be negligible compared with the bias. The first result is "ideal" but somewhat unrealistic in that a more or less complete knowledge of the design density is required. The second result avoids this, but unfortunately the design density has to be very smooth. The authors do not know if the excessive smoothness of the design is in fact a necessary condition.

We begin with a somewhat unrealistic setting. Let $A_h(x, t)$, $0 < h \leqslant 1$, be a family of kernels such that

$$\text{(10.1)} \qquad A_h(x, t)\, \omega(x) \, , \; 0 < h \leqslant 1, \quad \text{is convolution-like of order } 1 \, .$$

For the general regression problem (1.1)–(1.6), the Nadaraya-Watson estimator is defined as

$$(10.2)\qquad f^{nh}_{NW}(t) = \frac{\frac{1}{n}\sum_{i=1}^{n} Y_i\, A_h(X_i, t)}{\frac{1}{n}\sum_{i=1}^{n} A_h(X_i, t)} , \qquad t \in [0, 1] ,$$

provided the denominator is nonnegative everywhere on $[0, 1]$. We shall not worry about misbehaving denominators.

Let $\mathfrak{S}^{nh}_{NW}$ be the part of the estimator due to the noise in the responses,

$$(10.3)\qquad \mathfrak{S}^{nh}_{NW}(t) = \frac{\frac{1}{n}\sum_{i=1}^{n} D_i\, A_h(X_i, t)}{\frac{1}{n}\sum_{i=1}^{n} A_h(X_i, t)} , \qquad t \in [0, 1] ,$$

and let $[\mathcal{A}_h f_o](t)$ be the continuous version of the pure-signal part of f^{nh}_{NW},

$$(10.4)\qquad [\mathcal{A}_h f_o](t) = \frac{\int_0^1 A_h(x, t)\, f_o(x)\, \omega(x)\, dx}{\int_0^1 A_h(x, t)\, \omega(x)\, dx} , \qquad t \in [0, 1] .$$

Note that here the denominator equals 1 since $A_h(x, t)\,\omega(x)$, $0 < h \leqslant 1$ is convolution-like of order 1.

(10.5) THEOREM. *Let $\alpha > 0$ and $f_o \in W^{1,\infty}(0,1)$. Under the assumptions (1.1)–(1.6) and (10.1),*

$$f^{nh}_{NW} = \mathcal{A}_h f_o + \mathfrak{S}^{nh}_{NW} + \delta^{nh} \quad \text{on} \quad [0, 1]$$

with, almost surely,

$$\limsup_{n\to\infty} \sup_{h\in\mathcal{H}_n(\alpha)} \frac{\|\,\delta^{nh}\,\|_\infty}{h\sqrt{(nh)^{-1}\log n}} < \infty .$$

See (14.6.2) for the definition of $\mathcal{H}_n(\alpha)$.

PROOF. Observe that $f^{nh}_{NW} = \mathbb{E}[\, f^{nh}_{NW} \mid \mathbb{X}_n\,] + \mathfrak{S}^{nh}_{NW}$, so that it suffices to bound $\delta^{nh} = \mathbb{E}[\, f^{nh}_{NW} \mid \mathbb{X}_n\,] - \mathcal{A}_h f_o$.

For fixed t, we have that $p = \mathbb{E}[\, f^{nh}_{NW}(t) \,|\, \mathbb{X}_n\,]$ is the solution to

$$(10.6)\qquad \begin{aligned} &\text{minimize} && \tfrac{1}{n}\sum_{i=1}^{n} A_h(X_i, t)\,|\, p - f_o(X_i)\,|^2 \\ &\text{subject to} && p \in \mathbb{R} . \end{aligned}$$

Unfortunately, this minimization problem has a solution if and only if $\sum_{i=1}^{n} A_h(X_i, t) > 0$, in which case the objective function is a nice quadratic function of $p \in \mathbb{R}$. Of course, this holds almost surely as $h \to 0$, $nh \to \infty$, by (10.1).

Next, still for fixed t, we observe that $q = [\mathcal{A}_h f_o](t)$ is the (unique) solution to

$$\text{(10.7)} \qquad \begin{array}{ll} \text{minimize} & \displaystyle\int_0^1 A_h(x,t)\,|\,q - f_o(x)\,|^2\,\omega(x)\,dx \\ \text{subject to} & q \in \mathbb{R}\;. \end{array}$$

Recall that Ω_n and Ω_o denote respectively the empirical distribution function of the design and the distribution function of the design density.

Let $\varepsilon \equiv p - q$. Then, the usual manipulations for the minimization problems (10.6) and (10.7) result in

$$\text{(10.8)} \quad \varepsilon^2 \int_0^1 A_h(x,t)\,d\Omega_o(x) = \varepsilon \int_0^1 A_h(x,t)\,\big(\,q - f_o(x)\,\big) \times \big(\,d\Omega_n(x) - d\Omega_o(x)\,\big)\;.$$

Of course, the left-hand side just equals ε^2.

For the right-hand side, write $q - f_o(x) = q - f_o(t) + f_o(t) - f_o(x)$, and observe that, by Theorem (14.6.13),

$$\int_0^1 A_h(x,t)\,\big(\,q - f_o(t)\,\big)\,\big(\,d\Omega_n(x) - d\Omega_o(x)\,\big) = \zeta^{nh}\,\big|\,q - f_o(t)\,\big|\;,$$

where

$$\lim_{n\to\infty}\;\sup_{h\in\mathcal{G}_n(\alpha)} \frac{|\,\zeta^{nh}\,|}{\sqrt{(nh)^{-1}\log n}} < \infty \quad \text{almost surely}\;.$$

We leave the bound $|\,q - f_o(t)\,| \leqslant c\,h$ uniformly in t as an exercise.

Next, observe that $h^{-1} A_h(x-t)\{\,f_o(x) - f_o(t)\,\}$, $0 < h \leqslant 1$, is convolution-like because f_o is Lipschitz. Then, by Theorem (14.6.13),

$$\sup_{x\in[0,1]} \left|\, h^{-1} \int_0^1 A_h(x-t)\{\,f_o(x) - f_o(t)\,\}\{\,d\Omega_n(x) - d\omega_o(x)\,\}\,\right| \leqslant \eta^{nh}\;,$$

with
$$\lim_{n\to\infty}\;\sup_{h\in\mathcal{G}_n(\alpha)} \frac{\eta^{nh}}{\sqrt{(nh)^{-1}\log n}} < \infty \quad \text{almost surely}.$$

With (10.8), we then get that

$$\sup_{t\in[0,1]} \big|\,\mathbb{E}[\,f^{nh}_{NW}(t)\,|\,\mathbb{X}_n\,] - [\mathcal{A}_h f_o](t)\,\big| \leqslant \delta\;,$$

where $\delta = \mathcal{O}\big(\,h\sqrt{(nh)^{-1}\log n}\,\big)$, almost surely, uniformly in $h \in \mathcal{H}_n(\alpha)$. Q.e.d.

We now consider what happens if the (awkward) appearance of the design density in condition (10.1) is "fixed". So, let the family of kernels

$$\text{(10.9)} \qquad B_h(x,t),\; 0 < h \leqslant 1 \quad \text{be convolution-like of order } m\;.$$

Define the associated Nadaraya-Watson operators by

$$(10.10)\qquad [\mathcal{N}_h f_o](t) = \frac{\displaystyle\int_0^1 B_h(x, t) f_o(x)\,\omega(x)\,dx}{\displaystyle\int_0^1 B_h(x, t)\,\omega(x)\,dx}, \quad t \in [0, 1].$$

Also, let

$$(10.11)\qquad [\mathcal{B}_h f_o](t) = \int_0^1 B_h(x, t) f_o(x)\,dx, \ t \in [0, 1].$$

(The denominator in (10.10) need not equal 1.) Denote the Nadaraya-Watson estimator based on the kernels B_h by g^{nh}_{NW}, and let $\mathfrak{T}^{nh}_{NW}$ be the part of g^{nh}_{NW} due to the noise $D_1, D_2, \cdots, D_n$.

(10.12) THEOREM. *Let $\alpha > 0$. Under the assumptions (1.1)–(1.6) and (10.9), with $f_o \in W^{m,\infty}(0,1)$ and $\omega \in W^{m,\infty}(0,1)$,*

$$g^{nh}_{NW} = \mathcal{B}_h f_o + \mathfrak{T}^{nh}_{NW} + \delta^{nh} \quad \text{on} \quad [0, 1],$$

where, almost surely,

$$\limsup_{n\to\infty} \sup_{h\in\mathcal{H}_n(\alpha)} \frac{\|\delta^{nh}\|_\infty}{h\left(h^m + \sqrt{(nh)^{-1}\log n}\right)} < \infty.$$

PROOF. If the family $B_h(x, t)$, $0 < h \leqslant 1$, is convolution-like of order m, then the family of kernels, defined by

$$A_h(x, t) = \frac{B_h(x, t)}{[\mathcal{B}_h \omega](t)}, \quad x, t \in [0, 1],$$

is such that $A_h(x, t)\,\omega(t)$, $0 < h \leqslant 1$, is convolution-like of order 1. As in (10.4), let $\mathcal{A}_h$ be the associated operator. Then, $\mathcal{A}_h = \mathcal{N}_h$, and so Theorem (10.5) implies that

$$g^{nh}_{NW} = \mathcal{N}_h f_o + \mathfrak{T}^{nh}_{NW} + \delta^{nh} \quad \text{on} \quad [0, 1]$$

with, almost surely,

$$\limsup_{n\to\infty} \sup_{h\in\mathcal{H}_n(\alpha)} \frac{\|\delta^{nh}\|_\infty}{h\sqrt{(nh)^{-1}\log n}} < \infty.$$

Hence, it suffices to show that $\|\mathcal{N}_h f_o - \mathcal{B}_h f_o\|_\infty = \mathcal{O}(h^{m+1})$. Note that $q = [\mathcal{N}_h f_o](t)$ and $r = [\mathcal{B}_h f_o](t)$ are the minimizers of

$$\int_0^1 B_h(x, t)\,|\,q - f_o(x)\,|^2\,\omega(x)\,dx \quad \text{and} \quad \int_0^1 B_h(x, t)\,|\,r - f_o(x)\,|^2\,\omega(t)\,dx.$$

Note the distinction $\omega(x)$ versus $\omega(t)$.

Let $\varepsilon = r - q$. The usual considerations regarding convex quadratic minimization problems lead to the equality

$$\varepsilon^2 \int_0^1 B_h(x, t)\,\omega(x)\,dx = \varepsilon \int_0^1 B_h(x, t)\left(r - f_o(x)\right)\left(\omega(x) - \omega(t)\right)dx.$$

Split the integral on the right in two. First,

$$\int_0^1 B_h(x,t)\,\bigl(r - f_o(t)\bigr)\,\bigl(\omega(x) - \omega(t)\bigr)\,dx = \bigl(r - f_o(t)\bigr)\int_0^1 B_h(x,t)\,\bigl(\omega(x) - \omega(t)\bigr)\,dx\;.$$

It is again an exercise to show that $\|\,\omega - \mathcal{B}_h\,\omega\,\|_\infty \leqslant c\,h^m\,\|\,\omega^{(m)}\,\|_\infty$ and likewise $\bigl|\,r - f_o(t)\,\bigr| \leqslant \|\,f_o - \mathcal{B}_h\,f_o\,\|_\infty \leqslant c\,h^m\,\|\,f_o^{(m)}\,\|_\infty$. Thus,

$$\Bigl|\int_0^1 B_h(x,t)\,\bigl(r - f_o(t)\bigr)\,\bigl(\omega(x)-\omega(t)\bigr)\,dx\Bigr| \leqslant c\,h^{2m}\,\|\,f_o^{(m)}\,\|_\infty\,\|\,\omega^{(m)}\,\|_\infty\;.$$

Finally, using Taylor approximations, we have

$$\int_0^1 B_h(x,t)\,\bigl(f_o(t) - f_o(x)\,\bigr)\,\bigl(\omega(x) - \omega(t)\bigr)\,dx = \sum_{j,\ell=1}^{m-1} c_{j,\ell}\int_0^1 B_h(x,t)\,(x-t)^{j+\ell}\,dx + \mathcal{O}\bigl(\,h^{m+1}\bigr) = \mathcal{O}\bigl(\,h^{m+1}\bigr)\;.$$

by assumption (10.9), where $c_{j,\ell} = \dfrac{f_o^{(j)}(t)}{j\,!}\,\dfrac{\omega^{(\ell)}(t)}{\ell\,!}$. Putting everything together, the result follows. Q.e.d.

(10.13) COROLLARY. *Under the conditions of Theorem* (10.12),

$$\limsup_{n\to\infty}\;\sup_{h\in\mathcal{H}_n(\alpha)}\;\frac{\|\,g^{nh}_{\mathit{NW}} - f_o\,\|_\infty}{h^m + \sqrt{(nh)^{-1}\log n}} < \infty$$

almost surely.

(10.14) EXERCISE. Prove the corollary.

(10.15) QUESTION. Can the condition $\omega \in W^{m,\infty}(0,1)$ in Theorem (10.11) be relaxed to mere Lipschitz continuity? The authors do not know.

(10.16) EXERCISE. Let $1 \leqslant p \leqslant \infty$, $m \geqslant 1$. Suppose $f \in W^{m,p}(0,1)$.
(a) Assuming (10.1), show that $\|\,f - \mathcal{A}_h\,f\,\|_p \leqslant c\,h^m\,\|\,f^{(m)}\,\|_p$.
(b) Assuming (10.9), show that $\|\,f - \mathcal{B}_h\,f\,\|_p \leqslant c\,h^m\,\|\,f^{(m)}\,\|_p$.

EXERCISES: (10.14), (10.16).

11. Additional notes and comments

Ad § 1 : The local polynomial estimators originated in theoretical work by STONE (1977) on nonparametric regression, culminating in the result that, under the "usual" nonparametric assumptions, the "usual" convergence rates apply and are best possible; see STONE (1982). As a practical method, it was proposed by CLEVELAND (1979) and CLEVELAND and DEVLIN (1988). See also FAN and GIJBELS (1996).

Regarding the conditions (1.7) on the kernel used for the local polynomial estimators, assuming compact support of A is neither helpful nor necessary. However, one could argue that the use of kernels with unbounded support defeats the purpose of local likelihood estimation. Well, argue.

The form of the local estimation problem (1.8) may seem a bit pedantic: Why not write it as

$$\text{(11.1)} \qquad \text{minimize} \quad \sum_{i=1}^{n} A_h(x - X_i) \left| Y_i - \sum_{k=0}^{m-1} a_k \, (x - X_i)^k \right|^2 \quad ?$$

The main reason is that we wish to keep the *definition* of the estimator separate from its *computation* since (11.1) immediately suggests a computational scheme. The other reason is that this suggested computational scheme is the worst imaginable way to do it; see § 19.7.

Finally, we note that it does not appear to be possible to get the usual bounds on the (unconditional) expected error without some special conditions. In our setup, it is the failure of Lemma (2.7) for the "full" expectations. It seems fixable for a *truncated* version of the problem,

$$\begin{aligned} &\text{minimize} \quad \tfrac{1}{n} \sum_{i=1}^{n} A_h(X_i - x) \, | \, p(X_i) - Y_i \, |^2 \\ &\text{subject to} \quad p \in \mathcal{P}_m \, , \quad \sup_{t \in [0,1]} \big| A\big(h^{-1}(t - x) \big) \, p(t) \big| \leqslant \log n \, . \end{aligned}$$

See, e.g., GYÖRFI, KOHLER, KRZYŻAK, and WALK (2002).

Ad § 9: For Nadaraya-Watson estimators based on a kernel K that is a bounded pdf (in particular, nonnegative) with $t \, K(t) \longrightarrow 0$ for $t \to \pm\infty$, ZIEGLER (2001) proves that, for an open interval $J \subset [0, 1]$, and for sequences a_n, b_n with $n \, a_n^2 \to \infty$, $b_n \to 0$,

$$\sup_{t \in J} \; \sup_{a_n < h < b_n} \sqrt{nh} \left| \mathbb{E}[f^{nh}(t)] - \frac{\int_0^1 K_h(x - t) \, f_o(x) \, \omega(x) \, dx}{\int_0^1 K_h(x - t) \, \omega(x) \, dx} \right| \leqslant C$$

with an explicitly specified constant. Note that he does not condition on the design when taking expectations. Also note that the order of the suprema can be interchanged. ZIEGLER (2001) does not interpret the Nadaraya-Watson estimator as the local polynomial estimator of degree 0.

17

Other Nonparametric Regression Problems

1. Introduction

The previous chapters dealt mostly with the Gauss-Markov model

$$(1.1) \qquad y_{in} = f_o(x_{in}) + d_{in} \;, \quad i = 1, 2, \cdots, n \;,$$

in the smooth space setting $f_o \in W^{m,2}(0,1)$ for some integer $m \geqslant 1$. Recall that in the Gauss-Markov model the noise $d_n = (d_{1,n}, \cdots, d_{n,n})^T$ satisfies

$$(1.2) \qquad \mathbb{E}[\, d_n \,] = 0 \;, \quad \mathbb{E}[\, d_n d_n^T \,] = \sigma^2 I \;.$$

We revert back to deterministic, asymptotically uniform designs, but for convenience we take

$$(1.3) \qquad x_{in} = \frac{i-1}{n-1} \;, \quad i = 1, 2, \cdots, n \;.$$

There is an obvious need to discuss what happens if any of these assumptions are violated. Thus, here we discuss the case where the smoothness condition $f_o \in W^{m,2}(0,1)$ is substantially weakened and the case in which the noise does not have finite variance or indeed does not even have an expectation. The case of heteroscedastic errors is considered briefly in § 7 and in great detail in Chapters 21 and 22.

We begin with the smoothness issue. In Chapter 13, the weakest smoothness condition considered was that the first derivative is square integrable, $f_o \in W^{1,2}(0,1)$. At the expense of a more involved analysis, this can be relaxed to the case where the derivative is integrable, $f_o \in W^{1,1}(0,1)$. Going a tiny step further, one arrives at the condition that f_o has finite total variation; i.e., $|\, f \,|_{TV} < \infty$. Here,

$$(1.4) \qquad |\, f \,|_{TV} \stackrel{\text{def}}{=} \sup \sum_{i=1}^{k} |\, f(t_i) - f(t_{i-1}) \,| \;,$$

in which the supremum is over all $k \in \mathbb{N}$ and over all t_i, with

$$0 = t_0 < t_1 < \cdots < t_k = 1 \;.$$

P.P.B. Eggermont, V.N. LaRiccia, *Maximum Penalized Likelihood Estimation*, Springer Series in Statistics, DOI 10.1007/b12285_6,

For the model (1.1)–(1.2), this suggests that f_o be estimated by solving

$$\text{(1.5)}\qquad \begin{aligned} &\text{minimize} && \tfrac{1}{n}\sum_{i=1}^{n} |\, f(x_{in}) - y_{in}\,|^2 + h\,|\, f\,|_{TV} \\ &\text{subject to} && |\, f\,|_{TV} < \infty\ . \end{aligned}$$

Denoting this estimator by f^{nh}, in § 3 we prove that, for $|\, f_o\,|_{TV} < \infty$,

$$\text{(1.6)}\qquad \mathbb{E}[\,\|\, f^{nh} - f_o\,\|^2\,] = \mathcal{O}\big(\,(n^{-1}\log n)^{2/3}\,\big)\ ,$$

provided $h \asymp (n^{-1}\log n)^{2/3}$. The powers of $\log n$ appear to be an artifact of the methods of proof employed in this text. Indeed, for errors having a finite exponential moment, VAN DE GEER (2000) proves (1.6) without them. Then, in comparison with the smooth case $f_o \in W^{1,2}(0,1)$, nothing is lost in the convergence rate. Of course, we only require second-order moments ... .

It should be noted that in (1.4)–(1.5) the functions f are supposed to be defined everywhere, and if two functions differ on a set of (Lebesgue) measure zero, they are assumed to be different. However, in the large-sample asymptotic version of (1.5),

$$\text{(1.7)}\qquad \begin{aligned} &\text{minimize} && \|\, f - f_o\,\|^2 + h\,|\, f\,|_{TV} \\ &\text{subject to} && |\, f\,|_{TV} < \infty\ , \end{aligned}$$

it is more convenient to identify two functions if they coincide almost everywhere. Thus, it makes sense to also consider a weak total variation

$$\text{(1.8)}\qquad |\, f\,|_{BV} \stackrel{\text{def}}{=} \inf\{\,|\,\varphi\,|_{TV} : \varphi = f \text{ a.e.}\,\}\ .$$

Thus, in (1.7), it is reasonable to replace $|\, f\,|_{TV}$ by $|\, f\,|_{BV}$. However, in (1.5), this is not permissible since then a solution would be $f(x_{in}) = y_{in}$ for all i and $f(x) = 0$ everywhere else. Note that in (1.8) the infimum is attained. Indeed, if $|\, f\,|_{TV} < \infty$, the function φ defined by

$$\text{(1.9)}\qquad \varphi(x) = \tfrac{1}{2}\,\{\,\limsup_{t\to x} f(t) + \liminf_{t\to x} f(t)\,\}\ ,\quad x \in [\,0\,,1\,]\ ,$$

will work. (The lim sup and lim inf exist since $|\, f\,|_{TV} < \infty$ implies that f is the difference of two increasing functions.)

We now return to the smooth setting, in which $f_o \in W^{m,2}(0,1)$ for some integer $m \geqslant 1$, and consider the case where the noise does not satisfy (1.1)–(1.2). Sadly, then it seems mandatory to assume that the d_{in} are independent. To make life a little easier, we assume that the d_{in} are in fact iid random variables with

$$\text{(1.10)}\qquad \text{median}(d_{1,n}) = 0\ .$$

Thus, $\mathbb{E}[\,|\, d_{1,n}\,|\,] = +\infty$ is permitted. Under these circumstances, estimating f_o by smoothing splines or kernel estimators does not give good results, and one estimates f_o by the solution to

$$(1.11)\qquad \begin{aligned} &\text{minimize} && \tfrac{1}{n} \sum_{i=1}^{n} |\, f(x_{in}) - y_{in} \,| + h^{2m} \, \| \, f^{(m)} \, \|^2 \\ &\text{subject to} && f \in W^{m,2}(0,1) \ . \end{aligned}$$

The estimator, denoted as f^{nh}, is again a polynomial spline of order $2m$. We refer to it as a *least-absolute-deviations* smoothing spline. In § 5, we prove that

$$(1.12)\qquad \mathbb{E}[\, \| \, f^{nh} - f_o \, \|_{m,h}^2 \,] = \mathcal{O}\big(n^{-2m/(2m+1)} \big) \ ,$$

provided $f_o \in W^{m,2}(0,1)$ and $h \asymp n^{-1/(2m+1)}$. In other words, the same convergence rate applies as in the Gauss-Markov model; recall § 13.4. This result takes some time to get used to.

Apart from the error bound (1.12), why would one think that (1.11) should provide good estimators ? To answer this question, let us consider the case in which the d_{in} are iid random variables distributed according to a two-sided exponential. The joint density is then given by

$$(1.13)\qquad f_{d_{1,n}, \cdots, d_{n,n}}(t_1, \cdots, t_n) = (2\lambda_o)^{-n} \exp\big(-\lambda_o^{-1} \sum_{i=1}^{n} |\, t_i \,| \big)$$

for some $\lambda_o > 0$, and the maximum penalized likelihood estimation problem for estimating f_o and λ_o becomes

$$(1.14)\qquad \text{minimize} \quad n \log \lambda + \lambda^{-1} \sum_{i=1}^{n} |\, f(x_{in}) - y_{in} \,| + h^{2m} \, \| \, f^{(m)} \, \|^2 \ .$$

This is easily transformed into (1.11). Thus, for the noise model (1.13), the problem (1.11) is in fact maximum penalized likelihood estimation. It turns out that the two-sided exponential distribution is a good stand-in for "arbitrary" heavy-tailed noise distributions, even though it still has moments of all orders and indeed a finite exponential moment.

We note that, in view of (1.10), the *robustness* of the estimator is an important issue. Alternative estimators may be developed from this point of view, leading to general penalized M-estimators, see COX (1981), given as the solution to

$$(1.15)\qquad \begin{aligned} &\text{minimize} && \tfrac{1}{n} \sum_{i=1}^{n} \gamma(\, f(x_{in}) - y_{in} \,) + h^{2m} \, \| \, f^{(m)} \, \|^2 \\ &\text{subject to} && f \in W^{m,2}(0,1) \end{aligned}$$

for general "contrast" functions γ. The choice $\gamma(x) = |\, x \,|$ corresponds to (1.11). Another choice is

$$(1.16)\qquad \gamma_\varepsilon(\, t \,) = |\, t \,|^2 - \big(\max\{\, |\, t \,| - \varepsilon \,,\, 0 \,\} \big)^2 \ ,$$

where $\varepsilon > 0$ is a fudge factor to be chosen appropriately, due to HUBER (1981). (This is just t^2 for $|\, t \,| < \varepsilon$ and straight lines with derivatives matching at $t = \pm\, \varepsilon$.)

Another approach is via the local polynomial approach of Chapter 16; i.e., for a given order r of the polynomials, the estimator of $f_o(x)$ is $f^{nh}(x) = p^{nh}(\,x\,;\,x\,)$, where $p = p^{nh}(\,\cdot\,;\,x\,)$ is the solution to

$$(1.17)\qquad \begin{array}{ll} \text{minimize} & \frac{1}{n}\sum_{i=1}^{n} A_h(x - x_{in})\,|\,p(x_{in}) - y_{in}\,| \\ \text{subject to} & p \text{ is a polynomial of order } r\ . \end{array}$$

Unfortunately, we shall not consider this further.

(1.18) EXERCISE. Repeat the development in § 4 for the problems (1.15)–(1.17). [The authors are hopeful that (1.15), (1.16), and (1.17) may be treated along the lines of § 4 as well, but no doubt there will be surprises.]

Finally, we discuss the case of heteroscedastic errors; that is, the case where (1.2) is replaced by

$$(1.19)\qquad \mathbb{E}[\,d_n\,] = 0 \quad\text{and}\quad \mathbb{E}[\,d_n d_n^T\,] = \sigma^2\, V_n\ ,$$

where V_n is a diagonal matrix with diagonal entries that are not all equal to each other. A nice example of this is nonparametric logistic regression. Of course, this is a bit of a misnomer, but it seems to indicate what we have in mind. In § 6, we consider smoothing estimation for nonparametric logistic regression and prove the "usual" convergence rates $\mathcal{O}\big(\,n^{-2m/(2m+1)}\,\big)$. Of course, maximum penalized likelihood estimation is an attractive alternative, but we do not do this here.

In the remainder of this chapter, the analytical aspects of these nonparametric regression problems are considered. The crucial feature is that they do not fit into the reproducing kernel Hilbert space framework but find their natural setting in Banach spaces. However, the reproducing kernel aspects are preserved to some extent, and in the general treatment we try to mimic the material of Chapter 13 on smoothing spline estimators as much as possible. In §§ 2 and 3, total-variation roughness penalization of nonparametric least-squares problems is considered. In §§ 4 and 5, least-absolute-deviations spline estimators are scrutinized. In § 6, we briefly discuss heteroscedasticity of the (independent) errors in the model (1.1) and study the particular instance of nonparametric logistic regression.

EXERCISE: (1.18).

2. Functions of bounded variation

The nonparametric regression problems discussed in the previous chapters dealt with the estimation of smooth functions in the presence of "nice" errors. In this section and the next, we deal with the estimation of *non-smooth* functions, but still with "nice" Gauss-Markov errors. Here, the nonsmoothness of f_o is interpreted to mean that f_o has bounded (total) variation. Recall that the total variation of a function f on $[0, 1]$ is

$$(2.1) \qquad |f|_{TV} = \sup \sum_{i=1}^{n-1} |f(t_{i+1}) - f(t_i)| ,$$

where the supremum is over all $n \in \mathbb{N}$, and all $0 \leqslant t_1 < t_2 < \cdots < t_n \leqslant 1$. Also, recall the weak version

$$(2.2) \qquad |f|_{BV} = \inf\{ |\varphi|_{TV} : \varphi = f \text{ a.e.} \} .$$

If we wish to explicitly show the interval in question, we write $|f|_{TV(0,1)}$ and $|f|_{BV(0,1)}$.

(2.3) EXERCISE. (a) Show that taking absolute values reduces the total variation; i.e., if $g = |f|$, then $|g|_{TV} \leqslant |f|_{TV}$ and $|g|_{BV} \leqslant |f|_{BV}$.
(b) Prove the triangle inequalities

$$|f+g|_{TV} \leqslant |f|_{TV} + |g|_{TV} \quad \text{and} \quad |f+g|_{BV} \leqslant |f|_{BV} + |g|_{BV} .$$

(2.4) EXERCISE. Let $1 \leqslant p \leqslant \infty$. Show that, for all $f \in W^{1,p}(0,1)$,

$$|f|_{BV} = \| f' \|_1 \leqslant \| f' \|_p .$$

(2.5) DEFINITION. (a) *The set* $TV(0,1)$ *consists of all functions* f *defined on* $[0, 1]$ *that satisfy* $|f|_{TV} < \infty$.
(b) *The set* $BV(0,1)$ *consists of all functions* f *defined on* $[0, 1]$ *that satisfy* $|f|_{BV} < \infty$.

(2.6) THEOREM. (a) *The set* $TV(0,1)$ *with the usual addition and scalar multiplication of functions is a vector space. It is a Banach space under the norm*

$$\| f \|_{TV} \stackrel{\text{def}}{=} |f(0)| + |f|_{TV} .$$

Moreover, $\| f g \|_{TV} \leqslant \| f \|_{TV} \| g \|_{TV}$ *for all* $f, g \in TV(0,1)$.
(b) *Likewise,* $BV(0,1)$ *is a Banach space under the norm*

$$\| f \|_{BV} \stackrel{\text{def}}{=} |f(0)| + |f|_{BV} ,$$

and $\| f g \|_{BV} \leqslant \| f \|_{BV} \| g \|_{BV}$ *for all* $f, g \in BV(0,1)$.

PROOF. The triangle inequalities for the norms $\| \cdot \|_{TV}$ and $\| \cdot \|_{BV}$ follow from Exercise (2.3)(b).

To prove the completeness of $TV(0,1)$, let $\{ f_j \}_j \subset TV(0,1)$ be a Cauchy sequence; i.e., $\| f_i - f_j \|_{TV} \longrightarrow 0$ for $i,j \to \infty$. It follows that $\{ f_j(0) \}_j$ is a Cauchy sequence, so $\lim_{j\to\infty} f_j(0)$ exists.

We show that $\lim_{j\to\infty} f_j(x)$ exists for every x. Let $0 < x \leqslant 1$. Clearly,

$$\begin{aligned} |\, f_i(x) - f_j(x) \,| &\leqslant |\, \{ f_i(x) - f_j(x) \} - \{ f_i(0) - f_j(0) \} \,| + |\, f_i(0) - f_j(0) \,| \\ &\leqslant \| f_i - f_j \|_{TV} \longrightarrow 0 \end{aligned}$$

for $i, j \to \infty$. Thus, $\{ f_j(x) \}_j$ is a Cauchy sequence, and the limit exists.

Now, define φ on $[\,0\,,1\,]$ by $\varphi(x) = \lim_{j\to\infty} f_j(x)$. We need to show that $|\,\varphi\,|_{TV} < \infty$ and that $\| f_j - \varphi \|_{TV} \longrightarrow 0$, but this is a standard exercise.

The proof for $BV(0,1)$ is only slightly more complicated. Suppose that $\{ f_j \}_j \subset BV(0,1)$ is Cauchy. Then, there exists a sequence $\{ \varphi_j \}_j$ in $TV(0,1)$ such that each φ_j coincides a.e. with f_j and $\{ \varphi_j \}_j$ is a Cauchy sequence in $TV(0,1)$. Q.e.d.

Later on we show that, for $1 \leqslant p \leqslant \infty$, the norms $\| f \|_p + |\, f \,|_{TV}$ are also norms for $TV(0,1)$, and similarly for $BV(0,1)$.

An equivalent definition of the *BV* norm may be based on Exercise (2.4). For $f \in W^{1,1}(0,1)$, we had

$$|\, f \,|_{TV} = \| f' \|_1 = \sup \int_0^1 f'(x)\, \varphi(x)\, dx \ ,$$

where the supremum is over all φ with $\| \varphi \|_\infty \leqslant 1$. However, this may be restricted to functions φ belonging to the set

$$\text{(2.7)} \qquad \mathcal{U}_\infty \stackrel{\text{def}}{=} \{\, \varphi \in W^{1,\infty}(0,1) : \| \varphi \|_\infty \leqslant 1 \ , \ \varphi(0) = \varphi(1) = 0 \,\} \ .$$

For such functions φ, we may perform integration by parts to see that

$$\int_0^1 f'(x)\, \varphi(x)\, dx = - \int_0^1 f(x)\, \varphi'(x)\, dx \ .$$

This leads to the following weak definition of the total variation semi-norm:

$$\text{(2.8)} \qquad |\, f \,|_{weakBV} \stackrel{\text{def}}{=} \sup \Bigl\{ \int_0^1 f(x)\, \varphi'(x)\, dx \;\Big|\; \varphi \in \mathcal{U}_\infty \Bigr\} \ .$$

Parenthetically, this characterization serves as the basis for the definition of the total variation of functions of more than one variable. See, e.g., Giusti (1984) or Ziemer (1989) for details. The following lemma is already in Giusti (1984), p. 29.

(2.9) Lemma. *If* $f \in BV(0,1)$, *then* $|\, f \,|_{weakBV} = |\, f \,|_{BV}$.

Proof. Let $\psi = f$ a.e. with ψ bounded and $|\,\psi\,|_{TV} = |\, f \,|_{BV}$.

We first prove the $\geqslant$ part. Let $\varepsilon > 0$. By the definition of the total-variation norm, there exist $0 \leqslant t_1 < t_2 < \cdots < t_n \leqslant 1$ such that

$$\sum_{i=1}^{n} |\, \psi(t_i) - \psi(t_{i-1}) \,| > |\,\psi\,|_{TV} - \varepsilon \ .$$

Without loss of generality, one may assume that consecutive terms of $\psi(t_i) - \psi(t_{i-1})$ have opposite signs; otherwise

$$|\psi(t_{i+1}) - \psi(t_i)| + |\psi(t_i) - \psi(t_{i-1})| = |\psi(t_{i+1}) - \psi(t_{i-1})|,$$

and the point t_i can be omitted.

Then, apart from the overall sign, the sum above equals

$$\psi(t_0) + 2\sum_{i=1}^{n-1}(-1)^i\,\psi(t_i) + (-1)^n\,\psi(t_n) = \int_0^1 \psi(t)\,d\Phi(t)$$

for a step function Φ with $\Phi(t) = \pm 1$ everywhere but $\Phi(0) = \Phi(1) = 0$. Thus, the integral may be approximated to arbitrary accuracy by integrals

$$\int_0^1 \psi(t)\,\varphi'(t)\,dt$$

with $\varphi \in \mathcal{U}_\infty$. It follows that there exists a $\varphi \in \mathcal{U}_\infty$ such that

$$\int_0^1 \psi(t)\,\varphi'(t)\,dt > \sum_{i=1}^{n-1} |\psi(t_i) - \psi(t_{i-1})| - \varepsilon\,.$$

Putting all of this together, we obtain that

$$\begin{aligned} |\psi|_{weakBV} &\geqslant \sum_{i=1}^{n-1} |\psi(t_i) - \psi(t_{i-1})| - \varepsilon \\ &> |\psi|_{TV} - 2\varepsilon = |f|_{BV} - 2\varepsilon\,. \end{aligned}$$

Since obviously $|f|_{weakBV} = |\psi|_{weakBV}$, the $\geqslant$ part follows.

Next, consider the $\leqslant$ part. Let $\varepsilon > 0$ and let $\varphi \in \mathcal{U}_\infty$ such that

$$|\psi|_{weakBV} \leqslant \int_0^1 \psi(x)\,\varphi'(x)\,dx + \varepsilon = -\int_0^1 \varphi(x)\,d\psi(x)\,dx + \varepsilon\,.$$

Here, the last integral is a Riemann-Stieltjes integral. Consequently, there exist $0 \leqslant t_1 < t_2 < \cdots < t_n \leqslant 1$ such that

$$-\int_0^1 \varphi(x)\,d\,\psi(x) \leqslant -\sum_{i=1}^{n} \varphi(t_i)\,\{\psi(t_i) - \psi(t_{i-1})\} + \varepsilon\,.$$

The sum on the right may be bounded by $\sum_{i=1}^{n} |\psi(t_i) - \psi(t_{i-1})| \leqslant |\psi|_{TV}$. Summarizing, we have shown that

$$|\psi|_{weakBV} \leqslant |\psi|_{TV} + 2\varepsilon = |f|_{BV} + 2\varepsilon\,.$$

The $\leqslant$ part of the lemma follows. Q.e.d.

To motivate what follows, consider the total-variation estimator of f_o, defined as the solution to

$$
\text{(2.10)} \qquad \begin{array}{ll} \text{minimize} & \frac{1}{n} \sum_{i=1}^{n} | f(x_{in}) - y_{in} |^2 + h \, | f |_{TV} \\ \text{subject to} & | f |_{TV} < \infty \, . \end{array}
$$

Thus, this is (still) maximum penalized likelihood estimation. It is clear that in the study of this estimator some basic facts about functions of bounded variation are needed. In particular, as in the roughness-penalized least-squares spline problem, we have to deal with sums

$$
\text{(2.11)} \qquad \frac{1}{n} \sum_{i=1}^{n} \varepsilon(x_{in}) \, d_{in} \; ,
$$

in which the d_{in} satisfy (1.2) and $\varepsilon \in TV(0,1)$ is a random function. This occupies us for the remainder of this section. In § 3, we deal with (2.11) proper.

Reproducing kernel Hilbert space tricks. It seems that we must now face the daunting task of coming up with integral representations for functions $f \in BV(0,1)$. This would seem doomed to fail, but fortunately we only need it for functions $f \in W^{1,1}(0,1)$, and that we have done already.

(2.12) LEMMA. *Let $f \in W^{1,1}(0,1)$. Then, for all $0 < h \leqslant 1$,*

$$
f(x) = \langle \, \mathfrak{R}_{1,h}(x, \cdot\,) \, , \, f \, \rangle_{1,h} \, , \qquad x \in [\, 0 \, , \, 1 \,] \, .
$$

PROOF. Note that

$$
\langle \, \mathfrak{R}_{1,h}(x, \cdot\,) \, , \, f \, \rangle_{1,h} = \int_0^1 \mathfrak{R}_{1,h}(x, t\,) \, f(\,t\,) \, dt + h^2 \int_0^1 \mathfrak{R}'_{1,h}(x, t\,) \, f'(\,t\,) \, dt
$$

is well-defined since f and f' are integrable and $\mathfrak{R}_{1,h}$ and $\mathfrak{R}'_{1,h}$ are bounded. Here, the prime denotes differentiation with respect to the first argument. Now, for $\lambda > 0$, define f_λ by

$$
\text{(2.13)} \qquad f_\lambda(x) = \langle \, \mathfrak{R}_{1,\lambda}(x, \cdot\,) \, , \, f \, \rangle_{L^2(0,1)} \, , \qquad x \in [\, 0 \, , \, 1 \,] \, .
$$

Then, since the family of kernels $\mathfrak{R}_{1,h}(x, t\,)$, $0 < h \leqslant 1$, is convolution-like of order 1, it is easy to show that $\| \, f_\lambda - f \, \|_{W^{1,1}(0,1)} \longrightarrow 0$ as $\lambda \longrightarrow 0$, and since $f_\lambda \in W^{m,2}(0,1)$, also for fixed h and all $x \in [\, 0 \, , \, 1 \,]$,

$$
f_\lambda(x) = \langle \, \mathfrak{R}_{1,h}(x, \cdot\,) \, , \, f_\lambda \, \rangle_{1,h} \longrightarrow \langle \, \mathfrak{R}_{1,h}(x, \cdot\,) \, , \, f \, \rangle_{1,h} \, .
$$

The lemma follows. Q.e.d.

Piecewise linear approximation. The goal is still to deal with the random sums (2.11). Lemma (2.12), in conjunction with *linear interpolation*, turns out to be the right tool.

Let $\mathcal{S}_1$ be the set of continuous functions on $(0,1)$, which are linear on each interval $[\,x_{i-1,n}\,,\,x_{in}\,]$,

$$\text{(2.14)} \qquad \mathcal{S}_1 \stackrel{\text{def}}{=} \left\{ f \in C[\,0\,,\,1\,] : \begin{array}{l} f \text{ linear on } [\,x_{in}\,,\,x_{i+1,n}\,] \\ \text{for } i = 1, \cdots, n-1 \end{array} \right\} .$$

In other words, $\mathcal{S}_1$ is the space of polynomial splines of order 2 with knots at the design points x_{in} ; see § 15.5. Recall that $x_{1,n} = 0$ and $x_{n,n} = 1$. Let the projection operator $\pi_n : BV(0,1) \longrightarrow \mathcal{S}_1$ be defined by the condition

$$\text{(2.15)} \qquad [\,\pi_n\, f\,](x_{in}) = f(x_{in}) \,, \quad i = 1, 2, \cdots, n \,.$$

Thus, for $i = 2, 3, \cdots, n$,

$$\text{(2.16)} \qquad \pi_n f(x) = f(x_{i-1,n})\,\Lambda_{i+1,n}(x) + f(x_{in})\,\Lambda_{i,n}(x)$$

for all x, $x_{i-1,n} \leqslant x \leqslant x_{in}$. Here, the basis functions are defined as $\Lambda_{i,n}(x) = \Lambda(\,nx - i\,)$, with $\Lambda(x) = (\,1 - |x|\,)_+$ the standard tent function.

(2.17) LEMMA. *For all functions $f \in TV(0,1)$,*

$$|\,\pi_n\, f\,|_{TV} \leqslant |\,f\,|_{TV} \quad \textit{and} \quad \|\,f - \pi_n\, f\,\|_\infty \leqslant |\,f\,|_{TV} \,.$$

(2.18) EXERCISE. Prove it.

Quadrature. As in the analysis of the smoothing spline problem, another useful result deals with approximating sums by integrals and vice versa.

(2.19) QUADRATURE LEMMA. *For quasi-uniform designs, there exist constants c_n such that, for all $f \in TV(0,1)$,*

$$c_n\,\|\,f\,\|_2 \leqslant \{\,\tfrac{1}{n}\sum_{i=1}^n |\,f(x_{in})\,|^2\,\}^{1/2} + \tfrac{1}{n}\,|\,f\,|_{TV}$$

and $c_n \to 1$ for $n \to \infty$.

PROOF. Assume that $f \in W^{1,1}(0,1)$. By the quasi-uniformity of the design, then

$$\text{(2.20)} \qquad \left|\,\tfrac{1}{n}\sum_{i=1}^n |\,f(x_{in})\,|^2 - \int_0^1 |\,f(x)\,|^2\,dx\,\right| \leqslant c\,n^{-1}\,\|\,(f^2)'\,\|_1 \,.$$

Of course, $\|\,(f^2)'\,\|_1 \leqslant 2\,\|\,f\,\|_\infty\,\|\,f'\,\|_1 \leqslant (\,\|\,f\,\|_\infty + \|\,f'\,\|_1\,)^2$, and so, by the Equivalence of Norms Lemma (2.26) below,

$$\|\,(f^2)'\,\|_1 \leqslant 4\,\{\,\|\,f\,\| + \|\,f'\,\|_1\,\}^2 \leqslant 8\,\{\,\|\,f\,\|^2 + \|\,f'\,\|_1^2\,\} \,.$$

Thus, for another constant c, the right-hand side of (2.20) is bounded by $c\,n^{-1}\,(\,\|\,f\,\|^2 + \|\,f'\,\|_1^2\,)$, and so, from (2.20), we obtain that

$$(\,1 - c\,n^{-1}\,)\,\|\,f\,\| \leqslant S_n + c\,n^{-1}\,\|\,f'\,\|_1^2 \,.$$

This is the lemma for $f \in W^{1,1}(0,1)$. If $f \in TV(0,1)$, then apply the lemma to f_λ, see (2.13), and take limits. Q.e.d.

(2.21) EXERCISE. Do the inequalities of Lemma (2.19) and Corollary (2.20) hold with $|\, f \,|_{TV}$ replaced by $|\, f \,|_{BV}$?

Equivalent norms. In the above, we needed equivalent norms for the space $TV(0,1)$. For $1 \leqslant p < \infty$, let

$$\| f \|_{p,TV} = \| f \|_p + |\, f \,|_{TV} \,, \tag{2.22}$$

$$\| f \|_{p,BV} = \| f \|_p + |\, f \,|_{BV} \,, \tag{2.23}$$

and, for $p = \infty$,

$$\| f \|_{p,TV} = \sup_{x \in [0,1]} |\, f(x) \,| + |\, f \,|_{TV} \,, \tag{2.24}$$

$$\| f \|_{p,BV} = \| f \|_\infty + |\, f \,|_{BV} \,. \tag{2.25}$$

(2.26) LEMMA [EQUIVALENCE OF NORMS]. *For all $1 \leqslant p \leqslant \infty$, the norms $\| \cdot \|_{p,TV}$ and $\| \cdot \|_{TV}$ are equivalent. In particular,*

$$\tfrac{1}{2} \| f \|_{BV} \leqslant \| f \|_{p,BV} \leqslant 2 \| f \|_{BV} \quad \textit{for all } f \in BV(0,1) \,,$$

and likewise for the norms $\| \cdot \|_{p,BV}$ and $\| \cdot \|_{BV}$.

PROOF. Obviously,

$$\| f \|_{p,TV} = \| f \|_p + |\, f \,|_{TV} \leqslant \| f \|_\infty + |\, f \,|_{TV} \,. \tag{2.27}$$

Let $\varepsilon > 0$, and choose $z \in (0,1)$ such that $\| f \|_\infty \leqslant |\, f(z) \,| + \varepsilon$. Then

$$\| f \|_\infty \leqslant |\, f(0) \,| + |\, f(z) - f(0) \,| + \varepsilon \leqslant |\, f(0) \,| + |\, f \,|_{TV} + \varepsilon \,.$$

Since $\varepsilon > 0$ is arbitrary, then $\| f \|_\infty \leqslant |\, f(0) \,| + |\, f \,|_{TV}$. With (2.27), this shows that $\| f \|_{p,TV} \leqslant 2 \| f \|_{TV}$.

For the reverse, observe that $|\, f(0) \,| \leqslant |\, f(x_{in}) \,| + |\, f \,|_{TV}$ for all i, so that upon averaging and appealing to the Quadrature Lemma (2.19),

$$|\, f(0) \,| \leqslant \tfrac{1}{n} \sum_{i=1}^{n} |\, f(x_{in}) \,| + |\, f \,|_{TV} \leqslant c_n \| f \|_1 + \left(1 + c_n \tfrac{1}{n} \right) |\, f \,|_{TV}$$

with $c_n \longrightarrow 1$. Letting $n \to \infty$ yields $|\, f(0) \,| \leqslant \| f \|_1 + |\, f \,|_{TV}$, and since $\| f \|_1 \leqslant \| f \|_p$, it follows that

$$\| f \|_{TV} \leqslant \| f \|_p + 2 \,|\, f \,|_{TV} \leqslant 2 \| f \|_{p,TV} \,.$$

The lemma follows. Q.e.d.

This concludes the preliminary treatment of the random sums (2.11).

Compact sets. Finally, we address the existence of the estimators. For the small-sample problem (1.5), a simple approach suffices, but for the large-sample asymptotic problem (1.7), it seems we must have recourse to compactness arguments. The sets

$$TV_{p,K} = \{\, f \in TV(0,1) : \| f \|_{p,TV} \leqslant K \,\} \,, \tag{2.28}$$

$$BV_{p,K} = \{\, f \in BV(0,1) : \| f \|_{p,BV} \leqslant K \,\} \,. \tag{2.29}$$

where K is any positive constant, are important. It turns out that these sets are weakly compact in $L^p(0,1)$ for $1 \leqslant p < \infty$. The tool to be used is the Fréchet-Kolmogorov theorem; see, e.g., YOSIDA (1980) or Theorem (10.4.18) in Volume I. To that end, we need the uniform continuity of the translation operator.

(2.30) LEMMA. *Let $f \in BV(0,1)$, and extend f periodically to all of $\mathbb{R}$. Let $1 \leqslant p < \infty$. Then, for all $h > 0$,*

$$\| f(\,\cdot\, + h) - f \|_p^p \leqslant h \cdot \big(| f |_{BV(0,1)} \big)^p \,.$$

PROOF. Let $\varphi = f$ almost everywhere. Obviously,

$$\int_0^1 | \varphi(x+h) - \varphi(x) |^p \, dx \leqslant \big(\, | \varphi |_{TV} \big)^{p-1} \int_0^1 | \varphi(x+h) - \varphi(x) | \, dx \,.$$

Let $h > 0$, and set $m = \lfloor 1/h \rfloor$. Then,

$$\begin{aligned} &\int_0^1 | \varphi(x+h) - \varphi(x) | \, dx \\ &\quad = \sum_{k=0}^{m} \int_{kh}^{(k+1)h} | \varphi(x+h) - \varphi(x) | \, 1\!\!1(\, x \in [\,0\,,\,1\,]) \, dx \\ &\quad = \int_0^h \sum_{k=0}^{m} | \varphi(kh + x + h) - \varphi(kh + x) | \, 1\!\!1(\, x \in [\,0\,,\,1\,]) \, dx \\ &\quad \leqslant h \, | \varphi |_{TV} \end{aligned}$$

since the integrand is bounded by $| \varphi |_{TV}$. Now, take the infimum over all admissible φ. Q.e.d.

(2.31) LEMMA. *Let $1 \leqslant p < \infty$. For any $K > 0$, the sets $TV_{p,K}$ and $BV_{p,K}$ are weakly compact subsets in $L^p(0,1)$.*

PROOF. First, we consider $BV_{p,K}$. The aforementioned Fréchet-Kolmogorov theorem shows that the closure of $BV_{p,K}$ is weakly compact. Note that the convexity of $BV_{p,K}$ implies that one need not distinguish between closedness and weak closedness; see, e.g., HOLMES (1975) or Corollary (10.4.9) in Volume I. So, it suffices to show that $BV_{p,K}$ is closed.

Let $\{ f_j \}_j \subset BV_{p,K}$ be convergent in $L^p(0,1)$. Thus, there exists a function $\psi \in L^p(0,1)$ such that $\lim_{j\to\infty} \| f_j - \psi \|_p = 0$. However, since

$$\| f_j \|_{BV} \leqslant 2 \| f_j \|_{p,BV} \leqslant 2K ,$$

it follows that $\| f_j \|_\infty \leqslant 2K$ for all j. This implies that the limit ψ is bounded almost everywhere.

Now, we use the characterization (2.8). Thus, for any $\varphi \in \mathcal{U}_\infty$,

$$\int_0^1 \psi_o(y)\,\varphi'(y)\,dy = \lim_{j\to\infty} \int_0^1 f_j(y)\,\varphi'(y)\,dy \leqslant \limsup_{j\to\infty} | f_j |_{BV} .$$

Taking the supremum over all $\varphi \in \mathcal{U}_\infty$, we obtain

$$\| \psi \|_{p,BV} \leqslant \limsup_{j\to\infty} \| f_j \|_{p,BV} \leqslant K ,$$

thus showing that $\psi \in BV_{p,K}$.

Now, consider the closedness of $TV_{p,K}$. Let $\{ f_j \}_j \subset TV_{p,K}$ converge to $\psi \in L^p(0,1)$. Then, $\{ f_j \}_j \subset BV_{p,K}$ as well. Consequently, by the closedness of $BV_{p,K}$, we have $\psi \in BV_{p,K}$. Thus, ψ coincides with a function φ for which $| \varphi |_{TV} = | \psi |_{BV}$ and then $\| \varphi \|_{p,TV} = \| \psi \|_{p,BV} \leqslant K$. The lemma follows. Q.e.d.

(2.32) EXERCISE. Show that the sets $TV_{\infty,K}$ and $BV_{\infty,K}$ are weak-$*$ compact in $L^\infty(0,1)$, viewed as the dual of $L^1(0,1)$.

EXERCISE: (2.3), (2.4), (2.18), (2.21), (2.32).

3. Total-variation roughness penalization

With the preliminaries out of the way, we now consider nonparametric least-squares estimation with total-variation penalization for the Gauss-Markov model (1.1)–(1.3) with

$$f_o \in TV(0,1) . \tag{3.1}$$

Recall that the total-variation penalized estimator of f_o is the solution to

$$\begin{aligned} &\text{minimize} \quad L_{nh}(f) \stackrel{\text{def}}{=} \tfrac{1}{n} \sum_{i=1}^{n} | f(x_{in}) - y_{in} |^2 + h \, | f |_{TV} \\ &\text{subject to} \quad f \in TV(0,1) , \end{aligned} \tag{3.2}$$

where $h > 0$ is the smoothing parameter. Any solution of (3.2) is denoted by f^{nh}. As usual, the questions of existence (yes), uniqueness (no), and convergence rates of the estimator must be addressed. To relieve the suspense, we prove for the unique, piecewise linear solution f^{nh} of (3.2) convergence rates under the additional assumption that

$$\text{(3.3)} \qquad \begin{aligned} &d_{1,n}, d_{2,n}, \cdots, d_{n,n} \text{ are iid with } \mathbb{E}[\, d_{1,n} \,] = 0 \text{ and} \\ &\mathbb{E}[\, |\, d_{1,n} \,|^{\kappa} \,] < \infty \text{ for some } \kappa > 3 \;. \end{aligned}$$

Indeed, in the approach of this section, we end up needing uniform bounds on pure-noise regression sums.

(3.4) THEOREM. *Let $f_o \in TV(0,1)$. Consider the model (1.1)–(1.3), and assume that the noise satisfies (3.3). Then, the total-variation penalized estimator f^{nh} satisfies*

$$\tfrac{1}{n} \sum_{i=1}^{n} |\, f^{nh}(x_{in}) - f_o(x_{in}) \,|^2 + h \,|\, f^{nh} \,|_{TV} \leqslant \zeta^{nh}$$

and

$$\|\, f^{nh} - f_o \,\|^2 + h \,|\, f^{nh} \,|_{TV} \leqslant \zeta^{nh}$$

with

$$\lim_{n\to\infty} \sup_{h \in \mathcal{H}_n(\alpha)} \frac{\zeta^{nh}}{h + ((nh)^{-1} \log n)^2} < \infty \quad \textit{almost surely}.$$

Uniqueness of the estimator. There is at least one source of nonuniqueness in (3.2): If the values $f(x_{in})$, $i = 1, 2, \cdots, n$, are uniquely determined, then any function that has the same function values at the x_{in} and is monotone in between the design points is also a solution. This is implied by the inequality for any function f defined on $[\, a\,,\, b\,]$,

$$|\, f \,|_{TV(a,b)} \geqslant |\, f(b) - f(a) \,| \;,$$

with equality if and only if f is monotone on $[\, a\,,\, b\,]$. But it is clear that the above is the only kind of nonuniqueness in (3.2). To avoid this nonuniqueness, we only consider continuous, piecewise linear functions with knots at the design points; in other words, functions belonging to the set $\mathcal{S}_1$, defined in (2.14). It should be noted that $\mathcal{S}_1 \subset W^{1,1}(0,1)$. The restricted minimization problem is then

$$\text{(3.5)} \qquad \begin{aligned} &\text{minimize} \quad L_{nh}(f) = \tfrac{1}{n} \sum_{i=1}^{n} |\, f(x_{in}) - y_{in} \,|^2 + h \,|\, f \,|_{TV} \\ &\text{subject to} \quad f \in \mathcal{S}_1 \;. \end{aligned}$$

Assuming existence and uniqueness, the solution of (3.5) is denoted by f^{nh}. Indeed, establishing the existence of solutions is easy. Note that

$$\text{(3.6)} \qquad |\, f \,|_{TV} = \sum_{i=1}^{n-1} |\, f(x_{i+1,n}) - f(x_{in}) \,| \quad \text{for all } f \in \mathcal{S}_1 \;,$$

so that the problem (3.5) reduces to

$$\text{(3.7)} \qquad \text{minimize} \quad \tfrac{1}{n} \,\|\, b - y_n \,\|^2 + h \,\|\, Db \,\|_1 \;,$$

where now $\|\cdot\|_p$ is the p norm on $\mathbb{R}^n$ and $D \in \mathbb{R}^{(n-1)\times n}$ is given by

$$(3.8) \qquad D_{i,j} = \begin{cases} 1 \ , & j = i \ , \\ -1 \ , & j = i+1 \ , \\ 0 \ , & \text{otherwise} \ . \end{cases}$$

Now the existence of solutions of (3.5) and (3.2) follows since (3.7) is a strongly convex minimization problem over $\mathbb{R}^n$ with a continuous objective function; see Theorem (9.4.14) in Volume I.

The large-sample asymptotic analogue of (3.2) is

$$(3.9) \qquad \begin{array}{ll} \text{minimize} & L_{\infty h}(f) \stackrel{\text{def}}{=} \| f - f_o \|^2 + h \, | f |_{TV} \\ \text{subject to} & f \in TV(0,1) \ , \end{array}$$

with the unique solution denoted by f_h.

(3.10) THEOREM. *The problems* (3.7) *and* (3.9) *have unique solutions.*

(3.11) EXERCISE. Prove the theorem. [Hint: For (3.9), consider a minimizing subsequence. This lies in some set $TV_{2,K}$, see (2.28), that is weakly compact by Lemma (2.30). Extract a weakly convergent subsequence with limit in $TV_{2,K}$, etc.]

Convergence rates. We move on to error bounds for f^{nh}. As in the smoothing spline problem, it turns out that the usual decomposition of the error into asymptotic bias and variance,

$$(3.12) \qquad f^{nh} - f_o = f_h - f_o + \{ f^{nh} - f_h \} \ ,$$

combined with the strong convexity of the objective function, constitutes somewhat of a detour, even if the strong convexity needs to be fixed.

(3.13) LEMMA. *For all $f \in \mathcal{S}_1$ and all $h > 0$,*

$$\text{(a)} \qquad \tfrac{1}{n} \sum_{i=1}^{n} | f(x_{in}) - f^{nh}(x_{in}) |^2 \leqslant L_{nh}(f) - L_{nh}(f^{nh}) \ ,$$

$$\text{(b)} \qquad \| f - f_h \|^2 \leqslant L_{\infty h}(f) - L_{\infty h}(f_h) \ .$$

(3.14) EXERCISE. Prove the lemma. (The lemma holds for all $f \in TV(0,1)$, but we only need it for piecewise linear functions.)

The bias. Despite the detour it constitutes, for later use it is a good idea to obtain asymptotic bounds on the bias. This proceeds as follows. Since f_h solves (3.9), then $L_{\infty h}(f_h) \leqslant L_{\infty h}(f_o)$, or

$$\| f_h - f_o \|^2 + h \, | f_h |_{TV} \leqslant h \, | f_o |_{TV} \ .$$

This proves the following lemma.

(3.15) LEMMA. *If $f_o \in TV(0,1)$, then for all $h > 0$*

$$| f_h |_{TV} \leqslant | f_o |_{TV} \quad , \quad \| f_h - f_o \| \leqslant \left(h \, | f_o |_{TV} \right)^{1/2} .$$

(3.16) EXERCISE. It is possible to get a better bound on the bias under very special conditions on f_o. By way of example, suppose that f_o is continuously differentiable and that $f_o{}'$ vanishes only on nonempty intervals and not at isolated points. Show that then $\| f_h - f_o \|_2 = \mathcal{O}(h)$ for $h \to 0$.

The asymptotic error. Here we set out to prove Theorem (3.4). In analogy with the treatment of the error in smoothing spline estimation, we use the strong convexity of the functional L_{nh}; see Lemma (3.13). However, a "fix" is in order. Note that $f_o \notin \mathcal{S}_1$ but $\pi_n f_o$ *does* belong to $\mathcal{S}_1$, so, by Lemma (3.13),

$$\begin{aligned} \tfrac{1}{n} \sum_{i=1}^{n} | f^{nh}(x_{in}) - f_o(x_{in}) |^2 \leqslant \tfrac{1}{n} \sum_{i=1}^{n} | d_{in} |^2 + h \, | \pi_n f_o |_{TV} \\ - \tfrac{1}{n} \sum_{i=1}^{n} | f^{nh}(x_{in}) - f_o(x_{in}) - d_{in} |^2 - h \, | f^{nh} |_{TV} \, . \end{aligned}$$

With $\varepsilon \equiv f^{nh} - f_o$, the usual manipulations yield

$$\tfrac{1}{n} \sum_{i=1}^{n} | \varepsilon(x_{in}) |^2 \leqslant \tfrac{1}{n} \sum_{i=1}^{n} \varepsilon(x_{in}) \, d_{in} + \tfrac{1}{2} h \, | f_o |_{TV} - \tfrac{1}{2} h \, | f^{nh} |_{TV} \, ,$$

where we also applied the inequality $| \pi_n f_o |_{TV} \leqslant | f_o |_{TV}$. Now, to both sides of the inequality, add $\tfrac{1}{2} h \, | \varepsilon |_{TV} = \tfrac{1}{2} h \, | f^{nh} - f_o |_{TV}$ to obtain

$$\text{(3.17)} \qquad \tfrac{1}{n} \sum_{i=1}^{n} | \varepsilon(x_{in}) |^2 + \tfrac{1}{2} h \, | \varepsilon |_{TV} \leqslant \tfrac{1}{n} \sum_{i=1}^{n} \varepsilon(x_{in}) \, d_{in} + h \, | f_o |_{TV} \, .$$

This inequality acts as the strong convexity inequality in $TV(0,1)$, even though the functional $| f |_{TV}$ is not strongly convex. Also note that the term $h \, | f_o |_{TV}$ on the right-hand side is not really troublesome in view of the bound for the asymptotic bias in Lemma (3.15).

(3.18) EXERCISE. Verify (3.17) by showing that

$$| f_o |_{TV} - | f^{nh} |_{TV} + | f^{nh} - f_o |_{TV} \leqslant 2 \, | f_o |_{TV} \, .$$

The final task is to give an appropriate bound for the random sum in the right-hand side of (3.17).

(3.19) LEMMA. *Under the assumption* (3.3) *on the noise, for all* $\lambda > 0$ *and* $\varepsilon \in TV(0,1)$,

$$\tfrac{1}{n} \sum_{i=1}^{n} \varepsilon(x_{in})\, d_{in} \leqslant \eta_{n,\lambda} \left\{ \| \varepsilon \| + \lambda \, | \varepsilon |_{TV} \right\} ,$$

where
$$\lim_{n \to \infty} \sup_{h \in \mathcal{H}_n(\alpha)} \frac{\eta^{n\lambda}}{\sqrt{(n\lambda)^{-1} \log n}} < \infty \quad \textit{almost surely}.$$

PROOF. Let $\delta = \pi_n \varepsilon$, the linear interpolant of $\varepsilon(x_{in})$, $i = 1, 2, \cdots, n$; see (2.16). Note that $|\delta|_{TV} \leqslant |\varepsilon|_{TV}$ and, by the quadrature result of Lemma (2.19), one also obtains that $\| \delta \|_2 \leqslant c_n \| \varepsilon \|_2 + c n^{-1/2} | \varepsilon |_{TV}$.

Since $\delta \in W^{1,1}(0,1)$, the integral representation of Lemma (2.12) applies and gives $\delta(x) = \langle \mathfrak{R}_{1,\lambda}(x, \cdot\,) \, , \, \delta \rangle_{1,\lambda}$, so that

$$\tfrac{1}{n} \sum_{i=1}^{n} \varepsilon(x_{in})\, d_{in} = \tfrac{1}{n} \sum_{i=1}^{n} \delta(x_{in})\, d_{in} = \langle \mathfrak{S}_{n,\lambda} \, , \, \delta \rangle_{1,\lambda} , \tag{3.20}$$

in which

$$\mathfrak{S}_{n,\lambda}(t) = \tfrac{1}{n} \sum_{i=1}^{n} d_{in} \, \mathfrak{R}_{1,\lambda}(x_{in}, t) \, , \qquad t \in [0, 1] \, . \tag{3.21}$$

It follows that

$$\tfrac{1}{n} \sum_{i=1}^{n} \varepsilon(x_{in})\, d_{in} \leqslant \| \mathfrak{S}_{n,\lambda} \| \, \| \delta \| + \lambda^2 \, \| \mathfrak{S}_{n,\lambda}' \|_{\infty} \, \| \delta' \|_1 \, .$$

Now, Theorem (14.6.12) gives the required bounds on $\| \mathfrak{S}_{n,\lambda} \| \leqslant \| \mathfrak{S}_{n,\lambda} \|_{\infty}$ and $\| \mathfrak{S}_{n,\lambda}' \|_{\infty}$, and of course $\| \delta \| \leqslant \| \varepsilon \|$ and $\| \delta' \| \leqslant \| \varepsilon \|_{TV}$. Q.e.d.

With the lemma in hand, we move on to the integrated squared error.

PROOF OF THEOREM (3.4). Substitute the bound of Lemma (3.19) into (3.17), and appeal to the Quadrature Lemma (2.19) to replace the integral with a sum. Then, for all $\lambda > 0$ and $h > 0$,

$$s_n^2 + \tfrac{1}{2} h \, | \varepsilon |_{TV} \leqslant \eta^{n\lambda} \left\{ s_n + \left(c_o n^{-1/2} + \lambda \right) | \varepsilon |_{TV} \right\} + h \, | f_o |_{TV} \, ,$$

in which

$$s_n^2 = \tfrac{1}{n} \sum_{i=1}^{n} | \varepsilon(x_{in}) |^2 \, .$$

Now, collect all of the terms involving $| \varepsilon |_{TV}$ on the left-hand side of the inequality. Then

$$s_n^2 + \nu \, | \varepsilon |_{TV} \leqslant \eta^{n\lambda} \, s_n + h \, | f_o |_{TV} \tag{3.22}$$

with $\nu = \frac{1}{2} h - \eta^{n\lambda} \left(c_o n^{-1/2} + \lambda \right)$. Assuming $nh^2 \longrightarrow \infty$, take λ as large as possible such that, say,

$$\nu \geqslant \tfrac{1}{4} h \, . \tag{3.23}$$

The choice

$$\lambda = c_1 \, n \, h^2 \, (\log n)^{-1} \tag{3.24}$$

for a suitably small constant c_1 will work (asymptotically, almost surely). Then, (3.22) yields the inequality

$$s_n^2 + \tfrac{1}{4} h \, | \, \varepsilon \, |_{TV} \leqslant \eta^{n\lambda} \, s_n + h \, | \, f_o \, |_{TV} \; . \tag{3.25}$$

Now, Exercise (13.4.10) implies the bound

$$s_n^2 + \tfrac{1}{4} h \, | \, \varepsilon \, |_{TV} \leqslant (\eta^{n\lambda})^2 + 2 \, h \, | \, f_o \, |_{TV} \; .$$

This is the inequality of the theorem. Q.e.d.

(3.26) EXERCISE. Fill in the missing details of the proof of Theorem (3.4).

(3.27) EXERCISE. Let $m \geqslant 1$, and consider the estimator f^{nh}, defined as the solution to

$$\begin{aligned} &\text{minimize} \quad \tfrac{1}{n} \sum_{i=1}^{n} | \, f(x_{in}) - y_{in} \, |^2 + h^{2m} \, | \, f^{(m)} \, |_{TV} \\ &\text{subject to} \quad f \in L^2(0,1) \; , \; f^{(m)} \in TV(0,1) \; . \end{aligned}$$

Show that $\mathbb{E}[\, \| \, f^{nh} - f_o \, \|^2 \,] = \mathcal{O}\big(n^{-2m/(2m+1)} \big)$ for $h \asymp n^{-1/(2m+1)}$ and likewise for the discrete error. Thus, there is no power of $\log n$ present. [Hint: Prove and use the inequality $\| \, f' \, \| \leqslant c_m \big(\, \| \, f \, \| + | \, f^{(m)} \, |_{TV} \, \big) \, .$]

(3.28) EXERCISE. Consider the partially linear model of § 13.8,

$$y_{in} = z_{in}^T \, \beta_o + f_o(x_{in}) + d_{in} \; , \quad i = 1, 2, \cdots, n \; ,$$

under the same assumptions as stated there, except that $f_o \in TV(0,1)$. Is it possible to get asymptotically normal estimators of β_o ? [The authors don't know. Recall that in § 13.8 we extensively used that the estimator of f_o was linear in the responses y_{in}. So in effect you are asked to reanalyze the problem (13.8.18) in the presence of a nice convex constraint $f \in \mathcal{C}$.]

EXERCISES: (3.11), (3.14), (3.16), (3.18), (3.26), (3.27), (3.28).

4. Least-absolute-deviations splines: Generalities

In the previous chapters, we considered the "standard" nonparametric regression problem with Gauss-Markov errors: uncorrelated, zero-mean random noise with finite variance. Here, we investigate what happens when the Gauss-Markov assumption fails miserably in the sense that

$$\mathbb{E}[\, | \, d_{in} \, | \,] = +\infty \; , \quad i = 1, 2, \cdots, n \; . \tag{4.1}$$

From a practical point of view, one should think of "$+\infty$" as merely "large". A typical example would be the case where most of the d_{in} are standard normals but a few of them (randomly) are distributed normally with mean 0 and standard deviation 100. So, the distribution of d_{in} would be a mixture of two normals; cf. HUBER (1981). Examples of this may be found in EFRON (1988).

In the presence of the catastrophe (4.1), it is customary to assume that

$$(4.2) \qquad d_{1,n}, d_{2,n}, \cdots, d_{n,n} \text{ are iid} \quad \text{and} \quad \text{median}[\, d_{1,n} \,] = 0.$$

In addition, we need to assume that the density p of the d_{in} is bounded away from zero on an interval around zero, i.e.,

$$(4.3) \qquad \inf \{\, p(t) : |\,t\,| \leqslant q \,\} \geqslant r \,,$$

for certain positive constants q and r. It is somewhat strange that one should have trouble with the better case where (4.3) is replaced by

$$(4.4) \qquad \mathbb{P}[\, d_{1,n} = 0 \,] = \alpha > 0 \,, \quad \mathbb{P}[\, |\, d_{1,n} \,| > q \,] = 1 - \alpha \,,$$

but in fact this is the case.

The standard interpretation of the model (1.1)-(4.1)-(4.2) is that one wishes to construct *robust* estimators; i.e., estimators that are not overly influenced by a few gross errors. In the parametric case, the *least-absolute-deviations* method has been found useful. That is,

$$(4.5) \qquad \text{minimize} \quad \tfrac{1}{n} \sum_{i=1}^{n} |\, f(x_{in}) - y_{in} \,| \quad \text{subject to} \quad f \in \mathcal{F} \,,$$

where $\mathcal{F}$ is the parametric family under consideration; cf. the discussion of robust estimation in Volume I, § 2.6. We note that (4.5) is maximum likelihood estimation when the errors have a two-sided exponential distribution. In the nonparametric case, the method of penalized least-absolute-deviations is natural, and this is the main topic of this section. Thus, for the model (1.1) with (4.2) and $f_o \in W^{m,2}(0,1)$, the estimator under consideration is the solution f^{nh} of

$$(4.6) \qquad \begin{aligned} &\text{minimize} \quad L_{nh}(f) \stackrel{\text{def}}{=} \tfrac{1}{n} \sum_{i=1}^{n} |\, f(x_{in}) - y_{in} \,| + h^{2m} \, \|\, f^{(m)} \,\|_2^2 \\ &\text{subject to} \quad f \in W^{m,2}(0,1) \,. \end{aligned}$$

The usual questions regarding (4.6) are considered: existence (yes) and uniqueness (no) of solutions, and asymptotic error bounds. The analysis follows parts of POLLARD (1991). See also VAN DE GEER (1990).

The existence of solutions is a nice exercise, following Chapter 10.

(4.7) EXERCISE. Fix $n \in \mathbb{N}$, $h > 0$. (a) Show that (4.6) admits a minimizing sequence: There exists a sequence $\{\, f_k \,\}_k \subset W^{m,2}(0,1)$ such that

$$\lim_{k \to \infty} L_{nh}(f_k) = \inf \{\, L_{nh}(f) : f \in W^{m,2}(0,1) \,\} \,.$$

(b) Show that $\{ f_k \}_k \subset W^{m,2}(0,1)$ is bounded.
(c) Show that $L_{nh}(\,\cdot\,)$ is weakly lower semi-continuous on $W^{m,2}(0,1)$.
(d) Show that there exists a subsequence $\{ \varphi_k \}_k$ of $\{ f_k \}_k$ that converges weakly in $W^{m,2}(0,1)$ to some element φ_o, and show that

$$L_{nh}(\varphi_o) = \lim_{k\to\infty} L_{nh}(f_k) \ .$$

(e) Conclude that (4.6) has a solution.

The lack of uniqueness is explained by the fact that the functional $L_{nh}(f)$ is not strictly convex. For the roughness penalization, we have

$$\| f^{(m)} \|_2^2 - \| \varphi^{(m)} \|_2^2 - 2 \left\langle \varphi^{(m)} , f^{(m)} - \varphi^{(m)} \right\rangle = \| f^{(m)} - \varphi^{(m)} \|_2^2 \ ,$$

but this vanishes if f and φ differ by a polynomial of degree $< m$. The other part of the functional does not help since $| f(x_{in}) - d_{in} |$ is a convex function of f but not strictly. How this leads to nonuniqueness is best illustrated by the case $m = 1$ and $n = 2$ (two data points). Even so, it also illustrates that for large n all solutions will be pretty close, and one need not worry about nonuniqueness.

(4.8) EXERCISE. Let $m = 1$, $n = 2$, $h = 1$, and, with $y_{1,2} = -1$, $y_{2,2} = 1$, consider (4.6).
(a) Denote the zero function by 0, and let $\ell(x) = 2x - 1$ be the linear interpolant of the data. Verify that

$$L_{nh}(0) = 2 \quad , \quad L_{nh}(\ell) = 4 \ .$$

(b) Verify that every solution of (4.6) is a linear function of the form

$$f(x) = (1 - x)\, f(0) + x\, f(1)$$

and that $-1 \leqslant f(0) \leqslant 0$, $0 \leqslant f(1) \leqslant 1$. [Hint: Fix $f(0)$ and $f(1)$, and solve (4.6) with these added constraints.]
(c) If f is a solution of (4.6), verify that $f(x) - 1 - f(0)$ and $f(x) + 1 - f(1)$ are solutions also.
(d) Conclude that the solution of (4.6) is not unique.

Consistency and convergence rates. It is clear that the niceties associated with least-squares problems do not apply to least absolute deviations. For example, we need the large-sample asymptotic problem. The first guess,

$$\text{(4.9)} \qquad \begin{array}{ll} \text{minimize} & \| f - f_o \|_1 + h^{2m} \| f^{(m)} \|_2^2 \\ \text{subject to} & f \in W^{m,2}(0,1) \ , \end{array}$$

is inspired, but quite wrong. See Exercise (4.30) as to why this is unfortunate. Why is it wrong? Ordinarily, one would assume that, for fixed smooth functions f,

$$\tfrac{1}{n}\sum_{i=1}^{n} |\, f(x_{in}) - y_{in} \,| \longrightarrow \int_0^1 \mathbb{E}[\, |\, f(x) - d_{1,n} \,| \,]\, dx \ ,$$

but the expected value equals $+\infty$. So this is no good. However, if the estimator under consideration is consistent, then $f^{nh}(x_{in}) \approx f_o(x_{in})$ for all i, and then, if d_{in} is large,

$$|\, f^{nh}(x_{in}) - f_o(x_{in}) - d_{in} \,| \approx |\, d_{in} \,| \ .$$

Thus, it is reasonable (mandatory, actually) to consider

$$(4.10) \qquad \tfrac{1}{n}\sum_{i=1}^{n} \{\, |\, f(x_{in}) - y_{in} \,| - |\, d_{in} \,| \,\} \longrightarrow_{\text{as}} \int_0^1 \mathfrak{L}(\, f(x) - f_o(x)\,)\, dx$$

for smooth deterministic f (and f_o), where

$$(4.11) \qquad \mathfrak{L}(\, t\,) = \mathbb{E}[\, |\, t \, - d_{1,n} \,| - |\, d_{1,n} \,| \,] \ .$$

(4.12) Exercise. Prove (4.10) as follows. (a) Use the method of bounded differences of McDiarmid (1989) and Devroye (1991), expounded in Volume I, Theorem (4.4.21), to show that

$$\mathbb{P}[\, |\, Q_n \,| > t \,] \leqslant 2\, \mathrm{e}^{-n\,t^2/2} \ ,$$

where Q_n is the quotient, with $\varepsilon = f - f_o$,

$$Q_n = \frac{\tfrac{1}{n}\sum_{i=1}^{n} \{\, |\, \varepsilon(x_{in}) - d_{in} \,| - |\, d_{in} \,| - \mathbb{E}[\, |\, \varepsilon(x_{in}) - d_{in} \,| - |\, d_{in} \,| \,] \,\}}{\Bigl(\tfrac{1}{n}\sum_{i=1}^{n} |\, \varepsilon(x_{in}) \,|^2 \Bigr)^{1/2}} \ .$$

(b) Conclude that $\mathbb{E}[\, |\, Q_n \,| \,] = \mathcal{O}(\, \sqrt{n^{-1}}\,)$ and $Q_n =_{\text{as}} \mathcal{O}(\, \sqrt{n^{-1}\log n}\,)$, and that (4.10) holds for smooth deterministic f and f_o.

The large-sample asymptotic problem is thus

$$(4.13) \qquad \begin{array}{ll} \text{minimize} & \Lambda_{\infty,h}(f) \\ \text{subject to} & f \in W^{m,2}(0,1) \ , \end{array}$$

where

$$(4.14) \qquad \Lambda_{\infty,h}(f) \stackrel{\text{def}}{=} \int_0^1 \mathfrak{L}(\, f(x) - f_o(x)\,)\, dx + h^{2m} \,\|\, f^{(m)} \,\|_2^2 \ .$$

We denote the solution of (4.13) by f_h. Of course, the questions of existence and uniqueness of f_h must be settled.

Strong convexity. As a first step in the analysis, we had better take a close look at the function $\mathfrak{L}(\, t\,)$. It is easy to show that $\mathfrak{L}$ is Lipschitz continuous:

(4.15) $$| \mathfrak{L}(t) - \mathfrak{L}(s) | \leqslant | t - s | \quad \text{for all } t, s \; .$$

In fact, $\mathfrak{L}$ is twice differentiable

(4.16) $$\mathfrak{L}'(t) = \int_{\mathbb{R}} \operatorname{sign}(t - x)\, p(x)\, dx \; , \quad \text{for all } t \; ,$$

(4.17) $$\mathfrak{L}''(t) = 2\, p(t) \; , \quad \text{a.e.}$$

(4.18) EXERCISE. Verify (4.15) through (4.17).

The conclusion is that $\mathfrak{L}''(t) \geqslant 0$ a.e., and so $\mathfrak{L}$ is convex. It is in fact strongly convex on an interval around 0.

(4.19) LEMMA. *Under the condition* (4.3), *for all* $| t | \leqslant q$ *and* $| s | \leqslant q$,

$$\mathfrak{L}(t) - \mathfrak{L}(s) - \mathfrak{L}'(s)(t - s) \geqslant r\,(t - s)^2 \; .$$

PROOF. Let $-q \leqslant s < t \leqslant q$. Since $\mathfrak{L}'' = 2\,p \in L^1(\mathbb{R})$, Taylor's theorem with exact remainder gives

$$\mathfrak{L}(t) - \mathfrak{L}(s) - \mathfrak{L}'(s)\,(t - s) = 2 \int_s^t (t - \tau)\, p(\tau)\, d\tau \; .$$

By the assumption (4.3), the right-hand side dominates

$$2\, r \int_s^t (t - \tau)\, d\tau = r\,(t - s)^2 \; .$$

The case $t < s$ goes similarly. Q.e.d.

Also note that $\mathfrak{L}(0) = 0$ (obviously) and that $\mathfrak{L}'(0) = 0$ by (4.16) and the zero-median assumption (4.2). Thus, Lemma (4.19) shows that $\mathfrak{L}(t) > 0 \iff t \neq 0$. Consequently, we have the following corollary.

(4.20) COROLLARY. *If* $f, f_o \in L^1(0,1)$, *then* $\int_{\mathbb{R}} \mathfrak{L}\big(f(x) - f_o(x) \big)\, dx \geqslant 0$, *with equality if and only if* $f = f_o$ *almost everywhere.*

The strong convexity result for $\mathfrak{L}$ has an analogue for $\Lambda_{\infty,h}$. Let

(4.21) $$B(\psi; \varrho) = \{\, f \in W^{m,2}(0,1) : \| f - \psi \|_\infty \leqslant \varrho \,\} \; ,$$

the ball around $\psi \in W^{m,2}(0,1)$ with radius ϱ in the uniform norm.

(4.22) LEMMA. *For all* f, $\varphi \in B(f_o; q)$,

$$\Lambda_{\infty,h}(f) - \Lambda_{\infty,h}(\varphi) - \delta\Lambda_{\infty,h}(\varphi\,; f - \varphi) \geqslant r\, \| f - \varphi \|_{m,h}^2 \; ,$$

where $\delta\Lambda_{\infty,h}$ *is the Gateaux variation of* $\Lambda_{\infty,h}$,

$$\delta\Lambda_{\infty,h}(\varphi\,;\varepsilon) = \big\langle\, \mathfrak{L}'\big(\,\varphi(x) - f_o(x)\,\big)\,,\,\varepsilon\,\big\rangle_{L^2(0,1)} + 2\,h^{2m}\,\big\langle\,\varphi^{(m)}\,,\,\varepsilon^{(m)}\,\big\rangle_{L^2(0,1)}\,.$$

(4.23) Exercise. Prove the lemma.

(4.24) Exercise. The finite-sample problem (4.6) is not strongly convex. However, show that for, all f, $\varphi \in W^{m,2}(0,1)$,

$$L_{nh}(f) - L_{nh}(\varphi) - \delta L_{nh}(\varphi\,; f - \varphi) \geqslant h^{2m}\,\|\,(f-\varphi)^{(m)}\,\|^2\,.$$

Convergence rates for f^{nh} are shown in the next section, but here we make the following crucial remark.

(4.25) Crucial Observation on Establishing Convergence Rates. It is clear that the local strong convexity of $\mathfrak{L}(f - f_o)$ is instrumental in establishing convergence rates, but its use would require that the estimators f^{nh} and the large-sample asymptotic estimator f_h eventually lie inside the ball $B(f_o; q)$; see (4.21). The simplest way around this problem is to impose this as a constraint on the solution. Thus, we consider the constrained small-sample problem

$$\text{(4.26)} \qquad \text{minimize} \quad L_{nh}(f) \quad \text{subject to} \quad f \in B(f_o; q)$$

and the constrained large-sample asymptotic problem

$$\text{(4.27)} \qquad \text{minimize} \quad \Lambda_{\infty,h}(f) \quad \text{subject to} \quad f \in B(f_o; q)\,.$$

We denote the solutions by φ^{nh} and φ_h, respectively. If we can show that

$$\text{(4.28)} \qquad \|\,\varphi^{nh} - f_o\,\|_\infty \longrightarrow_{\text{as}} 0\,,$$

then φ^{nh} and f^{nh} eventually coincide, and likewise for φ_h and f_h. Thus, we start with proving convergence rates for φ^{nh} and φ_h and take it from there. Lest we forget, we have the following.

(4.29) Exercise. Show the existence and uniqueness of φ_h and the existence of φ^{nh}. See Theorem (10.4.14) in Volume I.

Finally, we present the long-awaited "unfortunate" exercise.

(4.30) Exercise. Suppose that f_o satisfies

$$f_o^{(\ell)}(0) = f_o^{(\ell)}(1) = 0\,, \quad m \leqslant \ell \leqslant 2m-1\,.$$

Show that $f = f_o$ is the solution of (4.9), provided $h^{2m}\,\|\,f_o^{(m)}\,\|_\infty \leqslant \frac{1}{2}$. [Hint: The Euler equations for (4.9) are

$$\tfrac{1}{2}\,\text{sign}(\,f(x) - f_o(x)\,) + h^{2m} f^{(2m)} = 0 \quad \text{in } (0,1)$$

plus appropriate boundary conditions.]

So, it is really a pity that (4.9) is not the large-sample asymptotic problem: If it were, then there would be no need for h to tend to 0, and the error bound $\mathcal{O}(n^{-1/2})$ could possibly be achieved! Of course, to take care of boundary corrections, a detour via § 13.4 would be required, but the point is moot.

EXERCISES: (4.7), (4.8), (4.12), (4.18), (4.23), (4.24), (4.29), (4.30).

5. Least-absolute-deviations splines: Error bounds

In this section, we prove convergence rates for the solution φ^{nh} of the problem (4.26), which we repeat here for convenience:

$$(5.1) \qquad \begin{aligned} &\text{minimize} && \tfrac{1}{n} \sum_{i=1}^{n} |\, f(x_{in}) - y_{in} \,| + h^{2m} \| \, f^{(m)} \, \|^2 \\ &\text{subject to} && f \in W^{m,2}(0,1) \; , \; \| \, f - f_o \, \|_\infty \leqslant q \; . \end{aligned}$$

As argued in Observation (4.25), the extra constraint $\| \, f - f_o \, \|_\infty \leqslant q$ is necessary. However, if $\| \, \varphi^{nh} - f_o \, \|_\infty \longrightarrow 0$, then the constraint is not active, and so φ^{nh} is also the unconstrained estimator f^{nh}.

When all is said and done, we get the usual mean integrated squared error bound.

(5.2) THEOREM. *If $f_o \in W^{m,2}(0,1)$, and the model (1.1), (4.2)–(4.3) applies, then, for asymptotically uniform designs,*

$$\mathbb{E}\Big[\, \| \, \varphi^{nh} - f_o \, \|_{m,h}^2 \Big] = \mathcal{O}\big(\, h^{2m} + (nh)^{-1} \, \big) \; ,$$

provided $h \to 0$, $nh \to \infty$, and likewise for the discrete version.

(5.3) COROLLARY. *Under the conditions of Theorem (5.2),*

$$\| \, f^{nh} - f_o \, \|_{m,h}^2 = \mathcal{O}_P\big(\, h^{2m} + (nh)^{-1} \, \big) \; ,$$

and likewise for the discrete version.

PROOF. By Lemma (13.2.11), we get

$$\mathbb{E}[\, \| \, f^{nh} - f_o \, \|_\infty^2 \,] = \mathcal{O}\big(\, h^{-1} \, (\, h^{2m} + (nh)^{-1} \,) \big) \; .$$

Thus, for $nh^2 \to \infty$, we have $f^{nh} = \varphi^{nh}$ in probability, and so, in probability,

$$\| \, f^{nh} - f_o \, \|_{m,h}^2 = \| \, f^{nh} - f_o \, \|_{m,h}^2 = \mathcal{O}\big(\, h^{2m} + (nh)^{-1} \, \big) \; . \qquad \text{Q.e.d.}$$

Asymptotic bias. Although there is no need for a bias/variance decomposition, it is useful to get some idea about the asymptotic bias.

Since φ_h solves (4.27), the Gateaux variation at φ_h in the direction $f_o - \varphi_h$ vanishes, that is $\delta\Lambda_{\infty,h}(\varphi_h; f_o - \varphi_h) = 0$, and so Lemma (4.22) implies that

$$(5.4) \qquad r\,\|\,\varphi_h - f_o\,\|^2 + h^{2m}\,\|\,(\,\varphi_h - f_o)^{(m)}\,\|^2 \leqslant \Lambda_{\infty,h}(f_o) - \Lambda_{\infty,h}(\varphi_h)\;.$$

The right-hand side may be rewritten as

$$-\int_0^1 \mathfrak{L}(\,\varphi_h(x) - f_o(x)\,)\,dx + h^{2m}\,\|\,f_o^{(m)}\,\|^2 - h^{2m}\,\|\,\varphi_h^{(m)}\,\|^2\;.$$

Since the first and last terms are negative, it follows that

$$(5.5) \qquad r\,\|\,\varphi_h - f_o\,\|^2 + h^{2m}\,\|\,(\,\varphi_h - f_o)^{(m)}\,\|^2 \leqslant h^{2m}\|\,f_o^{(m)}\,\|^2\;.$$

Consequently,

$$\|\,f_o^{(m)} - \varphi_h^{(m)}\,\| \leqslant \|\,f_o^{(m)}\,\| \quad\text{and}\quad \|\,f_o - \varphi_h\,\|^2 \leqslant r^{-1}\,h^{2m}\,\|\,f_o^{(m)}\,\|^2\;.$$

We have proved the following theorem.

(5.6) ASYMPTOTIC BIAS THEOREM. *Let $m \in \mathbb{N}$. If $f_o \in W^{m,2}(0,1)$, then*

$$\|\,f_o - \varphi_h\,\| \leqslant r^{-1/2}\,h^m\,\|\,f_o^{(m)}\,\|_2 \quad\text{and}\quad \|\,\varphi_h^{(m)}\,\| \leqslant 2\,\|\,f_o^{(m)}\,\|\;.$$

(5.7) COROLLARY. *Let $m \geqslant 1$. If $f_o \in W^{m,2}(0,1)$, then*

$$\|\,\varphi_h - f_o\,\|_\infty = \mathcal{O}(\,h^{m-1/2}\,)\;,$$

and $\varphi_h = f_h$ for all positive h small enough.

PROOF. Lemma (13.2.11) does the trick. Q.e.d.

(5.8) EXERCISE. (a) Assume that $f_o \in W^{2m,2}(0,1)$ and satisfies

$$f_o^{(\ell)}(0) = f_o^{(\ell)}(1) = 0\;, \quad \ell = m, m+1, \cdots, 2m-1\;.$$

Show that $\|\,(\varphi_h - f_o)^{(m)}\,\| \leqslant h^m\,\|\,f_o^{(m)}\,\|$ and

$$\|\,f_o - \varphi_h\,\| \leqslant r^{-1/2}\,h^{2m}\,\|\,f_o^{(2m)}\,\|_2\;.$$

(b) What happens if only $f_o^{(\ell)}(0) = f_o^{(\ell)}(1) = 0$, $\ell = m, m{+}1, \cdots, m{+}k$ for some k with $1 \leqslant k < 2m-1$?

Asymptotic variation. We start with the (strong) convexity inequality of Exercise (4.24) with $\varphi = \varphi^{nh}$ and $f = f_o$. Writing $\varepsilon \equiv \varphi^{nh} - f_o$, then

$$(5.9) \qquad h^{2m}\,\|\,\varepsilon^{(m)}\,\|^2 \leqslant \Lambda_{n,h}(f_o) - \Lambda_{n,h}(\varphi^{nh})\;,$$

while Lemma (4.22) gives that

$$r\,\|\,\varepsilon\,\|_{m,h}^2 \leqslant \Lambda_{\infty,h}(\,\varepsilon\,) - \Lambda_{\infty,h}(0) - \delta\Lambda_{\infty,h}(\,0\,;\varepsilon) = \Lambda_{\infty,h}(\,\varepsilon\,)\;.$$

Adding these two inequalities yields

$$(5.10) \qquad r\,\|\,\varepsilon\,\|_{m,h}^2 \leqslant S_n + h^{2m}\,\|\,(f_o)^{(m)}\,\|^2\;,$$

with

$$S_n = \int_{\mathbb{R}} \frac{1}{n} \sum_{i=1}^{n} \{ |t| - |\varepsilon(x_{in}) - t| \} \{ dP_i(t) - dP(t) \} , \tag{5.11}$$

in which

$$P_i(t) = 1\!\!1(d_{in} < t) , \quad P(t) = \mathbb{E}[1\!\!1(d_{in} < t)] . \tag{5.12}$$

Of course, $P(t)$ is just the distribution function of the d_{in}.

Reproducing kernel Hilbert space tricks. To get a handle on S_n, we must resort to reproducing kernel Hilbert space tricks and even plain integration by parts. As usual, the crucial quantity is the error for a kernel estimator applied to pure noise,

$$\begin{aligned} \mathfrak{S}_{n\lambda}(y, t) &= \frac{1}{n} \sum_{i=1}^{n} \mathfrak{R}_{1,\lambda}(x_{in}, y) \{ P_i(t) - P(t) \} , \\ \mathfrak{M}_{n\lambda}(y) &= \sup_{|t| \leqslant q} | \mathfrak{S}_{n\lambda}(y, t) | , \quad \mathfrak{N}_{n\lambda}(y) = \sup_{|t| \leqslant q} | \mathfrak{S}'_{n\lambda}(y, t) | , \end{aligned} \tag{5.13}$$

with $\mathfrak{R}_{1,h}(x, y)$ the reproducing kernel for $W^{1,2}(0,1)$ and with differentiation with respect to the first argument.

(5.14) LEMMA. $S_n \leqslant c \| \varepsilon \|_{m,\lambda} (\| \mathfrak{M}_{n,\lambda} \| + \lambda \| \mathfrak{N}_{n,\lambda} \|)$.

The proof is given in the next section. Below, we establish the bound for deterministic λ, $\lambda \to 0$, $n\lambda \to \infty$,

$$\mathbb{E}[\| \mathfrak{M}_{n,\lambda} \|^2 + \lambda^2 \| \mathfrak{N}_{n,\lambda} \|^2] \leqslant c\,(n\lambda)^{-1} . \tag{5.15}$$

For now, substitute this bound on S_n into the strong convexity inequality (5.10). For $\lambda = h \to 0$ and $nh \longrightarrow \infty$, this proves Theorem (5.2).

(5.16) EXERCISE. Make sure that, apart from proving (5.15), we really did prove Theorem (5.2).

PROOF OF (5.15). First, the change of variables $u_i = P(d_{in})$ reduces everything to Uniform(0, 1) random variables.

(5.17) LEMMA. *Let $u_1, u_2, \cdots, u_n$ be iid uniform(0, 1) random variables. For $0 \leqslant x \leqslant 1$, $0 \leqslant y \leqslant 1$, define*

$$\Psi_{n,\lambda}(y, p) = \frac{1}{n} \sum_{i=1}^{n} \mathfrak{R}_{1,\lambda}(x_{in}, y) \{ 1\!\!1(u_i \leqslant p) - p \} , \tag{5.18}$$

and let I_q be the interval $I_q = (P(-q) , P(q))$. Then,

$$\mathfrak{M}_{n\lambda}(y) = \sup_{p \in I_q} | \Psi_{n,\lambda}(y, p) | \quad \textit{and} \quad \mathfrak{N}_{n\lambda}(y) = \sup_{p \in I_q} | \Psi'_{n,\lambda}(y, p) | ,$$

with differentiation with respect to the first variable.

Now, finding an appropriate bound on the expectations of $\| \mathfrak{M}_{n\lambda} \|^2$ and $\| \mathfrak{N}_{n\lambda} \|^2$ may be done using the McDiarmid (1989) method of bounded differences. See Devroye (1991) or § 4.4 in Volume I. For fixed y, define

$$\psi_n \equiv \psi_n(u_1, u_2, \cdots, u_n) = \sup_{p \in I_q} \Bigl| \tfrac{1}{n} \sum_{i=1}^{n} \mathfrak{R}_{1,\lambda}(x_{in}, y) \bigl(1\!\!1(u_i \leqslant p) - p \bigr) \Bigr| .$$

With the notation $[u_{(i)}, w] = (u_1, \cdots, u_{i-1}, w, u_{i+1}, \cdots, u_n)$, in which u_i is replaced by w, we have, for all $v, w \in [0, 1]$,

$$| \psi_n[u_{(i)}, v] - \psi_n[u_{(i)}, w] | \leqslant \sup_{p \in I_q} \tfrac{1}{n} | \mathfrak{R}_{1,\lambda}(x_{in}, y) | \, | 1\!\!1(v < p) - 1\!\!1(w < p) | \leqslant \tfrac{1}{n} | \mathfrak{R}_{1,\lambda}(x_{in}, y) | .$$

(Note that, by the triangle inequality for the sup norm,

$$\Bigl| \sup_{p \in I_q} | F(p) | - \sup_{p \in I_q} | G(p) | \Bigr| \leqslant \sup_{p \in I_q} | F(p) - G(p) |$$

for all functions F and G on I_q.) Now, let

$$r_n^2 = n^{-2} \sum_{i=1}^{n} | \mathfrak{R}_{1,\lambda}(x_{in}, y) |^2 .$$

Then, McDiarmid's Lemma (4.4.21) gives that, for all $t > 0$,

$$\mathbb{P}[\psi_n > r_n t] \leqslant 2 \exp(-\tfrac{1}{2} t^2) ,$$

and it follows that $\mathbb{E}[| \psi_n |^2] \leqslant c r_n^2$. Since $r_n^2 \leqslant c_1 (nh)^{-1}$, this says that

$$\mathbb{E}[| \mathfrak{M}_{n\lambda}(y) |^2] \leqslant c_2 (nh)^{-1} \quad \text{uniformly in } y .$$

The bound on $\mathbb{E}[\| \mathfrak{M}_{n\lambda} \|^2]$ follows by integration. In a similar fashion, one obtains the bound on $\mathbb{E}[\lambda^2 \| \mathfrak{N}_{n\lambda} \|^2]$. Q.e.d.

(5.19) Exercise. Consider the partially linear model of § 13.7,

$$y_{in} = z_{in}^T \beta_o + f(x_{in}) + d_{in} , \quad i = 1, 2, \cdots, n ,$$

under the same assumptions as stated there, except that it is assumed that the noise components $d_{1,n}, d_{2,n}, \cdots, d_{nn}$ are iid and satisfy (4.2) and (4.3). Consider estimating β_o by the solution to

$$\begin{aligned} &\text{minimize} \quad \tfrac{1}{n} \sum_{i=1}^{n} | z_{in}^T \beta + f(x_{in}) - y_{in} | + h^{2m} \| f^{(m)} \|^2 \\ &\text{subject to} \quad \beta \in \mathbb{R}^d , \ f \in W^{m,2}(0,1) . \end{aligned}$$

Is it possible to get asymptotically normal estimators of β_o?

Exercises: (5.7), (5.16), (5.19).

6. Reproducing kernel Hilbert space tricks

Here we prove Lemma (5.14). The difference with the "usual" proofs lies in the combinatorial-like aspects, leading to (6.5) and (6.7) below.

PROOF OF LEMMA (5.14). As noted already, we must do two integrations by parts, one in the variable x and one in t. First, plain integration by parts in t gives

$$S_n = \int_{\mathbb{R}} \tfrac{1}{n} \sum_{i=1}^{n} s(x_{in}, t) \{ P_i(t) - P(t) \} \, dt \,, \tag{6.1}$$

where

$$s(x, t) = \operatorname{sign}(t - \varepsilon(x)) - \operatorname{sign}(t) \tag{6.2}$$

with $\varepsilon \equiv \varphi^{nh} - f_o$. Since $\varepsilon \in B(0, q)$, then $s(x, t) = \pm 2$ on the crooked bowtie region in the εt plane, bounded by the lines $\varepsilon = t$, $t = 0$, and $\varepsilon = \pm q$ and vanishes everywhere else. So, $s(x, t)$ vanishes for $|t| > q$. Consequently, the integration in (6.1) is only over $|t| < q$. Now, replace the function $s(x, t)$ by its piecewise linear interpolant in x,

$$s_n(x, t) \stackrel{\text{def}}{=} [\pi_n s(\cdot, t)](x) \,. \tag{6.3}$$

First, we digress. Observe that, for fixed x, the function values of $s_n(x, t)$ lie between -2 and 2 and equal 0 otherwise. In view of the discussion above, it follows that

$$\int_{|t|<q} |s_n(x, t)| \, dt \leqslant 2 \, | \pi_n \varepsilon(x) | \,. \tag{6.4}$$

Also, observe that, by the special design (1.3),

$$s_n'(y, t) \stackrel{\text{def}}{=} \frac{\partial s_n(y, t)}{\partial y} \tag{6.5}$$

equals $\pm 2(n-1)$ whenever it is nonzero, which is precisely when t lies between $\varepsilon(x_{in})$ and $\varepsilon(x_{i-1,n})$, and $y \in (x_{i-1,n}, x_{in})$. Let $\omega_{i,n}$ denote the interval bounded by $\varepsilon(x_{in})$ and $\varepsilon(x_{i-1,n})$. Then,

$$s_n'(y, t) = \sum_{i=2}^{n} \pm 2(n-1) \, 1\!\!1\big(t \in \omega_{i,n} \big) \, 1\!\!1\big(y \in (x_{i-1,n}, x_{in}) \big) \tag{6.6}$$

for the appropriate choices of the $\pm$ signs. Then, for $y \in (x_{i-1,n}, x_{in})$,

$$\begin{aligned} \int_{|t|<q} |s_n'(y, t)| \, dt &= \int_{\omega_{i,n}} |s_n'(y, t)| \, dt = 2(n-1) \, |\omega_{i,n}| \\ &= 2(n-1) \, | \varepsilon(x_{in}) - \varepsilon(x_{i-1,n}) | = 2 \, | (\pi_n \varepsilon)'(y) | \,. \end{aligned}$$

To summarize, we get the nice bound

$$\int_{|t|<q} |s_n'(y, t)| \, dt = 2 \, | (\pi_n \varepsilon)'(y) | \,. \tag{6.7}$$

We pick up the thread. The next step is the reproducing kernel trick. Observe that $s_n(\,\cdot\,,t\,) \in W^{1,1}(0,1)$ for fixed t. Thus, for all y and fixed t,

$$s_n(y,t\,) = \langle\, s_n(\,\cdot\,,t\,)\,,\, \mathfrak{R}_{1,\lambda}(y,\,\cdot\,)\,\rangle_{m,\lambda} \,. \tag{6.8}$$

Then, keeping in mind that the integration in (6.1) is over $|\,t\,| < q$ only,

$$S_n = \int_{|\,t\,|<q} \langle\, s_n(\,\cdot\,,t\,)\,,\, \mathfrak{S}_{n\lambda}(y,\,\cdot\,)\,\rangle_{m,\lambda} \,, \tag{6.9}$$

with $\mathfrak{S}_{n\lambda}(y,t\,)$ as in (5.13). Then we may write $S_n = U_n + W_n$ with

$$U_n = \int_{|\,t\,|<q} \int_0^1 \mathfrak{S}_{n\lambda}(y,t\,)\, s_n(y,t\,)\, dy\, dt$$

and

$$W_n = \int_{|\,t\,|<q} \int_0^1 \mathfrak{S}'_{n\lambda}(y,t\,)\, s_n(y,t\,)\, dy\, dt \,,$$

with differentiation with respect to the first argument.

We proceed to bound U_n and W_n. Interchanging the order of integration and bounding using the functions $\mathfrak{M}_{n\lambda}$ and $\mathfrak{N}_{n\lambda}$ from (5.13) gives

$$U_n \leqslant \int_0^1 M(y)\, \mathfrak{M}_{n\lambda}(y)\, dy \quad \text{and} \quad W_n \leqslant \int_0^1 N(y)\, \mathfrak{N}_{n\lambda}(y)\, dy \,,$$

where

$$M(y) = \int_{|\,t\,|<q} |\, s_n(y,t\,)\,|\, dt \leqslant 2 \big|\, [\,\pi_n \varepsilon\,](y)\, \big|$$

and

$$N(y) = \int_{|\,t\,|<q} |\, s'_n(y,t\,)\,|\, dt \leqslant 2 \big|\, [\,(\pi_n)' \varepsilon\,](y)\, \big| \,,$$

in which the bounds (6.5) and (6.6) were used. Putting it all together yields

$$\begin{aligned} S_n &\leqslant \langle\, |\, \pi_n \varepsilon\,|\,,\, \mathfrak{M}_{n\lambda}\,\rangle + \lambda^2 \langle\, |\,(\pi_n \varepsilon)'\,|\,,\, \mathfrak{N}_{n\lambda}\,\rangle \\ &\leqslant \|\, \pi_n \varepsilon\,\|_{1,\lambda} \big\{\, \|\, \mathfrak{M}_{n\lambda}\,\| + \lambda\, \|\, \mathfrak{N}_{n\lambda}\,\|\, \big\} \,, \end{aligned}$$

the last step by Cauchy-Schwarz and obvious bounding. Q.e.d.

7. Heteroscedastic errors and binary regression

In the previous section, we discussed some drastic violations of the Gauss-Markov conditions for the model (1.1)–(1.2). Here, the mildest and most common case is considered, where the d_{in} are independent random variables with zero means and *unequal* variances,

$$\mathbb{E}[\, d_n\,] = 0 \quad \text{and} \quad \mathbb{E}[\, d_n d_n^T\,] = \sigma^2\, V_n \,. \tag{7.1}$$

Here, V_n is an unknown diagonal matrix with bounded entries:

$$(7.2)\qquad \text{There exists a constant } K \text{ such that, for all } n\,, \quad (V_n)_{i,i} \leqslant K\;, \quad i = 1, 2, \cdots, n\;.$$

This goes by the name of *heteroscedasticity*. A nice example of this is nonparametric binary regression, discussed later in this section.

How would one estimate f_o in the model (1.1), (7.1)–(7.2)? Here, we assume that, for some integer $m \geqslant 1$,

$$(7.3)\qquad f_o \in W^{m,2}(0,1)\;,$$

and that the x_{in} are given by (1.3).

If V_n is a known function of f_o, then one could take as the estimator the solution of the penalized *weighted* least-squares problem

$$(7.4)\qquad \begin{array}{ll} \text{minimize} & \dfrac{1}{n}\displaystyle\sum_{i=1}^{n} \frac{|\, f(x_{in}) - y_{in}\,|^2}{\{\, V_n(f)\,\}_{i,i}} + h^{2m} \,\|\, f^{(m)}\,\|^2 \\ \text{subject to} & f \in W^{m,2}(0,1)\;. \end{array}$$

Considering that even the parametric nonlinear least-squares estimation problem is troublesome, we shall not discuss this further.

Under the circumstances, the only viable alternative is to ignore the heteroscedasticity initially and solve the usual roughness-penalized least-squares estimation problem

$$(7.5)\qquad \begin{array}{ll} \text{minimize} & \frac{1}{n}\displaystyle\sum_{i=1}^{n} |\, f(x_{in}) - y_{in}\,|^2 + h^{2m} \,\|\, f^{(m)}\,\|^2 \\ \text{subject to} & f \in W^{m,2}(0,1)\;. \end{array}$$

Then, estimate the variances by spline smoothing on the squared residuals, see § 22.4, and then solve

$$(7.6)\qquad \begin{array}{ll} \text{minimize} & \frac{1}{n}\displaystyle\sum_{i=1}^{n} \frac{|\, f(x_{in}) - y_{in}\,|^2}{\widehat{\sigma}^2(x_{in})} + h^{2m} \,\|\, f^{(m)}\,\|^2 \\ \text{subject to} & f \in W^{m,2}(0,1)\;. \end{array}$$

The existence and uniqueness of the solution f^{nh} of (7.5) and (7.6) have been adequately treated in § 13.3. Also, for (7.5), the usual convergence rates apply. Since the estimator of (7.5) achieves the optimal rate of convergence, all that can be gained by considering (7.6) is a smaller constant. In simulation results, this is hardly noticeable. However, the gains in the confidence bands are much more dramatic; see the simulations in § 23.7.

(7.7) THEOREM. *Let* $m \geqslant 1$ *and* $f_o \in W^{m,2}(0,1)$. *Considering the model* (1.1), (7.1) *under the assumptions* (7.2), (7.3), *and* (1.3), *the solution* f^{nh}

of (7.5) *satisfies*

$$\mathbb{E}[\, \| f^{nh} - f_o \|_{m,h}^2 \,] = \mathcal{O}\big(n^{-2m/(2m+1)} \big) \; ,$$

provided $h \asymp n^{-1/(2m+1)}$ *(deterministically). Moreover,*

$$\| f^{nh} - f_o \|_\infty = \mathcal{O}_P\big((n^{-1} \log n)^{2m/(2m+1)} \big) \; ,$$

provided $h \asymp (n^{-1} \log n)^{1/(2m+1)}$ *(deterministically).*

The proof of Theorem 7.7 is simple. Indeed, in comparison with § 13.4, the only complication is that the variances of the noise are not all equal.

(7.8) EXERCISE. Prove Theorem (7.7).

This concludes the general discussion of heteroscedasticity. Of course, it is open to criticism since we dealt with the problem by ignoring it.

Shifting gears somewhat, we now consider nonparametric binary regression and treat it as a special case of heteroscedasticity. In binary regression, the data y_{in} for any design x_{in} , $i = 1, 2, \cdots, n$, satisfy:

$$\text{(7.9)} \qquad \begin{array}{l} \text{The } y_{in} \text{ are independent and} \\ \left.\begin{array}{l} \mathbb{P}[\, y_{in} = 1 \,] = p_o(x_{in}) \\ \mathbb{P}[\, y_{in} = 0 \,] = 1 - p_o(x_{in}) \end{array}\right\} \text{ for } i = 1, 2, \cdots, n \; , \end{array}$$

where p_o is smooth; in particular,

$$\text{(7.10)} \qquad p_o \in W^{m,2}(0,1) \quad \text{for an integer } m \geqslant 1 \; .$$

Moreover, p_o obviously satisfies $0 \leqslant p_o(x) \leqslant 1$ for all $x \in [\, 0 \, , \, 1 \,]$. In many applications, it may seem reasonable to assume that p_o is increasing, but it is usually safer *not* to assume it.

The model (7.9) may be phrased as a special case of (1.1) and (7.1),

$$\text{(7.11)} \qquad y_{in} = p_o(x_{in}) + d_{in} \; , \quad i = 1, 2, \cdots, n \; ,$$

with $d_{in} = y_{in} - p_o(x_{in})$. One verifies that the d_{in} are independent and that $d_n = (\, d_{1,n}, d_{2,n}, \cdots, d_{n,n})^T$ satisfies

$$\text{(7.12)} \qquad \mathbb{E}[\, d_n \,] = 0 \; , \quad \mathbb{E}[\, d_n d_n^T \,] = \Sigma \; ,$$

where Σ is a diagonal matrix with

$$\text{(7.13)} \qquad \Sigma_{i,i} = p_o(x_{in}) \{\, 1 - p_o(x_{in}) \,\} \; , \quad i = 1, 2, \cdots, n \; .$$

Thus, the variances are not all the same.

As a nonparametric estimator of p_o, we consider the roughness-penalized least-squares spline estimator, the solution to the problem

$$\text{(7.14)} \qquad \begin{array}{ll} \text{minimize} & \frac{1}{n} \sum_{i=1}^{n} | p(x_{in}) - y_{in} |^2 + h^{2m} \| p^{(m)} \|^2 \\ \text{subject to} & p \in W^{m,2}(0,1) \,, \\ & 0 \leqslant p(x) \leqslant 1 \text{ for all } x \in [\, 0\,,\,1\,] \,. \end{array}$$

Theorem (7.7) implies convergence rates for this estimator.

(7.15) THEOREM. *Let $m \geqslant 1$ and $p_o \in W^{m,2}(0,1)$. Under the assumption (7.9), the solution p^{nh} of (7.14) satisfies*

$$\mathbb{E}[\, \| p^{nh} - p_o \|_{m,h}^2 \,] = \mathcal{O}(\, n^{-2m/(2m+1)} \,) \,,$$

provided $h \asymp n^{-1/(2m+1)}$. Moreover,

$$\| p^{nh} - p_o \|_\infty = \mathcal{O}_P(\, (n^{-1} \log n)^{2m/(2m+1)} \,) \,,$$

provided $h \asymp (n^{-1} \log n)^{1/(2m+1)}$.

For nonparametric binary regression, alternative estimators suggest themselves. The maximum penalized likelihood problem for estimating p_o is

$$\text{(7.16)} \qquad \begin{array}{ll} \text{minimize} & L_n(p) + R_h(p) \\ \text{subject to} & p \in W^{m,2}(0,1) \,,\ 0 \leqslant p(x) \leqslant 1 \text{ on } [\, 0\,,\,1\,] \,, \end{array}$$

with

$$\text{(7.17)} \qquad L_n(p) = \tfrac{1}{n} \sum_{i=1}^{n} \{\, y_{in} \log p(x_{in}) + (1 - y_{in}) \log(1 - p(x_{in})) \,\} \,,$$

and $R_h(p)$ the penalization functional depending on the smoothing parameter h. The choices of Chapter 13 and § 3 suggest themselves, as does Good's roughness penalization. The smoothed maximum likelihood approach as in EGGERMONT and LARICCIA (1995), see also § 1.2 in Volume I, has a surprising explicit solution; see the exercise below.

(7.18) EXERCISE. Let A_h be a suitable nonnegative boundary kernel. Show that the solution of the smoothed maximum likelihood estimation problem,

$$\begin{array}{ll} \text{minimize} & SL_{nh}(\, p \,) \\ \text{subject to} & 0 \leqslant p(x) \leqslant 1 \text{ on } [\, 0\,,\,1\,] \,, \end{array}$$

where

$$SL_{nh}(\, p \,) = -\tfrac{1}{n} \sum_{i=1}^{n} \{\, y_{in} \,[\, A_h \log p \,](x_{in}) - (\, 1 - y_{in} \,)\,[\, A_h \log(\, 1 - p \,) \,](x_{in}) \,\} \,,$$

is given by the Nadaraya-Watson estimator

$$p^{nh}(x) = \frac{\frac{1}{n}\sum_{i=1}^{n} y_{in}\, A_h(x_{in}, x)}{\frac{1}{n}\sum_{i=1}^{n} A_h(x_{in}, x)} .$$

In the definition of SL_{nh}, we used the notation, for nice functions f,

$$[A_h f](x) = \int_0^1 A_h(x, \tau)\, f(\tau)\, d\tau , \qquad x \in [0, 1] .$$

(7.19) EXERCISE. Investigate the weighted least-squares problem

$$\begin{aligned} &\text{minimize} && \frac{1}{n}\sum_{i=1}^{n} \frac{|\,p(x_{in}) - y_{in}\,|^2}{p(x_{in})\,(1 - p(x_{in}))} + h^{2m}\,\|\,p^{(m)}\,\|^2 \\ &\text{subject to} && p \in W^{m,2}(0,1) , \\ & && 0 \leqslant p(x) \leqslant 1 \text{ for all } x \in [0, 1] , \end{aligned}$$

for smooth binary regression. Compared with (7.4), we now have a convex problem (again), which should help matters.

EXERCISES: (7.8), (7.18), (7.19).

8. Additional notes and comments

Ad § 1: COX (1981) gives the complete analysis of the problem (1.16) for smooth functions γ based on Green's functions, i.e., "our" reproducing kernels, for $m = 2$. TSYBAKOV (1996) shows optimal rates of pointwise convergence of the estimators (1.17) when $f_o^{(r)}$ is Lipschitz continuous (in fact, for the problem

$$\text{minimize} \quad \frac{1}{n}\sum_{i=1}^{n} A_h(x - x_{in})\,\Phi\big(p(x_{in}) - y_{in}\big) \quad \text{subject to} \quad p \in \mathcal{P}_m ,$$

where Φ is any convex function).

Ad § 3: The bivariate case is treated in KOENKER and MIZERA (2004).

Ad §§ 4, 5: ROUSSEEUW (1992) introduced and made a case for the least-median method

$$\begin{aligned} &\text{minimize} && \text{median}\{\,|\,f(x_{in}) - y_{in}\,| \;:\; i = 1, 2, \cdots, n\,\} \\ &\text{subject to} && f \in \mathcal{F} , \end{aligned} \tag{8.1}$$

at least for the linear model $Y = X\beta + \varepsilon$.

The condition (4.3) already appears in TRUONG (1989), who constructs his estimators by local medians, carefully choosing the neighborhoods over which the medians are taken. When (4.4) holds, one should get better results than when (4.3) holds.

Nonparametric median smoothing goes way back; see, e.g., GEBSKI and MCNEIL (1984) and TRUONG (1989). BELITSER and VAN DE GEER (2000) study an interesting *on-the-fly* method. This is especially useful for the online analysis of time series; see, e.g., ARCE, GRABOWSKI, and GALLAGHER (2000) and FRIED, EINBECK, and GATHER (2007), and references therein.

The bound (5.15) may also be shown by martingale methods. This is based on the example in SHORACK (2000), p. 333. The first observation is that, for u a uniform $(0,1)$ random variable,

$$u(p) \stackrel{\text{def}}{=} \frac{1\!\!1(\,u < p\,) - p}{1-p} \ , \quad 0 \leqslant p < 1 \ , \tag{8.2}$$

is a continuous-time martingale in the sense that, for $0 < s < p < 1$,

$$\mathbb{E}\Big[\, u(p) \;\Big|\; 1\!\!1(\,u < s\,)\Big] = u(s) \ . \tag{8.3}$$

This may be seen as follows. For $0 < s < p < 1$, we have

$$\begin{aligned} \mathbb{E}\Big[\, 1\!\!1(\,u < p\,) \;\Big|\; 1\!\!1(\,u < s\,)\Big] &= \begin{cases} \quad 1 \quad , & \text{if } u < s \ , \\ \dfrac{p-s}{1-s} \ , & \text{if } u \geqslant s \ , \end{cases} \\ &= 1\!\!1(\,u < s\,) + \frac{p-s}{1-s}\, 1\!\!1(\,u \geqslant s\,) \ . \end{aligned}$$

The last line is just a convenient way to summarize the two cases. Since $1\!\!1(\,u \geqslant s\,) = 1 - 1\!\!1(u < s\,)$, elementary manipulations give (8.3).

Recall that $u_1, u_2, \cdots, u_n$ are independent uniform$(0,1)$ random variables. Then, for all $y \in (0,1)$,

$$\Big|\, \tfrac{1}{n} \sum_{i=1}^{n} \mathfrak{R}_{1,\lambda}(x_{in}, y)\,\big(\, 1\!\!1(u_i < p) - p\,\big)\,\Big|^2 \ , \quad p \in I_q \ ,$$

is a submartingale with respect to the σ-fields $\mathcal{F}_p$, $p \in I_q$, in which $\mathcal{F}_p$ is generated by the event $\big\{\, u_1 < p\,,\, u_2 < p\,, \cdots ,\, u_n < p\,\big\}$. This follows from the fact that a convex function of a martingale is a submartingale. See Lemma (4.4.4) in Volume I for the discrete case.

Now, Doob's submartingale inequality gives that

$$\begin{aligned} \mathbb{E}\Big[\, \sup_{p \in I_q} \Big|\, \tfrac{1}{n} \sum_{i=1}^{n} \mathfrak{R}_{1,\lambda}(x_{in}, y)\, u_i(p)\,\Big|^2\,\Big] &\leqslant 4\,\mathbb{E}\Big[\,\Big|\, \tfrac{1}{n} \sum_{i=1}^{n} \mathfrak{R}_{1,\lambda}(x_{in}, y)\, u_i(p_o)\,\Big|^2\,\Big] \\ &= \mathcal{O}\big(\,(nh)^{-1}\,\big) \ . \end{aligned}$$

Here p_o is the right endpoint of I_q. Again, the bound (5.15) follows.

Ad § 7: Maximum penalized likelihood for the problem (7.16)–(7.17) with the penalization $h^2 \| (\sqrt{p})' \|^2 + h^2 \| (\sqrt{1-p})' \|^2$ was treated in EGGERMONT and LARICCIA (2003).

18

Smoothing Parameter Selection

1. Notions of optimality

We start the treatment of the selection of the smoothing parameter in nonparametric regression with a discussion of optimality criteria. In the remainder of this chapter, we discuss their implementation for linear least-squares problems, in particular for smoothing spline estimators, the polynomial sieve, and local polynomials.

We are mostly concerned with plain function estimation but also pay some attention to estimating derivatives. Innocent as this little paragraph sounds, we are already prejudicing the proceedings as will become clear.

We consider the standard Gauss-Markov problem in which we wish to estimate the smooth function f_o from the data $y_n = \bigl(y_{1,n}, y_{2,n}, \cdots, y_{n,n} \bigr)^T$ according to the model

$$y_n = r_n f_o + d_n \; , \tag{1.1}$$

where $r_n f_o = \bigl(f_o(x_{1,n}), f_o(x_{2,n}), \cdots, f_o(x_{n,n}) \bigr)^T$. The components of $d_n = (d_{1,n}, d_{2,n}, \cdots, d_{n,n})^T$ are assumed to be uncorrelated, zero-mean random variables with common variance σ^2

$$\mathbb{E}[\, d_n \,] = 0 \quad , \quad \mathbb{E}[\, d_n \, d_n^T \,] = \sigma^2 I \; . \tag{1.2}$$

Here, typically, σ^2 is unknown. The heteroscedastic case

$$\mathbb{E}[\, d_n \,] = 0 \quad , \quad \mathbb{E}[\, d_n \, d_n^T \,] = \sigma^2 \, V^2 \; , \tag{1.3}$$

with V a diagonal matrix, is discussed also. At times, we assume that d_n is a multivariate normal,

$$d_n \sim \text{Normal}\bigl(0 \, , \, \sigma^2 \, I \bigr) \; . \tag{1.4}$$

In (1.1), the x_{in} are deterministic design points. No assumptions are made regarding their being asymptotically uniformly distributed over $[\, 0 \, , \, 1 \,]$.

Note that in the formulation above, the objective is to estimate the function f_o ; that is, to estimate the graph

$$\mathcal{G}_o = \Bigl\{ \, \bigl(x, f_o(x) \bigr) \; : \; x \in [\, 0 \, , \, 1 \,] \, \Bigr\} \; .$$

P.P.B. Eggermont, V.N. LaRiccia, *Maximum Penalized Likelihood Estimation*,
Springer Series in Statistics, DOI 10.1007/b12285_7,

Consequently, below, when we say "Let f^{nh} be an estimator of f_o depending on the smoothing parameter h", then

$$\mathcal{G}^{nh} = \Big\{ \, \big(x, f^{nh}(x) \big) \, : x \in [\, 0 \, , 1 \,] \, \Big\}$$

is an estimator of $\mathcal{G}_o$. The point is that the one *scalar* parameter h has to do the job for all $x \in [\,0\,,1\,]$. The alternative is to estimate $\mathcal{G}_o$ by

$$\Big\{ \, \big(x, f^{n,h(x)}(x) \big) \, : x \in [\, 0 \, , 1 \,] \, \Big\} \, ,$$

where h is allowed to change with x; e.g., in a piecewise constant fashion. For small sample sizes, choosing a possibly different h for each x seems a bit too ambitious. In § 7, we briefly come back to this. Another way in which we prejudice the proceedings is by assuming that the "order" of the estimator is fixed. Nevertheless, usually it is obvious how the order may be chosen as well.

As we were saying, let f^{nh} be an estimator of f_o depending on a scalar smoothing parameter h. To keep the discussion focused, consider the quintic smoothing spline estimators of the Wood Thrush Data Set of Brown and Roth (2004) in Figure 12.7.1. Obviously, the smoothing parameter determines what the estimator will look like. The question then is to decide what the estimator *should* look like. Phrased differently, what is needed is a quantitative notion of the optimality of the estimator. Once such a notion has been established, one can attempt to construct procedures that achieve the optimal result.

In nonparametric regression, three strains of optimality of the smoothing parameter are doing the rounds, of which minimum mean squared error and the optimality criterion of Akaike (1973) are related. The third one is uniquely concerned with data fitting and seems somewhat at odds with the idea of estimating f_o (and σ^2). We shall discuss each one briefly before diving in the remainder of the chapter.

No doubt, the standard objective in nonparametric regression is to estimate the regression function f_o. This leads to the quantitative notion of optimality of the smoothing parameter as the one that realizes the minimum of the mean squared error

$$\text{(1.5)} \qquad \min_{h>0} \, \| \, f^{nh} - f_o \, \|^2_{L^2(0,1)} \; ,$$

but the discrete version

$$\text{(1.6)} \qquad \min_{h>0} \frac{1}{n} \sum_{i=1}^{n} | \, f^{nh}(x_{in}) - f_o(x_{in}) \, |^2$$

is more tractable. In this chapter, we let $\langle \, \cdot \, , \, \cdot \, \rangle$ denote the Euclidean inner product on $\mathbb{R}^n$,

$$\text{(1.7)} \qquad \langle \, a \, , \, b \, \rangle = \sum_{i=1}^{n} a_i b_i \quad \text{for} \quad a, \, b \in \mathbb{R}^n \; ,$$

and define the Euclidean norm $\| \, a \, \|$ by way of $\| \, a \, \|^2 = \langle \, a \, , \, a \, \rangle$.

Although it is merely a matter of emphasis, note that we are not so much interested in the optimal h as in the associated estimator. The approach to smoothing parameter selection is to estimate the error $\| r_n f_o - f^{nh} \|^2$ and minimize this estimated error as a function of h.

Choosing the smoothing parameter by minimizing an estimator of the (discrete) squared L^2 error typically leads to undersmoothing and "rough" estimators of the regression function. However, it should be realized that this is not a fault of the particular selection procedure but rather of the optimality criterion under consideration. We will briefly explore remedial criteria along the lines of

$$\text{(1.8)} \qquad \min_{h>0} \; \| f^{nh} - f_o \|^2 + \alpha \, \| (f^{nh} - f_o)'' \|^2_{L^2(0,1)}$$

for a suitable fudge factor $\alpha > 0$, possibly depending on h but not on the data. In fact, α is a universal constant, but each experimenter must choose one. It is clear that the introduction of derivatives presents its own problems. However, frequently the experimenter wishes to estimate f_o' or f_o'', in which case presumably a good optimality criterion would be

$$\text{(1.9)} \qquad \min_{h>0} \; \| (f^{nh} - f_o)^{(k)} \|^2_{L^2(0,1)}$$

for the appropriate value of k. So, one way or another, one must come to terms with estimating derivatives.

To finish this discussion, we note that (1.6) immediately suggests the use of other norms or distance measures. It would be nice if one could estimate the (weighted) uniform error

$$\text{(1.10)} \qquad \max_{1 \leqslant i \leqslant n} \; \{ w_o(x_{in}) \}^{-1/2} \, | \, f^{nh}(x_{in}) - f_o(x_{in}) \, | \; ,$$

say for $w_o^2(x) = \mathbb{V}\mathrm{ar}[\, f^{nh}(x) \,]$, and then minimize this estimator. The problem is to properly balance the bias and variance of the estimator. Unfortunately, the authors do not know how to do this.

The second optimality criterion may be described as the model estimation approach. In the model (1.1)–(1.2), we wish to estimate both f_o and σ^2. This goes by the name of the information criterion of AKAIKE (1973), or AIC. The starting point is the model (1.1)–(1.4) with Gaussian noise. In this case, the data y_n have a normal distribution,

$$\text{(1.11)} \qquad y_n \sim \mathrm{Normal}(\, r_n f_o \, , \, \sigma^2 I \,) \; .$$

Let ϕ_o denote its density. The objective of the approach of AKAIKE (1973) is to see how well one can estimate this "model" density. Let f^{nh} and σ^2_{nh} be estimators of f_o and σ^2, and let ϕ^{nh} denote the normal pdf with mean $r_n f^{nh}$ and variance σ^2_{nh}. Now, a measure of the accuracy of the proposed model ϕ^{nh} is the Kullback-Leibler divergence $\mathrm{KL}(\, \phi_o \, , \, \phi^{nh} \,)$. In general, for arbitrary pdfs φ and ψ on $\mathbb{R}^n$,

$$\text{(1.12)} \qquad \mathrm{KL}(\varphi\,,\,\psi\,) = \int_{\mathbb{R}^n} \Bigl\{ \varphi(x) \log \frac{\varphi(x)}{\psi(x)} + \psi(x) - \varphi(x) \Bigr\} \, dx \ ,$$

with integration with respect to Lebesgue measure on $\mathbb{R}^n$. (The integrand is measurable and nonnegative, so $\mathrm{KL}(\varphi, \psi)$ is well-defined if we admit $+\infty$ as a possible value.) Note that the Kullback-Leibler divergence is invariant under monotone transformation of x and that it is more sensitive to tail behavior than the L^1 distance; see § 1.3 in Volume I. Now, for the normal pdfs at hand, $\mathrm{KL}(\,\phi_o\,,\,\phi^{nh}\,) = \mathit{AOC}(h)$, where

$$\text{(1.13)} \qquad \mathit{AOC}(h) \stackrel{\text{def}}{=} \frac{\|\, r_n(\,f^{nh} - f_o\,)\,\|^2}{2\sigma_{nh}^2} + n\, \mathit{BE}(\,\sigma_{nh}^2\,,\,\sigma^2\,)$$

with
$$\mathit{BE}(\,\sigma_{nh}^2\,,\,\sigma^2\,) = \log \frac{\sigma_{nh}^2}{\sigma^2} + \frac{\sigma^2}{\sigma_{nh}^2} - 1 \ .$$

Note that $\mathit{BE}(\,\sigma_{nh}^2\,,\,\sigma^2\,)$ is nonnegative, so that $\mathit{AOC}(h)$ makes sense, whether the normality assumption (1.4) holds or not. Thus, Akaike's Optimality Criterion as it applies to the estimation problem for the model (1.1)–(1.2) is to choose h as the solution to

$$\text{(1.14)} \qquad \text{minimize} \quad \mathit{AOC}(h) \quad \text{over} \quad h > 0 \ ,$$

and one can envision selection procedures based on estimators of (1.13); see § 18.6. In fact, this is not what is done. Instead, the optimality criterion (1.14) is replaced by

$$\text{(1.15)} \qquad \text{minimize} \quad \mathbb{E}[\,\mathrm{KL}(\phi_o\,,\,\phi^{nh}\,)\,] \quad \text{over } h > 0 \ ,$$

and some rather strange goings-on are going on to partially compute this expectation under the normality assumption (1.4). Although this relies even more on normality, the final result makes sense generally, so this does not appear to be an issue.

Of course, other measures of the difference between ϕ_o and ϕ^{nh} suggest themselves, such as the L^1 distance, but do not seem very tractable.

The final class of optimality criteria centers on the qualitative notion of selecting a model that best fits the data. The standard setting for this point of view is that of a collection of models of increasing "complexity", which one may think of as a sieve as in Chapter 15. Thus, we have a nested sequence of finite-dimensional subspaces or compact subsets of $L^2(0,1)$,

$$\text{(1.16)} \qquad C_1 \subset C_2 \subset \cdots \subset C_m \subset \cdots \subset L^2(0,1) \ ,$$

and the goal is to identify the optimal subspace C_m as well as the optimal element $f \in C_m$ that realizes the best data fit. Quantitatively, the objective is to realize the minimum residual sum of squares (what else?)

$$\text{(1.17)} \qquad \min_m \min_{f \in C_m} \|\, y_n - r_n f\,\|^2 \ ,$$

but, of course, at the same time one wants to avoid overfitting. There are various ways to implement this. The standard way is to penalize for the complexity of the model. For finite-dimensional subspaces, this is measured as the dimension of C_m. Thus, one selects m to achieve the minimum in

$$(1.18) \qquad \min_{m} \min_{f \in C_m} \| y_n - r_n f \|^2 + R(\dim C_m) ,$$

where R is some increasing function. (In fact, $R(t) = c\, t$ for a suitable constant c is the standard choice.) Phrased this way, one gets the distinct impression that one is confusing end and means: Obviously, (1.18) is the means for computing the optimal m, but the quantitative purpose of it is unclear. By way of example, in this context it is hard to find a satisfactory answer to the complaint that this estimator, too, is undersmoothed. The funny thing is that, nevertheless, the scheme (1.18) fits nicely into the minimum mean squared error notion of optimality by interpreting (1.18) as an *upper bound* on the actual error.

To the authors, data-splitting schemes, briefly discussed in § 12.7, seem more in line with the data-fitting view. They should be especially effective when considering nonlinear estimators or non-L^2 norms of the estimation error. Unfortunately, complexity penalization and data splitting will not be considered further.

At the other end of the data-splitting "spectrum" is the leave-one-out approach. Consider again the data-fitting point of view and the objective (1.17). It is useful to explicitly exhibit the objective function,

$$\sum_{i=1}^{n} | y_{in} - f(x_{in}) |^2 .$$

The best fit is obtained by connecting the dots, but this obviously defeats the purpose. Instead, to avoid connecting the dots, for each i, estimate the value $f(x_{in})$ on the basis of all the data except y_{in} by solving

$$(1.19) \qquad \text{minimize} \quad \sum_{\substack{j=1 \\ j \neq i}}^{n} | y_{jn} - f(x_{jn}) |^2 \quad \text{subject to} \quad f \in C_m .$$

Denote the estimator by $f_{\text{-}i}^{nm}$, and now determine m by solving

$$(1.20) \qquad \text{minimize} \quad \sum_{i=1}^{n} | y_{in} - f_{\text{-}i}^{nm}(x_{in}) |^2 \quad \text{subject to} \quad m \geqslant 1 .$$

So, we are trying to predict the "missing" data y_{in} by $f_{\text{-}i}^{nh}$ as well as possible. Thus, here one wishes to construct the model with the best predictive results. This appears to be different from the minimum squared error criterion (1.6) but in fact is remarkably similar, see §§ 3 and 4.

The leave-one-out method is the standard cross-validation approach and is even applicable to spline smoothing. For homoscedastic errors, ultimately this leads to coordinate-free cross-validation (GCV). For heteroscedastic errors, things are a little more complicated; see § 7.

In the above, we stated our views on how one should go about constructing selection procedures. In the next sections, we do the actual work first for the smoothing spline estimators. Thus, the vector $r_n f^{nh}$ is equal to the solution $b = b^{nh}$ of the problem

$$\text{(1.21)} \qquad \begin{aligned} &\text{minimize} && \| b - y_n \|^2 + nh^{2m} \langle a , Ta \rangle \\ &\text{subject to} && Ta = Sb , \end{aligned}$$

with T and S as in § 19.5. It follows that

$$\text{(1.22)} \qquad r_n f^{nh} = R_{nh} \, y_n ,$$

where

$$\text{(1.23)} \qquad R_{nh} = (I + nh^{2m} M)^{-1}$$

with $M = S^T T^{-1} S$. Consequently, the estimation process is described by the matrix $R_{nh} \in \mathbb{R}^{n \times n}$, with n the order of the matrix and h the smoothing parameter. When the order of spline matters, the notation is expanded to R_{nmh} or R_{nrh}. It follows from (1.23) that

(1.24) R_{nh} and $I - R_{nh}$ are symmetric and semi-positive-definite.

The representation (1.22) and properties (1.23) apply to the polynomial-sieve estimator in the form

$$\text{(1.25)} \qquad r_n \, p^{nr} = P_{nr} y_n$$

with P_{nr} a discrete projection operator onto the polynomials of order $\leqslant r$. In general, the treatment of splines applies to sieved least-squares estimators as well. For local polynomial estimators, the situation is somewhat different and we comment on it at the end of the chapter. For local polynomials, following Fan and Gijbels (1995a), we also study the minimizers of the asymptotic (global) pointwise and local errors and show that they are related by a universal fudge factor. Since the local error can be estimated quite reliably, this leads to stable methods for estimating the global pointwise error. We do not pay much attention to plug-in methods.

When considering the "other" regression problems or least-squared problems with constraints, the niceties associated with linearity disappear and other ideas, such as data splitting, have to come into play. Unfortunately, again, we shall not pursue this.

2. Mallows' estimator and zero-trace estimators

In this section, we derive Mallows' procedure for selecting the smoothing parameter. This procedure is based on an unbiased estimator of the mean squared error. Unfortunately, it depends on the unknown variance σ^2. However, this dependence may be eliminated by the construction of zero-trace estimators (Li, 1986).

We consider estimators f^{nh} of f_o in the model (1.1)–(1.2) that are linear in the data y_n. Thus

$$r_n f^{nh} = R_{nh} y_n \ , \tag{2.1}$$

where $R_{nh} \in \mathbb{R}^{n \times n}$. The optimal h is defined to be any solution of

$$\text{minimize} \quad \| \, y_o - R_{nh} y_n \, \|^2 \quad \text{subject to} \quad h \geqslant 0 \ , \tag{2.2}$$

where, for notational convenience, we let

$$y_o = r_n f_o \ . \tag{2.3}$$

For all nonparametric estimators except the ones based on sieves of finite-dimensional subspaces, the hat matrix also depends on the order m. Then, the hat matrix may be expressed as R_{nmh}, and the objective is to choose h and m to achieve the minimum in

$$\begin{aligned} &\text{minimize} \quad \| \, y_o - R_{nmh} \, y_n \, \|^2 \\ &\text{subject to} \quad h \geqslant 0 \ , \ m \in \mathbb{N} \ . \end{aligned} \tag{2.4}$$

Apart from a more cumbersome notation, this causes no extra problems, so we will just work with (2.2). At the appropriate times, reference to (2.4) will be made.

To approximate the optimal h in a rational manner, we must first estimate the loss function. Here, for notational convenience, set

$$y_o = r_n f_o \ . \tag{2.5}$$

In the typical squared error frame of mind, we write

$$\| \, y_o - R_{nh} \, y_n \, \|^2 = \| \, y_o \, \|^2 + \| \, R_{nh} \, y_n \, \|^2 - 2 \, \langle \, y_o \, , \, R_{nh} \, y_n \, \rangle \tag{2.6}$$

(inner product on $\mathbb{R}^n$) and note that in the final analysis we wish to minimize this over h. The first term, $\| \, y_o \, \|^2$, is unknown, but since it does not depend on h, we may ignore it. The second term, $\| \, R_{nh} y_n \, \|^2$, can be computed exactly, and so only the last term must be estimated. Obviously, $\langle \, y_n \, , \, R_{nh} \, y_n \, \rangle$ comes to mind, but then the right-hand side of (2.6) equals

$$\| \, y_n - R_{nh} \, y_n \, \|^2 + \| \, y_o \, \|^2 - \| \, y_n \, \|^2 \ ,$$

which is no good: Minimizing the last expression over all h will result in $R_{nh} \, y_n = y_n$; i.e., we are definitely overfitting. To more accurately pinpoint the inadequacy of the estimator $\langle \, y_n \, , \, R_{nh} \, y_n \, \rangle$, let us compute the expectations of the quantities in question: We have

$$\mathbb{E}[\, \langle \, y_n \, , \, R_{nh} \, y_n \, \rangle \,] = \langle \, y_o \, , \, R_{nh} \, y_o \, \rangle + \sigma^2 \, \mathrm{trace}(\, R_{nh} \,) \ , \tag{2.7}$$

while

$$\mathbb{E}[\, \langle \, y_o \, , \, R_{nh} \, y_n \, \rangle \,] = \langle \, y_o \, , \, R_{nh} \, y_o \, \rangle \ . \tag{2.8}$$

Thus, accounting for the difference,

$$\mathfrak{M}(h) = \| \, y_o \, \|^2 + \| \, R_{nh} \, y_n \, \|^2 - 2 \, \langle \, y_n \, , \, R_{nh} \, y_n \, \rangle + 2 \, \sigma^2 \, \mathrm{trace}(\, R_{nh} \,)$$

is an unbiased estimator of the loss in (2.1), which may be rewritten as

$$(2.9) \qquad \mathfrak{M}(h) = \| \, y_n - R_{nh} \, y_n \, \|^2 + 2 \, \sigma^2 \, \mathrm{trace}(\, R_{nh} \,) - n \, \sigma^2 \; .$$

This estimator is due to MALLOWS (1972), who refers to it as the C_L criterion. Note that it is based on y_n being an unbiased estimator of y_o.

(2.10) EXERCISE. Verify that $\mathfrak{M}(h)$ is indeed an unbiased estimator of $\| \, r_n f_o - r_n f^{nh} \, \|^2$.

Of course, the reference to the variance σ^2 is unfortunate, which immediately suggests that we estimate it. Note, however, that any estimator must be judged by its effect on the choice of (h and) the resulting estimator f^{nh}. In the next two sections, we derive some estimators of the variance in a roundabout way, that suggest other estimators. The various estimators are given in (5.15) and (6.8)–(6.11). See also Exercises (2.17) and (6.14). Here, as a way around this problem, we explore the zero-trace estimators of LI (1986), which eliminate the dependence on σ^2, apparently without estimating it. The idea behind zero-trace estimators is that since we have two estimators for y_o, viz. $R_{nh} y_n$ *and* y_n, it is natural to consider linear combinations of the two, say

$$(2.11) \qquad S_{nh\alpha} \, y_n \stackrel{\text{def}}{=} \alpha R_{nh} y_n + (1-\alpha) y_n = \{ \, \alpha R_{nh} + (1-\alpha) \, I \, \} \, y_n \; ,$$

and to see what may be done with them. Then, in (2.9), replacing R_{nh} with $S_{nh\alpha}$ results in

$$(2.12) \quad \mathfrak{M}(\alpha, h) = \alpha^2 \, \| \, y_n - R_{nh} \, y_n \, \|^2 \\ - 2 \, \alpha \, \sigma^2 \, \mathrm{trace}(\, I - R_{nh} \,) + 2 \, \sigma^2 \, \mathrm{trace}(\, I \,) - n \, \sigma^2 \; .$$

Now, a miracle happens. By taking α to be the solution of the equation

$$- \, 2 \, \alpha \, \sigma^2 \, \mathrm{trace}(\, I - R_{nh} \,) + 2 \, \sigma^2 \, \mathrm{trace}(\, I \,) = 0 \; ,$$

the unknown σ^2 is effectively eliminated! Thus, taking $\alpha = \alpha_o$ with

$$(2.13) \qquad \alpha_o = \{ \, \tfrac{1}{n} \, \mathrm{trace}(\, I - R_{nh} \,) \, \}^{-1} \; ,$$

we obtain,

$$(2.14) \qquad \mathfrak{M}(\alpha_o, h) = \frac{\| \, y_n - R_{nh} \, y_n \, \|^2}{\{ \, \frac{1}{n} \, \mathrm{trace}(\, I - R_{nh} \,) \, \}^2} - n \, \sigma^2$$

as an estimator of $\| \, y_o - \{ \, (\, 1 - \alpha_o \,) \, y_n + \alpha_o \, R_{nh} \, y_n \, \} \, \|^2$. The functional $\mathfrak{M}(\alpha_o, h)$ is in fact the generalized cross-validation (GCV) functional of GOLUB, HEATH, and WAHBA (1979), but their motivation is quite different; see § 4.

For smoothing splines, it turns out that by minimizing this over h we arrive at a value of α_o that is very close to 1. Thus, the original idea of approximation/estimation of the mean squared error is intact. Note that. for all intents and purposes, the unknown σ^2 has disappeared, but it is not at all clear that we actually estimated it. However, see Exercise (2.17) below as well as Exercise (4.19). LI (1986) refers to the resulting estimator as a zero-trace estimator (nil-trace, actually), which is more than apt.

(2.15) EXERCISE. In the above, why not solve

$$-2\,\alpha\,\sigma^2\,\text{trace}(\,I - R_{nh}\,) + 2\,\sigma^2\,\text{trace}(\,I\,) - n\,\sigma^2 = 0\ ?$$

(a) Show that the solution of this equation is given by $\alpha_1 = \frac{1}{2}\,\alpha_o$.
(b) Show that

$$\mathfrak{M}(\alpha_1, h) = \frac{\|\,y_n - R_{nh}\,y_n\,\|^2}{\{\,\frac{2}{n}\,\text{trace}(\,I - R_{nh}\,)\,\}^2}\ ,$$

which has the same minimizers as $\mathfrak{M}(\alpha_o, h)$.
(c) Why can this approach and (2.14) not *both* be any good as estimators of the error $\|\,y_o - R_{nh}\,y_n\,\|^2$?

(2.16) EXERCISE. Another idea is to minimize $\mathfrak{M}(\alpha, h)$ over α. Verify that

$$\min_{\alpha}\ \mathfrak{M}(\alpha, h) = -\sigma^4\,\frac{(\,\text{trace}(\,I - R_{nh}\,)\,)^2}{\|\,y_n - R_{nh}\,y_n\,\|^2} + n\,\sigma^2\ .$$

Thus, minimizing this over h is equivalent to minimizing $\mathfrak{M}(\alpha_o, h)$. LI (1985) studies the above from the point of view of Stein estimators.

(2.17) EXERCISE. Did we or did we not estimate the variance in deriving the zero-trace estimator $\mathfrak{M}(\alpha_o, h)$? Show that the expressions for $\mathfrak{M}(h)$ and $\mathfrak{M}(\alpha_o, h)$ coincide if σ^2 is replaced by the estimator

$$\widehat{\sigma}^{\,2}_{nh} = \tfrac{1}{n}\,\|\,y_n - R_{nh}\,y_n\,\|^2 \cdot \frac{1 - \frac{2}{n}\,\text{trace}(\,R_{nh}\,)}{\{\,\frac{1}{n}\,\text{trace}(\,I - R_{nh}\,)\,\}^2}\ .$$

(Since trace$(\,R_{nh}\,) \asymp (nh)^{-1}$, the trailing fraction just about equals 1.)

(2.18) EXERCISE. Consider the polynomial-sieve estimator $p = p^{n,r}$ of § 15.2. Here, r is the order of the polynomial. Derive the Mallows estimator of the error

$$\tfrac{1}{n}\sum_{i=1}^{n}\,|\,p^{n,r}(x_{in}) - f_o(x_{in})\,|^2\ ,$$

and show that the zero-trace version takes the form

$$GCV_{\text{poly}}(\,r\,) = \frac{\frac{1}{n}\sum_{i=1}^{n}\,|\,p^{n,r}(x_{in}) - y_{in}\,|^2}{\big(\,1 - \frac{r}{n}\,\big)^2}\ -\ \sigma^2\ .$$

We refer to this as GCV for selecting the order of the polynomial estimator.

EXERCISES: (2.10), (2.15), (2.16), (2.17), (2.18).

3. Leave-one-out estimators and cross-validation

In the next two sections, we develop good estimators for the error

$$\| y_o - R_{nh} y_n \|^2 \tag{3.1}$$

that do not refer to the variance of the noise. We remind the reader and ourselves that the development should also apply to estimators that depend on extra parameters; cf. (2.4).

The starting point is again the identity

$$\| y_o - R_{nh} y_n \|^2 = \| y_o \|^2 + \| R_{nh} y_n \|^2 - 2 \langle y_o , R_{nh} y_n \rangle , \tag{3.2}$$

and we must estimate the third term. Since R_{nh} is symmetric, we have

$$\langle y_o , R_{nh} y_n \rangle = \langle R_{nh} y_o , y_n \rangle , \tag{3.3}$$

and we must estimate $R_{nh} y_o$. As in § 2, the obvious first attempt would be to take $R_{nh} y_n$ (or maybe even y_n) but this leads to Mallows' estimator. Actually, what is needed is an estimator for $\langle R_{nh} y_o , y_n \rangle$ that has expected value $\langle R_{nh} y_o , y_o \rangle$, or as close to it as we can get. This would or could be the case if, for each i, the estimator of $[R_{nh} y_o]_i$ were independent of y_{in}, so that $[R_{nh} y_o]_i$ should be estimated on the basis of

$$y_{1,n} , \cdots , y_{i-1,n} , y_{i+1,n} , \cdots , y_{n,n} \tag{3.4}$$

only; i.e., with y_{in} left out. This is the classical leave-one-out method. Thus, consider the spline estimation problem (13.1.7), omitting the ℓ-th datum ($1 \leqslant \ell \leqslant n$),

$$\begin{aligned} &\text{minimize} && \tfrac{1}{n} \sum_{\substack{i=1 \\ i \neq \ell}}^{n} | f(x_{in}) - y_{in} |^2 + h^{2m} \| f^{(m)} \|^2 \\ &\text{subject to} && f \in W^{m,2}(0,1) , \end{aligned} \tag{3.5}$$

where $\| f^{(m)} \|^2$ is the squared $L^2(0,1)$ norm. The solution is a polynomial spline of order $2m$ on the partition

$$x_{1,n} < x_{2,n} < \cdots < x_{\ell-1,n} < x_{\ell+1,n} < \cdots < x_{n,n} .$$

The unfortunate thing is that the partition keeps changing. Of course, a spline on the partition above is also a spline on the partition with the knot $x_{\ell,n}$ added in. Thus, if $f_{[\ell]}$ is the solution of (3.5), then $b_{[\ell]} = r_n f_{[\ell]}$ is the solution to

$$\begin{aligned} &\text{minimize} && \sum_{\substack{i=1 \\ i \neq \ell}}^{n} | b_i - y_{in} |^2 + nh^{2m} \langle a , Ta \rangle \\ &\text{subject to} && Ta = Sb . \end{aligned} \tag{3.6}$$

It is useful to write

$$\sum_{\substack{i=1 \\ i \neq \ell}}^{n} | b_i - y_{in} |^2 = \| I_\ell (b - y_n) \|^2 , \tag{3.7}$$

where I_ℓ is the $n \times n$ identity matrix, with the ℓ-th diagonal element set equal to zero. Note that

$$I_\ell = I - e_\ell \, e_\ell^T \, , \tag{3.8}$$

where $e_1, e_2, \cdots, e_n$ is the standard basis for $\mathbb{R}^n$. Thus, (3.6) may be rewritten as

$$\begin{aligned} &\text{minimize} \quad \| \, I_\ell (\, b - y_n \,) \, \|^2 + nh^{2m} \, \langle \, a \, , \, Ta \, \rangle \\ &\text{subject to} \quad Ta = Sb \, . \end{aligned} \tag{3.9}$$

We proceed to solve (3.9). The normal equations are

$$(\, I_\ell + \lambda \, M \,) \, b = I_\ell \, y_n \, , \tag{3.10}$$

with $\lambda = nh^{2m}$ and $M = S^T T^{-1} S$. Note that M is semi-positive-definite. Now, it is possible of course to solve (3.10) for every single ℓ and thus arrive at the estimator for $\langle \, y_n \, , \, R_{nh} \, y_o \, \rangle$, but it is more fruitful to perform the following elementary manipulations.

Using (3.8), the normal matrix may be written as

$$I_\ell + \lambda \, M = I + \lambda \, M - e_\ell e_\ell^T \, , \tag{3.11}$$

which differs from $I + \lambda \, M$ by a matrix of rank one. Then, by the following lemma, its inverse differs from $(\, I + \lambda \, M \,)^{-1}$ by a matrix of rank one as well.

(3.12) Lemma (Sherman-Morrison-Woodbury). (a) *Let $u, w \in \mathbb{R}^n$ such that $w^T u \neq 1$. Then $I - uw^T$ is invertible, and*

$$(\, I - uw^T)^{-1} = I + \frac{uw^T}{1 - w^T u} \, .$$

(b) *Let $U, W \in \mathbb{R}^{n \times m}$ such that $I - W^T U$ is invertible. Then, $I - UW^T$ is invertible and*

$$(\, I - UW^T)^{-1} = I + U(\, I - W^T U \,)^{-1} \, W^T \, .$$

Proof. We only prove part (a). Consider the system of equations

$$(I - uw^T) \, x = y \, ,$$

and solve for x in terms of y. Multiplying out the left-hand side shows that

$$x = y + t \, u \, , \quad \text{with } t = w^T x \, .$$

Then, premultiplication with w^T gives

$$w^T x = w^T y + (w^T u) \, w^T x \, ,$$

which may be solved for $w^T x$. The final result is that x may be written as

$$x = y + \frac{uw^T}{1 - w^T u} \, y \, ,$$

from which the inverse may be read off. Q.e.d.

(3.13) EXERCISE. (a) Prove part (b) of Lemma (3.12).
(b) Formulate and prove the lemma for a matrix of the form $A - uw^T$.

Returning to (3.11), with Lemma (3.12) it follows that

$$(3.14)\qquad \begin{aligned}(I_\ell + \lambda M)^{-1} &= (I - R_{nh}\, e_\ell e_\ell^T)^{-1}(I + \lambda M)^{-1}\\ &= \Bigl(I + \frac{R_{nh}\, e_\ell e_\ell^T}{1 - e_\ell^T R_{nh} e_\ell}\Bigr) R_{nh}\;.\end{aligned}$$

Recall that $(I + \lambda M)^{-1} = R_{nh}$. Then, $e_\ell^T R_{nh} e_\ell = [R_{nh}]_{\ell,\ell}$, the ℓ-th element on the main diagonal of R_{nh}.

Consequently, the solution $b_{[\ell]}$ of (3.10) satisfies

$$(3.15)\qquad b_{[\ell]} = \left(I + \frac{R_{nh}\, e_\ell\, e_\ell^T}{1 - [R_{nh}]_{\ell,\ell}}\right) R_{nh}\, I_\ell\, y_n\;.$$

After some elementary manipulations, this gives

$$(3.16)\qquad b_{[\ell]} = R_{nh}\, y_n + \frac{R_{nh}\, e_\ell\, e_\ell^T (R_{nh}\, y_n - y_n)}{1 - [R_{nh}]_{\ell\ell}}\;.$$

(3.17) EXERCISE. Verify (3.16).

The estimator (3.16) gives the leave-one-out estimator for $[R_{nh}\, y_o]_\ell$,

$$(3.18)\qquad [b_{[\ell]}]_\ell = [R_{nh}\, y_n]_\ell - \frac{[R_{nh}]_{\ell,\ell}}{1 - [R_{nh}]_{\ell,\ell}}\,[y_n - R_{nh}\, y_n]_\ell\;.$$

Returning to the scene of the crime, the estimator for $\langle\, y_n\,,\, R_{nh}\, y_o\,\rangle$ is then

$$(3.19)\qquad \sum_{\ell=1}^{n} [b_{[\ell]}]_\ell\, y_{\ell,n} = \langle\, y_n\,,\, R_{nh}\, y_n\,\rangle - \langle\, y_n\,,\, D_{nh}(y_n - R_{nh}\, y_n)\,\rangle\;,$$

where

$$(3.20)\qquad D_{nh} = \sum_{\ell=1}^{n} \frac{[R_{nh}]_{\ell\ell}}{1 - [R_{nh}]_{\ell\ell}}\, e_\ell\, e_\ell^T$$

is a diagonal matrix with diagonal entries $[R_{nh}]_{\ell\ell}/(1 - [R_{nh}]_{\ell\ell})$. Finally, the estimator for the squared error $\|\, y_o - R_{nh}\, y_n\,\|^2$ may then be written as

$$(3.21)\qquad \begin{aligned}CV_I(h) = \|\, y_n - R_{nh}\, y_n\,\|^2 + 2\,\langle\, y_n\,,\, D_{nh}(y_n - R_{nh}\, y_n)\,\rangle&\\ + \|\, y_o\,\|^2 - \|\, y_n\,\|^2&\;.\end{aligned}$$

Of course, this is not an estimator since it still depends on the unknown y_o, but not in a way that hurts us. So, a good way to select h ought to be by minimizing $CV_I(h)$ over $h > 0$.

The estimator $CV_I(h)$ above is in fact not the usual cross-validation functional. The usual version is related to the estimation of the residual sum of squares. It sets itself the goal of "predicting" each datum $y_{\ell n}$ based on the other data. Thus, the functional in question is

$$CV_{II}(h) = \sum_{\ell=1}^{n} \left| [y_n]_\ell - [b_{[\ell]}]_\ell \right|^2 , \tag{3.22}$$

and h is chosen by minimizing this. This may be rewritten as

$$CV_{II}(h) = \| (I + D_{nh})(y_n - R_{nh} y_n) \|^2 . \tag{3.23}$$

(3.24) EXERCISE. Verify the expressions (3.23).

There are many other estimators of the error $\| y_o - R_{nh} y_n \|^2$. By way of example, it would seem equally reasonable to interpret $[b_{[\ell]}]_\ell$ as an estimator of $[y_o]_\ell$ rather than $[y_n]_\ell$, so that, estimating $[y_o]_\ell$ by $[R_{nh} y_n]_\ell$, a good estimator of $\| y_o - R_{nh} y_n \|^2$ should be

$$\frac{1}{n} \sum_{\ell=1}^{n} \left| [b_{[\ell]}]_\ell - [R_{nh} y_n]_\ell \right|^2 ,$$

which takes the form

$$CV_{III}(h) \stackrel{\text{def}}{=} \| D_{nh} (y_n - R_{nh} y_n) \|^2 . \tag{3.25}$$

So, a good criterion for choosing h should be to minimize $CV_{III}(h)$.

(3.26) EXERCISE. The interpretation of $[b_{[\ell]}]_\ell$ above may be applied to the estimation of $\langle y_o , R_{nh} y_n \rangle$ in (3.2). Show that, apart from constants, the error (3.1) is estimated by the gruesome-looking

$$CV_{IV}(h) = \| D_{nh} (y_n - R_{nh} y_n) \|^2 - \| R_{nh} y_n - D_{nh} (y_n - R_{nh} y_n) \|^2 .$$

Thus, minimizing this over h yields yet another selection procedure for h.

An unanswered question is whether these various functionals $CV_{??}(h)$ have useful minimizers. However, the interest is really in the coordinate-free versions of them, derived in the next section, and there we do indeed address the issue.

EXERCISES: (3.13), (3.17), (3.24), (3.26).

4. Coordinate-free cross-validation (GCV)

The leave-one-out idea of the previous section is of course intimately tied up with the particular way in which the data are represented ("measured" if you will). Equivalently, it is closely connected with the standard basis $e_1, e_2, \cdots, e_n$ for $\mathbb{R}^n$ since the leave-one-out approach is encoded in the leave-one-out matrix $I_\ell = I - e_\ell e_\ell^T$, or more suggestive of leaving-one-out,

$$I_\ell = \sum_{\substack{i=1\\ i\neq\ell}}^{n} e_i\, e_i^T \ . \tag{4.1}$$

The nice thing about bases for finite-dimensional vector spaces is that they may be changed at will. Thus, one can replace the standard basis by any other basis, and this is what we are going to do. However, in view of the least-squares problems, we must restrict our bases to be orthonormal. In fact, there is a preferred orthonormal basis for which the leave-one-out estimator is particularly simple and which shows that the resulting generalized cross-validation functional

$$\mathit{GCV}(h) = \frac{\|\, y_n - R_{nh}\, y_n \,\|^2}{\{\, \mathrm{trace}(\, I - R_{nh}\,)\,\}^2} \tag{4.2}$$

is the usual cross-validation function $\mathit{CV}_{III}(h)$ in the preferred basis. This makes generalized cross-validation in fact coordinate-free. Similar simplifications apply to the other $\mathit{CV}_{??}(h)$ functionals.

So, let $u_1, u_2, \cdots, u_n$ be any orthonormal basis for $\mathbb{R}^n$, and consider the leave-one-out procedure in this coordinate system. Analogous to (4.1), let

$$U_\ell = I - u_\ell\, u_\ell^T \ , \tag{4.3}$$

and consider the analogue of (3.9),

$$\begin{aligned} &\text{minimize} && \|\, U_\ell\,(\, b - y_n\,)\,\|^2 + nh^2 \,\langle\, a\,,\, Ta\,\rangle \\ &\text{subject to} && Ta = Sb \ . \end{aligned} \tag{4.4}$$

Denoting the solution of (4.4) again by $b_{[\ell]}$, one verifies that

$$b_{[\ell]} = \left(\, I + \frac{R_{nh}\, u_\ell\, u_\ell^T\, R_{nh}}{1 - u_\ell^T\, R_{nh}\, u_\ell}\,\right) R_{nh}\, U_\ell\, y_n \ , \tag{4.5}$$

with R_{nh} the same old hat matrix, and provided that

$$u_\ell^T\, R_{nh}\, u_\ell < 1 \ , \quad \ell = 1, 2, \cdots, n \ . \tag{4.6}$$

(4.7) Exercise. Verify that $0 < u_\ell^T\, R_{nh}\, u_\ell \leqslant 1$ for $\ell = 1, 2, \cdots, n$.

Consequently,

$$b_{[\ell]} = R_{nh}\, y_n + \frac{R_{nh}\, u_\ell\, u_\ell^T\,(\, R_{nh}\, y_n - y_n\,)}{1 - u_\ell^T\, R_{nh}\, u_\ell} \ , \tag{4.8}$$

and the analogue of $\mathit{CV}_I(h)$ is then

$$\mathit{CV}_I(h\,;U) \;=\; \|\, y_n \;-\; R_{nh}\, y_n\,\|^2 \;+\; 2\,\langle\, y_n\,,\, D_U(\, I \;-\; R_{nh}\,)\, y_n\,\rangle \ , \tag{4.9}$$

where we dropped the uninteresting terms $\|\, y_o\,\|^2 - \|\, y_n\,\|^2$, and where

$$D_U = \sum_{k=1}^{n} \frac{u_\ell^T\, R_{nh}\, u_\ell}{1 - u_\ell^T\, R_{nh}\, u_\ell}\, u_\ell\, u_\ell^T \ . \tag{4.10}$$

In $CV_I(h\,;U)$, the dependence on $U = (\,u_1\,|\,u_2\,|\;\cdots\;|\,u_n)$ is explicitly noted.

The question is whether anything is to be gained by choosing U different from the identity matrix. Ideally, we should or could choose U such that the h that minimizes $CV_I(h\,;U)$ subject to $h > 0$ results in the smallest possible error $\|\,y_o - R_{nh}\,y_n\,\|$, but this leads to a vicious circle. A second idea would be to solve the problem

$$(4.11)\qquad \begin{array}{ll} \text{minimize} & \mathbb{E}[\,|\,CV_I(h\,;U) - \|\,y_o - R_{nh}\,y_n\,\|^2\,|\,] \\ \text{subject to} & U \text{ orthogonal}\,. \end{array}$$

A simple calculation yields

$$(4.12)\quad CV_I(h\,;U) - \|\,y_o - R_{nh}\,y_n\,\|^2 = \\ -\,2\,\langle\, d_n\,,\,R_{nh}\,y_n\,\rangle + 2\,\langle\, y_n\,,\,D_U\,(\,I - R_{nh}\,)\,y_n\,\rangle\;,$$

and the derivation of the expectation in (4.11) "follows". Without stating the explicit result, it is obvious that the answer depends not only on the unknown y_o but also on various measures of the size of $D_U(I - R_{nh})$. This implies that solving the problem (4.11) is impossible, and it might not be such a bad idea to minimize the size of the matrix D_U. Since D_U is a diagonal matrix, all measures of its size relate to the largest diagonal element. Thus, we consider the problem of minimizing

$$(4.13)\qquad \|\,D_U\,\| = \max\Bigl\{\,\Bigl|\,\frac{u_\ell^T\,R_{nh}\,u_\ell}{1 - u_\ell^T\,R_{nh}\,u_\ell}\,\Bigr|\;:\;\ell = 1, 2, \,\cdots\,, n\,\Bigr\}$$

over all orthogonal matrices U. In view of Exercise (4.7), the matrix U that minimizes $\|\,D_U\,\|$ is the same one that solves

$$(4.14)\qquad \text{minimize}\quad \max\bigl\{\,u_\ell^T\,R_{nh}\,u_\ell \;:\; \ell = 1, 2, \,\cdots\,, n\,\bigr\}\;.$$

Solving (4.14) appears not to be easy, but in fact it is. We have

$$(4.15)\qquad \max_\ell\; u_\ell^T\,R_{nh}\,u_\ell \geqslant \tfrac{1}{n}\,\text{trace}(\,U^T R_{nh}\,U\,) = \tfrac{1}{n}\,\text{trace}(\,R_{nh}\,)\;,$$

with equality only if, for all $\ell = 1, 2, \,\cdots\,, n$,

$$(4.16)\qquad u_\ell^T\,R_{nh}\,u_\ell = \tfrac{1}{n}\,\text{trace}\,R_{nh}\;.$$

Assuming that an orthogonal matrix U exists for which this holds, then $\|\,D_U\,\|$ is minimal for this choice and the optimal D_U would equal

$$(4.17)\qquad D_U = \frac{\text{trace}(\,R_{nh}\,)}{\text{trace}(\,I - R_{nh}\,)}\;I\;.$$

The corresponding expression for $CV_I(h\,;U)$ would be

$$(4.18)\quad CFCV_I(h) = \|\,y_n - R_{nh}\,y_n\,\|^2 + 2\,\frac{\langle\, y_n\,,\,y_n - R_{nh}\,y_n\,\rangle}{\text{trace}(\,I - R_{nh}\,)}\;\text{trace}(\,R_{nh}\,)\;.$$

(Here CFCV stands for Coordinate-Free Cross-Validation.) At this point, it is not really important whether the orthogonal matrix with the advertised properties exists or not. However, an explicit solution is possible, as shown in Exercise (4.20).

(4.19) EXERCISE. It is instructive to compare the expression for $CFCV_I(h)$ with the expression (3.9) for Mallows' estimator $\mathfrak{M}(h)$. Show that if we estimate σ^2 in $\mathfrak{M}(h)$ by

$$\widehat{\sigma}^2 = \frac{\langle\, y_n \,,\, y_n - R_{nh}\, y_n \,\rangle}{\text{trace}(\, I - R_{nh}\,)} \;,$$

then $\mathfrak{M}(h) = CFCV_I(h)$. (Note that the estimator $\widehat{\sigma}^2$ is more than reasonable: It is always positive and $\mathbb{E}[\,\widehat{\sigma}^2\,] > \sigma^2$, which may not be a bad thing considering the purpose for which it is used.)

(4.20) EXERCISE. A matrix $C \in \mathbb{R}^{n\times n}$ is said to be *circulant* if

$$C_{ij} = a_{i-j} \;, \quad i,j = 1,2,\,\cdots,n \;,$$

where the a_i satisfy $a_{-i} = a_{n-i}$ for $i = 1,2,\,\cdots,n$.

(a) Let F be the discrete $n \times n$ Fourier transform matrix; that is,

$$F_{k+1,\ell+1} = \tfrac{1}{\sqrt{n}} \exp(-2\pi i k\ell/n\,) \;, \quad k,\ell = 0,1,\,\cdots,n-1 \;.$$

Then F is unitary and the columns of the discrete Fourier matrix are (the) eigenvectors of circulant matrices.

(b) Let $V \in \mathbb{R}^{n\times n}$ be an orthogonal matrix and $\Lambda \in \mathbb{R}^{n\times n}$ be a diagonal matrix such that $R_{nh} = V\,\Lambda\, V^T$ is the eigenvalue decomposition of R_{nh}. Take $U = V\,F$, and show that $U^*\,R_{nh}$ is a circulant matrix.

(d) Conclude that $[\,U^*\,R_{nh}\,U\,]_{\ell\ell} = u_\ell^*\,R_{nh}\,u_\ell$ is constant in ℓ.

(e) The above is the complex version. If this bothers you, modify it so that everything is real.

Procedures of this kind go by the name of "generalized" cross-validation. The previous exercise shows that generalized cross-validation is the same as "ordinary" cross-validation in a preferred coordinate system. Since the final result is coordinate-free, the authors are tempted to call it coordinate-free cross-validation.

It is now an easy exercise to derive the coordinate-free analogues of the various cross-validation functions. We list them here:

$$\text{(4.21)}\quad CFCV_I(h) = \|\, R_{nh}\, y_n \,\|^2 + 2\,\frac{\langle\, y_n \,,\, y_n - R_{nh}\, y_n \,\rangle}{\text{trace}(\, I - R_{nh}\,)}\;\text{trace}(\, R_{nh}\,) \;,$$

$$\text{(4.22)}\qquad CFCV_{II}(h) = GCV(h) = \frac{\|\, y_n - R_{nh}\, y_n \,\|^2}{\{\,\text{trace}(\, I - R_{nh}\,)\,\}^2} \;,$$

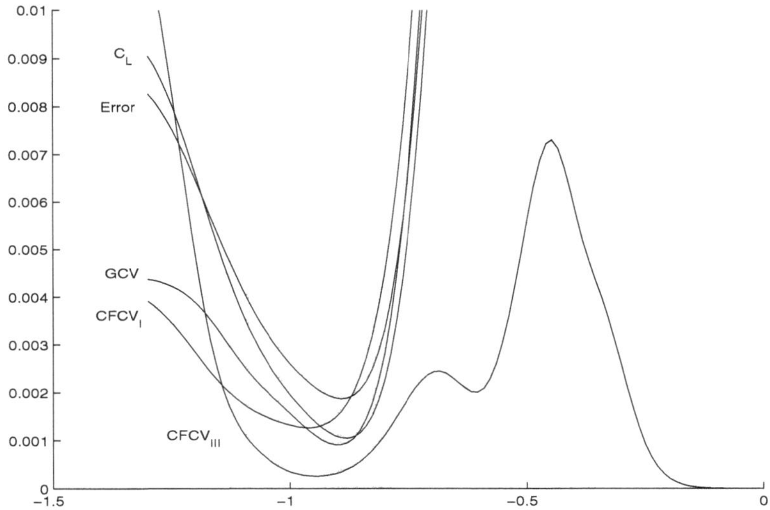

Figure 4.1. Graphs of the various estimators of $\frac{1}{n}\| y_o - R_{nh}\, y_n \|^2$ (the "Error") in cubic smoothing splines for the example $[y_o]_i = \sin(2\,\pi\, x_{in}^5)$ on the standard deterministic grid on $[\,0\,,\,1\,]$. Here, $n = 100$ and $\sigma = 0.1$. The horizontal axis is the logarithm (base 10) of h. C_L refers to Mallows' estimator with the exact σ^2 (see § 2). Usually(?), the "Error" curve and the "GCV" curve are much closer over the whole range of h. Except for the $CFCV_{III}$ curve, all curves are U-shaped over the whole range of h values indicated.

but CV_{III} has two coordinate-free versions,

$$(4.23)\qquad CFCV_{III}'(h) = \frac{\{\,\mathrm{trace}(\,R_{nh}\,)\,\}^2}{\{\,\mathrm{trace}(\,I - R_{nh}\,)\,\}^2}\,\|\, y_n - R_{nh}\, y_n \,\|^2\;,$$

$$(4.24)\qquad CFCV_{III}(h) = \frac{\|\, R_{nh}\,\{\, y_n - R_{nh}\, y_n \,\}\,\|^2}{\{\,\mathrm{trace}(\,I - R_{nh}\,)\,\}^2}\;.$$

(4.25) EXERCISE. Verify these formulas.

(4.26) REMARK. Actually, the expression for $CFCV_{III}$ is interesting but also somewhat in doubt, so here we give a "derivation". Analogous to the expression for $CV_I(h\,;\,U)$, one obtains $CV_{III}(h\,;\,U) = \|\, D_U\,(\,y_n - R_{nh}y_n\,)\,\|^2$. Now, if $U = V$ is the matrix of eigenvectors of R_{nh} as in Exercise (4.20)(b), then

$$\begin{aligned} D_U\,(\,y_n - R_{nh}y_n\,) &= U\,(\,I - \Lambda\,)^{-1}\,\Lambda\, U^T(\,y_n - R_{nh}\, y_n\,) \\ &= U\,(\,I - \Lambda\,)^{-1}\, U^T U \Lambda\, U^T(\,y_n - R_{nh}\, y_n\,) \\ &= U\,(\,I - \Lambda\,)^{-1}\, U^T R_{nh}\,(\,y_n - R_{nh}\, y_n\,)\;. \end{aligned}$$

So,

$$\| D_U (y_n - R_{nh} y_n) \|^2 = \sum_{\ell=1}^{n} \left| \frac{u_\ell^T R_{nh} (y_n - R_{nh} y_n)}{1 - u_\ell^T R_{nh} u_\ell} \right|^2 .$$

This holds for a particular coordinate system, but now we change it again to get the coordinate-free version.

All these derivations seem reasonable, but the question is of course whether any one of them works. The answer is yes, but some tweaking may be required. In Figure 4.1, we show the graphs of various $CFCV_{??}(h)$ functionals as well as the actual error $\| y_o - R_{nh} y_n \|^2$ for a typical (?) data set. The functionals $CFCV_I(h)$ and $GCV(h)$ seem to be well-behaved, but the global minimum of the functional $CFCV_{III}(h)$ is not acceptable, and we must choose a local minimum. Unfortunately, we shall not pursue this further.

EXERCISES: (4.7), (4.19), (4.20), (4.25).

5. Derivatives and smooth estimation

In this section, we discuss the selection of the smoothing parameter for the optimal estimation of derivatives, including the smooth estimation of the regression function, allowing for some trade-off between the optimal mean squared error and smoothness.

To focus our attention, we start with the estimation of the first derivative of the regression function in the model (1.1)–(1.3). In the context of spline estimation, the estimator for f_o' is $(f^{nh})'$, the derivative of the estimator for f_o. Optimality of the estimator is taken to be in the sense of minimum squared error,

$$\min_{h>0} \| (f^{nh})' - f_o' \|_{L^2(0,1)}^2 , \tag{5.1}$$

or its discrete counterpart. Thus, to construct rational selection procedures, we must estimate the squared error. A minor/major annoyance here is that there are no unbiased estimators of f_o', not even at the design points. Consequently, one would have to worry about the effect of the bias on selecting h. To avoid this, it seems reasonable to replace derivatives with difference quotients. For the first derivative, the difference quotients are implemented with a matrix $\nabla \in \mathbb{R}^{(n-1)\times(n-1)}$ such that, for all $b \in \mathbb{R}^n$,

$$[\nabla b]_i = \frac{b_{i+1} - b_i}{x_{i+1,n} - x_{in}} , \quad i = 1, 2, \cdots, n-1 . \tag{5.2}$$

Then, the optimality criterion (5.1) is replaced by

$$\min_{h>0} \| \nabla(r_n f^{nh} - r_n f_o) \|^2 . \tag{5.3}$$

It stands to reason that estimating $\nabla r_n f_o$ is not much different from estimating $r_n(f_o')$.

From here on, everything is smooth sailing. It seems obvious that a Mallows-like estimator for the error must be based on $\| \nabla(y_n - R_{nh} y_n) \|^2$. Recalling the notation $y_o = r_n f_o$ and the symmetry of R_{nh}, then

$$(5.4) \quad \mathbb{E}[\, \| \nabla(y_n - R_{nh}\, y_n) \|^2 \,] = \| \nabla(y_o - R_{nh}\, y_o) \|^2 + \sigma^2 \operatorname{trace}(\nabla^T \nabla \, (I - R_{nh})^2)$$

and

$$(5.5) \quad \mathbb{E}[\, \| \nabla(y_o - R_{nh}\, y_n) \|^2 \,] = \| \nabla(y_o - R_{nh}\, y_o) \|^2 + \sigma^2 \operatorname{trace}(\nabla^T \nabla \, R_{nh}^2) \, ,$$

so that

$$(5.6) \qquad \mathfrak{M}_\nabla(h) = \| \nabla(y_n - R_{nh}\, y_n) \|^2 + 2\,\sigma^2 \operatorname{trace}(\nabla^T \nabla \, R_{nh})$$

is an unbiased estimator of the error, apart from the fact that we dropped the term $-\sigma^2 \operatorname{trace}(\nabla^T \nabla)$.

(5.7) EXERCISE. Verify (5.4)–(5.5) and that $\mathfrak{M}_\nabla(h)$ has its advertised property.

It is now very tempting to jump the gun and take

$$(5.8) \quad \mathit{CFCV}_{\nabla,I}(h) = \| \nabla(y_n - R_{nh}\, y_n) \|^2 + 2\, \frac{\langle\, y_n \,,\, y_n - R_{nh}\, y_n \,\rangle}{\operatorname{trace}(I - R_{nh})} \operatorname{trace}(\nabla^T \nabla \, R_{nh})$$

as the analogue of $\mathit{CFCV}_I(h)$, and apparently we have given in. Apart from everything else, the good point about (5.8) is that both terms are positive.

(5.9) EXERCISE. The derivation of $\mathit{CFCV}_{\nabla,I}(h)$ via leave-one-out estimation is not obvious. (a) Consider the leave-one-out procedure for estimating $\langle\, y_o \,,\, \nabla_n^T \nabla_n R_{nh}\, y_n \,\rangle = \langle\, R_{nh}\, \nabla_n^T \nabla_n\, y_o \,,\, y_n \,\rangle$. By interpreting $b_{[\ell]}$ of (3.16) as an estimator of $R_{nh}\, y_o$, show that the leave-one-out estimator of the inner product $\langle\, \nabla_n\, y_o \,,\, \nabla_n R_{nh}\, y_n \,\rangle = \langle\, R_{nh} \nabla_n^T \nabla_n\, y_o \,,\, y_n \,\rangle$ is given by

$$\sum_{\ell=1}^{n} [\, R_{nh}\, \nabla_n^T \nabla_n\, R_{nh}^{-1} b_{[\ell]} \,]_\ell \, y_{\ell,n} \, ,$$

so that, apart from terms that do not depend on h, the error in (5.3) is estimated by

$$\mathit{CV}_{\nabla,I}(h) = \| \nabla(y_n - R_{nh}\, y_n) \|^2 + 2 \, \langle\, y_n \,,\, \mathfrak{D}_{nh}\, (y_n - R_{nh}\, y_n) \,\rangle \, ,$$

where $\mathfrak{D}_{nh}$ is a diagonal matrix with entries

$$[\, \mathfrak{D}_{nh} \,]_{\ell\ell} = \frac{[\, R_{nh} \nabla^T \nabla \,]_{\ell\ell}}{[\, I - R_{nh} \,]_{\ell\ell}} \, , \qquad \ell = 1, 2, \cdots, n \, .$$

(b) Repeat the above for an arbitrary orthonormal basis U. Does there exist an orthonormal basis U for which both $U^T R_{nh} U$ and $U^T R_{nh} \nabla^T \nabla U$ have constant main diagonals? If so, then $CFCV_{\nabla,I}(h)$ is the coordinate-free version. [The authors do not know the answer.]

(c) Returning to part (a), we may also interpret $b_{[\ell]}$ as an estimator for y_o itself. Show that

$$\sum_{\ell=1}^{n} y_{\ell,n} [\, R_{nh} \nabla_n^T \nabla_n \, b_{[\ell]} \,]_\ell$$

is an estimator for the inner product and that

$$CV_{\nabla,II}(h) = -\| \nabla R_{nh} \, y_n \|^2 + 2 \langle \, y_n \, , \, \mathcal{D}_{nh} \, (y_n - R_{nh} \, y_n \,) \, \rangle$$

is an estimator for the error, where $\mathcal{D}_{nh}$ is a diagonal matrix with entries

$$[\, \mathcal{D}_{nh} \,]_{\ell\ell} = \frac{[\, R_{nh} \nabla^T \nabla \, R_{nh}]_{\ell\ell}}{[\, I - R_{nh} \,]_{\ell\ell}} \, , \qquad \ell = 1, 2, \cdots, n \, .$$

(d) The coordinate-free version is then

$$CFCV_{\nabla,II}(h) = -\| \nabla R_{nh} \, y_n \|^2 + 2 \, \frac{\text{trace}(\, R_{nh} \nabla_n^T \nabla_n R_{nh} \,)}{\text{trace}(\, I - R_{nh} \,)} \langle \, y_n \, , \, y_n - R_{nh} \, y_n \, \rangle \, ,$$

but does it correspond to leave-one-out in a particular basis?

The exercise above casts some doubt on the analogue of the GCV functional as coordinate-free cross-validation. However, it is possible to construct zero-trace estimators: Replacing R_{nh} by $\alpha \, R_{nh} + (1 - \alpha) \, I$ in (5.6), we get

$$\mathfrak{M}_{\nabla,\alpha}(h) = \alpha^2 \, \| \, \nabla(\, y_n - R_{nh} \, y_n \,) \, \|^2 + 2 \, \sigma^2 \, \text{trace}(\, \nabla^T \nabla \{ \, I - \alpha(\, I - R_{nh} \,) \, \} \,) \, .$$

Setting the first trace equal to zero and solving for α gives $\alpha = \alpha_o$, with

$$\alpha_o = \frac{\text{trace}(\, \nabla^T \nabla \,)}{\text{trace}(\, (\, I - R_{nh} \,) \nabla^T \nabla \,)} \, ,$$

so that the zero-trace estimator $\mathfrak{M}_{\nabla,\alpha_o}(h)$ equals $GCV_\nabla(h)$, where

$$(5.10) \qquad \frac{GCV_\nabla(h)}{\{ \, \text{trace}(\, \nabla^T \nabla \,) \, \}^2} = \frac{\| \, \nabla(\, y_n - R_{nh} \, y_n \,) \, \|^2}{\text{trace}(\, \nabla(\, I - R_{nh} \,) \nabla^T \,)} \, \frac{\sigma^2}{\text{trace}(\, \nabla^T \nabla \,)} \, .$$

This is the analogue of the GCV functional for plain function estimation.

(5.11) EXERCISE. Verify that

$$\min_\alpha \mathfrak{M}_{\nabla,\alpha}(h) = - \frac{\sigma^4 \, \text{trace}(\, \nabla(\, I - R_{nh} \,) \nabla^T \,)}{\| \, \nabla(\, y_n - R_{nh} \, y_n \,) \, \|^2} + 2 \, \sigma^2 \, \text{trace}(\, \nabla^T \nabla \,) \, ;$$

cf. Exercise (2.15) and Stein estimators. So, minimizing this over h gives the same minimizing h as minimizing the zero-trace estimator $\mathfrak{M}_{\nabla,o}$.

The above actually takes care of estimating derivatives of all orders, modulo the potential trouble caused by replacing derivatives with difference quotients. The only difference is in the definition (5.2) of the matrix ∇. Also, in the above, note that the treatment also works for orthogonal projectors: We only used that

$$r_n f^{nh} = R_{nh}\, y_n \ , \tag{5.12}$$

with R_{nh} symmetric and all eigenvalues between 0 and 1. Of course, one also has to choose the order of the spline estimators to be able to differentiate the spline estimator and to get good estimation results; cf. (2.4).

We said above that the unavailability of unbiased estimators of f_o' at the design points is a nuisance. Of course, as RICE (1986b) proposed, one can construct difference quotients $D_n\, y_n$ that are quite accurate. All that is really needed is that

$$\max_{1\leqslant i\leqslant n} \| D_n y_o - r_n(f_o') \|_\infty = \mathcal{O}(n^{-1}) \ , \tag{5.13}$$

which is easy to do if $f_o \in W^{1,2}(0,1)$.

As an estimator of $r_n(f_o')$, we may take the derivative of the spline estimator, denoted as $R_{nh}' y_n$, and one verifies that

$$\mathfrak{M}_D \stackrel{\text{def}}{=} \| D_n\, y_n - R_{nh}' y_n \|^2 + 2\,\sigma^2 \operatorname{trace}(\, D_n^T\, R_{nh}' \,) \tag{5.14}$$

is an unbiased estimator of $\| r_n(f_o') - R_{nh}' y_n \|^2 + \sigma^2 \operatorname{trace}(\, D_n^T D_n \,)$. Now, σ^2 may be estimated as above or also as

$$\widehat{\sigma}^2 = \tfrac{1}{2(n-1)} \sum_{i=1}^{n-1} |\, y_{i+1,n} - y_{in} \,|^2 \ . \tag{5.15}$$

RICE (1986b) goes on to analyze this procedure.

To finish this section, we now return to the plain function estimation problem. It seems clear that we now have the means to implement trade-offs between accuracy and smoothness of the function estimators in the form of the optimality criterion

$$\min_{h>0} \| f^{nh} - f_o \|_{L^2(0,1)}^2 + \alpha \, \| (f^{nh})'' - (f_o)'' \|_{L^2(0,1)}^2 \tag{5.16}$$

for a suitable fudge factor α, depending on the goals of the experimenter. Be that as it may, true to form, we replace the above by a discrete version, including the replacement of the second derivative by a difference quotient,

$$\min_{h>0} \| r_n f^{nh} - r_n f_o \|^2 + \lambda \, \| \Delta(\, r_n f^{nh} - r_n f_o \,) \|^2 \ , \tag{5.17}$$

where Δ is the second order difference operator

$$[\Delta b\,]_i = \frac{[\,\nabla b\,]_{i+1} - [\,\nabla b\,]_i}{x_{i+2,n} - x_{in}} \ , \qquad i = 1, 2, \cdots, n-2 \ . \tag{5.18}$$

Reasonable estimators of the objective functions in (5.17) are then

$$(5.19)\quad \underset{\text{smooth}}{\mathit{CFCV}}(h) = \| y_n - R_{nh}\, y_n \|^2 + \lambda \| \Delta(y_n - R_{nh}\, y_n) \|^2 + 2\,\frac{\langle y_n\,,\, y_n - R_{nh}\, y_n \rangle}{\operatorname{trace}(I - R_{nh})} \operatorname{trace}\bigl(R_{nh}(I + \lambda \Delta^T \Delta) \bigr)$$

and

$$(5.20)\quad \underset{\text{smooth}}{\mathit{GCV}}(h) = \frac{\| y_n - R_{nh}\, y_n \|^2}{\{ \operatorname{trace}(I - R_{nh}) \}^2} + \frac{\lambda}{n}\,\frac{\operatorname{trace}(\Delta^T \Delta) \| \Delta(y_n - R_{nh}\, y_n) \|^2}{\{ \operatorname{trace}(\Delta(I - R_{nh})\Delta^T) \}^2} .$$

In the last expression, one must keep track of the constant factors (and we hope we did it right).

EXERCISES: (5.7), (5.9), (5.11).

6. Akaike's optimality criterion

In the previous sections, the optimality criterion for function estimation was minimum squared error. In a different view of the world, the objective is the optimal estimation of the whole model (1.1)–(1.2) under the normality assumption (1.4). Thus, both $r_n f_o$ and σ^2 must be estimated. To define optimality, let ϕ_o be the n-dimensional normal density with mean $y_o = r_n f_o$ and variance $\sigma^2 I$. With f^{nh} and σ^2_{nh} estimators of the regression function and the variance in the model (1.1)–(1.4) both depending on the same smoothing parameter h, let ϕ^{nh} be the normal density with mean $r_n f^{nh}$ and variance $\sigma^2_{nh} I$. The notion of optimality is now that ϕ^{nh} should be as close as possible to ϕ_o, quantified as realizing the minimum in

$$(6.1)\qquad \text{minimize} \quad \mathrm{KL}(\phi_o\,,\, \phi^{nh}) \quad \text{over} \quad h > 0 ,$$

with KL the Kullback-Leibler distance; see (1.12). The above is the specialization to the regression model (1.1)–(1.4) of the criterion by which AKAIKE (1973) judges optimality. Assuming a (linear) estimator

$$(6.2)\qquad r_n f^{nh} = R_{nh}\, y_n ,$$

for some matrix $R_{nh} \in \mathbb{R}^{n \times n}$, recall that $2\,\mathrm{KL}(\phi_o\,,\, \phi^{nh}) = \mathit{AOC}(h)$, where

$$(6.3)\qquad \mathit{AOC}(h) = \frac{\| y_o - R_{nh}\, y_n \|^2}{\sigma^2_{nh}} + n \left\{ \log \frac{\sigma^2_{nh}}{\sigma^2} + \frac{\sigma^2}{\sigma^2_{nh}} - 1 \right\} .$$

The functional $\mathit{AOC}(h)$ makes sense as a measure of the adequacy of the model (1.1)–(1.2), regardless of the normality assumption (1.4), so we take

Akaike's Optimality Criterion to be that of achieving the minimum of the $\mathit{AOC}(h)$ functional; i.e., ideally, we select h to achieve the minimum in

$$\text{(6.4)} \qquad \text{minimize} \quad \mathit{AOC}(h) \quad \text{over} \quad h > 0 \; .$$

Again, we remind the reader that this can also be used to estimate the order m of the smoothing spline or local polynomial estimator; cf. (2.4).

Now, the task at hand is to construct estimators of $\mathit{AOC}(h)$ and to select the smoothing parameter h by minimizing this estimator over $h > 0$.

From what we have learned in the previous sections, the GCV functional is a pretty good estimator of the squared error. In fact, keeping track of the constant terms, the correct approximation is

$$\text{(6.5)} \qquad \| \, y_o - R_{nh} \, y_n \, \|^2 \approx \frac{\| \, y_n - R_{nh} \, y_n \, \|^2}{\{ \, \frac{1}{n} \text{trace}(\, I - R_{nh} \,) \, \}^2} - n \, \sigma^2 \; ,$$

so that the GCV approximation to $\mathit{AOC}(h)$ is

$$\text{(6.6)} \qquad \mathit{AOC}_{\mathit{GCV}}(h) \stackrel{\text{def}}{=} \frac{\| \, y_n - R_{nh} \, y_n \, \|^2}{\sigma_{nh}^2 \, \{ \, \frac{1}{n} \text{trace}(\, I - R_{nh} \,) \, \}^2} + n \, \log \frac{\sigma_{nh}^2}{\sigma^2} - n \; .$$

Here, as a reward for clean living, the need to estimate the term $n \, (\sigma / \sigma_{nh})^2$ in (6.3) has disappeared.

(6.7) EXERCISE. Indeed, how would you go about estimating $n \, (\sigma / \sigma_{nh})^2$ using the data? [The authors have no idea.]

Note that in the above we have not committed ourselves to a choice for σ_{nh}^2. Several choices come to mind; e.g., for $\alpha = 0, 1$,

$$\text{(6.8)} \qquad \sigma_{nh,1+\alpha}^2 = \frac{\frac{1}{n} \, \| \, y_n - R_{nh} \, y_n \, \|^2}{\{ \, \frac{1}{n} \, \text{trace}(\, (\, I - R_{nh} \,)^2 \,) \, \}^\alpha} \; ,$$

$$\text{(6.9)} \qquad \sigma_{nh,3+\alpha}^2 = \frac{\frac{1}{n} \, \langle \, y_n \, , \, y_n - R_{nh} \, y_n \, \rangle}{\{ \, \frac{1}{n} \, \text{trace}(\, I - R_{nh} \,) \, \}^\alpha} \; ,$$

and

$$\text{(6.10)} \qquad \sigma_{nh,5}^2 = \frac{\frac{1}{n} \, \| \, y_n - R_{nh} \, y_n \, \|^2}{\frac{1}{n} \, \text{trace}(\, I - R_{nh} \,)} \; ,$$

but none of them are *compelling* in this context. (In § 8, we list the origins of these estimators.)

An intriguing choice arises by realizing that we wish to achieve the optimum in the criterion (6.4), so picking σ_{nh}^2 to be the minimizer of $\mathit{AOC}_{\mathit{GCV}}(h)$ seems to make sense. Some simple calculus gives the minimizing σ_{nh}^2 as

$$\text{(6.11)} \qquad \sigma_{nh,o}^2 = \frac{\frac{1}{n} \, \| \, y_n - R_{nh} \, y_n \, \|^2}{\{ \, \frac{1}{n} \, \text{trace}(\, I - R_{nh} \,) \, \}^2} \; ,$$

and then the minimum of $\mathit{AOC}_{GCV}(h)$ equals

$$(6.12) \qquad \mathit{AOC}_{GCV,0}(h) = n \log \frac{\frac{1}{n} \| y_n - R_{nh} y_n \|^2}{\{ \frac{1}{n} \operatorname{trace}(I - R_{nh}) \}^2} .$$

With these choices (GCV approximation, "minimum" variance), this model estimation procedure gives the same choices of h as the GCV procedure.

(6.13) REMARK. The plot thickens even more if we go back to the original objective (6.4). As already observed above, the choice of the estimator σ^2_{nh} is somewhat arbitrary. However, since the objective is to minimize $\mathit{AOC}(h)$, choose σ^2_{nh} by minimization. This gives $\sigma^2_{nh} = \sigma^2_{\min}$ with

$$\sigma^2_{\min} = \tfrac{1}{n} \| y_o - R_{nh} y_n \|^2 + \sigma^2 ,$$

and then $\mathit{AOC}(h) = \mathit{AOC}_{\min}(h)$ with

$$\mathit{AOC}_{\min}(h) = \log(\tfrac{1}{n} \| y_o - R_{nh} y_n \|^2 + \sigma^2) - \log \sigma^2 ,$$

so that in this interpretation, Akaike's Optimality Criterion and the minimum squared error criterion (1.5) give identical "optimal" values of h. Then, it is not surprising that there are rational implementations of both optimality criteria that give the same answers.

(6.14) EXERCISE. In (12.2.24) in the context of smoothing spline estimation, we made the rather outlandish suggestion to estimate the variance by the maximum penalized likelihood estimator

$$\sigma^2_{nh,6} = \tfrac{1}{n} \| y_n - r_n f^{nh} \|^2 + h^{2m} \| (f^{nh})^{(m)} \|^2 .$$

Show that

$$\sigma^2_{nh,6} = \tfrac{1}{n} \langle y_n , y_n - R_{nh} y_n \rangle = \sigma^2_{nh,3} ,$$

with R_{nh} as in (1.23) and § 3.

The above is the *unauthorized* version of the information criterion of AKAIKE (1973). The *authorized* way is to approximate/estimate $\mathit{AOC}(h)$ by

$$(6.15) \qquad n \log \sigma^2_{nh} + \mathbb{E}\Big[\frac{\| y_o - R_{nh} y_n \|^2}{\sigma^2_{nh}} \Big] + \mathbb{E}\Big[\frac{n \sigma^2}{\sigma^2_{nh}} \Big]$$

for suitable deterministic approximations to the expectations. Now, before we can compute the expectation, we must make a choice for σ^2_{nh}. In this setting, the universal choice is $\sigma^2_{nh,1}$ of (6.8) with $\alpha = 0$. Since the expectations depend on the unknown f_o, the simplifying assumption

$$(6.16) \qquad y_o = R_{nh} y_o$$

is also made. This assumption is based on the fact that, for the optimal h, the squared error is much smaller than the variance of the noise, so the bias must be much smaller also. Of course, the proof is in the pudding: The final result must be sensible and lead to good estimators of the regression function (it does). Under the assumption (6.16), then

(6.17) $$\sigma_{nh}^2 = \tfrac{1}{n} \| d_n - R_{nh} d_n \|^2$$

and

(6.18) $$\| y_o - R_{nh} y_n \|^2 = \| R_{nh} d_n \|^2$$

are positive-definite quadratic forms in $d_n \sim \text{Normal}(0, \sigma^2 I)$. Thus, a reasonable approximation to $AOC(h)$ is

(6.19) $$AIC(h) \stackrel{\text{def}}{=} n \log \sigma_{nh}^2 + \text{Quot} + \text{Recip} ,$$

where

$$\text{Quot} = \mathbb{E}\Big[\frac{\| R_{nh} d_n \|^2}{\| d_n - R_{nh} d_n \|^2} \Big] \quad , \quad \text{Recip} = \mathbb{E}\Big[\frac{n \sigma^2}{\| d_n - R_{nh} d_n \|^2} \Big] .$$

These expectations may be computed exactly (well, sort of) by means of the celebrated "differentiation under the integral" technique. To state the results, we need the eigenvalue decomposition of $(I - R_{nh})^T (I - R_{nh})$,

(6.20) $$(I - R_{nh})^T (I - R_{nh}) = U \Sigma U^T ,$$

where $U \in \mathbb{R}^{n \times n}$ is orthogonal and Σ is diagonal with diagonal elements $s_1 \geqslant s_2 \geqslant \cdots \geqslant s_n \geqslant 0$. Let r_i, $i = 1, 2, \cdots, n$, be the diagonal elements of the matrix $U^T R_{nh}^T R_{nh} U$. Then,

(6.21) $$\text{Recip} = \tfrac{1}{2} n \int_0^\infty \prod_{i=1}^n (1 + t s_i)^{-1/2} \, dt$$

and

(6.22) $$\text{Quot} = \tfrac{1}{2} \int_0^\infty \Big\{ \prod_{i=1}^n (1 + t s_i)^{-1/2} \Big\} \sum_{i=1}^n r_i (1 + t s_i)^{-1} \, dt .$$

(6.23) Exercise. In this exercise, we step though the process of computing the expectations "Quot" and "Recip". We begin with "Recip"and assume that $n \geqslant 3$.

(a) Using the eigenvalue decomposition of $(I - R_{nh})^T (I - R_{nh})$, show that

$$\| d_n - R_{nh} d_n \|^2 = \sigma^2 \varepsilon^T \Sigma \varepsilon ,$$

where $\varepsilon = \sigma^{-1} U d_n \sim \text{Normal}(0, I)$.

(b) Now, consider the function

$$F(x) = \mathbb{E}\Big[\frac{\exp(-\frac{1}{2} x \varepsilon^T \Sigma \varepsilon)}{\varepsilon^T \Sigma \varepsilon} \Big] , \quad x > -1/s_1 .$$

Then, $F(0)$ is what we want. Show that

$$-2 F'(x) = \mathbb{E}\big[\exp(-\tfrac{1}{2} x \varepsilon^T \Sigma \varepsilon) \big] .$$

(This is an exercise in advanced calculus.)
(c) Show that

$$-2\,F'(x) = \int_{\mathbb{R}^n} \exp\Bigl(-\tfrac{1}{2}\,\varepsilon^T (\,I + x\,\Sigma\,)\,\varepsilon \Bigr)\, d\mu(\varepsilon) = \prod_{i=1}^{n} (\,1 + x\,s_i\,)^{-1/2} \ ,$$

where $d\mu(\varepsilon) = (2\pi)^{-n/2}\, d\varepsilon_1\, d\varepsilon_2 \,\cdots\, d\varepsilon_n$.
(d) Show that

$$F(0) = -\int_0^{\infty} F'(x)\, dx \ ,$$

and draw your conclusions.
(e) The treatment of "Quot" requires some extra work. Show that

$$\|\, R_{nh}\, d_n \,\|^2 = \sigma^2\, \varepsilon^T A\, \varepsilon \ ,$$

with $A = U\, R_{nh}^T R_{nh} U^T$, and ε as in part (a), and U as in (6.20). Observe that A is diagonal and denote the diagonal elements by r_i, $i = 1, 2, \cdots, n$.
(f) Consider the function

$$G(x) = \mathbb{E}\left[\, \frac{\varepsilon^T A\, \varepsilon}{\varepsilon^T\, \Sigma\, \varepsilon}\, \exp\bigl(-\tfrac{1}{2}\, x\, \varepsilon^T\, \Sigma\, \varepsilon \,\bigr) \right] \ , \qquad x > -1/s_1 \ .$$

Show that

$$-2\, G'(x) = \mathbb{E}\Bigl[\, \varepsilon^T A\, \varepsilon\, \exp\bigl(-\tfrac{1}{2}\, x\, \varepsilon^T\, \Sigma\, \varepsilon \,\bigr) \Bigr] \ .$$

(g) Show that

$$-2\, G'(x) = \bigl\{\, \det(\, I + x\,\Sigma\,)\, \bigr\}^{-1/2} \int_{\mathbb{R}^n} \delta^T B\, \delta\, \exp\bigl(-\tfrac{1}{2}\, \|\, \delta \,\|^2\,\bigr)\, d\mu(\delta) \ ,$$

where $d\mu(\delta) = (2\pi)^{-n/2}\, d\delta_1\, d\delta_2 \,\cdots\, d\delta_n$ and

$$B = (\, I + x\,\Sigma\,)^{-1/2} A (\, I + x\,\Sigma\,)^{-1/2} \ ,$$

so that $\quad -2\, G'(x) = \text{trace}(\, B\,) \cdot \prod_{i=1}^{n} \, (1 + x s_i)^{-1/2}$ and, of course,

$$\text{trace}(\, B\,) = \sum_{i=1}^{n} \, r_i\, (\,1 + x\, s_i\,)^{-1} \ .$$

(h) Complete the solution as for "Recip".

Computing the integrals for "Quot" and "Recip" is a tedious exercise, even using numerical integration software packages, unless you are interested in that sort of thing. To get some feeling for these integrals, it is instructive to consider the case where R_{nh} is an orthogonal projector with rank denoted by p. Then, the diagonal elements of $U^T R_{nh}^T R_{nh} U$ are just the eigenvalues of $R_{nh}^T R_{nh}$, so

$$r_i = \begin{cases} 1\ , & j = 1, 2, \cdots, p\ , \\ 0\ , & j = p+1, \cdots, n\ , \end{cases} \tag{6.24}$$

and $s_i = 1 - r_i$ for all i. So, the sum in (6.22) equals p, and then

$$\text{(6.25)} \qquad \text{Recip} = \tfrac{1}{2}\, n \int_0^\infty (1+t)^{-(n-p)/2}\, dt = \frac{n}{n-p-2} \,,$$

$$\text{(6.26)} \qquad \text{Quot} = \tfrac{1}{2}\, p \int_0^\infty (1+t)^{-(n-p)/2}\, dt = \frac{p}{n-p-2} \,.$$

Now, since $p = \text{trace}(R_{nh})$ when R_{nh} is an orthogonal projector with rank p, then we may approximate $AIC(h)$ by $AIC_o(h)$, where

$$\text{(6.27)} \qquad AIC_o(h) = n \log \sigma_{nh}^2 + 1 + \frac{2\,\text{trace}(R_{nh}) + 2}{\text{trace}(I - R_{nh}) - 2} \,.$$

The suggestion is now to take $AIC_o(h)$ as an approximation to $AIC(h)$, even when R_{nh} is not an orthogonal projector. It seems reasonable though to restrict its use to the case of symmetric R_{nh} with eigenvalues between zero and one.

(6.28) REMARK. Actually, it is not clear that (6.27) is above reproach. If $R_{nh} = P$ is an orthogonal projector with rank p, one could equally well argue that $p = \text{trace}\big(P^T P\big)$ and $n - p = \text{trace}\big((I-P)^T (I-P)\big)$, and so we get the approximation to $AIC(h)$

$$AIC_p(h) = n \log \sigma_{nh}^2 + \frac{2\,\text{trace}\big(R_{nh}^T R_{nh}\big) + 2}{\text{trace}\big((I - R_{nh})^T (I - R_{nh})\big) - 2} \,.$$

Thus, one might apply this formula if R_{nh} is not a projector.

(6.29) EXERCISE. For the case where R_{nh} is an orthogonal projector and $d_n \sim \text{Normal}(0, \sigma^2 I)$, show directly that

$$\mathbb{E}\Big[\frac{\|R_{nh}\, d_n\|^2}{\|d_n - R_{nh}\, d_n\|^2}\Big] = \mathbb{E}\Big[\|R_{nh}\, d_n\|^2\Big] \cdot \mathbb{E}\Big[\frac{1}{\|d_n - R_{nh}\, d_n\|^2}\Big] \,.$$

EXERCISES: (6.7), (6.14), (6.23), (6.29).

7. Heterogeneity

In this section, we briefly consider two kinds of heterogeneity, to wit, heteroscedastic errors as in (2.4) and heterogeneity of scale, as discussed briefly in § 15.6. Heteroscedastic errors may be handled in a straightforward manner. Heterogeneity of scale appears to be different. As before, we concentrate on smoothing splines.

First, we discuss the C_L method of MALLOWS (1972) and the GCV method for selecting the smoothing parameter h (and the order m) for the smoothing spline in the model (1.1) with heteroscedastic errors (1.3),

$$\mathbb{E}[\, d_n \,] = 0 \quad , \quad \mathbb{E}[\, d_n \, d_n^T \,] = \sigma^2 \, V \ , \tag{7.1}$$

with V a diagonal matrix.

We start out by assuming that V is known but that σ^2 is not. Now, there are two ways to proceed, depending on whether one ignores or incorporates the heteroscedasticity of the noise into the estimator. In the first case, the estimator is as before,

$$r_n f^{nh} = R_{nh} \, y_n \ , \tag{7.2}$$

with the hat matrix

$$R_{nh} = \bigl(I + nh^{2m} \, M \bigr)^{-1} \ , \tag{7.3}$$

and $M = S^T T^{-1} S$ as before. Observe that R_{nh} is symmetric. Then, as in § 2, one verifies that

$$\mathbb{E}[\, \| \, y_o - R_{nh} \, y_n \, \|^2 \,] = \| \, y_o - R_{nh} \, y_o \, \|^2 + \sigma^2 \, \mathrm{trace}\bigl(V \, R_{nh}^2 \bigr) \ , \tag{7.4}$$

where we used the property $\mathrm{trace}(\, R_{nh}^T V \, R_{nh} \,) = \mathrm{trace}(\, V \, R_{nh}^2 \,)$ and

$$\mathbb{E}[\, \| \, y_n - R_{nh} \, y_n \, \|^2 \,] = \| \, y_o - R_{nh} \, y_o \, \|^2 + \sigma^2 \, \mathrm{trace}\bigl(V \, (\, I - R_{nh} \,)^2 \bigr) \ , \tag{7.5}$$

so that

$$\mathfrak{M}_V(\, h \,) = \| \, y_n - R_{nh} \, y_n \, \|^2 + 2 \, \sigma^2 \, \mathrm{trace}\bigl(V \, R_{nh} \bigr) - \sigma^2 \, \mathrm{trace}\bigl(V \bigr) \tag{7.6}$$

is an unbiased estimator of $\mathbb{E}[\, \| \, y_o - R_{nh} \, y_n \, \|^2 \,]$.

Similarly to the way it was done in previous sections, one derives the zero-trace estimator of $\mathbb{E}[\, \| \, y_o - R_{nh} \, y_n \, \|^2 \,]$. The final result is that

$$GCV(h) = \frac{ \{ \mathrm{trace}(\, V \,) \}^2 \, \| \, y_n - R_{nh} \, y_n \, \|^2 }{ \{ \mathrm{trace}\bigl(V \, (\, I - R_{nh} \,) \bigr) \}^2 } - \sigma^2 \, \mathrm{trace}(\, V \,) \ . \tag{7.7}$$

If one wishes to incorporate the heteroscedasticity into the estimator, things get a little more involved. The estimation problem (1.21) must then be replaced by

$$\begin{aligned} &\text{minimize} && \| \, V^{-1/2} (\, b - y_n \,) \, \|^2 + nh^{2m} \, \langle \, a \, , \, T \, a \, \rangle \\ &\text{subject to} && T \, a = S \, b \ . \end{aligned} \tag{7.8}$$

The solution is given by

$$b^{nh} = R_{Vnh} \, y_n \ , \tag{7.9}$$

with the hat matrix

$$R_{Vnh} = \bigl(V + nh^{2m} \, M \bigr)^{-1} V \ , \tag{7.10}$$

and M as before. Observe that R_{Vnh} is not symmetric in general.

We again work out the details of the C_L and zero-trace estimators. One derives that

$$(7.11)\quad \mathbb{E}[\, \| \, y_o - R_{Vnh}\, y_n \, \|^2 \,] = \| \, y_o - R_{Vnh}\, y_o \, \|^2 + \sigma^2 \operatorname{trace}(\, R_{Vnh}^T V R_{Vnh} \,)$$

and that

$$(7.12)\quad \mathbb{E}[\, \| \, y_n - R_{Vnh}\, y_n \, \|^2 \,] = \| \, y_o - R_{Vnh}\, y_o \, \|^2 + \sigma^2 \operatorname{trace}(\, (\, I - R_{Vnh}\,)^T V (\, I - R_{Vnh}\,)) \;,$$

so that in this case the MALLOWS (1972) functional

$$(7.13)\quad \mathfrak{M}_V(\, h \,) = \| \, y_n - R_{Vnh}\, y_o \, \|^2 + 2\,\sigma^2 \operatorname{trace}(\, V R_{Vnh} \,) - \sigma^2 \operatorname{trace}(\, V \,)$$

is an unbiased estimator of $\mathbb{E}[\, \| \, y_o - R_{Vnh}\, y_n \, \|^2 \,]$. Note that

$$(7.14)\quad V R_{Vnh} = V (\, V + n h^{2m} M \,)^{-1} V$$

is symmetric positive-definite again, and so is $I - V R_{Vnh}$.

Finally, the zero-trace estimator of the mean squared error is then

$$(7.15)\quad GCV_V(\, h \,) = \frac{\{\, \operatorname{trace}(\, V \,)\,\}^2 \, \| \, y_n - R_{Vnh}\, y_n \, \|^2}{\{\, \operatorname{trace}(\, V\, (\, I - R_{Vnh}\,)\,)\,\}^2} - \sigma^2 \operatorname{trace}(\, V \,) \;.$$

Considering what happens for parametric models, it seems clear that the estimator $R_{Vnh}\, y_n$, with h chosen by the GCV_V method, is better than the estimator $R_{nh}\, y_n$, with h chosen by the GCV method of (7.6). It would be nice if this could be quantified.

When the variance V is not known, then it must be estimated. See § 22.4 for some approaches. So, if $\widehat{V}$ is an estimator of the variance, then the $GCV_{\widehat{V}}$ method suggests itself. The only drawback is that this method would require larger sample sizes to offset the extra variability due to having to estimate the variances. Clearly, one can think of two- or even three-stage estimators. First, use the residuals $y_n - R_{nH}\, y_n$ to estimate V, and then use (7.7) with the estimated V.

Heterogeneity of scale was discussed briefly in § 15.6. Although heterogeneity of scale may be offset by heteroscedasticity of the noise and nonuniformity of the design, in general in nonparametric regression it requires smoothing parameters that vary with location. Now, one can envision choosing a possibly different smoothing parameter at each location, but then the paucity of available data near each point becomes an issue. So, it is more prudent to choose them in a piecewise fashion. The hardest part about this is choosing the pieces; i.e., to partition the interval $[\, 0\,,\, 1\,]$ into a small number of intervals on each of which the smoothing parameter is to be constant. However, the application itself may suggest a natural partition; see, e.g., the discussion in BROWN, EGGERMONT, LARICCIA, and ROTH (2008). Alternatively, a way around this, and also a way around the pointwise protocol, is to view the smoothing parameter as a function of bounded variation and to consider the problem

$$(7.16)\qquad \begin{array}{ll} \text{minimize} & \frac{1}{n}\sum_{i=1}^{n} |\, f(x_{in}) - y_{in}\,|^2 + \|\, f^{(m)}\,\|^2_{L^2(h^{2m})} + \lambda\,|\,h\,|_{BV} \\ \text{subject to} & f \in W^{m,2}(0,1)\,,\; h \in BV(0,1)\,,\; h > 0\,, \end{array}$$

where now the scalar λ must be chosen in a data-driven way. Compare this with the problem (15.6.20). If only the authors knew how to solve this!

However, one might indeed determine $h = H$ in a pointwise fashion, as outlined by CUMMINS, FILLOON, and NYCHKA (2001) and discussed below, and then smooth the function H by solving

$$(7.17)\qquad \begin{array}{ll} \text{minimize} & \frac{1}{n}\sum_{i=1}^{n} |\, h(x_{in}) - H(x_{in})\,|^2 + \lambda\,|\,h\,|_{BV} \\ \text{subject to} & h \in BV(0,1)\,,\; h > 0\,. \end{array}$$

(7.18) EXERCISE. Solve the problem (7.16) from both the theoretical and practical points of view. Conduct a simulation study for a critical set of test regression functions, and tell the authors about it.

We now discuss how CUMMINS, FILLOON, and NYCHKA (2001) solve the problem of choosing the smoothing parameter in a pointwise fashion. Let x_{in} be a design point in the model (1.1)–(1.2). Let f^{nh} be the (smoothing spline) estimator of f_o. The goal is to find the solution of

$$(7.19)\qquad \text{minimize} \quad |\, f^{nh}(x_{in}) - f_o(x_{in})\,|^2 \quad \text{subject to} \quad h > 0\,.$$

The task at hand is to estimate the objective function, and we proceed by way of the MALLOWS (1972) and zero-trace approach. With (2.1), we have

$$(7.20)\qquad f^{nh}(x_{in}) = e_i^T R_{nh}\, y_n\,,$$

where $e_1, e_2, \cdots, e_n$ is the standard basis of $\mathbb{R}^n$. Then,

$$(7.21)\qquad f_o(x_{in}) - f^{nh}(x_{in}) = e_i^T(\, y_o - R_{nh}\, y_n\,)\,.$$

From the standard computations

$$\mathbb{E}[\,|\, e_i^T(\, y_o - R_{nh}\, y_n\,)\,|^2\,] = |\, e_i^T(\, I - R_{nh}\,)\, y_o\,|^2 + \sigma^2\,[\,(R_{nh})^2\,]_{i,i}\,,$$

$$\mathbb{E}[\,|\, e_i^T(\, y_n - R_{nh}\, y_n\,)\,|^2\,] = |\, e_i^T(\, I - R_{nh}\,)\, y_o\,|^2 + \sigma^2\,[\,(\, I - R_{nh}\,)^2\,]_{i,i}\,,$$

it follows that

$$(7.22)\qquad \mathfrak{M}_i(h) = |\, e_i^T(\, I - R_{nh}\,)\, y_n\,|^2 + 2\,\sigma^2\,[\, R_{nh}\,]_{i,i} - \sigma^2$$

is an unbiased estimator of $\mathbb{E}[\,|\, e_i^T(\, y_o - R_{nh}\, y_n\,)\,|^2\,]$. The corresponding zero-trace estimator is then readily seen to be

$$(7.23)\qquad GCV_i(h) = \frac{|\, e_i^T(\, I - R_{nh}\,)\, y_n\,|^2}{\{\, 1 - [\, R_{nh}\,]_{i,i}\,\}^2} - \sigma^2\,.$$

(7.24) THE LOCAL GCV METHOD. So, the local GCV method for choosing the smoothing parameter in a pointwise fashion is: *For* $i = 1, 2, \cdots, n$, *choose* $h = H_i$ *as the minimizer of* $GCV_i(h)$ *over* $h > 0$.

(7.25) REMARK. CUMMINS, FILLOON, and NYCHKA (2001) actually suggest this local GCV method for bias reduction, which is important for the construction of honest pointwise confidence intervals. The implications for the construction of simultaneous confidence intervals (confidence *bands*) are not clear, especially when it comes to obtaining accurate critical values. See Chapter 22 and the simulations in Chapter 23.

(7.26) REMARK. The computational issues of implementing the local GCV method are benign, as noted by CUMMINS, FILLOON, and NYCHKA (2001). Compared with the global GCV method, where one must store

$$\sum_{i=1}^{n} | e_i^T (I - R_{nh}) y_n |^2 \Big/ \Big\{ \sum_{i=1}^{n} (1 - [R_{nh}]_{i,i}) \Big\}^2$$

for a fine grid of all possible values of h, now one must store

$$e_i^T (I - R_{nh}) y_n \quad \text{and} \quad 1 - [R_{nh}]_{i,i}$$

for $i = 1, 2, \cdots, n$, and all values of h. This merely requires a suitably large store of memory. In 2001, this might still have been problematic; eight years later, it barely deserves mentioning.

The CUMMINS, FILLOON, and NYCHKA (2001) scheme may also be applied when choosing the smoothing parameter in a piecewise fashion. Assuming we have a suitable partition of the interval $[0, 1]$ into intervals $\mathcal{I}_1, \mathcal{I}_2, \cdots, \mathcal{I}_J$, we may construct diagonal matrices $W_j \in \mathbb{R}^{n \times n}$ such that

$$[W_j]_{i,i} = 1 \quad \text{if} \quad x_{in} \in \mathcal{I}_j \tag{7.27}$$

and $= 0$ otherwise. Then, the goal is to choose $h = H_j$ as the solution to

$$\text{minimize} \quad \| W_j (y_o - R_{nh} y_n) \|^2 \quad \text{over} \quad h > 0 . \tag{7.28}$$

One readily derives the MALLOWS (1972) estimators as well as the zero-trace estimators. The zero-trace estimators take the form

$$GCV_{W_j}(h) = \frac{ \{ \text{trace}(W_j) \}^2 \, \| W_j (I - R_{nh}) y_n \|^2 }{ \{ \text{trace}(W_j (I - R_{nh})) \}^2 } , \tag{7.29}$$

where we dropped the $-\sigma^2 \, \text{trace}(W_j)$ term. Note the similarity with (7.7).

(7.30) EXERCISE. Derive the unbiased MALLOWS (1972) estimator of $\mathbb{E}[\| W_j (y_o - R_{nh} y_n) \|^2]$, and use it to verify the zero-trace estimator.

EXERCISES: (7.18), (7.30).

8. Local polynomials

We now discuss the selection of the smoothing parameter in local polynomial estimation. Although the language is about choosing the smoothing parameter h when the order m of the local polynomials is given, one may view the smoothing parameter to be (h, m); cf. (2.4).

The model under consideration is (16.1.5)–(16.1.6), repeated here for convenience,

$$Y_i = f_o(X_i) + D_i \ , \quad i = 1, 2, \cdots, n \ , \tag{8.1}$$

with the conditional covariance matrix V,

$$V \stackrel{\text{def}}{=} \mathbb{E}[\, \mathbb{D}_n \mathbb{D}_n^T \,|\, \mathbb{X}_n \,] \ , \tag{8.2}$$

where $\mathbb{D}_n = (\, D_1, \, D_2, \cdots, \, D_n \,)^T$ and $\mathbb{X}_n = (\, X_1, \, X_2, \cdots, \, X_n \,)$. Thus, V is a diagonal matrix with

$$V_{ii} = \sigma^2(X_i) \ , \quad i = 1, 2, \cdots, n \ . \tag{8.3}$$

We assume that the function $\sigma^2(x)$ is continuous.

Recall that the local polynomial estimator of f_o in the model (8.1)–(8.3) is obtained as follows. Fix $x \in [\,0\,,\,1\,]$, and for $h > 0$ determine the solution $p = p^{nh}(\,\cdot\,; x)$ of the problem

$$\begin{aligned} &\text{minimize} \quad \sum_{i=1}^{n} w_{in}(x) \, | \, p(X_i) - Y_i \, |^2 \\ &\text{subject to} \quad p \in \mathcal{P}_m \ , \end{aligned} \tag{8.4}$$

where

$$w_{in}(x) = \frac{A_h(x - X_i)}{\sum_{j=1}^{n} A_h(x - X_j)} \ , \qquad i = 1, 2, \cdots, n \ . \tag{8.5}$$

Here, the kernel A satisfies (16.1.7), and it is assumed that the denominator in (8.5) is positive. The estimator of $f_o(x)$ is then defined as

$$f^{nh}(x) = p^{nh}(x\,; x) \ . \tag{8.6}$$

As discussed in § 2, the optimality criterion for choosing the smoothing parameter is to minimize the error, so first we must decide how to measure it. For equally spaced design points, the choice we have made throughout this chapter (the sum of the squared error at the design points) is quite reasonable. For random designs, this choice, now denoted as

$$\mathit{RPW}(h) = \tfrac{1}{n} \sum_{i=1}^{n} | \, f^{nh}(X_i) - f_o(X_i) \, |^2 \tag{8.7}$$

and referred to as the random pointwise error, still makes sense.

Theoretically, see § 16.2, the local squared error is much nicer than the pointwise error,

$$(8.8) \qquad LE(x;h) = \sum_{i=1}^{n} w_{in}(x)\,|\,p^{nh}(X_i\,;x) - f_o(X_i)\,|^2 \ ,$$

as is the global random version

$$(8.9) \qquad RLE(h) = \tfrac{1}{n} \sum_{i=1}^{n} LE(X_i\,;h) \ .$$

Here, both measures of the error are considered. However, note that even though one can select the smoothing parameter in a pointwise manner, following FAN and GIJBELS (1995a), the authors are leaning towards using one global smoothing parameter.

The local error. As advertised, for the local error we follow § 2 to the letter. The first item is to formulate the problem in the language of vector spaces. Consider the local error at $x = X_i$, a design point. Define the diagonal matrix $W_i \in \mathbb{R}^{n\times n}$ with diagonal entries $[W_i]_{jj} = \{\, w_{jn}(X_i)\,\}^{1/2}$, and, for nice functions f, let

$$(8.10) \qquad r_n f = \big(\, f(X_1), f(X_2), \cdots, f(X_n)\big)^T \ .$$

Finally, define the vector space

$$(8.11) \qquad W_i\, r_n \mathcal{P}_m \overset{\text{def}}{=} \{\, W_i\, r_n p \,:\, p \in \mathcal{P}_m \,\} \ .$$

The local polynomial estimation problem may now be formulated as

$$(8.12) \qquad \text{minimize} \quad \|\, q - W_i \mathbb{Y}_n \,\|^2 \quad \text{subject to} \quad q \in W_i\, r_n \mathcal{P}_m \ ,$$

where $\mathbb{Y}_n = (Y_1, Y_2, \cdots, Y_n)^T$. Its solution is given by $q = Q_i\, W_i\, \mathbb{Y}_n$, where $Q_i \in \mathbb{R}^{n\times n}$ is the orthogonal projector onto $W_i\, r_n \mathcal{P}_m$.

We now proceed to Mallows' estimator of the (normalized) local error, which may be written as

$$(8.13) \qquad LE(x;h) = \|\, Q_i\, W_i\, \mathbb{Y}_n - W_i\, \mathbb{Y}_o \,\|^2 \ ,$$

with $\mathbb{Y}_o = r_n f_o$. The required expectations (conditional on the design) are

$$(8.14) \qquad \mathbb{E}[\, LE(x;h) \,|\, \mathbb{X}_n \,] = \|\,(\, Q_i - I\,)\, W_i\, \mathbb{Y}_o \,\|^2 + \operatorname{trace}\big(\, Q_i^2\, W_i^2\, V \,\big)$$

and

$$(8.15) \qquad \mathbb{E}[\, \|\,(\, Q_i - I\,)\, W_i\, \mathbb{Y}_n \,\|^2 \,|\, \mathbb{X}_n \,] = \|\,(\, Q_i - I\,)\, W_i\, \mathbb{Y}_o \,\|^2 + \operatorname{trace}\big(\,(\, Q_i - I\,)^2\, W_i^2\, V \,\big) \ .$$

It follows that

$$(8.16) \qquad \mathfrak{M}_{LE}(x\,;h) \overset{\text{def}}{=} \|\,(\, I - Q_i\,)\, W_i\, \mathbb{Y}_n \,\|^2 + 2\operatorname{trace}\big(\, Q_i\, W_i^2\, V\big) - \operatorname{trace}(\, W_i^2\, V)$$

is an unbiased estimator of the local error. Of course, the fact that the variance matrix does not have a constant diagonal causes trouble. However, the error variance is continuous, so we have the approximations (recall that Q_i is an orthogonal projector)

$$\text{trace}(\,Q_i\,W_i^2\,V\,) \approx \sigma^2(X_i)\,\text{trace}(\,Q_i\,W_i^2\,)\;,$$
$$\text{trace}(\,(\,Q_i - I\,)^2\,W_i^2\,V\,) \approx \sigma^2(X_i)\,\text{trace}(\,(\,I - Q_i\,)\,W_i^2\,)\;.$$

Since $\text{trace}(\,W_i^2\,) = 1$, then

$$\text{(8.17)}\quad M_{LE}(X_i\,;h) \stackrel{\text{def}}{=} \|\,(\,I - Q_i\,)\,W_i\,\mathbb{Y}_n\,\|^2 + 2\,\sigma^2(X_i)\,\text{trace}(\,Q_i\,W_i^2\,) - \sigma^2(X_i)$$

is an (approximately) unbiased estimator of the local error, and

$$\text{(8.18)}\qquad M_{RLE}(h) \stackrel{\text{def}}{=} \tfrac{1}{n}\sum_{i=1}^{n} M_{LE}(X_i\,;h)$$

is an (approximately) unbiased estimator of the global local error $RLE(h)$.

(8.19) EXERCISE. Verify (8.14)–(8.15), and if $\sigma^2(x)$ is a constant function, that $M_{LE}(h)$, as given by (8.18), is an unbiased estimator of $LE(x\,;h)$ *conditional on the design* and likewise for $M_{RLE}(h)$. [Hint for (8.14): The orthogonality of Q_i comes in.]

Next, we consider the possibility of zero-trace estimators. Thus, replace the estimators $Q_i\,W_i\,\mathbb{Y}_n$ by $\alpha\,W_i\,\mathbb{Y}_n + (1-\alpha)\,Q_i\,W_i\,\mathbb{Y}_n$, or equivalently, replace each $Q_i\,W_i$ by $\alpha\,W_i + (1-\alpha)\,Q_i\,W_i$. Then, the choice

$$\text{(8.20)}\qquad (\,1-\alpha\,)^{-1} = \text{trace}(\,(\,I - Q_i\,)\,W_i^2\,)$$

gives the zero-trace estimator

$$\text{(8.21)}\qquad GCV_{LE}(X_i\,;h) = \frac{\|\,(\,I - Q_i\,)\,W_i\,\mathbb{Y}_n\,\|^2}{\Big\{\,\text{trace}(\,(\,I - Q_i\,)\,W_i^2\,)\,\Big\}^2} - \sigma^2(X_i)$$

corresponding to $M_{LE}(h)$. Then, without further ado, we arrive at the zero-trace estimator of $RLE(h)$,

$$\text{(8.22)}\qquad GCV_{RLE}(h) = \tfrac{1}{n}\sum_{i=1}^{n} GCV_{LE}(X_i\,;h)\;.$$

Here, we give in to temptation and call this the generalized cross-validation estimator of the squared error. However, it is not clear that this may be derived as ordinary cross-validation in a favored coordinate system, analogous to § 4, even apart from the nonconstancy of $\sigma^2(x)$.

(8.23) EXERCISE. Fill in the details of the derivation of (8.21)–(8.22).

(8.24) EXERCISE. (a) Show that $\mathfrak{M}_{LE}(X_i; h)$ is an unbiased estimator of

$$\mathbb{E}\Big[\,\|\,W_i\,\mathbb{Y}_o - Q_i(\alpha_i)\,W_i\,\mathbb{Y}_n\,\|^2\,\Big]\;,$$

where $Q_i(\alpha) = \alpha I + (1-\alpha) Q_i$ and α_i is given via

$$1 - \alpha_i = \frac{\text{trace}(W_i^2 V)}{\text{trace}((I - Q_i) W_i^2 V)} .$$

(b) Show that

$$\mathfrak{GCV}_{LE}(X_i; h) \stackrel{\text{def}}{=} (1 - \alpha_i)^2 \, \| (I - Q_i) W_i \mathbb{Y}_n \|^2 - \text{trace}(W_i^2 V)$$

is the zero-trace version of $\mathfrak{M}_{LE}(X_i; h)$; see (8.16).

(8.25) EXERCISE. Assume that the error variance is constant: $\sigma^2(x) = \sigma^2$ for all $x \in [0, 1]$. Show that

$$GCV_{RLE\text{-const}}(h) = \frac{\frac{1}{n} \sum_{i=1}^{n} \| (I - Q_i) W_i \mathbb{Y}_n \|^2}{\Big\{ \frac{1}{n} \sum_{i=1}^{n} \text{trace}((I - Q_i) W_i^2) \Big\}^2} - \sigma^2$$

is the zero-trace estimator corresponding to $M_{RLE}(h)$, which may now be written as

$$M_{RLE\text{-const}}(h) = \frac{1}{n} \sum_{i=1}^{n} \| (I - Q_i) W_i \mathbb{Y}_n \|^2 + \\ 2 \sigma^2 \, \text{trace}\Big(\frac{1}{n} \sum_{i=1}^{n} Q_i W_i^2 \Big) - \sigma^2 .$$

(8.26) EXERCISE. Work out the details of the leave-one-out cross-validation, analogous to § 3, for $LE(X_i; h)$. Note that X_i is a design point. Is there a coordinate-free version? [The authors do not know.]

The pointwise error. We now come to the selection of the smoothing parameter for the pointwise error. First, for a "fixed" design point X_i, consider the error in the estimator of $f_o(X_i)$. Then,

$$(8.27) \qquad f^{nh}(X_i) = p^{nh}(X_i; X_i) = [P_i \mathbb{Y}_n]_i = e_i^T P_i \mathbb{Y}_n ,$$

where e_i is the i-th unit vector in the standard basis for $\mathbb{R}^n$ and the matrix P_i is given by

$$(8.28) \qquad P_i = W_i^\dagger Q_i W_i ,$$

in which $W_i^\dagger \in \mathbb{R}^{n \times n}$ is again diagonal with

$$(8.29) \qquad (W_i^\dagger)_{jj} = \begin{cases} 1/w_{jn}(X_i) , & \text{if} \quad w_{jn}(X_i) > 0 , \\ 0 , & \text{otherwise} . \end{cases}$$

Consequently, the operator P_i is a projector, but unless A is the uniform density, not an *orthogonal* projector.

Thus, the random pointwise error may be written as

$$(8.30) \qquad RPW(h) = \frac{1}{n} \sum_{i=1}^{n} | e_i^T (\mathbb{Y}_o - P_i \mathbb{Y}_n) |^2 .$$

Proceeding along Mallows' way, we have

$$\mathbb{E}[\,RPW(h)\,|\,\mathbb{X}_n\,] \approx \tfrac{1}{n}\sum_{i=1}^{n}\Bigl\{\,|\,e_i^T(I-P_i)\,\mathbb{Y}_o\,|^2+\sigma^2(X_i)\,\mathrm{trace}(\,P_i^T\,e_i\,e_i^T\,P_i\,)\,\Bigr\},$$

so that, since $\mathrm{trace}(\,P_i^T\,e_i\,e_i^T\,P_i\,) = \mathrm{trace}(\,e_i^T\,P_i\,P_i^T\,e_i\,) = (\,P_i\,P_i^T)_{ii}$,

$$\text{(8.31)}\quad \mathbb{E}[\,RPW(h)\,|\,\mathbb{X}_n\,] = \tfrac{1}{n}\sum_{i=1}^{n}\Bigl\{\,|\,e_i^T(I-P_i)\,\mathbb{Y}_o\,|^2+\sigma^2(X_i)\,(\,P_i\,P_i^T)_{ii}\,\Bigr\}.$$

As always, the estimator of $RPW(h)$ is based on the $e_i^T(\,I-P_i\,)\,\mathbb{Y}_n$, for which we obtain

$$\text{(8.32)}\quad \mathbb{E}\bigl[\,|\,e_i^T(\,I-P_i\,)\,\mathbb{Y}_n\,|\,\mathbb{X}_n\,|^2\,\bigr] = |\,e_i^T(\,P-I\,)\,\mathbb{Y}_o\,|^2 + \sigma^2(X_i)\,\mathrm{trace}\bigl(\,(\,I-P_i\,)^T\,e_i\,e_i^T\,(\,I-P_i\,)\bigr)\,,$$

and the trace equals $e_i^T\,(\,I-P_i\,)(\,I-P_i\,)^T\,e_i = 1 - 2\,(\,P_i\,)_{ii} + (\,P_i\,P_i^T)_{ii}$. It follows that

$$\text{(8.33)}\quad M_{RPW}(h) \stackrel{\mathrm{def}}{=} \tfrac{1}{n}\sum_{i=1}^{n}\Bigl\{\,|\,e_i^T(\,I-P_i\,)\,\mathbb{Y}_n\,|^2 + 2\,\sigma^2(X_i)\,(\,P_i\,)_{ii} - \sigma^2\,\Bigr\}$$

is an (approximately) unbiased estimator of $\mathbb{E}[\,RPW(h)\,|\,\mathbb{X}_n\,]$.

Continuing fearlessly toward zero-trace estimators, in (8.33) replace the projectors P_i by $(\,\alpha_i\,I + (1-\alpha_i)\,P_i\,)$, and choosing α_i such that

$$(\,1-\alpha_i\,)^{-1} = \bigl(\,1-(\,P_i\,)_{ii}\,\bigr)$$

then gives the zero-trace estimator

$$\text{(8.34)}\qquad GCV_{RPW}(h) = \tfrac{1}{n}\sum_{i=1}^{n}\frac{|\,e_i^T(\,I-P_i\,)\,\mathbb{Y}_n\,|^2}{|\,1-(\,P_i\,)_{ii}\,|^2} \;-\; \sigma^2(X_i)\,.$$

So now one has to worry about the denominators being well-behaved, and of course also about whether this actually works or not.

(8.35) EXERCISE. Verify the formula for $GCV_{RPW}(h)$.

(8.36) EXERCISE. If the error variance is constant, show that

$$GCV_{RPW\text{-const}}(h) = \frac{\tfrac{1}{n}\sum_{i=1}^{n}|\,e_i^T(\,I-P_i\,)\,\mathbb{Y}_n\,|^2}{\Bigl\{\,\tfrac{1}{n}\sum_{i=1}^{n}\bigl(\,1-(\,P_i\,)_{ii}\,\bigr)\,\Bigr\}^2} - \sigma^2$$

is the zero-trace estimator of

$$M_{RPW\text{-const}}(h) = \tfrac{1}{n}\sum_{i=1}^{n}\Bigl\{\,|\,e_i^T(\,I-P_i\,)\,\mathbb{Y}_n\,|^2 + 2\,\sigma^2\,(\,P_i\,)_{ii} - \sigma^2\,\Bigr\}\,.$$

EXERCISES: (8.19), (8.23), (8.24), (8.25), (8.26), (8.35), (8.36).

9. Pointwise versus local error, again

In this section, we investigate the issue of pointwise versus local errors from the asymptotic viewpoint. In particular, following FAN and GIJBELS (1995a), we show that, for local polynomial estimators of even order (odd degree), the asymptotically optimal smoothing parameters for the (global) pointwise error and the (global) local error are related by a universal fudge factor and the same holds for the GCV approximation to the global local error.

First, we discuss the optimal estimator for the "continuous" error, which attempts to choose the smoothing parameter h so as to minimize the global pointwise error

$$(9.1)\qquad GPW(h) = \| f^{nh} - f_o \|^2 = \int_0^1 | f^{nh}(x) - f_o(x) |^2 \, dx$$

and relate it to the optimal smoothing parameter for the global normalized local error

$$(9.2)\qquad GNLE(h) = \int_0^1 NQ_h(p^{nh} - f_o) \, dx \; ,$$

where $NQ_h(p^{nh} - f_o)$ is shorthand for the normalized quadratic form

$$(9.3)\quad NQ_h(p^{nh} - f_o) = \frac{\displaystyle\int_0^1 A_h(x-t) \, \big| p^{nh}(t;x) - f_o(t) \big|^2 \, \omega(t) \, dt}{\displaystyle\int_0^1 A_h(x-t) \, \omega(t) \, dt} \; .$$

Recall that $\omega(t)$ denotes the design density.

The starting point is the asymptotic behavior of the expected pointwise error. Let $x \in [h, 1-h]$. From Lemma (16.6.38), the variance, conditioned on the design, satisfies

$$(9.4)\quad \mathbb{E}[\, | f^{nh}(x) - f_h(x) |^2 \mid \mathbb{X}_n \,] = \frac{c_o \, \sigma^2(x)}{\omega(x)} \, (nh)^{-1} + o(\,(nh)^{-1}\,) \; ,$$

with c_o given by (16.6.36). Now, let

$$(9.5)\qquad \Phi(\sigma, \omega) = \int_0^1 \frac{\sigma^2(x)}{\omega(x)} \, dx \; .$$

Then, integrating both sides of (9.4) over $[0, 1]$, one gets

$$(9.6)\qquad \mathbb{E}[\, \| f^{nh} - f_h \|^2 \mid \mathbb{X}_n \,] = c_o \, \Phi(\sigma, \omega) \, (nh)^{-1} + o(\,(nh)^{-1}\,) \; .$$

This holds, e.g., if (9.4) holds uniformly over $x \in [0, 1]$, but the actual conditions suggest themselves.

For the bias, Lemma (16.6.15) and Theorem (16.7.7) tell us that

$$(9.7)\qquad f_h(x) - f_o(x) = \frac{f_o^{(m)}(x)}{m!} \, \varrho(\, \lambda(x;h)\,) \, h^m + o(\, h^m \,)$$

for a suitable function ϱ not depending on f_o or the design density, and $\lambda(x\,;h) = 0 \vee \min(\,x/h\,,\,1\,,\,(1-x)/h\,)$, the hip-roof function. Thus, the asymptotic behavior of the global bias is

$$\text{(9.8)} \qquad \|\, f^{nh} - f_o \,\|^2 = \frac{\varrho^2(1)\, h^{2m}}{(\,m\,!\,)^2}\, \|\, f_o^{(m)} \,\|^2 + o(\, h^{2m}\,)\ .$$

(One verifies that the contributions from the intervals $[\,0\,,\,h\,]$ and $[\,1-h\,,\,1\,]$ may be ignored, so that effectively $\lambda(x\,;h) = 1$ always.) Note that for odd-order local polynomials (m odd) this is misleading since in this case $\varrho(1) = 0$. For even m, we have $\varrho(1) > 0$; see § 16.9.

Thus, for even m, the asymptotic minimizer of $\mathbb{E}[\,GPW(h)\,|\,\mathbb{X}_n\,]$ is

$$\text{(9.9)} \qquad h_{GPW} = \left(\frac{c_o}{2\,m} \left\{ \frac{m\,!}{\varrho(1)} \right\}^2 \frac{\Phi(\sigma,\omega)}{\|\, f_o^{(m)} \,\|^2}\, \frac{1}{n} \right)^{1/(2m+1)} .$$

Now, let us repeat this development for the global normalized local error $GNLE(h)$. Note that, asymptotically,

$$\mathbb{E}[\,NQ_h(\,p^{nh} - f_o\,)\,|\,\mathbb{X}_n\,] \approx \{\,\omega(x)\,\}^{-1}\, \mathbb{E}[\,Q_h(\,p^{nh} - f_o\,)\,|\,\mathbb{X}_n\,]\ ,$$

where Q_h is defined in (16.3.1), and, by Corollary (16.3.9), that

$$\mathbb{E}[\,Q_h(\,p^{nh} - f_o\,)\,|\,\mathbb{X}_n\,] \approx \mathbb{E}[\,Q_h(\,\pi^{nh} - f_o\,)\,|\,\mathbb{X}_n\,]\ .$$

Then, Lemma (16.6.25) implies that, for the bias part of the local error,

$$\text{(9.10)} \qquad NQ_h(\,p_h - f_o\,) = c_1\, h^{2m}\, \frac{|\, f_o^{(m)}(x)\,|^2}{(\,m\,!\,)^2} + o(\, h^{2m}\,)\ ,$$

where c_1 is given by (16.6.24) and, upon integration,

$$\text{(9.11)} \qquad \int_0^1 NQ_h(\,p_h - f_o\,)\, dx = c_1\, h^{2m}\, \frac{\|\, f_o^{(m)} \,\|^2}{(\,m\,!\,)^2} + o(\, h^{2m}\,)\ .$$

For the variance part, we have by Lemma (16.6.40)

$$\text{(9.12)} \quad \mathbb{E}[\,NQ_h(\,p^{nh} - p_h\,)\,|\,\mathbb{X}_n\,] = \frac{\text{trace}(\,\mathbb{A}_u^{-1}\mathbb{B}_u\,)}{nh} \cdot \frac{\sigma^2(x)}{\omega(x)} + o(\,(nh)^{-1}\,)\ ,$$

so that again upon integration,

$$\text{(9.13)} \qquad \mathbb{E}\Bigl[\, \int_0^1 NQ_h(\,p^{nh} - p_h\,)\, dx \,\Bigm|\, \mathbb{X}_n \Bigr] \approx \frac{\text{trace}(\,\mathbb{A}_u^{-1}\mathbb{B}_u\,)}{nh}\, \Phi(\sigma,\omega)\ .$$

Combining (9.11) and (9.13), it follows that $\mathbb{E}[\,GNLE(h)\,|\,\mathbb{X}_n\,]$ behaves like

$$\text{(9.14)} \qquad c_1\, h^{2m}\, \|\, f_o^{(m)} \,\|^{\,2} + \Phi(\sigma,\omega)\, \text{trace}(\,\mathbb{A}_u^{-1}\,\mathbb{B}_u\,)\,(nh)^{-1}\ ,$$

and so its asymptotic minimizer is given by

$$\text{(9.15)} \qquad h_{GNLE} = \left(\frac{(\,m\,!\,)^2}{2\,m}\, \frac{\text{trace}(\,\mathbb{A}_u^{-1}\,\mathbb{B}_u\,)}{c_1}\, \frac{\Phi(\sigma,\omega)}{\|\, f_o^{(m)} \,\|^2}\, \frac{1}{n} \right)^{1/(2m+1)} .$$

Table 8.1. Values of the universal factors $F_{\text{Local}}(A,m)$ and $F_{IRSC}(A,m)$ (for the Epanechnikov kernel).

m	2	4	6	8
Local	0.9221	0.9755	0.9882	0.9931
IRSC	0.8941	0.8718	0.8819	0.8932

The point of it all is to see how these two asymptotic minimizers differ. We may relate them by

$$h_{GPW} = \mathbb{F}_{\text{Local}}(A;m)\, h_{GNLE} \ , \tag{9.16}$$

with the ugly but *universal* fudge factor

$$\mathbb{F}_{\text{Local}}(A;m) = \left(\frac{\{\,\varrho(1)\,\}^2}{c_o\, c_1} \operatorname{trace}(\,\mathbb{A}_u^{-1}\,\mathbb{B}_u\,) \right)^{1/(2m+1)} \ , \tag{9.17}$$

which only depends on known things such as the kernel A and the order m of the local polynomials. For the Epanechnikov kernel, $A(x) = \frac{3}{4}\,(1-x^2\,)_+$, the values of $F_{\text{Local}}(A,m)$ are noted in Table 8.1.

Of course, the global normalized local error is not known, but it indicates that if we approximate $GNLE(h)$ by the $GCV_{RLE}(h)$ functional of (8.21), then the same ought to apply. Denoting the asymptotic minimizer of $GCV_{RLE}(h)$ by $h_{GCV\text{-}RLE}$, one would expect that

$$h_{GPW} = \mathbb{F}_{\text{Local}}(A;m)\, h_{GCV\text{-}RLE} \ , \tag{9.18}$$

with the same fudge factor as before. This is indeed the case.

Outline of the proof of (9.18). Instead of $GCV_{RLE}(h)$, we consider the version

$$\mathfrak{GCV}_{RLE}(h) \stackrel{\text{def}}{=} \tfrac{1}{n} \sum_{i=1}^{n} \mathfrak{GCV}_{LE}(X_i;h) \ ; \tag{9.19}$$

cf. Exercise (8.24). By that construction,

$$\mathbb{E}[\,\mathfrak{GCV}_{LE}(X_i;h)\,|\,\mathbb{X}_n\,](\,1-\alpha_i\,)^2\,\|\,(\,I-Q_i\,)\,W_i\,\mathbb{Y}_o\,\|^2 + \text{stuff}\ ,$$

where

$$\text{stuff} = (\,1-\alpha_i\,)^2 \operatorname{trace}\bigl(\,(\,I-Q_i\,)\,W_i^2\,V\,\bigr) - \operatorname{trace}\bigl(\,W_i^2\,V\,\bigr)\ .$$

Here, α_i is given by way of

$$1-\alpha_i = \frac{\operatorname{trace}\bigl(\,W_i^2\,V\bigr)}{\operatorname{trace}\bigl(\,(\,I-Q_i\,)\,W_i^2\,V\bigr)}\ . \tag{9.20}$$

Note that $1 - \alpha_i > 1$. Now, an uninspiring computation yields

$$\text{(9.21)} \qquad \text{stuff} = (1 - \alpha_i)\, \text{trace}(\, Q_i W_i^2 V\,) \;,$$

so that

$$\text{(9.22)} \quad \mathbb{E}[\, \mathfrak{GCV}_{LE}(X_i; h) \mid \mathbb{X}_n \,] = \\ (1 - \alpha_i)^2 \, \| \, (I - Q_i)\, W_i \, \mathbb{Y}_o \|^2 + (1 - \alpha_i)\, \text{trace}(\, Q_i W_i^2 V\,) \;.$$

A comparison with

$$\text{(9.23)} \quad \mathbb{E}\bigl[\, \| \, W_i \, \mathbb{Y}_o - Q_i \, W_i \, \mathbb{Y}_n \, \|^2 \;\bigm|\; \mathbb{X}_n \,\bigr] = \\ \| \, (I - Q_i)\, W_i \, \mathbb{Y}_o \, \|^2 + \text{trace}(\, Q_i W_i^2 V\,)$$

shows that

$$\text{(9.24)} \qquad 1 - \alpha_i \leqslant \frac{\mathbb{E}[\, \mathfrak{GCV}_{LE}(X_i; h) \;\mid\; \mathbb{X}_n \,]}{\mathbb{E}[\, \| \, W_i \, \mathbb{Y}_o - Q_i \, W_i \, \mathbb{Y}_n \, \|^2 \;\mid\; \mathbb{X}_n \,]} \leqslant (1 - \alpha_i)^2 \;.$$

It follows that

$$\text{(9.25)} \quad 1 - \beta_o \leqslant \frac{\mathbb{E}[\, \mathfrak{GCV}_{RLE}(h) \mid \mathbb{X}_n \,]}{\mathbb{E}\left[\frac{1}{n} \sum_{i=1}^{n} \| \, W_i \, \mathbb{Y}_o - Q_i \, W_i \, \mathbb{Y}_n \, \|^2 \;\middle|\; \mathbb{X}_n \right]} \leqslant (1 - \beta_1)^2 \;,$$

where $\beta_o = \max_{1 \leqslant i \leqslant n} \; \alpha_i$ and $\beta_1 = \min_{1 \leqslant i \leqslant n} \; \alpha_i$. Now, for the moment, assume that

$$\text{(9.26)} \qquad \max_{1 \leqslant i \leqslant n} |\alpha_i| = \mathcal{O}(\,(nh)^{-1}\,) \;.$$

Then, since the denominator in (9.24) is a pretty good approximation of the global normalized local error $GNLE(h)$, which has the asymptotic behavior (9.14), the inequalities (9.25) show that the minimizers of the asymptotic expressions for $\mathbb{E}[\, \mathfrak{GCV}_{RLE}(h) \mid \mathbb{X}_n \,]$ and $\mathbb{E}[\, GNLE(h) \mid \mathbb{X}_n \,]$ are the same. In self-explanatory (?) notation,

$$\text{(9.27)} \qquad h_{\mathfrak{GCV}\text{-}RLE} = h_{GNLE} \;.$$

Together with (9.16), this shows (9.18), if one may assume (we will) that

$$\text{(9.28)} \qquad h_{\mathfrak{GCV}\text{-}RLE} = h_{GCV\text{-}RLE} \;.$$

We still need to show the asymptotic bound (9.26) on the α_i. Of course, $\text{trace}(\, Q_i \, W_i^2 \, V\,)$ is the variance part of $\mathbb{E}[\, \| \, W_i \, \mathbb{Y}_o - Q_i \, W_i \, \mathbb{Y}_n \, \|^2 \mid \mathbb{X}_n \,]$, see (9.23), so by (9.12)

$$\text{(9.29)} \qquad \text{trace}(\, Q_i \, W_i^2 \, V\,) = \frac{\text{trace}(\, \mathbb{A}_u^{-1} \, \mathbb{B}_u \,)}{nh} \cdot \frac{\sigma^2(X_i)}{\omega(X_i)} + o(\,(nh)^{-1}\,) \;.$$

Now, since $\text{trace}(\, W_i^2 \, V\,) \approx \sigma^2(X_i)$, the definition (9.20) implies that

$$\alpha_i = \frac{\text{trace}(\, \mathbb{A}_u^{-1} \, \mathbb{B}_u \,)}{nh} \cdot \frac{\sigma^2(X_i)}{\omega(X_i)} + o(\,(nh)^{-1}\,) \;,$$

and (9.26) holds. Q.e.d.

(9.30) EXERCISE. Fill in the precise details of the proof of (9.27) above and show (9.28). [Actually, all of this may be an unrewarding exercise in view of the following remark.]

(9.31) REMARK. In all of the above, we cavalierly ignored the difference between the minimizer of the asymptotic expected value of the global pointwise error and the asymptotics of the minimizer of the global pointwise error, and likewise for the normalized local error. Theoretically, in view of The Missing Theorem (12.7.33), this seems permissible, although a careful analysis is in order (but not by the authors). The whole development is reminiscent of the situation for smoothing splines, where CRAVEN and WAHBA (1979) show that the minimizer of the expected GCV score behaves in the right way.

(9.32) EXERCISE. One might view $RPW(h)$ of (8.7) as a discretization of the weighted global pointwise error, accounting for the design density,

$$WGPW_{DESIGN}(h) = \| f^{nh} - f_o \|^2_{L^2(\omega)} = \int_0^1 | f^{nh}(x) - f_o(x) |^2 \, \omega(x) \, dx \; .$$

After all, one would expect f^{nh} to be much more accurate in regions with a high density of design points and weight the pointwise error accordingly. Repeat the development above for this measure of the error. In particular, show that the fudge factor $F_{\text{Local}}(A, m)$ stays the same.

The previous development follows FAN and GIJBELS (1995a), with one or two differences. Instead of the statistic $GCV_{RLE}(x\,;h)$, see (8.22), they use the rather perplexing RSC statistic, which in our notation reads

$$RSC(x\,;h) = \frac{(\,1 + m\,(nh)^{-1}\,c_o\,)\, \|\,(\,I - Q\,)W\mathbb{Y}_n\,\|^2}{\text{trace}(\,(\,I - Q\,)W^2\,)} \,, \tag{9.33}$$

where c_o is given by (16.6.36) (actually, with $\mathbb{A}_u^{-1}\mathbb{B}_u\mathbb{A}_u^{-1}$ replaced by $\mathbb{A}^{-1}\mathbb{B}\mathbb{A}^{-1}$.) Also, they do an accurate approximation of the integral

$$IRSC(h) = \int_0^1 RSC(x;h)\,dx$$

instead of using a random sum. Proceeding just as we do (well, actually, it is the other way around), they relate the asymptotic minimizer of the $IRSC(h)$ statistic to the minimizer of the global pointwise error by way of the universal fudge factor, $F_{IRSC}(A, m)$. Some of the values are reproduced in Table 8.1. Perplexing as the choice of the IRSC functional is, there is more to it than meets the eye, apparently; see § 23.3.

(9.34) PLUG-IN METHODS. We briefly describe the plug-in method that is suggested by (9.9) and (9.18). Obtain a pilot estimator of the smoothing parameter by minimizing $GCV_{RLE}(h)$ over h. Then, using this pilot estimator, estimate $f_o^{(m)}$ and $\| f_o^{(m)} \|^2$. Likewise, this pilot estimator can be used to estimate the design density by the kernel estimator. The variance $\sigma^2(X_i)$ can be taken as $| f^{nh}(X_i) - Y_i |^2$ (or possibly as a smoothed version), and $\Phi(\sigma, \omega)$ is then estimated, e.g., by way of

$$\frac{1}{n} \sum_{i=1}^{n} \sigma^2(X_i)/\omega(X_i) .$$

By plugging these estimators into (9.9), the asymptotically optimal smoothing parameter h_{GPW} is determined.

Since asymptotically this is equivalent to the pilot estimator, the difference only shows up for (small) sample problems. The difference appears to be that the resulting smoothing parameter has a smaller variance than the pilot estimator of the smoothing parameter, see FAN and GIJBELS (1995a), but it is not obvious that this implies smaller $L^2(0,1)$ errors, even asymptotically. We shall not belabor the point.

EXERCISES: (9.30), (9.32).

10. Additional notes and comments

Ad § 1: The point of view that one should choose an optimality criterion and then design a selection procedure to come as close to optimality as (practically) possible is not universally accepted. By way of example, the literature is fraught with complaints that one popular selection procedure (GCV) results in estimators that are too rough (undersmoothed). The cure is to design an optimality criterion that incorporates smoothness, e.g., along the lines of (1.8), and then design a procedure accordingly.

For selecting the smoothing parameter for sieved estimators, we mention POLYAK and TSYBAKOV (1991).

Ad §§ 3 and 4: The leave-one-out idea has a venerable history in statistics. The generalized cross-validation procedure originated with GOLUB, HEATH, and WAHBA (1979), who credit ALLEN (1974) for the leave-one-out idea in this context. In the first edition already, EUBANK (1999) observes that "generalized" cross-validation is "plain" cross-validation in a favored coordinate system, which makes it in fact coordinate-free cross-validation.

There exist several modifications of the GCV functional that are aimed at avoiding the undersmoothing of GCV alluded to before. First, write

$$n^2 GCV(h) = \frac{\| y_n - R_{nh} y_n \|^2}{(1-t)^2} , \quad \text{where} \quad t = \tfrac{1}{n} \operatorname{trace}(R_{nh}) .$$

Now, t is presumed to be small, so the approximations

$$(1-t)^{-2} \approx (1-2t)^{-1} \approx (1+t)(1-t)^{-1} \approx 1+2t$$

are valid. Then we arrive at the modifications of RICE (1984b), AKAIKE (1970), SHIBATA (1981), and HURVICH, SIMONOFF, and TSAI (1995),

$$\frac{\| y_n - R_{nh}\, y_n \|^2}{1 - \frac{2}{n}\,\mathrm{trace}(R_{nh})}\,, \tag{10.1}$$

$$\frac{1 + \frac{1}{n}\,\mathrm{trace}(R_{nh})}{1 - \frac{1}{n}\,\mathrm{trace}(R_{nh})}\, \| y_n - R_{nh}\, y_n \|^2\,, \tag{10.2}$$

$$\{ 1 + \tfrac{2}{n}\,\mathrm{trace}(R_{nh}) \}\, \| y_n - R_{nh}\, y_n \|^2\,, \tag{10.3}$$

$$\log\big(\tfrac{1}{n} \| y_n - R_{nh}\, y_n \|^2 \big) + 1 + \frac{2\,\mathrm{trace}(R_{nh}) + 2}{n - \mathrm{trace}(R_{nh}) - 2}\,. \tag{10.4}$$

The second one goes by the name of Akaike's final prediction error (FPE).

For the case of orthogonal projectors, SHIBATA (1981) observed that, asymptotically, all of these variations are asymptotically equivalent. This also seems to be the case for our spline estimators if we believe The Missing Theorem (12.7.33) since $\frac{1}{n}\,\mathrm{trace}(R_{nh}) \asymp (nh)^{-1}$ for $nh \to \infty$. So, in the author's opinion, as a cure for undersmoothing this must be deemed doubtful.

Ad § 5: The criterion (5.16) is an attempt at an optimality criterion for more visually pleasing estimators that fit in well with standard approaches. MARRON and TSYBAKOV (1995) explore criteria along the lines of the minimum Skorokhod distance (a.k.a. dog-leash metric) between the graphs of f_o and the estimator.

Ad § 6: Although implicit in the work of AKAIKE (1973), the motivation of the AIC setup as that of estimating the model (1.1)–(1.4) is due to HURVICH and TSAI (1989). Having immersed ourselves first into the minimum L^2 error criterion, the unauthorized version of AIC as given here seems quite natural. The authorized version of the AIC for nonparametric regression is from HURVICH, SIMONOFF, and TSAI (1995), who use results of JONES (1986, 1987) for the computation of "Recip" and "Quot" in (6.20)–(6.21). The computations (6.24)–(6.25) were first shown by SUGIURA (1978) using χ^2 approximations (which are exact for the case of orthogonal projectors). See also HURVICH and TSAI (1989).

The observation regarding the exact equivalence of the GCV and the (unauthorized) AIC procedure based on (6.11) is new. For the orthogonal projection case, the asymptotic equivalence of the AIC procedure based on (6.26) and the GCV procedure was observed by SHIBATA (1981).

The pros and cons of some of the variance estimators are summarized by CARTER and EAGLESON (1992). BUCKLEY, EAGLESON, and SILVERMAN (1988) has a nice discussion on the construction of estimators of the variance that are optimal with respect to mean squared error $\mathbb{E}[\,|\,\widehat{\sigma}^2 - \sigma^2\,|^2\,]$.

In this chapter, interest is in the use of these estimators for the estimation of the error $\mathbb{E}[\, \| y_o - R_{nh}\, y_n \|^2 \,]$, which is not necessarily the same thing. The origins of the various estimators are listed below.

$\sigma_{nh,1}$: Used in AIC by SUGIURA (1978), HURVICH and TSAI (1989);

$\sigma_{nh,2}$: BUCKLEY, EAGLESON, and SILVERMAN (1988);

$\sigma_{nh,3}$: See Exercise (6.14);

$\sigma_{nh,4}$: ANSLEY, KOHN, and THARM (1990);

$\sigma_{nh,5}$: WAHBA (1983);

$\sigma_{nh,6}$: Rather strange, but the motivation is there.

See also the estimator of RICE (1986b) in (5.15) and the estimator of Exercise (2.17), implicit in LI (1986).

We have not mentioned the actual AIC method of AKAIKE (1973): For a sample $X_1, X_2, \cdots, X_n$ of the random variable X having pdf $f(x;\theta)$, indexed by a parameter θ belonging to a sieve of finite-dimensional subspaces (nested here for convenience) $\Theta_1 \subset \Theta_2 \subset \cdots \subset \Theta_m \subset \cdots$, the optimal "dimension" m is defined as the solution to

$$\text{(10.5)} \qquad \begin{aligned} &\text{minimize} && -2 \sum_{i=1}^{n} \log f(X_i;\widehat{\theta}_m) + 2 \dim(\Theta_m) \\ &\text{subject to} && m \geqslant 1 , \end{aligned}$$

where $\widehat{\theta}_m$ is the maximum likelihood estimator of θ assuming the model $\theta \in \Theta_m$. This arises after several (Taylor series) approximations that are not quite valid in nonparametric regression when m can be large. See HURVICH and TSAI (1989) and HURVICH, SIMONOFF, and TSAI (1995). STONE (1977, 1979) and NISHII (1984) prove the equivalence of AIC and (plain) cross-validation for model selection in parametric regression. Here, cross-validation refers to replacing $\widehat{\theta}_m$ in $\log f(X_i;\widehat{\theta}_m)$ in (10.1) by the maximum likelihood estimator with the datum X_i left out.

Ad § 7: For local polynomial estimation, the decision to choose h in a pointwise fashion seems easy to make, but doing it piecewise seems more "robust", as advocated by FAN and GIJBELS (1995a). Nevertheless, FAN and GIJBELS (1995b) suggest choosing h *and* the order m in a strictly pointwise fashion. Their method may be interpreted as a version of the pointwise Mallows estimator of the error, with estimated variance. SEIFERT and GASSER (1996, 2000) observe that the pointwise asymptotically optimal choice of the smoothing parameter is likely to fail since there will always be points for which there are too few design points used in the local estimation. For the local linear estimator in the form of Exercise (8.27), they suggest replacing S^2 by $S^2 + \alpha^2$ for a suitable parameter $\alpha > 0$ and observe that the particular choice of α does not seem to matter much.

Ad § 8: Fan, Gijbels, Hu, and Huang (1996) determine the leading terms of the asymptotic expression of $\mathbb{E}\big[\, | f^{nh}(x) - f_o(x) |^2 \,\big|\, \mathbb{X}_n \big]$ by way of a detailed study of the normal equations for the least-squares problem (1.7) in the implementation (16.11.1).

19

Computing Nonparametric Estimators

1. Introduction

In this chapter, we discuss the computation of the various nonparametric regression estimators encountered in the previous chapters (with kernel estimators being superseded by local polynomials).

For the most part, the estimators under consideration are solutions to quadratic minimization problems and the discussion centers on the construction of the normal equations. In particular, this applies to smoothing spline, polynomial, and local polynomial estimators. For smoothing splines, we discuss the classical approach via spline interpolation. The resulting systems of linear equations are easily solved by standard numerical methods, but we should point out problems with the numerical stability when design points are very close together. (Exact replications are not an issue.) The way around the numerical instability by means of Kalman filtering is discussed in the next chapter. For polynomial and local polynomial estimation, the use of discretely orthogonal polynomials is advocated.

The "other" estimation problems, total-variation penalized least-squares and roughness penalized least-absolute-deviations, are more troublesome. We discuss their formulation as finite-dimensional convex minimization problems and show that their dual formulations are positive-definite quadratic minimization problems with box constraints. We develop efficient "active constraint set" methods for their solution.

2. Cubic splines

In this section, we discuss the computation of the least-squares cubic spline estimator. It turns out to be useful to start with cubic spline interpolation so as to get a representation of splines in terms of their function values.

P.P.B. Eggermont, V.N. LaRiccia, *Maximum Penalized Likelihood Estimation*,
Springer Series in Statistics, DOI 10.1007/b12285_8,

We begin with the definition of a spline function on a collection of adjacent intervals of the real line. The intervals are determined by the knots

$$(2.1) \qquad t_1 < t_2 < \cdots < t_n \, .$$

(2.2) DEFINITION. *Let $s \geqslant 2$ be an integer. A function f is a polynomial spline of order s (or degree $s-1$) on the partition $t_1 < t_2 < \cdots < t_n$ if it satisfies*

(a) $f \in C^{s-2}(t_1, t_n)$ *and*

(b) f *is a polynomial of degree* $\leqslant s-1$ *on each interval* (t_i, t_{i+1}) *for* $i = 1, 2, \cdots, n-1$.

If, in addition, the order $s = 2m$ is even and f satisfies

(c) $f^{(k)}(t_1) = f^{(k)}(t_n) = 0 \, , \quad k = m, m+1, \cdots, 2m-1 \, ,$

then f is called a natural polynomial spline.

Practically, interest is in the choice $s = 4$, in which case one speaks of cubic splines. We also have an interest in quintic and septic splines. The case $s = 2$ is much less interesting: A spline of order 2 is just a piecewise linear, continuous function (a polygonal function). For $s = 1$, the definition above does not work.

(2.3) EXERCISE. Let $s \geqslant 3$. Show that the derivative of a spline of order s is a spline of order $s-1$. (Accordingly, we may define a spline of order 1 as the derivative of a spline of order 2. In other words, a spline of order 1 is a piecewise constant step function.)

We next introduce the spline interpolation problem, but only for splines of even order. (Odd-order interpolating splines are odd for more than one reason; see below.) Roughly speaking, the issue is to determine a spline function f of predetermined order $2m$ such that

$$(2.4) \qquad f(x_{in}) = y_{in} \, , \quad i = 1, 2, \cdots, n \, .$$

Here, the y_{in} are the data and the x_{in} are the design points, which are assumed to satisfy

$$(2.5) \qquad x_{1,n} < x_{2,n} < \cdots < x_{n,n} \, .$$

It is customary to take the design points as the knots for the spline interpolator, but it is useful to introduce two additional knots, $x_{o,n}$ and $x_{n+1,n}$, chosen more or less arbitrarily such that

$$(2.6) \qquad x_{o,n} < x_{1,n} \quad , \quad x_{n,n} < x_{n+1,n} \, .$$

At these additional points, there are no data.

(2.7) THE NATURAL SPLINE INTERPOLATION PROBLEM. *Let $m \geqslant 1$ be an integer. The problem is to determine a spline of even order $2m$ on the partition $x_{o,n} < x_{1,n} < x_{2,n} < \cdots < x_{n,n} < x_{n+1,n}$ such that*

$$f(x_{in}) = y_{in} \;, \quad i = 1, 2, \cdots, n \;, \tag{2.8}$$

and

$$f^{(k)}(x_{o,n}) = f^{(k)}(x_{n+1,n}) = 0 \;, \quad k = m, m+1, \cdots, 2m-1 \;. \tag{2.9}$$

The conditions (2.8) and (2.9) are the interpolation and boundary conditions, respectively.

It is possible to define the natural interpolation problem without reference to the arbitrary design points $x_{o,n}$ and $x_{n+1,n}$ as follows.

(2.10) ALTERNATIVE FORMULATION OF (2.7). Let $m \geqslant 1$. Determine a spline of order $2m$ on the partition $x_{1,n} < x_{2,n} < \cdots < x_{n,n}$ such that

$$f(x_{in}) = y_{in}, \quad i = 1, 2, \cdots, n \;, \tag{2.11}$$

and

$$f^{(k)}(x_{1,n}) = f^{(k)}(x_{n,n}) = 0 \;, \quad k = m, m+1, \cdots, 2m-1 \;. \tag{2.12}$$

Note that in the alternative formulation there are still $2m$ boundary conditions: The two interpolation conditions in (2.11),

$$f(x_{1,n}) = y_{1,n} \;, \quad f(x_{n,n}) = y_{n,n} \;, \tag{2.13}$$

are also boundary conditions. The advantage of the formulation (2.7) is that all of the design points are treated the same.

(2.14) EXERCISE. Show that the solutions of (2.7) and the alternative formulation (2.10) coincide on the interval $(x_{1,n}, x_{n,n})$.

It turns out that the solution of the spline interpolation problem exists and is unique for $n \geqslant 2m$. While this may be shown along the lines of § 13.3, here we *compute* the solution for some relevant values of m.

(2.15) COMPUTING NATURAL CUBIC INTERPOLATION SPLINES. Our starting point is the formulation (2.7). The obvious approach is to represent the cubic spline on $(x_{in}, x_{i+1,n})$ as

$$f(t) = f_i(t) = a_i + b_i(x_{i+1,n} - t) + c_i(x_{i+1,n} - t)^2 + d_i(x_{i+1,n} - t)^3$$

and to use the various conditions of the interpolation problem to set linear equations for the unknown coefficients; see, e.g., REINSCH (1967).

The following is a slight variation on this theme. By Exercise (2.3), the function f'' is a spline of order 2; i.e., it is a polygonal function with knots at the x_{in}. Abbreviating

(2.16) $$a_i = f''(x_{in}) \,, \quad i = 0, 1, \cdots, n+1 \,,$$

we may represent f'' on $(x_{in}, x_{i+1,n})$ as

(2.17) $$f''(t) = f_i''(t) = a_i\, r_i\,(x_{i+1,n} - t) + a_{i+1}\, r_i\,(t - x_{in}) \,,$$

where $r_i = (x_{i+1,n} - x_{in})^{-1}$. In this representation, f'' is continuous for any choice of $a_0, a_1, \cdots, a_{n+1}$.

The natural boundary conditions imply that $a_0 = a_{n+1} = 0$ as well as $a_0 - a_1 = a_n - a_{n+1} = 0$ or

(2.18) $$a_0 = a_1 = 0 \quad , \quad a_n = a_{n+1} = 0 \,.$$

Twice integrating (2.17) gives

(2.19) $$f(t) = f_i(t) = \tfrac{1}{6}\, a_i\, r_i\,(x_{i+1,n} - t)^3 + \tfrac{1}{6}\, a_{i+1}\, r_i\,(t - x_{in})^3 + p_i(t)$$

for polynomials p_i of degree $\leqslant 1$ yet to be determined. It seems reasonable to represent the p_i as in (2.17),

(2.20) $$p_i(t) = c_i\, r_i\,(x_{i+1,n} - t) + c_{i+1}\, r_i\,(t - x_{in}) \,.$$

Then, $\frac{1}{6}\, a_i\, r_i^{-2} + c_i = f_i(x_{in})$, and (2.19) may be written as

(2.21) $$f_i(t) = \; b_i\, r_i\,(x_{i+1,n} - t) + b_{i+1}\, r_i\,(t - x_{in}) + \tfrac{1}{6}\, r_i^{-2}\, \{ a_i\, P(r_i\,(x_{i+1,n} - t)) + a_{i+1}\, P(\, r_i\,(t - x_{in}\,)) \} \,,$$

where

(2.22) $$P(t) = t^3 - t \,.$$

Since $P(0) = P(1) = 0$, the b_i may be interpreted as the function values of the spline f,

(2.23) $$b_i = f(x_{in}) \,, \quad i = 0, 1, \cdots, n+1 \,.$$

Again, in this form, f is continuous for any choice of the a_i and the b_i. Of course, $b_i = y_{in}$ for $i = 1, 2, \cdots, n$, so that only b_0 and b_{n+1} are unknown.

We must still ensure that f' is continuous. This requires that the f_i' fit smoothly together,

(2.24) $$f_{i-1}'(x_{in}) = f_i'(x_{in}) \,, \quad i = 1, 2, \cdots, n \,,$$

which yields the equations

(2.25) $$\tfrac{1}{6}\, a_{i-1}\, \varrho_{i-1} + \tfrac{1}{3}\, a_i\,(\varrho_{i-1} + \varrho_i) + \tfrac{1}{6}\, a_{i+1}\, \varrho_i = r_{i-1}\, b_{i-1} - (r_{i-1} + r_i)\, b_i + r_i\, b_{i+1}$$

for $i = 1, 2, \cdots, n$, where

(2.26) $$\varrho_i = r_i^{-1} = x_{i+1,n} - x_{in} \,.$$

The above is the defining system of equations for the interpolating spline. Note that there are n equations and, in view of (2.18), there are n unknown coefficients: Besides the $n-2$ unknown a_i, there are also the unknown b_0 and b_{n+1}.

We now show that (2.25) has a unique solution. Let

$$a = (a_2, a_3, \cdots, a_{n-1})^T, \quad b = (b_1, b_2, \cdots, b_n)^T. \tag{2.27}$$

The equations (2.25), for $i = 2, 3, \cdots, n-1$, result in the linear system

$$T\,a = S\,b, \tag{2.28}$$

where $T \in \mathbb{R}^{(n-2)\times(n-2)}$ and $S \in \mathbb{R}^{(n-2)\times n}$ are tridiagonal matrices. The matrix T is defined by

$$T = \begin{bmatrix} T_{2,2} & T_{2,3} & 0 & \cdots & \cdots & \cdots & 0 \\ T_{3,2} & T_{3,3} & T_{3,4} & 0 & \cdots & \cdots & 0 \\ 0 & T_{4,3} & T_{4,4} & T_{4,5} & 0 & \cdots & 0 \\ \vdots & \ddots & \ddots & \ddots & \ddots & & \vdots \\ \vdots & & \ddots & \ddots & \ddots & \ddots & \vdots \\ 0 & \cdots & \cdots & 0 & T_{n-2,n-3} & T_{n-2,n-2} & T_{n-2,n-1} \\ 0 & \cdots & \cdots & \cdots & 0 & T_{n-1,n-2} & T_{n-1,n-1} \end{bmatrix}. \tag{2.29}$$

For S, the same indexing scheme applies, but the first and last columns

$$[\,S_{2,1} \quad 0 \quad \cdots \quad 0\,]^T \quad \text{and} \quad [\,0 \quad \cdots \quad 0 \quad S_{n-1,n}\,]^T$$

must be added. Then,

$$T_{i,j} = \begin{cases} \frac{1}{6}\,\varrho_{i-1}, & j = i-1, \quad 3 \leqslant i \leqslant n-1, \\ \frac{1}{3}\,(\varrho_{i-1} + \varrho_i), & j = i, \quad 2 \leqslant i \leqslant n-1, \\ \frac{1}{6}\,\varrho_i, & j = i+1, \quad 2 \leqslant i \leqslant n-2, \end{cases} \tag{2.30}$$

and $= 0$ otherwise, and for $i = 2, 3, \cdots, n-1$,

$$S_{i,j} = \begin{cases} r_{i-1}, & j = i-1, \\ -r_{i-1} - r_i, & j = i, \\ r_i, & j = i+1, \end{cases} \tag{2.31}$$

and $= 0$ otherwise. Now, since the matrix T is strongly diagonally dominant and hence nonsingular, see, e.g., Kincaid and Cheney (1991), it follows that the solution of the system (2.28) exists and is unique. The remaining two equations in (2.25) (for $i = 1$ and $i = n$) serve to determine b_0 and b_n.

(2.32) THE VARIATIONAL FORMULATION OF SPLINE INTERPOLATION. As a bridge to cubic smoothing splines, we take a brief look at the problem

$$
\begin{aligned}
&\text{minimize} \quad \int_{x_{0,n}}^{x_{n+1,n}} |\, f^{(m)}(t)\,|^2\, dt \\
(2.33) \qquad &\text{subject to} \quad f \in W^{m,2}(x_{0,n}\,,\, x_{n+1,n})\;, \\
&\qquad\qquad f(x_{in}) = y_{in}\;, \quad i = 1, 2, \cdots, n\;.
\end{aligned}
$$

The surprise is that this is just the spline interpolation problem. Of course, these two formulations are so heavily intertwined that it is hard to say which problem came first; see HOLLADAY (1957), SCHOENBERG (1973), AHLBERG, NILSON, and WALSH (1967), and DE BOOR (1978).

(2.34) THEOREM. *The solutions of the natural spline interpolation problem (2.7) and the problem (2.33) coincide.*

PROOF. The proof is essentially computational. The Euler equations for (2.33), see § 10.2, are

$$
\begin{aligned}
&(-1)^m f^{(2m)} + \sum_{i=1}^{n} \lambda_i\, \delta(\,\cdot - x_{in}) = 0 \quad \text{on } (\,x_{0,n}\,,\, x_{n+1,n}\,)\;, \\
(2.35) \qquad &f(x_{in}) = y_{in}\;, \quad i = 1, 2, \cdots, n\;, \\
&f^{(k)}(x_{0,n}) = f^{(k)}(x_{n+1,n}) = 0\;, \quad k = m, m+1, \cdots, 2m-1\;.
\end{aligned}
$$

Here, the λ_i are the Lagrange multipliers and $\delta(\,\cdot\,)$ is the Dirac delta function (or the point mass at 0). See Volume I, § 10.6.

Now, proceed to solve the Euler equations. The differential equation says that f is a spline of order $2m$. The interpolation and boundary conditions make f the natural interpolating spline. Q.e.d.

Note that the scheme (2.33) does not apply to odd-order splines, so they are odd indeed. An alternative proof of Theorem (2.34) is given in the next exercise, which amounts to *verifying* that the natural interpolating spline satisfies the necessary and sufficient conditions for a solution of (2.33).

(2.36) EXERCISE. (a) Verify that the solution f_s of the natural spline interpolation problem (2.7) satisfies

$$
\begin{aligned}
&\int_{x_{0,n}}^{x_{n+1,n}} f_s^{(m)}(t)\,\{\,\varphi^{(m)}(t) - f^{(m)}(t)\,\}\, dt = 0 \\
(2.37) \qquad &\text{for all } \varphi \in W^{m,2}(x_{0,n}\,,\, x_{n+1,n}) \text{ that satisfy the} \\
&\text{interpolation conditions (2.8)}\,.
\end{aligned}
$$

(b) Conclude that f_s solves (2.33). In other words, (2.7) and (2.33) are the "same". [Hint: Convexity.]

To finish this section, we now formulate the periodic cubic spline problem, but leave its treatment as an exercise.

(2.38) THE PERIODIC SPLINE INTERPOLATION PROBLEM. Let $m \geqslant 1$.
(a) Determine a spline of order $2m$ on the partition

$$x_{1,n} < x_{2,n} < \cdots < x_{n,n} < x_{n+1,n}$$

such that $f(x_{in}) = y_{in}$, $i = 1, 2, \cdots, n$, and

$$f^{(k)}(x_{1,n}) = f^{(k)}(x_{n+1,n}) , \quad k = 0, 1, \cdots, 2m-2 .$$

The solution f is a periodic spline of order $2m$.
(b) Determine the variational formulation of the periodic cubic spline interpolation problem.

EXERCISES: (2.4), (2.14), (2.36), (2.38).

3. Cubic smoothing splines

Here we discuss the computation of the least-squares cubic spline estimator. It makes heavy use of the representation of cubic splines in terms of its function values, derived in § 2.

The task at hand is to solve the problem (13.1.7). In (13.1.7), we assume that the design points $x_{1,n}$ and $x_{n,n}$ are boundary points, which causes complications. So, as in § 2, we add extra design points without data and reformulate (13.1.7) as

$$\text{(3.1)} \qquad \begin{aligned} &\text{minimize} \quad \tfrac{1}{n} \sum_{i=1}^{n} | f(x_{in}) - y_{in} |^2 + h^4 \int_{x_{0,n}}^{x_{n+1,n}} | f''(t) |^2 \, dt \\ &\text{subject to} \quad f \in W^{m,2}(x_{0,n} \, , \, x_{n+1,n}) \, . \end{aligned}$$

Before going on, we need to show that this is equivalent to the original problem (13.1.7).

(3.2) EXERCISE. Let f be the solution of (3.1), and define $\varphi(t) = f(t)$, $x_{1,n} \leqslant t \leqslant x_{n,n}$. Show that φ is the solution of (13.1.7) for $m = 2$.

Solving (3.1) again starts with the Euler equations, which read as

$$\text{(3.3)} \qquad \alpha^2 f^{(4)} + \sum_{i=1}^{n} (f(x_{in}) - y_{in}) \, \delta(\cdot - x_{in}) = 0 \quad \text{on } (x_{0,n} \, , \, x_{n+1,n}) \, ,$$

$$\text{(3.4)} \qquad f^{(k)}(x_{0,n}) = f^{(k)}(x_{n+1,n}) = 0 \, , \quad k = 2, 3 \, ,$$

where $\alpha^2 = nh^4$ and $\delta(\cdot)$ the Dirac delta function. The differential equation (3.3) again says that f is a cubic spline function. The boundary conditions (3.4) imply that it is a *natural* cubic spline. So, it may be represented as in (2.21) by

$$(3.5) \qquad f(t) = f_i(t)\,, \quad x_{in} \leqslant t \leqslant x_{i+1,n}\,, \quad i = 1, 2, \cdots, n\,,$$

where

$$(3.6) \quad f_i(t) = b_i\, r_i\,(x_{i+1,n} - t) + b_{i+1}\, r_i\,(t - x_{in}) + \frac{1}{6}\, r_i^{-2}\,\{\, a_i\, P(\, r_i\,(x_{i+1,n} - t)) + a_{i+1}\, P(\, r_i\,(t - x_{in}\,))\}\,,$$

in which $P(t) = t^3 - t$, and the a_i and b_i may be interpreted as

$$(3.7) \qquad a_i = f''(x_{in}) \;,\quad b_i = f(x_{in})\,, \quad i = 1, 2, \cdots, n\,.$$

Also, recall that

$$(3.8) \qquad a_0 = a_1 = 0\,, \quad a_n = a_{n+1} = 0\,,$$

and that the vectors a and b,

$$(3.9) \qquad a = (a_2, a_3, \cdots, a_{n-1})^T\,, \quad b = (b_1, b_2, \cdots, b_n)^T\,,$$

are connected by the equations

$$(3.10) \qquad Ta = Sb\,,$$

with the matrices S and T as in (2.30)–(2.31).

Of course, the solution of (3.1) is not just any natural cubic spline. The differential equation (3.3) says that $f^{(4)}$ is a sum of delta functions with the appropriate weights. Hence, $f^{(3)}$ has jumps at the x_{in} with these same weights, or

$$(3.11) \quad \alpha^2\,\{\, f_i^{(3)}(x_{in}) - f_{i-1}^{(3)}(x_{in})\,\} + f(x_{in}) = y_{in}\,, \quad i = 1, 2, \cdots, n\,.$$

Carefully checking, one obtains that (3.11) is equivalent to

$$(3.12) \qquad \alpha^2\, S^T a + b = y_n$$

for the data vector y_n.

The two equations (3.10) and (3.12) uniquely determine the least-squares spline. Since T is positive-definite, see Exercise (3.14) below, then (3.10) implies that $a = T^{-1} S\, b$ and then (3.12) gives the equation for b,

$$(3.13) \qquad (\, I + \alpha^2\, S^T\, T^{-1}\, S\,)\, b = y_n\,.$$

The argument may be repeated: Since $S^T\, T^{-1}\, S$ is semi-positive-definite, then (3.13) has a unique solution, and so $a = T^{-1} S\, b$ is uniquely determined as well. Thus, the cubic spline exists and is unique.

(3.14) EXERCISE. (a) Verify that, for f given by (3.5),

$$\langle\, a\,, T\, a\,\rangle = \int_{x_{1,n}}^{x_{n,n}} |\, f''(\,t\,)\,|^2\, dt\,.$$

(b) Conclude that T is positive-definite.

(3.15) NUMERICAL ASPECTS. Regarding the numerical solution of (3.13), the following seems to be the authorized version. First, use Cholesky factorization to compute a bidiagonal, lower-triangular matrix $L \in \mathbb{R}^{(n-2)\times(n-2)}$ such that $T = L\,L^T$. Next, compute the QR factorization of the matrix

$$\begin{bmatrix} L^T \\ --- \\ \alpha\,S^T \end{bmatrix} = Q\,R\ ,$$

with $Q \in \mathbb{R}^{(2n-2)\times(n-2)}$ having orthogonal columns and $R \in \mathbb{R}^{(n-2)\times(n-2)}$ upper-triangular and nonsingular, and solve (3.10) in the form

$$R^T R\,a = S\,y$$

using back substitution. See KINCAID and CHENEY (1991) for some of the details.

(3.16) EXERCISE. (a) Consider the metroscedastic cubic spline smoothing problem with known variances

$$\text{minimize} \quad \tfrac{1}{n}\sum_{i=1}^{n} \frac{|\,f(x_{in}) - y_{in}\,|^2}{\sigma_{in}^2} + h^4\,\|\,f''\,\|^2$$

(over all $f \in W^{2,2}(0,1)$). Let the vectors a and b be as in (3.7), and let the matrices T and S be as in (2.30)–(2.31). Define the diagonal $n \times n$ matrix V by $V_{ii} = \sigma_{in}$, $i = 1, 2, \cdots, n$. Show that a and b satisfy

$$T\,a = S\,b \quad , \quad S^T a + V^{-2}\,b = V^{-2}\,y_n\ .$$

(b) Consider the smoothing spline problem (3.1) with homoscedastic noise. Assume that there are independent replications; i.e., there are k distinct design points $t_{j,n}$, $j = 1, 2, \cdots, k$, with multiplicities r_{jn} ,

$$r_{jn} = \#\{\, i : x_{in} = t_{jn}\,\}\ .$$

Show that (3.1) may be formulated similar to (a), and work out the computational details.

This concludes the discussion of cubic smoothing splines. We briefly mention the periodic problem as well.

(3.17) THE PERIODIC SMOOTHING SPLINE PROBLEM. This reads as

$$\begin{aligned} &\text{minimize} && \tfrac{1}{n}\sum_{i=1}^{n} |\,f(x_{in}) - y_{in}\,|^2 + h^4 \int_{x_{1,n}}^{x'_{n+1,n}} |\,f''(\,t\,)\,|^2\,dt \\ &\text{subject to} && f \in W^{m,2}(\,x_{0,n}\,,\,x'_{n+1,n}\,) \\ & && f^{(k)}(x_{1,n}) = f^{(k)}(x'_{n+1,n})\ , \quad k = 0, 1, 2, \cdots, 2m-1\ . \end{aligned}$$

(3.18) EXERCISE. Show that the solution to the problem (3.17) may be computed from

$$(3.19) \qquad T_{\text{per}} a = S_{\text{per}} b \quad , \quad \alpha^2 \, S_{\text{per}}^T a + b = y_n \ ,$$

but with T, S, a, and b all different from the natural case. First,

$$(3.20) \qquad a = (\, a_1, a_2, \cdots, a_n \,)^T \ , \quad b = (\, b_1, b_2, \cdots, b_n \,)^T \ ,$$

with $a_i = f''(x_{in})$, $b_i = f(x_{in})$ for all i. Second, with

$$r_i = (\, x_{i+1,n} - x_{in} \,)^{-1} \quad , \quad \varrho_i = x_{i+1,n} - x_{in} \ ,$$
$$r_0 = r_n \quad , \quad \varrho_0 = \varrho_n \ ,$$

and following the indexing scheme (2.29), the matrices $T_{\text{per}} \in \mathbb{R}^{n \times n}$ and $S_{\text{per}} \in \mathbb{R}^{n \times n}$ are defined by

$$(3.21) \qquad [T_{\text{per}}]_{i,j} = \begin{cases} \frac{1}{6}\, \varrho_{i-1} & , \quad j = i-1 \ , \quad 2 \leqslant i \leqslant n \ , \\ \frac{1}{3}\, (\, \varrho_{i-1} + \varrho_i \,) & , \quad j = i \quad , \quad 1 \leqslant i \leqslant n \ , \\ \frac{1}{6}\, \varrho_i & , \quad j = i+1 \ , \quad 1 \leqslant i \leqslant n-1 \ , \end{cases}$$

and

$$(3.22) \qquad [T_{\text{per}}]_{1,n} = [T_{\text{per}}]_{n,1} = \tfrac{1}{6}\, \varrho_n \ ,$$

and all other elements equal to 0, and

$$(3.23) \qquad [S_{\text{per}}]_{i,j} = \begin{cases} r_{i-1} & , \quad j = i-1 \ , \quad 2 \leqslant i \leqslant n \ , \\ -\, r_{i-1} - r_i \ , & \quad j = i \quad , \quad 1 \leqslant i \leqslant n \ , \\ r_i & , \quad j = i+1 \ , \quad 2 \leqslant 1 \leqslant i \leqslant n-1 \ , \end{cases}$$

and

$$(3.24) \qquad [S_{\text{per}}]_{1,n} = [S_{\text{per}}]_{n,1} = r_n \ ,$$

and, again, all other elements equal to 0.

EXERCISES: (3.2), (3.14), (3.16), (3.18).

4. Relaxed boundary cubic splines

We now discuss the relaxed boundary splines of OEHLERT (1992) in the "cubic" case; i.e., the solution to

$$(4.1) \quad \text{minimize} \quad \tfrac{1}{n} \sum_{i=1}^{n} |\, f(x_{in}) - y_{in} \,|^2 + h^4 \int_{x_{0,n}}^{x_{n+1,n}} w(\,t\,)\,|\, f''(\,t\,)\,|^2 \, dt$$

over the relevant class of functions on $(\, x_{0,n} \,,\, x_{n+1,n})$. Following OEHLERT (1992), we refer to the solution of (4.1)–(4.3) as a spline, but it is not a polynomial spline. However, its computation is remarkably similar to that of natural cubic splines. We assume that

$$(4.2) \qquad x_{0,n} < 0 = x_{1,n} < x_{2,n} < \cdots < x_{n,n} = 1 < x_{n+1,n} \; .$$

Interest is in the choice of OEHLERT (1992),

$$(4.3) \qquad w(t) = [\, t\,(1-t\,)\,]^2 \; , \quad x_{1,n} = 0 \leqslant t \leqslant 1 = x_{n,n} \; ,$$

and periodically with period 1 outside $(0,1)$. Again, this provides the solution of the problem we are actually interested in. The algorithm to be developed works for general measurable, nonnegative functions w for which $1/w$ is integrable, and we present it as such. Thus, it may be used when the smoothing parameter varies with location, as in § 18.7. Of course, (4.3) is not covered, but the scheme will be seen to apply to (4.3) nevertheless.

(4.4) EXERCISE. Let f be the solution of (4.1)–(4.3). Define $\varphi(x) = f(x)$, $x \in (\, x_{1,n} \, , \, x_{n+1,n}\,)$. Prove that φ solves the "original" problem (13.6.3) for $m = 2$. [Hint: Check the computational scheme to follow.]

The computational scheme presented here is similar to that of §§ 2 and 3 and starts with the equations for (4.1),

$$(4.5) \qquad \begin{aligned} & h^4\, D^2\, w\, D^2 f + \tfrac{1}{n} \sum_{i=1}^{n} (\, f(x_{in}) - y_{in}\,)\, \delta(\,\cdot\, - x_{in}\,) = 0 \; , \\ & [\, w\, D^2 f\,](x_{0,n}) = [\, w\, D^2 f\,](x_{n+1,n}) = 0 \; , \\ & [\, D\, w\, D^2 f\,](x_{0,n}) = [\, D\, w\, D^2 f\,](x_{n+1,n}) = 0 \; , \end{aligned}$$

where D denotes the differentiation operator and $D^k\, w\, D^2 f$ should be interpreted as $D^k(\, w\, D^2 f\,)$. The differential equation holds on $(x_{0,n}\, , \, x_{n+1,n}\,)$.

The differential equation says that $D^2\, w\, D^2 f$ is a sum of delta functions. Consequently, $w\, D^2 f$ is piecewise linear and continuous. Since $1/w$ is integrable, then f'' is integrable, and f' and f must be continuous.

Now, backtracking a bit, we have that $w\, f''$ is a polygonal function and may be represented as

$$(4.6) \qquad w(\,t\,)\, f''(\,t\,) = a_i\, r_i\, (\, x_{i+1,n} - t\,) + a_{i+1}\, r_i\, (\, t - x_{i+1,n}\,)$$

for $x_{in} \leqslant t \leqslant x_{i+1,n}$, where $a_i = [\, w\, f''\,](x_{in})$, $i = 1, 2 \cdots , n$.

The boundary conditions imply, cf. (2.18),

$$(4.7) \qquad a_0 = a_1 = 0 \quad , \quad a_n = a_{n+1,n} = 0 \; .$$

Next, we divide by w and integrate twice to obtain

$$(4.8) \quad f(\,t\,) = f_i(\,t\,) = a_i\, \widetilde{V}_i(\,t\,) + a_{i+1}\, \widetilde{W}_i(\,t\,) + p_i(\,t\,) \; , \quad x_{in} \leqslant t \leqslant x_{i+1,n} \; ,$$

where the p_i are polynomials of degree $\leqslant 1$ and

$$(4.9)\qquad \begin{aligned} \widetilde{V}_i(t) &= \int_{x_{in}}^{t} r_i(x_{i+1,n} - s)(t - s)\,d\mu(s) , \\ \widetilde{W}_i(t) &= \int_{t}^{x_{i+1,n}} r_i(s - x_{in})(s - t)\,d\mu(s) , \end{aligned}$$

in which $d\mu(s) = [w(s)]^{-1}\,ds$. If $1/w$ is integrable, then all of these integrals are finite.

Now, similar to the way (2.21) was obtained from (2.19), we obtain that

$$(4.10)\qquad f_i(t) = a_i\,V_i(t) + a_{i+1}\,W_i(t) + b_i\,r_i(x_{i+1,n} - t) + b_{i+1}\,r_i(t - x_{in}) ,$$

where

$$(4.11)\qquad \begin{aligned} V_i(t) &= \widetilde{V}_i(t) - r_i(t - x_{in})\,\widetilde{V}_i(x_{i+1,n}) , \\ W_i(t) &= \widetilde{W}_i(t) - r_i(x_{i+1,n} - t)\,\widetilde{W}_i(x_{in}) . \end{aligned}$$

Again, the b_i may be interpreted as the values of f,

$$(4.12)\qquad b_i = f(x_{in}) , \quad i = 1, 2, \cdots, n .$$

Thus, the function f as given by (4.8)–(4.10) is continuous for any choice of the coefficients a_i and b_i.

We still need to enforce that f' is continuous by requiring that

$$(4.13)\qquad f_{i-1}'(x_{in}) = f_i'(x_{in}) , \quad i = 1, 2, \cdots, n .$$

This leads to

$$a_{i-1}\,V_{i-1}'(x_{in}) + a_i\,W_{i-1}'(x_{in}) + r_{i-1}(b_i - b_{i-1}) = a_i\,V_i'(x_{in}) + a_{i+1}\,W_i'(x_{in}) + r_i(b_{i+1} - b_i)$$

for $i = 1, 2, \cdots, n$, or

$$(4.14)\qquad T_w\,a = S\,b ,$$

with $S \in \mathbb{R}^{(n-2)\times n}$ given by (2.31), and $T_w \in \mathbb{R}^{(n-2)\times(n-2)}$ is a tridiagonal matrix with its elements indexed as in (2.29) and given by

$$(4.15)\qquad (T_w)_{i,j} = \begin{cases} V_{i-1}'(x_{in}) , & j = i-2,\ 3 \leqslant i \leqslant n-1 , \\ W_{i-1}'(x_{in}) - V_i'(x_{in}) , & j = i ,\quad 2 \leqslant i \leqslant n-1 , \\ W_i'(x_{in}) , & j = i+1,\ 2 \leqslant i \leqslant n-2 , \end{cases}$$

and $= 0$ otherwise. Finally,

$$(4.16)\qquad a = (a_2, a_3, \cdots, a_{n-1})^T , \quad b = (b_1, b_2, \cdots, b_n)^T .$$

As in the case of cubic spline interpolation, we would like to conclude that T_w is nonsingular (e.g., by showing that T_w is diagonally dominant), but that is not apparent. However, observe that for f given by (4.8)–(4.10), keeping (4.7) in mind,

$$\langle a , T_w a \rangle = \int_{x_{1,n}}^{x_{n,n}} w(t) \, | \, f''(t) \, |^2 \, dt \; . \tag{4.17}$$

(Note the integration bounds!) It follows that T_w is positive-definite, and that is good enough.

Finally, using the Euler equations once more yields that

$$\alpha^2 \, S^T a + b = y_n \; , \tag{4.18}$$

$$(T_w + \alpha^2 \, S \, S^T) \, a = S \, y_n \; , \tag{4.19}$$

where $\alpha^2 = nh^4$. Thus, (4.9) is the defining system of equations for the relaxed boundary cubic spline, with b determined from (4.18). Since (4.19) has the same structure as in ordinary cubic splines, we have one more reason to keep referring to relaxed boundary "splines".

(4.20) Exercise. A simple formulation of the interpolation problem associated with the least-squares problem of this section is not obvious, but a variational formulation is: Derive a scheme for the computation of the solution of

$$\text{minimize} \quad \int_{x_{0,n}}^{x_{n+1,n}} w(t) \, | \, f''(t) \, |^2 \, dt$$

$$\text{subject to} \quad f(x_{in}) = y_{in} \; , \quad i = 1, 2 \cdots , n \; ,$$

where the minimization is over the appropriate class of functions.

Of course, in the case (4.3), the function $1/w$ is not integrable. Nevertheless, one verifies that all the integrals defining T_w are finite. Thus, even for the case (4.3), the system (4.19) is a correct way to compute them.

The computations are somewhat unpleasant. However,

$$\begin{aligned} &V_i'(x_{in}) = - r_i \, \widetilde{V}_i(x_{i+1,n}) \; , \quad W_i'(x_{i+1,n}) = r_i \, \widetilde{W}_i(x_{in}) \; , \\ &V_i'(x_{i+1,n}) = \widetilde{V}_i'(x_{i+1,n}) - r_i \, \widetilde{V}_i(x_{i+1,n}) \; , \\ &W_i'(x_{in}) = \widetilde{W}_i'(x_{in}) + r_i \, \widetilde{W}_i(x_{in}) \; , \end{aligned} \tag{4.21}$$

which leads to $V_i'(x_{i+1,n}) = - W_i'(x_{in})$ and

$$\begin{aligned} V_i'(x_{in}) &= - \int_{x_{in}}^{x_{i+1,n}} \frac{(x_{i+1,n} - s)^2}{(x_{i+1,n} - x_{in})^2} \, d\mu(s) \; , \\ W_i'(x_{i+1,n}) &= \int_{x_{in}}^{x_{i+1,n}} \frac{(s - x_{in})^2}{(x_{i+1,n} - x_{in})^2} \, d\mu(s) \; , \\ V_i'(x_{i+1,n}) &= \int_{x_{in}}^{x_{i+1,n}} \frac{(x_{i+1,n} - s)(s - x_{in})}{(x_{i+1,n} - x_{in})^2} \, d\mu(s) \; . \end{aligned} \tag{4.22}$$

Since $d\mu(s) = [\, t\,(1 - t\,)\,]^{-2}\, ds$, these integrals can be evaluated in closed form using partial fraction decomposition. Here, a computer algebra package is helpful. The required accuracy can be achieved by combining symbolic methods with variable-precision arithmetic. This is a costly process, but it gets the job done.

A simple way to avoid all of this is given in the following exercise. Theoretically, the resulting approximation errors appear to be negligible compared with the usual error bounds. Unfortunately, to do the comparison in the small-sample case, one still has to compute the original integrals ... .

(4.23) Exercise. The following approximation scheme prevents all those nasty integrals in (3.22) from popping up. Compute the averages of w over each of the relevant intervals,

$$w_n(\,t\,) = w_i = r_i \int_{x_{in}}^{x_{i+1,n}} w(s)\, ds \ , \quad t \in (\, x_{in}\, ,\, x_{i+1,n}\,),$$

for $i = 1, 2, \cdots, n-1$.
(a) Show that the minimizer of (4.1), with w replaced by w_n, may be computed via (4.14)–(4.19), in which T_w is given by (2.30) with the ϱ_i replaced by $\varrho_i = (\, x_{i+1,n} - x_{in}\,)/w_i$ for $i = 1, 2, \cdots, n-1$.
(b) Do the bounds of Theorem (13.6.4) still apply? [The authors conjecture that the answer is yes.]

Exercises: (4.4), (4.20), (4.23).

5. Higher-order smoothing splines

In this section, we consider the computation of smoothing splines of arbitrary even order, interest being in the orders six and eight. We take a limited point of view in that we only consider the computation of the smoothed data or, equivalently, the values of the smoothing splines at the design points. Determining the splines themselves is an additional problem. The goal is again to derive a system of equations with simple matrices. It turns out that the general structure is the same as for cubic splines, but rather than tridiagonal matrices, we will end up with banded matrices, with the number of diagonals depending on the order of the splines.

As in the treatment of cubic splines, we begin with natural spline interpolation. Let $m \geqslant 1$. The starting point is the variational formulation

$$\begin{aligned} &\text{minimize} && \int_{x_{0,n}}^{x_{n+1,n}} |\, f^{(m)}(\,t\,)\,|^2\, dt \\ (5.1)\qquad &\text{subject to} && f \in W^{m,2}(x_{0,n}, x_{n+1,n}) \ , \\ & && f(x_{in}) = y_{in} \ , \quad i = 1, 2, \cdots, n \ . \end{aligned}$$

The Euler equations for this problem read

$$(5.2)\qquad \begin{aligned} &(-1)^m f^{(2m)}(t) = \sum_{i=1}^{n} a_i\, \delta(t - x_{in}) \quad \text{on } (x_{0,n}, x_{n+1,n}) , \\ &f^{(k)}(x_{0,n}) = f^{(k)}(x_{n+1,n}) = 0 , \quad k = m, m+1, \cdots, 2m-1 , \\ &f(x_{in}) = y_{in} , \quad i = 1, 2, \cdots, n . \end{aligned}$$

Here, the a_i are the Lagrange multipliers and δ is the Dirac delta function.

We proceed to "solve" the Euler equations. The differential equation implies that f is a spline of order $2m$ on the knot sequence

$$x_{0,n} < x_{1,n} < \cdots < x_{n,n} < x_{n+1,n}$$

and so is a polynomial of order $2m$ on each of the intervals $[x_{in}, x_{i+1,n}]$,

$$(5.3)\qquad f(t) = f_i(t) = \sum_{\ell=0}^{2m-1} b_{i,\ell}\, (t - x_{in})^{\ell} , \quad x_{in} \leqslant t \leqslant x_{i+1,n} .$$

Since $f \in C^{2m-2}(x_{0,n}, x_{n+1,n})$, the pieces fit together in the form

$$(5.4)\qquad f_{i-1}^{(k)}(x_{in}) = f_i^{(k)}(x_{in}) , \quad \begin{aligned} &i = 1, 2, \cdots, n, \\ &k = 0, 1, \cdots, 2m-2 . \end{aligned}$$

The natural boundary conditions

$$(5.5)\qquad f^{(k)}(x_{0,n}) = f^{(k)}(x_{n+1,n}) = 0 , \quad k = m, m+1, \cdots, 2m-1 ,$$

and the interpolation conditions

$$(5.6)\qquad f(x_{in}) = y_{in} , \quad i = 1, 2, \cdots, n ,$$

round out the set of equations. A quick count reveals that in (5.3) there are $2m(n+1)$ unknowns ($2m$ per interval) and $(2m-1)n + 2m + n$ equations. Thus, there are as many equations as unknowns. However, this does not quite say that there is a unique solution to these equations. Moreover, although the system of equations may be derived in a mechanical fashion, it is not easy to discern an overall structure in them.

Hence, we travel an alternate route to solving the Euler equations, ultimately leading to a banded system of equations and encountering the B-splines of Schoenberg (1973) along the way.

The approach is straightforward: Integrate the differential equation in (5.2) a total of $2m$ times. First, integrate m times from $x_{0,n}$ to t. Then,

$$(5.7)\qquad (-1)^m f^{(m)}(t) = p(t) + \sum_{i=1}^{n} a_i \frac{(t - x_{in})_+^{m-1}}{(m-1)!} ,$$

where $p(t)$ is a polynomial of degree $\leqslant m-1$. To apply the boundary conditions at $x_{0,n}$, note that $(-1)^m f^{(m)}(t) = p(t)$ on $[x_{0,n}, x_{1,n}]$, so

$$(5.8)\qquad f^{(m+k)}(x_{0,n}) = p^{(k)}(x_{0,n}) = 0 , \quad k = 0, 1, \cdots, m-1 .$$

Consequently, the polynomial p vanishes everywhere. Then the boundary conditions at $t = x_{n+1,n}$ yield the equations

(5.9) $$\sum_{i=1}^{n} a_i \frac{(x_{n+1,n} - x_{in})^k}{k!} = 0 , \quad k = 0, 1, \cdots, m-1 .$$

Second, integrate (5.7) for a total of m times from t to $x_{n+1,n}$. So

(5.10) $$f(t) = P(t) + \sum_{i=1}^{n} a_i V(t, x_{in}) ,$$

where $P(t)$ is another polynomial of degree $\leqslant m-1$ and

(5.11) $$V(t, x_{in}) = \int_0^{x_{n+1,n}} \frac{(s-t)_+^{m-1}}{(m-1)!} \frac{(s-x_{in})_+^{m-1}}{(m-1)!} \, ds .$$

Note that the integral is really over $t < s < x_{n+1,n}$.

(5.12) EXERCISE. Show (5.11). [Hint: Let $k \geqslant 0$. For a nice function f on $[a, b]$, where $a \leqslant t \leqslant b$, show that

$$\int_t^b \int_s^b \frac{(r-s)^k}{k!} f(r)\, dr\, ds = \int_t^b \frac{(r-t)^{(k+1)}}{(k+1)!} f(r)\, dr$$

by interchanging the order of integration.]

Now, restrict (5.10) to the design points. Then,

(5.13) $$y_{in} = P(x_{in}) + \sum_{j=1}^{n} a_j V(x_{in}, x_{j,n}) , \quad i = 1, 2, \cdots, n .$$

Observe that (5.9)–(5.13) constitute $n+m$ linear equations in the n unknown coefficients a_i and the m unknown coefficients needed to determine the polynomial P; e.g., in the form

(5.14) $$P(t) = \sum_{k=0}^{m-1} p_k \frac{(x_{n+1,n} - t)^k}{k!} .$$

(The reason for this "strange" form will become apparent shortly.)

It is useful to write the system (5.9) and (5.13)–(5.14) in matrix vector notation,

(5.15) $$\begin{bmatrix} V & X \\ X^T & 0 \end{bmatrix} \begin{bmatrix} a \\ p \end{bmatrix} = \begin{bmatrix} y_n \\ 0 \end{bmatrix} ,$$

with 0 denoting matrices and vectors of the appropriate dimensions with all elements zero,

(5.16) $$a = (a_1, a_2, \cdots, a_n)^T , \quad p = (p_0, p_1, \cdots, p_{m-1})^T ,$$

and $V \in \mathbb{R}^{n \times n}$, $X \in \mathbb{R}^{n \times m}$ given by

$$(5.17)\qquad \begin{aligned} V_{i,j} &= V(x_{in}, x_{j,n}) \,, & i,j &= 1,2,\cdots,n\,, \\ X_{i,k} &= \frac{(x_{n+1,n} - x_{in})^k}{k!} \,, & i &= 1,2,\cdots,n, \\ & & k &= 0,1,\cdots,m-1 \end{aligned} \quad .$$

The system (5.15) has a unique solution, as shown below, but the matrix V is dense: All of its coefficients are nonzero, so the solution would cost $\mathcal{O}(n^3)$ floating-point operations, see, e.g., KINCAID and CHENEY (1991). Moreover, the system is not well-conditioned. The reason for the bad conditioning of (5.15) is that the rows of V are nearly linearly dependent. Fortunately, this may be used to advantage in obtaining a better system, both in terms of conditioning and structure. The easiest way to see this is by introducing the celebrated B-splines of SCHOENBERG (1973).

First, consider the polynomials $(t - x_{j,n})^m$, $j = 1,2,\cdots,n$. We begin with the identity

$$(5.18)\qquad \sum_{i=j}^{j+m} \frac{(t - x_{in})^m}{\omega_j'(x_{in})} = (-1)^m \,,$$

for $j = 1,2,\cdots,n-m$. Here,

$$(5.19)\qquad \omega_j(t) = (t - x_{j,n})(t - x_{j+1,n}) \cdots (t - x_{j+m,n}) \,,$$

so that

$$(5.20)\qquad \omega_j'(x_{in}) = \prod_{\substack{\ell=j \\ \ell \neq i}}^{j+m} (x_{in} - x_{\ell,n}) \,.$$

The identity (5.18) looks rather formidable but easily follows by relating it to Lagrange interpolation.

(5.21) EXERCISE. Prove (5.18) as follows. Let $1 \leqslant j \leqslant n - m$ and fix t.
(a) Consider the polynomial $(t - s)^m$ as a function of s. Show that

$$(t - s)^m = \sum_{i=j}^{j+m} (t - x_{in})^m \prod_{\substack{\ell=j \\ \ell \neq i}}^{j+m} \frac{s - x_{\ell,n}}{x_{in} - x_{\ell,n}} \,.$$

(b) Differentiate both sides m times with respect to s.
[Hint: For (a), the right-hand side is the Lagrange form of an interpolating polynomial of degree $\leqslant m$.]

Now, differentiate (5.18) any number of times with respect to t. Then, for $j = 1,2,\cdots,n-m$ and $k = 0,1,\cdots,m-1$,

$$(5.22)\qquad \sum_{i=j}^{j+m} \frac{(t - x_{in})^k}{\omega_j'(x_{in})} = 0 \quad \text{for all } t \,.$$

So far, so good, but this is not quite what we want: $(t - x_{in})^{m-1}$ must be replaced by $(t - x_{in})_+^{m-1}$, and the scaling is different as well. Let

$$M_j(t) = \sum_{i=j}^{j+m} \frac{(t - x_{in})_+^{m-1}}{(m-1)!\; \omega_j'(x_{in})} \;, \qquad j = 1, 2, \cdots, n-m \;. \tag{5.23}$$

The functions $(-1)^m\,(m-1)!\,M_j(t)$ are the B-splines of Schoenberg (1973). They depend on the knot sequence and the order m, although the notation does not reflect this. We need a few elementary properties. Observe that $M_j(t) = 0$ for $t \leqslant x_{j,n}$ because then all terms in the sum (5.23) vanish and also for $t \geqslant x_{j+m,n}$, by (5.22). This shows the first part of

$$\begin{aligned} &\text{support}(M_j) \subset (x_{j,n} \,,\, x_{j+m,n}) \;, \quad j = 1, 2, \cdots, n-m \;, \\ &(-1)^m\, M_j(t) > 0 \quad \text{on } (x_{j,n}, x_{j+1,n}) \;. \end{aligned} \tag{5.24}$$

The second part may be shown by observing that on $[\, x_{j,n} \,,\, x_{j+1,n} \,]$ the sum in (5.23) consists of just one term, so

$$(m-1)!\, M_j(t) = \frac{(t - x_{j,n})^{m-1}}{\omega_j'(x_{in})} \qquad \text{on } (x_{j,n}, x_{j+1,n}) \;,$$

and $(-1)^m\,\omega_j'(x_{j,n}) > 0$. (In fact, $(-1)^m\, M_j(t) > 0$ on $(x_{j,n}, x_{j+m,n})$, but never mind.)

How does this relate to the system (5.15)? Define $D_m \in \mathbb{R}^{n\times n}$ as

$$[\, D_m\, a\,]_j = \sum_{i=j}^{(j+m)\vee n} \frac{a_j}{\omega_j'(x_{in})} \;, \qquad j = 1, 2, \cdots, n \;,$$

or, equivalently, for $i = 1, 2, \cdots, n$,

$$(D_m)_{i,j} = \begin{cases} 1/\omega_j'(x_{in}) \;, & j = i, \cdots, (j+m)\vee n \;, \\ \quad 0 \;, & \text{otherwise} . \end{cases} \tag{5.25}$$

Then, taking linear combinations of rows and columns of V gives

$$[\, D_m V D_m^T \,]_{i,j} = \int_{x_{0,n}}^{x_{n+1,n}} M_i(t)\, M_j(t)\, dt \;, \quad i,j = 1, 2, \cdots, n \;, \tag{5.26}$$

and this is a banded matrix,

$$[\, D_m V D_m^T \,]_{i,j} = 0 \quad \text{for} \quad |\, i - j \,| \geqslant m \;. \tag{5.27}$$

So, $D_m V D_m^T$ has $2m-1$ nonzero diagonals (the main diagonal and $m-1$ codiagonals on either side).

The transformation above applied to the system (5.15) leads to

$$\begin{bmatrix} D_m V D_m^T & | & D_m X \\ ---- & & -- \\ X^T D_m^T & | & 0 \end{bmatrix} \begin{bmatrix} c \\ -- \\ p \end{bmatrix} = \begin{bmatrix} D_m\, y \\ -- \\ 0 \end{bmatrix} , \tag{5.28}$$

with $c = (D_m^T)^{-1}a$. Note that

$$(5.29) \qquad D_m X = \begin{bmatrix} 0 \\ -- \\ \Sigma \end{bmatrix} ,$$

with $\Sigma \in \mathbb{R}^{m \times m}$ nonsingular. (For the first $n-m$ rows, (5.26) applies. Moreover, since D_m is nonsingular and X has independent columns, then the matrix Σ *must* be nonsingular.)

In view of the special structure of $D_m X$ and $X^T D_m^T$ and the right-hand side, a further partitioning in (5.28) is in order. Partition c according to

$$(5.30) \qquad c^T = [\, d^T \,|\, e^T \,] ,$$

with $d \in \mathbb{R}^{n-m}$ and $e \in \mathbb{R}^m$. Now, the last block of equations of (5.29) says that $X^T D_m^T c = 0$, and this implies that $\Sigma e = 0$ or

$$(5.31) \qquad e = 0 .$$

Then, the first $n-m$ equations (of the first block of equations) of (5.28) read as

$$(5.32) \qquad T d = S y ,$$

with T the $(n-m)\times(n-m)$ principal minor of $D_m V D_m^T$ and $S y$ the first $n-m$ rows of $D_m y$. Equivalently, $T \in \mathbb{R}^{(n-m)\times(n-m)}$ is given by

$$(5.33) \qquad T_{i,j} = [\, D_m V D_m^T \,]_{i,j} , \quad i,j = 1,2,\cdots,n-m ,$$

and $S \in \mathbb{R}^{(n-m)\times n}$ by

$$(5.34) \qquad S_{i,j} = [\, D_m \,]_{i,j} , \quad i = 1,2,\cdots,n-m, \; j = 1,2,\cdots,n .$$

Now, (5.32) essentially solves the interpolation problem since we can retrieve the vector a from $a = D_m^T c$. However, a small refinement is helpful. (The details are outlined in Exercise (5.39).) Note that in terms of the vector a, the statement (5.31) says that

$$(5.35) \qquad [\, D_m^{-T} a \,]_i = 0 , \quad i = n-m+1, \cdots, n ,$$

where D_m^{-T} is shorthand for $(D_m^T)^{-1}$. This implies that

$$(5.36) \qquad a_i = 0 , \quad j = n-m+1, \cdots, n ,$$

and we have

$$(5.37) \qquad d = \Delta_m^{-T} \widetilde{a} ,$$

with Δ_m the $(n-m)\times(n-m)$ leading principal minor of D_m, and

$$(5.38) \qquad \widetilde{a} = (a_1, a_2, \cdots, a_{n-m})^T .$$

It follows that $\widetilde{a} = \Delta_m^{-1} d$.

(5.39) EXERCISE. (a) Let $R \in \mathbb{R}^{n \times n}$ be a nonsingular upper-triangular matrix. Partition R as

$$R = \left[\begin{array}{c|c} U & V \\ \hline 0 & W \end{array} \right] .$$

Show that U and W are nonsingular and that

$$U^{-1} = \left[\begin{array}{c|c} U^{-1} & -U^{-1} V W^{-1} \\ \hline 0 & W^{-1} \end{array} \right] .$$

(b) Show that (5.35) implies (5.36), and derive (5.37).

Thus, the Lagrange multipliers, a_i, are completely determined by the vector d; i.e., by $[(D_m^T)^{-1} a]_i$, $i = 1, 2, \cdots, n-m$. Moreover, the vector d is uniquely determined by the system (5.32) since the matrix T is symmetric and (strictly) positive-definite.

(5.40) THEOREM. *T is positive-definite.*

PROOF. Let $a \in \mathbb{R}^{n-m}$. Then,

$$a^T T a = \int_{x_{0,n}}^{x_{n+1,n}} \Big| \sum_{i=1}^{n-m} a_i M_i(t) \Big|^2 dt \geqslant 0 , \tag{5.41}$$

so that T is semi-positive-definite. Now, if $a^T T a = 0$, then

$$\sum_{i=1}^{n-m} a_i M_i(t) = 0 \quad \text{for all } t$$

and, in particular, it vanishes on $(x_{1,n}, x_{2,n})$, but the only M_i that does not vanish on this interval is $M_1(t)$. Thus, $a_1 = 0$. Now, proceed to the next interval $(x_{2,n}, x_{3,n})$, and conclude that $a_2 = 0$. By induction, all a_i vanish. Q.e.d.

It turns out that the matrix T is well-conditioned compared with the original system (5.15). The "bad" part has been isolated as the computation of $S y$. However, the matrices T and S may still be badly scaled if some of the design points are much closer together than most of them.

The final step in the solution of the interpolation problem is the computation of the coefficients p of the polynomial P in (5.28), which may be done by solving the first block of (5.28), which reads

$$\left[\begin{array}{c} 0 \\ \hline \Sigma \end{array} \right] p = D_m y - D_m V D_m c . \tag{5.42}$$

Thus, the interpolating spline is given as

$$
(5.43)\qquad
\begin{aligned}
f^{(m)}(t) &= \sum_{i=1}^{n-m} d_i \, M_i(t) \ , \\
f(t) &= P(t) + \sum_{i=1}^{n-m} d_i \int_{x_{0,n}}^{x_{n+1,n}} M_i(t) \, \frac{(s-t)_+^{m-1}}{(m-1)!} \, ds \ ,
\end{aligned}
$$

(5.44) EXERCISE. Verify (5.43).

With the interpolation problem well-understood, we move on to the smoothing spline problem,

$$
(5.45)\qquad
\begin{aligned}
&\text{minimize} \quad \tfrac{1}{n} \sum_{i=1}^{n} |\, f(x_{in}) - y_{in} \,|^2 + h^{2m} \int_{x_{0,n}}^{x_{n+1,n}} |\, f^{(m)}(t) \,|^2 \, dt \\
&\text{subject to} \quad f \in W^{m,2}(x_{0,n}, x_{n+1,n}) \ .
\end{aligned}
$$

The Euler equations read as

$$
(5.46)\qquad
\begin{aligned}
&(-h^2)^m \, f^{(2m)}(t) + \tfrac{1}{n} \sum_{i=1}^{n} (\, f(x_{in}) - y_{in} \,) \, \delta(t - x_{in}) = 0 \ , \\
&f^{(k)}(x_{0,n}) = f^{(k)}(x_{n+1,n}) = 0 \ , \qquad k = m, m+1, \cdots, 2m-1 \ ,
\end{aligned}
$$

where the differential equation holds on $(x_{0,n}, x_{n+1,n})$. As in the interpolation problem, the differential equation implies that f is a polynomial spline of order $2m$, and the boundary conditions imply it is a natural spline. Thus, f may be represented as in (5.10), with

$$
(5.47)\qquad a_i = -(nh^{2m})^{-1} (\, f(x_{in}) - y_{in} \,) \ , \quad i = 1, 2, \cdots, n \ .
$$

Following the same development, the function f may be represented as in (5.43), where the d_i satisfy the system

$$
(5.48)\qquad Td = Sb \ ,
$$

in which d is related to the a_i via (5.37)–(5.38), and

$$
(5.49)\qquad b = \bigl(\, f(x_{1,n}), f(x_{2,n}), \cdots, f(x_{n,n}) \,\bigr)^T \ .
$$

Now, in view of (5.34), with $\lambda = nh^{2m}$, the equation (5.47) implies that

$$
(5.50)\qquad \lambda \, S^T d + b - y_n = 0 \ .
$$

The systems (5.48) and (5.50) are the familiar ones and lead to

$$
(5.51)\qquad (\, T + \lambda \, S \, S^T \,) \, d = S y_n \ ,
$$

after which b may be determined from (5.50). Thus, formally we get the same kind of system as in the cubic spline case. Note that the matrix in the system (5.51) has $2m+1$ nonzero diagonals (S is upper-triangular with $m+1$ nonzero diagonals).

This concludes the discussion of smoothing splines of even order. There are still a few gaps that should be filled in, most prominent of which is the (stable) computation of the matrices T and S. More importantly, the numerical stability of solving the system of equations is suspect, especially for smoothing splines of high order and very nonuniform designs (but Kalman filtering nicely avoids that). For exact replications, see Exercise (3.16)(b).

(5.52) EXERCISE. Show that the system (5.50)–(5.51) implies that

$$\left(I + nh^{2m} \, S^T T^{-1} S \right) b = y_n$$

so that (Euclidean vector norms) $\| b \| \leqslant \| y_n \|$ since the matrix T^{-1} is positive-definite and then $S^T T^{-1} S$ is semi-positive-definite.

(5.53) EXERCISE. Work out the details for the problem (5.45) with independent replications and with heteroscedastic noise, as in Exercise (3.16).

EXERCISES: (5.12), (5.21), (5.39), (5.44), (5.52), (5.53).

6. Other spline estimators

In this section, we consider the computation of least-absolute-deviations smoothing splines and total-variation-penalized least-squares estimators as discussed in Chapter 17. These two estimators are similar in that the objective functions are given as the sum of a squared discrete ℓ^2 norm and a discrete ℓ^1 norm and that their *duals* are standard least-squares problems with " box constraints", for which we develop an efficient active constraint set method.

Total-variation penalization. Of the two, the simplest problem is the total-variation penalization,

$$\begin{aligned} &\text{minimize} \quad \tfrac{1}{n} \sum_{i=1}^{n} | f(x_{in}) - y_{in} |^2 + h \, | f |_{TV} \\ &\text{subject to} \quad f \in TV(0,1) \; . \end{aligned} \tag{6.1}$$

By § 17.3, this is equivalent to

$$\text{minimize} \quad L(z) \stackrel{\text{def}}{=} \tfrac{1}{2} \| z - y_n \|^2 + \alpha \| D z \|_1 \; , \tag{6.2}$$

in which $\alpha = \frac{1}{2} nh$, $z_i = f(x_{in})$, $i = 1, 2, \cdots, n$, and $D \in \mathbb{R}^{(n-1) \times n}$ is the difference matrix,

$$D_{i,j} = \begin{cases} -1 \, , & i = j \, , \\ 1 \, , & i = j - 1 \, , \\ 0 \, , & \text{otherwise} \, . \end{cases} \tag{6.3}$$

Of course, the solution of (6.2) exists and is unique. The difficulty in *computing* the solution is that the objective function is not differentiable, so it is difficult to consider gradient-type methods. However, at the price of additional variables and constraints, it can be transformed into a smooth problem with linear inequality constraints. Let $s = |\, Dz \,|$ componentwise. Then, surely, $s \geqslant Dz$ and $s \geqslant -Dz$. Moreover, in the minimization problem (over z and s)

$$
\begin{aligned}
&\text{minimize} \quad \mathcal{L}(z,s) \stackrel{\text{def}}{=} \tfrac{1}{2} \|\, z - y_n \,\|^2 + \alpha \langle\, 1\!\!1 \,,\, s \,\rangle \\
&\text{subject to} \quad s \geqslant Dz \;,\; s \geqslant -Dz \;,
\end{aligned}
\tag{6.4}
$$

where $1\!\!1 \in \mathbb{R}^{n-1}$ is the vector of all ones, the minimizing s is as small as possible (componentwise); i.e., $s = |\, Dz \,|$.

(6.5) EXERCISE. Show that (6.4) has a unique solution z^*, s^*, and that then $s^* = |\, Dz^* \,|$.

Note that (6.4) is a convex optimization problem and that the objective function is now even differentiable, so "standard" optimization methods apply. The only drawback is the doubling of the number of variables. This can be corrected by considering the dual problem, which turns out to be

$$
\begin{aligned}
&\text{maximize} \quad \Lambda(\theta) \stackrel{\text{def}}{=} -\tfrac{1}{2} \|\, D^T\theta - y_n \,\|^2 + \tfrac{1}{2} \|\, y_n \,\|^2 \\
&\text{subject to} \quad -\alpha\, 1\!\!1 \leqslant \theta \leqslant \alpha\, 1\!\!1 \;.
\end{aligned}
\tag{6.6}
$$

Despite appearances, (6.6) is again a convex *minimization* problem, the solution of which exists and is unique. Moreover, the dual problem is equivalent to the primal problem (6.4) or (6.2) since the objective function is strongly convex and the constraint set satisfies the weak/strong Slater conditions

$$
\{\, s \in \mathbb{R}^{n-1} \;\big|\; s > Dz \;\text{ and }\; s > -Dz \,\} \neq \emptyset \;;
\tag{6.7}
$$

see, e.g., HIRIART-URRUTY and LEMARÉCHAL (1993) or § 9.4 in Volume I, in particular the *constraint qualification* (9.3.4). We formulate it as a theorem.

(6.8) THEOREM (TV DUALITY). *Let z^* and θ^* be the solutions of* (6.2) *and* (6.6). *Then, for all z and all θ satisfying $-2\,\alpha\, 1\!\!1 \leqslant \theta \leqslant -2\,\alpha\, 1\!\!1$,*

$$
\Lambda(\theta) \leqslant \Lambda(\theta^*) = L(z^*) \leqslant L(z) \;.
$$

Moreover, $z^* = y_n - D^T\theta^*$.

(6.9) EXERCISE. Prove the theorem. [Hint: Use the Lagrange multiplier theorem, see, e.g., Theorem (9.3.5) and take into account that the Lagrange multipliers for the inequalities $s_i \geqslant (Dz)_i$ and $s_i \geqslant -(Dz)_i$ are not unrelated.]

Computing the TV dual. In the case above, the duality is easily verified. Setting up and proving the "general" case requires work; see, e.g., HIRIART-URRUTY and LEMARÉCHAL (1993). However, in the present case, one can compute the dual problem rather than having to guess it.

First, rewrite the constraints in (6.4) as $Dz - s \leqslant 0$ and $-Dz - s \leqslant 0$, and introduce the *Lagrangian*

$$\mathfrak{L}(z, s\,; \lambda, \mu) \stackrel{\text{def}}{=} \mathcal{L}(z, s) + \langle\, \lambda\,,\, Dz - s\,\rangle + \langle\, \mu\,,\, -Dz - s\,\rangle\ . \tag{6.10}$$

Here, λ, $\mu \in \mathbb{R}^{n-1}$ are the dual variables (the Lagrange multipliers), and $\lambda \geqslant 0$, $\mu \geqslant 0$. The problem dual to (6.4) is

$$\text{maximize} \quad \mathfrak{K}(\lambda, \mu) \quad \text{subject to} \quad \lambda \geqslant 0\ ,\ \mu \geqslant 0\ , \tag{6.11}$$

in which the objective function is defined as

$$\mathfrak{K}(\lambda, \mu) \stackrel{\text{def}}{=} \min\{\, \mathfrak{L}(z, s\,; \lambda, \mu) : z \in \mathbb{R}^n\ , s \in \mathbb{R}^{n-1} \,\}\ . \tag{6.12}$$

This minimum value may in fact be explicitly determined. First, rewrite the Lagrangian as

$$\mathfrak{L}(z, s\,; \lambda, \mu) = \tfrac{1}{2}\,\|\, z - y_n \,\|^2 + \langle\, \alpha\,\mathbb{1} - \lambda - \mu\,,\, s\,\rangle + \langle\, D^T(\,\lambda - \mu\,)\,,\, z\,\rangle\ .$$

It is clear that the minimum over s (no constraints!) yields $-\infty$, unless $\alpha\,\mathbb{1} - \lambda - \mu = 0$, so

$$\mathfrak{K}(\lambda, \mu) = -\infty\ , \qquad \lambda + \mu \neq \alpha\,\mathbb{1}\ . \tag{6.13}$$

Now consider the case $\lambda + \mu = \alpha\,\mathbb{1}$. Then, the Lagrangian equals

$$\mathfrak{L}(z, s\,; \lambda, \mu) = \tfrac{1}{2}\,\|\, z - y_n \,\|^2 + \langle\, D^T(\,\lambda - \mu\,)\,,\, z\,\rangle\ . \tag{6.14}$$

Setting the gradient equal to zero gives $z = y_n - D^T(\,\lambda - \mu\,)$, and so the minimum of $\mathfrak{L}(z, s\,; \lambda, \mu)$ equals

$$\begin{aligned}\mathfrak{K}(\lambda, \mu) &= -\tfrac{1}{2}\,\|\, D^T(\lambda - \mu) \,\|^2 + \langle\, D^T(\,\lambda - \mu\,)\,,\, y_n\,\rangle \\ &= -\tfrac{1}{2}\,\|\, D^T(\,\lambda - \mu\,) - y_n \,\|^2 + \tfrac{1}{2}\,\|\, y_n \,\|^2\ .\end{aligned}$$

To summarize,

$$\mathfrak{K}(\lambda, \mu) = -\tfrac{1}{2}\,\|\, D^T(\,\lambda - \mu\,) - y_n \,\|^2 + \tfrac{1}{2}\,\|\, y_n \,\|^2 \quad \text{if} \quad \lambda + \mu = \alpha\,\mathbb{1}\ , \tag{6.15}$$

and $= -\infty$ otherwise. In the maximization problem (6.11), we obviously need not worry about the $-\infty$ values. So the dual problem is

$$\begin{aligned}&\text{maximize} \quad -\tfrac{1}{2}\,\|\, D^T(\,\lambda - \mu\,) - y_n \,\|^2 + \tfrac{1}{2}\,\|\, y_n \,\|^2 \\ &\text{subject to} \quad \lambda \geqslant 0\ ,\ \mu \geqslant 0\ ,\ \lambda + \mu = \alpha\,\mathbb{1}\ .\end{aligned} \tag{6.16}$$

Finally, the transformation $\theta = \lambda - \mu$ suggests itself and leads to the constraints on θ,

$$-\alpha\,\mathbb{1} \leqslant \theta \leqslant \alpha\,\mathbb{1}\ , \tag{6.17}$$

after which the problem (6.16) reduces to (6.6).

Least-absolute-deviations. The treatment of the least-absolute-deviations problem goes similarly but is a bit more complicated due to the possible nonuniqueness of the estimator. The penalization involving higher-order derivatives only makes matters worse, so we restrict attention to the cubic case. Recall the problem

$$(6.18)\qquad \begin{array}{ll} \text{minimize} & \frac{1}{n}\sum_{i=1}^{n} |\, f(x_{in}) - y_{in} \,| + h^4 \,\|\, f'' \,\|^2 \\ \text{subject to} & f \in W^{2,2}(0,1)\ , \end{array}$$

along with its equivalent extended formulation

$$(6.19)\qquad \begin{array}{ll} \text{minimize} & \frac{1}{n}\sum_{i=1}^{n} |\, f(x_{in}) - y_{in} \,| + h^4 \,\|\, f'' \,\|^2 \\ \text{subject to} & f \in W^{2,2}(x_{o,n}, x_{n+1,n})\ , \end{array}$$

where $x_{o,n} < x_{1,n}$ and $x_{n+1,n} > x_{n,n}$ but are otherwise arbitrary.

At one point or another, we must come to terms with the possible nonuniqueness of the solution of (6.18), and we might as well do it now. In § 17.4, it was observed that solutions of (6.18) differ at most by a linear function. To formalize this, write any $f \in W^{2,2}(0,1)$ as

$$(6.20)\qquad f(x) = \varphi(x) + f(0)\,(\,1-x\,) + f(1)\,x\ , \quad x \in [\,0\,,\,1\,]\ ,$$

where $\varphi(0) = \varphi(1) = 0$ and, of course, $\varphi \in W^{2,2}(0,1)$. The set of these functions φ is usually denoted as

$$(6.21)\qquad W_o^{2,2}(0,1) \stackrel{\text{def}}{=} \big\{\, \varphi \in W^{2,2}(0,1) \,\big|\, \varphi(0) = \varphi(1) = 0 \,\big\}\ .$$

Then, the problem (6.19) is equivalent to

$$(6.22)\qquad \begin{array}{ll} \text{minimize} & \frac{1}{n}\sum_{i=1}^{n} |\, \varphi(x_{in}) + p(x_{in}) - y_{in} \,| + h^4 \,\|\, \varphi'' \,\|^2 \\ \text{subject to} & \varphi \in W_o^{2,2}(0,1)\ ,\ p \in \mathcal{P}_2\ . \end{array}$$

Recall that $\mathcal{P}_2$ is the set of all polynomials of order $\leqslant 2$ (degree $\leqslant 1$).

Now, the solution of (6.22) is given by a pair (φ, p), in which φ is unique, since the functional $\varphi \longmapsto \|\, \varphi'' \,\|^2$ is strongly convex on $W_o^{2,2}(0,1)$. This is the content of the next exercise.

(6.23) Exercise. (a) Show the Poincaré inequality,

$$\|\, \varphi'' \,\| \geqslant \pi^2 \,\|\, \varphi \,\| \quad \text{for all} \quad \varphi \in W_o^{2,2}(0,1)\ .$$

(b) Show that $\varphi \longmapsto \|\, \varphi'' \,\|^2$ is strongly convex on $W_o^{2,2}(0,1)$. [Hint: For part (a), note that φ'' is an arbitrary element of $L^2(0,1)$. Express it in terms of the sines, $\sin(\,\pi k x\,)$, $k = 1, 2, \cdots$, which forms a complete orthogonal system for $L^2(0,1)$, and compute φ, keeping the boundary conditions in mind. Take it from there.]

We are now ready for the computation of solutions of (6.19). First, we derive a finite-dimensional formulation, followed by the switch to (6.22). The first observation is that the solutions of (6.19) are cubic splines. This is most readily seen from the Euler equations: The function f solves (6.19) if and only if it satisfies

$$
\begin{aligned}
&2\,h^4\, f^{(4)} + \sum_{i=1}^{n} \operatorname{sign}(\, f(x_{in}) - y_{in}\,)\, \delta(\,\cdot - x_{in}) = 0 \;, \\
&f^{(k)}(x_{0,n}) = f^{(k)}(x_{n+1,n}) = 0 \;, \quad k = m, m+1, \cdots, 2m-1 \;,
\end{aligned} \tag{6.24}
$$

where the differential equation holds on the interval $(\, x_{0,n} \,,\, x_{n+1,n}\,)$ and the sign function is defined in the usual way,

$$
\operatorname{sign}(\,t\,) = \begin{cases} 1 & , \quad t > 0 \;, \\ -1 & , \quad t < 0 \;, \\ \in [-1\,,\,1] \;, & \quad t = 0 \;. \end{cases} \tag{6.25}
$$

In other words, $\operatorname{sign}(\,0\,)$ is determined by "circumstances".

As in § 2, the differential equation says that f is a cubic spline function and so may be represented as in (2.21) with $a_i = f''(x_{in})$, $b_i = f(x_{in})$, $i = 1, 2, \cdots, n$, and

$$
a_0 = a_1 = 0 \;, \quad a_n = a_{n+1} = 0 \;. \tag{6.26}
$$

Setting $a = (a_2, a_3, \cdots, a_{n-1})^T$ and $b = (b_1, b_2, \cdots, b_n)^T$, we have the by now usual equation

$$
T\,a = S\,b \;, \tag{6.27}
$$

with the matrices S and T as in (2.29)–(2.31).

Now, using Exercise (3.14), the problem (6.19) is equivalent to

$$
\begin{aligned}
&\text{minimize} \quad \|\, b - y\,\|_1 + \tfrac{1}{2}\,\alpha^2 \,\langle\, a\,,\, T\,a\,\rangle \\
&\text{subject to} \quad T\,a = S\,b \;,
\end{aligned} \tag{6.28}
$$

where $\alpha^2 = 2\,n\,h^4$. This is the finite-dimensional computational problem. Now, incorporate the switch to (6.22) by writing

$$
b = X\,\gamma + \beta_{\text{ext}} \;, \tag{6.29}
$$

where β_{ext} is the "extended" β, i.e.,

$$
\beta_{\text{ext}} = \bigl[\, 0 \mid \beta^T \mid 0 \,\bigr]^T , \tag{6.30}
$$

with $\beta \in \mathbb{R}^{n-2}$, and

$$
X = \bigl[\, 1\!\!1 \mid x \,\bigr] \in \mathbb{R}^{n \times 2} \;, \tag{6.31}
$$

in which $x = (\,x_{1,n}, x_{2,n}, \cdots, x_{n,n}\,)^T$. Note that $SX = 0$. (This is the analogue of $f''(x) = 0$ for all linear functions f.) Then, with b as in (6.29), we have $S\,b = \Sigma\,\beta$, with Σ equal to the matrix S with the first and last columns removed. Thus, $\Sigma \in \mathbb{R}^{(n-2)\times(n-2)}$ is tridiagonal and nonsingular.

Eliminating the constraint $T\,a = S\,b = \Sigma\,\beta$, the problem (6.28) now translates to

$$\text{(6.32)}\qquad \begin{array}{ll} \text{minimize} & \|\,\beta_{\text{ext}} + X\,\gamma - y\,\|_1 + \tfrac{1}{2}\,\alpha^2\,\langle\,\beta\,,\,M\beta\,\rangle \\ \text{subject to} & \gamma \in \mathbb{R}^2\ ,\ \beta_{\text{ext}} \text{ and } \beta \text{ as in (6.30)}\ , \end{array}$$

where

$$\text{(6.33)}\qquad M = \Sigma^T\,T^{-1}\,\Sigma\ .$$

We proceed to calculate the dual problem. First, getting rid of the ℓ^1 norm leads to the problem

$$\text{(6.34)}\qquad \begin{array}{ll} \text{minimize} & \langle\,\mathbb{1}\,,\,s\,\rangle + \tfrac{1}{2}\,\alpha^2\,\langle\,\beta\,,\,M\beta\,\rangle \\ \text{subject to} & s \geqslant \beta_{\text{ext}} + X\,\gamma - y\ , \\ & s \geqslant -\beta_{\text{ext}} - X\,\gamma + y\ . \end{array}$$

(6.35) Exercise. Show that (6.34) and (6.32) are equivalent.

Next, we determine the Lagrangian for (6.34). Partitioning the dual variables the way we did for β_{ext},

$$\text{(6.36)}\qquad \lambda_{\text{ext}} = \begin{bmatrix} \lambda_1 \\ --- \\ \lambda \\ --- \\ \lambda_n \end{bmatrix}\ ,\qquad \mu_{\text{ext}} = \begin{bmatrix} \mu_1 \\ --- \\ \mu \\ --- \\ \mu_n \end{bmatrix}\ ,$$

we have $\langle\,\lambda_{\text{ext}} - \mu_{\text{ext}}\,,\,\beta_{\text{ext}}\,\rangle = \langle\,\lambda - \mu\,,\,\beta\,\rangle$, so that the Lagrangian may be written as

$$\text{(6.37)}\quad \mathfrak{L}(\,\beta, s\,;\,\lambda, \mu\,) = \langle\,\mathbb{1} - \lambda_{\text{ext}} - \mu_{\text{ext}}\,,\,s\,\rangle - \langle\,\gamma\,,\,X^T\,(\,\lambda_{\text{ext}} - \mu_{\text{ext}}\,)\,\rangle + \langle\,\lambda - \mu\,,\,\beta\,\rangle + \tfrac{1}{2}\,\alpha^2\,\langle\,\beta\,,\,M\,\beta\,\rangle - \langle\,\lambda_{\text{ext}} - \mu_{\text{ext}}\,,\,y\,\rangle\ .$$

We now must compute

$$\text{(6.38)}\qquad \mathfrak{K}(\lambda, \mu) = \min_{\beta,\gamma,s}\ \mathfrak{L}(\,\beta, \gamma, s\,;\,\lambda_{\text{ext}}, \mu_{\text{ext}})\ .$$

Considering that there are no constraints, the minimization over s and γ yields $-\infty$ whenever $\mathbb{1} - \lambda_{\text{ext}} - \mu_{\text{ext}} \neq 0$ and/or $X^T(\,\lambda_{\text{ext}} - \mu_{\text{ext}}\,) \neq 0$. So, consider the case where these two quantities indeed vanish. Setting the gradient with respect to β equal to 0 gives $\beta = -\alpha^{-2}\,M^{-1}\,(\,\lambda - \mu\,)$, and then

$$\mathfrak{K}(\,\lambda, \mu) = -(2\,\alpha^2)^{-1}\,\langle\,\lambda - \mu\,,\,M^{-1}\,(\,\lambda - \mu\,)\,\rangle - \langle\,\lambda_{\text{ext}} - \mu_{\text{ext}}\,,\,y\,\rangle\ .$$

Thus, except for some cleaning up, the dual problem is

$$
(6.39)\quad
\begin{aligned}
&\text{maximize} && -(2\alpha^2)^{-1}\langle \lambda-\mu, M^{-1}(\lambda-\mu)\rangle \\
& && \qquad -\langle \lambda_{\text{ext}}-\mu_{\text{ext}}, y\rangle \\
&\text{subject to} && 1\!\!1-\lambda_{\text{ext}}-\mu_{\text{ext}}=0,\ \lambda_{\text{ext}}\geqslant 0,\ \mu_{\text{ext}}\geqslant 0, \\
& && X^T(\lambda_{\text{ext}}-\mu_{\text{ext}})=0.
\end{aligned}
$$

Now, the constraints $X^T(\lambda_{\text{ext}}-\mu_{\text{ext}})=0$ are used to eliminate $\lambda_1-\mu_1$ and $\lambda_n-\mu_n$, resulting in

$$
(6.40)\qquad \langle \lambda_{\text{ext}}-\mu_{\text{ext}}, y\rangle = \langle \lambda-\mu, \widetilde{y}\rangle,
$$

where $\widetilde{y}\in\mathbb{R}^{n-2}$ is defined as

$$
(6.41)\qquad \widetilde{y}_i = y_i - y_1\,\frac{x_{n,n}-x_{in}}{x_{n,n}-x_{1,n}} - y_n\,\frac{x_{in}-x_{1,n}}{x_{n,n}-x_{1,n}}
$$

for $i=2,3,\cdots,n-1$. Note the similarity with (6.20).

(6.42) EXERCISE. Show that the constraint $X^T(\lambda_{\text{ext}}-\mu_{\text{ext}})=0$ together with the partitioning (6.36) implies (6.40)–(6.41). [Hint: Redefine X in (6.31) as $X=[\,\ell\mid r\,]$, where $\ell,\ r\in\mathbb{R}^n$ with

$$
\ell_i = \frac{x_{n,n}-x_{in}}{x_{n,n}-x_{1,n}}\,, \qquad r_i = \frac{x_{in}-x_{1,n}}{x_{n,n}-x_{1,n}}\,.
$$

Then, $X\gamma$ still represents all polynomials of degree $\leqslant 1$. Now, the constraint $X^T(\lambda_{\text{ext}}-\mu_{\text{ext}})=0$ takes a very simple form.]

Returning to (6.39), with (6.40), the objective function now only depends on $\lambda-\mu$, and the constraints may be particularized to them. Setting $\theta=-\alpha^{-2}(\lambda-\mu)$, the constraints then are $-\alpha^{-2}\,1\!\!1\leqslant\theta\leqslant\alpha^{-2}\,1\!\!1$, but now the vector $1\!\!1$ is an element of $\mathbb{R}^{n-2}$. So, finally, the dual problem is

$$
(6.43)\quad
\begin{aligned}
&\text{maximize} && -\tfrac{1}{2}\langle\theta, M^{-1}\theta\rangle + \langle\theta,\widetilde{y}\rangle \\
&\text{subject to} && -\alpha^{-2}\,1\!\!1\leqslant\theta\leqslant\alpha^{-2}\,1\!\!1.
\end{aligned}
$$

The tie-in between the primal and dual problems is that

$$
(6.44)\qquad a=\Sigma^{-T}\theta \quad,\quad T\,a=\Sigma\,\beta\,.
$$

So, once the dual problem has been solved, we determine a and then β and β_{ext}. Then, the linear part $X\gamma$ is determined as the (not necessarily unique) solution to

$$
(6.45)\qquad \text{minimize}\quad \|X\gamma-y+\beta_{\text{ext}}\|_1 \quad\text{subject to}\quad \gamma\in\mathbb{R}^2\,.
$$

This is a simple linear regression problem where the usual least-squares estimator is replaced by the least-absolute-deviations estimator.

EXERCISES: (6.5), (6.9), (6.23), (6.35), (6.42).

7. Active constraint set methods

In this section, we develop efficient methods for the solution of the dual problems (6.6) and (6.43)-(6.45). The methods are active constraint set methods, prototypical of which is the NNLS procedure of LAWSON and HANSON (1995), see also BJÖRCK (1996), for the nonnegatively constrained least-squares problem

$$\text{(7.1)} \qquad \text{minimize} \quad \| Ax - b \|^2 \quad \text{subject to} \quad x \geqslant 0 ,$$

with $A \in \mathbb{R}^{n \times m}$, $b \in \mathbb{R}^n$, and A having independent columns. The inequality constraint $x \geqslant 0$ is to be interpreted componentwise. If A indeed has independent columns, then the solution of (7.1) is unique (solutions always exist).

We first give an outline of "our" interpretation of the NNLS algorithm and then specialize to the problems of interest. The idea of active constraint set methods is to guess which constraints in (7.1) are "active". Suppose that x^* is the solution of (7.1), and define Π^* as $\Pi^* = \Pi(x^*)$, where, for any $x \in \mathbb{R}^m$,

$$\text{(7.2)} \qquad \Pi(x) = \{ \, 1 \leqslant j \leqslant m \mid x_j = 0 \, \} .$$

The constraints $x_j \geqslant 0$ for which $x_j^* = 0$ are "active". The point is that the "inactive" constraints do not matter, but of course the problem is figuring out which ones they are.

(7.3) EXERCISE. Let x^* be the solution of (7.1). Show that x^* is also the unique solution of

$$\text{minimize} \quad \| Ax - b \|^2 \quad \text{subject to} \quad x_j = 0 \quad \text{for all} \quad j \in \Pi^* .$$

So, we start by guessing x^* and Π^*; e.g., $x^{[0]} = \mathbb{1}$ and $\Pi^{[0]} = \emptyset$. There might be cases where better choices suggest themselves, but that is really a separate issue. Of course, these guesses are not right, and we must make better guesses, in an iterative manner.

We now describe the general step of the algorithm. Suppose we have a guess $x^{[k]}$ for the solution of (7.1), and let $\Pi^{[k]} = \Pi(x^{[k]})$. Now, compute $z^{[k+1]}$, the solution to

$$\text{(7.4)} \qquad \begin{aligned} &\text{minimize} && \| Ax - b \|^2 \\ &\text{subject to} && x_j = 0 \quad \text{for all} \quad j \in \Pi(x^{[k]}) . \end{aligned}$$

Note that we may eliminate the x_j with $x_j = 0$, so that we must solve

$$(7.5) \qquad \begin{aligned} [\, A^T(A x - b) \,]_j &= 0 \,, \quad j \notin \Pi^{[k]} \,, \\ x_j &= 0 \,, \quad j \in \Pi^{[k]} \,. \end{aligned}$$

Does $z^{[k+1]}$ solve the original problem (7.1)? The first thing to verify is whether $z^{[k+1]}$ satisfies $z^{[k+1]} \geqslant 0$. If this does not hold, then it obviously is not the solution of (7.1). Even if it did, that still does not mean it is.

First, suppose that $z^{[k+1]}$ does not satisfy the constraints. Can we use $z^{[k+1]}$ to make a better guess for the solution than $x^{[k]}$? Indeed, we obtain a new and improved guess by moving from $x^{[k]}$ toward $z^{[k+1]}$ until we hit the boundary. Thus, we take

$$(7.6) \qquad x^{[k+1]} = x^{[k]} + t^{[k]} \, (\, z^{[k+1]} - x^{[k]} \,) \,,$$

where $t^{[k]}$ is given by

$$(7.7) \qquad t^{[k]} = \min \left\{ \frac{x_j^{[k]}}{z_j^{[k+1]} - x_j^{[k]}} \;\middle|\; z_j^{[k+1]} - x_j^{[k]} < 0 \right\} .$$

Note that if $x_j^{[k]} = 0$ then $j \in \Pi^{[k]}$, and so $z_j^{[k+1]} = 0$ as well. The new active constraint set is then $\Pi^{[k+1]} = \Pi(x^{[k+1]})$.

How do we know that we have an improved guess? Observe that surely $\| \, Az^{[k+1]} - b \, \|^2 \lneqq \| \, A x^{[k]} - b \, \|^2$, so that, by the convexity of $\| A x - b \, \|^2$,

$$(7.8) \qquad \| \, A x^{[k+1]} - b \, \|^2 \lneqq \| \, A x^{[k]} - b \, \|^2 \,.$$

So, the objective function is strictly decreasing. Also, note that the index set $\Pi^{[k+1]}$ has at least one more element than $\Pi^{[k]}$. Now, increment k by 1 and start over with (7.4). Then, since $\Pi^{[k]}$ can only grow so big, eventually we arrive at the situation where $z^{[k+1]}$ does satisfy the constraints, $z^{[k+1]} \geqslant 0$, and we define

$$(7.9) \qquad x^{[k+1]} = z^{[k+1]} \quad \text{if} \quad z^{[k+1]} \geqslant 0 \,.$$

Now, we need to find out whether $x^{[k+1]}$ solves (7.1): We must verify whether $x^{[k+1]}$ satisfies the linear complementarity conditions,

$$(7.10) \qquad \begin{aligned} & A^T(A x^{[k+1]} - b) \geqslant 0 \quad , \quad x^{[k+1]} \geqslant 0 \,, \\ & x_j^{[k+1]} \, [\, A^T(A x^{[k+1]} - b) \,]_j = 0 \,, \quad j = 1, 2, \cdots, m \,. \end{aligned}$$

see Hiriart-Urruty and Lemaréchal (1993) or Example (9.2.10) in Volume I. One verifies that the complementary part is satisfied by virtue of (7.5). So, it boils down to whether

$$(7.11) \qquad [\, A^T(A x^{[k+1]} - b) \,]_j \geqslant 0 \quad \text{for all} \quad j \in \Pi^{[k+1]} \,.$$

If this holds, then $x^{[k+1]}$ solves (7.1), and we are done. If it fails, say for some $\ell \in \Pi^{[k]}$, then the active constraint $x_\ell = 0$ should not have been active. To see why, define $x(t) = x^{[k+1]} + t\, e_\ell$, where e_ℓ is the ℓ-th unit vector in the common basis for $\mathbb{R}^m$, and observe that

$$\frac{d}{dt} \| A\, x(t) - b \|^2 \Big|_{t=0} = 2\,[\, A^T(\, A\, x^{[k+1]} - b\,)\,]_\ell < 0 \,.$$

Now, for $t^* > 0$ but small enough, $x(t^*)$ satisfies the constraints, and

$$\| A\, x(t^*) - b \|^2 \lneqq \| A\, x(0) - b \|^2 = \| A\, x^{[k+1]} - b \|^2 \,.$$

Thus, by moving slightly away from the boundary, we get a smaller value of the objective function. This shows that the constraint $x_\ell = 0$ should not have been active.

So, we must remove (at least one of) the indices ℓ from $\Pi^{[k]}$ for which $[\, A^T(\, A\, x^{[k+1]} - b\,)\,]_\ell < 0$. Thus, e.g., define

$$\Pi^{[k+1]} = \Pi^{[k]} \setminus \{\, \ell \,\} \,. \tag{7.12}$$

Now again we increment k by one and resume at (7.4).

Does this procedure converge? The answer is yes. In fact, as we now show, the procedure terminates after a finite number of steps. Note that the active constraint set $\Pi^{[k]}$ sometimes gets bigger and sometimes gets smaller. The one "invariant" is that the objective function is strictly decreasing,

$$\| A\, x^{[k+1]} - b \|^2 \lneqq \| A\, x^{[k]} - b \|^2 \,, \qquad k = 1, 2, \cdots \,. \tag{7.13}$$

Also note that every so often we have a solution of a problem of the form (7.5). By (7.13), then we cannot revisit this problem with the same active constraint set $\Pi^{[k]}$. Since there are only finite many such problems/active constraint sets, the procedure must terminate in a finite number of steps. On a practical note, there are 2^m distinct possible sets $\Pi^{[k]}$, so the number of iterations could be prohibitively large, but the property (7.13) appears to rule out visits to many configurations $\Pi^{[k]}$.

We do not give a formal statement of the procedure above but instead specialize it to the solution of "our" dual problems.

Total-variation penalization. It is useful to write the problem (6.6) as

$$\begin{array}{ll} \text{minimize} & \tfrac{1}{2} \langle\, x\,,\, Mx \,\rangle + \langle\, D\, y_n \,,\, x \,\rangle \\ \text{subject to} & -\alpha\, \mathbb{1} \leqslant x \leqslant \alpha\, \mathbb{1} \,, \end{array} \tag{7.14}$$

where $M = DD^T$ is the negative discrete Laplacian. In the computations to follow, we need the Cholesky factorization of the matrix M,

$$M = LL^T \,, \tag{7.15}$$

with L lower triangular and bidiagonal. So,

$$M = \begin{bmatrix} 2 & -1 & & & \\ -1 & 2 & -1 & & \\ & -1 & \ddots & \ddots & \\ & & \ddots & \ddots & -1 \\ & & & -1 & 2 \end{bmatrix} , \quad L = \begin{bmatrix} d_1 & & & & \\ \ell_2 & d_2 & & & \\ & \ell_3 & d_3 & & \\ & & \ddots & \ddots & \\ & & & \ell_n & d_n \end{bmatrix} ,$$

with zero components not indicated.

We briefly highlight the salient points when specializing the NNLS algorithm to solving (7.14). (a) The initial guess we use is

$$x^{[0]} = 0 \,, \quad \Pi^{[0]} = \emptyset \,. \tag{7.16}$$

(b) When guessing the active constraints in (7.14), it obviously makes sense to treat them in pairs. The active constraint set $\Pi(x)$ is then organized so that, for $j = 1, 2, \cdots, m$,

$$\begin{aligned} j \in \Pi(x) &\iff x_j = +\alpha \,, \\ -j \in \Pi(x) &\iff x_j = -\alpha \,. \end{aligned} \tag{7.17}$$

(c) The computation of the solution of the system,

$$\begin{aligned} &\text{minimize} && \tfrac{1}{2} \langle\, x \,,\, Mx \,\rangle - \langle\, x \,,\, D\, y_n \,\rangle \\ &\text{subject to} && x_j = +\alpha \quad \text{for} \quad j \in \Pi^{[k]} \,, \\ & && x_j = -\alpha \quad \text{for} \quad -j \in \Pi^{[k]} \,, \end{aligned} \tag{7.18}$$

amounts to solving the linear equations

$$\begin{aligned} [\, Mx \,]_j = [\, D\, y_n \,]_j \,, &\quad j \notin \Pi^{[k]} \cup (\, -\Pi^{[k]} \,) \,, \\ x_j = +\alpha \,, &\quad j \in \Pi^{[k]} \,, \\ x_j = -\alpha \,, &\quad -j \in \Pi^{[k]} \,. \end{aligned} \tag{7.19}$$

Eliminating the constrained x_j then yields a system

$$M^{[k]}\, x = q^{[k]} \,, \tag{7.20}$$

where $q^{[k]}$ is the appropriate right-hand side and $M^{[k]}$ is obtained from M by deleting the rows and columns with indices appearing in $\Pi^{[k]}$. So then $M^{[k]}$ is block-diagonal, with each block being a negative discrete Laplacian of smaller order,

$$M^{[k]} = \begin{bmatrix} M_1 & & & \\ & M_2 & & \\ & & \ddots & \\ & & & M_p \end{bmatrix} . \tag{7.21}$$

One verifies that the Cholesky factorizations of the M_i are then given as

$$M_i = L_i L_i^T \ , \tag{7.22}$$

where L_i is a leading principal minor of the Cholesky factor L; see (7.15). So the factorization (7.15) covers all the occurrences of (7.22).

(d) The computation of $x^{[k+1]}$ if $z^{[k+1]}$ does not satisfy the constraints $-\alpha \mathbb{1} \leqslant z^{[k+1]} \leqslant \alpha \mathbb{1}$ goes the same way as in (7.6)-(7.7), with the obvious modifications. We did add a twist though: For each block of equations (7.20), we compute the solution $x^{[k+1],i}$ and move from the old $x^{[k],i}$ toward the new $x^{[k+1],i}$ until we hit the boundary. Because the M_i are tridiagonal, this is done without reference to the other blocks.

(e) The necessary and sufficient conditions for a minimum of (7.14) are

$$\begin{aligned} &M\,x^{[k+1]} - D\,y_n \geqslant 0 \quad , \quad -\alpha \mathbb{1} \leqslant x^{[k+1]} \leqslant \alpha \mathbb{1} \ , \\ &\big(x_j^{[k+1]} + s_j\,\alpha \big)\,[\,M\,x^{[k+1]} - D\,y_n\,]_j = 0 \ , \quad j = 1, 2, \cdots, m \ , \end{aligned} \tag{7.23}$$

where $s_j = +1$ if $j \in \Pi^{[k]}$, $s_j = -1$ if $-j \in \Pi^{[k]}$, and $s_j = 0$ otherwise. A curious point is that the optimality check is *always* successful; i.e., as soon as we reach a feasible solution, it is optimal. This appears to be closely related to the fact that the inverse of M (and hence the inverse of every principal minor of M) is positive componentwise.

All of this leads to the MATLAB implementation listed in Appendix 5.

Least-absolute-deviations smoothing splines. Comparison with the dual formulation (6.6) of the total-variation penalization problem reveals that the dual least-absolute-deviations problem (6.43) is different only in the system matrix. Whereas in (6.6) we had the negative discrete Laplacian, the system matrix in (6.43) is $M^{-1} = \Sigma^{-1}\,T\,\Sigma^{-T}$. Thus, the previous algorithm will work, except that the problems (7.4) must be replaced by

$$\begin{aligned} &\text{minimize} \quad \tfrac{1}{2}\,\langle\, x \,,\, M^{-1}\,x \,\rangle - \langle\, x \,,\, \widetilde{y} \,\rangle \\ &\text{subject to} \quad \Pi_k\,x = \alpha\,\mathbb{1} \ . \end{aligned} \tag{7.24}$$

Here, with e_j the j-th unit vector of the common basis for $\mathbb{R}^m$, the matrix Π_k consists of the rows $\pm\, e_j^T$ for $j \in \Pi^{[k]} \cup \big(-\Pi^{[k]} \big)$, such that the constraints in (7.24) and (6.43) coincide. Thus $\Pi_k \in \mathbb{R}^{p \times m}$, where p is the number of active constraints; i.e., the number of elements in $\Pi^{[k]}$.

Using Lagrange multipliers, we may then compute the solution of (7.24) by solving the system of linear equations,

$$\begin{aligned} M^{-1}\,x + \Pi_k^T\,\lambda &= \widetilde{y} \ , \\ \Pi_k\,x \qquad\quad &= 0 \ . \end{aligned} \tag{7.25}$$

With $a = \Sigma^{-T}\,x$, this system becomes

$$\left[\begin{array}{c|c} T & \Sigma\,\Pi_k^T \\ \hline \Pi_k\,\Sigma^T & 0 \end{array}\right] \left[\begin{array}{c} x \\ \hline \lambda \end{array}\right] = \left[\begin{array}{c} \widetilde{y} \\ \hline \alpha\,\mathbb{1} \end{array}\right] . \tag{7.26}$$

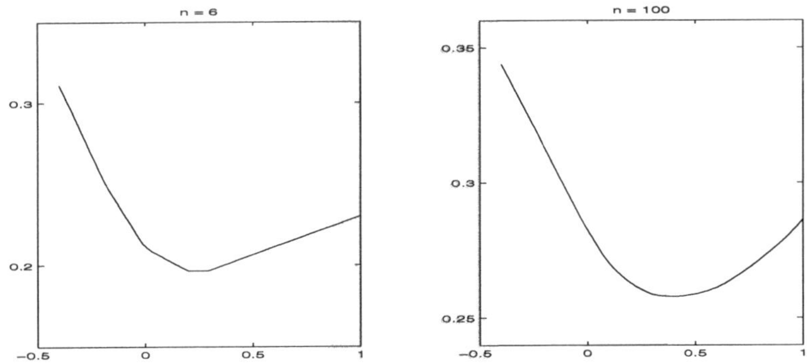

Figure 6.1. Graphs of the objective function in (6.33) for $y_{in} = \sin(2\pi x_{in}^6)$ and $\widetilde{b} = 0$ as a function of d for $n = 6$ and $n = 100$. Here, c is the median of $z_{in} - d\,x_{in} : i = 1, 2, \cdots .n$.

The coefficient matrix is not positive-definite, but it does have the nice factorization (note the minus sign)

$$(7.27) \qquad \begin{bmatrix} L & 0 \\ \Pi_k \Sigma^T L^{-T} & U_k^T \end{bmatrix} \begin{bmatrix} L^T & L^{-1} \Sigma \Pi_k^T \\ 0 & -U_k \end{bmatrix} ,$$

where $T = L L^T$ is the Cholesky factorization of T and $U_k \in \mathbb{R}^{p \times p}$ is upper triangular and may be obtained via the QR factorization,

$$(7.28) \qquad Q_k U_k = L^{-1} \Sigma \Pi_k^T .$$

Here, $Q \in \mathbb{R}^{m \times p}$ has orthonormal columns, so that $Q^T Q = I \in \mathbb{R}^{p \times p}$. Now, computing this QR factorization should be "easy" in that the matrix $L^{-1} \Sigma \Pi_k^T$ is the same as for the previous k, with one column added or removed. So, one should be able to efficiently *update* or *downdate* the factorization. (Updating is easy.) However, we shall leave this for the experts; see Björck (1996) and references therein. (In our implementation, we just slugged it out.)

To round it out, we give an exercise on how to compute the solution of (6.45).

(7.29) Exercise. Note that in (6.45) we may write $[X\gamma]_i = c + d\,x_{in}$ for all i. Then, once the optimal slope d has been determined, the intercept c equals

$$c = \operatorname{median}\big(y_{in} - (\beta_{\text{ext}})_i - d\,x_{in} \; : \; i = 1, 2, \cdots, n \big) .$$

So, the objective function in (6.45) may be viewed as a function of only the slope d. A graph of an example is shown in Figure 6.1: It is a piecewise linear convex function. For the typical application to LAD splines, the graph is very close to a V function (two straight lines). The graph immediately suggests an interesting way to solve the problem (6.45): Start with the interval $(-\infty, \infty)$, and note that the slopes of the tangent lines at the endpoints have opposite signs. Now, determine the intersection of these two tangent lines, and determine the slope of the tangent line at the intersection. Proceed with the interval for which the slopes of the tangent lines at the endpoints have opposite signs, etc. Moreover, it has a built-in stopping criterion; i.e., when the intersection of the tangent lines lies on the graph itself. Show that this algorithm stops in a finite number of steps.

EXERCISES: (7.3), (7.29).

8. Polynomials and local polynomials

The computational challenges of polynomial estimators are quite different from those encountered so far. The main hurdle is that the common basis for the polynomials is no good for numerical computations. Instead, one should work with orthogonal polynomials.

Consider the polynomial-sieve estimation problem

$$\text{(8.1)} \qquad \text{minimize} \quad \tfrac{1}{n} \sum_{i=1}^{n} |\, p(x_{in}) - y_{in} \,|^2 \quad \text{subject to} \quad p \in \mathcal{P}_r \ ,$$

where $\mathcal{P}_r$ is the customary vector space of all polynomials of order $\leqslant r$ (degree $\leqslant r-1$). We assume that the design points lie in the interval $[\,0\,,\,1\,]$ but make no assumptions on their distribution. How would one go about computing the solution?

A definitely bad idea is to represent the solution p as

$$\text{(8.2)} \qquad p(x) = \sum_{\ell=0}^{r-1} \varpi_\ell \, x^\ell \ , \quad x \in [\,0\,,\,1\,]$$

and solve for the coefficients $\varpi = (\varpi_0\,,\varpi_1\,,\cdots,\varpi_{r-1})^T$. The resulting normal equations for (8.1) are then $G_{rn}\varpi = \zeta$, with

$$\text{(8.3)} \qquad \begin{aligned} [\, G_{rn} \,]_{\ell,m} &= \tfrac{1}{n} \sum_{i=1}^{n} {x_{in}}^{\ell+m-2} \ , \\ \zeta_\ell &= \tfrac{1}{n} \sum_{i=1}^{n} y_{in} \, {x_{in}}^{\ell-1} \ , \quad \ell, m = 1, 2, \cdots, r-1 \ . \end{aligned}$$

The practical difficulty is that the Gram matrix G_{rn} is badly conditioned, even for small r: For the standard uniform design (13.2.2), as $n \to \infty$, we have that $G_{rn} \longrightarrow \mathfrak{H}_r$, the $r \times r$ Hilbert matrix,

$$\text{(8.4)} \qquad [\, \mathfrak{H}_r \,]_{\ell,m} = (\, \ell + m - 1\,)^{-1} \ , \quad \ell, m = 1, 2, \cdots, r \ ,$$

which is notoriously badly conditioned even for small values of m ($m = 5$ is bad already). For arbitrary asymptotically uniform designs, it is even worse.

A much better idea is to represent the solution of (8.1) in terms of the orthogonal polynomials associated with the discrete inner product

(8.5) $$\langle\!\langle\, f\,,g\,\rangle\!\rangle = \tfrac{1}{n}\sum_{i=1}^{n} f(x_{in})\,g(x_{in}) \;.$$

Let $P_0\,, P_1\,,\cdots, P_{n-1}$ be these orthogonal polynomials. They are uniquely determined by the conditions

(8.6) $$\begin{gathered} P_\ell(x) \text{ is a polynomial in } \ x \ \text{ of exact degree } \ \ell\;, \\ \langle\!\langle\, P_\ell\,, P_m\,\rangle\!\rangle = \begin{cases} 1 \;, & \text{if } \ \ell = m\;, \\ 0 \;, & \text{otherwise}\;. \end{cases} \end{gathered}$$

The solution of (8.1) is then given as

(8.7) $$p(x) = \sum_{\ell=0}^{r-1} p_\ell\, P_\ell(x)\;, \quad x\in[\,0\,,1\,]\;,$$

with

(8.8) $$p_\ell = \langle\!\langle\, y_n\,, P_\ell\,\rangle\!\rangle\;, \quad \ell = 1,2,\cdots,n-1\;,$$

where, with a slight abuse of notation,

(8.9) $$\langle\!\langle\, y_n\,, P_\ell\,\rangle\!\rangle = \tfrac{1}{n}\sum_{i=1}^{n} y_{in}\, P_\ell(x_{in})\;.$$

(8.10) EXERCISE. (a) Why are $P_0\,, P_1\,,\cdots, P_{n-1}$ *all* of the orthogonal polynomials associated with $\langle\!\langle\,\cdot\,,\cdot\,\rangle\!\rangle$?
(b) Verify (8.7)–(8.8) using (8.5).
(c) Show that $\frac{1}{n}\sum_{i=1}^{n} |\,p(x_{in})\,|^2 \leqslant \frac{1}{n}\sum_{i=1}^{n} |\,y_{in}\,|^2$.

This is nice, but how does one determine the P_k? In essence, it boils down to the three-point recurrence relation for orthogonal polynomials. For $k = 1,2,\cdots,n-1$, and suitable constants $a_k\,, b_k$ and c_k, one has

(8.11) $$P_{k+1}(x) = \big(\,a_k\,x + b_k\,\big)\,P_k(x) - c_k\,P_{k-1}(x)\;,$$

(8.12) $$P_0(x) = 1\;, \quad P_1(x) = (x-\overline{x})/s\;,$$

where $\overline{x}$ is the mean of the x_{in} and s^2 the mean of the $|\,x_{in}-\overline{x}\,|^2$.

A proof of the recurrence relation follows. The first observation is that $P_\ell\,$, $\ell = 0,1,\cdots,k$, is a basis for $\mathcal{P}_k$. The next step is to note that since P_{k+1} and P_k have exact degrees $k+1$ and k, there exists a constant a_k such that $P_{k+1}(x) - a_k x\,P_k(x)$ has degree $\leqslant k$. So, there exist coefficients δ_ℓ such that

(8.13) $$P_{k+1}(x) - a_k x\,P_k(x) = \sum_{\ell=0}^{k} \delta_\ell\,P_\ell(x)\;.$$

Now let us take the inner product with P_m , $m = 1, 2, \cdots, k-2$. Then,

$$\delta_m \langle\langle P_m , P_m \rangle\rangle = \langle\langle P_{k+1} , P_m \rangle\rangle - a_k \langle\langle Q_k , P_m \rangle\rangle ,$$

where $Q_k(x) \equiv x\, P_k(x)$. Now, $\langle\langle P_{k+1} , P_m \rangle\rangle = 0$ and

$$\langle\langle Q_k , P_m \rangle\rangle = \langle\langle P_k , Q_m \rangle\rangle = 0$$

since Q_m is a polynomial of degree $m+1 < k$. Consequently, $\delta_\ell = 0$ for $\ell = 0, 1, \cdots, k-2$, and what is left in (8.13) is the three-point recurrence relation, which may be written as (8.11).

Now, to compute the coefficients in (8.11), we take inner products again. From (8.11), taking inner products with P_k gives

$$\langle\langle P_{k+1} , P_k \rangle\rangle = \langle\langle a_k\, Q_k + b_k\, P_k , P_k \rangle\rangle - c_k \langle\langle P_{k-1} , P_k \rangle\rangle , \tag{8.14}$$

and in view of the orthonormality, then

$$b_k = -a_k \langle\langle Q_k , P_k \rangle\rangle . \tag{8.15}$$

Likewise, taking inner products with P_{k-1} yields

$$\langle\langle P_{k+1} , P_{k-1} \rangle\rangle = \langle\langle a_k\, Q_k + b_k\, P_k , P_{k-1} \rangle\rangle - c_k \langle\langle P_{k-1} , P_{k-1} \rangle\rangle , \tag{8.16}$$

from which we obtain

$$c_k = a_k \langle\langle Q_k , P_{k-1} \rangle\rangle . \tag{8.17}$$

Noting that

$$\begin{aligned} a_{k-1} \langle\langle Q_k , P_{k-1} \rangle\rangle &= \langle\langle P_k , a_{k-1}\, x P_{k-1} \rangle\rangle \\ &= \langle\langle P_k , (a_{k-1}\, x + b_{k-1})\, P_{k-1} - c_{k-1}\, P_{k-2} \rangle\rangle \\ &= \langle\langle P_k , P_k \rangle\rangle = 1 , \end{aligned}$$

as a refinement, one would obtain

$$c_k = a_k / a_{k-1} . \tag{8.18}$$

Either way, the polynomial P_{k+1} has been determined up to the factor a_k , which may be determined from the orthonormality requirement

$$\langle\langle P_{k+1} , P_{k+1} \rangle\rangle = 1 . \tag{8.19}$$

As a final comment, we note that in (8.11) it seems reasonable to replace $a_k\, x + b_k$ with $a_k\, (\, x - \overline{x}\,) + b_k$ and make the corresponding changes to the computational formulas. Note that besides computing the recurrence relation, we also computed the orthogonal polynomials at the design points.

In infinite-precision arithmetic, this works for $k = 1, 2, \cdots, n-1$, since the vector space of all polynomials restricted to the n distinct design points has dimension n, and so any (orthogonal) basis has n elements. In finite-precision arithmetic, it works quite well up to a point. By way of illustration, it is instructive to compute the matrices $U_r \in \mathbb{R}^{n \times r}$,

$$[\, U_r\,]_{i,\ell} = P_\ell(x_{in}) , \quad i = 1, 2, \cdots, n, \;\; \ell = 1, 2, \cdots, r , \tag{8.20}$$

Table 7.1. Values of $\| \frac{1}{n} U_r^T U_r - I_{r \times r} \|_2$ for $r = j\sqrt{n}$, $j = 3, 4, \cdots, 8$. The design points were $((i-1)/(n-1))^{3/2}$, $i = 1, 2, \cdots, n$. The results are acceptable up to $r \approx 6\sqrt{n}$. (The computations were done in double precision.)

r	$3\sqrt{n}$	$4\sqrt{n}$	$5\sqrt{n}$	$6\sqrt{n}$	$7\sqrt{n}$	$8\sqrt{n}$
$n = 100$	3.3e-14	4.3e-12	3.9e-09	2.6e-05	0.84	1.00
$n = 400$	1.8e-13	3.7e-12	2.3e-09	7.4e-06	0.13	0.99
$n = 1600$	5.6e-13	2.3e-11	1.3e-08	3.6e-05	0.45	1.00
$n = 6400$	1.5e-12	3.1e-11	1.7e-08	4.4e-05	0.50	1.00

that in exact arithmetic satisfy

$$\frac{1}{n} U_r^T U_r = I_{r \times r} , \tag{8.21}$$

where $I_{r \times r} \in \mathbb{R}^{r \times r}$ is the identity matrix. In finite-precision arithmetic, this is no longer true. To illustrate the severity, in Table 7.1 we show the discrepancy $\| \frac{1}{n} U_r^T U_r - I_{r \times r} \|_2$ for various values of r depending on the design size n. The results indicate that for the design in question one can compute polynomial regression estimators of degree $\approx 6\sqrt{n}$. For the design $x_{in} = ((i-1)/(n-1))^2$, it is lowered to about $5\sqrt{n}$, and for the uniform design it goes up to $7\sqrt{n}$. For the random design with uniform density, the cutoff appears to be $5\sqrt{n}$. Considering that the "optimal" r satisfies $r \ll \sqrt{n}$, we may consider this problem solved.

The case of local polynomial estimation differs significantly from the computations for the polynomial sieve. Rather than one big minimization problem, now many small ones must be solved. Moreover, only polynomials of low degree are involved.

We recall the problem. Fix $x \in [0, 1]$, let $m \geqslant 0$ be the order of the local polynomial, and choose A to be a reasonable nonnegative kernel satisfying the conditions (16.1.7). Let $h > 0$ be the smoothing parameter. We must compute $p(x)$, where p is the solution to

$$\text{minimize} \quad \sum_{i=1}^{n} w_i \, | p(X_i) - Y_i |^2 \quad \text{subject to} \quad p \in \mathcal{P}_m , \tag{8.22}$$

in which

$$w_i = A_h(x - X_i) \Big/ \sum_{i=1}^{n} A_h(x - X_i) , \quad i = 1, 2, \cdots, n . \tag{8.23}$$

Note that $w_i \geqslant 0$, but we assume that at least one of them is strictly positive.

Even though the order of the polynomials is low, for the same reasons as before, it is still a good idea to use orthogonal polynomials. The inner product is now

$$(8.24) \qquad \langle\!\langle\, f\,,g\,\rangle\!\rangle = \sum_{i=!}^{n} w_i\, f(X_i)\, g(X_i)\ .$$

We denote the associated orthogonal polynomials by $P_o(t), P_1(t), \cdots$, $P_{r-1}(t)$, where r is the number of positive w_i (assuming the X_i are distinct). The polynomials may be generated by (8.15), (8.17), and (8.19), with the initialization

$$(8.25) \qquad P_0(t) = 1 \quad , \quad P_1(t) = (t - \widetilde{X}\,)/S\ ,$$

where

$$(8.26) \qquad \widetilde{X} = \sum_{i=1}^{n} w_i X_i \quad , \quad S^2 = \sum_{i=1}^{n} w_i\, |\, X_i - \widetilde{X}\,|^2\ .$$

(8.27) EXERCISE. For $m = 1$, show that

$$p(x) = \sum_{i=1}^{n} w_i Y_i + S^{-2}\,(x - \widetilde{X}\,) \sum_{i=1}^{n} w_i Y_i(X_i - \widetilde{X}\,)\ .$$

Note that the first sum is the Nadaraya-Watson estimator.

EXERCISES: (8.10), (8.27).

9. Additional notes and comments

Ad § 1: Unfortunately, we omitted consideration of constrained least-squares splines. This is quite complicated, even for simple nonnegativity constraints, since there may be extra breakpoints besides the design points; see, e.g., ELFVING and ANDERSSON (1988). However, as an approximation, one may consider the usual splines, which are piecewise polynomial on each interval $[\,x_{in}, x_{i+1,n}\,]$, constrained to be nonnegative or isotone on each interval, see WANG and LI (2008), or one may just take the regular smoothing spline and make it monotone using the pool-adjacent-violators algorithm of KRUSKAL (1964) in the style of EGGERMONT and LARICCIA (2000a), see Volume I, Chapter 6.

Ad § 5: The development here shows how the B-splines *might* have arisen. The standard reference for B-spline computations is DE BOOR (1978). The procedure to compute the smoothing spline is outlined by ANSELONE and LAURENT (1968). An explicit algorithm for quintic spline interpolation is provided by HERRIOT and REINSCH (1976). For spline interpolation of arbitrary order, see HERRIOT and REINSCH (1973) and, of course, the next chapter.

Ad § 7: In a strange turn of events, for large-scale problems, Madsen and Nielsen (1993) solve problems like (6.6) by transforming it back into the original (6.2) and using "interior point" methods.

Ad § 8: The first concise and sound source on discrete orthogonal polynomials for polynomial regression was Forsythe (1957).

20

Kalman Filtering for Spline Smoothing

1. And now, something completely different

In this chapter, we give an account of some Bayesian aspects of spline smoothing for nonparametric regression. The Bayesian view leads to concepts that do not arise in the distinctly non-Bayesian presentation of the previous chapters. Indeed, novel solutions appear to which non-Bayesians cannot object; e.g., the various developments culminating in the *Kalman filter* for computing spline estimators of arbitrary order m. The efficient computation of the GML and GCV functionals by way of the Kalman filter should not be forgotten. Things already get dicey when justifying the GML method for choosing the smoothing parameter, but it must be admitted that this seems to work very well (see Chapter 23). However, the authors draw the line at Bayesian confidence bands for the unknown regression function.

We recall the non-Bayesian model for nonparametric regression,

$$(1.1) \qquad y_{in} = f_o(x_{in}) + d_{in} \ , \quad i = 1, 2, \cdots, n \ ,$$

where $0 < x_{1,n} < x_{2,n} < \cdots < x_{n,n} < 1$ are the design points (in this chapter, it is convenient to *exclude* the endpoints 0 and 1 from being design points), the vector $d_n = (d_{1,n}, d_{2,n}, \cdots, d_{n,n})^T$ contains the noise, and f_o is a smooth function on $[\,0\,,\,1\,]$, say $f_o \in W^{m,2}(0,1)$ for some integer $m \geqslant 1$. We are exclusively concerned with the case of iid normal errors,

$$(1.2) \qquad d_n \sim \text{Normal}(\,0\,, \sigma^2 I\,) \ ,$$

with σ^2 unknown. Our interest is in the smoothing spline estimator, defined as the solution to

$$(1.3) \qquad \begin{array}{ll} \text{minimize} & \frac{1}{n} \sum\limits_{i=1}^{n} |\, f(x_{in}) - y_{in} \,|^2 + h^{2m} \,\|\, f^{(m)} \,\|^2 \\ \text{subject to} & f \in W^{m,2}(0,1) \ . \end{array}$$

P.P.B. Eggermont, V.N. LaRiccia, *Maximum Penalized Likelihood Estimation*,
Springer Series in Statistics, DOI 10.1007/b12285_9,

To keep things as simple as possible in this introduction, we consider the spline interpolator with exact data; i.e., the solution to

$$\text{(1.4)}\qquad \begin{array}{ll} \text{minimize} & \| f^{(m)} \|^2 \\ \text{subject to} & f \in W^{m,2}(0,1) \ , \ f(x_{in}) = y_{in} \ , \ i = 1, 2, \cdots, n \ . \end{array}$$

There is an implicit shift in emphasis here. Whereas for (1.3) we were mostly interested in estimating $f_o(x_{in})$, $i = 1, 2, \cdots, n$, for the spline interpolation problem the emphasis is on estimating the function over the whole interval, $f_o(t)$, $t \in [0, 1]$.

The basic tenet of Bayesian estimation is to prescribe a prior distribution on the parameter to be estimated (f_o in our case) and then to compute its posterior distribution given the data. The parameter may then be estimated as the conditional mean or conditional mode or some such thing. In a finite-dimensional setting, this may be implemented without conceptual difficulties (see § 2). For the spline interpolation problem (1.4), things are complicated by the fact that we are dealing with infinite-dimensional objects. The customary way to prescribe a prior distribution on $f_o \in W^{m,2}(0,1)$ is by interpreting it to be a section of the *sample path* of the stochastic process

$$\text{(1.5)}\qquad X(t) = P(t) + \kappa\, \Phi(t) \ , \quad t \geqslant 0 \ ,$$

in which κ is unknown with

$$\text{(1.6)}\qquad P(t) = \sum_{k=0}^{m-1} p_k\, t^k \quad \text{and} \quad \Phi(t) = \int_0^t (t - \tau)^{m-1}\, dW(\tau) \ .$$

Here, $W(t)$, $t \geqslant 0$, is the standard Wiener process and the p_k are zero-mean random variables with a "diffuse" normal prior, independent of the Wiener process; that is,

$$\text{(1.7)}\qquad (p_0, p_1, \cdots, p_{m-1})^T \sim \text{Normal}(0, \xi^2 I) \ ,$$

and the diffusivity arises by letting $\xi \to \infty$.

Actually, the above sounds persuasive, but it is in effect nonsense. In particular, with probability 1, the sample paths of the Gaussian process (1.5) do not, repeat do not, lie in $W^{m,2}(0,1)$, so (1.5) does not constitute a prior distribution on $W^{m,2}(0,1)$. But then saying that the regression $f_o \in W^{m,2}(0,1)$ is a sample path of the Gaussian process in question does not make sense. Getting slightly ahead of ourselves, the correct interpretation is that the basic estimation problem for the Gaussian process (1.5) (estimate a sample path, given some points on the path) has the same solution as the smoothing spline problem for the standard nonparametric regression problem. In this chapter, we work the connection between the estimation problem for the Gaussian process and the spline smoothing for nonparametric regression.

We shall not attempt to make complete sense out of (1.5)–(1.6); e.g., what a Wiener process is and how one integrates it. For this, see, e.g.,

SHORACK (2000). However, the following must be noted. First, in (1.5), the process $X(t)$, $t \geqslant 0$, is zero-mean *Gaussian*, meaning that, for any $0 \leqslant t_1 < t_2 < \cdots < t_n$, the random variables $X(t_1), X(t_2), \cdots, X(t_n)$ are jointly normal. Specifically,

$$(1.8)\qquad \bigl(X(t_1), X(t_2), \cdots, X(t_n) \bigr)^T \sim \text{Normal}\bigl(0 \,,\, \xi^2 T T^T + \kappa^2 V \bigr) \,,$$

with $V \in \mathbb{R}^{n \times n}$ defined by

$$(1.9)\qquad V_{i,j} = \int_0^1 (t_i - \tau)_+^{m-1} (t_j - \tau)_+^{m-1} \, d\tau \;, \quad i,j = 1, 2, \cdots, n \,,$$

and $T \in \mathbb{R}^{n \times m}$ defined by

$$(1.10)\qquad T_{i,k} = t_i^k \;, \quad i = 1, 2, \cdots, n, \; k = 0, 1, \cdots, m-1 \,.$$

Thus, the stochastic process (1.5) is completely characterized by the fact that it is a zero-mean Gaussian process with covariance function

$$(1.11)\qquad \mathbb{E}[\, X(t)\, X(s) \,] = \xi^2 \sum_{k=0}^{m-1} t^k s^k + \kappa^2 \int_0^1 (t - \tau)_+^{m-1} (s - \tau)_+^{m-1} \, d\tau$$

for all $s, t \geqslant 0$.

Second, our interest is in the basic problem of estimating the *sample path* given a finite number of points on it. Specifically, for $t \in [\,0\,,\,1\,]$ fixed, the problem is to

$$(1.12)\qquad \begin{array}{ll} \text{estimate} & X(t) \\ \text{given the data} & y_i = X(t_i) \;, \; i = 1, 2, \cdots, n \,, \end{array}$$

where $t_1, t_2, \cdots, t_n$, are strictly positive "design" points. Obviously, this problem may be solved by determining the conditional distribution of $X(t)$, which, by Bayes' rule, may be related to the joint distributions of $\bigl(X(t_1), X(t_2), \cdots, X(t_n) \bigr)^T$ and

$$(1.13)\qquad w = \bigl(X(t), X(t_1), X(t_2), \cdots, X(t_n) \bigr)^T \,.$$

This is precisely the information provided by the interpretation of Gaussian processes. The "optimal" estimator is then the conditional mean,

$$(1.14)\qquad \widehat{X}(t) = \mathbb{E}[\, X(t) \mid y_1, y_2, \cdots, y_n \,] \,.$$

Of course, since we are dealing with multivariate normal random variables, the conditional mean is equal to the conditional mode. It is also equal to the constrained mode, or constrained maximum likelihood estimator, defined as the solution to

$$(1.15)\qquad \begin{array}{ll} \text{minimize} & u^T \bigl(\xi^2 T T^T + \kappa^2 V_t \bigr)^{-1} u \\ \text{subject to} & Au = y \,, \end{array}$$

where V_t is the covariance matrix of w, see (1.13), and $A \in \mathbb{R}^{n \times (n+1)}$ is such that if $u = (u_0, u_1, \cdots, u_n)^T$, then $Au = (u_1, u_2, \cdots, u_n)^T$. As already stated, all of these approaches yield the same answer, but the authors find the constrained maximum likelihood formulation to be more transparent.

It is crucial to the whole development that (1.15) is equivalent to the problem

$$\text{(1.16)}\qquad \begin{array}{ll} \text{minimize} & \xi^{-2} \sum_{k=0}^{m} |\, f^{(k)}(0)\,|^2 + \kappa^{-2} \,\|\, f^{(m)} \,\|^2 \\ \text{subject to} & f \in W^{m,2}(0,1)\ ,\ f(t_i) = y_{in}\ ,\ i = 1,2,\cdots,n\ , \end{array}$$

in the sense that $f(t) = \widehat{X}(t)$, $t > 0$, is the solution of (1.16) and vice versa. Then, in the diffusivity limit, we obtain the plain spline interpolation problem. Establishing the equivalence requires a bit of work, to put it mildly, but it is just a standard exercise for people versed in stochastic processes. Indeed, it follows from the deep connection between stochastic processes and reproducing kernel Hilbert spaces, as pioneered by LOÈVE (1948) and PARZEN (1961, 1967); see § 3.

Having made some operational sense of (1.5), we immediately transform it into two more manageable models. In the first rewrite, the process (1.5), once discretized, has a representation as an m-th-order *autoregressive* model, as follows. First, let us extend the process $X(t)$, $t \geqslant 0$, to the whole line. For any function $\Phi(t)$, $t \geqslant 0$, define φ_+ by

$$\text{(1.17)}\qquad \Phi_+(t) = \begin{cases} \Phi(t)\ , & t \geqslant 0\ , \\ 0\ , & t < 0\ , \end{cases}$$

and then set

$$\text{(1.18)}\qquad Y(t) = P(t) + \kappa\, \Phi_+(t)\ , \quad -\infty < t < \infty\ .$$

Then, obviously $X(t) = Y(t)$, $t \geqslant 0$. Now, there exist deterministic constants λ_{ik} depending on the design points and the order of the spline only (with $\lambda_{i,0} = 1$), such that

$$\text{(1.19)}\qquad \begin{array}{l} \sum_{k=0}^{m} \lambda_{ik}\, Y(t_{i-k}) = \Psi_i\ , \quad i = 1,2,\cdots,n\ , \\ Y(t_i) = P(t_i)\ , \quad i = 0,-1,-2,\cdots,-(m-1)\ , \end{array}$$

with

$$\text{(1.20)}\qquad \Psi_i = \kappa \int_0^{t_i} \sum_{i=0}^{k} \lambda_{ik}\, (t_{i-k} - \tau)_+^{m-1}\, dW(\tau)\ , \quad i = 1,2,\cdots,n\ .$$

It is straightforward to show that the process Ψ_i, $i = 1,2,\cdots,n$, is again Gaussian and to compute its covariance structure. In particular, one verifies that the process Ψ_i has limited memory in the sense that

$$\text{(1.21)}\qquad \mathbb{E}[\,\Psi_i\, \Psi_j\,] = 0 \quad \text{for } |\, i - j\,| \geqslant m\ ;$$

see Exercise (1.38). Now, from (1.19), one may deduce the distribution of $G = \big(Y(t_{-m+1}),\, Y(t_{-m+2}),\, \cdots,\, Y(t_n) \big)^T$, which may be put to good use in solving the basic estimation problems for (1.19); e.g., the estimation problem may be formulated as

$$\text{(1.22)} \qquad \begin{array}{ll} \text{estimate} & Y(t) \\ \text{given the data} & y_i = Y(t_i)\ ,\ i = 1, 2, \cdots, n\ . \end{array}$$

In § 4, we work out the data-smoothing problem with noisy data.

In the second rewrite, the model (1.5) has a *state-space* formulation. The starting point is the initial value problem for the stochastic differential equation

$$\text{(1.23)} \qquad \begin{array}{ll} X^{(m)}(t) = \nu(t)\ , & t > 0\ , \\ X^{(k)}(0) \ = k!\, p_k\ , & k = 0, 1, \cdots, m-1\ , \end{array}$$

with the p_k as in (1.6) and where $\nu(t)$ is scaled white noise; i.e., formally

$$\text{(1.24)} \qquad \nu(t) = \kappa\, \frac{dW(t)}{dt}\ .$$

Again, we interpret (1.24) as a Gaussian process with $\mathbb{E}[\,\nu(t)\,] = 0$ and

$$\text{(1.25)} \qquad \mathbb{E}[\,\nu(s)\,\nu(t)\,] = \begin{cases} \kappa^2\ , & t = s\ , \\ 0\ , & \text{otherwise}\,. \end{cases}$$

Writing (1.23) as a system of first-order differential equations gives what is called the state-space formulation,

$$\text{(1.26)} \qquad \begin{array}{ll} \mathbb{S}'(t) = A\,\mathbb{S}(t) + N(t)\ , & t > 0\ , \\ \mathbb{S}(0) = (\,\cdots, k!\, p_k\,, \cdots)^T\ , & \text{(see (1.23))}\ , \end{array}$$

where $\mathbb{S}(t)$ is the state at time t and $N(t)$ is the noise driving the system,

$$\text{(1.27)} \qquad \begin{aligned} \mathbb{S}(t) &= \big(\,X(t),\ X'(t),\ X''(t),\ \cdots,\ X^{(m-1)}(t)\,\big)^T\ , \\ N(t) &= \big(\,0, 0, \cdots, 0,\ \nu(t)\,\big)^T\ , \end{aligned}$$

and $A \in \mathbb{R}^{m\times m}$ is the fixed matrix, cf. (1.15),

$$\text{(1.28)} \qquad A = \begin{pmatrix} o & I \\ 0 & o^T \end{pmatrix}\ ,$$

with I the $(m-1)\times(m-1)$ identity matrix and o the zero vector in $\mathbb{R}^{m-1}$.

Starting at time s, the state at time t may then be expressed as

$$\text{(1.29)} \qquad \mathbb{S}(t) = Q(t\,|\,s)\,\mathbb{S}(s) + \mathbb{U}(t\,|\,s)\ ,$$

where the transition matrix $Q(t\,|\,s) \in \mathbb{R}^{m\times m}$ is upper triangular with

$$\text{(1.30)} \qquad [\,Q(t\,|\,s)\,]_{k,\ell} = \frac{(t-s)^{\ell-k}}{(\ell-k)!}\ , \qquad \ell = k, k+1, \cdots, m\ ,$$

and

$$\mathbb{U}(t \mid s) = \int_s^t Q(t \mid \tau) N(\tau) \, d\tau \,. \tag{1.31}$$

See Exercise (1.37) for some of the details. Note that the expression for $\mathbb{U}(t \mid s)$ can be simplified considerably due to the special form of $N(t)$. Again, $\mathbb{U}(t \mid s)$, $t > s$, is to be interpreted as a vector-valued Gaussian process with a suitable covariance structure.

It follows that for a fixed sequence of design points $t_0 < t_1 < \cdots < t_n$,

$$\mathbb{S}(t_i) = Q(t_i \mid t_{i-1}) \, \mathbb{S}(t_{i-1}) + \mathbb{U}(t_i \mid t_{i-1}) \,, \quad i = 1, 2, \cdots, n \,, \tag{1.32}$$

and what stands out here is that

$$\mathbb{S}(t_0) \,, \, \mathbb{U}(t_1 \mid t_0) \,, \mathbb{U}(t_2 \mid t_1) \,, \cdots , \mathbb{U}(t_n \mid t_{n-1}) \quad \text{are independent} \,. \tag{1.33}$$

The covariance matrix $\Sigma_i = \mathbb{E}\big[\, \mathbb{U}(t_i \mid t_{i-1}) \, \mathbb{U}(t_i \mid t_{i-1})^T \big]$ is given by

$$[\, \Sigma_i \,]_{k,\ell} = \frac{(\, t_i - t_{i-1} \,)^{2m-k-\ell+1}}{(2m - k - \ell + 1) \, (m-k)! \, (m - \ell)!} \tag{1.34}$$

for $k, \, \ell = 1, 2, \cdots .m$.

(1.35) EXERCISE. Verify (1.34).

Now, traditionally, the basic estimation problem for the state-space model (1.29) is one of prediction, typically in the form

$$\begin{aligned} &\text{estimate} \quad \mathbb{S}(t_{i+1}) \\ &\text{given} \qquad M(t_j) \, \mathbb{S}(t_j) = Y_j \,, \; j = 1, 2, \cdots, i \,, \end{aligned} \tag{1.36}$$

for some fixed known matrix function $M(t) \in \mathbb{R}^{\ell \times m}$ with linearly independent rows. The case that interests us is where $M(t) \, \mathbb{S}(t)$ is equal to the first component of $\mathbb{S}(t)$. Now, the estimation problem (1.36) is *not* equivalent to (1.12), but it turns out that the solution of (1.36) by way of the Kalman filter is quite relevant to the solution of (1.12). Also note that the diffuse prior on the stochastic process corresponds to a diffuse initial condition in the state-space model.

Of course, we are not really interested in spline interpolation but in the smoothing spline problem corresponding to noisy data. However, we can modify the stochastic process (1.5) by adding an independent stochastic process describing the noise, so $(\, X(t) \,, \, \varepsilon(t) \,)$, $t \geqslant 0$, and observe points on the sample path of $Y(t) = X(t) + \varepsilon(t)$. For the state-space formulation, obviously, the state $\mathbb{S}(t)$ is extended by the extra component $\varepsilon(t)$, which is independent of everything else.

We work out the details of the equivalence of the basic estimation problem for Gaussian processes with noisy data and the roughness-penalized least-squares estimation problem in § 3. In § 4, the autoregressive formulation is taken up. State-space models and various aspects of Kalman filtering, including the issue of diffuse priors, are discussed in §§ 5 through 8. Finally, in § 9, the algorithmic details for smoothing splines are summarized. Some of these details actually confuse the issue, so in § 2 we consider a simple penalized least-squares problem and study its Bayesian interpretation. This also leads to the Bayesian estimation of the smoothing parameter, the so-called GML procedure. There, we also get a first whiff of "diffuse" priors.

Finally, an exercise on various ways to estimate a multivariate normal random variable given linear information, and two exercises on the derivation of the autoregressive and state-space models for spline smoothing.

(1.37) EXERCISE. (a) Let X and Y be zero-mean, jointly normal, multivariate random variables with X nondegenerate; i.e., $\mathbb{V}\mathrm{ar}[\,X\,]$ is a positive-definite matrix. Then, $\mathbb{E}[\,Y \mid X\,] = \mathbb{C}\mathrm{ov}[\,Y, X\,]\,\{\,\mathbb{V}\mathrm{ar}[\,X\,]\,\}^{-1}\,X$.
(b) Let $Y \sim \mathrm{Normal}(\,0\,,\,V\,)$ for some positive-definite matrix $V \in \mathbb{R}^{n \times n}$. Let $A \in \mathbb{R}^{m \times n}$ have independent rows. Then,

$$\mathbb{E}[\,Y \mid AY = b\,] = VA^T(AVA^T)^{-1}\,b\ ,$$

which is also the solution to the constrained maximum likelihood problem

$$\text{minimize} \quad y^T V^{-1} y \quad \text{subject to} \quad Ay = b\ .$$

(1.38) EXERCISE. The autoregressive model (1.19) is based on polynomial interpolation. (a) For distinct t_i, define

$$L(\,t\,) = \sum_{k=1}^{m} p(\,t_{i-k})\,\Lambda_{i,k}(\,t\,) \quad \text{where} \quad \Lambda_{i,k}(\,t\,) = \prod_{\substack{\ell=1 \\ \ell \neq k}}^{m} \frac{t - t_{i-\ell}}{t_{i-k} - t_{i-\ell}}\ .$$

Show that if $p(\,t\,)$ is a polynomial in t of degree $\leqslant m-1$, then $p(\,t\,) = L(\,t\,)$ for all t.
(b) Define $\lambda_{i,k} = -\Lambda_{i,k}(\,t_i)$ for all i, k for which it makes sense, and show that, for $i \geqslant m$,

$$X(\,t_i) + \sum_{k=1}^{m} \lambda_{ik}\, X(\,t_{i-k}) = \Psi_i \quad \text{where} \quad \Psi_i = \kappa \int_0^{t_i} P_i(\tau)\, dW(\tau)\ ,$$

in which $\quad P_i(\tau) = (\,t_i - \tau\,)_+^{m-1} + \sum_{k=1}^{m} \lambda_{ik}\,(\,t_{i-k} - \tau\,)_+^{m-1}\ .$

(c) Finally, show that $P_i(\tau) = 0$ for $\tau \leqslant t_{i-m}$, which implies that

$$\mathbb{E}[\,\Psi_i\,\Psi_j\,] = 0\ , \quad |\,i - j\,| \geqslant m\ .$$

[Hint for (a): Show that $L(\,t\,)$ is a polynomial of degree $\leqslant m - 1$ and that $L(\,t_{i-k}) = p(\,t_{i-k})$, $k = 1, 2, \cdots, m$. Now, count the number of zeros of the polynomial $p(\,t\,) - L(\,t\,)$, and compare this with the degree of

$p(t) - L(t)$. Conclude that $p(t) - L(t) = 0$ everywhere. For more on Lagrange interpolation; see, e.g., KINCAID and CHENEY (1991).]

(1.39) EXERCISE. The transition matrix $Q(t \mid s)$ arises as the fundamental matrix for the system (1.23) as follows.

(a) Show that the homogeneous initial value problem

$$\begin{aligned} \mathbb{S}'(t) &= A\,\mathbb{S}(t)\,, \quad t > s\,, \\ \mathbb{S}(s) &= b\,, \end{aligned}$$

has the solution $\mathbb{S}(t) = \Phi(t \mid s)\, b$, where

$$\Phi(t \mid s) = \sum_{k=0}^{m-1} \frac{(t-s)^k A^k}{k!}\,.$$

[Actually, for arbitrary square matrices A, the expression for $\Phi(t \mid s)$ would be a (convergent) infinite series, which is usually denoted by the exponential $\exp\bigl((t-s)A\bigr)$. It just so happens that for the matrix A at hand we have $A^m = O$.]

(b) Show that, for all t, τ, and s,

$$\Phi(t \mid s) = \Phi(t \mid \tau)\,\Phi(\tau \mid s) \quad \text{and} \quad [\,\Phi(t \mid s)\,]^{-1} = \Phi(s \mid t)\,.$$

(c) To solve the inhomogeneous initial value problem

$$\begin{aligned} \mathbb{S}'(t) &= A\,\mathbb{S}(t) + N(t)\,, \quad t > s\,, \\ \mathbb{S}(s) &= b\,, \end{aligned}$$

assume that $N(t)$ is a deterministic, integrable vector-valued function. Now, try the *ansatz* $\mathbb{S}(t) = \Phi(t \mid s)\, Y(t)$ for a function Y yet to be determined. Show that this gives the equation for Y,

$$\Phi(t \mid s)\, Y'(t) = N(t)\,, \quad t > s\,,$$

and solve it. This should give

$$\mathbb{S}(t) = \Phi(t \mid s)\, b + \int_s^t \Phi(t \mid \tau)\, N(\tau)\, d\tau\,.$$

[This technique of solving the inhomogeneous problem is called "variation of constants". The constant that is allowed to vary is the vector b in $\Phi(t \mid s)\, b$, the solution of the homogeneous problem; see, e.g., SÁNCHEZ (1968). Now, the variation of constants formula holds when $N(t)$ is white noise; see, e.g., ØKSENDAL (2003), Chapter VI, Step 4.]

(d) If p is a polynomial of degree $\leqslant m - 1$ and $\Pi(t)$ is defined as

$$\Pi(t) = \bigl(p(t), p'(t), p''(t), \cdots, p^{(m-1)}(t)\bigr)^T\,,$$

then $\Pi(t) = \Phi(t \mid s)\, \Pi(s)$. Prove this directly, without recourse to differential equations. [Hint: Taylor's theorem.]

EXERCISES: (1.35), (1.37), (1.38), (1.39).

2. A simple example

We recall that the basic mode of operation in Bayesian estimation is to prescribe a prior distribution on the parameter (finite- or infinite-dimensional) to be estimated and then compute or estimate the posterior distribution of the parameter, given the data. The parameter may then be estimated as the posterior mean or the posterior mode. The authorized application of this idea to spline smoothing is complicated by several factors that obscure the underlying ideas. So, it is useful to consider the following simple example and do a Bayesian mock-up of the smoothing spline problem.

Consider the usual model for noisy data,

$$Y = B + \varepsilon \ , \tag{2.1}$$

where $Y \in \mathbb{R}^n$ contains the observations, $B \in \mathbb{R}^n$ is the parameter we wish to estimate, and $\varepsilon \in \mathbb{R}^n$ is the noise, assumed to have iid components,

$$\varepsilon \sim \text{Normal}(\, 0 \, , \, \sigma^2 \, I \,) \ , \tag{2.2}$$

with σ^2 unknown. Regarding B, we assume a normal prior,

$$B \sim \text{Normal}(\, 0 \, , \, \kappa^2 \, V \,) \ , \tag{2.3}$$

with V positive-definite and nonsingular and κ^2 unknown as well. In addition, we assume that

$$B \quad \text{and} \quad \varepsilon \quad \text{are independent} \ . \tag{2.4}$$

It is useful to think of an example like $B = \big(\, f_o(\, t_1), \, f_o(\, t_2), \, \cdots \, , \, f_o(\, t_n) \,\big)^T$, where f_o is a nice function, and $V_{i,j} = i \wedge j$.

The interpretation of this model is as follows. First, B is drawn from the advertised multivariate normal distribution. Denote this realization by b_o. Next, independently, the noise ε is generated from its normal distribution and the observations $Y = b_o + \varepsilon$ are made. The goal is now to estimate b_o, given the data Y. Note that σ^2 and κ^2 are nuisance parameters.

Assuming σ and κ are known, we can estimate B (or should we say b_o) by either the conditional mean $\mathbb{E}[\, B \,|\, Y \,]$ or the conditional mode; i.e., the solution to the maximum *a posteriori* likelihood problem. (Per Exercise (1.37), these two estimators are equal.) To compute either one of them, the distribution of B conditioned on Y is required. Using Bayes' formula, we have for the conditional probability density function

$$f_{B\,|\,Y}(\, b \,|\, y \,) = \frac{f_{Y\,|\,B}(\, y \,|\, b \,)\, f_B(b)}{f_Y(y)} \ , \tag{2.5}$$

and so

$$f_{B\,|\,Y}(\, b \,|\, y \,) \ \propto \ \exp\Big(- \frac{\|\, b - y \,\|^2}{2\sigma^2} - \frac{b^T V^{-1} b}{2\kappa^2} \Big) \ , \tag{2.6}$$

with the constant of proportionality not depending on b. Consequently, the maximum *a posteriori* likelihood problem is given by

$$\text{(2.7)} \qquad \begin{array}{ll} \text{minimize} & \| b - Y \|^2 + \lambda^{-2} b^T V^{-1} b \\ \text{subject to} & b \in \mathbb{R}^n , \end{array}$$

where $\lambda = \kappa / \sigma$. Then, the conditional mode of B given Y is

$$\text{(2.8)} \qquad \widehat{B} = (I + \lambda^{-2} V^{-1})^{-1} Y ,$$

and this equals $\mathbb{E}[B \mid Y]$.

We now see a first complication: What if V is only semi-positive-definite? Then, obviously, the above does not work. However, note that if V is indeed positive-definite, the final answer can be rewritten by way of

$$\text{(2.9)} \qquad (I + \lambda^{-2} V^{-1})^{-1} = \lambda^2 V (I + \lambda^2 V)^{-1} = I - (I + \lambda^2 V)^{-1} .$$

For later reference, it is useful to define

$$\text{(2.10)} \qquad R_{n\lambda} = I - (I + \lambda^2 V)^{-1} .$$

This notation is similar to the smoothing spline case of (18.1.23). Note that the last inverse exists even if V is merely semi-positive-definite. Thus,

$$\text{(2.11)} \qquad \widehat{B} = R_{n\lambda} Y .$$

We now consider the Bayesian way of choosing the fudge factor λ in (2.8). Actually, the idea is to choose $\kappa = \lambda \sigma$ in the model (2.2)–(2.5) and to do it by way of maximum *a posteriori* likelihood. Obviously, we have

$$\text{(2.12)} \quad Y \sim \text{Normal}(0 , \sigma^2 (I + \lambda^2 V)) = \text{Normal}(0 , \sigma^2 (I - R_{n\lambda})^{-1}) ,$$

so that the negative log-likelihood of σ and λ given $Y = y$ equals, apart from terms not depending on λ and σ,

$$\text{(2.13)} \quad L(\sigma, \lambda \mid y) = \frac{y^T (I - R_{n\lambda}) y}{2 \sigma^2} + \tfrac{n}{2} \log(\sigma^2) + \tfrac{1}{2} \log\det(I - R_{n\lambda})^{-1} .$$

Now, minimizing over σ^2 gives

$$\text{(2.14)} \qquad \widehat{\sigma}^2 = \tfrac{1}{n} y^T (I - R_{n\lambda}) y ,$$

so that apart from constant terms, the negative log-likelihood as a function of λ alone is given by

$$\text{(2.15)} \qquad L^*(\lambda \mid y) = \tfrac{1}{2} n \log K^*(\lambda \mid y) ,$$

$$\text{where} \qquad K^*(\lambda \mid y) = \frac{y^T (I - R_{n\lambda}) y}{\{ \det(I - R_{n\lambda}) \}^{1/n}} ,$$

and a factor $\frac{1}{n}$ was dropped in the numerator. It hardly needs mentioning (well, it does) that a comparison with the GCV procedure is intriguing. More on this in § 9.

Thus, the smoothing parameter λ is chosen by minimizing $K^*(\lambda \mid y)$, and then σ is determined by (2.14). We will have more on $M^*(\lambda \mid y)$ under diffuse priors later, but first an exercise on designs with independent replications.

(2.16) EXERCISE. Consider the model (2.1) with independent replications. Thus, let $t_1 \leqslant t_2 \leqslant \cdots \leqslant t_n$ be the design points with replications, resulting in J *distinct* design points, $x_1 \leqslant x_2 \leqslant \cdots \leqslant x_J$. Let n_j be the multiplicity of x_j, i.e., the number of indices in the set $I_j = \{ i \mid t_i = x_j \}$. Define the $J \times J$ diagonal matrix S by

$$S_{jj} = n_j^{-1/2} \ , \quad j = 1, 2, \cdots, J \ .$$

The model under consideration is then transformed into

$$Y_i = \mathbb{B}(t_i) + \varepsilon_i \ , \quad i = 1, 2, \cdots, n \ ,$$

where $\mathbb{B} = \big(\mathbb{B}(x_1), \mathbb{B}(x_2), \cdots, \mathbb{B}(x_n) \big)^T$, and we assume that $\mathbb{B}$ and ε are independent, with

$$\mathbb{B} \sim \text{Normal}(0, \kappa^2 V) \quad , \quad \varepsilon \sim \text{Normal}(0, \sigma^2 I_{n \times n}) \ .$$

We assume that V is known but σ^2 and κ^2 are not.

Finally, define $\overline{\mathbb{Y}}$ and the pure error by

$$\overline{\mathbb{Y}}_j = \frac{1}{n_j} \sum_{i \in I_j} Y_i \quad , \quad \delta = \sum_{j=1}^{J} \sum_{i \in I_j} | Y_i - \overline{\mathbb{Y}}_j |^2 \ .$$

(a) For the purpose of estimating $\mathbb{B}$, show that it suffices to consider the (much smaller) system $\overline{Y} = \mathbb{B} + S\eta$, where η is independent of $\mathbb{B}$ and $\eta \sim \text{Normal}(0, \sigma^2 I_{J \times J})$. In particular, show that

$$\mathbb{E}[\, \mathbb{B} \mid Y \,] = \mathbb{E}[\, \mathbb{B} \mid \overline{Y} \,] = S \, R_{J,\lambda} \, S^{-1} \overline{Y} \ ,$$

in which $R_{J,\lambda} = I - (I + \lambda^{-1} S V^{-1} S)^{-1}$ and $\lambda = \kappa^2 / \sigma^2$.

(b) Show that the negative log-likelihood $L^*(\lambda \mid Y)$ of λ alone is given by

$$\tfrac{1}{n} L^*(\lambda \mid Y) = 1 + \log\Big(\frac{2\pi}{n} \{ \delta + \overline{Y}^T S^{-1} (I - R_{J,\lambda}) S^{-1} \overline{Y} \} \Big) - \frac{1}{n} \log\det(I - R_{J,\lambda}) \ .$$

Diffuse priors. We now briefly discuss the case of what one euphemistically calls "diffuse" priors. In the previous example, we gave a Bayesian interpretation of the penalized least-squares problem (2.6). Of course, there the penalization $b^T V^{-1} b$ was positive-definite, so what does one do if the penalization is only semi-positive-definite as in the cubic smoothing spline problem (19.3.13)?

So, let us start with a semi-positive-definite matrix $M \in \mathbb{R}^{n \times n}$ with rank $n - m$. (Think of the case $m = 2$.) The penalized least-squares problem of interest is

$$\text{(2.17)} \qquad \begin{array}{ll} \text{minimize} & \| \, b - y \, \|^2 + \lambda^2 \, b^T M \, b \\ \text{subject to} & b \in \mathbb{R}^n \; . \end{array}$$

The goal is to construct a Bayesian model for this, in particular by approximating the matrix M with positive-definite matrices M_ξ. Thus, (2.17) is approximated by the problems

$$\text{(2.18)} \qquad \begin{array}{ll} \text{minimize} & \| \, b - y \, \|^2 + \lambda^2 \, b^T M_\xi \, b \\ \text{subject to} & b \in \mathbb{R}^n \end{array}$$

for suitable positive-definite matrices M_ξ, with $M_\xi \to M$ as $\xi \to \infty$. Of course, a Bayesian interpretation of (2.18) is in the bag.

Note that the solution b of (2.17) is given as $b = \mathfrak{R}_{n\lambda} \, y$, where

$$\text{(2.19)} \qquad \mathfrak{R}_{n\lambda} = \big(\, I + \lambda^2 \, M \, \big)^{-1} \; .$$

The following computational approach to the Bayesian estimation of λ seems to be the simplest. Consider the eigenvalue-eigenvector factorization of M in the form

$$\text{(2.20)} \qquad M = U \, \Lambda \, U^T \; ,$$

with orthogonal matrix $U \in \mathbb{R}^{n \times n}$ and diagonal matrix $\Lambda \in \mathbb{R}^{n \times n}$. The eigenvalues and eigenvectors may be arranged such that

$$\text{(2.21)} \qquad \Lambda = \begin{pmatrix} \Lambda_{11} & O \\ O & O \end{pmatrix} \quad , \quad U = (\, U_{11} \quad U_{12} \,) \; ,$$

where $\Lambda_{11} \in \mathbb{R}^{(n-m) \times (n-m)}$ is diagonal and positive-definite, the O denote zero matrices of suitable dimensions, and $U_{11} \in \mathbb{R}^{n \times (n-m)}$, $U_{12} \in \mathbb{R}^{n \times m}$. Note that the columns of U_{11} resp. U_{12} form an orthogonal basis for the range resp. the null space of M. Now, define

$$\text{(2.22)} \qquad \Lambda_\xi = \begin{pmatrix} \Lambda_{11} & O \\ O & \xi^{-2} \, I_m \end{pmatrix} \quad , \quad M_\xi = U \, \Lambda_\xi \, U^T \; ,$$

in which I_m is the identity matrix in $\mathbb{R}^{m \times m}$. Then, M_ξ is positive-definite and

$$\text{(2.23)} \qquad M_\xi^{-1} = M^\dagger + \xi^2 \, U_{12} U_{12}^T \; ,$$

where $M^\dagger = U_{11} \Lambda_{11}^{-1} U_{11}^T$ is the Moore-Penrose inverse of M.

Now, the Bayesian model may be stated as

$$\text{(2.24)} \qquad Y = U_{11} B + U_{12} C + \varepsilon \; ,$$

where $B \in \mathbb{R}^{n-m}$, $C \in \mathbb{R}^m$, and $\varepsilon \in \mathbb{R}^n$ are independent multivariate random variables with

$$\text{(2.25)} \qquad B \sim \text{Normal}\big(\, 0 \, , \, \kappa^2 \, \Lambda_{11}^{-1} \, \big) \quad , \quad C \sim \text{Normal}\big(\, 0 \, , \, \xi^2 \, I \, \big) \; ,$$

and ε the usual noise, as in (2.2). Here, $\kappa = \sigma/\lambda$. Note that $U_{12}C$ in (2.24) corresponds to $p(t)$ in (1.1)–(1.5).

(2.26) EXERCISE. (a) Verify (2.23) and that $Y \sim \text{Normal}(0, M_\xi^{-1}+\sigma^2 I)$.
(b) Verify that the model (2.24)–(2.25) leads to the problem (2.18), where b is to be interpreted as (the estimator of) $U(B^T \,|\, C^T)^T$.
(c) Verify that $U_{11}^T Y \sim \text{Normal}(0, \kappa^2 \Lambda_{11}^{-1} + \sigma^2 I)$.

How does this help in the Bayesian estimation of the parameter λ in the problem (2.18)? It seems clear that estimating λ given Y leads to trouble as $\xi \to \infty$. So, it makes sense to filter the contribution of C by conditioning on $U_{11}^T Y$. From the previous exercise, the maximum likelihood estimator of κ given $U_{11}^T Y = z$ is the minimizer of

$$\tfrac{1}{2}\, z^T \big(\kappa^2 \Lambda_{11}^{-1} + \sigma^2 I\big)^{-1} z + \log\det\big(2\pi\,(\kappa^2 \Lambda_{11}^{-1} + \sigma^2 I)\big)^{1/2} .$$

Now, set $\kappa = \sigma/\lambda$, and note that

$$\det\big(2\pi\,(\sigma^2\lambda^{-2}\Lambda_{11}^{-1} + \sigma^2 I)\big) = (2\pi\sigma^2)^{n-m} \det(A_\lambda) ,$$

where $A_\lambda = \lambda^{-2}\Lambda_{11}^{-1} + I$. So, the problem with σ factored out is

$$\text{(2.27)}\qquad \begin{aligned} &\text{minimize} && \frac{z^T A_\lambda^{-1} z}{2\sigma^2} + \tfrac{1}{2}(n-m)\log(2\pi\sigma^2) - \tfrac{1}{2}\log\det(A_\lambda^{-1}) \\ &\text{subject to} && \lambda > 0 ,\ \sigma > 0 . \end{aligned}$$

Minimizing first over σ^2 then gives

$$\widehat{\sigma}^2 = \frac{1}{n-m}\, z^T A_\lambda^{-1} z$$

and so leads to the problem (with a factor 2 ignored)

$$\text{(2.28)}\qquad \text{minimize}\quad (n-m)\log\left(\frac{\frac{2\pi}{n-m}\, z^T A_\lambda^{-1} z}{\big(\det(A_\lambda^{-1})\big)^{1/(n-m)}}\right) + n - m .$$

This is the Bayesian way of choosing λ. We retained the factors/terms $n-m$ since in spline smoothing the order of the smoothing spline, corresponding to m here, must also be chosen.

It is useful to rewrite the above in a coordinate-free way. First, using the trick (2.9), we may write

$$A_\lambda^{-1} = (\lambda^{-2}\Lambda_{11}^{-1} + I)^{-1} = I - (I + \lambda^2 \Lambda_{11})^{-1} .$$

Then

$$\text{(2.29)}\qquad \det(A_\lambda^{-1})^{-1} = \prod_{j=1}^{n-m}\Big(1 - \frac{1}{1+\lambda^2\mu_j}\Big) ,$$

where $\mu_1 > \mu_2 > \cdots > \mu_{n-m} > 0$ are the positive eigenvalues of the original matrix M, see (2.17), the other ones being equal to 0. For these positive eigenvalues, obviously $(1+\lambda^2\mu_j)^{-1} < 1$, so that

$$1-(1+\lambda^2\mu_j)^{-1}\ , \quad j=1,2,\cdots,n-m\ ,$$

are all of the positive eigenvalues of $I-(1+\lambda^2 M)^{-1}$. With the notation (2.19) in mind, we may write this suggestively as

$$\text{(2.30)} \qquad \det_+(I-\mathfrak{R}_{n\lambda}) = \prod_{j=1}^{n-m}\Bigl(1-\frac{1}{1+\lambda^2\mu_j}\Bigr)\ .$$

As far as $z^T A_\lambda^{-1} z$ is concerned, write $z=U_{11}^T y$ and observe that

$$\begin{aligned}\text{(2.31)} \qquad U_{11}^T A_\lambda^{-1} U_{11} &= U_{11}^T\,(I-(I+\lambda^2\Lambda_{11})^{-1})\,U_{11}\\ &= \begin{pmatrix}U_{11} & U_{12}\end{pmatrix}\begin{pmatrix}I-(I+\lambda^2\Lambda_{11})^{-1} & O\\ O & O\end{pmatrix}\begin{pmatrix}U_{11}^T\\ U_{12}^T\end{pmatrix}\\ &= I-(I+\lambda^2 M)^{-1} = I-\mathfrak{R}_{n\lambda}\ .\end{aligned}$$

(2.32) EXERCISE. Verify this formula. [Hint: Note that the O matrix on the diagonal may be written as $I-I$.]

Putting all of this together, we find that the Bayesian way to estimate λ in (2.18) is to

$$\text{minimize} \quad (n-m)\,\log\left(\frac{2\pi\, y^T(I-\mathfrak{R}_{n\lambda})\,y/(n-m)}{\{\det_+(I-\mathfrak{R}_{n\lambda})\}^{1/(n-m)}}\right)+n-m\ .$$

If the "order" m is fixed beforehand, then the problem is equivalent to

$$\text{(2.33)} \qquad \text{minimize} \quad \frac{y^T(I-\mathfrak{R}_{n\lambda})\,y}{\{\det_+(I-\mathfrak{R}_{n\lambda})\}^{1/(n-m)}}\ .$$

When specialized to spline smoothing, this method goes by the name GML (generalized maximum likelihood).

(2.34) EXERCISE. Work out the details of the diffuse prior for the model with replications discussed in Exercise (2.16).

EXERCISES: (2.16), (2.26), (2.32), (2.34).

3. Stochastic processes and reproducing kernels

In this section, we work out the connection between stochastic processes and reproducing kernel Hilbert spaces. The tangible result is that the basic estimation problem for a Gaussian stochastic process with noisy data is equivalent to a roughness-penalized least-squares problem. However, to take matters one step at a time, the case of exact data is considered first, leading to a "generalized" spline interpolation problem.

The basic estimation problem for Gaussian processes. We start with a scalar Gaussian stochastic process $X(t)$, $t \geqslant 0$, satisfying

$$\text{(3.1)} \qquad \mathbb{E}[X(t)] = 0 \quad , \quad \mathbb{E}[X(s)\,X(t)] = K(s,t) \,, \quad t, s \geqslant 0 \,.$$

Here, K is any *positive-definite function* with finite diagonal, i.e.,

$$\text{(3.2)} \qquad K(s,s) < \infty \quad \text{for all } s \geqslant 0 \,,$$

and for all $k \in \mathbb{N}$, all $0 < \tau_1 < \tau_2 < \cdots < \tau_k$ and all $x_1, x_2, \cdots, x_k \in \mathbb{R}$, one must have

$$\text{(3.3)} \qquad \sum_{i,j=1}^{k} K(\tau_i, \tau_j)\, x_i\, x_j > 0$$

unless $x_i = 0$, $i = 1, 2, \cdots, k$. The standard problem for the stochastic process (3.1) is to

$$\text{(3.4)} \qquad \begin{array}{ll} \text{estimate} & X(t) \\ \text{given the data} & y_i = X(t_i) \,, \quad i = 1, 2, \cdots, n \,, \end{array}$$

for a fixed $t \geqslant 0$. Here, $t_1, t_2, \cdots, t_n$ are strictly positive "design" points. We define the best estimator to be the one that solves

$$\text{(3.5)} \qquad \begin{array}{l} \text{minimize} \quad \mathbb{E}\big[\, | X(t) - \widehat{X} \,|^2 \;\big|\; X(t_i) = y_i \,,\; i = 1, 2, \cdots, n \big] \\ \text{over all estimators } \widehat{X} \,. \end{array}$$

It is well-known that the best estimator is the conditional expectation

$$\text{(3.6)} \qquad \widehat{X} = \mathbb{E}\big[X(t) \;\big|\; X(t_i) = y_i \,,\; i = 1, 2, \cdots, n \big]$$

and that it is a linear combination of the (observed) $X(t_i)$, $i = 1, 2, \cdots, n$. Consequently, the problem (3.5) is equivalent to

$$\text{(3.7)} \qquad \begin{array}{ll} \text{minimize} & \mathbb{E}\Big[\Big| X(t) - \sum\limits_{i=1}^{n} w_i\, X(t_i) \Big|^2 \Big] \\ \text{subject to} & w_1, w_2, \cdots, w_n \in \mathbb{R} \,. \end{array}$$

The span of the stochastic process. The process (3.1), or should we say the estimator (3.6), generates a Hilbert space as follows. First, let $\mathcal{L}(X)$ be the vector space of all finite linear combinations of values of $X(t)$, $t \geqslant 0$. Thus, a random variable Y belongs to $\mathcal{L}(X)$ if there exists an integer $k \in \mathbb{N}$, weights $w_1, w_2, \cdots, w_k \in \mathbb{R}$, and design points $\tau_1, \tau_2, \cdots, \tau_k$, such that

$$\text{(3.8)} \qquad Y = \sum_{j=1}^{k} w_j\, X(\tau_j) \,.$$

On $\mathcal{L}(X)$, an inner product may be defined by

(3.9) $$\langle Z , Y \rangle_X = \mathbb{E}[ZY] , \quad Z, Y \in \mathcal{L}(X) .$$

(3.10) EXERCISE. Show that the inner product satisfies the usual properties, viz. for every scalar $a \in \mathbb{R}$ and all $W, Y, Z \in \mathcal{L}(X)$,

(a) $\langle Y , Z \rangle_X = \langle Z , Y \rangle_X$;
(b) $\langle Y + aZ , W \rangle_X = \langle Y , W \rangle_X + a \langle Z , W \rangle_X$;
(c) $\langle Y , Y \rangle_X \geqslant 0$; and
(d) $\langle Y , Y \rangle_X = 0$ if and only if $Y = 0$ (almost surely).

(3.11) EXERCISE. (a) Note that $\langle X(s) , X(t) \rangle_X = K(s, t)$

(b) and that if $Y = \sum_{j=1}^{k} w_j X(\tau_j)$ and $Z = \sum_{\ell=1}^{m} u_\ell X(s_\ell)$, then

$$\langle Y , Z \rangle_X = \sum_{j=1}^{k} \sum_{\ell=1}^{m} K(\tau_j, s_\ell)\, w_j\, u_\ell .$$

Now, the inner product generates a norm by the usual recipe,

(3.12) $$\| Y \|_X = \{ \langle Y , Y \rangle_X \}^{1/2} , \quad Y \in \mathcal{L}(X) .$$

With the norm in place, we can talk about Cauchy sequences: A sequence $\{ Y_k \}_{k \in \mathbb{N}}$ is a Cauchy sequence in $\mathcal{L}(X)$ if

(3.13) $$\lim_{k,\ell \to \infty} \| Y_k - Y_\ell \|_X = 0 .$$

Of course, not every Cauchy sequence in $\mathcal{L}(X)$ need converge, but the *completion* of $\mathcal{L}(X)$ can be constructed in the usual manner. We denote the completion by $L^2(X)$. There are actually some subtleties here analogous to the construction of the reproducing kernel Hilbert space below, but we shall not dwell on them.

(3.14) REMARK. Do we really need the completion of $\mathcal{L}(X)$? Well, maybe not. Note, however, that the completion consists of all random variables $Z_1, \cdots , Z_k$ for which problems like

$$\text{estimate} \quad X(t) \quad \text{given} \quad Z_i = z_i , \quad i = 1, 2, \cdots , k ,$$

make sense and can be solved as a limit of "basic" problems.

Reproducing kernel Hilbert spaces, again. The Hilbert space $L^2(X)$ of random variables may now be transformed into a reproducing kernel Hilbert space of functions by the canonical map $X(t) \longmapsto K(t, \cdot)$. In fact, to every *random variable* $Y \in L^2(X)$, one associates a *function* in the reproducing kernel Hilbert space by using the recipe

(3.15) $$f(t) = \mathbb{E}[Y X(t)] , \quad t \geqslant 0 .$$

However, we prefer the more pedestrian course and essentially repeat the construction of $L^2(X)$ as follows.

Let $\mathcal{L}(K)$ be the linear span of the set of functions $\{ K(t, \cdot) : t \geqslant 0 \}$. Thus, a function f belongs to $\mathcal{L}(K)$ if there exists an integer $k \in \mathbb{N}$, weights $w_1, w_2, \cdots, w_k \in \mathbb{R}$, and design points $\tau_1, \tau_2, \cdots, \tau_k \geqslant 0$, such that

$$f(s) = \sum_{j=1}^{k} w_j K(\tau_j, s) , \quad s \geqslant 0 . \tag{3.16}$$

On $\mathcal{L}(K)$, an inner product may be defined by way of the recipe

$$\langle K(t, \cdot), K(s, \cdot) \rangle_K = K(t, s) , \quad t, s \geqslant 0 . \tag{3.17}$$

One verifies that this is an inner product in the sense of Exercise (3.10). Then, we may define a norm in the usual manner,

$$\| f \|_K = \{ \langle f, f \rangle_K \}^{1/2} . \tag{3.18}$$

(3.19) Exercise. Show that $\mathcal{L}(\mathcal{K})$ is a reproducing kernel inner product space; i.e., verify that, for all $f \in \mathcal{L}(K)$,

$$f(s) = \langle f, K(s, \cdot) \rangle_K , \quad s \geqslant 0 .$$

Now, the completion of $\mathcal{L}(K)$ may be constructed as usual ("limits" of Cauchy sequences in $\mathcal{L}(K)$), but it is expedient to do it differently to ensure that we end up with a *functional completion* in the sense of Aronszajn (1950). We need some preparatory results.

(3.20) Lemma. *Let $\{ f_k \}_k \subset \mathcal{L}(\mathcal{K})$ be a Cauchy sequence. Then, the sequence $\{ f_k(s) \}_k \subset \mathbb{R}$ converges for each $s \geqslant 0$.*

Proof. Note that, by Exercise (3.19),

$$| f_k(s) - f_\ell(s) | = | \langle f_k - f_\ell , K(s, \cdot) \rangle_K | \leqslant \| K(s \cdot) \|_K \| f_k - f_\ell \|_K .$$

Of course, $\| K(s \cdot) \|_K^2 = K(s, s) < \infty$, and it follows that $\{ f_k(s) \}_k \subset \mathbb{R}$ is a Cauchy sequence, and $\lim_{k \to \infty} f_k(s)$ exists. Q.e.d.

(3.21) Lemma. *Let $\{ f_k \}_k \subset \mathcal{L}(K)$ be a Cauchy sequence, and suppose that $f_k(s) \longrightarrow 0$ for all $s \geqslant 0$. Then $\| f_k \|_K \longrightarrow 0$.*

Proof. Note that $\langle f_k , K(s \cdot) \rangle_K = f_k(s) \longrightarrow 0$ for all s. Then, for a fixed m, consider f_m. It may be represented by a finite linear combination,

$$f_m = \sum_{i=1}^{n} w_{im} K(t_{im}, \cdot) .$$

Then, for m (and hence n) fixed but $k \to \infty$,

$$\langle f_m , f_k \rangle_K = \sum_{i=1}^{n} w_{im} \langle f_k , K(t_{im}, \cdot) \rangle_K \longrightarrow 0 .$$

Now, let $\varepsilon > 0$ be arbitrary. Since $\{ f_k \}_k$ is Cauchy, there exists an m such that

$$\| f_m - f_k \|_K < \varepsilon \quad \text{for all} \quad k \geqslant m .$$

Then, for this fixed m,

$$\begin{aligned} \| f_m \|_K^2 &= \langle f_m , f_k \rangle_K + \langle f_m , f_m - f_k \rangle_K \\ &\leqslant \varepsilon + \| f_m \|_K \| f_m - f_k \|_K \leqslant (1 + C) \varepsilon \end{aligned}$$

if we choose k large enough. Here, $C = \sup_k \| f_k \|_K < \infty$ since $\{ f_k \}_k$ is Cauchy. It follows that there exists a sequence $\{ m_\ell \}_\ell \subset \mathbb{N}$ such that

$$\| f_{m_\ell} \|_K \longrightarrow 0 .$$

Again, since $\{ f_k \}_k$ is Cauchy, then $\| f_k \|_K \longrightarrow 0$. Q.e.d.

Now, construct the completion of $\mathcal{L}(K)$ as follows.

(3.22) DEFINITION. The vector space $\mathcal{H}(K)$ consists of all $\varphi : \mathbb{R} \to \mathbb{R}$ for which there exists a Cauchy sequence $\{ f_k \}_k \subset \mathcal{L}(K)$ such that

(a) $$\varphi(s) = \lim_{k \to \infty} f_k(s) \quad \text{for all} \quad s \geqslant 0 .$$

The collection of Cauchy sequences satisfying (a) is denoted by $\mathcal{P}(\varphi)$. An inner product $\langle \cdot , \cdot \rangle_1$ is defined on $\mathcal{H}(K)$ by

(b) $$\langle \varphi , \psi \rangle_1 = \lim_{m,k \to \infty} \langle f_k , g_m \rangle_1 ,$$

where $\{ f_k \}_k \in \mathcal{P}(\varphi)$ and $\{ g_k \}_k \in \mathcal{P}(\psi)$. Finally, the norm $\| \cdot \|_1$ is defined by $\| \varphi \|_1 = \{ \langle \varphi , \varphi \rangle_1 \}^{1/2}$.

(3.23) REMARK. The definition (3.22)(b) does not depend on the particular choices of $\{ f_k \}_k \in \mathcal{P}(\varphi)$ and $\{ g_k \}_k \in \mathcal{P}(\psi)$: If, for $i = 1, 2$, we have $\{ f_{k,i} \}_k \in \mathcal{P}(\varphi)$ and $\{ g_{k,i} \}_k \in \mathcal{P}(\psi)$, then

$$\begin{aligned} &\Big| \langle f_{k,1} , g_{m,1} \rangle_1 - \langle f_{k,2} , g_{m,2} \rangle_2 \Big| \\ &\qquad \leqslant \Big| \langle f_{k,1} - f_{k,2} , g_{m,1} \rangle_1 \Big| + \Big| \langle f_{k,2} , g_{m,1} - g_{m,2} \rangle_2 \Big| \\ &\qquad \leqslant \| f_{k,1} - f_{k,2} \|_1 \| g_{m,1} \|_1 + \| f_{k,2} \|_1 \| g_{m,1} - g_{m,2} \|_1 \\ &\qquad \leqslant C \{ \| f_{k,1} - f_{k,2} \|_1 + \| g_{m,1} - g_{m,2} \|_1 \} \end{aligned}$$

because Cauchy sequences are bounded. Now, since $\| f \|_1 = \| f \|_K$ for all $f \in \mathcal{L}(K)$, then, by Lemma (3.21), the last expression tends to 0 as k and m tend to ∞.

After these preparations, we are ready.

(3.24) THEOREM. *The space $\mathcal{H}(K)$ with the inner product $\langle \cdot , \cdot \rangle_1$ is a reproducing kernel Hilbert space with reproducing kernel $K(s, t)$.*

PROOF. It is easy to show that $\langle \cdot , \cdot \rangle_1$ is indeed an inner product. In fact, that is the content of Remark (3.23). If one checks it directly, then the only curious property is the following. If $\varphi(s) = 0$ for all s, is then $\| \varphi \|_1 = 0$? Lemma (3.21) says that this is indeed so. It is also easy to show that $\mathcal{H}(K)$ is complete and that $\mathcal{L}(K)$ is dense in $\mathcal{H}(K)$.

Finally, we show that $\mathcal{H}(K)$ indeed has a reproducing kernel. Note that $K(s, \cdot) \in \mathcal{L}(K)$ for all s, so that $K(s, \cdot) \in \mathcal{H}(K)$ as well. Now, let $\varphi \in \mathcal{H}(K)$ and let $\{ f_k \}_k \in \mathcal{P}(\varphi)$. Then, for all s,

$$\varphi(s) = \lim_{k \to \infty} f_k(s) = \lim_{k \to \infty} \langle f_k , K(s \cdot) \rangle_K = \langle \varphi , K(s \cdot) \rangle_1 ,$$

the last equality by the definition of the inner product $\langle \cdot , \cdot \rangle_1$. Q.e.d.

(3.25) NOTATION. From now on, we denote the inner product $\langle \cdot , \cdot \rangle_1$ on $\mathcal{H}(K)$ as just $\langle \cdot , \cdot \rangle_K$. Note that they already coincided on $\mathcal{L}(K)$.

We now return to the estimation problem (3.7). Since

$$\mathbb{E}\Big[\Big| X(t) - \sum_{i=1}^{n} w_i X(t_i) \Big|^2 \Big] = \| K(t, \cdot) - \sum_{i=1}^{n} w_i K(t_i, \cdot) \|_K^2 ,$$

it is apparent that the problem (3.7) is equivalent to

$$\text{(3.26)} \qquad \begin{aligned} &\text{minimize} \quad \| K(t, \cdot) - \sum_{i=1}^{n} w_i K(t_i, \cdot) \|_K^2 \\ &\text{subject to} \quad w_1, w_2, \cdots , w_n \in \mathbb{R} . \end{aligned}$$

(3.27) EXERCISE. The problem (3.26) suggests the following approximation problem: For fixed $f \in \mathcal{H}(K)$ and $t \geqslant 0$,

$$\begin{aligned} &\text{minimize} \quad \Big| f(t) - \sum_{i=1}^{n} w_i f(t_i) \Big|^2 \\ &\text{subject to} \quad w_1, w_2, \cdots , w_n \in \mathbb{R} . \end{aligned}$$

This problem asks one to express $f(t)$ in terms of the given function values $f(t_i)$, $i = 1, 2, \cdots , n$. It seems clear that this is not a useful problem since one must rule out the "best" choice $w_1 = f(t)/f(t_1)$ and all other w_i equal to 0 (if indeed $f(t_1) \neq 0$). Here is a way to enforce that: Show that the problem (3.26) is equivalent to the minimax problem, for fixed $t \geqslant 0$,

$$\begin{aligned} &\text{minimize} \quad \max\Big\{ \Big| f(t) - \sum_{i=1}^{n} w_i f(t_i) \Big|^2 : \| f \|_K \leqslant 1 \Big\} \\ &\text{subject to} \quad w_1, w_2, \cdots , w_n \in \mathbb{R} . \end{aligned}$$

The equivalent "generalized" spline interpolation problem. The exercise above is cute, but we really want to get an interpolation problem. The starting point is the constrained maximum likelihood formulation of (3.4), which we already partially explored in § 1, see (1.12)–(1.15). A more detailed treatment goes as follows.

The random vector $(X(t), X(t_1), X(t_2), \cdots X(t_n))^T$ is normally distributed with covariance matrix $V_t \in \mathbb{R}^{(n+1)\times(n+1)}$ given by

$$[V_t]_{i,j} = K(t_i, t_j) , \quad i, j = 0, 1, \cdots, n , \tag{3.28}$$

where we have set $t_0 = t$. So, the constrained maximum likelihood estimation problem is

$$\begin{aligned} &\text{minimize} \quad w^T V_t^{-1} w \\ &\text{subject to} \quad w_i = y_i \ , \ i = 1, 2, \cdots, n . \end{aligned} \tag{3.29}$$

Here, $w = (w_0, w_1, \cdots, w_n)^T$. Since V_t is positive-definite, the solution exists and is unique. Note that we treat all of the components of w as unknowns.

Now, let us try to find a solution of (3.29) in the form $w = V_t u$. Then, $w^T V_t^{-1} w = u^T V_t u$ and

$$u^T V_t u = \sum_{i,j=0}^{n} u_i u_j K(t_i, t_j) = \Big\| \sum_{i=0}^{n} u_i K(t_i, \cdot) \Big\|_K^2 . \tag{3.30}$$

Also, for $1 \leqslant j \leqslant n$,

$$w_j = \sum_{i=0}^{n} u_i K(t_i, t_j) . \tag{3.31}$$

Consequently, the problem (3.29) is equivalent to

$$\begin{aligned} &\text{minimize} \quad \Big\| \sum_{i=0}^{n} u_i K(t_i, \cdot) \Big\|_K^2 \\ &\text{subject to} \quad \sum_{i=0}^{n} u_i K(t_i, t_j) = y_j \ , \quad j = 1, 2, \cdots, n . \end{aligned} \tag{3.32}$$

Note that the constraints may be written as

$$\Big\langle \sum_{i=0}^{n} u_i K(t_i, \cdot) , K(t_j, \cdot) \Big\rangle_K = y_j \ , \quad j = 1, 2, \cdots, n .$$

Thus, (3.32) is equivalent to

$$\begin{aligned} &\text{minimize} \quad \| f \|_K \\ &\text{subject to} \quad f \in \operatorname{span}\{ K(t_j, \cdot) : j = 0, 1, 2, \cdots, n \} \\ &\qquad\qquad\quad \langle f , K(t_i, \cdot) \rangle_K = y_i \ , \quad i = 1, 2, \cdots, n . \end{aligned} \tag{3.33}$$

(Pay close attention to the ranges for the integers i and j.) Written in the form above, the geometry of the problem gives us the existence and uniqueness of the solution f and that $f \in \operatorname{span}\{ K(t_i, \cdot) : i = 1, 2, \cdots, n \}$.

But then the problems (3.29) and (3.33) are equivalent to

$$\text{(3.34)} \qquad \begin{array}{ll} \text{minimize} & \| f \|_K \\ \text{subject to} & f \in \mathcal{H}(K) \\ & f(t_i) = y_i \,, \quad i = 1, 2, \cdots, n \,. \end{array}$$

We refer to this as a "generalized" spline interpolation problem. The solution exists and is unique, and is of the form

$$\text{(3.35)} \qquad f = \sum_{i=1}^{n} u_i \, K(t_i, \cdot)$$

for suitable u_i, and then $w = (f(t), y_1, y_2 \cdots, y_n)^T$ (for that original, fixed t) is the solution of the constrained maximum likelihood estimation problem (3.29).

Noisy data. Of course, we are interested in the smoothing problem, not in interpolation problems. So, what can be done when the data are noisy? A clean way of incorporating noisy data into the basic estimation problem for a Gaussian process is to make it a part of the process. This approach makes even more sense for the state-space models of the next section. (Indeed, it was a suggestion by KALMAN (1960) himself.) Thus, we consider the vector-valued process

$$\text{(3.36)} \qquad \mathbb{X}(t) = \begin{pmatrix} X(t) \\ \varepsilon(t) \end{pmatrix} , \quad t \geqslant 0 ,$$

where $X(t)$ and $\varepsilon(t)$ are independent Gaussian processes. Then, the process $\mathbb{X}(t)$, $t \geqslant 0$, has the covariance structure

$$\text{(3.37)} \qquad \mathbb{E}[\,\mathbb{X}(t)\,\mathbb{X}(t)^T\,] = \mathbb{K}(s, t) \,, \quad s, t \geqslant 0 \,,$$

where $\mathbb{K}(s, t) \in \mathbb{R}^{2\times 2}$ is given by

$$\text{(3.38)} \qquad \mathbb{K}(s, t) = \begin{bmatrix} \kappa^2 K(s, t) & 0 \\ 0 & \sigma^2 G(s, t) \end{bmatrix} .$$

Here, $G(s, t) = 1$ if $s = t$ and $= 0$ otherwise, and K is positive-definite. The parameters κ and σ are new here and are needed for a proper enunciation of the theory.

The basic estimation problem for the process $\mathbb{X}(t)$, $t \geqslant 0$, is to

$$\text{(3.39)} \qquad \begin{array}{ll} \text{estimate} & \mathbb{X}(t), \mathbb{X}(t_1), \mathbb{X}(t_2), \cdots, \mathbb{X}(t_n) \\ \text{given the data} & y_i = \mathbb{1}^T \, \mathbb{X}(t_i) \,, \quad i = 1, 2, \cdots, n \,, \end{array}$$

for a fixed $t \geqslant 0$. Here, $\mathbb{1} = (1, 1)^T$. However, we prefer the formulation of (3.39) as the constrained maximum likelihood estimation problem

$$\text{(3.40)} \qquad \begin{array}{ll} \text{minimize} & w^T \, \mathbb{V}_t^{-1} \, w \\ \text{subject to} & \mathbb{A} w = y \,. \end{array}$$

Here, w is a potential estimator of

$$\mathcal{X} = \left(\mathbb{X}(t)^T, \mathbb{X}(t_1)^T, \mathbb{X}(t_2)^T, \cdots, \mathbb{X}(t_n)^T \right)^T ,$$

and $\mathbb{V}_t = \mathbb{E}[\mathcal{X}\mathcal{X}^T]$ is a block matrix with 2×2 blocks,

$$(3.41) \qquad [\mathbb{V}_t]_{ij} = \mathbb{K}(t_i, t_j), \quad i,j = 0,1,\cdots,n .$$

Finally, $\mathbb{A}$ is block-diagonal with diagonal blocks $\mathbb{1}^T \in \mathbb{R}^{1\times 2}$. In short,

$$(3.42) \quad \mathbb{V}_t = \begin{bmatrix} \mathbb{K}_{00} & \mathbb{K}_{01} & \cdots & \mathbb{K}_{0n} \\ \mathbb{K}_{10} & \mathbb{K}_{11} & \cdots & \mathbb{K}_{1n} \\ \vdots & \vdots & \ddots & \vdots \\ \mathbb{K}_{n0} & \mathbb{K}_{n1} & \cdots & \mathbb{K}_{nn} \end{bmatrix}, \quad \mathbb{A} = \begin{bmatrix} O & \mathbb{1}^T & & \\ O & & \mathbb{1}^T & \\ \vdots & & & \ddots \\ O & & & & \mathbb{1}^T \end{bmatrix},$$

where $\mathbb{K}_{ij} = \mathbb{K}(t_i, t_j)$.

(3.43) EXERCISE. Show that $\mathbb{V}_t$ is positive-definite, so that the solution of the problem (3.40) exists and is unique.

We shall not bother with the construction of $\mathcal{L}(\mathbb{X})$ or $L^2(\mathbb{X})$ and proceed straightaway to the reproducing kernel Hilbert space. It is apparent that $\mathcal{L}(\mathbb{K})$ consists of 2×2 diagonal matrices, but it seems more reasonable to construct $\mathcal{L}(\mathcal{K})$, where $\mathcal{K}$ is the diagonal of $\mathbb{K}$,

$$(3.44) \qquad \mathcal{K}(s,t) = \left(\kappa^2 K(s,t), \sigma^2 G(s,t) \right), \quad s,t \geqslant 0 .$$

Obviously, we dealt with the first component already since if K is a positive-definite function, so is $\kappa^2 K$. Now, $\sigma^2 G$ is positive-definite also. Consequently, we may construct the reproducing kernel Hilbert spaces $\mathcal{H}(\kappa^2 K)$ and $\mathcal{H}(\sigma^2 G)$ with inner products defined via the usual recipes for s, $t \geqslant 0$,

$$(3.45) \qquad \begin{aligned} \langle K(s,\cdot), K(t,\cdot) \rangle_{\kappa^2 K} &= \kappa^{-2} K(s,t), \\ \langle G(s,\cdot), G(t,\cdot) \rangle_{\sigma^2 G} &= \sigma^{-2} G(s,t). \end{aligned}$$

One verifies that these definitions are instances of (3.17). It should be noted that $\mathcal{H}(\kappa^2 K) = \mathcal{H}(K)$ as sets, but their intrinsic inner products and norms differ by a constant,

$$(3.46) \qquad \langle f, g \rangle_{\kappa^2 K} = \kappa^{-2} \langle f, g \rangle_K \quad \text{for all } f, g \in \mathcal{H}(K).$$

It is useful to observe that if $g = \sum_{i=1}^{n} a_i G(t_i, \cdot)$ for real numbers a_i, then

$$(3.47) \qquad \| g \|^2_{\sigma^2 G} = \sigma^{-2} \sum_{i=1}^{n} | a_i |^2 .$$

(3.48) EXERCISE. Verify (3.46) and (3.47). [Hint for (3.46): First, show it for $f,\ g \in \mathcal{L}(K)$ and then by a limiting process for all $f,\ g \in \mathcal{H}(K)$.]

Now, we may define $\mathcal{H}(\mathcal{K})$ as the Cartesian product

$$\mathcal{H}(\mathcal{K}) = \mathcal{H}(\kappa^2 K) \otimes \mathcal{H}(\sigma^2 G) \ , \tag{3.49}$$

or, in detail,

$$\mathcal{H}(\mathcal{K}) = \{ \, (f,g) \ : \ f \in \mathcal{H}(\kappa^2 K) \, , \ g \in \mathcal{H}(\sigma^2 G) \, \} \ ,$$

with inner product

$$\langle \, (f,g) \, , \, (\varphi,\psi) \, \rangle_{\mathcal{H}(\mathcal{K})} = \langle \, f \, , \, \varphi \, \rangle_{\kappa^2 K} + \langle \, g \, , \, \psi \, \rangle_{\sigma^2 G} \ . \tag{3.50}$$

(3.51) THEOREM. *$\mathcal{H}(\mathcal{K})$, with the inner product $\langle \, \cdot \, , \, \cdot \, \rangle_{\mathcal{H}(\mathcal{K})}$, is a reproducing kernel Hilbert space with reproducing kernel $\mathcal{K}(s,\, t)$.*

(3.52) EXERCISE. Prove it! [Hint: Copy the construction of $\mathcal{H}(K)$ and the proof of Theorem (3.24), but apply the required cosmetic changes.]

The equivalent "generalized" smoothing spline problem. The purpose of the construction of $\mathcal{H}(\mathcal{K})$ is to show the equivalence of the basic estimation problem for the process $\mathbb{X}(t)$, $t \geqslant 0$, to a roughness-penalized least-squares problem. Thus, consider the problem (3.40) and attempt a solution w in the form $w = \mathbb{V}_t \, u$, where $u = (u_0^T, u_1^T, \cdots, u_n^T)^T$ with $u_i^T = (u_{i1}, u_{i2})$. Then, $w^T \mathbb{V}_t^{-1} w = u^T \mathbb{V}_t \, u$ and

$$u^T \mathbb{V}_t \, u = \sum_{i,j=0}^{n} u_i^T \, \mathbb{K}(t_i \, , \, t_j) \, u_j = \| \sum_{i=0}^{n} u_i^T \, \mathcal{K}(t_i \, , \, \cdot \,) \, \|_{\mathcal{H}(\mathcal{K})}^2 \ . \tag{3.53}$$

Also, $\mathbb{A}\, w = y$ is equivalent to $\mathbb{A}\mathbb{V}_t \, u = y$, and

$$\begin{aligned}(\mathbb{A}\mathbb{V}_t \, u)_i = \mathbb{1}^T \, (\mathbb{V}_t \, u)_i = [\, (\mathbb{V}_t \, u)_i \,]^T \, \mathbb{1} = \\ \{ \sum_{j=0}^{n} u_j^T \, \mathbb{K}(t_i \, , \, t_j) \, \} \, \mathbb{1} = \sum_{j=0}^{n} u_j^T \, \mathcal{K}(t_i \, , \, t_j) \ .\end{aligned} \tag{3.54}$$

Now, $v = \sum_{i=0}^{n} u_i^T \, \mathcal{K}(t_i \, , \, \cdot \,)$ is just a generic element in

$$\mathcal{S}_t = K_t + G_t \ , \tag{3.55}$$

where

$$\begin{aligned} K_t &= \operatorname{span}(\, K(t \, , \, \cdot \,), K(t_1, \, \cdot \,), \, \cdots \, , K(t_n, \, \cdot \,) \,) \ , \\ G_t &= \operatorname{span}(\, G(t \, , \, \cdot \,), G(t_1, \, \cdot \,), \, \cdots \, , G(t_n, \, \cdot \,) \,) \ . \end{aligned} \tag{3.56}$$

So, the problem (3.40) is equivalent to

$$\begin{aligned} &\text{minimize} && \| v \|_{\mathcal{H}(\mathcal{K})} \\ &\text{subject to} && v \in \mathcal{S}_t \,, \\ & && \mathbb{1}^T v(t_i) = y_i \,, \quad i = 1, 2, \cdots, n, \end{aligned}$$

or, writing $v = (f, g)$ with $f \in \mathcal{K}_t$ and $g \in \mathcal{G}_t$,

$$\begin{aligned} &\text{minimize} && \kappa^{-2} \| f \|_K^2 + \sigma^{-2} \| g \|_G^2 \\ &\text{subject to} && f \in \mathcal{K}_t \,, \; g \in \mathcal{G}_t \,, \\ & && f(t_i) + g(t_i) = y_i \,, \quad i = 1, 2, \cdots, n \,. \end{aligned} \tag{3.57}$$

Unfortunately, we are not done yet. Moreover, it is clear that g must be treated differently from f if we are to reach a roughness-penalized least-squares problem.

First, pretend that g is known, and write the constraints as

$$\langle K(t_i, \cdot), f \rangle_K = y_i - g(t_i) \,, \quad i = 1, 2, \cdots, n \,.$$

Now, similar to the generalized spline interpolation case, we may extend the minimization in (3.57) to *all* $f \in \mathcal{H}(K)$.

Second, pretending f is known and writing the constraints as

$$\langle G(t_i, \cdot), g \rangle_G = y_i - f(t_i) \,, \quad i = 1, 2, \cdots, n \,,$$

we see that we may *restrict* the minimization over $g \in \mathcal{G}_t$ to

$$g \in \operatorname{span}\big(G(t_1, \cdot), G(t_2, \cdot), \cdots, G(t_n, \cdot) \big) \,.$$

So, we may parametrize g as

$$g(s) = \sum_{i=1}^{n} a_i \, G(t_i, s) \,, \quad s \geqslant 0 \,, \tag{3.58}$$

for arbitrary real numbers a_i, and then $g(t_j) = a_j$ for $j = 1, 2, \cdots, n$. Thus, the constraints in (3.58) take the form

$$a_i = y_i - f(t_i) \,, \quad i = 1, 2, \cdots, n \,.$$

Moreover,

$$\| g \|_G^2 = \sum_{i=1}^{n} | y_i - f(t_i) |^2 \,.$$

Putting it all together, we may write the problem (3.57) as

$$\begin{aligned} &\text{minimize} && \sigma^{-2} \sum_{i=1}^{n} | y_i - f(t_i) |^2 + \kappa^{-2} \| f \|_K^2 \\ &\text{subject to} && f \in \mathcal{H}(K) \,, \end{aligned} \tag{3.59}$$

and then $f(t)$ is the estimator of $X(t)$, the first component of $\mathbb{X}(t)$, in (3.39). This is the "generalized" smoothing spline problem.

Noisy data, continued. The development so far has been focused on the basic estimation problem (3.39) for the Gaussian process with noisy observations, (3.36)–(3.38), and the equivalence with the roughness-penalized least-squares problem (3.59). Here, we rework the whole thing from the point of view of data smoothing, viz.

$$\text{(3.60)} \qquad \begin{array}{ll} \text{estimate} & X(t_i)\ , \quad i = 1, 2, \cdots, n \\ \text{given the data} & y_i = X(t_i) + \varepsilon(t_i)\ , \quad i = 1, 2, \cdots, n\ . \end{array}$$

We also reorganize things a bit. Let

$$\mathfrak{X} = \big(X(t_1), X(t_2), \cdots, X(t_n) \big)^T ,$$

and

$$\mathfrak{E} = \big(\varepsilon(t_1), \varepsilon(t_2), \cdots, \varepsilon(t_n) \big)^T .$$

Then,

$$\text{(3.61)} \qquad \begin{bmatrix} \mathfrak{X} \\ \mathfrak{E} \end{bmatrix} \sim \text{Normal} \left(0, \begin{bmatrix} \kappa^2 \mathcal{V} & O \\ O & \sigma^2 I \end{bmatrix} \right)$$

with

$$\text{(3.62)} \qquad \mathcal{V} = \begin{bmatrix} K_{11} & K_{12} & \dots & K_{1n} \\ K_{21} & K_{22} & \dots & K_{2n} \\ \vdots & \vdots & \ddots & \vdots \\ K_{n1} & K_{n2} & \dots & K_{nn} \end{bmatrix} ,$$

where $K_{ij} = K(t_i, t_j)$ for all i, j.

The constrained maximum likelihood estimation problem is then

$$\begin{array}{ll} \text{minimize} & \kappa^{-2} u^T \mathcal{V}^{-1} u + \sigma^{-2} e^T e \\ \text{subject to} & y = u + e\ , \ u \in \mathbb{R}^n\ , \ e \in \mathbb{R}^n\ . \end{array}$$

Here, u is the potential estimator of $\mathfrak{X}$ and e refers to the error $\mathfrak{E}$. Eliminating e yields the discrete formulation

$$\text{(3.63)} \qquad \begin{array}{ll} \text{minimize} & \sigma^{-2} \sum\limits_{i=1}^{n} |\, u_i - y_i \,|^2 + \kappa^{-2} u^T \mathcal{V}^{-1} u \\ \text{subject to} & u \in \mathbb{R}^n\ . \end{array}$$

In the next section, we arrive at this formulation via autoregressive models.

Now, as before, we may transfer (3.63) to the reproducing kernel Hilbert space setting by attempting a solution of (3.63) in the form $u = \mathcal{V} w$. Then, $u^T \mathcal{V}^{-1} v = w^T \mathcal{V} w$ and

$$\text{(3.64)} \qquad w^T \mathcal{V} w = \Big\| \sum_{i=1}^{n} w_i K(t_i, \cdot) \Big\|_K^2 ,$$

and, after a few necessary shenanigans, the problem (3.63) is seen to be equivalent to (3.59).

(3.65) EXERCISE. Formulate in what precise sense the problems (3.63) and (3.59) are equivalent. (And prove it.)

(3.66) EXERCISE. One could and should argue that we really ought to have developed the theory for vector-valued stochastic processes with jointly Gaussian components, say $\mathcal{X}(t)$, $t \geqslant 0$, where $\mathcal{X}(t) \in \mathbb{R}^p$ for some integer p, with $\mathbb{E}[\mathcal{X}(t)] = 0$, and $\mathbb{E}[\mathcal{X}(s)\mathcal{X}(t)^T]$ a suitable symmetric positive-definite matrix function. Well, go to it!

EXERCISES: (3.10), (3.11), (3.19), (3.27), (3.43), (3.48), (3.52), (3.65), (3.66).

4. Autoregressive models

We now consider what happens when the Gaussian process under consideration is equivalent to an autoregressive process driven by Gaussian noise with short-range dependence. Interest is centered on the computational aspects of the basic estimation problem for the stochastic process or, more precisely, on the data-smoothing problem.

Thus, consider the Gaussian process

(4.1) $$\mathbb{X}(t) = \begin{bmatrix} X(t) \\ \varepsilon(t) \end{bmatrix} , \qquad t \geqslant 0 ,$$

with independent Gaussian components. The problem is to

(4.2) $$\begin{array}{ll} \text{estimate} & X(t_i),\ i = 1, 2, \cdots, n \\ \text{given the data} & y_i = X(t_i) + \varepsilon(t_i) , \quad i = 1, 2, \cdots, n . \end{array}$$

Now, suppose that there exist deterministic coefficients $\lambda_{i,j}$, depending only on the design points $t_1, t_2, \cdots, t_n$, with $\lambda_{i,0} \neq 0$, and processes q_i, $i = 0, -1, -2, \cdots, -m+1$, and ψ_i, $i = 1, 2, \cdots, n$, such that

(4.3) $$\begin{aligned} &\sum_{k=0}^{m} \lambda_{ik}\, X(t_{i-k}) = \psi_i , \quad i = 1, 2, \cdots, n , \\ &X(t_i) = q_i , \quad i = 0, -1, -2, \cdots, -m+1 . \end{aligned}$$

We assume that the q_i have a diffuse prior,

(4.4) $$(q_0, q_{-1}, \cdots, q_{-m+1})^T \sim \text{Normal}(0, \xi^2 \Sigma) ,$$

with $\xi \to \infty$ and Σ some fixed, positive-definite covariance matrix. Last but not least, we assume that the ψ_i, $i = 1, 2, \cdots, n$, form a zero-mean Gaussian process with short-range dependence; i.e., ψ_i and ψ_j are independent for $|i - j| \geqslant m$. Assume that there exists a fixed $T \in \mathbb{R}^{n \times n}$ and an unknown parameter κ such that

(4.5) $$\mathbb{E}[\psi_i \psi_j] = \kappa^2 T_{ij} , \quad i, j = 1, 2, \cdots, n .$$

Then $T_{ij} = 0$ for $|i - j| \geqslant m$. So, T has $2m + 1$ nonzero diagonals.

Of course, one can write any stochastic process in autoregressive form, but the question is what one would hope to gain by doing so. The limited memory of the noise process would be one such thing. (Ideally, one would like the ψ_i to be independent and the λ_{ik} not to depend on i, but in our context that is too much to hope for.)

We proceed to solve the data-smoothing problem (4.2) by considering its formulation as a constrained maximum likelihood problem (again). First, we need the joint distribution of

$$(4.6) \quad X_n = \bigl(X(t_{-m+1}), \, \cdots, X(t_{-1}), X(t_0), X(t_1), \, \cdots, X(t_n) \bigr)^T .$$

Write (4.3) in matrix-vector notation as

$$(4.7) \qquad \mathfrak{L} X_n = q_{(m)} + \psi_{(n)} ,$$

where $\mathfrak{L} \in \mathbb{R}^{(n+m)\times(n+m)}$ and

$$(4.8) \qquad \begin{aligned} q_{(m)} &= \bigl(q_{-m+1}, \cdots, q_{-1}, q_0, 0, 0, \cdots, 0 \bigr)^T , \\ \psi_{(n)} &= \bigl(0, \cdots, 0, 0, \psi_1, \psi_2, \cdots, \psi_n \bigr)^T . \end{aligned}$$

Note that $\mathfrak{L}$ is a banded lower-triangular matrix and that its main diagonal has only nonzero elements on it. Thus, $\mathfrak{L}$ is nonsingular. It is useful to partition $\mathfrak{L}$ as

$$(4.9) \qquad \mathfrak{L} = \begin{bmatrix} I & O \\ \mathfrak{L}_{2,1} & \mathfrak{L}_{2,2} \end{bmatrix} , \quad \text{with} \quad \mathfrak{L}_{2,2} \in \mathbb{R}^{n\times n} .$$

The distribution of X_n is now easy to determine. Obviously,

$$(4.10) \qquad \mathfrak{L} X_n \sim \text{Normal}(0, \mathcal{V}) ,$$

where

$$(4.11) \qquad \mathcal{V} = \begin{bmatrix} \xi^2 \Sigma & O \\ O & \kappa^2 T \end{bmatrix} ,$$

and then $X_n \sim \text{Normal}(0, \mathfrak{L}^{-1} \mathcal{V} \mathfrak{L}^{-T})$. So, the constrained maximum likelihood estimation problem is

$$\begin{aligned} &\text{minimize} \quad w^T \mathfrak{L}^T \mathcal{V}^{-1} \mathfrak{L} w + \sigma^{-2} e^T e \\ &\text{subject to} \quad w + e = y , \end{aligned}$$

or equivalently

$$(4.12) \qquad \begin{aligned} &\text{minimize} \quad \sigma^{-2} \sum_{i=1}^{n} | w_i - y_i |^2 + w^T \mathfrak{L}^T \mathcal{V}^{-1} \mathfrak{L} w \\ &\text{subject to} \quad w \in \mathbb{R}^n . \end{aligned}$$

Finally, in view of (4.11), we have the diffuse limits

$$\mathcal{V}^{-1} = \begin{bmatrix} \xi^{-2}\,\Sigma^{-1} & O \\ O & \kappa^{-2}\,T^{-1} \end{bmatrix} \longrightarrow \begin{bmatrix} O & O \\ O & \kappa^{-2}\,T^{-1} \end{bmatrix} \tag{4.13}$$

and

$$\mathfrak{L}^T\,\mathcal{V}^{-1}\,\mathfrak{L} \longrightarrow \kappa^{-2}\,\mathfrak{L}_{2,2}^T\,T^{-1}\,\mathfrak{L}_{2,2}\ . \tag{4.14}$$

This leads to the final form

$$\begin{aligned} &\text{minimize} \quad \sigma^{-2}\sum_{i=1}^{n} |\,b_i - y_i\,|^2 + \kappa^{-2}\,a^T\,T\,a \\ &\text{subject to} \quad T\,a = S\,b\ , \end{aligned} \tag{4.15}$$

where $S = \mathfrak{L}_{2,2}$.

(4.16) Exercise. Let $\alpha = \sigma^2/\kappa^2$. Show that the solution of (4.15) may be computed by solving $(\,T + \alpha\,S\,S^T\,)\,a = S\,y$ and then $b = y - \alpha\,S^T a$.

Of course, we wish to work out the details for spline smoothing. However, all that is required is the computation of the matrices S and T, which we already did in Exercise (1.38). Finally, one should note the similarity with the computational scheme developed in §§19.3 and 19.5.

Exercise: (4.16).

5. State-space models

We continue to study the data-smoothing problem for a Gaussian process with noisy observations. We are particularly interested in what happens when miraculously there is a state-space formulation of the stochastic process under scrutiny, as in (1.7)–(1.29).

Thus, consider again the process

$$\mathbb{X}(\,t\,) = \begin{pmatrix} X(\,t\,) \\ \varepsilon(\,t\,) \end{pmatrix}\ , \qquad t \geqslant 0\ , \tag{5.1}$$

where $X(\,t\,)$ and $\varepsilon(\,t\,)$ are independent zero-mean Gaussian processes, as before; see (3.36)–(3.38).

Now, suppose that there is a state-space model formulation of the stochastic process $X(\,t\,)$, $t \geqslant 0$; cf. (1.26). Then, per Exercise (1.39), we may obtain a discrete-time formulation in the form

$$\mathbb{S}(\,t_{i+1}) = Q(\,t_{i+1}\,|\,t_i)\,\mathbb{S}(\,t_i) + \mathbb{U}(\,t_{i+1})\ , \quad i = 0, 1, 2, \cdots\ , \tag{5.2}$$

where $\mathbb{S}(\,t\,) \in \mathbb{R}^p$ for all t, the matrices $Q(\,t\,|\,s\,)$ are deterministic, and

$$\mathbb{S}(\,t_0), \mathbb{U}(\,t_1), \mathbb{U}(\,t_2), \mathbb{U}(\,t_3), \cdots \ \text{are independent normals}\,. \tag{5.3}$$

Of course, we assume that they have mean zero and set

$$(5.4)\qquad \Sigma_0 = \mathbb{E}[\,\mathbb{S}(t_0)\,\mathbb{S}(t_0)^T\,]\ , \quad \Sigma_i = \mathbb{E}[\,\mathbb{U}(t_i)\,\mathbb{U}(t_i)^T\,]\ ,\ i = 1, 2, \cdots\ .$$

The basic assumption is that the stochastic process and the state space model are equivalent in the sense that, almost surely,

$$(5.5)\qquad [\,\mathbb{S}(t_i)\,]_1 = X(t_i)\ , \quad i = 0, 1, 2, \cdots\ .$$

The noise process $\varepsilon(t_i)$ must be added, but is kept separate. The significance of the state-space model lies in the independence of the noise $\mathbb{U}(t_i)$, $i = 1, 2, \cdots$, that drives it.

(5.6) REMARK. Starting the application to spline smoothing early, note that $[\,\mathbb{S}(0)\,]_k = k\,!\,p_k\ ,\ k = 0, 1, \cdots, m-1$. So, a diffuse prior on $\mathbb{S}(0)$ (or $\mathbb{S}(t_0)$ below) is the same as the diffuse prior in (1.7).

The state-space formulation of the data-smoothing problem is then

$$(5.7)\qquad \begin{array}{ll} \text{estimate} & \bigl(\,\mathbb{S}(t_i)\,,\,\varepsilon(t_i)\,\bigr)\ ,\ i = 0, 1, 2, \cdots, n\ , \\ \text{given the data} & y_i = e_1^T\,\mathbb{S}(t_i) + \varepsilon(t_i)\ ,\ i = 1, 2, \cdots, n\ , \end{array}$$

with e_1 the first element of the standard basis for $\mathbb{R}^p$. Note that $\mathbb{S}(t_0)$ must be estimated as well, although there are no direct data involving it. Also, there is no need to estimate the $\varepsilon(t_i)$.

Of course, we are interested in the maximum likelihood formulation. Let

$$(5.8)\qquad \begin{aligned} \mathcal{S} &= \bigl(\,\mathbb{S}(t_0)^T, \mathbb{S}(t_1)^T, \cdots, \mathbb{S}(t_n)^T\bigr)^T\ , \\ \mathcal{E} &= \bigl(\,\varepsilon(t_0), \varepsilon(t_1), \cdots, \varepsilon(t_n)\bigr)^T\ . \end{aligned}$$

Then, $\mathcal{S}$ and $\mathcal{E}$ are independent, and

$$(5.9)\qquad \mathcal{E} \sim \text{Normal}\bigl(\,0\,,\,\sigma^2\,I\,\bigr)\ .$$

Regarding $\mathcal{S}$, (5.3) and (5.4) tell us that

$$(5.10)\qquad \mathbb{L}\,\mathcal{S} \sim \text{Normal}\bigl(\,0\,,\,\mathbb{D}\,\bigr)\ ,$$

where $\mathbb{D} = \text{block-diag}\bigl(\,\Sigma_0, \Sigma_1, \cdots, \Sigma_n\,\bigr)$ and

$$(5.11)\qquad \mathbb{L} = \begin{bmatrix} I & & & & \\ -Q_{1|0} & I & & & \\ & -Q_{2|1} & I & & \\ & & \ddots & \ddots & \\ & & & -Q_{n|n\text{-}1} & I \end{bmatrix}\ ,$$

in which $Q_{i|i\text{-}1} = Q(\,t_i\,|\,t_{i-1}\,)$. Then, apparently,

$$(5.12)\qquad \mathcal{S} \sim \text{Normal}\bigl(\,0\,,\,\mathbb{L}^{-1}\,\mathbb{D}\,\mathbb{L}^{-T}\,\bigr)\ ,$$

and the constrained maximum likelihood formulation of (5.7) is

$$\begin{aligned} &\text{minimize} \quad s^T \mathbb{L}^T \mathbb{D}^{-1} \mathbb{L}\, s + \sigma^{-2} e^T e \\ &\text{subject to} \quad s \in \mathbb{R}^{(n+1)p} \,, \quad e \in \mathbb{R}^n \\ &\qquad\qquad s_{i,1} + e_i = y_i \,, \quad i = 1, 2, \cdots, n \,. \end{aligned}$$

Note that s has one more block than e has elements. After the elimination of e, we then get the problem

$$(5.13) \qquad \begin{aligned} &\text{minimize} \quad \sigma^{-2} \sum_{i=1}^{n} |\, s_{i,1} - y_i \,|^2 + s^T \mathbb{L}^T \mathbb{D}^{-1} \mathbb{L}\, s \\ &\text{subject to} \quad s \in \mathbb{R}^{(n+1)p} \,. \end{aligned}$$

Now, the problems (4.11) and (5.13) may be identified with each other, if s is partitioned as $s = (\, s_0^T, s_1^T, s_2^T, \cdots, s_n^T \,)^T$, and interpreted as $s_{i,1} = w_i$ for $i = 1, 2, \cdots, n$.

(5.14) EXERCISE. It is perhaps useful to observe that the problem (5.13) may be equivalently formulated as

$$\text{minimize} \quad \|\, s_0 \,\|^2_{\Sigma_o^{-1}} + \sum_{i=1}^{n} \Big\{ \|\, s_i - Q_{i|i\text{-}1}\, s_{i\text{-}1} \,\|^2_{\Sigma_i^{-1}} + \frac{|\, e_1^T s_i - y_{in} \,|^2}{\sigma^2} \Big\} \,.$$

Verify this! We will return to this in Exercise (6.27).

The computational issues are now easily settled, sort of. It is useful to introduce the matrix $\mathbb{B} \in \mathbb{R}^{n \times (n+1)m}$ by

$$(5.15) \qquad \mathbb{B} = \begin{bmatrix} 0 & e_1^T & & & \\ 0 & & e_1^T & & \\ \vdots & & & \ddots & \\ 0 & & & & e_1^T \end{bmatrix} ,$$

where $e_1 = (\,1, 0, \cdots, 0\,)^T \in \mathbb{R}^m$. Then, $\sum_{i=1}^{n} |\, s_{i,1} - y_i \,|^2 = \|\, \mathbb{B}\, s - y \,\|^2$, with the obvious definition of y. The problem (5.13) is then equivalent to

$$(5.16) \qquad \begin{aligned} &\text{minimize} \quad \sigma^{-2} \,\|\, \mathbb{B}\, s - y \,\|^2 + s^T \mathbb{L}^T \mathbb{D}^{-1} \mathbb{L}\, s \\ &\text{subject to} \quad s \in \mathbb{R}^{(n+1)\, m} \,. \end{aligned}$$

The normal equations for this problem are

$$(5.17) \qquad \big(\sigma^{-2} \mathbb{B}^T \mathbb{B} + \mathbb{L}^T \mathbb{D}^{-1} \mathbb{L}\big)\, s = \mathbb{B}^T y \,,$$

which may be solved using Cholesky factorization. This requires that the matrix product $\mathbb{L}^T \mathbb{D}^{-1} \mathbb{L}$ be multiplied out explicitly. Fortunately, it is a block-tridiagonal matrix, so this is not a full matrix multiplication. Also, the matrix $\mathbb{B}^T \mathbb{B}$ is really simple (so as not to require explicit computation).

Looking ahead to the application to spline smoothing, the problem with all of this is that it seems to require the inverses of the matrices Σ_i. Since these matrices are scaled (and permuted) versions of the Hilbert matrices, see (19.8.4), this may lead to trouble even for splines of moderate order m.

Also note that (5.17) involves the solution of an $(n+1)\,m$ by $(n+1)\,m$ system of equations but that the data is only n-dimensional. We can put this to good use by noting that

$$(5.18) \qquad \left(\sigma^{-2}\,\mathbb{B}^T\mathbb{B} + \mathbb{L}^T\mathbb{D}^{-1}\mathbb{L}\right)^{-1}\mathbb{B}^T = \mathbb{A}\,\mathbb{B}^T\left(\sigma^{-2}\,I + \mathbb{B}\,\mathbb{A}\,\mathbb{B}^T\right)^{-1},$$

where $\mathbb{A} = \mathbb{L}^{-1}\,\mathbb{D}\,\mathbb{L}^{-T}$.

Now, a system of size n by n must be solved, which is much better, but it seems to require the explicit computation of $\mathbb{L}^{-1}\,\mathbb{D}\,\mathbb{L}^{-T}$, which is *not* a clever thing to do. So, we are stuck! Actually, in the next section, KALMAN (1960) comes to the rescue, although initially it is not obvious that he does.

(5.19) EXERCISE. Verify (5.18). [Hint: View $\mathbb{B}^T\mathbb{B}$ as a low-rank perturbation of the matrix $\mathbb{A}$, and use the Sherman-Morrison-Woodbury Lemma (18.3.12).]

(5.20) EXERCISE (DIFFUSE PRIORS). In the model (5.1)–(5.4), consider the case of the diffuse initial state, $\Sigma_0 = k^2\,I$, $k \to \infty$. Denote the solution of (5.16) by s^k. Show that $s^o = \lim_{k\to\infty} s^k$ exists and that $s = s^0$ solves

$$\begin{array}{ll} \text{minimize} & \sigma^{-2}\,\|\,\mathbb{B}\,s - y\,\|^2 + s^T\,\mathbb{L}^T\,\mathbb{D}_o\,\mathbb{L}\,s \\ \text{subject to} & s \in \mathbb{R}^{(n+1)\,m}\,, \end{array}$$

where $\mathbb{D}_o = \text{block-diag}(\,O\,,\,\Sigma_1^{-1}\,,\,\Sigma_2^{-1}\,,\,\cdots\,,\,\Sigma_n^{-1}\,)$. [Hint: The normal equations are the simplest route. The hardest part is to show that the inverse $\left(\mathbb{B}^T\,\mathbb{B} + \mathbb{L}^T\,\mathbb{D}_o\,\mathbb{L}\right)^{-1}$ exists. Then, in fact, for n fixed, one gets the bound $\|\,s^k - s^o\,\| = \mathcal{O}(\,k^{-2}\,)$.]

EXERCISES: (5.14), (5.19), (5.20).

6. Kalman filtering for state-space models

In the previous section, data smoothing for state-space models was studied from the point of view of maximum *a posteriori* likelihood estimation. Ultimately, this requires the Cholesky factorization of the normal equations. The only criticism one can level against this development (if such it may be called and only after viewing the alternative) is that the Cholesky factorization lacks a probabilistic (Bayesian ?) interpretation. Enter the Kalman filter, which nicely provides the missing pieces. Actually, the Kalman filter was designed mostly for prediction and filtering problems, but of course all of these problems are "equivalent" in a broad sense, as discussed in § 3.

In this section, we derive the Kalman prediction algorithm. In § 7, it is related to the data-smoothing problem. Recall the discrete-time model

(5.2)–(5.5) for the partially observed stochastic process $X(t)$,

$$(6.1)\qquad \begin{aligned} \mathbb{S}(t_{i+1}) &= Q_{i+1|i}\,\mathbb{S}(t_i) + \mathbb{U}(t_{i+1})\,, \quad i = 0, 1, \cdots, n-1\,, \\ y(t_{i+1}) &= e_1^T\,\mathbb{S}(t_{i+1}) + \varepsilon(t_{i+1})\,, \end{aligned}$$

with the design points $t_0 < t_1 < \cdots < t_n$, deterministic transition matrices $Q_{i|j} = Q(t_i \mid t_j)$, and $\mathbb{S}(t_0)$, $\mathbb{U}(t_1)$, $\mathbb{U}(t_2)$, $\cdots$, $\mathbb{U}(t_n)$ independent, zero-mean multivariate Gaussian random variables with

$$(6.2)\quad \Sigma_0 = \mathbb{E}[\,\mathbb{S}(t_0)\,\mathbb{S}(t_0)^T\,]\,,\quad \Sigma_i = \mathbb{E}[\,\mathbb{U}(t_i)\,\mathbb{U}(t_i)^T\,]\,,\ i = 1, 2, \cdots, n\,.$$

Also, $\varepsilon(t_1)$, $\varepsilon(t_2)$, $\cdots$, $\varepsilon(t_n)$, are independent Normal$(0, \sigma^2)$ random variables, independent of everything else.

The *prediction* problem for the discrete-time model (6.1) is

$$(6.3)\qquad \begin{array}{ll} \text{estimate} & \mathbb{S}(t_{i+1}) \\ \text{given the data} & y(t_1),\, y(t_2),\, \cdots,\, y(t_i)\,. \end{array}$$

Note that, there is no direct information at $t = t_{i+1}$, which explains why it is a prediction problem. In the *filtering* problem, there is such information:

$$(6.4)\qquad \begin{array}{ll} \text{estimate} & \mathbb{S}(t_i) \\ \text{given the data} & y(t_1),\, y(t_2),\, \cdots,\, y(t_i)\,. \end{array}$$

Note that this is not quite the same as the data-smoothing problem in that only the current state must be estimated. The previous states are deemed to be uninteresting.

The Kalman predictor provides a recursive solution to the prediction problem, but in doing so it also solves the filtering problem. To explain the notion of a recursive solution, let $\mathbb{S}_{i|j}$ denote the state at time $t = t_i$ given the information $y(t_1)$, $y(t_2)$, $\cdots$, $y(t_j)$, or

$$(6.5)\qquad \mathbb{S}_{i|j} = \mathbb{S}(t_i) \,\big|\, y(t_1),\, y(t_2),\, \cdots,\, y(t_j)\,.$$

The Kalman predictor determines the distribution of $\mathbb{S}_{i+1|i}$ in terms of the distribution of $\mathbb{S}_{i|i-1}$ and, of course, the known distributions of the driving noise $\mathbb{U}(t_{i+1})$ and $\varepsilon(t_i)$. Since everything in sight is Gaussian, this means that we only need to be concerned with the (conditional) means and variances. With that in mind, we have that $\mathbb{S}_{i|j}$ is normal, with the appropriate conditional mean and variance, denoted as

$$(6.6)\qquad \mathbb{S}_{i|j} \sim \text{Normal}(\,\widehat{\mathbb{S}}_{i|j}\,,\, \mathbb{V}_{i|j}\,)\,.$$

We move on to the derivation of the Kalman predictor by way of the Kalman filter. In the Gaussian setting, the "best" predictor of $\mathbb{S}(t_{i+1})$ is

$$(6.7)\qquad \widehat{\mathbb{S}}_{i+1|i} = \mathbb{E}\big[\,\mathbb{S}_{i+1|i}\,\big]\,.$$

Since the model (6.1) only involves linear transformations of Gaussians, it follows that

$$\widehat{\mathbb{S}}_{i+1|i} = Q_{i+1|i}\, \widehat{\mathbb{S}}_{i|i} \tag{6.8}$$

and that

$$\mathbb{V}_{i+1|i} = Q_{i+1|i}\, \mathbb{V}_{i|i}\, Q^T_{i+1|i} + \Sigma_{i+1} \,, \tag{6.9}$$

where $\mathbb{V}_{i|j} = \mathbb{V}\mathrm{ar}[\, \mathbb{S}_{i|j}\,]$; see (6.6). Thus, we may concentrate on the filtering problem. It seems advantageous to compute the (conditional) expectation of $\mathbb{S}_{i|i}$ by means of the linearly constrained maximum likelihood estimation problem. It is apparent that the random variables $\mathbb{S}_{i|i\text{-}1}$ and $\varepsilon(\,t_i\,)$ are independent, with $\varepsilon(\,t_i\,) \sim \mathrm{Normal}(\,0\,,\, \sigma^2\,)$ and

$$\mathbb{S}_{i|i\text{-}1} \sim \mathrm{Normal}\Big(\, \widehat{\mathbb{S}}_{i|i\text{-}1}\,,\, \mathbb{V}_{i|i\text{-}1}\,\Big)\,. \tag{6.10}$$

Then, $s = \widehat{\mathbb{S}}_{i|i}$ solves the constrained maximum likelihood problem

$$\begin{aligned} &\text{minimize} \quad \| \, s - \widehat{\mathbb{S}}_{i|i\text{-}1} \, \|^2_{\mathbb{V}^{-1}_{i|i\text{-}1}} + \sigma^{-2}\, e^2 \\ &\text{subject to} \quad y(\,t_i\,) = e_1^T s + e \,, \end{aligned} \tag{6.11}$$

where we used the notation $\|\, a \,\|^2_B = a^T B\, a$. Now, eliminate e and apply the substitution $\delta = s - \widehat{\mathbb{S}}_{i|i\text{-}1}$ to obtain the problem

$$\text{minimize} \quad \|\, \delta \,\|^2_{\mathbb{V}^{-1}_{i|i\text{-}1}} + \sigma^{-2}\, |\, e_1^T \delta - \widetilde{y}_{i|i\text{-}1}\, |^2 \,, \tag{6.12}$$

where $\widetilde{y}_{i|i\text{-}1} = y(\,t_i\,) - e_1^T\, \widehat{\mathbb{S}}_{i|i\text{-}1}$. Denoting its solution by $\widehat{\delta}$, we obtain

$$\widehat{\mathbb{S}}_{i|i} = \widehat{\mathbb{S}}_{i|i\text{-}1} + \widehat{\delta}\,. \tag{6.13}$$

The problem (6.12) is straightforward to solve: The normal equations are

$$(\, \mathbb{V}^{-1}_{i|i\text{-}1} + \sigma^{-2}\, e_1\, e_1^T\,)\, \delta = \sigma^{-2}\, e_1\, \widetilde{y}_{i|i\text{-}1} \,, \tag{6.14}$$

so that the solution $\widehat{\delta}$ is given by

$$\widehat{\delta} = \big(\, \mathbb{V}^{-1}_{i|i\text{-}1} + \sigma^{-2}\, e_1\, e_1^T\,\big)^{-1} \sigma^{-2}\, e_1\, \widetilde{y}_{i|i\text{-}1} \,. \tag{6.15}$$

Using the Sherman-Morrison-Woodbury Lemma (18.3.12), one shows that this equals

$$\widehat{\delta} = \mathbb{V}_{i|i\text{-}1}\, e_1\, \big(\, \sigma^2 + e_1^T\, \mathbb{V}_{i|i\text{-}1}\, e_1\,\big)^{-1}\, \widetilde{y}_{i|i\text{-}1} \,. \tag{6.16}$$

Putting it all together, we then get

$$\widehat{\mathbb{S}}_{i|i} = \widehat{\mathbb{S}}_{i|i\text{-}1} + L_{i|i\text{-}1}\, \big(\, y(\,t_i\,) - e_1^T\, \widehat{\mathbb{S}}_{i|i\text{-}1}\,\big)\,. \tag{6.17}$$

Here,

$$L_{i|i\text{-}1} = \mathbb{V}_{i|i\text{-}1}\, e_1\, \big(\, \sigma^2 + e_1^T\, \mathbb{V}_{i|i\text{-}1}\, e_1\,\big)^{-1}\,. \tag{6.18}$$

The next step is to compute the variance of $\mathbb{S}_{i|i}$. To that end, an expression for

$$(6.19) \qquad \widetilde{\mathbb{S}}_{i|i} = \mathbb{S}_{i|i} - \widehat{\mathbb{S}}_{i|i}$$

is needed. Obviously, from (6.17), and using (6.1) to rewrite $y(t_i)$, we obtain

$$(6.20) \qquad \widetilde{\mathbb{S}}_{i|i} = (I - L_{i|i-1}\, e_1^T)\, \widetilde{\mathbb{S}}_{i|i-1} + L_{i|i-1}\, \varepsilon(t_i) \,.$$

Since $\varepsilon(t_i)$ is independent of $\mathbb{S}_{i|i-1}$, then

$$(6.21) \qquad \mathbb{V}_{i|i} = (I - L_{i|i-1}\, e_1^T)\, \mathbb{V}_{i|i-1}\, (I - L_{i|i-1}\, e_1^T)^T + \sigma^2 L_{i|i-1}\, L_{i|i-1}^T \,.$$

With (6.8), a recursive expression for the variances $\mathbb{V}_{i+1|i}$ follows.

Thus, to summarize, the Kalman filter reads as follows:

$$(6.22) \qquad \widehat{\mathbb{S}}_{i|i} = Q_{i|i-1}\, \widehat{\mathbb{S}}_{i-1|i-1} + L_{i|i-1}\left(y(t_i) - e_1^T\, Q_{i|i-1}\, \widehat{\mathbb{S}}_{i-1|i-1} \right) ,$$

$$(6.23) \qquad \mathbb{V}_{i|i} = (I - L_{i|i-1}\, e_1^T)\, \mathbb{V}_{i|i-1}\, (I - L_{i|i-1}\, e_1^T)^T + \sigma^2 L_{i|i-1}\, L_{i|i-1}^T ,$$

in which

$$(6.24) \qquad \mathbb{V}_{i|i-1} = Q_{i|i-1}\, \mathbb{V}_{i-1|i-1}\, Q_{i|i-1}^T + \Sigma_i$$

and

$$(6.25) \qquad L_{i|i-1} = \mathbb{V}_{i|i-1}\, e_1 \left(\sigma^2 + e_1^T\, \mathbb{V}_{i|i-1}\, e_1 \right)^{-1} .$$

Apart from two exercises below, this concludes the treatment of the Kalman filter. It is clear that the Kalman filter by itself does not solve the smoothing spline problem. Something extra is needed, which in the context of (general purpose) Kalman filtering goes by the name of *fixed interval smoothing*. A nice exposition of this may be found in SAGE and MELSA (1971), § 8.3. In the next section, we take a different route. Then, our interest is mostly in (6.23)–(6.25), which after introducing

$$(6.26) \qquad M_{i+1|i} \stackrel{\text{def}}{=} Q_{i+1|i}\, L_{i|i-1}\, e_1^T$$

may be rearranged as

$$(6.27) \qquad \begin{aligned} D_{i|i-1} &= \sigma^2 + e_1^T\, \mathbb{V}_{i|i-1}\, e_1 \,, \\ M_{i+1|i} &= Q_{i+1|i}\, \mathbb{V}_{i|i-1}\, e_1\, D_{i\,|\,i-1}^{-1} e_1^T \,, \\ K_{i+1|i} &= Q_{i+1|i} - M_{i+1|i} \,, \\ \mathbb{V}_{i+1|i} &= K_{i+1|i}\, \mathbb{V}_{i|i-1}\, K_{i+1}^T + \sigma^2 M_{i+1|i}\, M_{i+1}^T + \Sigma_{i+1} \,. \end{aligned}$$

(6.28) EXERCISE. Verify this! Note that $L_{i|i-1}\, L_{i\,|\,i-1}^T = L_{i|i-1}\, e_1^T\, e_1\, L_{i\,|\,i-1}^T$.

(6.29) EXERCISE (ORTHOGONAL PROJECTIONS). (a) The derivation of the Kalman filter above is short and sweet, but in fact it is not the whole

story. The tip of the iceberg is that $\widetilde{y}_{i|i\text{-}1}$, see (6.12), is orthogonal to $y(t_1), y(t_2), \cdots, y(t_{i-1})$, in the sense that $\mathbb{E}[\, \widetilde{y}_{i|i\text{-}1}\, y(t_j)\,] = 0$ for all j with $1 \leqslant j \leqslant i-1$. Verify this.
(b) While we are at it, verify that $\mathbb{V}\mathrm{ar}[\,\widetilde{y}(t_i)\,] = \sigma^2 + e_1^T\, \mathbb{V}_{i|i\text{-}1}\, e_1$. [Hint for (a): This is rather obvious upon viewing $e_1^T\, \mathbb{S}_{i|i\text{-}1}$ as the orthogonal projection of $e_1^T\, \mathbb{S}(t_i)$ onto $\mathcal{Y}_{i-1} = \mathrm{span}\{\, y(t_1), y(t_2), \cdots, y(t_{i-1})\,\}$. See Exercise (1.37).]

(6.30) Exercise. Consider the problem of Exercise (5.14) with the following recursive solution. For $i = 1, 2, \cdots, n$, define

$$\mathcal{L}_i(s) = \|\, s_0 \,\|^2_{\Sigma_o^{-1}} + \sum_{j=1}^{i} \left\{ \|\, s_j - Q_{j|j\text{-}1}\, s_{j-1} \,\|^2_{\Sigma_j^{-1}} + \frac{|\, e_1^T s_j - y_{jn}\,|^2}{\sigma^2} \right\},$$

$$\mathcal{L}_{i+1/2}(s) = \mathcal{L}_i(s) + \|\, s_{i+1} - Q_{i+1|i}\, s_i \,\|^2_{\Sigma_{i+1}^{-1}}.$$

(a) Let $s^{i-1/2} \in \mathbb{R}^{(i+1)m}$ be the minimizer of $\mathcal{L}_{i-1/2}(s)$ and $s^i \in \mathbb{R}^{(i+1)m}$ the minimizer of $\mathcal{L}_i(s)$. Show that

$$[\, s^{i-1/2}\,]_{i+1} = \widehat{\mathbb{S}}_{i|i\text{-}1} \quad \text{and} \quad [\, s^i\,]_{i+1} = \widehat{\mathbb{S}}_{i|i}, \qquad i = 1, 2, \cdots, n.$$

(b) Derive the recursions for $s^{i-1/2}$ and s^i in the present context. [This is the (?) interpretation of the Kalman filter in computational linear algebra terms.]

Exercises: (6.28), (6.29), (6.30).

7. Cholesky factorization via the Kalman filter

In this section, we develop a data-smoothing algorithm for state-space models based on the Kalman filter due to Kohn and Ansley (1989). As a bonus, this also provides an efficient way to compute the leverage values of the regression problem and hence the computation of the GCV functional. (See also Exercise (8.15). For the computation of the GML statistic, see Exercise (8.14).) As a result, the emphasis here is on the estimation of the original stochastic process, not on estimating the complete states.

The starting point is the following observation. Let $\mathcal{S}$ and y denote the collective states and data,

$$\begin{aligned} \mathcal{S} &= \big(\, \mathbb{S}(t_0)^T, \mathbb{S}(t_1)^T, \cdots, \mathbb{S}(t_n)^T \,\big)^T, \\ y &= \big(\, y(t_1), y(t_2), \cdots, y(t_n)\,\big)^T, \end{aligned} \tag{7.1}$$

as in §§ 4, 5, and 6. The objective is to estimate $\mathcal{S}$ given the data y, and since everything in sight is a zero-mean Gaussian, the estimator of the states is $\widehat{\mathcal{S}} = \mathbb{E}[\, \mathcal{S} \mid y\,]$, and the estimator of the original stochastic process is $\mathbb{B}\,\widehat{\mathcal{S}} = \mathbb{E}[\,\mathbb{B}\,\mathcal{S}\,|\,y\,]$, computable as

$$
(7.2)\qquad \begin{aligned} \widehat{\mathcal{S}} &= \mathbb{C}\mathrm{ov}[\,\mathcal{S}, y\,]\,\{\,\mathbb{V}\mathrm{ar}[\,y\,]\,\}^{-1}\,y\;, \\ \mathbb{B}\,\widehat{\mathcal{S}} &= \mathbb{C}\mathrm{ov}[\,\mathbb{B}\,\mathcal{S}, y\,]\,\{\,\mathbb{V}\mathrm{ar}[\,y\,]\,\}^{-1}\,y\;; \end{aligned}
$$

see Exercise (1.37). Note that, in the notation of § 6, the i-th block of $\widehat{\mathcal{S}}$ corresponds to $\widehat{\mathbb{S}}_{i|n}$. Also, the customary notation for $\mathbb{B}\,\widehat{\mathbb{S}}$ is $\widehat{y}$.

(7.3) EXERCISE. (a) Compute $\mathbb{C}\mathrm{ov}[\,\mathcal{S}, y\,]$ and $\mathbb{V}\mathrm{ar}[\,y\,]$, and show that

$$
\widehat{\mathcal{S}} = \mathbb{A}\,\mathbb{B}^T\,(\,\sigma^2\,I + \mathbb{B}\,\mathbb{A}\,\mathbb{B}^T\,)^{-1}\,y\;,
$$

where $\mathbb{A} = \mathbb{L}^{-1}\,\mathbb{D}\,\mathbb{L}^{-T}$. (Where have we seen this before?)
(b) Show that $\mathbb{B}\,\widehat{\mathcal{S}} = y - \sigma^2\,(\,\sigma^2\,I + \mathbb{B}\,\mathbb{A}\,\mathbb{B}^T\,)^{-1}\,y$, or equivalently

$$
\widehat{y} = y - \sigma^2\,\{\,\mathbb{V}\mathrm{ar}[\,y\,]\,\}^{-1}\,y\;.
$$

The observation that gets us going is that the "innovations"

$$
(7.4)\qquad \widetilde{y}(\,t_i\,) = y(\,t_i\,) - e_1^T\,\widehat{\mathbb{S}}_{i|i\text{-}1}\;, \quad i = 1, 2, \cdots, n\;,
$$

are independent and that $\widehat{\mathbb{S}}_{i|i\text{-}1}$ is a linear combination of the "preceding" data $y(\,t_1\,)$, $y(\,t_2\,)$, $\cdots$, $y(\,t_{i-1}\,)$. So, $\widetilde{y}(\,t_i\,)$ is a linear combination of $y(\,t_1\,)$, $y(\,t_2\,)$, $\cdots$, $y(\,t_{i-1}\,)$, $y(\,t_i\,)$, and conversely $y(\,t_i\,)$ is a linear combination of $\widetilde{y}(\,t_1\,)$, $\widetilde{y}(\,t_2\,)$, $\cdots$, $\widetilde{y}(\,t_{i-1}\,)$, $\widetilde{y}(\,t_i\,)$. All of this shows that there exists a lower-triangular matrix $L_K \in \mathbb{R}^{n\times n}$ such that

$$
(7.5)\qquad y = L_K\,\widetilde{y}\;,
$$

and the diagonal elements of L_K are all equal to 1. It follows that

$$
(7.6)\qquad \mathbb{V}\mathrm{ar}[\,y\,] = L_K\,D_K\,L_K^T\;,
$$

with $D_K = \mathbb{V}\mathrm{ar}[\,\widetilde{y}\,]$ a diagonal matrix with $[\,D_K\,]_{i,i} = D_{i|i\text{-}1}$ of (6.27),

$$
(7.7)\qquad [\,D_K\,]_{i,i} = \sigma^2 + e_1^T\,\mathbb{V}_{i|i\text{-}1}\,e_1\;.
$$

Of course, the Kalman filter provides for $\widetilde{y}$ and D_K and in so doing provides the Cholesky factorization of $\mathbb{V}\mathrm{ar}[\,y\,]$. We may compute the matrix L_K more or less explicitly, as follows. It is useful to consider the "state innovations"

$$
(7.8)\qquad \check{y}(\,t_i\,) = e_1\,y(\,t_i\,) - \widehat{\mathbb{S}}_{i|i\text{-}1}\;, \quad i = 1, 2, \cdots, n\;.
$$

Note that $e_1^T\,\check{y}(\,t_i\,) = \widetilde{y}(\,t_i\,)$. Now, a recursion for the state innovations may be obtained using the Kalman filter (6.22)–(6.25). First,

$$
(7.9)\qquad \begin{aligned} \check{y}(\,t_{i+1}\,) &= e_1\,y(\,t_{i+1}\,) - Q_{i+1|i}\,\widehat{\mathbb{S}}_{i|i} \\ &= e_1\,y(\,t_{i+1}\,) - Q_{i+1|i}\,\{\widehat{\mathbb{S}}_{i|i\text{-}1} + L_{i|i\text{-}1}\,\widetilde{y}(\,t_i\,)\}\;. \end{aligned}
$$

Then, writing $\widehat{\mathbb{S}}_{i|i\text{-}1} = e_1\, y(t_i) - \overset{\vee}{y}(t_i)$ and $L_{i|i\text{-}1}\, \widetilde{y}(t_i) = L_{i|i\text{-}1}\, e_1^T\, \overset{\vee}{y}(t_i)$, we obtain the recursion

$$
\begin{aligned}
(7.10)\quad \overset{\vee}{y}(t_{i+1}) - Q_{i+1|i}\,(I - L_{i|i\text{-}1}\, e_1^T)\, \overset{\vee}{y}(t_i) &= \\
e_1\, y(t_{i+1}) - Q_{i+1|i}\, e_1\, y(t_i)\,, &\quad i = 0, 1, \cdots, n-1\,.
\end{aligned}
$$

Of course, for $i = 0$, this needs to be adjusted. In matrix-vector formulation, this may be written as

$$
(7.11)\qquad \Lambda\, \overset{\vee}{y} = \mathbb{L}\, \mathbb{B}^T y\,,
$$

with $\mathbb{L}$ and $\mathbb{B}$ as in (5.11) and (5.15) and Λ a block-lower-triangular matrix,

$$
(7.12)\qquad \Lambda = \begin{bmatrix} I & & & \\ -K_{1|0} & I & & \\ & \ddots & \ddots & \\ & & -K_{n|n\text{-}1} & I \end{bmatrix},
$$

with the $K_{i+1|i}$ of (6.27),

$$
(7.13)\qquad K_{i+1|i} = Q_{i+1|i}\,(I - L_{i|i\text{-}1}\, e_1^T)\,.
$$

It follows that $\overset{\vee}{y} = \Lambda^{-1}\, \mathbb{L}\mathbb{B}^T y$, and so, since $\widetilde{y} = \mathbb{B}\, \overset{\vee}{y}$, then

$$
(7.14)\qquad \widetilde{y} = \mathbb{B}\, \Lambda^{-1}\, \mathbb{L}\, \mathbb{B}^T y\,.
$$

Thus

$$
(7.15)\qquad L_K^{-1} = \mathbb{B}\, \Lambda^{-1}\, \mathbb{L}\, \mathbb{B}^T \quad,\quad L_K^{-T} = \mathbb{B}\, \mathbb{L}^T\, \Lambda^{-T}\, \mathbb{B}^T\,,
$$

and both are easy to apply to a vector.

Traces. When applying all of this to spline smoothing and the concomitant smoothing parameter selection via the GCV procedure, in view of Exercise (7.3)(b), we must compute trace$(\{\,\mathbb{V}\mathrm{ar}[y]\,\}^{-1})$. Note that for the GML procedure of § 2, the determinant is needed, but this is easy:

$$
(7.16)\quad \det(\{\,\mathbb{V}\mathrm{ar}[y]\,\}^{-1}) = \det(\{D_K\}^{-1}) = \prod_{i=1}^{n}\,(\sigma^2 + e_1^T\, \mathbb{V}_{i|i\text{-}1}\, e_1)^{-1}\,.
$$

For the trace, apparently the diagonal elements of $\{\,\mathbb{V}\mathrm{ar}[y]\,\}^{-1}$ are needed. Using (7.6) and (7.15), this may be done as follows. First, let

$$
(7.17)\qquad z = \{\,\mathbb{V}\mathrm{ar}[y]\,\}^{-1}\, y
$$

and note that

$$
(7.18)\qquad \mathbb{V}\mathrm{ar}[z] = \{\,\mathbb{V}\mathrm{ar}[y]\,\}^{-1}\,.
$$

Now, the vector z can be computed using (7.15) and, as it turns out, so can the variances. With (7.6), we have

$$
(7.19)\qquad z = L_K^{-T}\, D_K^{-1}\, \widetilde{y}\,,
$$

where $\widetilde{y}$ may be computed by the Kalman filter. Moreover, the components of $\widetilde{y}$ are independent, and $\mathbb{V}\mathrm{ar}[\widetilde{y}] = D_K$. Next, use (7.15) so

$$z = \mathbb{B}\,\mathbb{L}^T \Lambda^{-T} \mathbb{B}^T D_K^{-1} \widetilde{y} \,. \tag{7.20}$$

Computing $u = \mathbb{B}^T D_K^{-1} \widetilde{y}$ is easy, and we will not dwell on it. The fun starts with the computation of $w = \Lambda^{-T} u$. Since Λ^T is block-bidiagonal, we may compute w by solving the system $\Lambda^T w = u$ recursively, starting with the last equation. This gives, see (7.12),

$$w_n = e_1 \,(D_K)^{-1}_{n,n} \, \widetilde{y}_n \,, \tag{7.21}$$

$$w_i = e_1 \,(D_K)^{-1}_{i,i} \, \widetilde{y}_i + K^T_{i+1|i} \, w_{i+1} \,, \quad i = n-1, n-2, \cdots, 2, 1 \,. \tag{7.22}$$

(7.23) EXERCISE. For later use, observe that w_i is a linear combination of $\widetilde{y}_j$, $j = i, i+1, \cdots, n$. Prove this using induction.

Next, we must compute $z = \mathbb{B}\,\mathbb{L}^T w$. This is straightforward. Again, starting at the end so we can merge it with (7.21)–(7.22), we get

$$\begin{aligned} z_n &= e_1^T \, w_n \,, \\ z_i &= e_1^T \, (w_i - Q^T_{i|i-1} \, w_{i+1}) \,, \quad i = n-1, n-2, \cdots, 2, 1 \,. \end{aligned} \tag{7.24}$$

This finishes the computation of z. It is now also clear how the variances of the components may be computed. From (7.22),

$$w_i - Q^T_{i|i-1} \, w_{i+1} = e_1 \,(D_K)^{-1}_{i,i} \, \widetilde{y}_i - (Q_{i|i-1} - K_{i+1|i})^T \, w_{i+1} \,,$$

and since $\widetilde{y}_i$ and w_{i+1} are independent by virtue of Exercise (7.23), then

$$\begin{aligned} \mathbb{V}\mathrm{ar}[\, w_i - Q^T_{i|i-1} \, w_{i+1} \,] = {} & e_1 \, e_1^T / (D_K)_{i,i} \, + \\ & (Q_{i|i-1} - K_{i+1|i})^T \, \mathbb{V}\mathrm{ar}[\, w_{i+1} \,] \, (Q_{i|i-1} - K_{i+1|i}) \,. \end{aligned} \tag{7.25}$$

The variances of the z_i, a.k.a. the diagonal elements of $\{\, \mathbb{V}\mathrm{ar}[\, y \,] \,\}^{-1}$, are then obtained as

$$\begin{aligned} &\mathbb{V}\mathrm{ar}[\, z_n \,] = e_1^T \, \mathbb{V}\mathrm{ar}[\, w_n \,] \, e_1 \qquad \text{and} \\ &\mathbb{V}\mathrm{ar}[\, z_i \,] = e_1^T \, \mathbb{V}\mathrm{ar}[\, w_i - Q^T_{i|i-1} \, w_{i+1} \,] \, e_1 \,, \quad i = 1, 2, \cdots, n-1 \,. \end{aligned} \tag{7.26}$$

Of course, we also need to compute $\mathbb{V}\mathrm{ar}[\, w_i \,]$, but similarly to the above, we get the recursive formula

$$\begin{aligned} &\mathbb{V}\mathrm{ar}[\, w_n \,] = e_1 \, e_1^T / (D_K)_{n,n} \qquad \text{and} \\ &\mathbb{V}\mathrm{ar}[\, w_i \,] = e_1 \, e_1^T / (D_K)_{i,i} + K^T_{i+1|i} \, \mathbb{V}\mathrm{ar}[\, w_{i+1} \,] \, K_{i+1|i} \,. \end{aligned} \tag{7.27}$$

The leverage values themselves are then obtained as $1 - \mathbb{V}\mathrm{ar}[\, z_i \,]$.

Thus, the leverage values may be computed via the Kalman filter! Note, however, that these are the Bayesian leverage values, if we call them such. The non-Bayesian leverage values would be the diagonal elements of the matrix $\left(I - \sigma^2 \left(L_K\, D_K\, L_K^T\right)^{-1}\right)^2$. It does not appear that these may be computed by the Kalman filter in $\mathcal{O}(n)$ operations.

EXERCISE: (7.23).

8. Diffuse initial states

In this section, we consider the application of the Kalman filter to state-space models when the initial state has a diffuse prior distribution. However, as in § 7, we restrict attention to the estimation of the original stochastic process. The consequences for the computation of the leverage values are briefly considered also.

Recall the discrete-time model (5.2)–(5.5) for the partially observed stochastic process $X(t)$,

$$\begin{aligned} \mathbb{S}(t_{i+1}) &= Q_{i+1|i}\, \mathbb{S}(t_i) + \mathbb{U}(t_{i+1})\ , \quad i = 0, 1, \cdots, n-1\ , \\ y(t_{i+1}) &= e_1^T\, \mathbb{S}(t_{i+1}) + \varepsilon(t_{i+1})\ , \end{aligned} \tag{8.1}$$

with the design points $0 < t_0 < t_1 < \cdots < t_n$, deterministic transition matrices $Q_{i|j} = Q(t_i \,|\, t_j)$, and $\mathbb{S}(t_0)$, $\mathbb{U}(t_1)$, $\mathbb{U}(t_2)$, $\cdots$, $\mathbb{U}(t_n)$ independent, multivariate zero-mean Gaussian random variables with variances given by

$$\Sigma_0 = \mathbb{E}[\,\mathbb{S}(t_0)\,\mathbb{S}(t_0)^T\,]\ , \quad \Sigma_i = \mathbb{E}[\,\mathbb{U}(t_i)\,\mathbb{U}(t_i)^T\,]\ ,\ i = 1, 2, \cdots, n\ . \tag{8.2}$$

Also, $\varepsilon(t_1)$, $\varepsilon(t_2)$, $\cdots$, $\varepsilon(t_n)$ are independent Normal$(0, \sigma^2)$ random variables, independent of everything else. The diffuse prior on the initial state is implemented as

$$\Sigma_0 = k^2 I\ , \quad k \to \infty\ . \tag{8.3}$$

Let $\mathcal{S} = (\mathbb{S}(t_0)^T, \mathbb{S}(t_1)^T, \cdots, \mathbb{S}(t_n)^T)^T$ be the compound states and let y denote the observations, $y = (y(t_1), y(t_2), \cdots, y(t_n))^T$. The goal is to compute the diffuse limit of the conditional expectations $\widehat{\mathcal{S}} = \mathbb{E}[\,\mathcal{S} \,|\, y\,]$ and $\mathbb{B}\,\widehat{\mathcal{S}} = \mathbb{E}[\,\mathbb{B}\,\mathcal{S} \,|\, y\,]$. We restrict attention to the latter and note that by Exercise (7.3)(b) we may write

$$\mathbb{B}\,\widehat{\mathcal{S}} = y - \sigma^2 \left\{ \mathbb{V}\mathrm{ar}[\,y\,] \right\}^{-1} y\ . \tag{8.4}$$

How should one go about computing the diffuse limit ? A naive idea is to take limits $(k \to \infty)$ in the Kalman filter. However, a cursory inspection reveals that all of the variances $\mathbb{V}_{i|i-1}$ and $\mathbb{V}_{i|i}$, $i = 1, 2, \cdots, n$, tend to "infinity". Indeed, in (8.4), the variance $\mathbb{V}\mathrm{ar}[\,y\,]$ tends to "infinity". Of course, we know from Exercise (5.20) that $\mathbb{E}[\,\mathbb{B}\,\mathcal{S} \,|\, y\,] = \mathbb{B}\,\mathbb{E}[\,\mathcal{S} \,|\, y\,]$ has a diffuse limit (and it is the solution to the maximum *a posteriori* likelihood problem). So, the right-hand side of (8.4) has a limit, and the question is if and how the Kalman filter can be used to compute it.

Here, we employ the brute force approach by directly computing the limit of $\{ \mathbb{V}\mathrm{ar}[y] \}^{-1}$.

Recall that $\mathbb{V}\mathrm{ar}[y] = \sigma^2 I + \mathbb{B}\mathbb{L}^{-1} \mathbb{D} \mathbb{L}^{-T}$. Since

$$\mathbb{D} = \text{block-diag}(k^2 I , \Sigma_1 , \Sigma_2 , \cdots , \Sigma_n) ,$$

then

$$\mathbb{V}\mathrm{ar}[y] = V_o + k^2 \ell \ell^T , \tag{8.5}$$

where

$$V_o = \sigma^2 I + \mathbb{B}\mathbb{L}^{-1} \mathbb{D}_o \mathbb{L}^{-T} \mathbb{B}^T \quad , \quad \ell = \mathbb{B} \mathbb{L}^{-1} \mathcal{E}_o , \tag{8.6}$$

with

$$\begin{aligned} \mathbb{D}_o &= \text{block-diag}(O , \Sigma_1 , \Sigma_2 , \cdots , \Sigma_n) , \\ \mathcal{E}_o &= (I , O , \cdots , O)^T . \end{aligned} \tag{8.7}$$

Note that $\mathcal{E}_o \in \mathbb{R}^{(n+1)m \times m}$.

Then, using the Sherman-Morrison-Woodbury Lemma (18.3.12),

$$\{ \mathbb{V}\mathrm{ar}[y] \}^{-1} = V_o^{-1} - k^2 V_o^{-1} \ell (I + k^2 \ell^T V_o^{-1} \ell)^{-1} \ell^T V_o^{-1} , \tag{8.8}$$

so that in the diffuse limit

$$\{ \mathbb{V}\mathrm{ar}[y] \}^{-1} \longrightarrow V_o^{-1} - V_o^{-1} \ell (\ell^T V_o^{-1} \ell)^{-1} \ell^T V_o^{-1} . \tag{8.9}$$

Then, the estimator of $\mathbb{B}\mathcal{S}$ in the diffuse limit equals

$$\widehat{y} = y - \sigma^2 \{ V_o^{-1} - V_o^{-1} \ell (\ell^T V_o^{-1} \ell)^{-1} \ell^T V_o^{-1} \} y . \tag{8.10}$$

(8.11) EXERCISE. Verify (8.8) and (8.9).

It is now easy to see how the Kalman filter may be used to compute $\widehat{y}$. Setting

$$\Sigma_o = 0 , \tag{8.12}$$

the Kalman filter computes the Cholesky factorization of V_o, so that computing $V_o^{-1} y$ is as in § 7. Computing $V_o^{-1} \ell$ goes the same way, and then the only extra work is computing $(\ell^T V_o^{-1} \ell)^{-1} z$ for $z = \ell^T V_o^{-1} y$. However, this is just solving an $m \times m$ system of equations.

Likewise, for the leverage values, or more specifically the diagonal elements of V_o^{-1}, as discussed in § 7, the Kalman filter may be used. Computing the diagonal elements of $V_o^{-1} \ell (\ell^T V_o^{-1} \ell)^{-1} \ell^T V_o^{-1}$ may be done by explicit computation: The matrix $V_o^{-1} \ell$ may be computed by the Kalman filter, and then computing the $m \times m$ matrix $\ell^T V_o^{-1} \ell$ as well as its inverse is a cinch.

(8.13) EXERCISE. Show that the diffuse estimator $\widehat{y}$ of (8.10) does not depend on the specific form of the diffuse limit (8.3). In particular, show that if $\{\Sigma_o(k)\}_k$ is a sequence of positive-definite but otherwise arbitrary matrices such that

$$\Omega_o = \lim_{k\to\infty} k^{-2}\,\Sigma_o(k)$$

exists and is positive-definite, then (8.10) holds. [Hint: With the Cholesky factorization $\Omega_o = R_o^T R_o$, replace the vector ℓ in (8.6) by $\ell = \mathbb{B}\,\mathbb{L}^{-1} R_o\,\mathcal{E}_o$ with $\mathcal{E}_o$ as is, and convince yourself that $V_o^{-1}\,\ell\,(\,\ell^T V_o^{-1}\,\ell\,)^{-1}\,\ell^T V_o^{-1}$ in (8.9) does *not* change when the old ℓ is replaced with the new one.]

(8.14) EXERCISE. The GML method for choosing the smoothing parameter was discussed in § 2. In the present context, its implementation requires the product of the positive eigenvalues of the matrix W_o, denoted as $\det_+(\,W_o\,)$, where $W_o \stackrel{\text{def}}{=} V_o^{-1} - V_o^{-1}\,\ell\,\big(\,\ell^T V_o^{-1}\,\ell\,\big)^{-1}\,\ell^T V_o^{-1}$. Prove that

$$\det_+(\,W_o\,) = \frac{\det(\,V_o^{-1}\,)\,\det(\,\ell^T\ell\,)}{\det(\,\ell^T V_o^{-1}\,\ell\,)}\;.$$

(This permits an efficient implementation by way of the Kalman filter.) A possible proof goes as follows.
(a) Show that W_o is semi-positive-definite.
(b) Prove that $\mathbb{W} \stackrel{\text{def}}{=} W_o + \ell\,(\,\ell^T\,\ell\,)^{-1}\,\ell^T$ has the same eigenvalues, with the same multiplicities, as W_o, except that the zero eigenvalues have been replaced by ones, so that $\det_+\big(\,W_o\,\big) = \det\big(\,\mathbb{W}\,\big)$.
(c) For arbitrary A, $B \in \mathbb{R}^{n\times m}$, show that $\det(\,I - A\,B^T\,) = \det(\,I - B^T A\,)$.
(d) Write $V_o\,\mathbb{W} = I - A\,B^T$ for the appropriate matrices A, $B \in \mathbb{R}^{n\times 2m}$, and apply (c). This gives the required result.
[Hint: For (b), note that $\mathcal{N}(\,W_o\,)$, the nullspace of W_o, consists of all vectors in $\mathcal{R}(\,\ell\,)$, the range of ℓ. Also, if x is orthogonal to $\mathcal{R}(\,\ell\,)$, then so are $W_o\,x$ and $\mathbb{W}\,x$. Part (c) is easy but requires a cheap (?) trick. Part (d) speaks for itself.]

(8.15) EXERCISE. The GCV method for choosing the smoothing parameter requires the computation of the trace of the matrix W_o defined in Exercise (8.14). Show that

$$\operatorname{trace}\big(\,W_o\,\big) = \operatorname{trace}\big(\,V_o^{-1}\,\big) - \operatorname{trace}\big(\,(\,\ell^T V_o^{-1}\ell\,)^{-1}\,\ell^T V_o^{-2}\,\ell\,\big)\;,$$

and note that the last trace applies to an m by m matrix, which just represents a fixed cost. Unfortunately, numerical stability issues seem to prohibit the direct evaluation. Instead, one may use that $x = (\,\ell^T V_o^{-1}\ell\,)^{-1}\,\ell^T V_o^{-2}\,\ell$ is the solution of the system

$$(\,\ell^T V_o^{-1}\ell\,)\,x = \ell^T V_o^{-2}\,\ell$$

which are the normal equations for the least-squares problem

$$\text{(8.16)} \qquad \text{minimize} \quad \big\|\, V_o^{-1/2}\,\ell\,x - V_o^{-1}(\,V_o^{-1/2}\,\ell\,)\,\big\|_F^2\;,$$

where $V_o^{-1/2} = D_K^{-1/2} L_K$, see (7.6). Here, for any $A \in \mathbb{R}^{n \times m}$, we denote the Frobenius norm of A by

$$\| A \|_F = \left(\text{trace}(A^T A) \right)^{1/2} ,$$

so that $\| A \|_F^2$ is the sum of squares of the elements of A. So, (8.16) is indeed a least-squares problem. It may be solved by way of the QR factorization, which costs $\mathcal{O}(n)$ operations; see BJÖRCK (1996).

EXERCISE: (8.11), (8.13), (8.14).

9. Spline smoothing with the Kalman filter

To wrap it all up, referring to the appropriate formulas from the previous sections, we summarize the algorithmic details for the computation of the smoothing spline

$$(9.1) \qquad \widehat{y} = y - \sigma^2 \{ V_o^{-1} - V_o^{-1} \ell \left(\ell^T V_o^{-1} \ell \right)^{-1} \ell^T V_o^{-1} \} y .$$

Although we keep referring to σ^2, one should take $\sigma^2 = 1$. Equivalently, replacing κ^2 in (3.38) with $\sigma^2 \kappa^2$ effectively eliminates σ^2, as evidenced by the equivalent smoothing spline problem (3.63).

The basic algorithmic building blocks implement matrix-vector operations for block-bidiagonal matrices. It tends to become messy when one realizes that these elementary pieces may be combined for (alleged) greater efficiency. The algorithms are presented in a pseudo programming language. The basic structure is the initialization followed by a **for** (or **do**) loop, with the direction from small to large consecutive values of the index parameter, but sometimes the direction is the other way around.

By way of an introductory example, for given vectors a and b, consider the computation of $u = \mathbb{L}\, a$ and $w = \Lambda^{-1} b$ for the matrices $\mathbb{L}$ and Λ of (5.11) and (7.12); see (7.15). This is accomplished by

$$(9.2) \qquad \begin{aligned} & u_o = a_o \ ; \ w_o = b_o \\ & \textbf{for} \quad i = 1 \quad \textbf{to} \quad n \quad \textbf{do} \\ & \quad u_i = a_i - Q_{i|i\text{-}1}\, a_{i-1} \\ & \quad w_i = b_i + K_{i|i\text{-}1}\, w_{i-1} \\ & \textbf{end} \end{aligned}$$

Then, $\mathbb{L}\, a = (w_1, w_2, \cdots, w_n)^T$ and likewise for $\Lambda^{-1} b$. Note that w_i can overwrite b_i; i.e., the line $w_i = b_i + K_{i|i\text{-}1}\, w_{i-1}$ can be replaced by $b_i = b_i + K_{i|i\text{-}1}\, b_{i-1}$. Also note that the computation of $w = \Lambda^{-1} b$ *has* to proceed in the order $w_o, w_1, w_2, \cdots$, whereas the order for $u = \mathbb{L}\, a$ is arbitrary. Things become interesting when the algorithm (9.2) is applied to the case $b = \mathbb{L}\,a$. Thus, the computation of $u = \Lambda^{-1}\,\mathbb{L}\,a$ is achieved by

$$
\begin{aligned}
&u_o = a_o \\
&\textbf{for} \quad i = 1 \quad \textbf{to} \quad n \quad \textbf{do} \\
&\qquad u_i = a_i - Q_{i|i\text{-}1}\, a_{i-1} \\
&\qquad u_i = u_i + K_{i|i\text{-}1}\, u_{i-1} \\
&\textbf{end}
\end{aligned}
\tag{9.3}
$$

The computation of $v = \mathbb{L}^T\,\Lambda^{-T}\,b$ proceeds similarly and involves overwriting the b_i as follows.

$$
\begin{aligned}
&v_n = b_n \\
&\textbf{for} \quad i = n-1 \quad \textbf{downto} \quad 0 \quad \textbf{do} \\
&\qquad b_i = b_i + K^T_{i+1\,|\,i}\, b_{i+1} \\
&\qquad v_i = b_i - Q^T_{i+1\,|\,i}\, b_{i+1} \\
&\textbf{end}
\end{aligned}
\tag{9.4}
$$

(9.5) EXERCISE. (a) Verify (9.4).
(b) If we wish to avoid overwriting the b_i, one can proceed as follows. Remembering the current w_i and the previous one, w_{i+1}, gives the algorithm

$$
\begin{aligned}
&w = b_n \ ; \ v_n = w \\
&\textbf{for} \quad i = n-1 \quad \textbf{downto} \quad 0 \quad \textbf{do} \\
&\qquad w_{\text{old}} = w \\
&\qquad w = b_i + K^T_{i+1\,|\,i}\, w_{\text{old}} \\
&\qquad v_i = w - Q^T_{i+1\,|\,i}\, w_{\text{old}} \\
&\textbf{end}
\end{aligned}
$$

Show that this works!

We are now ready for the computation of $\widehat{y}$. It suffices to calculate

$$
\widehat{\delta} = \{\, V_o^{-1} - V_o^{-1}\,\ell\,(\,\ell^T\,V_o^{-1}\,\ell\,)^{-1}\,\ell^T\,V_o^{-1}\,\}\,y
\tag{9.6}
$$

since $\widehat{y} = y - \sigma^2\,\widehat{\delta}$. The first step is the computation of the Cholesky factorization of V_o,

$$
V_o = L_K\, D_K\, L_K^T \ ,
\tag{9.7}
$$

with D_{κ} and L_{κ}^{-1} given by (7.7) and (7.15). Here, the work has been done in (6.27): With $(D_{\kappa})_{i,i} \equiv D_{i+1|i}$, we get

$$
\begin{aligned}
&\mathbb{V}_{0|-1} = 0 \; ; \; \mathbb{D}_{0|-1} = \sigma^2 \\
&\textbf{for} \quad i = 0 \quad \textbf{to} \quad n-1 \quad \textbf{do} \\
&\quad M_{i+1|i} = Q_{i+1|i} \, \mathbb{V}_{i|i-1} \, e_1 \, D_{i\,|\,i-1}^{-1} e_1^T \, , \\
&\quad K_{i+1|i} = Q_{i+1|i} - M_{i+1|i} \, , \\
&\quad \mathbb{V}_{i+1|i} = K_{i+1|i} \, \mathbb{V}_{i|i-1} \, K_{i+1|i}^T + \sigma^2 M_{i+1|i} \, M_{i+1|i}^T + \Sigma_{i+1} \, , \\
&\quad D_{i+1|i} = \sigma^2 + e_1^T \, \mathbb{V}_{i+1|i} \, e_1 \, , \\
&\textbf{end}
\end{aligned}
\tag{9.8}
$$

With the factorization (9.7), we may rewrite $\widehat{\delta}$ as

$$
\widehat{\delta} = L_{\kappa}^{-T} D_{\kappa}^{-1/2} \{ I - \psi (\psi^T \psi)^{-1} \psi^T \} D_{\kappa}^{-1/2} L_{\kappa}^{-1} y \, , \tag{9.9}
$$

with $\psi = D_{\kappa}^{-1/2} L_{\kappa}^{-1} \ell$, and the fun begins. We must compute

$$
\widetilde{y} = L_{\kappa}^{-1} y \qquad \text{and} \qquad \widetilde{\ell} = L_{\kappa}^{-1} \ell \, .
$$

The identification (7.15) for L_{κ} now readily gives a suitable algorithm, see (9.3), were it not for the fact that $\ell = \mathbb{B} \, \mathbb{L}^{-1} \mathcal{E}_o$ must be computed as well. Here, we bite the bullet and put everything together in one loop,

$$
\begin{aligned}
&\ell_o = 0 \; ; \; y_o = 0 \\
&L_o = I \in \mathbb{R}^{m \times m} \; ; \; U_o = 0 \in \mathbb{R}^{m \times (m+1)} \\
&\textbf{for} \quad i = 1 \quad \textbf{to} \quad n \quad \textbf{do} \\
&\quad L_i = Q_{i|i-1} \, L_{i-1} \; ; \; \ell_i = e_1^T L_i \\
&\quad U_i = e_1 [\, y_i \,|\, \ell_i \,] - Q_{i|i-1} \, e_1 [\, y_{i-1} \,|\, \ell_{i-1} \,] \\
&\quad U_i = U_i + K_{i|i-1} \, U_{i-1} \\
&\quad z_i = e_1^T \, U_i \, D_{i|i-1}^{-1/2} \\
&\textbf{end}
\end{aligned}
\tag{9.10}
$$

Now, $D_{\kappa}^{-1/2} L_{\kappa}^{-1} y$ and $\psi = D_{\kappa}^{-1/2} L_{\kappa}^{-1} \ell$ can be extracted from z. So then

$$
\eta \stackrel{\text{def}}{=} \left(I - \psi (\psi^T \psi)^{-1} \psi^T \right) D_{\kappa}^{-1/2} L_{\kappa}^{-1} y \tag{9.11}
$$

can be computed by brute force since computing $\psi^T \psi$ costs (exactly) $n m^2$ floating-point operations and solving a system of equations with the coefficient matrix $\psi^T \psi$ costs $\mathcal{O}(m^3)$ operations. Finally, the computation of

$\widehat{\delta} = L_K^{-T} D_K^{-1/2} \eta$ follows analogously to (9.4),

$$
(9.12)\qquad
\begin{aligned}
& w_n = e_1 \eta_n \; ; \; v_n = w_n \\
& \textbf{for} \quad i = n-1 \quad \textbf{downto} \quad 1 \quad \textbf{do} \\
& \quad w_i = e_1 \eta_i - K_{i+1\,|\,i}^T w_{i+1} \\
& \quad v_i = w_i - Q_{i+1\,|\,i}^T w_{i+1} \\
& \quad \widehat{\delta}_i = e_1^T v_i \\
& \textbf{end}
\end{aligned}
$$

The trace of V_o^{-1} may be computed by way of the algorithm

$$
(9.13)\qquad
\begin{aligned}
& v_n = w_n \\
& \mathbb{V}\mathrm{ar}[\, w_n \,] = e_1 e_1^T / (\, D_K \,)_{n,n} \\
& T_n = e_1^T \mathbb{V}\mathrm{ar}[\, w_n \,]\, e_1 \\
& \textbf{for} \quad i = n-1 \quad \textbf{downto} \quad 1 \quad \textbf{do} \\
& \quad b_i = (\, Q_{i\,|\,i-1} - K_{i+1\,|\,i} \,)\, e_1 \\
& \quad T_i = T_{i+1} + b_i^T \, \mathbb{V}\mathrm{ar}[\, w_{i+1} \,]\, b_i \\
& \quad \mathbb{V}\mathrm{ar}[\, w_i \,] = e_1 e_1^T / (\, D_K \,)_{i,i} + K_{i+1\,|\,i}^T \mathbb{V}\mathrm{ar}[\, w_{i+1} \,]\, K_{i+1\,|\,i} \\
& \textbf{end}
\end{aligned}
$$

Then, $T_1 = \mathrm{trace}(V_o^{-1})$. The trace of the updated matrix, due to the diffuse prior, may then be computed as per Exercise (8.15). Note that the diagonal elements of V_o^{-1} are computed by algorithm (9.13) in the form:

$$
(9.14)\qquad
[\, V_o^{-1} \,]_{i,i} = \begin{cases} T_n \quad , & i = n \, , \\ T_i - T_{i+1} \, , & i = 1, 2, \cdots, n-1 \, . \end{cases}
$$

The diagonal elements of $V_o^{-1} - V_o^{-1} \ell \, (\, \ell^T V_o^{-1} \ell \,)^{-1} \ell^T V_o^{-1}$ may then be determined by explicitly computing the update.

Finally, the computation of the product of the positive eigenvalues of the matrix $V_o^{-1} - V_o^{-1} \ell \, (\, \ell^T V_o^{-1} \ell \,)^{-1} \ell^T V_o^{-1}$ may be done by way of Exercise (8.14). The computation of $\det(\, V_o^{-1} \,)$ is easy: It is just the product of the $(D_K)_{i.i}$. However, because of finite-precision arithmetic issues, since we actually need $(\, \det(\, V_o^{-1} \,) \,)^{1/(n-m)}$, one should compute this as

$$
(\, \det(\, V_o^{-1} \,) \,)^{1/(n-m)} = \prod_{i=1}^{n} \Big((\, [\, D_K \,]_{i.i} \,)^{1/(n-m)} \Big)
$$

(note the parentheses) or as

$$
(9.15)\qquad
(\, \det(\, V_o^{-1} \,) \,)^{1/(n-m)} = \exp\Big(\tfrac{1}{n-m} \sum_{i=1}^{n} \log(\, [\, D_K \,]_{i,i} \,) \Big) \; .
$$

(9.16) VERIFYING THE IMPLEMENTATION. The algorithms of this section are easy to implement, but there remains the nagging problem of verifying that it has been done correctly. One way is to implement the autoregressive model of § 4 by brute force numerical linear algebra (construct the matrices in question explicitly) and compare the results. Apart from roundoff errors, they should be equal. One tell-tale instance is the limiting case where the "variance" κ^2 in (1.5) tends to 0. In that case, the estimator is just the least-squares polynomial estimator of degree $\leqslant m-1$. For cubic smoothing splines, the standard method of § 19.3 may be used and compared with the Kalman filter implementation, including the computations of the GCV and GML functionals.

EXERCISE: (9.5).

10. Notes and comments

Ad § 1: The section title is from Monty Python's Flying Circus and seems an apt description of the Bayesian statistics associated with the use of splines.

The reason for this chapter is the approach to the computation of spline estimators using Kalman filtering, as pioneered by WEINERT and KAILATH (1974). WEINERT and SIDHU (1978), WEINERT, BYRD, and SIDHU (1980), WECKER and ANSLEY (1983), ANSLEY AND KOHN (1985, 1987), and KOHN and ANSLEY (1985, 1989) carried the torch further. Of course, all of this presumes the Bayesian model (1.5)-(1.6) for spline smoothing, which originated with KIMELDORF and WAHBA (1970a, 1970b).

There are two ways of handling the polynomial parts. The first one, which seems to fit in best with the Bayesian approach, is by way of noninformative or diffuse priors; see WAHBA (1978) and also O'HAGAN (1978). The second one is to view them as deterministic parameters and deal with them using maximum likelihood estimation; see WECKER and ANSLEY (1983). They cite difficulties in the diffuse prior approach when using the WEINERT and SIDHU (1978) and WEINERT, BYRD, and SIDHU (1980) setup. These difficulties were resolved by DE JONG (1991).

The real reason for this chapter is the advocacy of Randy Eubank; see EUBANK and WANG (2002), EUBANK, HUANG, and WANG (2003), and EUBANK (2005).

Ad § 2: The GML procedure originated with BARRY (1983, 1986) and WAHBA (1985).

Ad § 3: The discussion of Gaussian processes largely follows PARZEN (1962). The identification of such processes with reproducing kernel Hilbert spaces is due to LOÈVE (1948). Its importance was realized by PARZEN (1961).

The equivalent "generalized" spline interpolation problem (3.39) is discussed by HILLE (1972).

The construction of the space $\mathcal{H}(K)$ follows the classic exposition in the field, ARONSZAJN (1950). Lemma (3.21) provides the verification of condition 2 of the theorem in ARONSZAJN (1950), Section 4 (bottom, p. 347). Definition (3.22) and Theorem (3.24) are a somewhat different rendition of that theorem with essentially the same proof. Strangely, some of the literature is remarkably silent about the need to verify ARONSZAJN's condition 2 in the construction of $\mathcal{H}(K)$, such as, e.g., MESCHKOWSKI (1962). HILLE (1972) essentially proves it under the extra condition that the reproducing kernel is (uniformly) continuous. However, *the* proof of ARONSZAJN's condition 2 is well-known; see, e.g. RITTER (2000) or BERLINET and THOMAS-AGNAN (2004).

The standard (and certainly the shortest) way of showing the equivalence of the generalized smoothing spline problem and the basic estimation problem for a Gaussian stochastic process is to compute the solutions of both of them and to exclaim: Look, they are the same! The authors tried hard to avoid this, but it is for the reader to decide whether they were successful and/or whether it was worth the effort.

Regarding the definition of a "best" estimator in (3.5), for Gaussian processes, all reasonable choices lead to the same estimator. The most general definition of optimality of an estimator is the one that realizes

$$\begin{aligned}&\text{minimize} \quad \mathbb{E}\big[\,\Lambda(\,X(\,t\,) - \widehat{X}\,) \,\big|\, X(\,t_i) = y_i\,,\ i = 1, 2, \cdots, n\,\big] \\ &\text{over all estimators } \widehat{X}\ ,\end{aligned}$$

where $-\Lambda(x)$ is any symmetric, strictly unimodal function of x.

Ad § 5: It is interesting to contrast the numerical linear algebra treatment of the normal equations (5.17), see OSBORNE and SÖDERKVIST (2004) and references therein, with the treatment of KOHN and ANSLEY (1989) developed in § 7.

Ad § 6: The Kalman filter originated with KALMAN (1960). What is striking about this paper is the steadfast geometric point of view, interpreting conditional expectations as orthogonal projections. Unfortunately, that still leaves some unappetizing formulas for digestion. It is curious that this seems to be due to the implicit incorporation of the Sherman-Morrison-Woodbury Lemma (18.3.12). The presentation of SAGE and MELSA (1971) follows the original one by Kalman quite closely, but they also present a derivation based on maximum *a posteriori* likelihood. The authors like the constrained likelihood approach the best. All of this is a reflection of the many equivalent ways of estimation for Gaussian processes. For a treatment of the continuous version of the Kalman filtering problem in the context of stochastic differential equations, see, e.g., ØKSENDAL (2003), Chapter VI.

Ad § 7: The version developed here, including the computation of the leverage values, was succinctly stated and proved to be correct by KOHN and ANSLEY (1989). We have concentrated our efforts on the Cholesky factorization, as in EUBANK and WANG (2002) and outlined in EUBANK (2005), where the more traditional derivations may be found also. This development, and also the ones in §§ 6 and 8, may be characterized as opportunistically going back and forth between the probabilistic model (in the guise of conditional expectations) and the constrained maximum likelihood (or the computational linear algebra) point of view. For the so-called square root form of the Kalman filter, see OSBORNE and PRVAN (1991).

Regarding the computation of the leverage values, we note that for the autoregressive model (4.15), or more specifically the solution method of Exercise (4.16), the purely "numerical" approach of HUTCHINSON and DE HOOG (1985) applies.

Ad § 8: The Kalman filter with diffuse initial state was succinctly stated and proved to be correct by DE JONG (1991). A nice alternative to the derivation here would be to filter out the bad part, along the lines of the GML method for choosing the smoothing parameter in the presence of a diffuse prior; see § 2. So, one would like to construct independent multivariate normal random variables y^o and y^k such that

$$y = y^o + y^k \,, \tag{10.1}$$

with

$$\mathbb{V}\mathrm{ar}[\, y^o \,] = V_o \quad , \quad \mathbb{V}\mathrm{ar}[\, y^k \,] = k^2 \,\ell\, \ell^T \,, \tag{10.2}$$

and instead of the conditional expectation $\mathbb{E}[\, \mathbb{B}\, \mathcal{S} \,|\, y \,]$ consider $\mathbb{E}[\, \mathbb{B}\, \mathcal{S} \,|\, y^o \,]$ and claim that

$$\lim_{k\to\infty} \mathbb{E}[\, \mathbb{B}\, \mathcal{S} \,|\, y \,] = \mathbb{E}[\, \mathbb{B}\, \mathcal{S} \,|\, y^o \,] \,. \tag{10.3}$$

Unfortunately, the authors do not know how to do this.

Ad § 9: The condensed version of the full Monty, including the computation of the leverage values, may be found in EUBANK, HUANG, and WANG (2003).

21

Equivalent Kernels for Smoothing Splines

1. Random designs

In this chapter, we return to smoothing splines of arbitrary order but now for general, nonparametric regression problems with random designs. One goal is indeed to rework parts of Chapter 13, but we implement it as a by-product of a more ambitious project regarding "asymptotically equivalent" (or just "equivalent") kernel approximations to smoothing splines. We interpret "equivalence" in the strict sense that *the (global) bias and variance of the difference between the equivalent kernel and smoothing spline estimators are asymptotically negligible compared with the bias and variance of either estimator.* (It would seem that under any weaker interpretation all "good" nonparametric regression estimators would be equivalent to one another.) An interesting twist in the "equivalent" kernel story is that the bias and the noise components require different representations for the "equivalence" to hold in the strict sense.

Now, the purpose of equivalent kernels is that one would like to use them to prove various facts concerning smoothing splines. Here, we use them to prove uniform error bounds for smoothing splines with random designs. In § 14.7, we already did this for the deterministic, asymptotically uniform design. In the next chapter, the equivalent kernels are used to construct strong approximations to the noise part of smoothing splines. This in turn is used to construct confidence bands for the regression function. There, we also take the first steps in establishing the asymptotic distribution theory of the uniform error by exhibiting appropriate limiting Gaussian processes. Since in general the processes in question are not stationary, we fall short of showing the actual asymptotic distribution. However, the Gaussian processes can be simulated and used to approximate the finite-sample distributions.

It will transpire that the equivalent kernels are the reproducing kernels of the natural reproducing kernel Hilbert spaces associated with the smoothing spline problem, with the proper inner products indexed by the

P.P.B. Eggermont, V.N. LaRiccia, *Maximum Penalized Likelihood Estimation*,
Springer Series in Statistics, DOI 10.1007/b12285_10,

smoothing parameter. Equivalently, they are the Green's functions for appropriate boundary value problems, depending on the design density and the smoothing parameter. The boundary requires special care only when making the transition to the equivalent convolution kernels of SILVERMAN (1984); see (1.24) below.

In this chapter, the following very general version of the nonparametric regression problem is considered. The data consist of the random sample $(X_1, Y_1), (X_2, Y_2), \cdots, (X_n, Y_n)$ of the bivariate random variable (X, Y) with $X \in [0, 1]$ almost surely. Assume that

$$f_o(x) = \mathbb{E}[\, Y \mid X = x\,] \tag{1.1}$$

exists and that, for some natural number m,

$$f_o \in W^{m,\infty}(0,1)\ . \tag{1.2}$$

For the definition of the Sobolev spaces $W^{m,p}(a,b)$, see (12.2.18).

Regarding the design, assume that

(1.3) $X_1, X_2, \cdots, X_n$ are independent and identically distributed, having a probability density function ω with respect to Lebesgue measure on $(0,1)$, that satisfies

$$\omega_1 \leqslant \omega(t) \leqslant \omega_2 \quad \text{for all} \quad t \in [0, 1] \tag{1.4}$$

for positive constants ω_1 and ω_2. We say that such designs (or design densities ω or functions in general) are *quasi-uniform*. (See Exercise (3.7) for a discussion of this assumption.)

With the random variable (X, Y), associate the noise D, defined by

$$D = Y - f_o(X)\ , \tag{1.5}$$

and set $D_i = Y_i - f_o(X_i)$, $i = 1, 2, \cdots, n$. Assume that

$$\sup_{x \in [0,1]} \mathbb{E}\big[\, |D|^\kappa \bigm| X = x\,\big] < \infty \quad \text{for some } \kappa > 2\ . \tag{1.6}$$

This is equivalent to $\sup\big\{\, \mathbb{E}\big[\, |Y|^\kappa \mid X = x\,\big] \bigm| x \in [0, 1]\,\big\} < \infty$ provided (1.2) holds. Of course, then

$$\sigma^2 \stackrel{\text{def}}{=} \sup_{x \in [0,1]} \sigma^2(x) < \infty\ , \tag{1.7}$$

where

$$\sigma^2(x) = \mathbb{E}[\, D^2 \mid X = x\,]\ .$$

Since we have occasion to condition on the design, it is useful to note that the random variables $Y_1, Y_2, \cdots, Y_n$ are independent *conditioned on the design*; see Exercise (1.9) below. It follows that, for $t, s \in [0, 1]$,

$$\mathbb{E}\big[\, D_1 D_2 \bigm| X_1 = t,\, X_2 = s\,\big] = 0 \quad \text{if} \quad t \neq s\ . \tag{1.8}$$

(1.9) Exercise. Show that the joint pdf of $Y_1 \,|\, X_1 = x_1$ and $Y_2 \,|\, X_2 = x_2$ satisfies

$$f_{Y_1, Y_2 \,|\, X_1, X_2}(y_1, y_2 \,|\, x_1, x_2) = f_{Y \,|\, X}(y_1 \,|\, x_1)\, f_{Y \,|\, X}(y_2 \,|\, x_2) \,,$$

and similarly for the joint pdf of $Y_i \,|\, X_i = x_i$, $i = 1, 2, \cdots, n$. [Hint: Start from the joint distribution of (X_1, Y_1) and (X_2, Y_2).]

We recall that the smoothing spline estimator is defined implicitly as the solution, denoted by $f = f^{nh}$, of the smoothing spline problem,

$$\text{(1.10)} \qquad \begin{aligned} &\text{minimize} \quad L_{nh}(f) \stackrel{\text{def}}{=} \tfrac{1}{n} \sum_{i=1}^{n} |\, f(X_i) - Y_i \,|^2 + h^{2m} \,\|\, f^{(m)} \,\|^2 \\ &\text{subject to} \quad f \in W^{m,2}(0,1) \,, \end{aligned}$$

where $\|\cdot\|$ denotes the $L^2(0,1)$ norm. The objective of this chapter is to show convergence rates for the estimator f^{nh} for both mean squared error and uniform error. With regard to mean squared error, the decision whether to condition on the design or not is easy: The unconditional expectations are based on the conditional ones, so we do them first and relegate the unconditional expectations to Exercise (4.9). Thus, under the assumptions (1.2)–(1.6), we get almost surely

$$\text{(1.11)} \qquad \mathbb{E}\Big[\, \tfrac{1}{n} \sum_{i=1}^{n} |\, f^{nh}(X_i) - f_o(X_i) \,|^2 \;\Big|\; \mathbb{X}_n \Big] = \mathcal{O}\big(\, h^{2m} + (nh)^{-1} \,\big) \,,$$

and likewise for $\mathbb{E}[\, \|\, f^{nh} - f_o \,\|^2 \,|\, \mathbb{X}_n \,]$. Here,

$$\text{(1.12)} \qquad \mathbb{X}_n \stackrel{\text{def}}{=} (X_1, X_2, \cdots, X_n)$$

is shorthand for the design.

The approach to uniform error bounds is by way of convolution-like kernel estimators/operators acting on "pure-noise" data and the "noiseless continuous" signal. Define the random sum and the smoothing operator,

$$\text{(1.13)} \qquad \mathfrak{S}_{\omega nh}(\,t\,) = \tfrac{1}{n} \sum_{i=1}^{n} D_i \, \mathfrak{R}_{\omega mh}(X_i, t\,) \,, \qquad t \in [\,0\,,\,1\,] \,,$$

$$\text{(1.14)} \qquad [\,\mathfrak{R}_{\omega mh} \, f\,](\,t\,) = \int_0^1 \mathfrak{R}_{\omega mh}(x, t\,)\, f(x)\, \omega(x)\, dx \,, \qquad t \in [\,0\,,\,1\,] \,,$$

where $\mathfrak{R}_{\omega mh}$ is the Green's function for a suitable boundary value problem; see (2.13). Also, for $\alpha > 0$, define the intervals of smoothing parameters,

$$\text{(1.15)} \qquad \mathcal{H}_n(\alpha) = \Big[\, \alpha \,\big(\, n^{-1} \log n \,\big)^{1-2/\kappa} \,,\, \tfrac{1}{2} \,\Big] \,.$$

(1.16) Equivalent Kernel Theorem. *Under the assumptions* (1.2) *to* (1.6) *on the model* (1.1), *the smoothing spline estimator* f^{nh} *of order* $2m$ *satisfies*

$$f^{nh}(\,t\,) = [\,\mathfrak{R}_{\omega mh} \, f_o\,](\,t\,) + \mathfrak{S}_{\omega nh}(\,t\,) + \varepsilon^{nh}(\,t\,) \,, \qquad t \in [\,0\,,\,1\,] \,,$$

where almost surely

$$\limsup_{n\to\infty} \sup_{h\in\mathcal{H}_n(\alpha)} \frac{\|\,\varepsilon^{nh}\,\|_{\omega mh}}{h^{2m}+(nh)^{-1}\log n} < \infty \; ,$$

$$\limsup_{n\to\infty} \sup_{h\in\mathcal{H}_n(\alpha)} \frac{\|\,\varepsilon^{nh}\,\|_{\infty}}{h^{-1/2}\,(\,h^{2m}+(nh)^{-1}\log n\,)} < \infty \; .$$

Note that the denominators are (roughly) the same size as the squared error $\mathbb{E}[\,\|\,f^{nh}-f_o\,\|^2\,]$, see (1.11). We also mention that here, and in the results below, the factor $\log n$ can be replaced by $\log(1/h)\vee\log\log n$ per EINMAHL and MASON (2005); see Remark (14.6.11). Since our interest is in $h \asymp n^{-1/(2m+1)}$ (random or deterministic), that causes no difficulties. The uniform-in-bandwidth character of this theorem (which admits *random* choices of the smoothing parameter) stands out.

The theorem above rolls (conditional) bias and variance into one. To prove the theorem, in the following two lemmas, the two are separated.

(1.17) LEMMA. *Under the assumptions of Theorem* (1.16),

$$\limsup_{n\to\infty} \sup_{h\in\mathcal{H}_n(\alpha)} \frac{\|\,f^{nh}-\mathbb{E}[\,f^{nh}\,|\,\mathbb{X}_n\,]-\mathfrak{S}_{\omega nh}\,\|_{\omega mh}}{(nh)^{-1}\log n} < \infty \; ,$$

$$\limsup_{n\to\infty} \sup_{h\in\mathcal{H}_n(\alpha)} \frac{\|\,f^{nh}-\mathbb{E}[\,f^{nh}\,|\,\mathbb{X}_n\,]-\mathfrak{S}_{\omega nh}\,\|_{\infty}}{h^{-1/2}(nh)^{-1}\log n} < \infty \; .$$

(1.18) LEMMA. *Under the assumptions of Theorem* (1.16),

$$\limsup_{n\to\infty} \sup_{h\in\mathcal{H}_n(\alpha)} \frac{\|\,\mathbb{E}[\,f^{nh}\,|\,\mathbb{X}_n\,]-\mathfrak{R}_{\omega mh}\,f_o\,\|_{\infty}}{h^{m-1/2}\sqrt{(nh)^{-1}\log n}} < \infty \; .$$

Obviously, the two lemmas imply the Equivalent Kernel Theorem (1.16). They also imply "equivalence" in the strict sense alluded to earlier.

(1.19) REMARK. We note here that the "equivalent" kernel estimator,

$$f^{nh}_{\text{equiv}}(\,t\,) \stackrel{\text{def}}{=} \frac{1}{n}\sum_{i=1}^{n} Y_i\,\mathfrak{R}_{\omega mh}(X_i,\,t\,) \; , \qquad t\in[\,0\,,\,1\,] \; ,$$

fails to be "equivalent" in the strict sense: It is fairly obvious that for the conditional squared bias one (only) gets

$$\Bigl|\,\frac{1}{n}\sum_{i=1}^{n} f_o(X_i)\,\mathfrak{R}_{\omega mh}(X_i,\,t\,)-[\,\mathfrak{R}_{\omega mh}\,f_o\,](\,t\,)\,\Bigr|^2 = \mathcal{O}\bigl(\,(nh)^{-1}\,\bigr) \; ,$$

(pointwise) which is not negligible compared with, say, $\mathbb{E}[\,|\,f^{nh}(\,t\,)-f_o(\,t\,)\,|^2\,]$. Thus, in f^{nh}_{equiv}, the randomness of the design is a major contributor to the error, whereas we know that in the smoothing spline estimator it is not.

On the other hand, the noise part, to wit $\mathfrak{S}_{\omega nh}(t)$ of (1.13), behaves just dandy; see Lemma (1.17). (Similar observations hold for global measures of the conditional bias and variance.) This argument does not apply to the Nadaraya-Watson-ized estimator

$$f^{nh}_{RK\text{-}NW}(t) = \frac{\frac{1}{n}\sum\limits_{i=1}^{n} Y_i\, \mathfrak{R}_{\omega mh}(X_i, t)}{\frac{1}{n}\sum\limits_{i=1}^{n} \mathfrak{R}_{\omega mh}(X_i, t)} \,, \qquad t \in [\,0\,,1\,]\;.$$

Indeed, it follows from Theorem (16.10.5) that this estimator is (asymptotically) "equivalent" to the pure-signal/pure-noise superposition

$$\mathfrak{R}_{\omega mh}\, f_o \;+\; \mathfrak{S}^{nh}_{RK\text{-}NW}$$

and then by Theorem (1.16) is "equivalent" to the smoothing spline estimator. Of course, for finite samples, there is a positive probability that the denominator is negative, which is a practical drawback. Finally, we note that, for deterministic designs, the estimator analogous to f^{nh}_{equiv} is "equivalent" to the smoothing spline estimator. See Exercise (1.31).

The Equivalent Kernel Theorem (1.16) allows us to prove uniform error bounds for smoothing splines. For this, we narrow the range of smoothing parameters from $\mathcal{H}_n(\alpha)$ to

$$\mathcal{G}_n(\alpha) = \Big[\; \alpha\,\big(n^{-1}\log n\,\big)^{1-2/\lambda}\,,\, \tfrac{1}{2}\;\Big]\;, \tag{1.20}$$

where $\lambda = \min(\,\kappa, 4\,)$. This way, the requirement $h \in \mathcal{G}_n(\alpha)$ implies that $h \gg (n^{-1}\log n)^{1/2}$, so that the error term $\varepsilon^{nh}(t)$ in Theorem (1.16) can be ignored.

(1.21) Uniform Error Theorem. *Consider the model* (1.1). *Under the conditions* (1.2) *through* (1.6), *the spline estimator of order* $2m$ *satisfies almost surely*

$$\limsup_{n\to\infty}\; \sup_{h\in\mathcal{G}_n(\alpha)} \; \frac{\|\, f^{nh} - f_o\,\|_\infty}{\sqrt{\,h^{2m} + (nh)^{-1}\,\log n}} < \infty\;.$$

Again, the uniform-in-bandwidth character of this theorem permits *random* choices of the smoothing parameter:

(1.22) Corollary. *Under the conditions of Theorem* (1.21), *if in* (1.6) *one assumes* $\kappa \geqslant 2 + 1/m$, *then in probability, resp. almost surely,*

$$\|\, f^{nh} - f_o\,\|_\infty = \mathcal{O}\big(\,(\,n^{-1}\log n)^{m/(2m+1)}\,\big)\;,$$

provided $h \asymp (n^{-1}\log n)^{1/(2m+1)}$ *in probability, resp. almost surely.*

(1.23) EXERCISE. Prove the corollary. [Hint: Note that the choice of h in the corollary indeed belongs to $\mathcal{G}_n(\alpha)$, so that Theorem (1.21) applies.]

In the remainder of this chapter, we prove Theorems (1.16) and (1.21), and in the process suitable versions of (1.11). We essentially repeat Chapter 13, but now for the random design regression problem outlined above. Thus, in the next section, we discuss the setting of reproducing kernel Hilbert spaces and the convolution-like properties of the reproducing kernels. The quadrature results of Chapter 13 are now replaced by suitable statements concerning reproducing kernel density estimation; see § 3. In §§ 4 and 5, bounds on the conditional mean squared error and uniform error (for random or deterministic h) are considered. All these results rely on the reproducing kernels being convolution-like. This is shown in § 6 using a theorem relating the collective boundedness of inverses of convolution-like integral operators of the second kind on L^p spaces, which is discussed in § 7. In § 8, we explore the boundary and interior behavior of smoothing spline estimators of order $2m$, especially when the regression function has extra smoothness; i.e., $f_o \in W^{2m,\infty}(0,1)$. In particular, we show "equivalent" representations of the estimators involving the reproducing kernels for the uniform design, even if the actual design is only quasi-uniform. Finally, in § 9, we show a precise version of: For smooth design densities ω and for all t away from the boundary, the smoothing spline estimator may be represented as in Theorem (1.16), with $\mathfrak{S}_{\omega nh}$ replaced by

$$S_{\omega mh}(t) = \tfrac{1}{n} \sum_{i=1}^{n} D_i \, \mathfrak{B}_{m,\lambda(t)}(X_i - t)/\omega(t) , \tag{1.24}$$

where $\mathfrak{B}_{m,h}$ is the convolution kernel/Green's function of MESSER and GOLDSTEIN (1993), defined in Lemma (14.7.12), and

$$\lambda(t) = h / (\omega(t))^{1/(2m)} . \tag{1.25}$$

This is shown in the following theorem, where the maximally useful smoothness condition on f_o is assumed, to wit

$$f_o \in W^{2m,\infty}(0,1) . \tag{1.26}$$

The region away from the boundary is defined as

$$\mathfrak{P}_h(k) = [\, k\, h \log(1/h) \, , \, 1 - k\, h \log(1/h) \,] . \tag{1.27}$$

(If the upper bound is smaller than the lower bound, we define $\mathfrak{P}_h(k) = \emptyset$.)

(1.28) THE SILVERMAN KERNEL THEOREM. *Under the assumptions* (1.2) *to* (1.6), *as well as* (1.26), *on the model* (1.1), *there exists a constant* γ_m *depending on* m *and* ω *only such that the smoothing spline estimator* f^{nh} *of order* $2m$ *satisfies, for all* $t \in [\,0\,,\,1\,]$,

$$f^{nh}(t) = \int_0^1 \mathfrak{B}_{m\lambda(t)}(x - t)\, f_o(x)\, dx + S_{\omega mh}(t) + \delta^{nh}(t) ,$$

where almost surely

$$\limsup_{n\to\infty} \sup_{h\in\mathcal{H}_n(\alpha)} \frac{\sup\{\,|\,\delta^{nh}(t)\,|\;:\; t\in\mathfrak{P}_h(\gamma_m)\,\}}{h^{2m+1}+h^{-1/2}(nh)^{-1}\,\log n} < \infty\,.$$

Here, we take $\sup\{\,\cdots\;|\; t\in\mathfrak{P}_h(\gamma_m)\,\}=0$ *if* $\mathfrak{P}_h(k)=\emptyset$.

Remark (1.19) may be repeated here, with $\mathfrak{R}_{\omega mh}$ replaced by the Silverman kernel. That is, the "estimator"

$$f_S^{nh} \stackrel{\text{def}}{=} \tfrac{1}{n}\sum_{i=1}^{n} Y_i\,\mathfrak{B}_{m,\lambda(t)}(\,X_i - t\,)/\omega(\,t\,)$$

is not "equivalent" to the smoothing spline estimator in the strict sense, not even away from the boundary. But, we have the following results for the variable kernel Nadaraya-Watson estimator,

$$f_{\text{S-NW}}^{nh}(\,t\,) \stackrel{\text{def}}{=} \frac{\frac{1}{n}\sum_{i=1}^{n} Y_i\,\mathfrak{B}_{m,\lambda(t)}(\,X_i-t\,)}{\frac{1}{n}\sum_{i=1}^{n} \mathfrak{B}_{m,\lambda(t)}(\,X_i-t\,)}\,. \tag{1.29}$$

(1.30) THE EQUIVALENT NADARAYA-WATSON ESTIMATOR. *Under the assumptions of Theorem* (1.28), *if in addition* $\omega\in W^{2m,\infty}(0,1)$, *then*

$$f^{nh} = f_{\text{S-NW}}^{nh}(\,t\,)+\eta^{nh}\,,$$

where almost surely

$$\limsup_{n\to\infty} \sup_{h\in\mathcal{H}_n(\alpha)} \frac{\sup\{\,|\,\eta^{nh}(t)\,|\;:\; t\in\mathfrak{P}_h(k)\,\}}{h^{2m+1}+h\,\sqrt{(nh)^{-1}\,\log n}} < \infty\,.$$

(1.31) EXERCISE. Let Ω be the distribution function corresponding to the quasi-uniform design density ω, and let

$$x_{in} = \Omega^{\text{inv}}\big(\tfrac{i}{n+1}\big)\,,\quad i=1,2,\cdots,n\,,$$

be the (deterministic) design points. Let f^{nh} be the smoothing spline estimator for the regression problem $y_{in}=f_o(x_{in})+d_{in}$, $i=1,2,\cdots,n$, and finally let

$$\nu_h(\,t\,)=\tfrac{1}{n}\sum_{i=1}^{n} f_o(x_{in})\,\mathfrak{R}_{\omega mh}(x_{in},\,t\,)\,,\quad t\in[\,0\,,\,1\,]\,.$$

Show that $\|\,\nu_h-\mathfrak{R}_{\omega mh}\,f_o\,\|_\infty=\mathcal{O}\big(\,(nh)^{-1}\,\big)$. [Hint: Check out the quadrature results of Lemma (13.2.27).]

EXERCISES: (1.9), (1.23), (1.31).

2. The reproducing kernels

Let $m \in \mathbb{N}$ be fixed. The smoothing spline estimator, denoted by f^{nh}, is defined as the solution of the minimization problem (1.10). As in Chapter 13, we must worry about the expressions $f(X_i)$ being well-defined, conditioned on X_i. We know from Lemma (13.2.10) that there exists a constant c_m such that, for all $0 < h \leqslant 1$, all $f \in W^{m,2}(0,1)$, and all $t \in [0,1]$,

$$(2.1) \qquad | f(t) | \leqslant c_m \, h^{-1/2} \, \| f \|_{mh}^2 \, .$$

Recall (§ 13.2) that

$$(2.2) \qquad \| f \|_{mh} \stackrel{\text{def}}{=} \{ \| f \|^2 + h^{2m} \| f^{(m)} \|^2 \}^{1/2} \, .$$

Of course, the inequality (2.1) is geared towards the uniform design. For the present "arbitrary" design, it is more appropriate to consider the weighted inner products

$$(2.3) \qquad \langle f , g \rangle_{\omega mh} = \langle f , g \rangle_{L^2((0,1),\omega)} + h^{2m} \langle f^{(m)} , g^{(m)} \rangle \, ,$$

where $\langle \cdot , \cdot \rangle$ is the usual $L^2(0,1)$ inner product and

$$(2.4) \qquad \langle f , g \rangle_{L^2((0,1),\omega)} = \int_0^1 f(t) \, g(t) \, \omega(t) \, dt \, .$$

The norms are then defined by

$$(2.5) \qquad \| f \|_{\omega mh} = \{ \langle f , f \rangle_{\omega mh} \}^{1/2} \, .$$

For special values of m, such as $m = 1$ or $m = k$, we write $\| \cdot \|_{\omega,1,h}$ and $\| \cdot \|_{\omega,k,h}$, and so on.

With the design density being bounded and bounded away from zero, see (1.4), it is obvious that the norms $\| \cdot \|_{mh}$ and $\| \cdot \|_{\omega mh}$ are equivalent uniformly in h. In particular, with the constants ω_1 and ω_2 as in (1.4), for all $f \in W^{m,2}(0,1)$,

$$(2.6) \qquad \omega_1 \| f \|_{mh}^2 \leqslant \| f \|_{\omega mh}^2 \leqslant \omega_2 \| f \|_{mh}^2 \, .$$

Then, the analogue of (2.1) holds as follows.

(2.7) LEMMA. *For quasi-uniform design densities ω, there exists a constant c_m such that, for all $0 < h \leqslant 1$, all $f \in W^{m,2}(0,1)$, and all $t \in [0,1]$,*

$$| f(t) | \leqslant c_m \, h^{-1/2} \, \| f \|_{\omega mh} \, .$$

Also, there exist constants $c_{k,k+1}$ such that

$$\| f \|_{\omega,k,h} \leqslant c_{k,k+1} \, \| f \|_{\omega,k+1,h}$$

for all $f \in W^{k+1,2}(0,1)$ and all $0 < h \leqslant 1$.

(2.8) EXERCISE. Prove (2.6) and Lemma (2.7). [Hint: Use the results for the uniform design from § 13.2.]

For later use, we quote the following multiplication result.

(2.9) MULTIPLICATION LEMMA. *There exists a constant c such that, for all elements f and $g \in W^{1,2}(0,1)$,*

$$\| f\, g \|_{L^1((0,1),\omega)} + h \, \| (f\, g)' \|_{L^1(0,1)} \leqslant c \, \| f \|_{\omega,1,h} \, \| g \|_{\omega,1,h} \; .$$

(2.10) EXERCISE. Prove Lemma (2.9). [Hint: Cauchy-Schwarz.]

Lemma (2.7) says that the linear functionals $f \mapsto f(t)$ are continuous in the $\| \cdot \|_{\omega m h}$-topology, so that $W^{m,2}(0,1)$ with the inner product $\langle \cdot , \cdot \rangle_{\omega m h}$ is a reproducing kernel Hilbert space; see ARONSZAJN (1950). Thus, for each t, there exists an element $\mathfrak{R}_{\omega m h t} \in W^{m,2}(0,1)$ such that

$$f(t) = \langle f \, , \mathfrak{R}_{\omega m h t} \rangle_{\omega m h}$$

for all $f \in W^{m,2}(0,1)$. Applying this to $\mathfrak{R}_{\omega m h t}$ itself gives the self-reproducing property

$$\mathfrak{R}_{\omega m h t}(s) = \langle \mathfrak{R}_{\omega m h t} \, , \mathfrak{R}_{\omega m h s} \rangle_{\omega m h} \; .$$

We define

$$\mathfrak{R}_{\omega m h}(t \, , s) = \mathfrak{R}_{\omega m h t}(s) = \mathfrak{R}_{\omega m h s}(t) \quad \text{for all } s, t \in [0 \, , 1] \; .$$

We summarize the above in a lemma.

(2.11) REPRODUCING KERNEL LEMMA. *Let $m \geqslant 1$. For quasi-uniform design densities ω, the spaces $W^{m,2}(0,1)$ with the inner products $\langle \cdot , \cdot \rangle_{\omega m h}$ are reproducing kernel Hilbert spaces: There exist functions $\mathfrak{R}_{\omega m h}$ on $[0 \, , 1] \times [0 \, , 1]$ such that $\mathfrak{R}_{\omega m h}(t \, , \cdot) \in W^{m,2}(0,1)$ for all $t \in [0 \, , 1]$, and for all $0 < h \leqslant 1$, $f \in W^{m,2}(0,1)$, and $t \in [0 \, , 1]$,*

$$f(t) = \langle f \, , \mathfrak{R}_{\omega m h}(t \, , \cdot) \rangle_{\omega m h} \; .$$

In addition,

$$\mathfrak{R}_{\omega m h}(t \, , t) = \| \mathfrak{R}_{\omega m h}(t \, , \cdot) \|_{\omega m h} \leqslant c_m \, h^{-1/2} \; ,$$

with the same constant c_m as in Lemma (2.7),

Now, the little information we have on the reproducing kernels suffices for obtaining useful bounds on random sums of the form

$$\tfrac{1}{n} \sum_{j=1}^{n} D_i \, f(X_i) \; ,$$

with $D_1, D_2, \cdots, D_n$ and $X_1, X_2, \cdots, X_n$ as in § 1 and $f \in W^{m,2}(0,1)$ random; i.e., depending on the D_i and X_i. Recall the definition of the pure-noise regression (or random) sums $\mathfrak{S}_{\omega nh}(t)$ of (1.13).

(2.12) LEMMA. *For every $f \in W^{m,2}(0,1)$, random or not,*

$$\Bigl|\, \tfrac{1}{n} \sum_{j=1}^{m} D_i\, f(X_i) \,\Bigr| \leqslant \|\, f \,\|_{\omega mh} \, \|\, \mathfrak{S}_{\omega nh} \,\|_{\omega mh} \; .$$

Moreover, under the assumptions (1.3) through (1.6), there exists a constant c_m not depending on h such that, for all deterministic h, $0 < h \leqslant 1$,

$$\mathbb{E}\bigl[\, \|\, \mathfrak{S}_{\omega nh} \,\|_{\omega mh}^2 \,\bigr] \leqslant c_m \, (nh)^{-1} \, ,$$

and almost surely

$$\mathbb{E}\bigl[\, \|\, \mathfrak{S}_{\omega nh} \,\|_{\omega mh}^2 \,\bigm|\, \mathbb{X}_n \,\bigr] \leqslant c_m \, (nh)^{-1} \; .$$

PROOF. The identity

$$\tfrac{1}{n} \sum_{i=1}^{n} D_i\, f(X_i) = \langle\, f \,,\, \mathfrak{S}_{\omega nh} \,\rangle_{\omega mh}$$

implies the first bound by way of Cauchy-Schwarz. For the expectations, using the self-reproducing property

$$\langle\, \mathfrak{R}_{\omega mh}(x, \,\cdot\,) \,,\, \mathfrak{R}_{\omega mh}(y, \,\cdot\,) \,\rangle_{\omega mh} = \mathfrak{R}_{\omega mh}(x,y) \, ,$$

we have

$$\|\, \mathfrak{S}_{\omega nh} \,\|_{\omega mh}^2 = n^{-2} \sum_{i,j=1}^{n} D_i\, D_j\, \mathfrak{R}_{\omega mh}(X_i, X_j) \; ,$$

so that, with (1.7)–(1.8) and Lemma (2.11),

$$\mathbb{E}\bigl[\, \|\, \mathfrak{S}_{\omega nh} \,\|_{\omega mh}^2 \,\bigm|\, \mathbb{X}_n \,\bigr] \leqslant \sigma^2\, n^{-2} \sum_{i=1}^{n} \mathfrak{R}_{\omega mh}(X_i, X_i) \leqslant c\, (nh)^{-1} \; .$$

Since this is a nonasymptotic bound for almost all designs, the expectation over the design then causes no problems. Q.e.d.

To obtain "better" results, we obviously need to better understand the reproducing kernels. The key to this is that reproducing kernels may be interpreted as the Green's functions for appropriate boundary value problems; see, e.g., MESCHKOWSKI (1962) or DOLPH and WOODBURY (1952). In the present case, $\mathfrak{R}_{\omega mh}(t\,, s)$ is the Green's function for

$$\text{(2.13)} \qquad \begin{aligned} &(-h^2)^m\, u^{(2m)} + \omega\, u = v \quad \text{on} \quad (0,1) \; , \\ &u^{(k)}(0) = u^{(k)}(1) = 0 \; , \quad k = m, \cdots, 2m-1 \; . \end{aligned}$$

For arbitrary design densities ω, this may still be troublesome, but it turns out that, for quasi-uniform designs, the boundary value problem (2.13) may be regarded as a compact perturbation of the case of the uniform design density, so that $\mathfrak{R}_{\omega mh}$ may be related to $\mathcal{R}_{mh}$, the Green's function for the uniform density, which we already "computed" in § 14.6.

In this chapter, all we need to know concerning the reproducing kernels is that they share many of the properties of true convolution kernels. In §§ 6 and 7, we prove the following theorem. For $\ell = 0, 1, \cdots, m$, let

$$\mathfrak{R}^{(\ell)}_{\omega mh}(t, s) = \frac{d^{\,\ell}}{ds^{\,\ell}} \mathfrak{R}_{\omega mh}(t, s) \tag{2.14}$$

denote the ℓ-th-order derivative of $\mathfrak{R}_{\omega mh}(t, s)$ with respect to s.

(2.15) DEFINITION. *We say the family of kernels* A_h, $0 < h < 1$, *on* $[0, 1] \times [0, 1]$ *is exponentially decaying if there exist positive constants* c *and* γ *such that, for all* $0 < h < 1$, *and for all* $s, t \in [0, 1]$,

$$|A_h(t, s)| \leqslant c\, h^{-1} \exp(-\gamma h^{-1} |s - t|) .$$

(2.16) THEOREM. *Let* ω *be a quasi-uniform design density; see* (1.4). *Then, for* $\ell = 0, 1, \cdots, m$, *the families*

$$h^{\ell} \mathfrak{R}^{(\ell)}_{\omega mh} , \quad 0 < h < 1 ,$$

are convolution-like in the sense of Definition (14.2.5) *and exponentially decaying in the sense of Definition* (2.15).

Since the reproducing kernels are convolution-like, the material of § 14.6 applies to the random sums $\mathfrak{S}_{\omega nh}$ of (1.13) and to its derivatives.

(2.17) THEOREM. *Under the assumptions* (1.3) *through* (1.6), *for* $\alpha > 0$, *and for* $\ell = 0, 1, \cdots, m$, *we have almost surely*

$$\limsup_{n\to\infty} \sup_{h \in \mathcal{H}_n(\alpha)} \frac{h^{\ell} \, \| \mathfrak{S}^{(\ell)}_{\omega nh} \|_{\infty}}{\sqrt{(nh)^{-1} \log n}} < \infty .$$

It turns out that we need a similar result for the $\| \cdot \|_{\omega mh}$ norm, which requires a result for the L^2 norm. The following is good enough for our purposes. Obviously, with $\mathfrak{S}^{(m)}_{\omega nh}$ denoting the m-th-order derivative of $\mathfrak{S}_{\omega nh}$, we have

$$\| \mathfrak{S}_{\omega nh} \|_{L^2((0,1),\omega)} \leqslant \| \mathfrak{S}_{\omega nh} \|_{\infty} \quad , \quad \| \mathfrak{S}^{(m)}_{\omega nh} \|_{L^2(0,1)} \leqslant \| \mathfrak{S}^{(m)}_{\omega nh} \|_{\infty} ,$$

and then Theorem (2.17) gives useful bounds for the $\| \cdot \|_{\omega mh}$ norm.

(2.18) COROLLARY. *Under the conditions of Theorem* (2.17), *we have*

$$\limsup_{n\to\infty} \sup_{h \in \mathcal{H}_n(\alpha)} \frac{\| \mathfrak{S}_{\omega nh} \|_{\omega mh}}{\sqrt{(nh)^{-1} \log n}} < \infty \quad \textit{almost surely}.$$

EXERCISES: (2.8), (2.10).

3. Reproducing kernel density estimation

As was the case for asymptotically uniform deterministic designs, we have a need for bounds on the difference between

$$\tfrac{1}{n} \sum_{i=1}^{n} |\, f(X_i)\,|^2 \quad \text{and} \quad \int_0^1 |\, f(x)\,|^2\, \omega(x)\, dx$$

for *random* functions f. Whereas for deterministic designs this is a question about quadrature, for random designs this leads to questions concerning the variance of reproducing kernel density estimators,

$$\tfrac{1}{n} \sum_{i=1}^{n} \mathfrak{R}_{\omega m h}(X_i, t)\,, \quad t \in [\,0\,,1\,]\,.$$

In this context, the bias is of no interest. Of course, the convolution-like structure of the reproducing kernels again comes to the fore. We prove the following almost sure result.

(3.1) THEOREM. *Assume* (1.3) *and* (1.4). *Then, for all* $f \in W^{m,2}(0,1)$, *deterministic or random,*

$$r_{nh}\, \|\, f\,\|_{\omega m h}^2 \leqslant \tfrac{1}{n} \sum_{j=1}^{n} |\, f(X_i)\,|^2 + h^{2m}\, \|\, f^{(m)}\,\|^2 \leqslant (\,1 + r_{nh}\,)\, \|\, f\,\|_{\omega m h}^2\,,$$

where $\displaystyle \limsup_{n\to\infty} \sup_{h\in\mathcal{H}_n(\alpha)} \frac{|\, r_{nh} - 1\,|}{\sqrt{(nh)^{-1}\log n}} = 0$ *almost surely*.

The proof follows from the next two lemmas. As in § 14.5, let Ω_o be the distribution function corresponding to the design density ω and let Ω_n be the empirical distribution function of the design $X_1, X_2, \cdots, X_n$. Then the reproducing kernel estimator may be written as

$$(3.2) \qquad \tfrac{1}{n} \sum_{i=1}^{n} \mathfrak{R}_{\omega m h}(X_i, t) = \int_0^1 \mathfrak{R}_{\omega m h}(x, t)\, d\Omega_n(x)\,, \quad t \in [\,0\,,1\,]\,,$$

with expected values

$$(3.3) \quad \mathbb{E}\Big[\, \tfrac{1}{n} \sum_{i=1}^{n} \mathfrak{R}_{\omega m h}(X_i, t)\,\Big] = \int_0^1 \mathfrak{R}_{\omega m h}(x, t)\, d\Omega_o(x)\,, \quad t \in [\,0\,,1\,]\,.$$

We will use Theorem (14.6.13), a somewhat weaker version of a result of EINMAHL and MASON (2005). Recall Definition (1.15) of $\mathcal{H}_n(\alpha)$. To prove Theorem (3.1), we start with simple "design sums".

(3.4) LEMMA. *Assume* (1.3) *through* (1.6). *Then, for all* $f \in W^{1,1}(0,1)$, *deterministic or random,*

$$\Big|\int_0^1 f(t)\,\{\, d\Omega_n(t) - d\Omega_o(t)\,\}\Big| \leqslant \zeta_{nh}\, \{\, \|\, f\,\|_{L^1((0,1),\omega)} + h\, \|\, f\,'\,\|_{L^1(0,1)}\,\}\,,$$

where $$\limsup_{n\to\infty} \sup_{h\in\mathcal{H}_n(\alpha)} \frac{\zeta_{nh}}{\sqrt{(nh)^{-1}\log n}} < \infty \qquad \textit{almost surely}.$$

PROOF. The reproducing kernel Hilbert space trick for $m = 1$ gives

$$f(t) = \langle f, \mathfrak{R}_{\omega,1,h}(t,\cdot)\rangle_{\omega,1,h} .$$

Then, by linearity and Fubini's theorem,

$$\int_0^1 f(t)\{d\Omega_n(t) - d\Omega_o(t)\} = \langle f, \delta^{nh}\rangle_{\omega,1,h} ,$$

where

$$\delta^{nh}(s) = \int_0^1 \mathfrak{R}_{\omega,1,h}(t,s)\{d\Omega_n(t) - d\Omega_o(t)\} , \quad s\in[0,1] .$$

This is the variance part of the pointwise error of the reproducing kernel estimator (3.2). Now, straightforward bounding gives

$$\langle f, \delta^{nh}\rangle_{L^2((0,1),\omega)} \leqslant \| f \|_{L^1((0,1),\omega)} \| \delta^{nh}\|_\infty ,$$
$$\langle f', (\delta^{nh})'\rangle_{L^2(0,1)} \leqslant \| f' \|_{L^1(0,1)} \| (\delta^{nh})'\|_\infty .$$

Consequently,

$$\langle f, \delta^{nh}\rangle_{\omega,1,h} \leqslant \Bigl\{ \| f \|_{L^1((0,1),\omega)} + h \| f' \|_{L^1(0,1)} \Bigr\} \times \Bigl\{ \| \delta^{nh}\|_\infty + h \| (\delta^{nh})'\|_\infty \Bigr\} ,$$

with, explicitly,

$$\delta^{nh} = \tfrac{1}{n}\sum_{i=1}^n \mathfrak{R}_{\omega,1,h}(X_i,\cdot) - \mathbb{E}[\mathfrak{R}_{\omega,1,h}(X_1,\cdot)] ,$$
$$h(\delta^{nh})' = \tfrac{1}{n}\sum_{i=1}^n h\mathfrak{R}'_{\omega,1,h}(X_i,\cdot) - h\,\mathbb{E}[\mathfrak{R}'_{\omega,1,h}(X_1,\cdot)] .$$

Both of these may be interpreted as the variance part of the uniform error of (reproducing) kernel estimators. Since the families $\mathfrak{R}_{\omega mh}(t,s)$ and $h\mathfrak{R}'_{\omega,1,h}(t,s)$ are convolution-like in the sense of Definition (14.2.5), an appeal to Theorem (14.6.13) clinches the deal. Q.e.d.

The following lemma gives the last step in the proof of Theorem (3.1).

(3.5) LEMMA. *Assume* (1.2) *through* (1.6). *Then, for all* $f, g \in W^{m,2}(0,1)$, *deterministic or random,*

$$\Bigl|\int_0^1 f(t)\,g(t)\{d\Omega_n(t) - d\Omega_o(t)\}\Bigr| \leqslant \eta_{nh} \| f \|_{\omega mh} \| g \|_{\omega mh} ,$$

where $$\limsup_{n\to\infty} \sup_{h\in\mathcal{H}_n(\alpha)} \frac{\eta_{nh}}{\sqrt{(nh)^{-1}\log n}} < \infty \quad \textit{almost surely}.$$

PROOF. From Lemma (3.4), we get the bound

$$\zeta_{nh} \left\{ \| f g \|_{L^1((0,1),\omega)} + h \| (f g)' \|_{L^1(0,1)} \right\}$$

with the requisite behavior of ζ^{nh}. Now, from Lemma (2.9),

$$\| f g \|_{L^1((0,1),\omega)} + h \| (f g)' \|_{L^1(0,1)} \leqslant c \| f \|_{\omega,1,h} \| g \|_{\omega,1,h} ,$$

and Lemma (2.7) gives $\| f \|_{\omega,1,h} \leqslant c \| f \|_{\omega mh}$, and likewise for g, all for appropriate constants c. Thus, η_{nh} satisfies $\eta_{nh} \leqslant c \zeta_{nh}$, and the lemma follows. Q.e.d.

(3.6) EXERCISE. Fill in the missing details of the proof of Theorem (3.1).

(3.7) EXERCISE/REMARK. (a) Does Theorem (3.1) hold when the design density is not bounded away from 0? (b) What if the design density is unbounded? [Comment: For (a), the inequality of Lemma (2.7) needs to hold. To get a glimpse of the trouble you (we) may be in, compare this with the relaxed boundary splines of OEHLERT (1992) discussed in § 13.6. So, the design density being bounded away from 0 appears to be an essential assumption. For (b), if the design density is bounded away from zero but possibly unbounded, Lemma (2.7) holds, but it is not clear whether the theorem holds. Of course, the theorem is not an end in itself: Unbounded design densities should not affect the convergence rates of the squared unweighted L^2 error $\| f^{nh} - f_o \|^2$ of the smoothing spline estimator, but its effect on the (weighted) $L^2((0,1),\omega)$ error is not obvious. So, is it merely a matter of convenience? (The authors do not know!)]

EXERCISES: (3.6), (3.7).

4. L^2 error bounds

We are now ready to prove bounds on $\| f^{nh} - f_o \|^2_{\omega mh}$ for the smoothing spline f^{nh}. The starting point is the convexity (in)equality for the smoothing spline problem (1.10). Let

$$\varepsilon \equiv f^{nh} - f_o \,. \tag{4.1}$$

Since the Gateaux variation of L_{nh} at its minimizer vanishes in every direction, the quadratic Taylor expansion of the objective function $L_{nh}(f)$ of (1.10) around its minimizer yields

$$\frac{1}{n} \sum_{i=1}^{n} | \varepsilon(X_i) |^2 + h^{2m} \| \varepsilon^{(m)} \|^2 = L_{nh}(f_o) - L_{nh}(f^{nh}) \,. \tag{4.2}$$

Now, again, simple quadratic Taylor expansion around f_o gives

$$
\begin{aligned}
(4.3)\qquad L_{nh}(f_o) - L_{nh}(f^{nh}) = -\tfrac{1}{n}\sum_{i=1}^{n} |\,\varepsilon(X_i)\,|^2 + \tfrac{2}{n}\sum_{i=1}^{n} D_i\,\varepsilon(X_i) \\
- h^{2m}\,\|\,\varepsilon^{(m)}\,\|^2 + 2\,h^{2m}\,\langle\, f_o^{(m)}\,,\, \varepsilon^{(m)}\,\rangle\,.
\end{aligned}
$$

Substitution into (4.2) gives

$$
\begin{aligned}
(4.4)\quad \tfrac{1}{n}\sum_{i=1}^{n} |\,\varepsilon(X_i)\,|^2 + h^{2m}\,\|\,\varepsilon^{(m)}\,\|^2 = \\
\tfrac{1}{n}\sum_{i=1}^{n} D_i\,\varepsilon(X_i) - h^{2m}\,\langle\, f_o^{(m)}\,,\,\varepsilon^{(m)}\,\rangle\,.
\end{aligned}
$$

(This is similar to the development in § 13.4.)

Now, with Lemma (2.12), Theorem (3.1), and Cauchy-Schwarz, one obtains, with $\mathfrak{S}_{\omega nh}$ as in (1.13),

$$
(4.5)\qquad r^{nh}\,\|\,\varepsilon\,\|_{\omega mh}^2 \leqslant \|\,\varepsilon\,\|_{\omega mh}\,\{\,\|\,\mathfrak{S}_{\omega nh}\,\|_{\omega mh} + h^m\,\|\, f_o^{(m)}\,\|\,\}\,,
$$

where we took the liberty of using $h^m\|\,\varepsilon^{(m)}\,\| \leqslant \|\,\varepsilon\,\|_{\omega mh}$. It follows that

$$
(4.6)\qquad r^{nh}\,\|\,\varepsilon\,\|_{\omega mh} \leqslant \|\,\mathfrak{S}_{\omega nh}\,\|_{\omega mh} + h^m\,\|\, f_o^{(m)}\,\|\,,
$$

and the following results emerge. Recall the definition (1.15) of $\mathcal{H}_n(\alpha)$.

(4.7) Theorem. *For the model* (1.1), *under the assumptions* (1.2) *through* (1.6), *for* $\alpha > 0$, *we have almost surely*

$$
\limsup_{n\to\infty}\ \sup_{h\in\mathcal{H}_n(\alpha)} \frac{\|\, f^{nh} - f_o\,\|_{\omega mh}}{\sqrt{h^{2m} + (nh)^{-1}\,\log n}} < \infty
$$

and

$$
\|\, f^{nh} - f_o\,\|_{\omega mh} = \mathcal{O}\big(\,(n^{-1}\log n)^{m/(2m+1)}\big) \quad \textit{almost surely},
$$

provided $h \asymp_{\text{as}} (n^{-1}\log n)^{1/(2m+1)}$ *(deterministic or random).*

Proof. This follows from (4.6) and Corollary (2.18). Q.e.d.

(4.8) Theorem. *For the model* (1.1), *under the assumptions* (1.2) *through* (1.6), *for* $n\to\infty$, $h \to 0$, *with* $nh\to\infty$,

$$
\mathbb{E}[\,\|\, f^{nh} - f_o\,\|_{\omega mh}^2 \;\big|\; \mathbb{X}_n\,] = \mathcal{O}\big(\,h^{2m} + (nh)^{-1}\,\big)\,,
$$

so that, for $h \asymp n^{-1/(2m+1)}$ *(deterministic),*

$$
\mathbb{E}[\,\|\, f^{nh} - f_o\,\|_{\omega mh}^2 \;\big|\; \mathbb{X}_n\,] = \mathcal{O}\big(\,n^{-2m/(2m+1)}\,\big)\,.
$$

(4.9) Exercise. (a) Prove Theorem (4.8). (b) Prove the same bound for the unconditional expectation of the discrete squared error

$$
SE = \tfrac{1}{n}\sum_{i=1}^{n} |\, f^{nh}(X_i) - f_o(X_i)\,|^2
$$

(c) and likewise for the unconditional expectation of the $\| f^{nh} - f_o \|^2_{\omega mh}$. [Hint: For part (b), one needs to apply a version of the Dominated Convergence Theorem. We have the nonasymptotic upper bound

$$\frac{1}{n} \sum_{i=1}^{n} | f^{nh}(X_i) - f_o(X_i) |^2 \leqslant \beta_n \stackrel{\text{def}}{=} \frac{1}{n} \sum_{i=1}^{n} | D_i |^2 + h^{2m} \| f_o^{(m)} \|^2 .$$

Of course, β_n is random, but well-behaved. This must be combined with a bound on $\mathbb{P}[r_{nh} \leqslant t]$ to obtain a suitable bound on

$$\mathbb{E}[\mathit{SE}\, \mathbb{1}(r_{nh} \leqslant t)] \leqslant \mathbb{E}[\beta_n \mathbb{1}(r_{nh} \leqslant t)] \leqslant \sqrt{ \mathbb{E}[\beta_n^2] \, \mathbb{P}[r_{nh} \leqslant t] }$$

for the appropriate value of t .]

(4.10) EXERCISE. Let $f_h \equiv \mathbb{E}[f^{nh} \,|\, \mathbb{X}_n]$. Under the assumptions of Theorem (4.7), show the bias-variance decomposition in the form

$$\limsup_{n\to\infty} \sup_{h \in \mathcal{H}_n(\alpha)} h^{-m} \| f_h - f_o \|_{\omega mh} < \infty$$

and almost surely

$$\limsup_{n\to\infty} \sup_{h \in \mathcal{H}_n(\alpha)} \sqrt{nh \, (\log n)^{-1}} \, \| f^{nh} - f_h \|_{\omega mh} < \infty .$$

[Hint: For the first part, the situation is as in Theorem (4.7), provided we view this as a special case of $\| f^{nh} - f_o \|_{\omega mh}$ with f^{nh} the solution of (1.10) when the noise satisfies $D_i = 0$, $i = 1, 2, \cdots, n$. Then, the (conditional) variance goes away. For the second part, assume $f_o \equiv 0$. Then the bias part goes away.]

EXERCISES: (4.9), (4.10).

5. Equivalent kernels and uniform error bounds

In this section, we prove Theorems (1.16) and (1.21). The main result is Lemma (1.17) since the bias is "easy". However, the "interior" behavior of the bias is interesting again.

Since the estimator f^{nh} is linear in the data, one sees that

(5.1) $$\varphi^{nh} \stackrel{\text{def}}{=} f^{nh} - \mathbb{E}[f^{nh} \,|\, \mathbb{X}_n]$$

is the solution to the "pure-noise" problem

(5.2) $$\begin{aligned} &\text{minimize} \quad \frac{1}{n} \sum_{i=1}^{n} | f(X_i) - D_i |^2 + h^{2m} \| f^{(m)} \|^2 \\ &\text{subject to} \quad f \in W^{m,2}(0,1) . \end{aligned}$$

The big result of this section is that $\varphi^{nh}(t) \approx \psi^{nh}(t)$, where $f = \psi^{nh}$ solves the C(ontinuous)-spline problem

$$\text{(5.3)}\qquad \begin{array}{ll} \text{minimize} & \| f \|^2_{L^2((0,1),\omega)} - \frac{2}{n} \sum_{i=1}^{n} D_i f(X_i) + h^{2m} \| f^{(m)} \|^2 \\ \text{subject to} & f \in W^{m,2}(0,1) . \end{array}$$

Interpreting the reproducing kernel $\mathfrak{R}_{\omega mh}$ as the Green's function for the boundary value problem (2.13), one sees that ψ^{nh} is given by

$$\text{(5.4)}\qquad \psi^{nh}(t) = \tfrac{1}{n} \sum_{i=1}^{n} D_i \, \mathfrak{R}_{\omega mh}(X_i, t) = \mathfrak{S}_{\omega nh}(t) ;$$

see (1.13). Therefore, it suffices to prove the following reformulation of Lemma (1.17). Recall the definition (1.15) of $\mathcal{H}_n(\alpha)$.

(5.5) THEOREM. *Under the assumptions* (1.2) *through* (1.6), *for* $\alpha > 0$, *we have almost surely*

$$\limsup_{n\to\infty} \sup_{h\in\mathcal{H}_n(\alpha)} \frac{\| \varphi^{nh} - \psi^{nh} \|_{\omega mh}}{(nh)^{-1} \log n} < \infty ,$$

$$\limsup_{n\to\infty} \sup_{h\in\mathcal{H}_n(\alpha)} \frac{\| \varphi^{nh} - \psi^{nh} \|_{\infty}}{h^{-1/2} (nh)^{-1} \log n} < \infty .$$

PROOF. Let $\varepsilon = \varphi^{nh} - \psi^{nh}$. Similar to the derivation of the inequality (4.4), one obtains quadratic inequalities for the discrete and the continuous spline problems (5.2) and (5.3). Adding these gives

$$\text{(5.6)}\qquad \| \varepsilon \|^2 + 2 h^{2m} \| \varepsilon^{(m)} \|^2 + \tfrac{1}{n} \sum_{i=1}^{n} | \varepsilon(X_i) |^2 = \text{rhs} ,$$

where, with $\theta = | \varphi^{nh} |^2 - | \psi^{nh} |^2 = (\varphi^{nh} + \psi^{nh}) \varepsilon$,

$$\text{rhs} = \int_0^1 \theta(t) \{ d\Omega_n(t) - d\Omega_o(t) \} .$$

Now, let $\alpha > 0$ be fixed. Then, the following statements hold uniformly in $h \in \mathcal{H}_n(\alpha)$. Using Lemma (3.5), one obtains almost surely

$$\text{rhs} = \mathcal{O}\big(\sqrt{(nh)^{-1} \log n} \,\big) \, \| \varepsilon \|_{\omega mh} \, \| \varphi^{nh} + \psi^{nh} \|_{\omega mh} ,$$

so that substitution into (5.6) yields almost surely

$$\| \varepsilon \|_{\omega mh} = \mathcal{O}\big(\sqrt{(nh)^{-1} \log n} \,\big) \, \| \varphi^{nh} + \psi^{nh} \|_{\omega mh} .$$

Now, $\| \varphi^{nh} + \psi^{nh} \|_{\omega mh} \leqslant \| \varphi^{nh} \|_{\omega mh} + \| \psi^{nh} \|_{\omega mh}$ and consequently, by Exercise (4.10) and its analogue for the C-spline estimator ψ^{nh},

$$\| \varphi^{nh} + \psi^{nh} \|_{\omega mh} = \mathcal{O}\big(\sqrt{(nh)^{-1} \log n} \,\big) \quad \text{almost surely} .$$

Thus, $\| \varepsilon \|_{\omega mh} = \mathcal{O}\big((nh)^{-1} \log n \big)$, and so, by Lemma (2.7), at the loss of a factor $h^{1/2}$, we obtain that $\| \varepsilon \|_\infty = \mathcal{O}\big(h^{-1/2} (nh)^{-1} \log n \big)$. The theorem has been proved. Q.e.d.

(5.7) EXERCISE. (a) Formulate and prove the analogue of Exercise (4.10) for the C-spline estimator ψ^{nh} of (5.3).
(b) Show that for $h \to 0$, $nh \to \infty$,

$$\mathbb{E}[\, \| \varphi^{nh} - \psi^{nh} \|_{\omega mh}^2 \,|\, \mathbb{X}_n \,] = \mathcal{O}\big((nh)^{-2} \big) \quad \text{almost surely}.$$

The above completes the proof of Lemma (1.17). Next, consider the bound on the bias, as expressed in Lemma (1.18). One way of approaching the problem is to show that the reproducing kernels $\mathfrak{R}_{\omega mh}$, $0 < h < 1$, are convolution-like of order m; see Definition (14.2.6). An alternative approach goes as follows. Note that $f_h = \mathbb{E}[\, f^{nh} \,|\, \mathbb{X}_n]$ is the solution of (1.10) with $D_i = 0$, $i = 1, 2, \cdots, n$; i.e., the solution of the discrete noiseless problem

$$(5.8) \qquad \begin{array}{ll} \text{minimize} & DN(f) \stackrel{\text{def}}{=} \frac{1}{n} \sum\limits_{i=1}^{n} |\, f(X_i) - f_o(X_i) \,|^2 + h^{2m} \| \, f^{(m)} \, \|^2 \\ \text{subject to} & f \in W^{m,2}(0,1) \; . \end{array}$$

The randomness in f_h is due to the randomness of the design. Now, let $\zeta_h = \mathfrak{R}_{\omega mh} f_o$, so ζ_h is the solution of the noiseless C-spline problem,

$$(5.9) \qquad \begin{array}{ll} \text{minimize} & CN(f) \stackrel{\text{def}}{=} \| \, f - f_o \, \|_{L^2((0,1),\omega)}^2 + h^{2m} \| \, f^{(m)} \, \|^2 \\ \text{subject to} & f \in W^{m,2}(0,1) \; . \end{array}$$

(5.10) EXERCISE. Prove the bound $\| \zeta_h - f_o \|_{\omega mh} \leqslant h^m \| \, f_o^{(m)} \, \|$.

(5.11) LEMMA. *Under the assumptions of Theorem (4.7), the solution ζ_h of (5.9) satisfies almost surely*

$$\limsup_{n \to \infty} \; \sup_{h \in \mathcal{H}_n(\alpha)} \frac{\| \zeta_h - f_h \|_\infty}{h^{m-1/2} \sqrt{(nh)^{-1} \log n}} < \infty \; .$$

PROOF. Let $\varepsilon = \zeta_h - f_h$. Similar to the derivation of (5.6), one obtains

$$(5.12) \qquad \| \, \varepsilon \, \|_{\omega mh}^2 + \tfrac{1}{n} \sum_{i=1}^{n} |\, \varepsilon(X_i) \,|^2 + 2\, h^{2m} \| \, \varepsilon^{(m)} \, \|^2 = \text{rhs} \; ,$$

where "rhs" is given by $\text{rhs} = DN(\zeta_h) - DN(f_h) + CN(f_h) - CN(\zeta_h)$. This simplifies to

$$\text{rhs} = \int_0^1 \theta(t) \, \{ \, d\Omega_n(t) - d\Omega_o(t) \, \} \; ,$$

with Ω_o and Ω_n as in Lemma (3.4), and

$$\begin{aligned}\theta(t) &= |\zeta_h(t) - f_o(t)|^2 - |f_h(t) - f_o(t)|^2 \\ &= \{\zeta_h(t) + f_h(t) - 2f_o(t)\}\,\varepsilon(t) .\end{aligned}$$

By Lemma (3.5), we get that

(5.13) $$\text{rhs} \leqslant \eta_{nh} \,\|\, \zeta_h + f_h - 2f_o \,\|_{\omega mh} \,\|\, \varepsilon \,\|_{\omega mh} ,$$

with almost surely, uniformly in $h \in \mathcal{H}_n(\alpha)$,

(5.14) $$\eta_{nh} = \mathcal{O}\big(\sqrt{(nh)^{-1} \log n}\,\big) .$$

Substituting (5.13) into (5.12) and omitting some (positive) terms on the left-hand side, we obtain

(5.15) $$\|\, \varepsilon \,\|_{\omega mh} \leqslant \eta_{nh} \,\|\, \zeta_h + f_h - 2f_o \,\|_{\omega mh} .$$

Of course, we have

$$\|\, \zeta_h + f_h - 2f_o \,\|_{\omega mh} \leqslant \|\, \varepsilon \,\|_{\omega mh} + 2 \,\|\, f_h - f_o \,\|_{\omega mh} ,$$

and so, from (5.15),

$$(1 - \eta_{nh}) \,\|\, \varepsilon \,\|_{\omega mh} \leqslant 2\,\eta_{nh} \,\|\, f_h - f_o \,\|_{\omega mh} .$$

Since $\eta_{nh} \longrightarrow 0$ almost surely, then

(5.16) $$\|\, \varepsilon \,\|_{\omega mh} \leqslant c\,\eta_{nh} \,\|\, f_h - f_o \,\|_{\omega mh} ,$$

with $c \longrightarrow 2$ almost surely.

Now, Exercise (4.10) provides us with the almost sure bound

(5.17) $$\|\, f_h - f_o \,\|_{\omega mh} = \mathcal{O}(\, h^m \,) ,$$

uniformly in $h \in \mathcal{H}_n(\alpha)$. (Note that there is still randomness in f_h due to the design.) It then follows from (5.16) that

$$\limsup_{n\to\infty} \sup_{h\in\mathcal{H}_n(\alpha)} \frac{\|\, \varepsilon \,\|_{\omega mh}}{h^m \sqrt{(nh)^{-1}\log n}} < \infty .$$

At the usual loss of a factor $h^{1/2}$, the required bound on $\|\, \varepsilon \,\|_\infty$ follows as well. Q.e.d.

THE PROOF OF THEOREM (1.16) now follows from Theorem (5.5) and Lemma (5.11). It is useful to rephrase Theorem (1.16) as

(5.18) $$f^{nh} = \mathfrak{R}_{\omega mh} f_o + \mathfrak{S}_{\omega nh} + \delta^{nh} \quad \text{on} \quad [\,0\,,1\,]$$

with $$\|\, \delta^{nh} \,\|_\infty = \mathcal{O}\big(\,\|\, \zeta_h - f_h \,\|_\infty + h^{-1/2}(nh)^{-1} \log n \,\big)$$

almost surely, uniformly in $h \in \mathcal{H}_n(\alpha)$.

Exercise (5.10) and Lemma (5.11) imply the following theorem.

(5.19) THEOREM. *Under the assumptions of Theorem* (4.7), *almost surely,*

$$\limsup_{n\to\infty} \sup_{h\in\mathcal{H}_n(\alpha)} \frac{\| f_h - f_o \|_\infty}{h^m + h^{-1/2}(nh)^{-1}\log n} < \infty \, .$$

(5.20) LEMMA. *Under the smoothness assumption* (1.2) *on* f_o *and the quasi-uniformity* (1.4) *of the design, there exists a constant* c *such that*

$$\| \zeta_h - f_o \|_\infty \leqslant c\, h^m \| f_o^{(m)} \|_\infty \, .$$

PROOF. One verifies that ζ_h is the solution to the differential equation

$$(-h^2)^m f^{(2m)} + \omega f = \omega f_o \quad \text{on } (0,1) \tag{5.21}$$

supplemented with the natural boundary conditions. Now, by assumption (1.2), we have $f_o \in W^{m,\infty}(0,1)$, so $f_o \in W^{m,2}(0,1)$. Then, cf. (2.13), the solution of (5.21) is given by

$$\zeta_h(t) = \int_0^1 \mathfrak{R}_{\omega mh}(t,s)\,\omega(s)\,f_o(s)\,ds = \big\langle \mathfrak{R}_{\omega mh}(t,\cdot),\, f_o \big\rangle_{L^2((0,1),\omega)} \, ,$$

and then

$$\zeta_h(t) = \big\langle \mathfrak{R}_{\omega mh}(t,\cdot),\, f_o \big\rangle_{\omega mh} - h^{2m} \big\langle \mathfrak{R}^{(m)}_{\omega mh}(t,\cdot),\, f_o^{(m)} \big\rangle_{L^2(0,1)} \, .$$

Since $\big\langle \mathfrak{R}_{\omega mh}(t,\cdot),\, f_o \big\rangle_{\omega mh} = f_o(t)$ and

$$\begin{aligned} \Big| \big\langle \mathfrak{R}^{(m)}_{\omega mh}(t,\cdot),\, f_o^{(m)} \big\rangle_{L^2(0,1)} \Big| &\leqslant \| \mathfrak{R}^{(m)}_{\omega mh}(t,\cdot) \|_{L^1((0,1),\omega)} \, \| f_o^{(m)} \|_\infty \\ &\leqslant c\, h^{-m} \| f_o^{(m)} \|_\infty \, , \end{aligned}$$

the last inequality by the convolution-likeness of $h^m \mathfrak{R}^{(m)}_{\omega mh}$, the lemma follows. Q.e.d.

(5.22) EXERCISE. Verify (5.12), (5.16), and (5.17).

Finally, we are ready to prove the uniform error bounds on the smoothing spline uniformly in the bandwidth over a wide (useful) range.

PROOF OF THEOREM (1.21). The starting point is the triangle inequality

$$\| f^{nh} - f_o \|_\infty \leqslant \| f^{nh} - f_h \|_\infty + \| f_h - f_o \|_\infty \tag{5.23}$$

with $f_h = \mathbb{E}[\, f^{nh} \,|\, \mathbb{X}_n \,]$. Now, recall the representation of Lemma (1.17),

$$f^{nh}(t) - f_h(t) = \mathfrak{S}_{\omega nh}(t) + \varepsilon^{nh}(t) \, ,$$

where $\mathfrak{S}_{\omega nh}$ is given by (1.13) and, almost surely, uniformly in $h \in \mathcal{H}_n(\alpha)$,

$$\| \varepsilon^{nh} \|_\infty = \mathcal{O}\big(h^{-1/2}(nh)^{-1} \log n \big) \, .$$

For $h \in \mathcal{G}_n(\alpha) \subset \mathcal{H}_n(\alpha)$, we may conclude that

$$\| \varepsilon^{nh} \|_\infty = o\Big(\sqrt{(nh)^{-1} \log n} \Big) .$$

Now, Theorem (2.17) gives

$$\| \mathfrak{S}_{\omega nh} \|_\infty = \mathcal{O}(\sqrt{(nh)^{-1} \log n}) \tag{5.24}$$

almost surely, uniformly in $h \in \mathcal{H}_n(\alpha)$. Thus, the bound on $\| \varepsilon^{nh} \|_\infty$ is negligible compared with the bound (5.24). Consequently, we have almost surely, uniformly in $h \in \mathcal{G}_n(\alpha)$,

$$\| f^{nh} - f_h \|_\infty \leqslant \| \mathfrak{S}_{\omega nh} \|_\infty + \| \varepsilon^{nh} \|_\infty = \mathcal{O}(\sqrt{(nh)^{-1} \log n}) .$$

Finally, Theorem (5.19) gives the bound, $\| f_h - f_o \|_\infty = \mathcal{O}(h^m)$ uniformly in $h \in \mathcal{G}_n(\alpha) \subset \mathcal{H}_n(\alpha)$. Substitution of these last two bounds into (5.23) yields the required result. Q.e.d.

EXERCISES: (5.7), (5.10), (5.22).

6. The reproducing kernels are convolution-like

In this section, we prove Theorem (2.16) and some related results. The main argument rests on a theorem concerning the uniform solvability of classes of convolution-like integral equations of the second kind on L^p spaces ("if it holds for $p = 2$, then it holds for all p"). This mostly falls under the heading of basic functional analysis. For a general reference, see, e.g., RIESZ and SZ-NAGY (1955).

We first outline the proof of Theorem (2.16), filling in the details later. Recall that $\mathfrak{R}_{\omega mh}$ is the Green's function of the boundary value problem

$$\begin{aligned} &(-h^2)^m \, u^{(2m)} + \omega \, u = v \quad \text{on } (0,1) , \\ &u^k(0) = u^{(k)}(1) = 0 , \quad k = m, \cdots, 2m-1 . \end{aligned} \tag{6.1}$$

Also recall that the case of the uniform design density was treated in § 14.7. In this case, the reproducing kernel $\mathcal{R}_{mh}$ is the Green's function of the boundary value problem (6.1) with $\omega = 1$ everywhere. Since we computed the solution more or less explicitly, *everything* is known about the kernels $\mathcal{R}_{mh}$. But, how do we get from $\mathcal{R}_{mh}$ to $\mathfrak{R}_{\omega mh}$? The answer is by way of integral equation methods.

Write the differential equation in (6.1) as

$$(-\lambda^2)^m \, u^{(2m)} + u = \widetilde{v} - \mathcal{M} \, u , \tag{6.2}$$

in which

$$\lambda = h \, \omega_{\text{low}}^{-1/(2m)} \quad , \quad \widetilde{v} = v / \omega_{\text{low}} \quad , \quad \text{where} \quad \omega_{\text{low}} = \tfrac{1}{2} \, \omega_1 \, .$$

Also, $\mathcal{M}$ is the operator of multiplication by the function M, so that $[\mathcal{M} u](t) = M(t) u(t)$ with

$$M(t) = (\omega(t) - \omega_{\text{low}})/\omega_{\text{low}} \,, \quad t \in [0, 1] \,. \tag{6.3}$$

Now, let f denote the right-hand side of (6.2), and assume that it is known. Then, the solution of (6.2) together with the boundary conditions of (6.1) is given by $u = T_\lambda f$, where $T_\lambda : L^2(0,1) \to L^2(0,1)$ is defined as

$$[T_\lambda g](t) = \int_0^1 \mathcal{R}_{m\lambda}(t, \tau) g(\tau) \, d\tau \,, \quad t \in [0, 1] \,. \tag{6.4}$$

Since the function f itself depends on u, then $u = T_\lambda f$ amounts to an integral equation of the second kind,

$$u + T_\lambda \mathcal{M} u = \omega_{\text{low}}^{-1} T_\lambda v \,. \tag{6.5}$$

Of course, the point is that (6.5) is equivalent to the boundary value problem (6.1).

(6.6) THEOREM. *For each $v \in L^2(0,1)$, the solutions of the boundary value problems* (6.1) *and* (6.5) *exist and are unique, and are equal. Moreover,*

$$\sup_{0<\lambda<1} \| (\mathcal{I} + T_\lambda \mathcal{M})^{-1} \|_2 \leqslant \tfrac{1}{2} + \frac{\omega_2}{\omega_1} \,.$$

Here, for $1 \leqslant p \leqslant \infty$, the norm of a linear operator $T : L^p(0,1) \to L^p(0,1)$ is defined as

$$\| T \|_p = \sup \{ \, \| T f \|_p \, : \, f \in L^p(0,1), \ \| f \|_p = 1 \, \} \,. \tag{6.7}$$

The main trick is now to infer the uniform invertibility of $\mathcal{I} + T_\lambda \mathcal{M}$ on $L^1(0,1)$, after which the rest is smooth sailing. The following theorem is proved in § 7.

(6.8) THEOREM. *For quasi-uniform design densities, see* (1.4), *there exists a constant C_1 such that, for all p, $1 \leqslant p \leqslant \infty$,*

$$\sup_{0<\lambda<1} \| (\mathcal{I} + T_\lambda \mathcal{M})^{-1} \|_p \leqslant C_1 \,.$$

PROOF OF THEOREM (2.16). Fix $s \in (0,1)$, and set $u = \mathfrak{R}_{\omega m h}(\,\cdot\,, s)$. Then, u is the solution to the boundary value problem with the right-hand side $v = \delta(\,\cdot\, - s)$, the point mass at s. Since $T_\lambda v = \mathcal{R}_{m\lambda}(\,\cdot\,, s)$, then u is the solution to

$$u + T_\lambda \mathcal{M} u = \omega_{\text{low}}^{-1} \mathcal{R}_{m\lambda}(\,\cdot\,, s) \,, \tag{6.9}$$

so that Theorem (6.8) gives

$$\| u \|_1 \leqslant C_1 \, \omega_{\text{low}}^{-1} \, \| \mathcal{R}_{m\lambda}(\,\cdot\,, s) \|_1 \leqslant C_2 \tag{6.10}$$

for a suitable constant C_2. This says that

$$\| \mathfrak{R}_{\omega mh}(\,\cdot\,, s) \|_1 \leqslant C_2 \quad \text{for all } s \in [\,0\,,1\,]\,.$$

Everything now follows from (6.10). First,

$$\| \mathcal{M}u \|_1 \leqslant \omega_{\text{low}}^{-1} \, \| w - \omega_{\text{low}} \|_\infty \, \| u \|_1 \leqslant \Bigl(2 \, \frac{\omega_2}{\omega_1} - 1 \Bigr) \, \| u \|_1 \leqslant C$$

for a suitable constant C. It follows that

$$|\, [\, T_\lambda \mathcal{M} u\,](\,t\,) \,| \leqslant \| \mathcal{R}_{m\lambda}(\,t\,,\,\cdot\,) \|_\infty \, \| \mathcal{M}u \|_1 \leqslant C \, \| \mathcal{R}_{m\lambda}(\,t\,,\,\cdot\,) \|_\infty \,.$$

Second, (6.9) may be rewritten as $u = \omega_{\text{low}}^{-1} \, \mathcal{R}_{m\lambda}(\,\cdot\,,\, s\,) - T_\lambda \, \mathcal{M} \, u$, from which we conclude that

$$|\, u(\,t\,) \,| \leqslant \omega_{\text{low}}^{-1} \, |\, \mathcal{R}_{m\lambda}(\,t\,,\, s\,) \,| + C \, \| \mathcal{R}_{m\lambda}(\,t\,,\,\cdot\,) \|_\infty \,.$$

Since the family $\mathcal{R}_{m\lambda}$, $0 < \lambda < 1$, is convolution-like, see Theorem (14.7.6), then $\| u \|_\infty \leqslant c_1 \, \lambda^{-1} \leqslant c \, h^{-1}$ or

$$\| \mathfrak{R}_{\omega mh}(\,\cdot\,, s) \|_\infty \leqslant c \, h^{-1} \quad \text{for all } s \in [\,0\,,1\,]\,.$$

For the BV-property, note that, for all f,

$$|\, T_\lambda \, f \,|_{BV} \leqslant \sup_{s \in [\,0\,,\,1\,]} \, |\, \mathcal{R}_{m\lambda}(\,\cdot\,,\, s\,) \,|_{BV} \, \| f \|_1 \,,$$

so that, again from (6.9),

$$|\, u \,|_{BV} \leqslant \omega_{\text{low}}^{-1} \, |\, \mathcal{R}_{m\lambda}(\,\cdot\,,\, s\,) \,|_{BV} + \sup_{s \in [\,0\,,\,1\,]} \, |\, \mathcal{R}_{m\lambda}(\,\cdot\,,\, s\,) \,|_{BV} \, \| \mathcal{M}u \|_1 \,,$$

and the bound $|\, u \,|_{BV} \leqslant c \, h^{-1}$ follows, or

$$|\, \mathfrak{R}_{\omega mh}(\,\cdot\,, s) \,|_{BV} \leqslant c \, h^{-1} \,.$$

Thus, the family of Green's functions $\mathfrak{R}_{\omega mh}(\,t\,,\, s\,)$, $0 < h < 1$, is convolution-like.

Now, let $1 \leqslant \ell \leqslant m$. After ℓ times differentiating both sides of (6.5), the derivations above may be repeated to show that the kernels

$$h^\ell \, \mathfrak{R}_{\omega mh}^{(\ell)}(\,t\,,\, s\,)$$

are convolution-like as well.

Apart from the exponential decay, this completes the proof. Q.e.d.

We finish this section with the proofs of Theorems (6.6) and the exponential decay, as well as a result on the dependence of the reproducing kernels on the smoothing parameter.

PROOF OF THEOREM (6.6). One verifies that the boundary value problem (6.1) constitutes the Euler equations for the problem

$$\begin{aligned} &\text{minimize} \quad \| u \|_{L^2(\omega)}^2 - 2 \langle\, u\,,\, v \,\rangle + h^{2m} \, \| u^{(m)} \|^2 \\ &\text{subject to} \quad u \in W^{m,2}(0,1) \;; \end{aligned}$$

see, e.g., TROUTMAN (1983) or Chapter 10 in Volume I. Here, for a weight function w, we write $L^2(w)$ to denote $L^2((0,1),w)$. Thus, for $v \in L^2(0,1)$, the solution u of (6.1) exists and is unique. Moreover, u satisfies

$$\| u \|^2_{L^2(\omega)} + h^{2m} \| u^{(m)} \|^2 = \langle u , v \rangle \leqslant \| u \|_{L^2(\omega)} \| v \|_{L^2(1/\omega)} .$$

It follows that

$$\| u \|_{L^2(\omega)} \leqslant \| v \|_{L^2(1/\omega)} ,$$

and then, by the assumption (1.4) on the design density,

$$\| u \| \leqslant \omega_1^{-1} \| v \| . \tag{6.11}$$

Now consider (6.5). To show that the solution is unique, let $u \in L^2(0,1)$ and suppose that $u + T_\lambda \mathcal{M} u = 0$. Then, $u = -T_\lambda \mathcal{M} u$, so that u satisfies

$$(-\lambda^2)^m u^{(2m)} + u = -\mathcal{M} u$$

together with the natural boundary conditions, but this implies that

$$(-h^2)^m u^{(2m)} + \omega u = 0$$

and consequently, see above, $u = 0$.

The existence of solutions of (6.5) now follows from the Fredholm alternative, see, e.g., RIESZ and SZ-NAGY (1955), since the integral operator $T_\lambda \mathcal{M} : L^2(0,1) \to L^2(0,1)$ is compact for each λ.

To continue, (6.5) implies that

$$u = \omega_{\text{low}}^{-1} (\mathcal{I} + T_\lambda \mathcal{M})^{-1} T_\lambda v ,$$

so that, in combination with (6.11),

$$\| (\mathcal{I} + T_\lambda \mathcal{M})^{-1} T_\lambda \|_2 \leqslant \omega_{\text{low}} / \omega_1 = \tfrac{1}{2} .$$

Then,

$$\| (\mathcal{I} + T_\lambda \mathcal{M})^{-1} T_\lambda \mathcal{M} \|_2 \leqslant \tfrac{1}{2} \| M \|_\infty \leqslant \tfrac{1}{2} (\omega_2 - \omega_{\text{low}}) / \omega_{\text{low}} ,$$

and finally, since $(\mathcal{I} + T_\lambda \mathcal{M})^{-1} = \mathcal{I} - (\mathcal{I} + T_\lambda \mathcal{M})^{-1} T_\lambda \mathcal{M}$, then

$$\| (\mathcal{I} + T_\lambda \mathcal{M})^{-1} \|_2 \leqslant 1 + (\omega_2 - \omega_{\text{low}}) / (2 \omega_{\text{low}}) .$$

This is the bound of the theorem.

As to the equivalence of (6.1) and (6.5), obviously, if u solves (6.1), then it also is a solution of (6.5). Conversely, if u solves (6.5), then

$$u = T_\lambda \left\{ \omega_{\text{low}}^{-1} v - \mathcal{M} u \right\} . \tag{6.12}$$

Since T_λ is the Green's function *operator* for (6.1) for the uniform design density, then (6.12) implies that

$$(-\lambda^2)^m u^{(2m)} + u = \omega_{\text{low}}^{-1} v - \mathcal{M} u \quad \text{on } (0,1)$$

together with the boundary conditions of (6.1). It follows that u solves (6.1) for the design density w. Q.e.d.

The decay of the Green's function. The proof of the exponential decay of the Green's function $\mathfrak{R}_{\omega m h}(t, s)$ rests on the following result. With λ as in (6.2), for $\gamma > 0$ define

$$a(t) = \lambda^{-1} \exp(-\gamma \lambda^{-1} |t|) . \tag{6.13}$$

Also, recall from Theorem (14.7.6) that

$$\left| \mathcal{R}_{m\lambda}(t, s) \right| \leqslant c_m \lambda^{-1} \exp(-k_m \lambda^{-1} |t - s|) . \tag{6.14}$$

(6.15) LEMMA. *For all* $0 < \gamma < k_m$,

$$\int_0^1 \left| \frac{a(\tau - s)}{a(t - s)} - 1 \right| \left| \mathcal{R}_{m\lambda}(t, \tau) \right| d\tau \leqslant \frac{2 c_m \gamma}{(k_m - \gamma) k_m} .$$

PROOF. First, for $|t| > |s|$, we have

$$0 \leqslant \frac{a(s)}{a(t)} - 1 = \exp\big(\gamma \lambda^{-1} (|t| - |s|)\big) - 1 \leqslant \exp\big(\gamma \lambda^{-1} |t - s|\big) - 1 .$$

For $|t| \leqslant |s|$, one obtains likewise

$$0 \leqslant 1 - \frac{a(s)}{a(t)} \leqslant 1 - \exp\big(-\gamma \lambda^{-1} |t - s|\big) \leqslant \exp\big(\gamma \lambda^{-1} |t - s|\big) - 1 .$$

It follows that the integral in the lemma is bounded by

$$c_m \lambda^{-1} \int_0^1 \left(e^{\gamma \lambda^{-1} |t - \tau|} - 1 \right) e^{-k_m \lambda^{-1} |t - \tau|} d\tau \leqslant$$

$$c_m \int_{-\infty}^{\infty} \left(e^{\gamma |\tau|} - 1 \right) e^{-k_m |\tau|} d\tau .$$

Now, multiply out the integrand and integrate. Q.e.d.

PROOF OF THE EXPONENTIAL DECAY OF $\mathfrak{R}_{\omega m h}$. Fix $s \in [0, 1]$. Then

$$v(t) = \mathfrak{R}_{\omega m h}(t, s) / a(t - s)$$

satisfies the equation, cf. (6.5),

$$v + T_{\lambda,a} \mathcal{M} v = b , \tag{6.16}$$

where $b(t) = \omega_{\text{low}}^{-1} \mathcal{R}_{m\lambda}(t, s) / a(t - s)$ and $T_{\lambda,a}$ is defined by

$$[T_{\lambda,a} g](t) = \int_0^1 \frac{a(\tau - s)}{a(t - s)} \mathcal{R}_{m\lambda}(t, \tau) g(\tau) d\tau .$$

Note that b is bounded uniformly in λ. Now, we may rewrite (6.16) as

$$v + T_\lambda \mathcal{M} v + (T_{\lambda,a} - T_\lambda) \mathcal{M} v = b$$

so that

$$v + \mathcal{E} v = (I + T_\lambda \mathcal{M})^{-1} b , \tag{6.17}$$

where $\mathcal{E} = (I + T_\lambda \mathcal{M})^{-1} (T_{\lambda,a} - T_\lambda) \mathcal{M}$. Now,

$$\| \mathcal{E} \|_\infty \leqslant \| (I + T_\lambda \mathcal{M})^{-1} \|_\infty \| (T_{\lambda,a} - T_\lambda) \|_\infty \| M \|_\infty \leqslant C \gamma$$

for a suitable constant C. This uses Theorem (6.8) with $p = \infty$ and Lemma (6.15) to bound $\| (T_{\lambda,a} - T_\lambda) \|_\infty$. Now, choose $\gamma \leqslant 1/(2C)$. Then, $\| \mathcal{E} \|_\infty \leqslant \frac{1}{2}$, so that the Banach contraction principle applied to (6.17) implies the inequality $\| v \|_\infty \leqslant \text{const} \| b \|_\infty$. Thus, v is bounded uniformly in λ. But this implies the exponential decay of $\mathfrak{R}_{\omega m h}(t , s)$. Q.e.d.

The pointwise modulus of continuity. Here, we consider the L^∞ modulus of continuity of $\mathfrak{R}_{\omega m h}(x, \cdot)$.

(6.18) LEMMA. *Let $m \geqslant 1$, and assume that ω is quasi-uniform. Then, there exist constants k, ℓ, and μ such that*

$$\big| \mathfrak{R}_{\omega m h}(x, t) - \mathfrak{R}_{\omega m h}(x, s) \big| \leqslant \frac{| t - s |}{h} \times \Big\{ \frac{k}{h} \exp\Big(-\frac{\mu | x - t |}{h} \Big) + \frac{k}{h} \exp\Big(-\frac{\mu | x - s |}{h} \Big) \Big\}$$

for all h, $0 < h \leqslant 1$, and all $x, t, s \in [0, 1]$, with $| t - s | \leqslant \ell h$.

The proof uses the fact that the kernels $\mathcal{R}_{mh}(x, t)$ satisfy the same bound.

(6.19) LEMMA. *Let $m \geqslant 1$, and assume that ω is quasi-uniform. Then, there exist positive constants k, ℓ and μ such that*

$$\big| \mathcal{R}_{mh}(x, t) - \mathcal{R}_{mh}(x, s) \big| \leqslant \frac{| t - s |}{h} \times \Big\{ \frac{k}{h} \exp\Big(-\frac{\nu | x - t |}{h} \Big) + \frac{k}{h} \exp\Big(-\frac{\nu | x - s |}{h} \Big) \Big\}$$

for all h, $0 < h \leqslant 1$ and all $x, t, s \in [0, 1]$, with $| t - s | \leqslant \ell h$.

(6.20) EXERCISE. (a) To fix the idea, first show the bound of Lemma (6.19) for the function $h^{-1} \cos(h^{-1} | x - t |) \exp(-h^{-1} | x - t |)$.
(b) Prove Lemma (6.19). [Hint: Use the (semi-) explicit representation of Lemma (14.7.11).]

PROOF OF LEMMA (6.18). For fixed t and s, let

$$u(x) = \mathfrak{R}_{\omega m h}(x, t) - \mathfrak{R}_{\omega m h}(x, s) .$$

Recall the definition (6.13) of the function $a(\cdot)$ and the bound (6.14) on $\mathcal{R}_{m\lambda}(\cdot, \cdot)$. Let $v(x) = u(x)/a(x - t)$. Then, v is the solution to the integral equation

$$v + T_{\lambda,a} \mathcal{M} v = b ,$$

with $T_{\lambda,a}$ as in equation (6.16), and

$$b(x) = \frac{\mathcal{R}_{m\lambda}(x, t) - \mathcal{R}_{m\lambda}(x, s)}{\omega_{\text{low}}\, a(x - t)} .$$

It follows from Lemma (6.19) that

$$|\, b(x) \,| \leqslant c \exp(r \,|\, x - t \,|/\lambda) \,|\, t - s \,|/\lambda ,$$

where $r = \gamma - \mu < 0$ for $\gamma < \mu$. So,

$$\|\, b \,\|_\infty \leqslant c \,|\, t - s \,|/\lambda .$$

Now, since the operator $\mathcal{I} + T_{\lambda,a}\mathcal{M}$ has a bounded inverse on $L^\infty(0,1)$ uniformly in λ, $0 < \lambda \leqslant 1$, it follows for yet another constant c that $\|\, v \,\|_\infty \leqslant c \,\|\, b \,\|_\infty$, which translates to

$$|\, u(x) \,| \leqslant c \, a(x - t) \,|\, t - s \,|/\lambda ,$$

and the lemma follows. Q.e.d.

The dependence on the smoothing parameter. Finally, we study the dependence of $\mathfrak{R}_{\omega m h}$ on the smoothing parameter h. The starting point is the behavior of the reproducing kernels for the uniform design.

(6.21) LEMMA. *Let $m \geqslant 1$. There exists a constant c such that, for all $h, \theta \in (0,1)$, all p, with $1 \leqslant p \leqslant \infty$, and $\ell = 0, 1, \cdots, m$,*

$$\sup_{s \in [\,0\,,\,1\,]} \Bigl\|\, h^\ell\, \mathcal{R}_{mh}^{(\ell)}(\,\cdot\,,\, s\,) - \theta^\ell\, \mathcal{R}_{m\theta}^{(\ell)}(\,\cdot\,,\, s\,) \Bigr\|_p \leqslant c\, h^{-1+1/p} \,\Bigl|\, 1 - \frac{h}{\theta} \,\Bigr| .$$

(6.22) EXERCISE. Prove the lemma as follows. Assume that $0 < h < \theta$. Recall the definition (14.5.12) of the one-sided exponential g_h.

(a) Show that $\|\, \theta\, g_\theta - h\, g_h \,\|_1 = \theta - h$ and $\|\, g_h - g_\theta \,\|_1 \leqslant 2 \,|\, 1 - h/\theta \,|$.
(b) Show that $\|\, g_h - g_\theta \,\|_\infty \leqslant 2 \,|\, h^{-1} - \theta^{-1} \,| = 2\, h^{-1} \,|\, 1 - h/\theta \,|$.
(c) Use Hölder's inequality to show that $\|\, g_h - g_\theta \,\|_p \leqslant 2\, h^{-1+1/p} \,|\, 1 - h/\theta \,|$.
(d) Observe that parts (a), (b), and (c) also apply to $g_h(-x)$, $x \in \mathbb{R}$.
(e) Prove Lemma (6.21) using the representation of Lemma (14.7.11) for the reproducing kernels $\mathcal{R}_{mh}$.
[Hint: In (a), for the first result, get rid of the absolute values. In (b), just use calculus to maximize $g_\theta(x) - g_h(x)$ over $x \geqslant 0$, checking the second derivative to make sure you have the maximum.]

(6.23) THEOREM. *For quasi-uniform design densities w, see (1.4), there exists a constant c such that for all $h, \theta \in (0,1)$ and p with $1 \leqslant p \leqslant \infty$, and $\ell = 0, 1, \cdots, m$,*

$$\sup_{s \in [\,0\,,\,1\,]} \Bigl\|\, h^\ell\, \mathfrak{R}_{\omega m h}(\,\cdot\,,\, s\,) - \theta^\ell\, \mathfrak{R}_{\omega m\theta}^{(\ell)}(\,\cdot\,,\, s\,) \Bigr\|_p \leqslant c\, h^{-1+1/p} \,\Bigl|\, 1 - \frac{h}{\theta} \,\Bigr| .$$

PROOF. It suffices to prove the cases $p = 1$ and $p = \infty$. For $s \in [0, 1]$, let $u = \mathfrak{R}_{\omega m h}(\cdot, s)$ and $v = \mathfrak{R}_{w m \theta}(\cdot, s)$, and set

$$\lambda = h\,\omega_{\text{low}}^{-1/(2m)} \quad \text{and} \quad \eta = \theta\,\omega_{\text{low}}^{-1/(2m)} . \tag{6.24}$$

As in (6.9), the functions u and v are the solutions to

$$(\mathcal{I} + T_\lambda \mathcal{M})\,u = \omega_{\text{low}}^{-1}\,\mathcal{R}_{m\lambda}(\cdot, s)\,, \quad (\mathcal{I} + \mathcal{T}_\eta \mathcal{M})\,v = \omega_{\text{low}}^{-1}\,\mathcal{R}_{m\eta}(\cdot, s)\,.$$

It follows that

$$\omega_{\text{low}} \cdot (u - v) = \text{first} + \text{second}\,, \tag{6.25}$$

with

$$\begin{aligned} \text{first} &= (\mathcal{I} + \mathcal{T}_\eta \mathcal{M})^{-1} \left\{ \mathcal{R}_{m\lambda}(\cdot, s) - \mathcal{R}_{m\eta}(\cdot, s) \right\}, \\ \text{second} &= \left\{ (\mathcal{I} + T_\lambda \mathcal{M})^{-1} - (\mathcal{I} + \mathcal{T}_\eta \mathcal{M})^{-1} \right\} \mathcal{R}_{m\lambda}(\cdot, s)\,. \end{aligned}$$

Everything is in place to bound the two terms. From Lemma (6.21), one obtains that for all p, $1 \leqslant p \leqslant \infty$,

$$\left\| \mathcal{R}_{m\lambda}(\cdot, s) - \mathcal{R}_{m\eta}(\cdot, s) \right\|_p \leqslant c\,\lambda^{-1+1/p} \left| 1 - \frac{\lambda}{\eta} \right| ,$$

and then, by Theorem (6.8), the same bound with a different constant applies to the "first" term in (6.25).

Regarding the second term, observe that

$$\begin{aligned} &(\mathcal{I} + T_\lambda \mathcal{M})^{-1} - (\mathcal{I} + \mathcal{T}_\eta \mathcal{M})^{-1} = \\ &\qquad\qquad (\mathcal{I} + T_\lambda \mathcal{M})^{-1} (\mathcal{T}_\eta - T_\lambda)\,\mathcal{M}\,(\mathcal{I} + \mathcal{T}_\eta \mathcal{M})^{-1} , \end{aligned}$$

so that, again with Theorem (6.8),

$$\left\| (\mathcal{I} + T_\lambda \mathcal{M})^{-1} - (\mathcal{I} + \mathcal{T}_\eta \mathcal{M})^{-1} \right\|_p \leqslant c \left\| \mathcal{T}_\eta - T_\lambda \right\|_p .$$

Now, by the analogue of Young's inequality, see Exercise (6.26) below,

$$\| T_\lambda - \mathcal{T}_\eta \|_p \leqslant c \sup_{x \in [0, 1]} \| \mathcal{R}_{m\lambda}(x, \cdot) - \mathcal{R}_{m\eta}(x, \cdot) \|_1 \leqslant c_2 \left| 1 - \frac{\lambda}{\eta} \right| .$$

Finally, by Hölder's inequality and then Theorem (14.7.6),

$$\| \mathcal{R}_{m\lambda}(x\,\cdot) \|_p \leqslant \| \mathcal{R}_{m\lambda}(x\,\cdot) \|_1^{1/p} \| \mathcal{R}_{m\lambda}(x\,\cdot) \|_\infty^{1-1/p} \leqslant c\,\lambda^{-1+1/p} .$$

Thus, the "second" term in (6.25) may be bounded as

$$|\, \text{second} \,| \leqslant c\,\lambda^{-1+1/p} \left| 1 - \frac{\lambda}{\eta} \right| .$$

Since $\lambda/\eta = h/\theta$ and $h = \omega_{\text{low}}\,\lambda$, the theorem follows. Q.e.d.

(6.26) EXERCISE. (a) Let $1 \leqslant p < \infty$, and define q by $(1/p) + (1/q) = 1$. Set $\mathcal{D}(x, s) = \mathcal{R}_{m\lambda}(x, s) - \mathcal{R}_{m\eta}(x, s)$. For $f \in L^p(0, 1)$, show that

$$\| (T_\lambda - T_\eta) f \|_p \leqslant \sup_{x \in [0,1]} \| \mathcal{D}(x, \cdot) \|_1^{1/q} \sup_{s \in [0,1]} \| \mathcal{D}(\cdot, s) \|_1^{1/p} \| f \|_p .$$

(b) Conclude that $\| T_\lambda - T_\eta \|_p \leqslant \sup_{x \in [0,1]} \| \mathcal{D}(x, \cdot) \|_1$. [Hint: For (a), inspect the proof of Young's inequality for convolutions.]

(6.27) EXERCISE. Recall the definition (1.13) of $\mathfrak{S}_{\omega nh}$. Show that there exists a constant c such that for all h and θ

$$\| \mathfrak{S}_{\omega nh} - \mathfrak{S}_{\omega n\theta} \|_\infty \leqslant c \left| 1 - \frac{h}{\theta} \right| \{ \| \mathfrak{S}_{\omega nh} \|_\infty + h^m \| (\mathfrak{S}_{\omega nh})^{(m)} \|_\infty \} .$$

[Hint: $\mathfrak{S}_{\omega nh}(t) - \mathfrak{S}_{\omega n\theta}(t) = \langle \mathfrak{S}_{\omega nh}, \mathfrak{R}_{\omega mh}(\cdot, t) - \mathfrak{R}_{\omega m\theta}(\cdot, t) \rangle_{\omega mh}$ and use that $\langle f, g \rangle \leqslant \| f \|_\infty \| g \|_1$.]

EXERCISES: (6.20), (6.22), (6.26), (6.27).

7. Convolution-like operators on L^p spaces

In this section, we prove Theorem (6.8) on the uniform solvability of classes of convolution-like integral equations on L^p spaces. To that effect, the aspects of convolution-like kernels that are emphasized differ somewhat from the ones used in the previous sections.

Throughout this section, let

(7.1) $\quad D \subset \mathbb{R}$ be a finite union of proper closed intervals,

(7.2) $\quad B \in L^1(\mathbb{R})$ be nonnegative, and

(7.3) $\quad \mu \in C([0, \infty))$ with $\mu(0) = 0$.

The results do not essentially depend on D being bounded, so (unions of) unbounded intervals are covered as well.

(7.4) DEFINITION. *Let $1 \leqslant p \leqslant \infty$. The class $\mathcal{F}_D(B, \mu)$ consists of all bounded linear operators $\mathcal{A} : L^p(D) \to L^p(D)$ for which there exists a measurable function A on $[0, 1] \times [0, 1]$ and a positive scalar h such that the following three statements hold:*

(a) $\quad | A(x, t) | \leqslant h^{-1} B(h^{-1} | x - t |), \quad x, t \in D,$

(b) $\quad \displaystyle\sup_{|x - z| \leqslant \delta} \int_D | A(x, t) - A(z, t) | \, dt \leqslant \mu(h^{-1}\delta), \quad 0 < h \leqslant 1,$

where the supremum is over all $x, z \in D$ with $|x - z| \leqslant \delta$ and

(c) $\quad \displaystyle \mathcal{A}f(x) = \int_D A(x, t) f(t) \, dt, \quad x \in D,$ *for all $f \in L^p(D)$.*

(7.5) REMARK. Note that the class $\mathcal{F}_D(B, \mu)$ does not depend on the particular value of p. In other words, if we extend the notation to $\mathcal{F}_{p,D}(B, \mu)$, then for all $1 \leqslant p \leqslant \infty$ we have $\mathcal{F}_{p,D}(B, \mu) = \mathcal{F}_{2,D}(B, \mu)$. This is due to the operators $\mathcal{A}$ being continuous, so they are uniquely determined once they are defined on, say, $L^\infty(D)$, a dense subset of $L^p(D)$ (in the $L^p(D)$-topology).

(7.6) DEFINITION. *A set $\mathcal{C}$ is a collection of convolution-like integral operators if there exist functions B and μ satisfying* (7.1)–(7.3), *such that*

$$\mathcal{C} \subset \mathcal{F}_D(B, \mu) .$$

(7.7) LEMMA. *The set $\mathcal{T} = \{ T_\lambda \mathcal{M} : 0 \leqslant \lambda \leqslant 1 \}$ with T_λ and $\mathcal{M}$ defined in* (6.3) *and* (6.4) *is a collection of convolution-like integral operators.*

PROOF. Theorem (2.16) provides property (a) of Definition (7.4). For property (b), note that for $z \leqslant x$, by Fubini,

$$\int_0^1 | \mathcal{R}_{mh}(x, t) - \mathcal{R}_{mh}(z, t) | \, dt \leqslant \int_z^x \int_0^1 \left| \frac{\partial}{\partial s} \mathcal{R}_{mh}(s, t) \right| ds \, dt .$$

Now, again Theorem (2.16) gives a bound for suitable constants c_m and β,

$$\left| \frac{\partial}{\partial s} \mathcal{R}_{mh}(s, t) \right| \leqslant c \, h^{-2} \exp(- \beta \, h^{-1} \, | s - t |) ,$$

for all s, t and $0 < h < 1$. It follows that

$$\int_0^1 | \mathcal{R}_{mh}(x, t) - \mathcal{R}_{mh}(z, t) | \, dt \leqslant c_1 \, (x - z) \, h^{-1}$$

for another constant c_1. Q.e.d.

(7.8) REMARK. For strict compliance, in Lemma (7.7), the range on λ should actually be $0 < \lambda < w_{\text{low}}^{-1}$, but we shall let it pass.

For operators $\mathcal{A}$ in $\mathcal{F}_D(B, e)$, we may define the spectrum with respect to $L^p(D)$ in the usual way by

(7.9) $\sigma_p(\mathcal{A}) = \{ \lambda \in \mathbb{C} : \lambda \mathcal{I} - \mathcal{A} \;$ has no bounded inverse on $L^p(D) \}$.

Thus, for $\lambda \notin \sigma_p(\mathcal{A})$, the solution of the equation $\lambda f - \mathcal{A} f = g$ exists and is unique for every $g \in L^p(0,1)$ and vice versa.

Now, the question is the following. Suppose we have a linear operator $\mathcal{A} \in \mathcal{F}_D(B, \mu)$ for given functions B and μ, and for a given $\lambda \in \mathbb{C}$ we are interested in the solvability of the equation

$$\lambda f - \mathcal{A} f = g$$

in $L^\infty(D)$ for (every) $g \in L^\infty(D)$. A not uncommon situation is that in the Hilbert space case ($p = 2$) it is "easy" to show that $\lambda \notin \sigma_2(\mathcal{A})$, and one wonders whether this is useful for establishing that it works for $p = \infty$ as well; i.e., whether $\lambda \notin \sigma_\infty(\mathcal{A})$. For convolution-like integral operators, this turns out to be true, but we want some sort of uniform boundedness of the inverses in the result above.

(7.10) THEOREM. *Let $\mathcal{C}$ be a class of convolution-like integral operators, let $\lambda \in \mathbb{C}$, and let $1 < p < \infty$. If $\lambda \notin \sigma_p(\mathcal{A})$ for all $\mathcal{A} \in \mathcal{C}$ and*

$$\sup_{\mathcal{A} \in \mathcal{C}} \| (\lambda \mathcal{I} - \mathcal{A})^{-1} \|_{L^p(D)} < \infty \,, \tag{7.11}$$

then $\lambda \notin \sigma_\infty(\mathcal{A})$ for all $\mathcal{A} \in \mathcal{C}$ and

$$\sup_{\mathcal{A} \in \mathcal{C}} \| (\lambda \mathcal{I} - \mathcal{A})^{-1} \|_{L^\infty(D)} < \infty \,. \tag{7.12}$$

There are two aspects to Theorem (7.10), to wit the existence of the inverses as well as their collective boundedness. The funny thing is that the knowledge that the inverses exist is of little use in showing their boundedness but the boundedness is quite helpful in showing the existence. Of course, it would seem that in this last setup we are begging the question, but one verifies that if (7.11) holds, then

$$\inf_{\mathcal{A} \in \mathcal{C}} \inf_{\substack{f \in L^p(D) \\ f \neq 0}} \frac{\| (\lambda \mathcal{I} - \mathcal{A}) f \|_p}{\| f \|_p} > 0 \,, \tag{7.13}$$

and vice versa if in addition the inverses exist.

(7.14) EXERCISE. Show that (7.12) implies (7.13) and that the converse holds if $\lambda \notin \sigma_p(\mathcal{A})$ for all $\mathcal{A} \in \mathcal{C}$.

Thus, the proof of Theorem (7.10) may be split into two unrelated parts.

(7.15) LEMMA. *Let $1 < p < \infty$, let $\mathcal{C}$ be a collection of convolution-like integral operators, and let $\lambda \in \mathbb{C}$. If there exists a positive constant c such that for all $\mathcal{A} \in \mathcal{C}$ and for all $f \in L^p(D)$*

$$\| \lambda f - \mathcal{A} f \|_{L^p(D)} \geqslant c \| f \|_{L^p(D)} \,, \tag{7.16}$$

then, for some $c_1 > 0$ and for all $\mathcal{A} \in \mathcal{C}$ and for all $f \in L^\infty(D)$,

$$\| \lambda f - \mathcal{A} f \|_{L^\infty(D)} \geqslant c_1 \| f \|_{L^\infty(D)} \,. \tag{7.17}$$

(7.18) LEMMA. *Let $1 < p < \infty$, and let $e \in C(\mathbb{R})$ with $e(0) = 0$. Then, for all $\mathcal{A} \in \mathcal{F}_D(B, e)$, we have $\sigma_\infty(\mathcal{A}) \subset \sigma_p(\mathcal{A})$.*

The case $p = 1$ is not covered by Lemma (7.18), but see Remark (7.28). The question of equality is not addressed either, but if in addition the operators are Fredholm integral operators, then it holds true. The operators of interest, $\mathcal{I}+T_\lambda\mathcal{M}$, are indeed Fredholm integral operators of the second kind. If their null spaces are trivial in $L^2(0,1)$, then they are trivial in all L^p spaces, and then by the Fredholm alternative each one of them must be invertible.

Using Lemmas (7.15) and (7.18), we can now prove Theorem (7.10).

PROOF OF THEOREM (7.10). Apparently, (7.11) implies (7.13) and (7.16), and thus also (7.17). From Lemma (7.18), we get that $\lambda \notin \sigma_\infty(\mathcal{A})$ for all $\mathcal{A} \in \mathcal{C}$ and thus $\lambda\mathcal{I} - \mathcal{A}$ is invertible on $L^\infty(D)$. Combined with (7.17) and Exercise (7.14), this gives (7.12). Q.e.d.

We now set out to prove the two lemmas. We provide the complete proof of Lemma (7.15) for the case where $\mathcal{C} \subset \mathcal{F}_D(B,\mu)$, with B exponentially decaying,

$$B(s,t) \leqslant c\, e^{-\beta|s-t|} , \quad s, t \in \mathbb{R} , \tag{7.19}$$

for positive constants c and β, and indicate what must be done in the general case.

PROOF OF LEMMA (7.15), ASSUMING (7.19). Clearly, the case $\lambda = 0$ is void. Then, we may as well assume that $\lambda = 1$.

The proof goes by way of contradiction. Suppose that (7.17) does not hold. Then there exist sequences $\{\mathcal{A}_n\} \subset \mathcal{C}$ and $\{f_n\}_n \subset L^\infty(D)$ such that, for all $n \in \mathbb{N}$,

$$\| f_n \|_{L^\infty(D)} = 1 \quad , \quad \| f_n - \mathcal{A}_n f_n \|_{L^\infty(D)} = \varepsilon_n , \tag{7.20}$$

with $\varepsilon_n \to 0$. Then, without loss of generality, we may take

$$\varepsilon_n \leqslant n^{-2} .$$

Note that, for each operator $\mathcal{A}_n$ with associated kernel $A_n(x,t)$, Definition (7.4) implies the existence of a scalar h_n such that parts (a) and (b) hold.

Observe that, by Definition (7.4)(b), it follows that

$$|\mathcal{A}_n f_n(x) - \mathcal{A}_n f_n(z)| \leqslant \mu(|x-z|/h_n) \tag{7.21}$$

for all n and all $x, z \in D$. To describe this succinctly, we say that

$$\{\mathcal{A}_n f_n\}_n \text{ is } \{h_n\}_n\text{-equi-uniformly continuous on } D . \tag{7.22}$$

Now, from (7.20), we have that $\| \mathcal{A}_n f_n \|_{L^\infty(D)} \geqslant 1 - n^{-2}$. This implies that $\varphi_n = \mathcal{A}_n f_n / \|\mathcal{A}_n f_n\|_{L^\infty(D)}$ satisfies

$$\| \varphi_n \|_{L^\infty(D)} = 1 \quad , \quad \| \varphi_n - \mathcal{A}_n \varphi_n \|_{L^\infty(D)} \leqslant c n^{-2} ,$$

where $c = \sup_n \| \mathcal{A}_n \|_{L^\infty(D)} \leqslant \| B \|_{L^1(\mathbb{R})}$, and of course

$$\{\varphi_n\}_n \text{ is } \{h_n\}_n\text{-equi-uniformly continuous on } D . \tag{7.23}$$

Since $\| \varphi_n \|_{L^\infty(D)} = 1$, there exist $\{ t_n \}_n \subset D$ with $| \varphi_n(t_n) | \geqslant \frac{2}{3}$, and then, by the $\{ h_n \}_n$-equi-uniform continuity, there exists a positive constant d such that

$$| \varphi_n(t) | \geqslant \tfrac{1}{2} \qquad \text{for all } t \in D\,, \quad |t - t_n| < h_n\, d\,.$$

Moreover, either $[t_n - h_n d\,, t_n] \subset D$ or $[t_n\,, t_n + h_n d] \subset D$, or both.

Now, with β as in (7.19), consider the functions

$$a_n(t) = \exp\Big(-\frac{\beta\,|t - t_n|}{nh_n} \Big)\,, \qquad t \in \mathbb{R}\,,$$

defined on $\mathbb{R}$ and hence, by restriction, also on D. One verifies that

$$\| a_n \|_{L^p(D)} \leqslant \| a_n \|_{L^p(\mathbb{R})} \leqslant c\,(nh_n)^{1/p}$$

for some positive constant c. Consequently,

$$\begin{aligned} \| a_n\varphi_n - a_n\mathcal{A}_n \varphi_n \|_{L^p(D)} &\leqslant \| a_n \|_{L^p(D)} \; \| \varphi_n - \mathcal{A}_n \varphi_n \|_{L^\infty(D)} \\ &\leqslant c\,(nh_n)^{1/p}\, n^{-2}\,. \end{aligned} \tag{7.24}$$

Also, since $| a_n(t) | \geqslant \exp\big(-\beta\, n^{-1} d \big)$ for $| t - t_n | \leqslant h_n\, d$, then

$$\| a_n\varphi_n \|_{L^p(D)}^p \geqslant \int_{\substack{t \in D \\ |t - t_n| \leqslant h_n d}} | \tfrac{1}{2}\, a_n(t) |^p \, dt \geqslant h_n\, d^*$$

for some $d^* > 0$ for all n. But then

$$\frac{\| a_n\varphi_n - a_n\mathcal{A}_n\varphi_n \|_{L^p(D)}}{\| a_n\varphi_n \|_{L^p(D)}} \leqslant \frac{c\,(nh_n)^{1/p}\, n^{-2}}{(d^*\, h_n)^{1/p}} \leqslant c_1\, n^{-1} \tag{7.25}$$

for a suitable constant c_1.

Now, if in (7.25) we can replace $a_n\mathcal{A}_n$ by $\mathcal{A}_n a_n$, then (7.25) shows that (7.16) does not hold and we are done. So, it suffices to show that, for all $f \in L^p(D)$,

$$\| a_n\mathcal{A}_n f - \mathcal{A}_n a_n f \|_{L^p(D)}^p \leqslant \eta_n\, \| a_n f \|_{L^p(D)}^p \tag{7.26}$$

with $\eta_n \to 0$ as $n \to \infty$.

For this, the expression on the left of (7.26) may be written as

$$e_n = \int_D \Big| \int_D A_n(x, t)\, r_n(x, t)\; a_n(t)\, f(t)\, dt \Big|^p \, dx\,,$$

where

$$r_n(x, t) = \frac{a_n(x)}{a_n(t)} - 1\,.$$

Now, as in the proof of Lemma (6.15),

$$\Big| \frac{a_n(x)}{a_n(t)} - 1 \Big| \leqslant \exp\Big(\frac{\beta\,|t - t_n|}{nh_n} \Big) - 1\,, \tag{7.27}$$

and then, with the exponential bound (7.19), we get

$$| A_n(x, t) r_n(x, t) | \leqslant h_n^{-1} \mathfrak{C}_n(h_n^{-1} | x - t |) ,$$

in which

$$\mathfrak{C}_n(x) = c \Big(\exp(\beta n^{-1} | x |) - 1 \Big) \exp(- \beta | x |) .$$

It follows that

$$e_n \leqslant \int_D \Big| \int_D h_n^{-1} \mathfrak{C}_n(h_n^{-1} | x - t |) a_n(t) f(t) \, dt \Big|^p dx ,$$

and then, by Young's inequality, $e_n^{1/p} \leqslant \| \mathfrak{C}_n \|_{L^1(\mathbb{R})} \| a_n f \|_{L^p(D)}$. Finally, as in the proof of Lemma (6.15), $\| \mathfrak{C}_n \|_{L^1(\mathbb{R})} \to 0$ for $n \to \infty$. This yields (7.26). Q.e.d.

(7.28) REMARK. Note that the proof above works for all p, $1 \leqslant p < \infty$.

(7.29) REMARK/EXERCISE. The general case of Theorem (7.15) may be proved as follows. For some $\alpha > 0$ and $\varrho > 0$, define the weight functions a_n as the solution to the convolution integral equation on the line,

$$a_n(x) - \varrho [\mathcal{A}_n a_n](x) = \big(1 + (nh_n)^{-1} | t - t_n | \big)^{-(1+\alpha)/p} , \quad x \in \mathbb{R} .$$

Now, choose $\varrho = (2 \| B \|_{L^1(\mathbb{R})})^{-1}$, so that, by the Banach contraction principle, the solution a_n exists and is unique. Now, retrace the proof of the exponential case, and at each step of the proof of the theorem decide what inequality is needed and prove it. Instead of Young's inequality, Hölder's inequality comes into play; see EGGERMONT and LUBICH (1991).

We now consider the proof of Lemma (7.18). A crucial role is played by the notion of strict convergence in $L^\infty(D)$.

(7.30) DEFINITION. *A sequence* $\{ f_n \}_n \subset L^\infty(D)$ *converges in the strict topology of* $L^\infty(D)$ *if the sequence is bounded and there exists a function* $f_o \in L^\infty(D)$ *such that, for every compact subset* K *of* $\mathbb{R}$,

$$\lim_{n \to \infty} f_n(x) = f_o(x) \quad \textit{uniformly in} \quad x \in K \cap D .$$

We abbreviate this last statement as $\lim_{n \to \infty} \| f_n - f_o \|_{L^\infty(K \cap D)} = 0$.

We next show that convolution-like integral operators are continuous in the strict topology and show an analogue of the Arzelà-Ascoli lemma.

(7.31) LEMMA (ANSELONE and SLOAN, 1990). *Let* D, B, *and* μ *satisfy* (7.1)–(7.3). *If* $\mathcal{A} \in \mathcal{F}_D(B, \mu)$ *and* $\{ f_n \}_n \subset L^\infty(D)$ *converges in the strict topology on* $L^\infty(D)$ *to some element* $f_o \in L^\infty(D)$, *then* $\{ \mathcal{A} f_n \}_n$ *converges in the strict topology to* $\mathcal{A} f_o$.

PROOF. If $\mathcal{A} \in \mathcal{F}_D(B, \mu)$, then parts (a) and (b) of Definition (7.4) hold for a suitable scaling parameter h. Without loss of generality, we may take $h = 1$. The proof now uses an old-fashioned "diagonal" argument. For each integer $j \in \mathbb{N}$, let

$$K(j) = D \cap [-j, j].$$

Let $m \in \mathbb{N}$ and $\varepsilon > 0$. We will show that

$$\text{(7.32)} \qquad \exists N \in \mathbb{N} \quad \forall n \in \mathbb{N}, n \geqslant N \;:\; \| \mathcal{A}f_n - \mathcal{A}f_o \|_{L^\infty(K(m))} < \varepsilon .$$

To this end, choose $\ell \in \mathbb{N}$ such that $\ell > m$ and

$$\sup_{x \in K(m)} \int_{D \setminus K(\ell)} | A(x, t) | \, dt < \varepsilon .$$

This is possible since $| A(x, t) | \leqslant B(x - t)$ and $| t - x | > \ell - m$ for $x \in K(m)$ and $t \notin K(\ell)$.

Next, choose $N \in \mathbb{N}$ such that, for all $n \geqslant N$,

$$\| f_n - f_o \|_{L^\infty(K(\ell))} < \varepsilon .$$

Now, in somewhat loose notation,

$$\begin{aligned} | \mathcal{A}f_n(x) - \mathcal{A}f_o(x) | &\leqslant \int_D B(x - t) \, | f_n(t) - f_o(t) | \, dt = \\ &\leqslant \Big(\int_{K(\ell)} + \int_{D \setminus K(\ell)} \Big) B(x - t) \, | f_n(t) - f_o(t) | \, dt , \end{aligned}$$

and we must bound the last two integrals. First, for all $x \in K(m)$,

$$\begin{aligned} &\int_{D \setminus K(\ell)} B(x - t) \, | f_n(t) - f_o(t) | \, dt \\ &\qquad \leqslant 2 \sup_n \| f_n \|_{L^\infty(D)} \int_{D \setminus K(\ell)} B(x - t) \, dt \leqslant 2 \varepsilon \sup_n \| f_n \|_{L^\infty(D)} \end{aligned}$$

(where we used that $\| f_o \|_{L^\infty(D)} \leqslant \sup_n \| f_n \|_{L^\infty(D)}$). Second, we have

$$\int_{K(\ell)} B(x - t) \, | f_n(t) - f_o(t) | \, dt \leqslant \varepsilon \int_{K(\ell)} B(x - t) \, dt \leqslant \varepsilon \, \| B \|_{L^1(\mathbb{R})} .$$

It follows that $\| \mathcal{A}f_n - \mathcal{A}f_o \|_{L^\infty(K(m))} \leqslant C \varepsilon$ for a suitable constant. This is (7.32) for a slightly different ε. Q.e.d.

(7.33) LEMMA (ANSELONE and SLOAN, 1990). *If $\{ f_n \}_n \subset L^\infty(D)$ is bounded and uniformly continuous on D, then it has a subsequence that is convergent in the strict topology on $L^\infty(D)$.*

PROOF. By the plain Arzelà-Ascoli lemma, there exists a subsequence, denoted by $\{ f_{n,1} \}_n$, that converges uniformly on $K(1)$ to a bounded function $f_{o,1}$. Next, select a subsequence $\{ f_{n,2} \}_n$ of $\{ f_{n,1} \}_n$ that converges uniformly on $K(2)$ to some bounded function $f_{o,2}$. Then $f_{o,1} = f_{o,2}$ on $K(1)$. Repeating this process, we obtain an infinite sequence of subsequences $\{ f_{n,j} \}_n$, $j = 1, 2, \cdots$, such that $\{ f_{n,j} \}_n$ is a subsequence of $\{ f_{n,j-1} \}_n$ and

$$f_{n,j} \longrightarrow f_{o,j} \quad \text{uniformly on} \quad K(j)$$

with each $f_{o,j}$ bounded. Also, $f_{o,j} = f_{o,\ell}$ on $K(\ell)$ for all $j \geqslant \ell$. This allows us to define f_o by

$$f_o(x) = f_{o,n}(x) \ , \quad x \in K(n) \ , \quad n = 1, 2, \cdots \ .$$

Obviously, $\| f_o \|_{L^\infty(D)} \leqslant \sup_n \| f_n \|_{L^\infty(D)}$, so $f_o \in L^\infty(D)$.

Finally, consider the sequence $\{ f_{n,n} \}_n$. One verifies that $f_{n,n} \longrightarrow f_o$ uniformly on $K(j)$ for each $j \in \mathbb{N}$. Q.e.d.

After these preliminaries, we can get down to business. To make life a little easier, assume that

$$B \in L^2(\mathbb{R}) \ . \tag{7.34}$$

PROOF OF LEMMA (7.18), ASSUMING (7.34). Let D, B, μ be as in (7.1)–(7.3), with B also satisfying (7.34). Let $\mathcal{A} \in \mathcal{F}_D(B, \mu)$, and assume without loss of generality that

$$| \mathcal{A}(x, t) | \leqslant B(x - t) \ , \quad x, t \in D \ .$$

Assume that $\lambda \notin \sigma_2(\mathcal{A})$. Let $g \in L^\infty(D)$, and consider the equation

$$f - \mathcal{A}f = g \ . \tag{7.35}$$

We must show that this equation has a solution. It is much nicer to consider the equation

$$u - \mathcal{A}\, u = \mathcal{A}\, g \tag{7.36}$$

since now the right-hand side is uniformly continuous, and if u_o solves this last equation, then $f_o = u_o + g$ solves the first equation, and vice versa.

Define $g_n(t) = g(t) \exp(-|t|/n)$, $t \in D$, and consider the equation

$$u - \mathcal{A}\, u = \mathcal{A}\, g_n \ . \tag{7.37}$$

Since $g_n \in L^2(D)$, this equation has a solution $u_n \in L^2(D)$, and then

$$| \mathcal{A}\, u_n(x) | \leqslant \| \mathcal{A}(x, \cdot\,) \|_{L^2(D)} \| u_n \|_{L^2(D)} \leqslant \| B \|_{L^2(\mathbb{R})} \| u_n \|_{L^2(D)} \ ,$$

so, by (7.34), we have $\mathcal{A}\, u_n \in L^\infty(D)$. Then, by Lemma (7.15), for some positive constant c,

$$\| u_n \|_{L^\infty(D)} \leqslant c \| u_n - \mathcal{A}u_n \|_{L^\infty(D)} = c \| g_n \|_{L^\infty(D)} \leqslant c \| g \|_{L^\infty(D)} \ .$$

So $\{u_n\}_n$ is a bounded subsequence of $L^\infty(D)$. But then $\{\mathcal{A} u_n\}_n$ is equi-uniformly continuous on D. Of course, so is $\{\mathcal{A} g_n\}_n$. From (7.37), it then follows that $\{u_n\}_n$ is itself equi-uniformly continuous.

By the "strict" Arzelà-Ascoli Lemma (7.33), the sequence $\{u_n\}_n$ has a subsequence that converges in the strict topology on $L^\infty(D)$ to some element $u_o \in L^\infty(D)$. Then, for that subsequence, by Lemma (7.31),

$$\mathcal{A} u_n \longrightarrow \mathcal{A} u_o \quad \text{strictly} ,$$

and of course $\mathcal{A} g_n \longrightarrow \mathcal{A} g$ strictly. It follows that $u_o = \mathcal{A} g + \mathcal{A} u_o$ belongs to $L^\infty(D)$. In other words, (7.37), and hence (7.36), has a solution. Lemma (7.15) provides for the boundedness as well as the uniqueness, so that the operator $\mathcal{I} - \mathcal{A}$ has a bounded inverse on $L^\infty(D)$. Q.e.d.

(7.38) REMARK/EXERCISE. To get rid of the assumption (7.34), one must add a truncation step. So, define

$$A_n(x,t) = A(x,t)\, 1\!\!1(\,|A(x,t)| \leqslant n\,) ,$$

and define the integral operator $\mathcal{A}_n$ accordingly. Then $\mathcal{A}_n \in \mathcal{F}_D(B, \mu)$. Instead of the integral equation (7.37), now consider $u - \mathcal{A}_n u = \mathcal{A} g_n$. The only extra result needed is: *If $\{u_n\}_n$ converges strictly to some element $u_o \in L^\infty(D)$, then $\{\mathcal{A}_n u_n\}_n$ converges strictly to $\mathcal{A} u_o$.*

This works since we may write

$$A_n(x,t) - A(x,t) = A(x,t)\, 1\!\!1(\,|A(x,t)| > n\,) ,$$

so that $|A_n(x,t) - A(x,t)| \leqslant B(x-t)\, 1\!\!1\big(B(x-t) > n\big)$. Then, since $B \in L^1(\mathbb{R})$,

$$\| \mathcal{A}_n - \mathcal{A} \|_p \leqslant \int_{\mathbb{R}} B(t)\, 1\!\!1\big(B(t) > n\big)\, dt \longrightarrow 0 \quad \text{as} \quad n \to \infty .$$

See, again, ANSELONE and SLOAN (1990).

For a different proof of the lemma, see EGGERMONT (1989).

EXERCISES: (7.5), (7.14), (7.29).

8. Boundary behavior and interior equivalence

In this section, we explore what happens when f_o is extra smooth and satisfies the natural boundary conditions associated with the smoothing spline problem (1.10); i.e.,

$$\text{(8.1)} \qquad f_o \in W^{2m,\infty} ,$$

$$\text{(8.2)} \qquad f_o^{(\ell)}(0) = f_o^{(\ell)}(1) = 0 , \quad j = m, \cdots, 2m-1 .$$

Even though now the bias is a lot smaller, it turns out that we still have "equivalence" in the strict sense. Of course, a much more interesting situation arises when (8.1) holds but (8.2) fails completely. In this case, the bias is a lot smaller only away from the boundary, and we show that we have strict equivalence away from the boundary.

We begin with the case where f_o is smooth and satisfies the natural boundary conditions.

(8.3) SUPER EQUIVALENT KERNEL THEOREM. *Under the assumptions* (1.2) *to* (1.6) *on the model* (1.1), *as well as* (8.1)–(8.2), *the smoothing spline estimator* f^{nh} *of order* $2m$ *satisfies*

$$f^{nh}(t) = [\mathfrak{R}_{\omega mh} f_o](t) + \mathfrak{S}_{\omega nh}(t) + \varepsilon^{nh}(t) , \quad t \in [0, 1] ,$$

where almost surely

$$\limsup_{n\to\infty} \sup_{h\in\mathcal{H}_n(\alpha)} \frac{\| \varepsilon^{nh} \|_{\omega mh}}{h^{4m} + (nh)^{-1} \log n} < \infty ,$$

$$\limsup_{n\to\infty} \sup_{h\in\mathcal{H}_n(\alpha)} \frac{\| \varepsilon^{nh} \|_{\infty}}{h^{4m} + h^{-1/2}(nh)^{-1} \log n} < \infty .$$

(8.4) EXERCISE. Prove the theorem. [Hint: The reformulation (5.18) of the Equivalent Kernel Theorem (1.16) should be useful. Also, compare this with the asymptotically uniform, deterministic design case in § 13.5.]

If the natural boundary conditions are not satisfied, then the bounds of the theorem above cannot hold. However, they do hold on the interior of $[0, 1]$ (i.e., on the set $\mathfrak{P}_h(k)$; see (1.27)) for a large enough constant k.

(8.5) INTERIOR SUPER EQUIVALENT KERNEL THEOREM. *Under the assumptions* (1.2) *to* (1.6) *on the model* (1.1), *as well as* (8.1) *but not necessarily* (8.2), *there exists a constant* γ_m, *depending on* m *and* ω *only, such that the smoothing spline estimator* f^{nh} *of order* $2m$ *satisfies*

$$f^{nh}(t) = [\mathfrak{R}_{\omega mh} f_o](t) + \mathfrak{S}_{\omega nh}(t) + \varepsilon^{nh}(t) , \quad t \in [0, 1] ,$$

where almost surely

$$\limsup_{n\to\infty} \sup_{h\in\mathcal{H}_n(\alpha)} \frac{\sup\{ | \varepsilon^{nh}(t) | : t \in \mathfrak{P}_h(\gamma_m) \}}{h^{4m} + (nh)^{-1} \log n} < \infty .$$

Here, we take $\sup\{ \cdots \mid t \in \mathfrak{P}_h(\gamma_m) \} = 0$ *if* $\mathfrak{P}_h(\gamma_m) = \emptyset$.

PROOF. Let $f_h = \mathbb{E}[f^{nh} \mid \mathbb{X}_n]$. From Theorem (5.5), we get the representation

$$f^{nh} = f_h + \mathfrak{S}_{\omega nh} + \eta^{nh} \quad \text{on } [0, 1]$$

with the bound

$$\limsup_{n\to\infty} \sup_{h\in\mathcal{H}_n(\alpha)} \frac{\| \eta^{nh} \|_{\infty}}{(nh)^{-1} \log n} < \infty$$

almost surely. Then, obviously,

$$f^{nh} = \mathfrak{R}_{\omega mh} f_o + \mathfrak{S}_{\omega nh} + \eta^{nh} + f_h - \mathfrak{R}_{\omega mh} f_o ,$$

and it suffices to get bounds on $f_h - \mathfrak{R}_{\omega mh} f_o$.

To cut down on some rather horrendous formulas, we introduce the operators T_n acting on smooth functions φ and deterministic, open subintervals J of $[0, 1]$ by way of

$$[T_n(\varphi, J)](t) = \int_J \mathfrak{R}_{\omega mh}(x, t)\, \varphi(x) \left(d\Omega_n(x) - d\Omega_o(x) \right) . \tag{8.6}$$

When $J = [0, 1]$, we just write $T_n\varphi$ or $[T_n\varphi](t)$. While we are at it, from Lemma (3.4), we get the bound, for all $t \in [0, 1]$,

$$\left| [T_n(\varphi, J)](t) \right| \leqslant \theta_{nh} \, \| \mathfrak{R}_{\omega mh}(\,\cdot\,, t)\, \varphi(\,\cdot\,) \|_{\omega,m,h,1}$$

(see (22.2.18) for the definition of the $\| \cdot \|_{\omega,m,h,1}$ norm), with

$$\theta_{nh} = \mathcal{O}\big(\sqrt{(nh)^{-1} \log n}\, \big) \quad \text{uniformly in } h \in \mathcal{H}_n(\alpha) . \tag{8.7}$$

Then, by the convolution-like properties of $\mathfrak{R}_{\omega mh}(x, t)$, $0 < h \leqslant 1$,

$$\left| [T_n(\varphi, J)](t) \right| \leqslant c\, \theta_{nh} \, \| \varphi \|_{\omega,m,h,\infty} \tag{8.8}$$

uniformly in $t \in [0, 1]$ (whether φ is random or not).

Recall that f_h is the solution to the noiseless smoothing spline problem (5.8). Thus, $u = f_h$ is the solution to the Euler equations

$$(-h^2)^m\, u^{(2m)} + \tfrac{1}{n} \sum_{i=1}^{n} \left(u(\,\cdot\,) - f_o(\,\cdot\,) \right) \delta(\,\cdot\, - X_i) = 0 \quad \text{on } (0,1) ,$$

$$u^{(\ell)}(0) = u^{(\ell)}(1) = 0 , \quad \ell = m, \cdots, 2m-1 .$$

The differential equation may be rewritten (in)formally as

$$(-h^2)^m\, u^{(2m)} + \omega\, u = \omega\, f_o + \left(f_o(\,\cdot\,) - u(\,\cdot\,) \right) \left(d\Omega_n(x) - d\Omega_o(x) \right) ,$$

so that f_h has the representation

$$f_h(t) = [\mathfrak{R}_{\omega mh} f_o](t) + [T_n(f_o - f_h)](t) ,$$

with T_n defined above. Finally, by linearity,

$$f_h - \mathfrak{R}_{\omega mh} f_o = T_n(f_o - \mathfrak{R}_{\omega mh} f_o) - T_n(f_h - \mathfrak{R}_{\omega mh} f_o) . \tag{8.9}$$

We now bound the two terms on the right. For the second term, we have

$$\| f_h - \mathfrak{R}_{\omega mh} f_o \|_{\omega,m,h,1} \leqslant \| f_h - \mathfrak{R}_{\omega mh} f_o \|_{\omega mh} = \mathcal{O}\big(h^m \sqrt{(nh)^{-1} \log n}\, \big)$$

almost surely so that, by (8.8),

$$[T_n(f_h - \mathfrak{R}_{\omega mh} f_o)](t) = \mathcal{O}\big(h^m (nh)^{-1} \log n \big)$$

uniformly in $t \in [0, 1]$ and $h \in \mathcal{H}_n(\alpha)$. This is negligible compared with the usual noise term $\mathcal{O}\big((nh)^{-1} \log n \big)$.

Now consider the first term of (8.9), $T_n(f_o - \mathfrak{R}_{\omega m h} f_o)$, and write

$$T_n(f_o - \mathfrak{R}_{\omega m h} f_o) = T_n(f_o - \mathfrak{R}_{\omega m h} f_o , \mathfrak{P}_h(k)) + T_n(f_o - \mathfrak{R}_{\omega m h} f_o , \mathcal{C}) ,$$

where $\mathcal{C} = [0, 1] \setminus \mathfrak{P}_h(k)$.

If k is large enough, by Lemma (8.10) below,

$$\sup_{t \in \mathfrak{P}_h(k)} | f_o(t) - \mathfrak{R}_{\omega m h} f_o(t) | = \mathcal{O}(h^{2m}) ,$$

and likewise for $\displaystyle \sup_{t \in \mathfrak{P}_h(k)} h \, | f_o'(t) - [(\mathfrak{R}_{\omega m h} f_o)'](t) |$. Then,

$$\begin{aligned} [T_n(f_o - \mathfrak{R}_{\omega m h} f_o , \mathfrak{P}_h(k))](t) &= \mathcal{O}\big(h^{2m} \sqrt{(nh)^{-1} \log n} \big) \\ &= \mathcal{O}\big(h^{4m} + (nh)^{-1} \log n \big) , \end{aligned}$$

almost surely, uniformly in $t \in [0, 1]$.

Finally, for the term $T_n(f_o - \mathfrak{R}_{\omega m h} f_o , \mathcal{C})$, we restrict t to $t \in \mathfrak{P}_h(2k)$. On the set $[0, 1] \setminus \mathfrak{P}_h(k)$, we only have the bound h^m for $f_o - \mathfrak{R}_{\omega m h} f_o$ and $h (f_o' - (\mathfrak{R}_{\omega m h} f_o)')$, but here the kernel comes to the rescue. For $t \in \mathfrak{P}_h(2k)$, we have the (exponential) bound

$$\begin{aligned} | \mathfrak{R}_{\omega m h}(x, t) | &\leqslant c\, h^{-1} \exp(-\kappa\, h^{-1} | x - t |) \\ &\leqslant c\, h^{-1} \exp(-\kappa\, k \log(1/h)) \leqslant c\, h^{2m} \end{aligned}$$

if k is large enough. The same bound holds for $h \, | \mathfrak{R}_{\omega m h}'(x, t) |$ (derivative with respect to x). So then we get the bound

$$[T_n(f_o - \mathfrak{R}_{\omega m h} f_o , \mathcal{C})](t) = \mathcal{O}\big(h^{2m} \sqrt{(nh)^{-1} \log n} \big) .$$

Of course, by the elementary inequality $2ab \leqslant a^2 + b^2$, this is of the order $h^{4m} + (nh)^{-1} \log n$. The corresponding bound on $[T_n(f_o - \mathfrak{R}_{\omega m h} f_o)](t)$ follows, and one checks that it holds uniformly in $t \in \mathfrak{P}_h(k)$. Q.e.d.

In the proof above, we used the following result, where we use the notation, for intervals $J \subset [0, 1]$,

$$\| f \|_{\omega,m,h,p,J} = \| f \|_{L^p(J, \omega)} + h^m \| f^{(m)} \|_{L^p(J)} .$$

(8.10) LEMMA. *Let $m \geqslant 1$ and $1 \leqslant p \leqslant \infty$. Assuming that the design density ω is quasi-uniform, there exist constants c and k such that, for all $f_o \in W^{2m,p}(0,1)$,*

$$\| f_o - \mathfrak{R}_{\omega m h} f_o \|_{\omega,m,h,p,\mathfrak{P}_h(k)} \leqslant c\, h^{2m} \| f_o \|_{W^{2m,p}(0,1)} .$$

PROOF. Let $\zeta_h = \mathfrak{R}_{\omega m h} f_o$. A slight perversion of the reproducing kernel Hilbert space trick shows that

$$\zeta_h(t) = f_o(t) - h^{2m} \langle \mathfrak{R}_{\omega m h}^{(m)}(\cdot , t) , f_o^{(m)} \rangle .$$

Integration by parts m times results in

$$\langle \mathfrak{R}^{(m)}_{\omega mh}(\,\cdot\,, t\,)\,,\, f_o^{(m)} \,\rangle = \int_0^1 f_o^{(m)}(x)\; \frac{\partial^m \mathfrak{R}_{\omega mh}(x, t\,)}{\partial x^m}\, dx$$

$$= -\sum_{\ell=1}^{m} (-1)^{\ell}\,\{\, N_\ell(1) - N_\ell(0)\,\} + (-1)^m \int_0^1 \mathfrak{R}_{\omega mh}(x, t\,)\, f_o^{(2m)}(x)\, dx \;,$$

where the $N_\ell(0)$ and $N_\ell(1)$ are the boundary terms, with

$$N_\ell(x) = \mathfrak{R}^{(m-\ell)}_{\omega mh}(x, t\,)\, f_o^{(m+\ell-1)}(x) \;.$$

One observes that first, since $0 \leqslant m+\ell-1 \leqslant 2m-1$,

$$(8.11) \qquad |\, f_o^{(m+\ell-1)}(x)\,| \leqslant c\,\|\, f_o \,\|_{W^{2m,p}(0,1)} \;,$$

and second, for all $t \in \mathfrak{P}_h(k)$, if k is large enough,

$$|\,\mathfrak{R}^{(m-\ell)}_{\omega mh}(0, t\,)\,| \leqslant c_m\, h^{-m+\ell-1} \exp\bigl(-\kappa_m\, t\,/h\,\bigr) \leqslant c \;.$$

Note that k being large enough only depends on c_m and κ_m. In other words, it only depends on the reproducing kernel and thus, ultimately, only on m and the design density. The same bound holds for $|\,\mathfrak{R}^{(m-\ell)}_{\omega mh}(1, t\,)\,|$, and it follows that

$$\Bigl|\, -\sum_{\ell=1}^{m} (-1)^{\ell}\,\{\, N_\ell(1) - N_\ell(0)\,\}\,\Bigr| \leqslant c\,\|\, f_o \,\|_{W^{2m,p}(0,1)} \;.$$

Of course, the bound

$$\Bigl\|\int_0^1 \mathfrak{R}_{\omega mh}(x, \,\cdot\,)\, f_o^{(2m)}(x)\, dx\,\Bigr\|_{L^p(0,1)} \leqslant$$

$$\sup_{t\in[\,0\,,\,1\,]} \|\,\mathfrak{R}_{\omega mh}(x, \,\cdot\,)\,\|_{L^1(0,1)}\, \|\, f_o^{(2m)} \,\|_{L^p(0,1)}$$

follows similarly to Young's inequality. Thus,

$$\Bigl|\,\langle \mathfrak{R}^{(m)}_{\omega mh}(\,\cdot\,, t\,)\,,\, f_o^{(m)} \,\rangle\,\Bigr| \leqslant c\,\|\, f_o \,\|_{W^{2m,p}(0,1)} \;,$$

and the lemma follows. Q.e.d.

(8.12) EXERCISE. Prove the inequality (8.11). [Hint: Check the proof of the case $p = 2$ in § 2.]

We are now ready for the superconvergence of the smoothing spline estimator in the interior of $[\,0\,,1\,]$, provided f_o is smooth enough.

(8.13) COROLLARY. *Under the conditions of Lemma* (8.10),

$$\sup_{h\in\mathcal{G}_n(\alpha)} \frac{\|\,\mathbb{E}[\, f^{nh} \,|\, \mathbb{X}_n\,] - f_o\,\|_{L^p(\mathfrak{P}_h(k))}}{h^{2m} + h^{-1/2}(nh)^{-1}\log n} < \infty \quad \textit{almost surely}.$$

(8.14) EXERCISE. Prove the corollary.

(8.15) EXERCISE. Formulate and prove the refinement of Lemma (8.10) when $f_o \in W^{m+\ell,p}(0,1)$ for $\ell = 1, 2, \cdots, m-1$.

EXERCISES: (8.4), (8.12), (8.14), (8.15).

9. The equivalent Nadaraya-Watson estimator

In this section, we study the approximate translation invariance of the reproducing kernels $\mathfrak{R}_{\omega mh}$ when the design density is smooth. This would be very much in the style of the original result of SILVERMAN (1984), except that, instead of approximating the reproducing kernels themselves, we approximate the reproducing kernel *estimators*. Ultimately, for very smooth design densities, this leads to the approximation of the smoothing spline estimator by a Nadaraya-Watson estimator with a local smoothing parameter, at least away from the boundary, as expressed in Theorem (1.30). The original variable kernel estimator of SILVERMAN (1984) fails to be "equivalent" in the strict sense.

Comparing with the uniform density. The first step is to replace the quasi-uniform design density by the uniform density. This works at the expense of a spatially dependent smoothing parameter, without any nasty boundary effects. The results are stated in Corollaries (9.16) and (9.19).

As always, the design density is assumed to be quasi-uniform in the sense of Definition (1.4), but, in addition, extra smoothness is required. Initially, we consider Lipschitz-continuous design densities; i.e., for some constant c,

(9.1) $$|\,\omega(x) - \omega(\,t\,)\,| \leqslant c\,|\,x - t\,|\;, \quad x, t \in [\,0\,,1\,]\;.$$

Let m be a positive integer, and let $h > 0$. Consider the boundary value problem

(9.2) $$\begin{aligned} &(-h^2)^m\,\psi^{(2m)} + \omega\,\psi = v \quad \text{on} \quad [\,0\,,1\,]\;, \\ &\psi^{(k)}(0) = \psi^{(k)}(1) = 0\;, \quad k = m, \cdots, 2m-1\;, \end{aligned}$$

where v is "nice". The cases of interest are

(9.3) $$v = \omega\,f \quad \text{with} \quad f \in L^p(0,1) \quad \text{for some } p \text{ with } 1 \leqslant p \leqslant \infty$$

and the case where v is a weighted sum of point masses,

(9.4) $$v(\,t\,) = \tfrac{1}{n} \sum_{i=1}^{n} D_i\,\delta(\,t\, - X_i)\;.$$

In general, the (X_i, D_i), $i = 1, 2, \cdots, n$, are random, but in this section randomness is mostly irrelevant.

We know that, for $f \in L^p(0,1)$, the solution of (9.2)–(9.3) is given by

$$\psi(t) = \int_0^1 \mathfrak{R}_{\omega mh}(x,t)\,\omega(x)\,f(x)\,dx\ , \quad t \in [0,1]\ , \tag{9.5}$$

and, for v given by (9.4), we have $\psi = \mathfrak{S}_{\omega nh}$, see (1.13),

$$\psi(t) = \tfrac{1}{n}\sum_{i=1}^{n} D_i\,\mathfrak{R}_{\omega mh}(X_i,t)\ , \quad t \in [0,1]\ . \tag{9.6}$$

Of course, the kernels $\mathfrak{R}_{\omega mh}$ are well-understood. However, "things" are easier for more explicitly convolution-like kernels, so here we wish to investigate the behavior of $\psi(t)$ around some point $x_o \in [0,1]$. Thus, in (9.2)–(9.3), replace ω with $\omega(x_o)$,

$$\begin{aligned}&(-h^2)^m\,\varphi^{(2m)} + \omega(x_o)\,\varphi = \omega(x_o)\,f \quad \text{on} \quad [0,1]\ ,\\ &\varphi^{(k)}(0) = \varphi^{(k)}(1) = 0\ , \quad k = m, \cdots, 2m-1\ .\end{aligned} \tag{9.7}$$

Rewriting the differential equation of (9.7) as $(-\lambda^2)^m\,\varphi^{(2m)} + \varphi = f$, where

$$\lambda = \lambda(x_o) = h \,/\, \bigl(\omega(x_o)\bigr)^{1/(2m)}\ , \tag{9.8}$$

we see that the solution of (9.7) for $f \in L^p(0,1)$ is given by $\varphi \equiv \varphi(\,\cdot\,|\,x_o)$,

$$\varphi(t\,|\,x_o) = \int_0^1 \mathcal{R}_{m\lambda}(x,t)\,f(x)\,dx\ . \tag{9.9}$$

If v is the sum of point masses (9.4), then we consider the boundary value problem

$$\begin{aligned}&(-h^2)^m\,\vartheta^{(2m)} + \omega(x_o)\,\vartheta = v \quad \text{on} \quad [0,1]\ ,\\ &\vartheta^{(k)}(0) = \vartheta^{(k)}(1) = 0\ , \quad k = m, \cdots, 2m-1\ ,\end{aligned} \tag{9.10}$$

the solution of which is given by

$$\vartheta(t\,|\,x_o) = \tfrac{1}{n}\sum_{i=1}^{n} D_i\,\mathcal{R}_{m\lambda}(X_i,t)\,/\,\omega(x_o)\ , \quad t \in [0,1]\ . \tag{9.11}$$

(9.12) THEOREM. *Assume that the design density is quasi-uniform and Lipschitz continuous. There exists a positive constant c such that if v satisfies (9.3), then, for all $t \in [0,1]$,*

$$\bigl|\,\psi(t) - \varphi(t\,|\,x_o)\,\bigr| \leqslant c\,\{\,|\,t - x_o\,| + h\,\}\,\|\,\psi - f\,\|_\infty\ .$$

If v satisfies (9.4), then

$$\bigl|\,\psi(t) - \vartheta(t\,|\,x_o)\,\bigr| \leqslant c\,\{\,|\,t - x_o\,| + h\,\}\,\|\,\psi\,\|_\infty\ .$$

PROOF. Consider the case where v satisfies (9.3). Let $\varepsilon \equiv \varphi(\,\cdot\,|\,x_o) - \psi$. From the boundary value problems (9.2) with $v = \omega f$, and (9.7), one derives that

$$(-h^2)^m \, \varepsilon^{(2m)} + \omega(x_o)\, \varepsilon = -\big(w - \omega(x_o) \big) \big(\psi - f \big) \quad \text{on} \quad [\,0\,,\,1\,]\;,$$
$$\varepsilon^{(k)}(0) = \varepsilon^{(k)}(1) = 0\;, \qquad k = m, \cdots, 2m-1\;,$$

so that, with λ as in (9.8), the solution is given by

$$\text{(9.13)} \qquad \varepsilon(\,t\,) = \int_0^1 \mathcal{R}_{m\lambda}(x,\,t\,)\, q(x, x_o) \big(\psi(x) - f(x) \big)\, dx$$

for $t \in [\,0\,,\,1\,]$, where $q(x,\,t\,) = \big(\omega(\,t\,) - \omega(x)\big)/\omega(\,t\,)$. By the quasi-uniformity and Lipschitz continuity of ω, we get $|\,q(x, x_o)\,| \leqslant c\,|\,x - x_o\,|$, so that

$$|\,\varepsilon(\,t\,)\,| \leqslant c\,\|\,\psi - f\,\|_\infty \int_0^1 |\,\mathcal{R}_{m\lambda}(x,\,t\,)\,|\,|\,x - x_o\,|\, dx\;.$$

Since the convolution-like properties of $\mathcal{R}_{m\lambda}$ imply that

$$\text{(9.14)} \qquad \int_0^1 |\,\mathcal{R}_{m\lambda}(x,\,t\,)\,|\,|\,x - \,t\,|\, dx \leqslant c\,h$$

for a suitable constant c, the elementary inequality

$$|\,x - x_o\,| \leqslant |\,x - \,t\,| + |\,t\, - x_o\,|$$

then implies that $|\,\varepsilon(\,t\,)\,| \leqslant c\,\|\,\psi - f\,\|_\infty \,\{\,h + |\,t\, - x_o\,|\,\}$. This concludes the case where v satisfies (9.3). The case where v is given by (9.4) follows similarly. Q.e.d.

(9.15) Exercise. (a) Verify (9.14).
(b) Fill in the details when v satisfies (9.4).

It is worthwhile to restate this explicitly for the cases where v is pure noise, see (9.4). It is useful to take $t = x_o$ or rather $x_o = t$.

(9.16) Corollary. *Under the conditions of Theorem* (9.12), *there exists a constant* c *such that, for all* $t \in [\,0\,,\,1\,]$ *and all* (X_i, D_i), $i = 1, 2, \cdots, n$,

$$\Big|\, \mathfrak{S}_{\omega n h}(\,t\,) - \tfrac{1}{n} \sum_{i=1}^{n} D_i\, \mathcal{R}_{m,\lambda(t)}(X_i,\,t\,)\,/\,\omega(\,t\,) \Big| \leqslant c\,h\;\|\,\mathfrak{S}_{\omega n h}\,\|_\infty\;,$$

with $\lambda(\,t\,) = h\,/\,\big(\omega(\,t\,)\big)^{1/(2m)}$.

The analogous result for the noiseless continuous signal follows similarly. Define the operator

$$\text{(9.17)} \qquad [\,\mathcal{R}_{m,\lambda(\,t\,)}\, f_o\,](\,t\,) = \int_0^1 \mathcal{R}_{m,\lambda(\,t\,)}(x,\,t\,)\, f_o(x)\, dx\;.$$

Note that the only reference to the design density is by way of the smoothing parameter $\lambda(\,t\,)$.

(9.18) COROLLARY. *Under the conditions of Theorem* (9.12), *there exists a constant* c *such that, for all* $t \in [0, 1]$ *and all* $f \in W^{m,\infty}(0,1)$,

$$\Big| [\mathfrak{R}_{\omega m h} f](t) - [\mathcal{R}_{m,\lambda(t)} f](t) \Big| \leqslant c h^{m+1} \| f_o^{(m)} \|_\infty .$$

PROOF. Theorem (9.12) says that

$$\Big| [\mathfrak{R}_{\omega m h} f](t) - [\mathcal{R}_{m\lambda(t)} f](t) \Big| \leqslant c h \| f - \mathfrak{R}_{\omega m h} f \|_\infty .$$

Now, with $\mathfrak{R}^{(m)}_{\omega m h}(x, t)$ denoting the derivative of $\mathfrak{R}_{\omega m h}(x, t)$ with respect to x,

$$\begin{aligned} [\mathfrak{R}_{\omega m h} f](t) &= \langle \mathfrak{R}_{\omega m h}(\cdot, t), f(\cdot) \rangle_{L^2((0,1),\omega)} \\ &= \langle \mathfrak{R}_{\omega m h}(\cdot, t), f(\cdot) \rangle_{\omega m h} - h^{2m} \langle \mathfrak{R}^{(m)}_{\omega m h}(\cdot, t), f^{(m)}(\cdot) \rangle \\ &= f(t) - h^{2m} \langle \mathfrak{R}^{(m)}_{\omega m h}(\cdot, t), f^{(m)}(\cdot) \rangle . \end{aligned}$$

Of course, by the convolution-like properties of $\mathfrak{R}_{\omega m h}(x, t)$, $0 < h \leqslant 1$,

$$\Big| \langle \mathfrak{R}^{(m)}_{\omega m h}(\cdot, t), f^{(m)}(\cdot) \rangle \Big| \leqslant \\ \| \mathfrak{R}^{(m)}_{\omega m h}(\cdot, t) \|_{L^1(0,1)} \| f^{(m)} \|_\infty \leqslant c h^{-m} \| f^{(m)} \|_\infty ,$$

and so $\| f - \mathfrak{R}_{\omega m h} f \|_\infty \leqslant c h^m \| f^{(m)} \|_\infty$. Q.e.d.

The story does not end here. We know from Lemma (8.10) and Corollary (8.13) that if the regression function is smooth enough, then the bias of smoothing spline estimators is a lot smaller in the interior than in the boundary region. So, we should like "equivalent" kernel approximations to reflect this, and indeed they do. We assume the maximally useful smoothness condition (1.26). Also, recall the definition (1.27) of the interior region $\mathfrak{P}_h(k)$.

(9.19) COROLLARY. *Assume the conditions of Theorem* (9.12). *Then, for a large enough constant* κ, *if* $f \in W^{2m,\infty}(0,1)$,

$$\sup_{t \in \mathfrak{P}_h(\kappa)} \Big| [\mathfrak{R}_{\omega m h} f](t) - [\mathcal{R}_{m,\lambda(t)} f](t) \Big| \leqslant c h^{2m+1} \| f^{(2m)} \|_\infty .$$

PROOF. From (9.13), we get for

$$\varepsilon(t) = \psi(t) - \varphi(t \,|\, t) = [\mathfrak{R}_{\omega m h} f](t) - [\mathcal{R}_{m,\lambda(t)} f](t)$$

that

$$\varepsilon(t) = \int_0^1 \mathcal{R}_{m,\lambda(t)}(x, t) \, q(x, t) \big(f(x) - [\mathfrak{R}_{\omega m h} f](x) \big) \, dx .$$

Recall that $|\,q(x,t)\,| \leqslant c\,|\,x-t\,|$. Split the integral into integrals over $\mathfrak{P}_h(k)$ and over $[\,0\,,\,1\,]\setminus\mathfrak{P}_h(k)$, with a constant k to be chosen large enough. Since by Lemma (8.10),

$$\|\, f - \mathfrak{R}_{\omega m h}\, f \,\|_{L^\infty(\mathfrak{P}_h(k))} \leqslant c\, h^{2m}\, \|\, f\,\|_{W^{2m,\infty}(0,1)}\;,$$

then

$$\begin{aligned}
&\Bigl|\int_{\mathfrak{P}_h(k)} \mathcal{R}_{m,\lambda(t)}(x,t)\, q(x,t)\,\bigl(\, f(x) - [\,\mathfrak{R}_{\omega m h}\, f\,](x)\,\bigr)\, dx\,\Bigr| \\
&\qquad \leqslant c\, h^{2m}\, \|\, f\,\|_{W^{2m,\infty}(0,1)} \int_{\mathfrak{P}_h(k)} |\,\mathcal{R}_{m,\lambda(t)}(x,t)\,|\,|\,q(x,t)\,|\, dx \\
&\qquad \leqslant c\, h^{2m}\, \|\, f\,\|_{W^{2m,\infty}(0,1)} \int_{[\,0\,,\,1\,]} |\,\mathcal{R}_{m,\lambda(t)}(x,t)\,|\,|\,x-t\,|\, dx \\
&\qquad \leqslant c\, h^{2m+1}\, \|\, f\,\|_{W^{2m,\infty}(0,1)}\;.
\end{aligned}$$

For the remaining integral over $[\,0\,,\,1\,]\setminus\mathfrak{P}_h(k)$, we use the exponential bound on $\mathcal{R}_{m,\lambda(t)}(x,t)$. For $x\in[\,0\,,\,1\,]\setminus\mathfrak{P}_h(k)$ and $t\in\mathfrak{P}_h(2k)$, we have $|\,x-t\,|\geqslant k\,h\,\log(1/h)$, and then, for k large enough,

$$|\,\mathcal{R}_{m,\lambda(t)}(x,t)\,| \leqslant c\, h^m\;.$$

Of course, $\|\, f - \mathfrak{R}_{\omega m h}\, f\,\|_\infty \leqslant c\, h^m\, \|\, f^{(m)}\,\|_\infty \leqslant c\, h^m\, \|\, f\,\|_{W^{2m,\infty}(0,1)}$, and the bound on $q(x,t)$ still applies, so that

$$\Bigl|\int_{[\,0\,,\,1\,]\setminus\mathfrak{P}_h(k)} \cdots\,\Bigr| \leqslant c\, h^{2m+1}\, \|\, f\,\|_{W^{2m,\infty}(0,1)}\;,$$

and that does it. Q.e.d.

The Silverman approximation. Now, we get to the explicit convolution-like structure of the "estimator". First, take t away from the boundary; i.e., for a suitable constant k, depending on the order of the smoothing spline, for $0<h<1$ we take $t\in\mathfrak{P}_h(k)$. At this point, recall the definition in Lemma (14.7.12) of the MESSER and GOLDSTEIN (1993) convolution kernel $\mathfrak{B}_{mh}$, the solution to the boundary value problem (14.7.17). Let

(9.20) $$\mathfrak{D}_{nh}(t) = \tfrac{1}{n}\sum_{i=1}^{n} D_i\,\mathfrak{B}_{m,\lambda(t)}(X_i - t)\,/\,\omega(t)\;,\quad t\in\mathfrak{P}_h(\gamma_m)\;,$$

with $\lambda(t) = h\,/\,\bigl(\omega(t)\bigr)^{1/(2m)}$; cf. (9.8).

(9.21) THEOREM. *Under the conditions of Theorem* (9.12), *with* $\mathfrak{B}_{mh}$ *as in Theorem* (14.7.12), *there exist positive constants* k *and* c *such that*

$$\sup_{t\in\mathfrak{P}_h(k)} \Bigl|\,\mathfrak{S}_{\omega nh}(t) - \mathfrak{D}_{nh}(t)\,\Bigr| \leqslant c\,h\,\bigl\{\,\|\,\mathfrak{S}_{\omega nh}\,\|_\infty + h^m\,\|\,(\mathfrak{S}_{\omega nh})^{(m)}\,\|_\infty\,\bigr\}\;.$$

We note that the factor h may not look very impressive but in fact it is because all of $\| \mathfrak{S}_{\omega nh} \|_\infty$, $\| \mathfrak{D}_{nh} \|_\infty$, and $h^m \| (\mathfrak{S}_{\omega nh})^{(m)} \|_\infty$ are $\mathcal{O}\big(\sqrt{(nh)^{-1} \log n}\, \big)$, so the approximation error is negligible compared with either of them.

PROOF OF THEOREM (9.21). It suffices to show the required bound on

$$\varepsilon_{nh}(t) \stackrel{\text{def}}{=} \tfrac{1}{n} \sum_{i=1}^{n} D_i\, b_{nh}(X_i, t) \,/\, \omega(t) \,,$$

where
$$b_{nh}(x, t) = \mathcal{R}_{m,\lambda(t)}(x, t) - \mathfrak{B}_{m,\lambda(t)}(x - t) \,.$$

Thus, b_{nh} represents the boundary corrections to the pure convolution kernel $\mathfrak{B}_{m\lambda(\cdot)}$. Then, since b_{nh} is a nice function, one obtains by way of the reproducing kernel Hilbert space trick that

$$\varepsilon_{nh}(t) = \big\langle\, \mathfrak{S}_{\omega nh} \,,\, b_{nh}(\,\cdot\,, t) \,\big\rangle_{\omega mh} \,/\, \omega(t) \,.$$

Since ω is bounded away from zero, we get

$$\big|\, \varepsilon_{nh}(t) \big| \leqslant c \,\big|\, \big\langle\, \mathfrak{S}_{\omega nh} \,,\, b_{nh}(\,\cdot\,, t) \,\big\rangle_{\omega mh} \,\big| \,,$$

and so, for another suitable constant c,

$$\big|\, \varepsilon_{nh}(t) \big| \leqslant c \big\{ \| \mathfrak{S}_{\omega nh} \|_\infty + h^m \| (\mathfrak{S}_{\omega nh})^{(m)} \|_\infty \big\} \times \\ \big\{ \| b_{nh}(\,\cdot\,, t) \|_{L^1(0,1)} + h^m \| b_{nh}^{(m)}(\,\cdot\,, t) \|_{L^1(0,1)} \big\} \,.$$

Here, $b_{nh}^{(m)}(x, t)$ denotes the m-th derivative with respect to x.

So, all that is left is to bound the norms of b_{nh}. From Lemmas (14.7.11) and (14.7.12), one gleans that there exist positive constants c_m and κ_m such that

$$\text{(9.22)} \qquad \big|\, b_{nh}(x, t) \,\big| \leqslant c_m\, h^{-1} \big\{ \exp(-\kappa_m\, h^{-1}\, (x + t)) + \\ \exp(-\kappa_m\, h^{-1}\, (2 - x - t)) \big\}$$

and likewise for $b_{nh}^{(m)}$. It follows that

$$\big\| \, b_{nh}(\,\cdot\,, t) \,\big\|_{L^1(0,1)} \leqslant c \exp\big(-\kappa_m\, h^{-1} \min(t, 1 - t)\big) \,.$$

So, for $t \in \mathfrak{P}_h(k)$, with a large enough constant k, this is bounded by h. The same bound applies to $\big\| \, b_{nh}^{(m)}(\,\cdot\,, t) \,\big\|_{L^1(0,1)}$. Q.e.d.

There is an analogue for the smoothing operators acting on the noiseless continuous signal. Define

$$\text{(9.23)} \qquad [\, \mathfrak{B}_{m,\lambda(t)}\, f_o \,](t) = \int_0^1 \mathfrak{B}_{m,\lambda(t)}(x, t)\, f_o(x)\, dx \,.$$

Then, we have the following continuous Silverman approximation. Since we need to be bounded away from the boundary anyway, we aim for the super equivalent interior kernel approximation.

(9.24) THEOREM. *Under the conditions of Theorem* (9.12), *there exist constants* c *and* k *such that, for all* $f_o \in W^{2m,\infty}(0,1)$ *and all* $t \in \mathfrak{P}_h(k)$,

$$\Big| [\mathcal{R}_{m,\lambda(t)} f_o](t) - [\mathfrak{B}_{m,\lambda(t)} f_o](t) \Big| \leqslant c h^{2m+1} \| f_o^{(2m)} \|_\infty .$$

(9.25) EXERCISE. Prove the theorem. [Hint: The proof of Theorem (9.21) should provide inspiration.]

The details are now in place for the proof of the Silverman Equivalent Kernel Theorem (1.28).

(9.26) EXERCISE. Prove Theorem (1.28).

(9.27) REMARK. It is somewhat surprising that the Silverman kernels are not symmetric. However, as the following result shows, there is some sort of symmetry in that $\lambda(t)$ may be replaced by $\lambda(x)$. Let

$$\mathfrak{L}_{nh}(t) = \tfrac{1}{n} \sum_{i=1}^{n} D_i \, \mathfrak{C}_{m,h}(X_i, t) , \tag{9.28}$$

where $\mathfrak{C}_{m,h}(x, t) = \mathfrak{B}_{m,\lambda(x)}(x - t)$.

(9.29) THEOREM. *Under the conditions of Theorem* (9.12), *if, in addition, the design density is Lipschitz continuous, then*

$$\sup_{t \in \mathfrak{P}_h(\gamma_m)} \big| \mathfrak{D}_{nh}(t) - \mathfrak{L}_{nh}(t) \big| \leqslant c h \{ \| S_{n,1,h} \|_\infty + h \| (S_{n,1,h})' \|_\infty \} ,$$

in which
$$S_{n,1,h}(t) = \tfrac{1}{n} \sum_{i=1}^{n} D_i \, \mathfrak{R}_{\omega,1,h}(X_i, t) .$$

Of course, the theorem says that we can work with the symmetric kernel $\frac{1}{2}\{ \mathfrak{C}_{m,h}(x, t) + \mathfrak{C}_{m,h}(t, x) \}$.

(9.30) EXERCISE. (a) Prove the theorem.
(b) Show that the family $\mathfrak{C}_{m,h}(x, t)$, $0 < h \leqslant 1$, is convolution-like.
(c) Show that the approximation of $\mathfrak{D}_{nh}$ by

$$\mathfrak{N}_{nh}(t) = \tfrac{1}{n} \sum_{i=1}^{n} D_i \, \frac{1}{\lambda(t)} \, \mathfrak{B}_{m,h}\left(\frac{X_i}{\lambda(X_i)} - \frac{t}{\lambda(t)} \right) , \quad t \in [0, 1] ,$$

does not work. [Hint: For (a), peek ahead to § 22.6. For (c), it is clear that the same proof cannot be modified. Indeed, if $x/\lambda(x)$ is constant over an open interval, then the approximation certainly fails since then $\mathbb{P}[X/\lambda(X) = c/h] > 0$ for a suitable value of c. This is reminiscent of material in STADTMÜLLER (1986).]

The equivalent Nadaraya-Watson estimator. The proof of Theorem (1.30) follows, but recall that the theorem requires the extra smoothness condition $\omega \in W^{2m,\infty}(0,1)$. Let

$$(9.31)\qquad \nu^{nh}(t) = \frac{\frac{1}{n}\sum_{i=1}^{n} D_i\, \mathfrak{B}_{m,\lambda(t)}(X_i, t)}{\frac{1}{n}\sum_{i=1}^{n} \mathfrak{B}_{m,\lambda(t)}(X_i, t)}\ , \qquad t \in [0, 1]\ ,$$

be the Nadaraya-Watson estimator acting on pure noise, and let $s^{nh}(t)$ be the corresponding estimator acting on the noiseless discrete signal. It is easy to show that, almost surely,

$$(9.32)\qquad \limsup_{n\to\infty}\ \sup_{h\in\mathcal{H}_n(\alpha)} \frac{\|\,\mathfrak{D}_{nh} - \nu^{nh}\,\|_\infty}{(nh)^{-1}\log n} < \infty\ .$$

In Theorem (16.10.12), we showed that

$$\eta_{nh} \stackrel{\text{def}}{=} s^{nh}(t) - [\,\mathfrak{B}_{m,\lambda(t)}\, f_o\,](t)$$

satisfies

$$(9.33)\qquad \sup_{t\,\in\mathfrak{P}_h(k)} \bigl|\,\eta_{nh}(t)\,\bigr| = \mathcal{O}\bigl(\,h^{2m+1}\,\bigr)\ .$$

Combining (9.32)–(9.33) along with Theorem(1.28) gives

$$\begin{aligned} f^{nh}(t) &= [\,\mathfrak{B}_{m,\lambda(t)}\, f_o\,](t) + \mathfrak{D}_{nh}(t) + \delta^{nh}(t) \\ &= s^{nh}(t) + \nu^{nh}(t) + \delta^{nh}(t) - \eta_{nh}(t) + \varepsilon_{nh}(t) \\ &= f^{nh}_{\text{S-NW}}(t) + \delta^{nh}(t) - \eta_{nh}(t) + \varepsilon_{nh}(t)\ , \end{aligned}$$

with the combined, almost sure bounds

$$\sup_{t\,\in\mathfrak{P}_h(k)} \bigl|\,\delta^{nh}(t) - \eta_{nh}(t) + \varepsilon_{nh}(t)\,\bigr| = \mathcal{O}\bigl(\,h^{2m+1} + h\,\sqrt{(nh)^{-1}\log n}\,\bigr)\ ,$$

uniformly in $h \in \mathcal{H}_n(\alpha)$.

(9.34) EXERCISE. Prove the bound (9.32).

EXERCISES: (9.15), (9.25), (9.30), (9.34).

10. Additional notes and comments

Ad § 1: An even more general nonparametric regression problem is considered by EFROMOVICH (1996): The random variable (X, Y) is supposed to have the joint pdf

$$f_{X,Y}(x,y) = p\bigl(\,y\,|\,f_o(x)\,\bigr)\,f_X(x)\ , \quad -\infty < x\,,\, y < \infty\ ,$$

where the (parametric) form of $p(\cdot\,|\,\cdot)$ is known. The objective is to estimate f_o. In this volume, we are (mostly) concerned with the additive model $p(\,y\,|\,f_o(x)\,) = \pi(\,y - f_o(x)\,)$, and now π need not be known (but it would help if it was).

Regarding equivalent kernels, it is not clear whether SILVERMAN (1984) had our strict interpretation of equivalence in mind, but certainly his results are not strong enough to draw such a conclusion. However, that has not stopped workers from using "equivalence"; see, e.g., LIN, WANG, WELSH, and CARROLL (2004), and many others. A notable exception is MA, CHIOU, and WANG (2006). Perhaps it is useful to state the result of SILVERMAN (1984). For the random design case, he showed in our notation that

$$h\,\mathfrak{R}_{\omega mh}(x,\,t\,) - h\,K_h(x,\,t\,|\,\mathbb{X}_n) = \mathcal{O}\big(\,h + h^{-1}n^{-1/2}\,\big)$$

uniformly in s and t away from the boundary of $[\,0\,,\,1\,]$. Here, K_h is the hat operator: The solution of the smoothing spline problem (1.10) has the representation

$$f^{nh}(\,t\,) = \tfrac{1}{n}\sum_{i=1}^{n} Y_i\,K_h(X_i\,,\,t\,|\,\mathbb{X}_n)\ .$$

Note that K_h depends on the design.

The earlier references on equivalent kernels all deal with deterministic designs; see SPECKMAN (1981) and COX (1981). Some later works, discussed later in this section, are COX (1984), MESSER (1991), MESSER and GOLDSTEIN (1993), NYCHKA (1995), and CHIANG, RICE, and WU (2001).

There are two aspects to the equivalent kernel setup. One aspect concerns the convolution-like properties of the reproducing kernel and the properties of the reproducing kernel estimator of the regression function. The other one deals with the accuracy of the reproducing kernel estimator as an approximation to the original smoothing spline estimator.

Regarding the first problem, attention is usually restricted to the mean squared error. For the uniform design density, COX (1981) computes the Green's function for (2.13) with periodic boundary conditions by means of Fourier series and then fixes the natural boundary conditions (for $m = 2$). MESSER and GOLDSTEIN (1993) determine the Green's function for (2.13) on the line by means of Fourier transform methods and then fix the natural boundary conditions on the finite interval, except that they ignore the interaction between the two boundary points. Since this interaction is of size $\exp(-1/h)$, saying that this causes a negligible error, even in the boundary region, is an understatement. See § 14.7 for the details.

For "arbitrary", smooth design densities ω, properties of the equivalent kernels were obtained by NYCHKA (1995) for $m = 1$ and CHIANG, RICE, and WU (2001) and ABRAMOVICH and GRINSHTEIN (1999) for $m = 2$ using the venerable WKB method, although only the latter explicitly mentions it. The WKB method applies to the boundary value problem

$$(10.1)\qquad \begin{aligned} &(-h^2)^m\, u^{(2m)} + \omega\, u = v\ , \quad \text{on the line}\ ,\\ &u^{(k)}(x) \to 0 \quad \text{for} \quad x \to \pm\infty\ , \quad \text{for} \quad k = m, \cdots, 2m-1\ , \end{aligned}$$

and deals with the asymptotic behavior of the solution as $h \to 0$. A classic reference for the WKB method is MATHEWS and WALKER (1979). One drawback of this approach is that the boundary behavior of the Green's function is inaccessible since the boundaries are pushed out to infinity. But then the approximations are only valid away from the boundary. To some extent, § 8 gets around this in that the reproducing kernel $\mathfrak{R}_{\omega mh}$ is compared with $\mathcal{R}_{mh}$, the kernel for the uniform design density, even near the boundary. The trouble arises only if one needs the local approximate translation invariance of $\mathcal{R}_{mh}$. We noted that ABRAMOVICH and GRINSHTEIN (1999) give convincing arguments that one should weight the penalization $\| f^{(m)} \|^2$ in spline smoothing. This leads to the differential equation

$$(-h^2)^m \frac{\partial^m}{\partial x^m}\Big(\,\kappa(x)\,\frac{\partial^m}{\partial x^m}\,u(x)\Big) + \omega(x)\,u(x) = v(x)$$

together with the natural boundary conditions. If κ is smooth, the WKB method applies again.

Regarding the accuracy of the equivalent kernel approximation, the approach of NYCHKA (1995) (for piecewise linear splines) and CHIANG, RICE, and WU (2001) (for cubic splines) is used in the proof of the Interior Super Equivalent Kernel Theorem (8.5). See also COX (1981).

Ad § 3: The title "Reproducing kernel density estimation" is misleading, to put it mildly, since we only considered the variance part of the density estimator. Worse, if $X_1, X_2, \cdots, X_n$ are iid with pdf ω, then

$$\mathbb{E}\Big[\,\tfrac{1}{n}\sum_{i=1}^{n} \mathfrak{R}_{\omega mh}(X_i, t)\,\Big] = 1 \quad \text{for all } t \in [\,0\,,\,1\,]\ ,$$

so that the reproducing kernel may be better described as an anti-kernel.

Ad § 4: In random design regression problems, there are two schools of thought: Should one condition on the design when obtaining error bounds or should one not? Considering that in simple linear regression one conditions on the design in order to apply the standard normal theory, this is what we do, too. Of course, for spline smoothing with random designs, there are theoretical and practical advantages to conditioning on the design; see Chapter 22. GYÖRFI, KOHLER, KRZYŻAK, and WALK (2002) do not condition on the design.

Ad §§ 5–6: These sections follow EGGERMONT and LARICCIA (2006a) and EGGERMONT and LARICCIA (2006b).

Ad § 7: This material constitutes the analogue of "pure" convolution equations on the line and the half-line. The case of the line is a consequence of Wiener's Lemma; see, e.g., DYM and MCKEAN (1972) or Volume I, Appendix 2, § 6. The case of the half-line was a major achievement of KREIN (1958). (In fact, KREIN (1958) goes much further by describing the solvability of pure convolution equations on the half-line in terms of the *index* of the operator.) The general case of the result in this section is from EGGERMONT and LUBICH (1991), § 3. For an approach based on Banach algebras, BARNES (1987) is a starting point. However, BARNES (1987) does not use the continuity properties (7.4)(b).

Ad § 9: The value of The equivalent Nadaraya-Watson Theorem (1.30) seems to be in the surprise: Even away from the boundary, the SILVERMAN (1984) estimator is *not* equivalent in the strict sense to the smoothing spline, but the variable-bandwidth Nadaraya-Watson estimator *is*.

22

Strong Approximation and Confidence Bands

1. Introduction

In the previous chapters, we have studied various aspects of nonparametric regression estimators: Convergence rates (mean squared errors, uniform errors), computation, and smoothing parameter selection. Of course, instead of convergence rates, one would like to have accurate, nonasymptotic error bounds, or better yet (asymptotic) confidence bands for the unknown regression function and the (asymptotic) distribution of various measures of the error. Now, confidence bands may be constructed based on the asymptotic distribution theory of the uniform error, but in this chapter we mostly use the distribution theory to justify the simulation of the distribution of the maximal error. The development is based on smoothing spline estimators of order $2m$ for the general regression problem (21.1.2)–(21.1.6) using a random smoothing parameter. However, the treatment extends to kernel estimators and Nadaraya-Watson estimators. Ordinarily, the reliance on smoothing spline estimators would only be explained by the authors' "Splines 'R Us" attitude, but actually, as discussed below, some real benefits accrue from this choice.

To set the stage, consider the general regression problem (21.1.2)–(21.1.6), written here in the signal-plus-noise form

$$Y_i = f_o(X_i) + D_i \ , \quad i = 1, 2, \cdots, n \ . \tag{1.1}$$

Let f^{nH} be the smoothing spline estimator of order $2m$ with the random smoothing parameter H. We are interested in constructing confidence bands for the unknown regression function, which likely will involve the asymptotic distribution theory of the uniform or weighted uniform error of the smoothing spline. The first question to be decided is whether the confidence bands should have constant width or not. In the constant-width case, the width is likely to be determined by the worst point, i.e., by the point where the variance of f^{nH} is largest, and so is likely to be too wide where the variance is (much) smaller. Thus, it seems reasonable to let the

P.P.B. Eggermont, V.N. LaRiccia, *Maximum Penalized Likelihood Estimation*, Springer Series in Statistics, DOI 10.1007/b12285_11, shape of the confidence bands be determined by the (estimated) variance. Hence, we take the asymptotic pivot value to be

$$\max_{t\in[0,1]} \frac{|\, f^{nH}(t) - f_o(t)\,|}{\sqrt{\mathbb{V}\mathrm{ar}[\, f^{nh}(t)\,]\big|_{h=H}}} .$$

Here, the variance $\mathbb{V}\mathrm{ar}[\, f^{nh}(t)\,]\big|_{h=H}$ is computed as $\mathbb{V}\mathrm{ar}[\, f^{nh}(t)\,]$ for deterministic smoothing parameter h, after which it is evaluated at the random value H of h. We refrain from taking the additional precautions of BERAN (1988) to guarantee that the resulting pointwise confidence intervals all have the same confidence levels.

In view of the previous chapter, we modify the above by conditioning on the design,

$$M_{0,n} = \max_{t\in[0,1]} \frac{|\, f^{nH}(t) - f_o(t)\,|}{\sqrt{\mathbb{V}\mathrm{ar}[\, f^{nh}(t) \,|\, \mathbb{X}_n\,]\big|_{h=H}}} . \tag{1.2}$$

Note that $\mathbb{V}\mathrm{ar}[\, f^{nh}(t) \,|\, \mathbb{X}_n\,]\big|_{h=H}$ may be obtained from the smoothing spline computation (although we only considered $t = X_i$, $i = 1, 2, \cdots, n$), and a suitable estimator of the variance function $\sigma^2(t)$, of (21.1.7).

For $0 < \alpha < 1$, a $100(1-\alpha)\%$ confidence band for the unknown regression function f_o is then given by

$$f^{nH}(t) \pm C_{\alpha,n,H} \sqrt{\mathbb{V}\mathrm{ar}[\, f^{nh}(t) \,|\, \mathbb{X}_n\,]\big|_{h=H}} , \qquad t \in [\,0\,,1\,] , \tag{1.3}$$

where the "constant" $C_{\alpha,n,H}$ satisfies

$$\mathbb{P}[\, M_{0,n} > C_{\alpha;n,H}\,] = \alpha . \tag{1.4}$$

Asymptotically, the value of $C_{\alpha,n,H}$ may be determined from the asymptotic distribution of $M_{0,n}$ (if it is known). In the uniform random-design case, we show, following KONAKOV and PITERBARG (1984), that for a suitable sequence $\{\,\ell_n\,\}_n$, for all x,

$$\mathbb{P}[\, \ell_n\,(\, M_{0,n} - \ell_n\,) < x\,] \longrightarrow \exp(-2\mathrm{e}^{-x}) , \tag{1.5}$$

under suitable regularity conditions on f_o, the variance function σ^2, and the smoothing parameter H. The condition on H (which depends on n) is that there exists a deterministic sequence $\{\,\gamma_n\,\}_n$ such that

$$\frac{H}{\gamma_n} - 1 = o_P\big(\,(\log n)^{-1}\big) \quad \text{and} \quad \gamma_n \asymp n^{-1/(2m+1)} . \tag{1.6}$$

In (1.5), we are tempted to say that

$$\ell_n = \sqrt{2\,\log(2\pi/(\Lambda H)\,)} \tag{1.7}$$

for a suitable, computable constant Λ, but in fact we prove it with H replaced by γ_n.

The difficulty with the asymptotic distribution approach is that the convergence in (1.5) is really slow. (In fact, for deterministic h, KONAKOV and PITERBARG (1984) provide a fix, but that spoils our buildup.) In § 5, we discuss the asymptotic distribution result. There, as is usual in this context, we first approximate $M_{0,n}$ by the maxima of a family of Gaussian processes, $\{ \mathcal{G}_{\omega,\gamma}(t) \}_\gamma$, indexed by the smoothing parameter $\gamma \equiv \gamma_n$ and the design density ω. This "convergence" to Gaussian processes is quite fast. However, the convergence of the distribution of

$$M_{\omega,\gamma} = \max_{t \in [0,1]} | \mathcal{G}_{\omega,\gamma}(t) | \tag{1.8}$$

is really slow; see KONAKOV and PITERBARG (1984). Thus, the asymptotic value of $C_{\alpha,n,H}$, obtained from (1.5) as

$$C_{\alpha,n,H} = \ell_n - \frac{1}{\ell_n} \log\Bigl(\frac{1}{2} \log \frac{1}{1-\alpha} \Bigr) , \tag{1.9}$$

must be validated for small sample sizes. Adding to the problem is our inability to find the analogue of (1.5) for general quasi-uniform designs.

However, we may obtain (more) accurate approximations of $C_{\alpha,n,H}$ even in the nonuniform case by simulating the distribution of $M_{\omega,\gamma}$ for the particular γ in question, viz. $\gamma = H$. This approach is justified in § 2 (for iid noise, independent of the design) and in § 5 (for the general regression model). Key is that the asymptotic distribution result (1.5) implies that $M_{0,n}$ may be approximated with accuracy $o\bigl((\log n)^{-1} \bigr)$ without affecting the asymptotic distribution. Now, using the approximation results of KOMLÓS, MAJOR, and TUSNÁDY (1976) and SAKHANENKO (1985), conditional on the design, we construct independent normal random variables $Z_1, Z_2, \cdots, Z_n$, independent of the smoothing parameter H, with $\mathbb{V}\mathrm{ar}[Z_i \,|\, \mathbb{X}_i] = \mathbb{V}\mathrm{ar}[D_i \,|\, \mathbb{X}_i]$ such that the smoothing spline estimator φ^{nH} in the model

$$Y_i = f_o(X_i) + Z_i \ , \quad i = 1, 2, \cdots, n \ , \tag{1.10}$$

(with the same random H) satisfies

$$f^{nH}(t) - \mathbb{E}[f^{nh} \,|\, \mathbb{X}_n] = \varphi^{nH}(t) - \mathbb{E}[\varphi^{nh} \,|\, \mathbb{X}_n] + \varepsilon_n \tag{1.11}$$

in distribution, with $\| \varepsilon_n \|_\infty = o\bigl((\log n)^{-1} \bigr)$ in probability (but the rate on $\| \varepsilon_n \|_\infty$ is typically much better). Note that $\varphi^{nH}(t) - \mathbb{E}[\varphi^{nh} \,|\, \mathbb{X}_n]$ is in fact the smoothing spline estimator for the pure-noise model

$$Y_i = Z_i \ , \quad i = 1, 2, \cdots, n \ ,$$

and so does not contain the unknown regression function, but the variances $\mathbb{V}\mathrm{ar}[D_i \,|\, X_i]$ must be estimated. Thus, ignoring the difference between $\mathbb{E}[\varphi^{nh} \,|\, \mathbb{X}_n]$ and f_o for now, we may simulate the distribution of $M_{0,n}$.

In the above, we stumbled on the bias versus variance issue. It is rather amazing that after imposing some added smoothness on the regression function f_o, the variance function $\sigma^2(t)$, and the design density $\omega(t)$, one may indeed ignore the conditional bias $\mathbb{E}[\varphi^{nh} \,|\, \mathbb{X}_n] - f_o(t)$. To see why, note that Corollary (21.8.13) implies that

$$(1.12) \quad \max_{t \in \mathfrak{P}_h(k)} \left| \mathbb{E}[\varphi^{nh}(t) \,|\, \mathbb{X}_n] - f_o(t) \right| = \mathcal{O}\big(h^{2m} + h^{-1/2}(nh)^{-1} \log n \big)$$

almost surely, uniformly over h in a wide range, where, as in (21.1.27),

$$(1.13) \qquad \mathfrak{P}_h(k) = \big[\, k\, h \log(1/h) \,,\, 1 - k\, h \log(1/h) \,\big]$$

for a large enough constant k. For our value of h (H or γ), we have

$$h^{2m} + h^{-1/2}(nh)^{-1} \log n \ll h^m + \big((nh)^{-1} \log n \big)^{1/2} \asymp \big((nh)^{-1} \log n \big)^{1/2} ,$$

so the pointwise bias is much smaller than the pointwise standard deviation of the estimator. Certainly, then the same is true for the uniform versions. While in the boundary region $[\,0\,,\,1\,] \setminus \mathfrak{P}_h(k)$, the bias and standard deviation are roughly of the same size, the maximum of the noise component of the estimator over $[\,0\,,\,1\,]$ occurs on $\mathfrak{P}_h(k)$ with probability tending to 1, so what happens on $[\,0\,,\,1\,] \setminus \mathfrak{P}_h(k)$ is mostly harmless. This is the advantage of using smoothing spline estimators alluded to before. For other estimators, such as the Nadaraya-Watson estimators studied by KONAKOV and PITERBARG (1984), one must find a way to undersmooth to make the bias disappear. Although theoretically the boundary region causes no problems, in practice one must employ boundary corrections as, e.g., in EUBANK and SPECKMAN (1993), or just be satisfied with confidence bands away from the boundary set.

In the remainder of this chapter, we go into the details of the outline above. In § 2, we consider the simplest possible case, where the noise is iid and independent of the design. We approximate the noise by normal noise using the KOMLÓS, MAJOR, and TUSNÁDY (1976) construction. We also deal with the random smoothing parameter. This leads to a simulation procedure for approximating the distribution of the noise component of the smoothing spline estimator. In § 3, we construct the actual confidence bands based on the simulation procedure. In effect, we spell out the effect of the (conditional) bias of the spline estimator. In § 4, we repeat the constructions of § 2 for the general regression model. The main difference is that the KOMLÓS, MAJOR, and TUSNÁDY (1976) result is replaced by the result of SAKHANENKO (1985). In §§ 5 and 6, we consider half of the asymptotic distribution theory: We show the convergence to the family of Gaussian processes for the general quasi-uniform design. Unfortunately, only for the uniform design do we get a nice stationary Gaussian process for which the asymptotic distribution theory is known. (For arbitrary Gaussian processes with almost surely continuous sample paths, the asymptotic distribution theory is not known.) In § 7, we consider something completely different: the (asymptotically) 100% confidence intervals of DEHEUVELS and MASON (2007) for Nadaraya-Watson estimators. (Of course, we concentrate on smoothing splines.) The development of §§ 6 and 7 allows us to come to terms with this fairly quickly.

2. Normal approximation of iid noise

In this section, we study the normal approximation of the noise component of smoothing spline estimators in the random-design regression problem, restricted to iid noise, independent of the design, and with a random (data-driven) smoothing parameter. In this setting, the ideas go back to STADTMÜLLER (1986), EUBANK and SPECKMAN (1993), and DIEBOLT (1993), but the implementation by way of the equivalent kernel approximation and the accompanying reproducing kernel Hilbert space trick make life really easy. See also WANG and YANG (2006) for the case of the spline sieve (§ 15.5). We note that, in combination with the strategy of conditioning on the design, the setting is just about equivalent to that of deterministic designs. The goal is to be able to simulate the maximum of the noise component of the smoothing spline. The construction of confidence bands for the regression function, which also involves the bias of the smoothing spline, is undertaken in § 3.

As said, consider the model

$$Y_i = f_o(X_i) + D_i \,, \quad i = 1, 2, \cdots \,. \tag{2.1}$$

For the design, assume that

$$\begin{aligned}&X_1, X_2, \cdots \text{ are independent and identically distributed,}\\ &\text{having a probability density function } \omega \text{ with respect to}\\ &\text{Lebesgue measure on } (0,1)\,,\end{aligned} \tag{2.2}$$

and that, for positive constants ω_1 and ω_2,

$$\omega_1 \leqslant \omega(t) \leqslant \omega_2 \quad \text{for all} \quad t \in [0,1]\,. \tag{2.3}$$

Regarding the noise, we require that

$$\begin{aligned}&D_1, D_2, \cdots \text{ are iid, independent of the design,}\\ &\text{with} \quad \mathbb{E}[D_1] = 0 \quad \text{and} \quad \mathbb{E}[\,|D_1|^4\,] < \infty\,.\\ &\text{We let} \quad \sigma^2 \stackrel{\text{def}}{=} \mathbb{E}[\,|D_1|^2\,]\,.\end{aligned} \tag{2.4}$$

Assuming a finite fourth moment does not seem like a severe constraint. In § 5, we discuss the case where the noise does depend on the design. Note that conditions on the regression function are not required for studying the random component of the estimator.

The normal approximation depends on the following strong approximation result of KOMLÓS, MAJOR, and TUSNÁDY (1976); see CSÖRGŐ and RÉVÉSZ (1981), Theorem 2.6.7.

(2.5) THEOREM (KOMLÓS, MAJOR, and TUSNÁDY, 1976). *Let $\kappa > 2$. Let $M(x)$, $x \geqslant 0$, be a positive continuous function such that $x^{-\kappa} M(x)$ is nondecreasing and $x^{-1} \log M(x)$ is nonincreasing. If D_1, D_2 $\cdots$ are iid* *mean zero random variables with* $\mathbb{E}[\,M(|\,D_1\,|)\,] < \infty$, *then there exist iid normally distributed random variables* $Z_1, Z_2, \cdots$ *with* $\mathbb{E}[\,Z_1\,] = 0$ *and* $\mathbb{V}\mathrm{ar}[\,Z_1\,] = \mathbb{V}\mathrm{ar}[\,D_1\,]$ *such that*

$$\mathbb{P}\left[\max_{1\leqslant i\leqslant n}\Bigl|\sum_{j=1}^{i}\{\,D_j - Z_j\,\}\Bigr| > t_n\right] \leqslant \frac{c\,n}{M(a\,t_n)}$$

for any t_n *satisfying* $M^{\mathrm{inv}}(n) \leqslant t_n \leqslant c_1\,(\,n\log n)^{1/2}$.

Here, the constants a, c, *and* c_1 *depend only on the distribution of* D_1.

Needless to say, we shall not prove this. Taking $M(x) = x^4$, we have that for noise satisfying (2.4), for all $t > 0$,

$$\mathbb{P}\left[\max_{1\leqslant i\leqslant n}\Bigl|\sum_{j=1}^{i}\{\,D_j - Z_j\,\}\Bigr| > (\,n\,t\,)^{1/4}\right] \leqslant \frac{k}{t}\,, \tag{2.6}$$

where k is a constant depending only on the distribution of D_1. Thus, we have the in-probability behavior

$$\max_{1\leqslant i\leqslant n}\Bigl|\sum_{j=1}^{i}\{\,D_j - Z_j\,\}\Bigr| = \mathcal{O}_P\bigl(\,n^{1/4}\,\bigr)\,. \tag{2.7}$$

In fact, it is easy to show that it is even $o\bigl(\,n^{1/4}\,\bigr)$ in probability since a finite fourth moment implies the finiteness of a slightly larger moment, as in the Improved Truncation Lemma (14.5.23). See Csörgő and Révész (1981).

The application of (2.7) to the normal approximation of our old acquaintance the random sum

$$\mathfrak{S}_{\omega nh}(x) = \tfrac{1}{n}\sum_{i=1}^{n} D_i\,\mathfrak{R}_{\omega mh}(X_i, x)\,, \quad x \in [\,0\,,1\,]\,, \tag{2.8}$$

is immediate. For iid normals $Z_1, Z_2, \cdots$, let

$$\mathfrak{N}_{nh}(\,t\,) = \tfrac{1}{n}\sum_{i=1}^{n} Z_i\,\mathfrak{R}_{\omega mh}(X_i, t\,)\,, \quad t \in [\,0\,,1\,]\,. \tag{2.9}$$

(2.10) Theorem. *Assuming* (2.2) *through* (2.4), *we may construct iid normals* $Z_1, Z_2, \cdots$, *with* $\mathbb{V}\mathrm{ar}[\,Z_1\,] = \mathbb{V}\mathrm{ar}[\,D_1\,]$, *such that*

$$\mathfrak{S}_{\omega nh} = \mathfrak{N}_{nh} + \varepsilon_{nh} \quad \textit{in distribution}\,,$$

with $\|\,\varepsilon_{nh}\,\|_\infty = \mathcal{O}\bigl(\,n^{-3/4}\,h^{-1}\,\bigr)$ *in probability, uniformly in* h, $0 < h \leqslant 1$.

Proof. At a crucial point in the proof, we have to rearrange the random design in increasing order. So, let $X_{1,n} \leqslant X_{2,n} \leqslant \cdots \leqslant X_{n,n}$ be the order statistics of the (finite) design $X_1, X_2, \cdots, X_n$, and let $D_{1,n}, D_{2,n}, \cdots, D_{n,n}$ and $Z_{1,n}, Z_{2,n}, \cdots, Z_{n,n}$ be the induced rearrangements of $D_1, D_2, \cdots, D_n$ and $Z_1, Z_2, \cdots, Z_n$.

Then, by (2.4),

$$\mathfrak{S}_{\omega nh}(x) = \tfrac{1}{n} \sum_{i=1}^{n} D_{in}\, \mathfrak{R}_{\omega mh}(X_{in}, x) =_{\mathrm{d}} \tfrac{1}{n} \sum_{i=1}^{n} D_i\, \mathfrak{R}_{\omega mh}(X_{in}, x) \; ,$$

and likewise

$$\mathfrak{N}_{nh}(x) =_{\mathrm{d}} \tfrac{1}{n} \sum_{i=1}^{n} Z_i\, \mathfrak{R}_{\omega mh}(X_{in}, x) \; , \quad x \in [\,0\,,\,1\,] \; .$$

We move on to the actual proof. Let $S_i = \sum_{j=1}^{i} \{\, D_{jn} - Z_j \,\}$. Summation by parts gives

$$\mathfrak{S}_{\omega nh}(x) - \mathfrak{N}_{nh}(x) =_{\mathrm{d}} \tfrac{1}{n} \sum_{i=1}^{n} \{\, D_{in} - Z_i \,\}\, \mathfrak{R}_{\omega mh}(X_{in}, x)$$

$$(2.11) \qquad = \tfrac{1}{n} \sum_{i=1}^{n-1} S_i \,\{\, \mathfrak{R}_{\omega mh}(X_{in}, x) - \mathfrak{R}_{\omega mh}(X_{i+1,n}, x) \,\} + \tfrac{1}{n}\, S_n\, \mathfrak{R}_{\omega mh}(X_{n,n}, x) \; .$$

In view of the obvious bound,

$$(2.12) \qquad \sum_{i=1}^{n-1} \bigl|\, \mathfrak{R}_{\omega mh}(X_{i,n}, x) - \mathfrak{R}_{\omega mh}(X_{i+1,n}, x) \,\bigr| \leqslant |\, \mathfrak{R}_{\omega mh}(\,\cdot\,, x)\,|_{BV}$$

(see the discussion of the BV semi-norm in § 17.2), the right-hand side of (2.11) may be bounded by

$$\tfrac{1}{n} \{\, \|\, \mathfrak{R}_{\omega mh}(\,\cdot\,, x)\,\|_{\infty} + |\, \mathfrak{R}_{\omega mh}(\,\cdot\,, x)\,|_{BV} \,\} \max_{1 \leqslant i \leqslant n} |\, S_i \,| \; ,$$

which by virtue of the convolution-like properties of the reproducing kernels $\mathfrak{R}_{\omega mh}$, $h > 0$, may be further bounded as $c\,(nh)^{-1} \max_i |\, S_i \,|$. If the Z_i are constructed as per Theorem (2.5), then (2.7) clinches the deal. Q.e.d.

Continuing, Theorem (2.10) provides for the normal approximation of the smoothing spline estimator with deterministic smoothing parameter. That is, with $Z_1, Z_2, \cdots$ constructed as per Theorem (2.5), consider the regression problem

$$(2.13) \qquad \widetilde{Y}_i = f_o(X_i) + Z_i \; , \quad i = 1, 2, \cdots, n \; .$$

Let φ^{nh} be the smoothing spline estimator of order $2m$ of f_o with smoothing parameter h. Finally, we want the approximation to hold uniformly over the range of smoothing parameters $\mathcal{H}_n(\alpha)$ of (21.1.15). With $\kappa = 4$ (fourth moment), this becomes

$$(2.14) \qquad \mathcal{F}_n(\alpha) = \Bigl[\, \alpha \,\bigl(n^{-1} \log n \bigr)^{1/2} , \tfrac{1}{2} \,\Bigr] \; .$$

(2.15) Normal Approximation for Smoothing Splines. *Under the conditions* (2.1) *through* (2.4), *the spline estimators* f^{nh} *and* φ^{nh} *satisfy*

$$f^{nh} = \varphi^{nh} + r_{nh} \quad \textit{in distribution},$$

where, for all $\alpha > 0$,

$$\limsup_{n\to\infty} \sup_{h\in\mathcal{F}_n(\alpha)} \frac{\|\, r_{nh} \,\|_\infty}{h^{-1}n^{-3/4} + h^{-1/2}\,(nh)^{-1}\log n} < \infty \quad \textit{in probability}.$$

(2.16) EXERCISE. Prove the theorem. [Hint: Obviously, the Equivalent Kernel Theorem (21.1.16) comes into play. Twice!]

Again, note that the result above allows random smoothing parameters. Thus, let $h = H_n$ be a random, data-driven smoothing parameter. Assume that the random sequence $\{ H_n \}_n$ behaves like a well-behaved deterministic sequence $\{ \gamma_n \}_n$ in the sense that

$$\frac{H_n}{\gamma_n} - 1 =_{\text{p}} o\big((\log n)^{-1} \big) \quad \text{and} \quad \gamma_n \asymp n^{-1/(2m+1)} . \tag{2.17}$$

This is a mild condition, which appears to hold for the GCV choice of the smoothing parameter; see The Missing Theorem (12.7.33). Note that DEHEUVELS and MASON (2004) succeed with the $o_p(1)$ bound for an *undersmoothed* Nadaraya-Watson estimator when obtaining the precise bound on the (weighted) uniform error; see (7.2). It is clear that all that is needed to accommodate random smoothing parameters is a bound on the modulus of continuity of $\mathfrak{S}_{\omega nh}$ as a function of h. Before stating the result, it is useful to introduce the notation, for $1 \leqslant p \leqslant \infty$ (but for $p = 1$ and $p = \infty$, mostly),

$$\|\, f \,\|_{\omega,m,h,p} \stackrel{\text{def}}{=} \|\, f \,\|_{L^p((0,1),\omega)} + h^m \,\|\, f^{(m)} \,\|_p \,, \tag{2.18}$$

for those functions f for which it makes sense. Here, $\|\cdot\|_p$ denotes the $L^p(0,1)$ norm and $\|\cdot\|_{L^p((0,1),\omega)}$ the weighted L^p norm.

(2.19) RANDOM-H LEMMA. *If the design density satisfies* (2.2), *then for all random* $h = H_n$ *and deterministic* γ_n,

$$\|\, \mathfrak{S}_{nH_n} - \mathfrak{S}_{n\gamma_n} \,\|_\infty \leqslant c \left| \frac{H_n}{\gamma_n} - 1 \right| \|\, \mathfrak{S}_{n\gamma} \,\|_{\omega,m,\gamma_n,\infty} \,.$$

PROOF. We drop the subscripts in H_n and γ_n. The reproducing kernel Hilbert space trick of Lemma (21.2.11) gives that

$$\mathfrak{S}_{nH}(x) - \mathfrak{S}_{n\gamma}(x) = \big\langle\, \Delta_{H\gamma}(\,\cdot\,,x)\,,\, \mathfrak{S}_{n\gamma} \,\big\rangle_{\omega m\gamma} \,,$$

where $$\Delta_{H\gamma}(t\,,x) = \mathfrak{R}_{\omega mH}(t\,,x) - \mathfrak{R}_{\omega m\gamma}(t\,,x) \,.$$

It follows that $\|\, \mathfrak{S}_{nH} - \mathfrak{S}_{n\gamma} \,\|_\infty \leqslant \|\, \mathfrak{S}_{n\gamma} \,\|_{\omega,m,\gamma,\infty} \,\|\, \Delta_{H\gamma}(\,\cdot\,,x) \,\|_{\omega,m,\gamma,1}$. Now, Theorem (21.6.23) tells us that

$$\sup_{x\in[\,0\,,1\,]} \|\, \Delta_{H\gamma}(\,\cdot\,,x) \,\|_{\omega,m,\gamma,1} \leqslant c \left| \frac{H}{\gamma} - 1 \right| ,$$

and we are done. Q.e.d.

(2.20) COROLLARY. *Under the conditions of* (2.2) *through* (2.4) *and condition* (2.17) *on* H,

$$\| \mathfrak{S}_{nH_n} - \mathfrak{S}_{n\gamma_n} \|_\infty =_{\rm p} o\big((n\gamma \log n)^{-1/2} \big) .$$

(2.21) EXERCISE. Prove the corollary. [Hint: Theorem (21.2.17).]

The results above allow us to decouple the random smoothing parameter from the noise and set the stage for the simulation of the confidence bands and the asymptotic distribution of the uniform error.

Let Z_1', Z_2', $\cdots$ be iid normals independent of the random smoothing parameter H with $\mathbb{E}[\,Z_1'\,] = 0$ and $\mathbb{V}\mathrm{ar}[\,Z_1'\,] = \mathbb{V}\mathrm{ar}[\,D_1\,]$, and let ψ^{nH} be the smoothing spline with the smoothing parameter H in the model

$$Y_i{}' = f_o(X_i) + Z_i' \ , \quad i = 1, 2, \cdots, n \ . \tag{2.22}$$

(2.23) THEOREM. *Assume the conditions* (2.2) *through* (2.4). *Then, with* f^{nH} *the smoothing spline estimator for the problem* (2.1) *with the random smoothing parameter* H *satisfying* (2.17), *and* ψ^{nH} *the smoothing spline estimator for the problem* (2.22), *we have*

$$f^{nH} = \psi^{nH} + \eta_{nH} \quad \textit{in distribution},$$

with $\| \eta_{nH} \|_\infty = \mathcal{O}\big((n\,\gamma_n^{3/2})^{-1} \log n + n^{-3/4}\,\gamma_n^{-1} \big)$ *in probability*.

PROOF. The proof is just a sequence of appeals to various theorems proved before. First, Lemma (21.1.17) says that

$$f^{nh}(x) - \mathbb{E}[\, f^{nh} \,|\, \mathbb{X}_n \,] = \mathfrak{S}_{\omega nh}(x) + e_{nh}(x) \ , \tag{2.24}$$

where, uniformly in $h \in \mathcal{F}_n(\alpha)$,

$$\| e_{nh} \|_\infty =_{\rm as} \mathcal{O}\big(h^{-1/2}\,(nh)^{-1} \log n \big) \ . \tag{2.25}$$

Thus, since $H \in \mathcal{F}_n(\alpha)$,

$$f^{nH} - \mathbb{E}[\, f^{nh} \,|\, \mathbb{X}_n \,]_{h=H} = \mathfrak{S}_{nH}(x) + e_{nH}(x) \ , \tag{2.26}$$

and the bound (2.25) applies to $\| e_{nH} \|_\infty$.

Next, by the Random-H Lemma (2.19), or more precisely by Corollary (2.20), under the assumption (2.17),

$$\| \mathfrak{S}_{nH} - \mathfrak{S}_{n\gamma} \|_\infty =_{\rm p} o\big((n\gamma \log n)^{-1} \big) \ . \tag{2.27}$$

Then, the normal approximation of Theorem (2.10) gives

$$\mathfrak{S}_{n\gamma} =_{\rm d} \mathfrak{N}_{n\gamma} + \varepsilon_{n\gamma} \ , \tag{2.28}$$

with $\| \varepsilon_{n\gamma} \|_\infty =_{\rm p} \mathcal{O}\big(n^{-3/4} h^{-1} \big)$.

Now, we need to reverse our steps. First, however, replace the Z_i, which are not (necessarily) independent of H, with iid normals Z_i' as in (2.22), and define

(2.29) $$\mathfrak{N}'_{nh}(x) = \tfrac{1}{n} \sum_{i=1}^{n} Z_i' \,\mathfrak{R}_{\omega m h}(X_i, x) \ .$$

So then, obviously, $\mathfrak{N}_{n\gamma} = \mathfrak{N}'_{n\gamma}$ in distribution (as Gaussian processes). Then, as in (2.27), we have

$$\| \mathfrak{N}'_{n\gamma} - \mathfrak{N}'_{nH} \|_\infty = o\big((n\gamma \log n)^{-1} \big) \ .$$

Finally, with the Equivalent Kernel Theorem again, we get

$$\mathfrak{N}'_{nH} = \psi_{nH} + e'_{nH} \ ,$$

with the bound (2.25) on $\| e'_{nH} \|_\infty$.

Collecting all the error terms, the theorem follows. Q.e.d.

(2.30) REMARK. The last result implies that the distribution of

(2.31) $$f^{nH} - \mathbb{E}[\, f^{nh} \,|\, \mathbb{X}_n \,]_{h=H}$$

can be simulated by way of

(2.32) $$\psi^{nH} - \mathbb{E}[\, \psi^{nH} \,|\, \mathbb{D}_n, \mathbb{X}_n \,] \ ,$$

the smoothing spline estimator for the problem (2.22), with $f_o = 0$. Here, $\mathbb{D}_n = (D_1, D_2, \cdots, D_n)$ is the collective noise, so in (2.32) the expectation is only over Z_1', Z_2', $\cdots$, Z_n'. In the same vein, the asymptotic distribution of (continuous) functionals of (2.31) can be based on the asymptotic distribution of the corresponding functionals of (2.32). Of course, all of this presumes that the approximation errors involved are negligible. It appears that this can be decided only after the asymptotic distribution itself has been determined. Indeed, the asymptotic distribution theory of $M_{0,n}$, see (1.2), shows that uniform errors of size $o\big((nh \log n)^{-1/2} \big)$ do not change the asymptotic distribution. Thus, we conclude that errors of that size will not adversely affect the simulated distribution for the finite-sample case. For some of the details, see § 5.

EXERCISES: (2.16), (2.21).

3. Confidence bands for smoothing splines

In this section, we construct confidence bands for the unknown regression function in the simple regression problem (2.1)–(2.4). The main difficulty is dealing with the bias. Asymptotically, all is well, even in the boundary region, but simulation experiments show that there are problems for small sample sizes. So, either one must stay away from the boundary or construct boundary corrections.

To proceed, some extra smoothness on the regression function, beyond the usual $f_o \in W^{m,\infty}(0,1)$, is required. We shall assume the maximally useful smoothness condition

$$f_o \in W^{2m,\infty}(0,1) \; . \tag{3.1}$$

Let f^{nH} be the smoothing spline estimator of order $2m$ with random smoothing parameter H for the problem (2.1)–(2.4). Further, let φ^{nH} denote the smoothing spline estimator of the same order $2m$ and with the same smoothing parameter H in the problem

$$Y_i' = f_o(X_i) + Z_i \; , \quad i = 1, 2, \cdots, n \; , \tag{3.2}$$

where $Z_1, Z_2, \cdots, Z_n$ are iid normals, conditioned on the design, independent of the smoothing parameter H, with $\mathbb{V}\text{ar}[\, Z_1 \,|\, \mathbb{X}_n \,] = \mathbb{V}\text{ar}[\, D_1 \,|\, \mathbb{X}_n \,]$. (For now, we ignore that the variance must be estimated. See, e.g., § 18.7, § 18.9, and the next section.) Finally, let $\zeta^{nH} = \varphi^{nH} - \mathbb{E}[\, \varphi^{nh} \,|\, \mathbb{X}_n \,]\big|_{h=H}$. Note that the variances are all equal: For all $t \in [\, 0\, , \, 1 \,]$,

$$\mathbb{V}\text{ar}[\, f^{nh}(\,t\,) \,|\, \mathbb{X}_n \,] = \mathbb{V}\text{ar}[\, \varphi^{nh}(\,t\,) \,|\, \mathbb{X}_n \,] = \mathbb{V}\text{ar}[\, \zeta^{nh}(\,t\,) \,|\, \mathbb{X}_n \,] \; .$$

Confidence bands. The construction of the confidence bands now proceeds as follows. It is based on ζ^{nh}, with h replaced by the random H but treated as if it were deterministic. Specifically, $\zeta^{nH} = \varphi^{nH} - \mathbb{E}[\, \varphi^{nh} \,|\, \mathbb{X}_n \,]\big|_{h=H}$. Let $0 < \alpha < 1$, and determine the "constant" $C_{\alpha,n,H}$ so that

$$\mathbb{P}\left[\max_{t \in [\,0\,,\,1\,]} \frac{|\, \zeta^{nH}(\,t\,) \,|}{\sqrt{\mathbb{V}\text{ar}[\, \zeta^{nh}(\,t\,) \,|\, \mathbb{X}_n \,]\big|_{h=H}}} > C_{\alpha,n,H} \;\Big|\; \mathbb{X}_n \right] = \alpha \; . \tag{3.3}$$

(Note that the constant $C_{\alpha,n,H}$ does not depend on the value of σ^2. Thus, in the simulation of the critical values, we may take $\sigma^2 = 1$.) Of course, the maximum over $[\, 0\, , \, 1 \,]$ is determined approximately as the maximum over a fine enough grid of points in $[\, 0\, , \, 1 \,]$. Alternatively, one might just consider the maximum over the design points. Apart from this glitch, we emphasize that the critical value $C_{\alpha,n,H}$ may be obtained by simulating the distribution of the random function under scrutiny. The confidence band for f_o is then

$$f^{nH}(\,t\,) \pm C_{\alpha,n,H} \sqrt{\mathbb{V}\text{ar}[\, \zeta^{nh}(\,t\,) \,|\, \mathbb{X}_n \,]\big|_{h=H}} \; , \qquad t \in A \; , \tag{3.4}$$

where either $A = [\, 0\, , \, 1 \,]$ or $A = \mathfrak{P}_H(k)$, with

$$\mathfrak{P}_h(k) = \big[\, k\, h \log(1/h) \, , \, 1 - k\, h \log(1/h) \,\big] \; , \tag{3.5}$$

for a large enough constant k and small enough h. The strange form of the boundary region is explained in § 21.8.

(3.6) REMARK. For kernel and Nadaraya-Watson estimators with a global smoothing parameter, the boundary region is $[0, H] \cup [1-H, 1]$ *provided* the kernel has compact support in $[-1, 1]$.

The question now is whether the boundary region $[0, 1] \setminus \mathfrak{P}(\gamma)$ causes problems. This depends on the accuracy of ζ^{nH} as an approximation of the error $f^{nH} - f_o$ and thus on the size of $f^{nH} - \mathbb{E}[f^{nh} \mid \mathbb{X}_n]\big|_{h=H}$. The following theorem states that asymptotically there are no problems.

(3.7) THEOREM. *Under the assumptions* (2.1)–(2.4), (2.17), *and* (3.1),

$$(a)\quad \mathbb{P}\Big[\max_{t \in \mathfrak{P}_H(k)} \frac{|f^{nH}(t) - f_o(t)|}{\sqrt{\mathbb{V}\mathrm{ar}[f^{nh}(t) \mid \mathbb{X}_n]\big|_{h=H}}} > C_{\alpha,n,H} \;\Big|\; \mathbb{X}_n\Big] \longrightarrow_{\mathrm{p}} \alpha ,$$

$$(b)\quad \mathbb{P}\Big[\max_{t \in [0,1]} \frac{|f^{nH}(t) - f_o(t)|}{\sqrt{\mathbb{V}\mathrm{ar}[f^{nh}(t) \mid \mathbb{X}_n]\big|_{h=H}}} > C_{\alpha,n,H} \;\Big|\; \mathbb{X}_n\Big] \longrightarrow_{\mathrm{p}} \alpha .$$

So all appears well. However, what the theorem fails to tell is that the convergence in part (a) is quite fast but that the convergence in (b) appears to be slow, due to what happens in the boundary region. This is made clear by the proof.

PROOF OF THEOREM (3.7). The proof relies on results regarding the asymptotic distribution of the quantities in question.

For part (a), the material of § 2 settles most of the questions. The two remaining issues are the bias and the behavior of the variance. It follows from Lemma (6.1) that $\mathbb{V}\mathrm{ar}[f^{nh}(t) \mid \mathbb{X}_n]\big|_{h=H} \asymp (nH)^{-1}$ in probability. From Theorem (21.5.19), it follows that

$$\| f_o - \mathbb{E}[f^{nh} \mid \mathbb{X}_n]\big|_{h=H} \|_\infty = \mathcal{O}\big(H^m + H^{-1/2} (nH)^{-1} \log n\big)$$

in probability. Taking into account that $H^m \asymp n^{-m/(2m+1)} \asymp (nH)^{-1/2}$, combining the two bounds above gives us

$$\varepsilon_{n,H} \stackrel{\mathrm{def}}{=} \frac{\| f_o - \mathbb{E}[f^{nh} \mid \mathbb{X}_n]\big|_{h=H} \|_\infty}{\sqrt{\mathbb{V}\mathrm{ar}[f^{nh}(t) \mid \mathbb{X}_n]\big|_{h=H}}} = \mathcal{O}\big(H^{m-\frac{1}{2}} \log n\big)$$

in probability. Then, it follows from (3.3) that

$$\mathbb{P}\Big[\max_{t \in \mathfrak{P}_H(k)} \frac{|f^{nH}(t) - f_o(t)|}{\sqrt{\mathbb{V}\mathrm{ar}[f^{nh}(t) \mid \mathbb{X}_n]\big|_{h=H}}} > C_{\alpha,n,H} - \varepsilon_{n,H} \;\Big|\; \mathbb{X}_n\Big] = \alpha .$$

Since the asymptotic distribution theory tells us that $C_{\alpha,n,H} = \ell_n - c_\alpha / \ell_n$ for a suitable c_α, with ℓ_n given by (1.7), then

$$C_{\alpha,n,H} - \varepsilon_{nH} = C_{\alpha,n,H} - \mathcal{O}\big(H^{m-\frac{1}{2}} \log n\big) = C_{\alpha',n,H} ,$$

with $\alpha' \longrightarrow \alpha$. This proves part (a).

For part (b), in Lemma (6.11), we show that the maximum over the boundary region $[0, 1] \backslash \mathfrak{P}(\gamma)$ is only $\mathcal{O}_P(\sqrt{\log\log n})$ in probability. Also, the asymptotic distribution theory shows that the maximum over $\mathfrak{P}(\gamma)$ itself is $\mathcal{O}_P(\sqrt{\log n})$ but not $o_P(\sqrt{\log n})$. The conclusion is that the maximum over the whole interval occurs on $\mathfrak{P}(\gamma)$ in probability, and then part (a) applies. Q.e.d.

So, the message is clear: On $\mathfrak{P}(\gamma)$, the empirical confidence bands should be close to the ideal ones, even for relatively small samples, and indeed, experiments bear this out. On $[0, 1] \setminus \mathfrak{P}(\gamma)$, we are relying on a term of size $\mathcal{O}_P(\sqrt{\log\log n})$ to be negligible compared with an $\mathcal{O}_P(\sqrt{\log n})$ term. While asymptotically this holds true, in the small-sample case the unmentioned constants mess things up.

Boundary corrections. It would be interesting to see whether boundary corrections salvage the situation in the boundary region. Again, asymptotically they do, and in a much less precarious way. After applying boundary correction, the conditional bias is essentially $\mathcal{O}_P(H^{2m})$ throughout the interval, including the boundary region. The only complications are that the variance in the boundary region increases and that the normal approximation of the noise must be justified for the boundary corrections. However, since the boundary correction, as described by the Bias Reduction Principle of EUBANK and SPECKMAN (1990b), see (13.5.4), is finite-dimensional, that *should* not cause problems. However, getting boundary corrections to work in practice is a nontrivial matter; see HUANG (2001). For overall bias corrections, see § 23.5.

(3.8) EXERCISE. Work out the confidence bands for the boundary corrections, and tell the authors about it!

EXERCISE: (3.8).

4. Normal approximation in the general case

We now study the normal approximation of the smoothing spline estimator in the general regression problem (21.1.1)–(21.1.6), culminating in Theorem (4.25), which allows us to simulate the confidence bands in question. Again, conditions on the regression function are not required for studying the random component of the estimator.

The ordered noise in the general case. Let us reexamine the noise in the general regression problem (21.1.1)–(21.1.6),

(4.1) $$Y = f_o(X) + D \ .$$

Since D and X are not (necessarily) independent, the distribution of D conditioned on X is a function of X. We denote it in standard fashion as $F_{D|X}(d\,|\,x)$. For fixed $x \in \mathbb{R}$, let $Q_{D|X}(d\,|\,x)$ be its inverse. So,

$$Q_{D|X}\big(F_{D|X}(d\,|\,x)\;\big|\;x\big) = d \quad \text{for all relevant } d \text{ and } x\ .$$

Then, with $U \sim \text{Uniform}(0,1)$, and U independent of X, we have

$$(4.2) \qquad D\,|\,X = Q_{D|X}(U\,|\,X) \quad \text{in distribution}.$$

As in the case where the noise was iid, independent of the design, we need to rearrange the design in increasing order. Thus, the iid data

$$(D_1, X_1),\ (D_2, X_2),\ \cdots,\ (D_n, X_n)$$

are rearranged as $(D_{1,n}, X_{1,n}),\ (D_{2,n}, X_{2,n}),\ \cdots,\ (D_{n,n}, X_{n,n})$ with the order statistics $X_{1,n} < X_{2,n} < \cdots < X_{n,n}$. Now, the data are no longer iid, but, conditioned on the design, the noise is still independent. An easy way to see this is that the distribution of $D_{i,n}$ conditioned on $X_{i,n}{=}x$ is still given by $F_{D|X}(d\,|\,x)$. In other words, it just depends on the *value* of the i-th order statistic and not on it *being* the i-th order statistic.

The normal approximation in the general case. We now are in need of the approximation

$$D_i\,|\,X_i \approx \sigma(X_i)\,Z_i\ ,$$

where, conditioned on the design, $Z_1, Z_2, \cdots$ are independent standard normal random variables and $\sigma^2(X) = \mathbb{V}\text{ar}[\,D\,|\,X\,]$. So, in effect, there are two flies in the ointment: The noise is independent but not identically distributed, and the variance of the noise is not constant. Both of them are taken care of by the following approximation result of SAKHANENKO (1985). See the introduction of ZAITSEV (2002) or SHAO (1995), Theorem B. Obviously, the theorem is well beyond the scope of this text.

(4.3) THEOREM (SAKHANENKO, 1985). *Let $p > 2$, and let $\{\,D_n\,\}_n$ be a sequence of independent, zero-mean random variables with*

$$M_{n,p} \overset{\text{def}}{=} \tfrac{1}{n} \sum_{i=1}^{n} \mathbb{E}[\,|\,D_i\,|^p\,] < \infty\ , \qquad n = 1, 2, \cdots\ .$$

Then, there exists a sequence of independent standard normal random variables $\{\,Z_n\,\}_n$ such that

$$\mathbb{P}\left[\ \max_{1\leqslant i\leqslant n} \left|\ \sum_{j=1}^{i} \{\,D_j - \sigma_j\,Z_j\,\}\right| > t\,n^{1/p}\ \right] \leqslant (A\,p)^p\,t^{-p}\,M_{n,p}\ ,$$

where $\sigma_j^2 = \mathbb{V}\text{ar}[\,D_j\,]$. Here, A is an absolute constant.

For $p = 4$, the theorem implies the in-probability bound,

$$\max_{1 \leqslant i \leqslant n} \Bigl| \sum_{j=1}^{i} \{ D_j - \sigma_j Z_j \} \Bigr| =_{\rm p} \mathcal{O}(n^{1/4}) , \tag{4.4}$$

analogous to (2.7). It even implies the bound for the triangular version, as stated in the following theorem.

(4.5) THEOREM. *Let $D_{i,n}$, $i = 1, 2, \cdots, n$, be a triangular array or row-wise independent random variables with*

$$\sup_{n \geqslant 1} \tfrac{1}{n} \sum_{i=1}^{n} \mathbb{E}[\, | D_{i,n} |^4 \,] < \infty .$$

Then, there exist standard normal random variables $Z_{i,n}$, $i = 1, 2, \cdots, n$, row-wise independent, such that

$$\max_{1 \leqslant i \leqslant n} \Bigl| \sum_{j=1}^{i} \{ D_{j,n} - \sigma_{j,n} Z_{j,n} \} \Bigr| =_{\rm p} \mathcal{O}(n^{1/4}) ,$$

where $\sigma_{j,n}^2 = \mathbb{V}\mathrm{ar}[\, D_{j,n} \,]$.

The application of Theorem (4.5) to the normal approximation of $\mathfrak{S}_{\omega nh}$ is again immediate. In fact, Theorem (2.10) applies almost as is since it only relies on the bound (2.7) resp. the bound of Theorem (4.5). Thus, for independent standard normals Z_1, Z_2, $\cdots$, define the random sum

$$\mathfrak{N}_{\sigma,nh}(t) = \tfrac{1}{n} \sum_{i=1}^{n} Z_i \, \sigma(X_i) \, \mathfrak{R}_{\omega mh}(X_i, t) , \quad t \in [\, 0 , 1 \,] . \tag{4.6}$$

(4.7) THEOREM. *Assuming* (21.1.2) *through* (21.1.6), *conditional on the design $\mathbb{X}_n$, there exist iid standard normals Z_1, Z_2, $\cdots$, such that*

$$\mathfrak{S}_{\omega nh} = \mathfrak{N}_{\sigma,nh} + \varepsilon_{nh} \quad \textit{in distribution} ,$$

with $\| \varepsilon_{nh} \|_\infty = \mathcal{O}(n^{-3/4} h^{-1})$ in probability, uniformly in h, $0 < h \leqslant 1$.

(4.8) EXERCISE. Prove the theorem. [Hint: See the construction of Theorem (2.10).]

We may apply Theorem (4.7) to the normal approximation of the smoothing spline estimator with deterministic smoothing parameter. Thus, with Z_1, Z_2, $\cdots$ constructed as per Theorem (4.5), consider the model

$$\widetilde{Y}_i = f_o(X_i) + \sigma(X_i) \, Z_i , \quad i = 1, 2, \cdots, n . \tag{4.9}$$

Let φ^{nh} be the smoothing spline estimator of order $2m$ of f_o with smoothing parameter h. Recall the definition (2.14) of $\mathcal{F}_n(\alpha)$.

(4.10) NORMAL APPROXIMATION FOR SMOOTHING SPLINES. *Under the conditions* (21.1.2) *through* (21.1.6), *the spline estimators f^{nh} and φ^{nh}*

satisfy

$$f^{nh} = \varphi^{nh} + r_{nh} \quad \textit{in distribution},$$

where, for all $\alpha > 0$,

$$\limsup_{n\to\infty} \sup_{h\in\mathcal{F}_n(\alpha)} \frac{\| r_{nh} \|_\infty}{h^{-1}n^{-3/4} + h^{-3/2}n^{-1}\log n} < \infty \quad \textit{in probability}.$$

(4.11) EXERCISE. Prove the theorem.

Random smoothing parameters may also be treated as before. Consider the random, data-driven smoothing parameter $h = H_n$. Assume that the random sequence $\{ H_n \}_n$ behaves like a deterministic sequence $\{ \gamma_n \}_n$ that itself behaves as expected in the sense that

$$\frac{H_n}{\gamma_n} - 1 =_{\mathrm{p}} o\big((\log n)^{-1} \big) \tag{4.12}$$

with

$$\gamma_n \asymp n^{-1/(2m+1)} . \tag{4.13}$$

To handle the random smoothing parameter, a bound on the modulus of continuity of $\mathfrak{S}_{\omega nh}$ as a function of h is needed. This we already did in Lemma (2.19).

(4.14) THEOREM. *Under the assumptions* (2.2) *through* (2.4), *for random* H *satisfying* (4.12)–(4.13),

$$\mathfrak{N}_{\sigma,nH} = \mathfrak{N}_{\sigma,n,\gamma_n} + \eta_{nH} \,,$$

with $\| \eta_{nH} \|_\infty = \mathcal{O}\big((n\gamma_n^{3/2})^{-1}\log n + n^{-3/4}\gamma_n^{-1} \big)$ *in probability*.

(4.15) REMARK. Without too much loss of information, we may summarize the bound on η_{nH} as

$$\| \eta_{nH} \|_\infty = \mathcal{O}\big(\big(n^{-m/(2m+1)} \big)^r \big)$$

with $r = \frac{3}{2} - 1/(4m)$. Since then $\frac{5}{4} \leqslant r < \frac{3}{2}$, the error is negligible compared with the size of $\| \mathfrak{N}_{\sigma,nH} \|_\infty$.

(4.16) COROLLARY. *For each* n, *let* $Z'_{1,n}, Z'_{2,n}, \cdots, Z'_{n,n}$ *be independent standard normal random variables independent of the noise* $\mathbb{D}_n$. *Then, under the assumptions of Theorem* (4.14),

$$\mathfrak{N}_{\sigma,nH} = \tfrac{1}{n} \sum_{i=1}^{n} Z'_{in}\, \sigma(X_{i,n})\, \mathfrak{R}_{\omega nH}(X_{i,n}, \cdot\,) + \eta_{nH}$$

in distribution, with

$$\| \eta_{nH} \|_\infty = \mathcal{O}\big((n\gamma_n^{3/2})^{-1}\log n + n^{-3/4}\gamma_n^{-1} \big) \quad \textit{in probability}.$$

(4.17) EXERCISE. Prove the theorem and corollary.

The final issue is the presence of the unknown variance function in the model (4.9). It is clear that it must be estimated and so some regularity of $\sigma^2(t)$ is required. We assume that

$$\sigma \in W^{1,\infty}(0,1) \quad \text{and} \tag{4.18}$$

$$\sigma \quad \text{is quasi-uniform in the sense of (21.1.4)}\,. \tag{4.19}$$

So, σ is Lipschitz continuous and bounded and bounded away from 0. Then, $\sigma^2 \in W^{1,\infty}(0,1)$ as well, and vice versa. If in addition

$$\sup_{x\in[\,0\,,\,1\,]} \mathbb{E}[\,D^6 \,|\, X = x\,] < \infty\;, \tag{4.20}$$

then one should be able to estimate σ^2 by an estimator s_{nH}^2 such that

$$\|\,\sigma^2 - s_{nH}^2\,\|_{\omega,1,\gamma_n} = \mathcal{O}\Bigl(\sqrt{\gamma_n^2 + (n\,\gamma_n)^{-1}\log n}\,\Bigr)\;,$$

which by (4.13) amounts to

$$\|\,\sigma^2 - s_{nH}^2\,\|_{\omega,1,\gamma_n} = \mathcal{O}\bigl(\,n^{-1/(2m+1)}\,\bigr) \quad \text{in probability}\,. \tag{4.21}$$

This is discussed below, but first we explore the consequences of (4.21).

It seems reasonable to extend the definition (4.6) to the case where σ is replaced with s_{nH},

$$\mathfrak{N}_{s,nh}(\,t\,) \stackrel{\text{def}}{=} \tfrac{1}{n}\sum_{i=1}^{n} Z_i\, s_{nH}(X_i)\,\mathfrak{R}_{\omega mh}(X_i,\,t\,)\;, \quad t\in[\,0\,,\,1\,]\,. \tag{4.22}$$

(4.23) THEOREM. *Under the assumptions* (2.2), (2.3), *and* (4.21), *with* γ_n *satisfying* (4.13),

$$\|\,\mathfrak{N}_{\sigma,n\gamma_n} - \mathfrak{N}_{s,n\gamma_n}\,\|_\infty = \mathcal{O}\bigl(\,\sqrt{n^{-1}\log n}\,\bigr) \quad \textit{in probability}.$$

PROOF. We abbreviate γ_n as just γ. Let $\varepsilon \equiv \sigma - s_{nH}$. Then, the object of interest is $\mathfrak{N}_{\varepsilon,nh}$. The reproducing kernel Hilbert space trick gives

$$|\,\mathfrak{N}_{\varepsilon,n\gamma}(\,t\,)\,| \leqslant \|\,\tfrac{1}{n}\sum_{i=1}^{n} Z_i\,\mathfrak{R}_{\omega,1,\gamma}(X_i,\,\cdot\,)\,\|_{\omega,1,\gamma,\infty}\;\|\,\varepsilon(\,\cdot\,)\,\mathfrak{R}_{\omega m\gamma}(\,\cdot\,,\,t\,)\,\|_{\omega,1,\gamma,1}\,.$$

Now, as always,

$$\|\,\tfrac{1}{n}\sum_{i=1}^{n} Z_i\,\mathfrak{R}_{\omega,1,\gamma}(X_i,\,\cdot\,)\,\|_{\omega,1,h,\infty} = \mathcal{O}\bigl(\,\sqrt{(n\gamma)^{-1}\log n}\,\bigr) \quad \text{almost surely}\,.$$

Also, by the Multiplication Lemma (21.2.9),

$$\|\,\varepsilon(\,\cdot\,)\,\mathfrak{R}_{\omega m\gamma}(X_i,\,\cdot\,)\,\|_{\omega,1,\gamma,1} \leqslant c\,\|\,\varepsilon\,\|_{\omega,1,\gamma}\,\|\,\mathfrak{R}_{\omega m\gamma}(\,\cdot\,,\,t\,)\,\|_{\omega,1,\gamma}\;,$$

and since by Corollary (4.46) below the condition (4.21) implies the same bound on $\|\,\sigma - s_{nH}\,\|_{\omega,1,\gamma}$, then

$$\|\,\varepsilon(\,\cdot\,)\,\mathfrak{R}_{\omega m\gamma}(X_i,\,\cdot\,)\,\|_{\omega,1,\gamma,1} = \mathcal{O}\bigl(\,\sqrt{\gamma + (n\gamma^2)^{-1}\log n}\,\bigr)\,.$$

Now, the product of the two bounds behaves as advertised since, by (4.13), $\gamma \asymp n^{-1/(2m+1)}$. Q.e.d.

Apart from the variance function estimation, we have reached the goal of this section. Recall that f^{nH} is the smoothing spline estimator or order $2m$ of f_o in the original regression problem (21.2.1) with random smoothing parameter H. Assume that

(4.24) φ^{nH} is the smoothing spline estimator of order $2m$ with the random smoothing parameter H for the problem

$$Y_i' = s_{nH}(X_i)\, Z_i' \ , \quad i = 1, 2, \cdots, n \ .$$

(4.25) THEOREM. *Under the assumptions* (21.2.1)–(21.2.4), (4.12)–(4.13), *and* (4.21), *we may construct independent standard normal random variables* $Z_1', Z_2', \cdots, Z_n'$, *independent of the noise* $\mathbb{D}_n$ *in* (21.2.1), *such that*

$$f^{nH} - \mathbb{E}[\, f^{nh} \,|\, \mathbb{X}_n \,]\big|_{h=H} = \varphi^{nH} + \varepsilon_{nH} \quad \textit{in distribution},$$

where

$$\|\, \varepsilon_{nH} \,\|_\infty = \mathcal{O}\big(\, n^{-3/4+1/(2m+1)} + n^{-1/2} \log n \,\big) \quad \textit{in probability}.$$

(4.26) REMARK. Note that the distribution of $\|\, \varphi^{nH} \,\|_\infty$ is amenable to simulation since everything is known. Also note that the bound on ε_{nH} vanishes compared with the optimal accuracy of the smoothing spline estimator, $\mathcal{O}\big(\, n^{-m/(2m+1)} \,\big)$ (give or take a power of $\log n$), since

$$-\tfrac{3}{4} + \tfrac{1}{2m+1} < -\tfrac{m}{2m+1} \quad \text{for} \quad m > \tfrac{1}{2} \ .$$

The error is even negligible compared with the superconvergence bound, $\mathcal{O}\big(\, n^{-2m/(4m+1)} \,\big)$, on the smoothing spline away from the boundary, as in § 21.8.

Estimating the variance function. We finish this section by constructing a (nonparametric) estimator of the variance function $\sigma^2(x)$, $x \in [\,0\,,\,1\,]$, for which the bound (4.21) holds. It seems clear that this must be based on the residuals of the smoothing spline estimator, but an alternative is discussed in Remark (4.48). In Exercise (4.47), the direct estimation of the standard deviation is discussed.

So, consider the data

$$V_i = |\, Y_i - f^{nH}(X_i) \,|^2 \ , \quad i = 1, 2, \cdots, n \ , \tag{4.27}$$

where f^{nH} is the smoothing spline estimator of order $2m$ for the original regression problem (21.2.1). Now, let ψ^{nH} be the solution for $h = H$ of the problem

$$\begin{aligned} &\text{minimize} \quad \tfrac{1}{n} \sum_{i=1}^{n} |\, \psi(X_i) - V_i \,|^2 + H^2 \,\|\, \psi' \,\|^2 \\ &\text{subject to} \quad \psi \in W^{1,2}(0,1) \ . \end{aligned} \tag{4.28}$$

Although one can envision the estimator ψ^{nh} for "arbitrary" smoothing parameter h, the estimator still depends on H by way of the data. The choice $h = H$ avoids a lot of issues.

(4.29) EXERCISE. Verify that ψ^{nH} is strictly positive on $[0, 1]$, so that there is no need to enforce the positivity of the variance estimator. (This only works for splines of order 2.)

(4.30) THEOREM. *Under the assumptions* (21.2.1)–(21.2.4) *and* (4.12)–(4.13),

$$\| \psi^{nH} - \sigma^2 \|_{\omega,1,\gamma} = \mathcal{O}(n^{-1/(2m+1)}) \quad \textit{in probability}.$$

PROOF. The proof appeals to most of the results on smoothing splines from Chapter 21. Recall the assumption (4.12)–(4.13) on H and $\gamma \equiv \gamma_n$. At this late stage of the game, one readily obtains that

$$\text{(4.31)} \qquad \| \psi^{nH} - \sigma^2 \|_{\omega,1,\gamma} \leqslant c \{ \gamma + \| \mathfrak{V}_{nH} \|_{\omega,1,\gamma} \} ,$$

where

$$\text{(4.32)} \qquad \mathfrak{V}_{nH}(t) = \tfrac{1}{n} \sum_{i=1}^{n} \{ V_i - \sigma^2(X_i) \} \mathfrak{R}_{\omega,1,H}(X_i, t) , \quad t \in [0, 1] .$$

(4.33) EXERCISE. Prove (4.31).

However, (4.32) is just the beginning of the trouble. We split $\mathfrak{V}_{nH}$ into three sums, conformal to the expansion

$$V_i - \sigma^2(X_i) = \{ D_i^2 - \sigma^2(X_i) \} + 2 D_i r_{nH}(X_i) + r_{nH}^2(X_i) ,$$

where $r_{nH} = f_o - f^{nH}$. Since

$$\text{(4.34)} \qquad \mathfrak{V}_{I,nH}(t) \stackrel{\text{def}}{=} \tfrac{1}{n} \sum_{i=1}^{n} \{ D_i^2 - \sigma^2(X_i) \} \mathfrak{R}_{\omega,1,H}(X_i, t)$$

is a "standard" random sum, we obtain

$$\text{(4.35)} \qquad \| \mathfrak{V}_{I,nH} \|_{\omega,1,\gamma} = \mathcal{O}(\sqrt{(n\gamma)^{-1} \log n}) \quad \text{in probability}.$$

Of course, this uses (4.20).

(4.36) EXERCISE. You guessed it: Prove (4.35).

For the third sum,

$$\text{(4.37)} \qquad \mathfrak{V}_{III,nH}(t) \stackrel{\text{def}}{=} \tfrac{1}{n} \sum_{i=1}^{n} r_{nH}^2(X_i) \mathfrak{R}_{\omega,1,H}(X_i, t) ,$$

straightforward bounding gives

(4.38) $$|\,\mathfrak{V}_{III,nH}(t)\,| \leqslant \|\, r_{nH} \,\|_\infty^2 \,\|\, S_{nH} \,\|_\infty \ ,$$

where $$S_{nh}(t) = \tfrac{1}{n} \sum_{i=1}^{n} |\,\mathfrak{R}_{\omega,1,h}(X_i, t)\,| \ .$$

Clearly,

(4.39) $$\|\, r_{nH} \,\|_\infty^2 = \mathcal{O}\big(\, (n\gamma)^{-1} \log n \,\big) \quad \text{in probability}\,.$$

Now,

$$S_{nh}(t) = \|\,\mathfrak{R}_{\omega,1,h}(\,\cdot\,, t\,)\,\|_{\omega,1,h} + \int_0^1 |\,\mathfrak{R}_{\omega,1,h}(x, t\,)\,| \,\{\, d\Omega_n(x) - d\Omega_o(x) \,\} \ ,$$

and Lemma (21.3.4) gives the bound

$$\int_0^1 |\,\mathfrak{R}_{\omega,1,h}(x, t\,)\,| \{\, d\Omega_n(x) - d\Omega_o(x) \,\} =$$
$$\mathcal{O}\big(\, \sqrt{(n\gamma)^{-1} \log n} \,\big) \cdot \|\, |\,\mathfrak{R}_{\omega,1,H}(\,\cdot\,, t\,)\,| \,\|_{\omega,1,\gamma,1} \ ,$$

uniformly in $h \in \mathcal{H}_n(\alpha)$. Since $\|\, |\,\mathfrak{R}_{\omega,1,\gamma}(\,\cdot\,, t\,)\,| \,\|_{\omega,1,\gamma,1} \asymp 1$, again uniformly in h, then $\|\, S_{nH} \,\|_\infty \asymp 1$ in probability, and it follows that

$$\|\, \mathfrak{V}_{III,nH} \,\|_{L^2((0,1),\omega)} = \mathcal{O}\big(\, (n\gamma)^{-1} \log n \,\big) \quad \text{in probability}\,.$$

The same bound applies to $\gamma \,\|\, (\,\mathfrak{V}_{III,nH}\,)' \,\|$, and so

(4.40) $$\|\, \mathfrak{V}_{III,nH} \,\|_{\omega,1,\gamma} = \mathcal{O}\big(\, (n\gamma)^{-1} \log n \,\big) \quad \text{in probability}\,.$$

(4.41) EXERCISE. Prove (4.39).

Finally, for

$$\mathfrak{V}_{II,nH}(t) = \tfrac{1}{n} \sum_{i=1}^{n} D_i \, r_{nH}(X_i)\, \mathfrak{R}_{\omega,1,H}(X_i, t\,) \ ,$$

the usual reproducing kernel Hilbert space trick works. This leads to

(4.42) $$|\,\mathfrak{V}_{II,nH}(t)\,| \leqslant \|\, \tfrac{1}{n} \sum_{i=1}^{n} D_i \,\mathfrak{R}_{\omega,1,\gamma}(X_i, \,\cdot\,) \,\|_{\omega,1,\gamma,\infty} \times$$
$$\|\, r_{nH}(\,\cdot\,)\, \mathfrak{R}_{\omega,1,\gamma}(\,\cdot\,, t\,) \,\|_{\omega,1,\gamma,1} \ .$$

The first factor is sufficiently covered by Theorem (14.6.12). For the second factor, the Multiplication Lemma (21.2.9) gives the bound

(4.43) $$\|\, r_{nH}(\,\cdot\,)\, \mathfrak{R}_{\omega,1,\gamma}(\,\cdot\,, t\,) \,\|_{\omega,1,\gamma,1} =_{\mathrm{p}} \mathcal{O}\big(\, \sqrt{(n\gamma^2)^{-1} \log n} \,\big) \ ,$$

so that (4.42) leads to the bound

$$\max_{t \in [\,0\,,\,1\,]} |\,\mathfrak{V}_{II,nH}(t)\,| =_{\mathrm{p}} \mathcal{O}\big(\, \gamma^{-1/2} (n\gamma)^{-1} \log n \,\big) \ .$$

The same derivation and bound applies to $\gamma\,|\,(\mathfrak{V}_{II,nH}\,)'(\,t\,)\,|$, so that

$$\|\,\mathfrak{V}_{II,nH}\,\|_{\omega,1,\gamma} = \mathcal{O}\big(\,\gamma^{-1/2}(n\gamma)^{-1}\log n\,\big)\ . \tag{4.44}$$

Combining (4.35), (4.40), and (4.44) leads to the in-probability bound

$$\|\,\mathfrak{V}_{nH}\,\|_{\omega,1,\gamma} = \mathcal{O}\big(\,\sqrt{(n\gamma)^{-1}\log n}\,\big) = \mathcal{O}\big(\,\sqrt{n^{-2/3}\log n}\,\big)\ . \tag{4.45}$$

Then, (4.31) proves the theorem. Q.e.d.

(4.46) COROLLARY. *Under the conditions of Theorem* (4.30),

$$\|\,\sqrt{\psi^{nH}} - \sigma\,\|_{\omega,1,\gamma} = \mathcal{O}\big(\,\sqrt{n^{-1/(2m+1)}\log n}\,\big) \quad \textit{in probability}.$$

PROOF. Define v by $v^2 = \psi^{nH}$ and $v \geqslant 0$. The corollary follows from

$$\|\,\sigma - v\,\| \leqslant k\,\|\,\sigma^2 - v^2\,\|\ ,$$

with $k = \max\ 1/\sigma(x)$, and $\|\,\sigma\,' - v\,'\,\| \leqslant \ell\,\|\,\sigma\,'\,v - v\,'\,v\,\|$, in which $\ell = \max\ 1/v(x)$, and then

$$\|\,\sigma\,' - v\,'\,\| \leqslant \ell\,\big(\,\|\,\sigma\,'\sigma - v\,'v\,\| + \|\,\sigma'(\,\sigma - v\,)\,\|\,\big)\ .$$

Since $\sigma\,'$ is bounded, the last term causes no problems. Also, since

$$\|\,\sigma^2 - v^2\,\|_\infty \leqslant c\,\gamma^{-1/2}\,\|\,\sigma^2 - v^2\,\|_{\omega,1,\gamma} \quad \text{in probability}\,,$$

one verifies that v is bounded away from 0 in probability. Then, the "constant" ℓ is bounded away from 0 in probability as well. The result follows. Q.e.d.

(4.47) EXERCISE. Here, we discuss another way of estimating $\sigma(\,t\,)$. The setting is as in (4.27). Consider the data

$$\mathcal{S}_i = |\,Y_i - f^{nH}(X_i)\,|\ , \quad i = 1, 2, \cdots, n\ ,$$

and consider the estimator θ^{nH} of $\sigma(x)$, defined as the solution of

$$\begin{aligned} &\text{minimize} \quad \tfrac{1}{n}\sum_{i=1}^{n}\,|\,\theta(X_i) - \mathcal{S}_i\,|^2 + H^2\,\|\,\theta\,'\,\|^2 \\ &\text{subject to} \quad \theta \in W^{1,2}(0,1)\ . \end{aligned}$$

Show that $\|\,\theta^{nH} - \sigma\,\|_{\omega,1,\gamma} = \mathcal{O}\big(\,n^{-1/(2m+1)}\big)$ in probability.

(4.48) REMARK/EXERCISE. An alternative to the use of the residuals (4.27) when estimating the variance function is to use the first- or second- (or higher) order differences

$$V_i = \frac{|\,(X_i - X_{i+1})(Y_{i-1} - Y_i) - (X_{i-1} - X_i)(Y_i - Y_{i+1})\,|^2}{(X_i - X_{i+1})^2 + (X_{i-1} - X_{i+1})^2 + (X_{i-1} - X_i)^2}\ ,$$

for $i = 2, 3, \cdots, n-1$, assuming the design has been arranged in increasing order; see, e.g., RICE (1986b) or BROWN and LEVINE (2007). One checks that if the regression function is linear, then $\mathbb{E}[\, V_i \,|\, \mathbb{X}_n \,]$ is a convex combination of the neighboring $\sigma^2(X_j)$,

$$\mathbb{E}[\, V_i \,|\, \mathbb{X}_n \,] = \sum_{j=-1}^{1} w_{i,j}\, \sigma^2(X_{i-j}) \ ,$$

for computable weights $w_{i,j}$ depending on the X_i only. For general smooth regression functions, the bias does not spoil things, or so one hopes. Of course, if the variance function is smooth, then one can consider the model

$$V_i = \sigma^2(X_i) + \nu_i \ , \quad i = 2, 3, \cdots, n-1 \ ,$$

where ν_i represents the noise (as well as the bias). Now, we may consider the usual nonparametric estimators, which leads to the usual error bounds on the estimator of $\sigma^2(\,t\,)$, conditioned on the design. For Nadaraya-Watson estimators with deterministic designs, see, e.g., BROWN and LEVINE (2007).

It seems much more interesting to consider the problem

$$V_i = \sum_{j=-1}^{1} w_{i,j}\, \sigma^2(X_{i-j}) + \nu_i \ , \quad i = 2, 3, \cdots, n-1 \ ,$$

and the resulting smoothing spline problem

$$\begin{aligned} &\text{minimize} && \tfrac{1}{n} \sum_{i=2}^{n-1} \Bigl|\, V_i - \sum_{j=-1}^{1} w_{i,j}\, f(X_{i-j}) \,\Bigr|^2 + h^{2m} \, \|\, f^{(m)} \,\|^2 \\ &\text{subject to} && f \in W^{m,2}(0,1) \ , \ f \geqslant 0 \ . \end{aligned}$$

Computationally, enforcing the nonnegativity constraint is not trivial, except in the piecewise linear case ($\, m = 1 \,$). Also, the V_i are correlated, and it is not clear if ignoring this causes problems. The authors do not know whether this actually works, either in theory or in practice. (Exercise!)

EXERCISES: (4.8), (4.11), (4.17), (4.29), (4.33), (4.36), (4.41), (4.47), (4.48).

5. Asymptotic distribution theory for uniform designs

In this section, we discuss the asymptotic distribution of the maximum error of smoothing spline estimators in the general regression problem (21.1.1)–(21.1.6). The way one goes about determining the asymptotic distribution of the maximum deviation in nonparametric function estimation is well-established practice; see, e.g., BICKEL and ROSENBLATT (1973), KONAKOV and PITERBARG (1984), STADTMÜLLER (1986), and EUBANK and SPECKMAN (1993). First, exhibit a limiting family of Gaussian processes, and second, either prove or scrounge the literature for useful results on such processes. For convolution kernel estimators, this works out nicely in that the limiting family consists of finite sections of a “fixed” stationary Gaussian process with almost surely continuous sample paths. Konakov and Piterbarg (1984) work this out for Nadaraya-Watson estimators (based on convolution kernels with compact support) in the general regression problem with a global smoothing parameter. They determine the asymptotic distribution of the maximum of the sections of the associated stationary Gaussian process with finite-range dependence,

$$\xi(t) \stackrel{\text{def}}{=} \int_{\mathbb{R}} K(x-t)\, dW(x) , \quad t > 0 ,$$

where $W(t)$ is the Wiener process on the line with $W(0) = 0$. Note that $\mathbb{E}[\, \xi(t)\, \xi(s)\,] = 0$ if $|\, t - s\, |$ is large enough since K has compact support. However, Pickands (1969) had shown long before that the finite-range aspect of the dependence is irrelevant.

We work out this approach for smoothing spline estimators, but the resulting family of Gaussian processes is stationary only in the case of uniform designs. Thus, we get the asymptotic distribution theory for uniform designs but fall short of the goal for arbitrary quasi-uniform designs. As a consequence, this justifies the simulation procedures of §§ 2 and 4 for uniform designs. For the simulation approach with arbitrary quasi-uniform designs, we optimistically call it corroborating evidence.

The treatment is based on the conditional normal approximations of § 4 and the reproducing kernel / Silverman kernel approximation of § 21.9. It appears to be necessary to require that the variance and the design density vary smoothly and that the regression function has extra smoothness. This is the usual conundrum: The extra smoothness leads to an asymptotic distribution theory, but not to more accurate estimators. Even though we consider the general regression problem with random designs, our approach resembles that of Eubank and Speckman (1993) in that we condition on the design, which is tantamount to considering deterministic designs.

Before digging in, it should be noted that the convergence to the asymptotic distribution is really slow. This is due to the slow convergence of the distribution of the maxima of the Gaussian processes, akin to the slow convergence to the asymptotic distribution of the maximum of an increasing number of independent standard normals. Konakov and Piterbarg (1984) prove the slow convergence and provide a fix, but we shall not explore this. As said, our main motivation is to justify the negligibility of the approximation errors in the simulation approach of §§ 2 and 4.

Let f^{nH} be the smoothing spline estimator of order $2m$ in the model (21.1.2) with the random smoothing parameter H. Although the above suggests that we should look at

$$\max_{t \in [0,1]} |\, f^{nH}(t) - f_o(t)\, | ,$$

we immediately switch to the scaled maximum error,

$$(5.1) \qquad M_{0,n} = \max_{t\in[0,1]} \frac{|\, f^{nH}(t) - f_o(t)\,|}{\sqrt{\mathbb{V}\mathrm{ar}[\, f^{nh}(t)\,|\,\mathbb{X}_n\,]\big|_{h=H}}} \;.$$

Recall that the variance $\mathbb{V}\mathrm{ar}[\, f^{nh}(t)\,|\,\mathbb{X}_n\,]$ is evaluated for deterministic h, after which the (random) H is substituted for h. In (5.1), we shall ignore the fact that the variance must be estimated. Note that $M_{0,n}$ is the maximum of a family of correlated standard normals. Also, it is interesting that the whole interval $[\,0\,,\,1\,]$ is considered, but it will transpire that the boundary region requires extra attention. Although this works asymptotically, it is probably a good idea to consider boundary corrections, as EUBANK and SPECKMAN (1993) did for kernel estimators (but we won't). Either way, some additional smoothness of the regression function is required. We will assume the maximally useful smoothness, but in fact $f_o \in W^{m+1,\infty}(0,1)$ would suffice.

(5.2) ASSUMPTIONS. Throughout this section, assume that (3.1), (21.1.3) through (21.1.5), (4.12)–(4.13), and (4.20) hold. In addition, assume that

$$f_o \in W^{2m,\infty}(0,1) \quad \text{and} \quad \sigma^2\,,\ \omega\ \in W^{1,\infty}(0,1)\ ,$$

so the variance function σ^2 and the density ω are Lipschitz continuous.

We now state the main results. Recall the MESSER and GOLDSTEIN (1993) kernel $\mathfrak{B}_m$ of § 14.7, and let

$$(5.3) \qquad \varpi(t) = \begin{cases} \big(\omega(t)\big)^{1/(2m)} \;, & t\in[\,0\,,\,1\,]\;, \\ 1 \;, & \text{otherwise}\;. \end{cases}$$

Thus, ϖ is a quasi-uniform function. Define the Gaussian process, indexed by the weight function ϖ and the smoothing parameter h,

$$(5.4) \qquad \mathbb{G}_{\varpi,h}(t) = \frac{\displaystyle\int_{\mathbb{R}} \mathfrak{B}_m\Big(\frac{x-t}{\varpi(ht)}\Big)\,dW(x)}{\displaystyle\Big\{\int_{\mathbb{R}} \Big|\,\mathfrak{B}_m\Big(\frac{x-t}{\varpi(ht)}\Big)\Big|^2\,dx\Big\}^{1/2}} \;, \qquad t\in\mathbb{R}\;,$$

where $W(t)$, $t\in\mathbb{R}$, is the standard Wiener process with $W(0)=0$. Let

$$(5.5) \qquad \mathbb{M}(\varpi,h) = \max_{0\leqslant t\leqslant(1/h)} \big|\,\mathbb{G}_{\varpi,h}(t)\,\big| \;.$$

(5.6) THEOREM. *Under the assumptions* (5.2),

$$M_{0,n} = \mathbb{M}(\varpi,\gamma) + \varepsilon_n \quad \textit{in distribution}\,,$$

with $\varepsilon_n = o\big((\log n)^{-1/2}\big)$ *in probability. Here, the parameter* $\gamma \equiv \gamma_n$ *comes to us by way of* $H \asymp \gamma$ *in the sense of* (4.12)–(4.13).

The case of the uniform design density, $\varpi(t) = 1\!\!1(t) = 1$ for all t, is interesting. Then, $\mathbb{G}_{\varpi,h}(t) = \mathbb{G}_{\mathbb{1},h}(t)$, $t \in \mathbb{R}$, does not depend on h and is a stationary Gaussian process with almost surely continuous sample paths. We then get the asymptotic distribution of $\mathbb{M}(1\!\!1, h)$ from KONAKOV and PITERBARG (1984), Theorem 3.2, and PICKANDS (1969) (to get rid of the compactly supported kernel condition). Let

$$(5.7) \qquad \Lambda_m = \frac{\| \mathfrak{B}_m' \|_{L^2(\mathbb{R})}}{\| \mathfrak{B}_m \|_{L^2(\mathbb{R})}} \quad \text{and} \quad \ell_n = \sqrt{2\log(2\pi\gamma_n/\Lambda_m)} \ .$$

(Evaluating Λ_m is probably best done in terms of the Fourier transforms, by way of Parseval's equality, and then by symbolic mathematics packages.)

(5.8) THEOREM. *Under the assumptions* (5.2) *and a uniform design,*

$$\mathbb{P}[\ell_n(M_{0,n} - \ell_n) < x] \longrightarrow \exp(-2\,\mathrm{e}^{-x}) \quad \textit{for all} \quad x\ .$$

The theorem permits perturbations of size $o(\ell_n^{-1}) = o((\log n)^{-1/2})$ in $M_{0,n}$, without affecting the limiting distribution. Thus, the approximation errors of $\mathfrak{S}_{\omega nh}$, which in §§2 and 4 we showed to be of size $o((nh\log n)^{-1/2})$, do not cause any harm.

For non-uniform designs, we are at a loss since the Gaussian processes $\mathbb{G}_{\varpi,h}$ are not stationary. The authors are unaware of results regarding the distribution of the maxima of sections of "arbitrary" nonstationary Gaussian processes. In the following theorem, we do get a comparison with the extreme case, in which the design density is replaced by its minimum, but the result does not completely justify the simulation approach for arbitrary quasi-uniform designs.

(5.9) THEOREM. *Under the assumptions* (5.2), *for all* x,

$$\limsup_{n\to\infty} \mathbb{P}\big[\ell_n'(M_{0,n} - \ell_n') > x\big] \leqslant 1 - \exp(-2\,\mathrm{e}^{-x})\ ,$$

where $\ell_n' = \sqrt{2\log(2\pi/(\gamma\varpi_{\min}\Lambda_m))}$. *Here,* Λ_m *is given by* (5.7).

PROOF. Define the interval $I = [\varpi_{\min}, \varpi_{\max}]$, where $\varpi_{\min}$ and $\varpi_{\max}$ are the minimum and maximum of $\varpi(t)$. Then, with $\theta 1\!\!1$ denoting the constant function with value θ everywhere,

$$\begin{aligned}
\max_{0\leqslant t\leqslant(1/\gamma)} |\mathbb{G}_{\varpi,\gamma}(t)| &\leqslant \max_{0\leqslant t\leqslant(1/\gamma)} \max_{\theta\in I} |\mathbb{G}_{\theta\mathbb{1},\gamma}(t)| \\
&= \max_{\theta\in I} \max_{0\leqslant t\leqslant(1/\gamma)} |\mathbb{G}_{\theta\mathbb{1},\gamma}(t)| =_{\mathrm{d}} \max_{\theta\in I} \max_{0\leqslant t\leqslant 1/(\theta\gamma)} |\mathbb{G}_{\mathbb{1},1}(t)| \\
&= \max_{0\leqslant t\leqslant 1/(\varpi_{\min}\gamma)} |\mathbb{G}_{\mathbb{1},1}(t)|\ ,
\end{aligned}$$

and the last quantity has the asymptotic distribution of Theorem (5.8), except that ℓ_n must be replaced by ℓ_n'. The lemma follows. Q.e.d.

In the remainder of this section, we outline the steps involved in obtaining the limiting family of Gaussian processes: Boundary trouble, bias considerations, the (equivalent) reproducing kernel approximation, getting rid of the random H, normal approximation, getting rid of the variance and (sort of) the design density, the Silverman kernel approximation, and Gaussian processes. The actual proofs are exhibited in the next section. First, we need some notation.

(5.10) NOTATION. For sequences of random variables $\{\mathbb{A}_n\}_n$ and $\{\mathbb{B}_n\}_n$, we write $\mathbb{A}_n \approx\!\approx \mathbb{B}_n$ if $\sqrt{\log n}\,\{\mathbb{A}_n - \mathbb{B}_n\} \longrightarrow 0$ in probability. We write $\mathbb{A}_n \approx\!\approx_d \mathbb{B}_n$, or $\mathbb{A}_n \approx\!\approx \mathbb{B}_n$ in distribution, if there exists a sequence of random variables $\{\mathbb{C}_n\}_n$ with $\mathbb{A}_n =_{\rm d} \mathbb{C}_n$ for all n and $\mathbb{C}_n \approx\!\approx \mathbb{B}_n$.

The conclusion of Theorem (5.6) may then be succinctly stated as

$$M_{0,n} \approx\!\approx_d \mathbb{M}(\varpi, \gamma)\,. \tag{5.11}$$

A final notation: For random, continuous functions f on $[0,1]$, and unions of intervals $A \subset [0,1]$, define

$$\mathcal{M}(f, A) \stackrel{\text{def}}{=} \sup_{t\in A} \frac{|f(t)|}{\sqrt{\operatorname{Var}[f(t)\,|\,\mathbb{X}_n]}}\,. \tag{5.12}$$

Note that $M_{0,n} = \mathcal{M}(f^{nh} - f_o, A)\big|_{h=H}$.

We now outline the various steps listed above.

Boundary trouble. It turns out that the region near the boundary of $[0,1]$ needs special attention, essentially because there the squared bias of the smoothing spline estimator is not negligible compared with the variance. Here, the region away from the boundary is taken to be $\mathfrak{P}(\gamma)$, with γ from (4.12)–(4.13), where

$$\mathfrak{P}(h) = \big[\,k\,h\log(1/h)\,, 1 - k\,h\log(1/h)\,\big] \tag{5.13}$$

for a large enough constant k. We take $\mathfrak{P}(h) = \emptyset$ if the upper bound is smaller than the lower bound. (Note the notational difference with (21.1.27).) The boundary region is then $[0,1] \setminus \mathfrak{P}(\gamma)$. We then write

$$M_{0,n} = \max\{\,M_{I,n}, M_{B,n}\,\}\,, \tag{5.14}$$

with the main, interior part $M_{I,n}$ (I for Interior)

$$M_{I,n} = \mathcal{M}(f^{nh} - f_o, \mathfrak{P}(\gamma))\big|_{h=H}\,, \tag{5.15}$$

and the allegedly negligible boundary part

$$M_{B,n} = \mathcal{M}(f^{nh} - f_o, [0,1]\setminus\mathfrak{P}(\gamma))\big|_{h=H}\,. \tag{5.16}$$

The claim is that $M_{B,n} = \mathcal{O}(\sqrt{\log\log n}\,)$ and $M_{I,n} \gg \sqrt{\log\log n}$ in probability, so that then

$$M_{0,n} = M_{I,n} \quad \text{in probability}. \tag{5.17}$$

Bias considerations. The next step is to show that the bias in the interior region is negligible. This is done by replacing $f^{nH} - f_o$ by

$$\varphi^{nH} = f^{nH} - \mathbb{E}[\, f^{nh} \,|\, \mathbb{X}_n \,]\big|_{h=H} \tag{5.18}$$

and showing that

$$M_{I,n} \approx\approx M_{II,n} \stackrel{\text{def}}{=} \mathcal{M}(\,\varphi^{nh}\,,\, \mathfrak{P}(\gamma)\,)\big|_{h=H}\;. \tag{5.19}$$

(Here, II is the Roman numeral 2.)

The equivalent kernel approximation, dealing with random $\boldsymbol{H}$**, and the normal approximation** are all rolled into one, even though they constitute distinct steps. We replace φ^{nH} by $\mathfrak{N}_{\sigma,n\gamma}$, where

$$\mathfrak{N}_{\sigma,n\gamma}(\,t\,) = \tfrac{1}{n} \sum_{i=1}^{n} Z_i\, \sigma(X_i)\, \mathfrak{R}_{\omega m\gamma}(X_i, t\,)\;, \quad t \in [\,0\,,1\,]\;, \tag{5.20}$$

in which $Z_1, Z_2 \cdots, Z_n$ are iid standard normals, conditioned on the design, as per the construction in Theorem (4.7). The task at hand is to show that

$$M_{II,n} \approx\approx M_{III,n} \stackrel{\text{def}}{=} \mathcal{M}(\,\mathfrak{N}_{\sigma,n\gamma}\,,\, \mathfrak{P}(\gamma)\,)\;. \tag{5.21}$$

Getting rid of the variance and design density. We now get rid of the variance function and the design density (but not completely). Let

$$v(\,t\,) = \big(\,\sigma(\,t\,)\, \sqrt{\omega(\,t\,)}\,\big)^{-1}\;, \quad t \in [\,0\,,1\,]\;. \tag{5.22}$$

and let $r(\,t\,) = \omega(\,t\,)^{-1/2}$. Then, one shows that $v(\,t\,)\, \mathfrak{N}_{\sigma,n\gamma}(\,t\,) \approx \mathfrak{N}_{r,n\gamma}(\,t\,)$ in the sense that

$$M_{III,n} \approx\approx M_{IV,n} \stackrel{\text{def}}{=} \mathcal{M}(\,\mathfrak{N}_{r,n\gamma}\,,\, \mathfrak{P}(\gamma)\,)\;. \tag{5.23}$$

The Silverman kernel approximation. Next, replace the reproducing kernel in the sum $\mathfrak{N}_{r,n\gamma}$ with the Silverman kernel. Let

$$\mathfrak{D}_{r,n\gamma}(\,t\,) = \tfrac{1}{n} \sum_{i=1}^{n} Z_{i,n}\, r(X_{i,n})\, \mathfrak{B}_{m,\lambda(t)}(X_{i,n} - t\,)\;, \quad t \in [\,0\,,1\,]\;, \tag{5.24}$$

with $\lambda(\,t\,) = \gamma/(\omega(\,t\,))^{1/(2m)}$; see § 21.9. We must show that

$$M_{IV,n} \approx\approx M_{V,n} \stackrel{\text{def}}{=} \mathcal{M}(\,\mathfrak{D}_{r,n\gamma}\,,\, \mathfrak{P}(\gamma)\,)\;. \tag{5.25}$$

(5.26) REMARK. At this point, the usefulness of replacing $\lambda(\,t\,)$ by $\lambda(X_{i,n})$ is not obvious, but we certainly may do so. Precisely, let

$$\mathfrak{L}_{n\gamma}(\,t\,) = \tfrac{1}{n} \sum_{i=1}^{n} Z_{i,n}\, r(X_{i,n})\, \mathfrak{C}_{m\gamma}(X_{i,n}, t\,)\;,$$

where

$$\mathfrak{C}_{m\gamma}(x, t\,) = \frac{1}{\lambda(\,t\,)}\, \mathfrak{B}_m\Big(\frac{x - t}{\lambda(x)}\Big)\;.$$

In $\mathfrak{C}_{m\gamma}(x, t)$, note the occurrence of both $\lambda(t)$ and $\lambda(x)$. Then, one can show that $M_{V,n} \approx\approx \mathcal{M}(\mathfrak{L}_{n\gamma}, \mathfrak{P}(\gamma))$.

Gaussian processes. Obviously, for $n = 1, 2, \cdots$, the processes $\mathfrak{N}_{r,n\gamma}(t)$, $t \in [0, 1]$, are Gaussian and finite-dimensional, but the dimension increases with n. Let

$$\mathfrak{K}_{\gamma}(t) = \int_0^1 \mathfrak{B}_{m,\lambda(t)}(x - t)\, dW(x) , \quad t \in [0, 1] , \tag{5.27}$$

where $W(t)$ is a standard Wiener process with $W(0) = 0$. Then, one shows that there exists a version of the Wiener process for which

$$M_{V,n} \approx\approx_d M_{VI,n} \stackrel{\text{def}}{=} \mathcal{M}(\mathfrak{K}_{\gamma}, \mathfrak{P}(\gamma)) . \tag{5.28}$$

The final step is to show that

$$M_{VI,n} \approx\approx \mathbb{M}(\varpi, \gamma) . \tag{5.29}$$

This involves some minor boundary trouble but is otherwise straightforward.

The proof of Theorem (5.6) then follows. In the next section, the proofs of the various steps are provided.

6. Proofs of the various steps

We now set out to prove the equivalences of the previous sections. Since some of the arguments for the boundary region and the interior are the same, the following results are offered first.

(6.1) LEMMA. *Under the assumptions* (5.2), *for* $h \to 0$, $nh \to \infty$,

$$\mathbb{E}[\, |\, \varphi^{nh}(t) - \mathfrak{S}_{\varpi nh}(t)\, |^2 \,|\, \mathbb{X}_n] = \mathcal{O}(h^{-1}(nh)^{-2}) ,$$

and $\mathbb{V}\mathrm{ar}[\varphi^{nh}(t) \,|\, \mathbb{X}_n] \asymp (nh)^{-1}$, *almost surely, uniformly in* $t \in [0, 1]$.

PROOF. By the reproducing kernel Hilbert space trick, we have, for all t and all h,

$$\mathbb{E}[\, |\, \varphi^{nh}(t) - \mathfrak{S}_{\varpi nh}(t)\, |^2 \,|\, \mathbb{X}_n] \leqslant c\, h^{-1}\, \mathbb{E}[\, \|\, \varphi^{nh} - \mathfrak{S}_{\varpi nh} \,\|_{\varpi mh}^2 \,|\, \mathbb{X}_n] .$$

Then, Exercise (21.5.7)(b) provides the almost sure bound $\mathcal{O}(h^{-1}(nh)^{-2})$. Since $\mathbb{V}\mathrm{ar}[\mathfrak{S}_{\varpi nh}(t) \,|\, \mathbb{X}_n] \asymp (nh)^{-1}$ whenever $h \to 0$, the lemma follows. Q.e.d.

The following factoid is used over and over.

(6.2) FACTOID. *Let* A_n *be subintervals of* $[0, 1]$, *let* $\{\varphi_n\}_n$, *and* $\{\psi_n\}_n$ *be sequences of random, zero-mean functions on* $[0, 1]$, *and let* $\{a_n\}_n$,

$\{ b_n \}_n$, $\{ v_n \}_n$, *and* $\{ w_n \}_n$ *be sequences of strictly positive numbers. If, uniformly in* $t \in A_n$,

$$| \varphi_n(t) - \psi_n(t) | = \mathcal{O}(a_n) \quad , \quad | \psi_n(t) | = \mathcal{O}(b_n) \; ,$$
$$\mathbb{E}[| \varphi_n(t) - \psi_n(t) |^2] = \mathcal{O}(w_n^2) \quad \text{and} \quad \mathbb{V}\mathrm{ar}[\psi_n(t)] \asymp v_n^2 \; ,$$

and $w_n = o(v_n)$, *then*

$$\sup_{t \in A_n} \left| \frac{\varphi_n(t)}{\sqrt{\mathbb{V}\mathrm{ar}[\varphi_n(t)]}} - \frac{\psi_n(t)}{\sqrt{\mathbb{V}\mathrm{ar}[\psi_n(t)]}} \right| = \mathcal{O}\Big(\frac{a_n \, v_n + b_n \, w_n}{v_n^2} \Big) \; .$$

(6.3) EXERCISE. Prove the factoid. [Hint: Note that $w_n = o(v_n)$ implies that $\mathbb{V}\mathrm{ar}[\varphi_n(t)] \asymp \mathbb{V}\mathrm{ar}[\psi_n(t)] \asymp v_n^2$. The triangle inequality in the form

$$\Big| \sqrt{ \mathbb{E}[| \varphi_n(t) |^2] } - \sqrt{ \mathbb{E}[| \psi_n(t) |^2] } \Big| \leqslant \sqrt{ \mathbb{E}[| \varphi_n(t) - \psi_n(t) |^2] }$$

comes into play as well.]

(6.4) LEMMA. *Let* K *be a positive integer. Let* $\{ A_n \}_n$ *be a sequence of deterministic subintervals of* $[0 , 1]$, *and let* $\varphi^{nh} = f^{nh} - \mathbb{E}[f^{nh} \mid \mathbb{X}_n]$. *Under the assumptions* (5.2),

$$\mathcal{M}(\varphi^{nh} , A_n) \big|_{h=H} \approx\approx \mathcal{M}(\mathfrak{N}_{\sigma, n\gamma} , A_n) \; .$$

PROOF. The proof consists of three parts: reproducing kernel approximation, dealing with the random H, and the normal approximation.

Reproducing kernel approximation. We must show that

$$\mathcal{M}(\varphi^{nh} , A_n)\big|_{h=H} \approx\approx \mathcal{M}(\mathfrak{S}_{nH} , A_n)\big|_{h=H} \; . \tag{6.5}$$

Now, Lemma (21.1.17), combined with the assumptions (4.12)–(4.13), provides us with the bound

$$\| \varphi^{nH} - \mathfrak{S}_{nH} \|_\infty = \mathcal{O}\big(\gamma^{-1/2} \, (n\gamma)^{-1} \log n \big) \quad \text{in probability} ,$$

and of course $\| \mathfrak{S}_{nH} \|_\infty = \mathcal{O}\big(\sqrt{(n\gamma)^{-1} \log n} \, \big)$ in probability. That takes care of the numerators in (6.5).

For the denominators, Lemma (6.1) gives us for all t and all h,

$$\mathbb{E}\big[\, | \varphi^{nh}(t) - \mathfrak{S}_{\omega nh}(t) |^2 \, \big| \, \mathbb{X}_n \big] = \mathcal{O}\big(h^{-1} \, (nh)^{-2} \big) \; ,$$

almost surely, and $\mathbb{V}\mathrm{ar}[\, \mathfrak{S}_{\omega nh}(t) \, | \, \mathbb{X}_n] \asymp (nh)^{-1}$. Then, Factoid (6.2) gives us

$$\Big| \mathcal{M}(\varphi^{nh} , A_n)\big|_{h=H} - \mathcal{M}(\mathfrak{S}_{nH} , A_n)\big|_{h=H} \Big| = \mathcal{O}\big(\, \gamma^{-1/2} (n\gamma)^{-1/2} \log n \, \big)$$

in probability. Since $\gamma \asymp n^{-1/(2m+1)}$ is at least of size $n^{-1/3}$, the right-hand side is at most $\mathcal{O}\big(n^{-1/6} \log n \big)$, thus showing (6.5).

Getting rid of the random H. We must show that

$$\mathcal{M}(\mathfrak{S}_{\omega nh}\,,\,A_n)\big|_{h=H} \approx\approx \mathcal{M}(\mathfrak{S}_{n\gamma}\,,\,A_n)\;. \tag{6.6}$$

For the difference in the numerators, Exercise (21.6.27) and (4.12)–(4.13) provide us with the in-probability bound

$$\|\,\mathfrak{S}_{nH} - \mathfrak{S}_{n\gamma}\,\|_\infty \leqslant c\,\|\,\mathfrak{S}_{n\gamma}\,\|_\infty \left|\,\frac{H}{\gamma} - 1\,\right| = \mathcal{O}\big(\,(n\gamma\,\log n)^{-1/2}\,\big)\;. \tag{6.7}$$

For the variances, let

$$\Delta(x,\,t\,) = \sigma^2(x)\,|\,\mathfrak{R}_{\omega m\gamma}(x,\,t\,) - \mathfrak{R}_{\omega mh}(x,\,t\,)\,|^2\;.$$

Then, a straightforward calculation gives

$$\begin{aligned} n\,\mathbb{E}\big[\;|\,\mathfrak{S}_{\omega nh}(\,t\,) - \mathfrak{S}_{n\gamma}(\,t\,)\,|^2\,\big|\,\mathbb{X}_n\,\big] &= \tfrac{1}{n}\sum_{i=1}^{n}\Delta(X_i,\,t\,) \\ &= \int_0^1 \Delta(x,\,t\,)\,\omega(x)\,dx + \mathrm{rem}\;, \end{aligned} \tag{6.8}$$

where the remainder is given by

$$\mathrm{rem} = \int_0^1 \Delta(x,\,t\,)\,\big(\,d\Omega_n(x) - d\Omega(x)\,\big)\;,$$

with Ω_n the empirical distribution function of the design. Lemma (21.3.4) implies

$$|\,\mathrm{rem}\,| \leqslant \|\,\Delta(\,\cdot\,,\,t\,)\,\|_{\omega,1,\gamma,1}\cdot\mathcal{O}\big(\,\sqrt{(n\gamma)^{-1}\log n}\,\big)\quad\text{almost surely}\,.$$

Now, since ω is bounded, by Theorem (21.6.23), then

$$\begin{aligned} \|\,\Delta(\,\cdot\,,\,t\,)\,\|_{L^1((0,1),\omega)} &\leqslant c\,\|\,\mathfrak{R}_{\omega mh}(\,\cdot\,,\,t\,) - \mathfrak{R}_{\omega m\gamma}(\,\cdot\,,\,t\,)\,\|^2_{L^2((0,1))} \\ &\leqslant c\,\gamma^{-1}\left|\,1-\frac{\gamma}{h}\,\right|^2\;. \end{aligned} \tag{6.9}$$

Now, note that (the prime denotes differentiation with respect to x)

$$\begin{aligned} \Delta'(x,\,t\,) = \big(\sigma^2(x)\big)'\,|\,\mathfrak{R}_{\omega mh}(x,\,t\,) - \mathfrak{R}_{\omega m\gamma}(x,\,t\,)\,|^2 + {}\\ \sigma^2(x)\,\big(\,(\,\mathfrak{R}_{\omega mh}(x,\,t\,) - \mathfrak{R}_{\omega m\gamma}(x,\,t\,)\,)^2\,\big)'\;. \end{aligned}$$

Since $\sigma^2 \in W^{1,\infty}(0,1)$, then

$$\begin{aligned} &\gamma\,\big\|\,\big(\sigma^2(\,\cdot\,)\big)'\,|\,\mathfrak{R}_{\omega mh}(\,\cdot\,,\,t\,) - \mathfrak{R}_{\omega m\gamma}(\,\cdot\,,\,t\,)\,|^2\,\big\|_1 \\ &\qquad\leqslant c\,\gamma\,\big\|\,\mathfrak{R}_{\omega mh}(\,\cdot\,,\,t\,) - \mathfrak{R}_{\omega m\gamma}(\,\cdot\,,\,t\,)\,\big\|_2^2 \;\leqslant\; c\,\gamma^{-1}\left|\,1-\frac{\gamma}{h}\,\right|^2\;. \end{aligned}$$

Likewise, with Cauchy-Schwarz,

$$\begin{aligned}\gamma \,\|\, \sigma^2(\cdot)\,\big(\,(\mathfrak{R}_{\omega m h}(\,\cdot\,,t\,) - \mathfrak{R}_{\omega m \gamma}(\,\cdot\,,t\,)\,)^2\,\big)'\,\|_1 &\\ \leqslant c\,\gamma \,\|\, \mathfrak{R}_{\omega m h}(\,\cdot\,,t\,) - \mathfrak{R}_{\omega m \gamma}(\,\cdot\,,t\,)\,\|_2 \;\times &\\ \|\, \mathfrak{R}'_{\omega m h}(\,\cdot\,,t\,) - \mathfrak{R}'_{\omega m \gamma}(\,\cdot\,,t\,)\,\|_2 &\\ \leqslant \; c\,\gamma^{-1}\,\Big|\, 1 - \frac{\gamma}{h}\,\Big|^2 \,. &\end{aligned}$$

So, the bound (6.9) applies to $\gamma \,\|\, \Delta'(\,\cdot\,,t\,)\,\|_{L^1((0,1))}$ as well. It follows that

$$|\,\mathrm{rem}\,| = \mathcal{O}\Big(\,\gamma^{-1}\,\Big|\,1 - \frac{\gamma}{h}\,\Big|^2\,\sqrt{(n\gamma)^{-1}\log n}\;\Big)\,.$$

But now observe that this bound is negligible compared with (the bound on) the leading term in (6.8).

The above shows that

$$\mathbb{E}[\,|\,\mathfrak{S}_{\omega n h}(\,t\,) - \mathfrak{S}_{n\gamma}(\,t\,)\,|^2 \,|\, \mathbb{X}_n\,] \leqslant c\,(n\gamma)^{-1}\,\Big|\,1 - \frac{\gamma}{h}\,\Big|^2\,.$$

Now substitute $h = H$. Then, (6.7) and the bound above imply by way of Factoid (6.3) that

$$\begin{aligned}\big|\,\mathcal{M}(\mathfrak{S}_{\omega n h}\,,\,A_n)\big|_{h=H} - \mathcal{M}(\mathfrak{S}_{n\gamma}\,,\,A_n)\,\big| &\\ \leqslant c\,\Big|\,\frac{\gamma}{H} - 1\,\Big|\,\sqrt{\log n} = o\big(\,(\log n)^{-1/2}\,\big)\,, &\end{aligned}$$

the last step by (4.12)–(4.13), and so (6.6) follows.

The normal approximation causes no problems. We must show that

$$(6.10)\qquad \mathcal{M}(\,\mathfrak{S}_{n\gamma}\,,\,A_n\,) \;\approx\!\approx\; \mathcal{M}(\,\mathfrak{N}_{\sigma,n\gamma}\,,\,A_n\,)\,.$$

Theorem (4.7) gives us the nice in-probability bound

$$\|\,\mathfrak{S}_{n\gamma} - \mathfrak{N}_{\sigma,n\gamma}\,\|_\infty = \mathcal{O}(\,n^{-3/4}\gamma^{-1}\,)\,,$$

and since

$$\mathbb{V}\mathrm{ar}[\,\mathfrak{S}_{n\gamma}(\,t\,)\,|\,\mathbb{X}_n\,] = \mathbb{V}\mathrm{ar}[\,\mathfrak{N}_{\sigma,n\gamma}(\,t\,)\,|\,\mathbb{X}_n\,] \asymp (n\gamma)^{-1}\,,$$

then Factoid (6.3) gives

$$|\,\mathcal{M}(\,\mathfrak{S}_{n\gamma}\,,\,A_n\,) - \mathcal{M}(\,\mathfrak{N}_{\sigma,n\gamma}\,,\,A_n\,)\,| = \mathcal{O}(\,(n\gamma^2)^{-1/4}\,)\,,$$

which is negligible compared with $(\log n)^{-1}$. Thus, (6.10) follows.

The lemma now follows from (6.5), (6.6), and (6.10). Q.e.d.

Next, we consider the result that $M_{0,n} = M_{I,n}$ in probability; i.e., that $M_{B,n}$ is negligible compared with $M_{I,n}$. The reason for this is that the boundary region $\mathfrak{P}(\gamma)$ is so short that nothing much of interest happens there. The next result is not meant to be sharp but is good enough for our purposes.

(6.11) LEMMA. *Let* $\{ A_n \}_n$ *be a sequence of subintervals of* $[0, 1]$. *If* $| A_n | = \mathcal{O}(\gamma \log n)$, *then, under the assumptions* (5.2),

$$\mathcal{M}(\varphi^{nh}, A_n)\big|_{h=H} = \mathcal{O}(\sqrt{\log\log n}) \quad \textit{in probability} .$$

PROOF. By Lemma (6.4), we have

$$\mathcal{M}(\varphi^{nh}, A_n) \approx\!\approx \mathcal{M}(\mathfrak{N}_{\sigma,n\gamma}, A_n) .$$

Below, we show that the function

$$\mathfrak{T}_{n\gamma}(t) = \mathfrak{N}_{\sigma,n\gamma}(t) \Big/ \sqrt{\mathbb{V}\mathrm{ar}[\mathfrak{N}_{\sigma,n\gamma}(t) \,|\, \mathbb{X}_n]} , \quad t \in [0, 1] ,$$

satisfies

$$\sup_{t, s \in [0,1]} \frac{| \mathfrak{T}_{n\gamma}(t) - \mathfrak{T}_{n\gamma}(s) |}{| t - s |} = \mathcal{O}(\gamma^{-1} \sqrt{\log\log n}) \tag{6.12}$$

in probability.

That being the case, we now cover the intervals A_n by enough points that the maximum of $\mathfrak{T}_{n\gamma}$ over these points is close enough to the supremum over all of A_n. Let $a_n = | A_n | / \gamma$ and $\delta = \gamma / b_n$, with $b_n \to \infty$ but yet to be specified. Then, $| t - s | \leqslant \delta$ implies that

$$| \mathfrak{T}_{n\gamma}(t) - \mathfrak{T}_{n\gamma}(s) | = \mathcal{O}(b_n^{-1} \sqrt{\log n}) \quad \text{in probability} .$$

Define

$$t_j = j\,\delta , \quad j = 0, 1, 2, \cdots , N - 1 ,$$

and set $t_N = 1$. Here, N denotes the smallest natural number exceeding $1/\delta$. Enumerate the t_j that belong to A_n by s_j, $j = 1, 2, \cdots , J$. Then, $J \leqslant | A_n |/\delta = a_n b_n$. In fact, J differs from $a_n b_n$ by at most 1. Then,

$$\| \mathfrak{T}_{n\gamma} \|_\infty \leqslant m_J + \mathcal{O}(b_n^{-1} \sqrt{\log n}) ,$$

with
$$m_J = \max_{1 \leqslant j \leqslant J} | \mathfrak{T}_{n\gamma}(s_j) | .$$

Now, it is apparent that

$$m_J = \mathcal{O}(\sqrt{\log J}) = \mathcal{O}(\sqrt{\log(a_n b_n)}) ,$$

and so

$$\| \mathfrak{T}_{n\gamma} \|_\infty = \mathcal{O}(b_n^{-1} \sqrt{\log n} + \sqrt{\log(a_n b_n)}) .$$

Now, take $b_n = \sqrt{\log n}$. Then, the first term is $\mathcal{O}(1)$ and the second is

$$\mathcal{O}(\sqrt{\log a_n + \log\log n}) = \mathcal{O}(\sqrt{\log a_n \vee \log\log n}) = \mathcal{O}(\sqrt{\log\log n}) .$$

So, all that is left is to prove (6.12). First, by Theorem (21.2.17),

$$\begin{aligned} | \mathfrak{N}_{\sigma,n\gamma}(t) - \mathfrak{N}_{\sigma,n\gamma}(s) | &\leqslant | t - s | \, \| \mathfrak{N}'_{\sigma,n\gamma} \|_\infty \\ &= \mathcal{O}(| t - s | \gamma^{-1} \sqrt{(n\gamma)^{-1} \log n}) \end{aligned} \tag{6.13}$$

almost surely, uniformly in t and s.

For the variances, one obtains by virtue of Lemma (21.6.18) that

$$\begin{aligned}\mathbb{E}[\,|\,\mathfrak{N}_{\sigma,n\gamma}(t) - \mathfrak{N}_{\sigma,n\gamma}(s)\,|^2\,|\,\mathbb{X}_n\,] \\ &\leqslant \frac{c}{n^2}\sum_{i=1}^{n}|\,\mathfrak{R}_{\omega m\gamma}(X_i,\,t) - \mathfrak{R}_{\omega m\gamma}(X_i,\,s)\,|^2 \\ &\leqslant \frac{c\,|\,t\,-\,s\,|^2}{n\,\gamma^2}\cdot\tfrac{1}{n}\sum_{i=1}^{n}|\,A_{n\gamma}(X_i,\,t\,,\,s\,)\,|^2\,,\end{aligned}$$

where

$$A_{n\gamma}(t) = \gamma^{-1}\Big\{\exp(-\kappa\,\gamma^{-1}|\,X_i\,-\,t\,|) + \exp(-\kappa\,\gamma^{-1}|\,X_i\,-\,s\,|)\Big\}$$

for a suitable positive constant κ. Now, by Theorem (14.6.13),

$$\sup_{t\,,\,s\in[\,0\,,\,1\,]}\ \tfrac{1}{n}\sum_{i=1}^{n}|\,A_{n\gamma}(X_i,\,t\,,\,s\,)\,|^2 = \mathcal{O}(\,\gamma^{-1}\,)$$

almost surely, so that

$$\mathbb{E}[\,|\,\mathfrak{N}_{\sigma,n\gamma}(t) - \mathfrak{N}_{\sigma,n\gamma}(s)\,|^2\,|\,\mathbb{X}_n\,] = \mathcal{O}\Big(\,\frac{|\,t\,-\,s\,|^2}{\gamma^2}\,\frac{1}{n\gamma}\,\Big)\,.$$

Together with (6.13), this implies the bound (6.12) (appealing again to Factoid (6.3)). Q.e.d.

We are now ready for the proofs of the various equivalences of § 5.

(6.14) Boundary trouble. Here we prove (5.17): $M_{0,n} = M_{I,n}$ in probability. Recall from Theorem (21.5.19) that, for $f_o \in W^{m,\infty}(0,1)$ and nice designs,

$$\|\,\mathbb{E}[\,f^{nh}\,|\,\mathbb{X}_n\,]\,\big|_{h=H} - f_o\,\|_\infty = \mathcal{O}(\,\gamma^m + \gamma^{-1/2}(n\gamma)^{-1}\log n\,) = \mathcal{O}(\,\gamma^m\,)$$

in probability. Since, by Lemma (6.1), $\mathbb{V}\mathrm{ar}[\,f^{nh}(\,t\,)\,|\,\mathbb{X}_n\,]\big|_{h=H} \asymp (n\gamma)^{-1}$, uniformly in t, then

$$M_{B,n} \leqslant \mathcal{M}(\,\varphi^{nh}\,,\,[\,0\,,\,1\,]\setminus\mathfrak{P}(\gamma)\,) + \text{rem}\,,$$

with

$$\text{rem} = \frac{\|\,\mathbb{E}[\,f^{nh}\,|\,\mathbb{X}_n\,] - f_o\,\|_\infty}{\sqrt{\mathbb{V}\mathrm{ar}[\,f^{nh}\,|\,\mathbb{X}_n\,]\big|_{h=H}}} = \frac{H^m}{(nH)^{-1/2}} = \mathcal{O}(\,1\,)\quad\text{in probability}\,.$$

Finally, Lemma (6.11) provides the bound

$$\mathcal{M}(\,\varphi^{nh}\,,\,[\,0\,,\,1\,]\setminus\mathfrak{P}(\gamma)\,) = \mathcal{O}(\,\sqrt{\log\log n}\,)\,,$$

so that $M_{B,n} = \mathcal{O}(\,\sqrt{\log\log n}\,)$ in probability.

Now, we need a lower bound on $M_{I,n}$. STONE (1982) has shown that for any estimator $\widehat{f}_n$ of f_o, the uniform error $\| \widehat{f}_n - f_o \|_\infty$ is at least of size $(n^{-1}\log n)^{m/(2m+1)}$, so that

$$\liminf_{n\to\infty} \inf_{h>0} \frac{\| f^{nh} - f_o \|_\infty}{(n^{-1}\log n)^{m/(2m+1)}} > 0 \quad \text{in probability} .$$

Since $(nH)^{-1/2} \asymp n^{-m/(2m+1)}$ in probability, then $M_{0,n}$ is at least of order $(\log n)^{m/(2m+1)}$. So $M_{B,n} \ll M_{I,n}$, or $M_{0,n} = M_{I,n}$ in probability. Q.e.d.

(6.15) BIAS CONSIDERATIONS. We must deal with the bias $f_o - \mathbb{E}[\, f^{nh} \mid \mathbb{X}_n \,]$ in the interior region; i.e., we must prove (5.19). Corollary (21.8.13) gives

$$\| \mathbb{E}[\, f^{nh} \mid \mathbb{X}_n \,] \,|_{h=H} - f_o \|_{L^\infty(\mathfrak{P}(\gamma))} = \mathcal{O}\big(\gamma^{2m} + \gamma^{-1/2}(n\gamma)^{-1}\log n \big)$$

in probability. This takes care of the difference in the numerators of $\mathcal{M}\big(f^{nh} - f_o , \mathfrak{P}(\gamma) \big)$ and $\mathcal{M}\big(f^{nh} - \mathbb{E}[\, f^{nh} \mid \mathbb{X}_n \,] , \mathfrak{P}(\gamma) \big)$.

Since the variances are the same, then, by Lemma (6.1),

$$\mathbb{V}\mathrm{ar}[\, f^{nh} \mid \mathbb{X}_n \,]\big|_{h=H} = \mathbb{V}\mathrm{ar}[\, \varphi^{nh} \mid \mathbb{X}_n \,]\big|_{h=H} \asymp (nH)^{-1} \asymp (n\gamma)^{-1}$$

in probability, and it follows that

$$M_{I,n} - M_{II,n} = \mathcal{O}\big(\gamma^{2m}(n\gamma)^{-1/2} + \gamma^{-1/2}(n\gamma)^{-1/2}\log n \big) ,$$

which decays fast enough to conclude that $M_{I,n} \approx\!\!\approx M_{II,n}$. Q.e.d.

(6.16) PROOF OF $M_{II,n} \approx\!\!\approx M_{III,n}$. Replacing φ^{nh} by $\mathfrak{N}_{\sigma,n\gamma}$ was done already in Lemma (6.4).

(6.17) PROOF OF $M_{III,n} \approx\!\!\approx M_{IV,n}$. We must show that in the sums $\mathfrak{N}_{\sigma,n\gamma}$, we may replace σ by (any) smooth quasi-uniform function without unduly affecting $\mathcal{M}\big(\mathfrak{N}_{\sigma,n\gamma} , \mathfrak{P}(\gamma) \big)$. Let v be a Lipschitz-continuous, quasi-uniform function, and let $r(t) = v(t)\,\sigma(t)$. Then, r is Lipschitz continuous and quasi-uniform as well. Let $\delta(t) = v(t)\,\mathfrak{N}_{\sigma n\gamma}(t) - \mathfrak{N}_{rn\gamma}(t)$. Observe that we may write

$$\delta(t) = \tfrac{1}{n} \sum_{i=1}^{n} Z_i\, q(X_i, t)\, \mathfrak{R}_{\omega m\gamma}(X_i, t) ,$$

where $q(x, t) = \big\{ v(t) - v(x) \big\}\, \sigma(x)$ for all $x, t \in [\,0 , 1\,]$.

Again, Lemma (21.3.4) implies that

$$|\, \delta(t) \,| = \| q(\,\cdot\,, t)\, \mathfrak{R}_{\omega m\gamma}(\,\cdot\,, t) \|_{\omega,1,\gamma,1} \cdot \mathcal{O}\big(\sqrt{(n\gamma)^{-1}\log n} \,\big)$$

almost surely, uniformly in t. The Lipschitz continuity of v yields that $|\, q(x, t) \,| \leqslant c \,|\, x - t \,|\, \| \sigma \|_\infty \leqslant c_1 \,|\, x - t \,|$, and so

$$\| q(\,\cdot\,, t)\, \mathfrak{R}_{\omega m\gamma}(\,\cdot\,, t) \|_{L^1(\omega)} \leqslant c_1 \int_0^1 |\, t - x \,|\, |\, \mathfrak{R}_{\omega m\gamma}(x, t) \,|\, dx \leqslant c_2\, \gamma .$$

The same argument yields

$$\gamma \, \| \, q(\,\cdot\,,t\,)\, \mathfrak{R}'_{\omega m\gamma}(\,\cdot\,,t\,) \, \|_{L^1(0,1)} \leqslant c\gamma \; ,$$

where the prime denotes differentiation with respect to the first argument. Finally, the Lipschitz continuity of v and σ shows that $\| \, q'(\,\cdot\,,t\,) \, \|_\infty \leqslant c$, uniformly in t, so that

$$\gamma \, \| \, q'(\,\cdot\,,t\,)\, \mathfrak{R}_{\omega m\gamma}(\,\cdot\,,t\,) \, \|_{L^1(0,1)} \leqslant c\gamma \; ,$$

all for the appropriate constants c. Thus, we get that

$$\| \, \delta \, \|_\infty = \mathcal{O}\big(\gamma \sqrt{(n\gamma)^{-1} \log n} \, \big) \; . \tag{6.18}$$

For the variances, observe that by the arguments above ,

$$\begin{aligned} \mathbb{E}[\, \delta^2(t) \mid \mathbb{X}_n \,] &= n^{-2} \sum_{i=1}^{n} \big| \, q(X_i, t)\, \mathfrak{R}_{\omega m\gamma}(X_i, t) \, \big|^2 \\ &\leqslant c n^{-2} \sum_{i=1}^{n} | \, t - X_i \, |^2 \, | \, \mathfrak{R}_{\omega m\gamma}(X_i, t) \, |^2 \\ &\leqslant \frac{c}{n} \Big\{ \int_0^1 | \, t - x \, |^2 \, | \, \mathfrak{R}_{\omega m\gamma}(x, t) \, |^2 \, \omega(x) \, dx + \text{rem} \Big\} \; , \end{aligned}$$

with

$$\text{rem} = \int_0^1 | \, t - x \, |^2 \, | \, \mathfrak{R}_{\omega m\gamma}(x, t) \, |^2 \, \{ \, d\Omega_n(x) - d\Omega_o(x) \, \} \; .$$

Again, by Lemma (21.3.4), one obtains the almost-sure bound

$$| \, \text{rem} \, | = \mathcal{O}\big(\sqrt{(n\gamma)^{-1} \log n} \, \big) \, \| \, (\, (t - \,\cdot\,)\, \mathfrak{R}_{\omega n\gamma}(\,\cdot\,, t\,) \,)^2 \, \|_{\omega,1,\gamma,1} \; .$$

Now, it is a straightforward exercise to show that

$$\| \, (\, (t - \,\cdot\,)\, \mathfrak{R}_{\omega n\gamma}(\,\cdot\,, t\,) \,)^2 \, \|_{\omega,1,\gamma,1} \leqslant c\gamma \; ,$$

and then "rem" is negligible. Thus, we obtain the almost-sure bound

$$\mathbb{E}[\, \delta^2(t) \mid \mathbb{X}_n \,] = \mathcal{O}\big(n^{-1} \gamma \big) \; .$$

Then, with (6.18) and the help of Factoid (6.2), we get that $M_{V,n} - M_{VI,n} = \mathcal{O}\big(\gamma \sqrt{\log n} \, \big)$, and the conclusion that $M_{V,n} \approx\approx M_{VI,n}$ follows. Q.e.d.

(6.19) PROOF OF (5.25): $M_{IV,n} \approx\approx M_{V,n}$. Here, we handle the Silverman kernel approximation. Let $\delta(t) = \mathfrak{N}_{r,n\gamma}(t) - \mathfrak{D}_{r,n\gamma}(t)$. The bound

$$\| \, \delta \, \|_\infty = \mathcal{O}\big(\gamma \sqrt{(n\gamma)^{-1} \log n} \, \big) \quad \text{almost surely} \tag{6.20}$$

follows from Theorem (21.9.21). For the variances, we get

$$\mathbb{E}[\, | \, \delta(t) \, |^2 \mid \mathbb{X}_n \,] = n^{-2} \sum_{i=1}^{n} | \, r(X_i)\, b_{n\gamma}(X_i, t) \, |^2 \; ,$$

where $b_{n\gamma}(x,t) = \mathfrak{R}_{\omega m\gamma}(x,t) - \mathfrak{B}_{m,\lambda(t)}(x,t)$. Since r is bounded, then

$$\mathbb{E}[\,|\,\delta(t)\,|^2 \mid \mathbb{X}_n\,] \leqslant c\,n^{-2} \sum_{i=1}^{n} |\,b_{n\gamma}(t)\,|^2 \;.$$

Now, (21.9.22) gives a bound on b_{nh} in terms of two exponentials. Thus, we must consider

$$S = (n\gamma)^{-1} \tfrac{1}{n} \sum_{i=1}^{n} \gamma^{-1} \exp\bigl(-\kappa\,\gamma^{-1}\,(X_i + t)\bigr) \;,$$

as well as the sum where $X_i + t$ is replaced by $2 - X_i - t$. It follows from standard arguments that

$$\begin{aligned} \tfrac{1}{n} \sum_{i=1}^{n} \gamma^{-1} \exp\bigl(-\kappa\,\gamma^{-1}\,(X_i + t)\bigr) &\leqslant c\,\gamma^{-1} \int_0^1 \exp\bigl(-\kappa(x+t)\bigr)\,dx \\ &\leqslant c_1\,\gamma^{-1} \exp(-\kappa\,\gamma^{-1}\,t) \end{aligned}$$

almost surely. Now, since $t \in \mathfrak{P}(\gamma)$, then $t \geqslant k\gamma \log(1/\gamma)$, and so

$$\exp(-\kappa\,\gamma^{-1}\,t) \leqslant \gamma^{k/\kappa} \leqslant \gamma^3$$

for k large enough. So then $S = \mathcal{O}\bigl(\gamma^2 (n\gamma)^{-1}\bigr)$ almost surely. The same bound applies to the other sum. It follows that

$$\mathbb{E}[\,|\,\delta(t)\,|^2 \mid \mathbb{X}_n\,] \leqslant c\,\gamma^2\,(n\gamma)^{-1} \;.$$

Together with (6.20), then $\mathcal{M}(\mathfrak{N}_{r,n\gamma}, \mathfrak{P}(\gamma)) \approx\approx \mathcal{M}(\mathfrak{D}_{r,n\gamma}, \mathfrak{P}(\gamma))$, again by way of Factoid (6.2). Q.e.d.

The pace now changes since now, Wiener processes enter.

(6.21) PROOF OF $M_{V,n} \approx\approx M_{VI,n}$. Recall that $\lambda(t) = \gamma/(\omega(t))^{1/(2m)}$; see § 21.9. With Ω_n the empirical distribution function of the design, for each n we may construct a standard Wiener process (conditional on the design) such that

$$W\Bigl(\Omega_n(X_{i,n})\Bigr) = W\Bigl(\frac{i}{n}\Bigr) = \frac{1}{\sqrt{n}} \sum_{j=1}^{i} Z_{j,n} \;, \quad i = 1, 2, \cdots, n \;.$$

Then, writing $\mathfrak{D}_{r,n\gamma}(t)$ in terms of the order statistics of the design,

$$\mathfrak{D}_{r,n\gamma}(t) = \tfrac{1}{n} \sum_{i=1}^{n} Z_{i,n}\, r(X_{i,n})\, \mathfrak{B}_{m,\lambda(t)}(X_{i,n}, t) \;,$$

we see that we may rewrite it as

$$\frac{1}{\sqrt{n}} \sum_{i=1}^{n} r(X_{i,n})\, \mathfrak{B}_{m,\lambda(t)}(X_{i,n}, t)\, \{\, W(\Omega_n(X_{i,n})) - W(\Omega_n(X_{i-1,n}))\, \} \;.$$

Here, $\Omega_n(X_{0,n}) = 0$. The last sum may then be interpreted as a Riemann-Stieltjes integral, so $\mathfrak{D}_{r,n\gamma}(t) = \frac{1}{\sqrt{n}} \mathfrak{K}_{n\gamma}(t)$ for all $t \in [0, 1]$, with

$$\text{(6.22)} \qquad \mathfrak{K}_{r,n\gamma}(t) = \int_0^1 r(x)\, \mathfrak{B}_{m,\lambda(t)}(x, t)\, dW(\Omega_n(x)) \ .$$

Obviously, $\mathcal{M}(\mathfrak{D}_{r,n\gamma}, \mathfrak{P}(\gamma)) = \mathcal{M}(\mathfrak{K}_{r,n\gamma}, \mathfrak{P}(\gamma))$.

We now wish to approximate $\mathfrak{K}_{r,n\gamma}(t)$ by the integral

$$\text{(6.23)} \qquad \mathfrak{K}_{r,\gamma}(t) = \int_0^1 r(x)\, \mathfrak{B}_{m,\lambda(t)}(x, t)\, dW(\Omega(x)) \ .$$

Note that we are integrating a deterministic function against white noise, so the full intricacies of the Itô integral do not enter into the picture. Nevertheless, see SHORACK (2000).

We prove that $\mathcal{M}(\mathfrak{K}_{r,n\gamma}, \mathfrak{P}(\gamma)) \approx \mathcal{M}(\mathfrak{K}_{r,\gamma}, \mathfrak{P}(\gamma))$. This depends on the modulus of continuity of the standard Wiener process,

$$\text{(6.24)} \qquad \lim_{\delta \to 0} \sup_{\substack{x,y \in [0,1] \\ |x-y| \leqslant \delta}} \frac{|W(x) - W(y)|}{\sqrt{2\delta \log(1/\delta)}} = 1 \quad \text{almost surely} \ ;$$

see, e.g., CSÖRGŐ and RÉVÉSZ (1981).

Now, integration by parts gives

$$\begin{aligned} \mathfrak{K}_{r,n\gamma}(t) - \mathfrak{K}_{r,\gamma}(t) &= \int_0^1 r(x)\, \mathfrak{B}_{m\lambda(t)}(x, t) \{ dW(\Omega_n(x)) - dW(\Omega(x)) \} \\ &= r(x)\, \mathfrak{B}_{m\lambda(t)}(x, t) \{ W(\Omega_n(x)) - W(\Omega(x)) \} \Big|_{x=0}^{x=1} \\ &\quad - \int_0^1 \{ W(\Omega_n(x)) - W(\Omega(x)) \} \frac{\partial}{\partial x} \{ r(x)\, \mathfrak{B}_{m\lambda(t)}(x, t) \} \ . \end{aligned}$$

Since $\Omega_n(0) = \Omega(0) = 0$ and $\Omega_n(1) = \Omega(1) = 1$, the boundary terms vanish.

By the Kolmogorov-Smirnov law of the iterated logarithm, see SHORACK (2000),

$$\limsup_{n \to \infty} \frac{\| \Omega_n - \Omega \|_\infty}{\sqrt{n^{-1} \log\log n}} = \sqrt{8} \quad \text{almost surely},$$

then (6.24) with $\delta = (8\, n^{-1} \log\log n)^{1/2}$ gives

$$\limsup_{n \to \infty} \frac{\| W(\Omega_n(\cdot)) - W(\Omega(\cdot)) \|_\infty}{n^{-1/4} (\log n)^{1/2} (\log\log n)^{1/4}} = \sqrt{8} \quad \text{almost surely}.$$

One checks that

$$\int_0^1 \left| \frac{\partial}{\partial x} \{ r(x)\, \mathfrak{B}_{m\lambda(t)}(x, t) \} \right| dx = \mathcal{O}(\gamma^{-1}) \ ,$$

and so, almost surely,

$$(6.25) \qquad \| \mathfrak{K}_{r,n\gamma} - \mathfrak{K}_{r,\gamma} \|_\infty = \mathcal{O}\big(\gamma^{-1} n^{-1/4} (\log n)^{1/2} (\log\log n)^{1/4} \big) .$$

Next, we need a bound on the difference between the variances (standard deviations, actually). Observe that

$$\begin{aligned} &\mathbb{V}\mathrm{ar}[\, \mathfrak{K}_{r,n\gamma}(t) \,|\, \mathbb{X}_n \,] - \mathbb{V}\mathrm{ar}[\, \mathfrak{K}_{r,\gamma}(t) \,|\, \mathbb{X}_n \,] \\ &\qquad\qquad = \int_0^1 \big| \, r(x) \, \mathfrak{B}_{m\lambda(t)}(x,\, t) \, \big|^2 \, \{ \, d\Omega_n(x) - d\Omega(x) \, \} \; . \end{aligned}$$

Since $h \,|\, r(x)\, \mathfrak{B}_{m,h}(x,\, t) \,|^2$, $0 < h \leqslant 1$, is a family of convolution-like kernels, then Lemma (21.3.4) implies the almost-sure bound

$$\mathbb{V}\mathrm{ar}[\, \mathfrak{K}_{r,n\gamma}(t) \,|\, \mathbb{X}_n \,] - \mathbb{V}\mathrm{ar}[\, \mathfrak{K}_{r,\gamma}(t) \,|\, \mathbb{X}_n \,] = \mathcal{O}\big(\gamma^{-1} \sqrt{(n\gamma)^{-1} \log n} \, \big) \; .$$

Since

$$\mathbb{V}\mathrm{ar}[\, \mathfrak{K}_{r,n\gamma}(t) \,|\, \mathbb{X}_n \,] \asymp \mathbb{V}\mathrm{ar}[\, \mathfrak{K}_{r,\gamma}(t) \,|\, \mathbb{X}_n \,] \asymp \gamma^{-1} \; ,$$

this implies that

$$\Big| \sqrt{\mathbb{V}\mathrm{ar}[\, \mathfrak{K}_{r,n\gamma}(t) \,|\, \mathbb{X}_n \,]} - \sqrt{\mathbb{V}\mathrm{ar}[\, \mathfrak{K}_{r,\gamma}(t) \,|\, \mathbb{X}_n \,]} \Big| = \mathcal{O}\big(\gamma^{-1/2} \sqrt{(n\gamma)^{-1} \log n} \, \big) \, .$$

In combination with $\| \mathfrak{K}_{r,n\gamma} \|_\infty = n^{1/2} \, \| \mathfrak{D}_{r,n\gamma} \|_\infty = \mathcal{O}\big(\sqrt{\gamma^{-1} \log n} \, \big)$, then

$$(6.26) \qquad \mathcal{M}\big(\mathfrak{K}_{r,n\gamma}, \mathfrak{P}(\gamma) \big) \approx\!\!\approx \mathcal{M}\big(\mathfrak{K}_{r,\gamma}, \mathfrak{P}(\gamma) \big)$$

follows from Factoid (6.2).

We now take a closer look at the process $\mathfrak{K}_{r,n}$; see (6.23). First, again by the scaling properties of the Wiener process,

$$\mathfrak{K}_{r,\gamma}(t) =_{\mathrm{d}} \int_0^1 \mathfrak{B}_{m,\lambda(t)}(\, x \,,\, t \,)\, r(x) \sqrt{\omega(x)} \, dW(x)$$

jointly for all $t \in [\, 0 \,,\, 1 \,]$. If we now choose $r(t) = 1/\sqrt{\omega(t)}$, $t \in [\, 0 \,,\, 1 \,]$, then $\mathfrak{K}_{r,\gamma}(t) = \mathfrak{K}_\gamma(t)$ in distribution, jointly for all t, where

$$\mathfrak{K}_\gamma(t) \stackrel{\mathrm{def}}{=} \int_0^1 \mathfrak{B}_{m,\lambda(t)}(\, x \,,\, t \,)\, dW(x) \; , \qquad t \in \mathfrak{P}(\gamma) \; .$$

Again, obviously $\mathcal{M}\big(\mathfrak{K}_{r,\gamma} \,,\, \mathfrak{P}(\gamma) \big) =_{\mathrm{d}} \mathcal{M}\big(\mathfrak{K}_\gamma \,,\, \mathfrak{P}(\gamma) \big)$.

Now extend the Wiener process to the whole line, but with $W(0) = 0$, and then approximate $\mathfrak{K}_\gamma$ by

$$(6.27) \qquad \mathfrak{M}_\gamma(t) = \int_{-\infty}^{\infty} \mathfrak{B}_{m,\lambda(t)}(\, x \,,\, t \,)\, dW(x) \; ,$$

where the design density ω is extended to the whole line by

$$(6.28) \qquad \omega(t) = 1 \; , \quad t \in \mathbb{R} \setminus [\, 0 \,,\, 1 \,] \; .$$

Of course, we must show

$$\mathcal{M}(\mathfrak{M}_\gamma, \mathfrak{P}(\gamma)) \approx\approx \mathcal{M}(\mathfrak{K}_\gamma, \mathfrak{P}(\gamma)) \; . \tag{6.29}$$

Note that the dependence of $\mathfrak{M}_\gamma$ on γ partially disappears since by a scaling of the integration variable

$$\begin{aligned} \mathfrak{M}_\gamma(t/\gamma) &= \frac{1}{\varpi(\gamma t)} \int_{\mathbb{R}} \mathfrak{B}\Big(\frac{x-t}{\varpi(\gamma t)}\Big)\, dW(\gamma x) \\ &=_{\mathrm{d}} \frac{\sqrt{\gamma}}{\varpi(\gamma t)} \int_{\mathbb{R}} \mathfrak{B}\Big(\frac{x-t}{\varpi(\gamma t)}\Big)\, dW(x) \; . \end{aligned}$$

Recall that ϖ was defined in (5.3). Then, jointly for all t,

$$\frac{\mathfrak{M}_\gamma(t/\gamma)}{\sqrt{\mathrm{Var}[\,\mathfrak{M}_\gamma(t/\gamma)\,|\,\mathbb{X}_n\,]}} = \mathbb{G}_{\varpi,\gamma}(t) \quad \text{in distribution}, \tag{6.30}$$

with $\mathbb{G}_{\varpi,h}$ defined in (5.4). Then, by chaining the equivalences,

$$\mathcal{M}(\mathfrak{D}_{r,n\gamma}, \mathfrak{P}(\gamma)) \approx\approx_d \max_{t\in\mathfrak{P}(\gamma)} |\,\mathbb{G}_{\varpi,\gamma}(t)\,| = \mathbb{M}(\varpi,\gamma)$$

and we are done, except that we must prove (6.29).

Since $\varpi(x) = 1$ for $x \notin [\,0\,,\,1\,]$, it suffices to show that

$$\sup_{t\in\mathfrak{P}(\gamma)} \Big|\int_1^\infty \mathfrak{B}_{m,\gamma}(x-t)\, dW(x)\Big| = \mathcal{O}(\gamma^{2m}) \quad \text{in probability} \; . \tag{6.31}$$

The same bound for the integral over $(-\infty, 0)$ follows then as well. Note that, for the ranges in question, we have $x - t > k\gamma \log(1/\gamma) > 0$, so the representation of Lemma (14.7.11) gives

$$\mathfrak{B}_m(x-t) = \sum_{k=1}^m a_k \exp(-b_k(x-t)) \; , \quad t < x \; , \tag{6.32}$$

for suitable complex constants a_k, b_k, with the b_k having positive real parts. Note that half of the terms are missing. Further, there exists a positive real number β such that

$$\mathrm{Re}\; b_k \geqslant \beta \; , \quad k = 1, 2, \cdots, m \; .$$

Now,

$$\int_1^\infty \mathfrak{B}_{m\gamma}(x-t)\, dW(x) = \sum_{k=1}^m a_k\, \gamma^{-1/2} \exp(b_k \gamma^{-1}(t-1))\, Z_{k,\gamma} \; ,$$

where, for all k,

$$Z_{k,\gamma} = \gamma^{-1/2} \int_1^\infty \exp(-b_k\gamma^{-1}(x-1))\, dW(x) \; .$$

Of course, the $Z_{k\gamma}$ are (complex) normals with mean 0 and

$$\mathbb{E}[\,|\,Z_{k\gamma}\,|^2\,|\,\mathbb{X}_n\,] = \frac{1}{2\,\mathrm{Re}\; b_k} \leqslant \frac{1}{2\beta} \; ,$$

and their distribution does not depend on γ. Now, for $t \in \mathfrak{P}(\gamma)$, we get

$$| \exp(-b_k(t-1)/\gamma) | \leqslant \exp(-\beta\, k \log(1/\gamma)) = \gamma^{k\beta} ,$$

so that, for a suitable constant c,

$$\begin{aligned} \Bigl| \int_1^\infty \mathfrak{B}_{m,\gamma}(x-t)\, dW(x) \Bigr| &\leqslant c\, \gamma^{k\beta-1/2} \sum_{k=1}^{m} | Z_{k\gamma} | \\ &= \mathcal{O}\bigl(\gamma^{k\beta-1/2} \bigr) \quad \text{in probability} . \end{aligned}$$

So, if we initially had chosen the constant k so large that $k\beta - 1/2 \geqslant 2m$, then we get the bound $\mathcal{O}(\gamma^{2m})$, and it follows that

$$\sup_{t \in \mathfrak{P}(\gamma)} (n\gamma)^{-1/2} \Bigl| \int_1^\infty \mathfrak{B}_{m\gamma}(x-t)\, dW(x) \Bigr| = \mathcal{O}(\gamma^{2m})$$

in probability, thus showing (6.31). The same bound applies to the integral over $(-\infty, 0)$. Q.e.d.

EXERCISE: (6.3).

7. Asymptotic 100% confidence bands

In the previous sections, we have discussed simulation procedures for constructing $100(1-\alpha)\%$ confidence bands for the unknown regression function in the general regression problem (21.1.1)–(21.1.6). These procedures were justified by way of the asymptotic distribution theory of the weighted maximal error (at least for uniform designs). The asymptotic distribution theory suggests what kind of approximation errors of the maximal errors may be considered negligible, even for small sample sizes. However, the slow convergence of the distribution of the maximal error, suitably scaled and centered, to the asymptotic distribution creates the need for simulation procedures. A different kind of simulation procedure comes out of the work of DEHEUVELS and MASON (2004, 2007) regarding the Nadaraya-Watson estimator for a general regression problem somewhat different from "ours". The analogue for nonparametric density estimation is treated in DEHEUVELS and DERZKO (2008). It seems that this approach is yet another way of avoiding the slow convergence in an asymptotic result. In this section, we give our slant to the result of DEHEUVELS and MASON (2007) and derive an analogous result (conditioned on the design, of course) for the smoothing spline estimator.

Consider the Nadaraya-Watson estimator in the general regression problem (21.1.1)–(21.1.6),

$$(7.1) \qquad f^{n,A,H}(t) = \frac{\frac{1}{n}\sum_{i=1}^{n} Y_i\, A_H(t - X_i)}{\frac{1}{n}\sum_{i=1}^{n} A_H(t - X_i)} \,, \qquad t \in [0,1] \,,$$

with a global, random smoothing parameter H satisfying

$$(7.2) \qquad \frac{H}{\gamma_n} - 1 = o_P(1) \quad \text{with} \quad \gamma_n \asymp n^{-1/(m+1)}$$

for a deterministic sequence $\{\gamma_n\}_n$. In (7.1), $A_h(t) = h^{-1}A(h^{-1}t)$ for all t, and A is a nice kernel of order m, as in (14.4.2)–(14.4.3).

The starting point is an application of the theorem of Deheuvels and Mason (2004), to the effect that under the conditions of the general regression problem (21.1.1)–(21.1.6),

$$(7.3) \qquad \limsup_{n\to\infty} \max_{t \in \mathfrak{P}(\gamma)} \frac{|\, f^{n,A,H}(t) - f_o(t)\,|}{\sqrt{\mathbb{V}\mathrm{ar}[\, f^{n,A,h}(t)\,]\big|_{h=H} \log(1/H)}} = C_1$$

in probability for a known constant C_1. Asymptotically, this gives a 100% confidence band for f_o,

$$(7.4) \qquad f^{n,A,H}(t) \pm C_1 \sqrt{\mathbb{V}\mathrm{ar}[\, f^{n,A,h}(t)\,]\big|_{h=H} \log(1/H)} \;, \qquad t \in \mathfrak{P}(\gamma) \,.$$

The practical implication of this is problematic, though, because the factor $\sqrt{\log(1/H)}$ is asymptotic. In fact, Deheuvels and Mason (2004) suggest approximations with better small-sample behavior, but we shall ignore that.

Now, if the random H satisfies (7.2), then the bias is negligible compared with the noise component of the estimator, so we are really dealing with $f^{n,A,H} - \mathbb{E}[\, f^{n,A,h}\,]\big|_{h=H}$. Now, symmetrization ideas lead to the alternative estimator of $f^{n,A,H} - \mathbb{E}[\, f^{n,A,h}\,]\big|_{h=H}$,

$$(7.5) \qquad s^{n,A,H}(t) = \frac{\frac{1}{n}\sum_{i=1}^{n} s_i\, Y_i\, A_H(t - X_i)}{\frac{1}{n}\sum_{i=1}^{n} A_H(t - X_i)} \,, \qquad t \in [0,1] \,,$$

where

(7.6) $s_1, s_1, \cdots, s_n$ are independent Rademacher random variables, independent of the design and the noise, with
$$\mathbb{P}[\, s_1 = -1\,] = \mathbb{P}[\, s_1 = 1\,] = \tfrac{1}{2} \,.$$

Then, the work of Deheuvels and Mason (2007) implies that

$$(7.7) \qquad \limsup_{n\to\infty} \max_{t \in \mathfrak{P}(\gamma)} \frac{|\, s^{n,A,H}(t)\,|}{\sqrt{\mathbb{V}\mathrm{ar}[\, s^{n,A,h}(t)\,]\big|_{h=H} \log(1/H)}} = C_1$$

in probability, with the same constant C_1 as in (7.3). Although it is not immediately obvious from comparing the two limit results, the asymptotic factors $\log(1/H)$ are just about the same for small samples as well. The net effect of this is that

$$\max_{t\in\mathfrak{P}(\gamma)} \frac{|\, f^{n,A,H}(t) - f_o(t)\,|}{\sqrt{\mathbb{V}\mathrm{ar}[\, f^{n,A,h}(t)\,]\big|_{h=H}}} \approx \max_{t\in\mathfrak{P}(\gamma)} \frac{|\, s^{n,A,H}(t)\,|}{\sqrt{\mathbb{V}\mathrm{ar}[\, s^{n,A,h}(t)\,]\big|_{h=H}}}$$

in a sense we will not make precise. At this stage, it is not clear that anything has been achieved since the estimator $s^{n,A,H}$ is *more* complicated that $f^{n,A,H}$. However, in the expression on the right, one may condition on the data (design and responses), and then it only depends on the random signs. In other words, it may be simulated. We then get the asymptotic 100% confidence band for $f_o(t)$, $t \in \mathfrak{P}(\gamma)$,

$$f^{n,A,H}(t) \pm (1+\delta)\, \mathfrak{W}_{nH} \sqrt{\mathbb{V}\mathrm{ar}[\, f^{n,A,h}(t)\,]\big|_{h=H}} \;, \tag{7.8}$$

where $\delta > 0$ is arbitrary, and

$$\mathfrak{W}_{nH} = \max_{t\in\mathfrak{P}(\gamma)} \frac{|\, s^{n,A,H}(t)\,|}{\sqrt{\mathbb{V}\mathrm{ar}[\, s^{n,A,H}(t)\,]}} \;. \tag{7.9}$$

The occurrence of δ is unfortunate, of course, but the distribution theory of Konakov and Piterbarg (1984) suggests that δ may be chosen depending on n and H. Alternatively, one may simulate the distribution to obtain the usual $100(1-\alpha)\%$ confidence bands.

The above is a most unauthorized interpretation of Deheuvels and Mason (2004, 2007), but it translates beautifully to smoothing spline estimators. The only real difference is that we condition on the design. Let f^{nh} be the smoothing spline estimator of order $2m$ for the general regression problem (21.1.1)–(21.1.6) with smoothing parameter h. Let s^{nh} be the smoothing spline estimator in the regression problem with the data

$$(\, X_i \,,\, s_i\, Y_i\,) \,, \quad i = 1, 2, \cdots, n \;, \tag{7.10}$$

with the Rademacher signs s_i as in (7.5)–(7.6). In §§ 5 and 6, we constructed a Gaussian process $\mathbb{G}_{\varpi,\gamma}(t)$, $t \in \mathfrak{P}(\gamma)$, such that

$$\max_{t\in[0,1]} \frac{|\, f^{nH}(t) - f_o(t)\,|}{\sqrt{\mathbb{V}\mathrm{ar}[\, f^{nh}(t)\,|\, \mathbb{X}_n\,]\big|_{h=H}}} \approx\!\approx_d \max_{t\in[0,1]} |\, \mathbb{G}_{\varpi,\gamma}(t)\,| \;, \tag{7.11}$$

$$\max_{t\in[0,1]} \frac{|\, s^{nH}(t)\,|}{\sqrt{\mathbb{V}\mathrm{ar}[\, s^{nh}(t)\,|\, \mathbb{X}_n\,]\big|_{h=H}}} \approx\!\approx_d \max_{t\in[0,1]} |\, \mathbb{G}_{\varpi,\gamma}(t)\,| \;, \tag{7.12}$$

so everything is dandy.

(7.13) Exercise. Check that we really did already show (7.11) and (7.12).

Of course, the conditional variance of s^{nh} is problematic, but it is reasonable to estimate it by conditioning on the Y_i as well,

$$\mathbb{V}\text{ar}[\, s^{nh}(t) \,|\, \mathbb{X}_n \,]\big|_{h=H} \approx \mathbb{V}\text{ar}[\, s^{nh}(t) \,|\, \mathbb{X}_n \,,\, \mathbb{Y}_n \,]\big|_{h=H} \,,$$

where $\mathbb{Y}_n = (Y_1, Y_2, \cdots, Y_n)^T$ represents the responses in (21.1.1). So, if we can prove the following lemma, then we are home free.

(7.14) LEMMA. *Under the assumptions* (5.2), *and* (7.6),

$$\max_{t \in [0,1]} \left| \frac{\mathbb{V}\text{ar}[\, s^{nh}(t) \,|\, \mathbb{X}_n \,]\big|_{h=H}}{\mathbb{V}\text{ar}[\, s^{nh}(t) \,|\, \mathbb{X}_n \,,\, \mathbb{Y}_n \,]\big|_{h=H}} - 1 \right| = \mathcal{O}_P\big(\sqrt{(n\gamma)^{-1} \log n} \,\big) \,.$$

It is clear that, combined with (7.12), the lemma implies that

$$\text{(7.15)} \qquad \max_{t \in \mathfrak{P}(\gamma)} \frac{|\, s^{nH}(t) \,|}{\sqrt{\mathbb{V}\text{ar}[\, s^{nh}(t) \,|\, \mathbb{X}_n \,,\, \mathbb{Y}_n \,]\big|_{h=H}}} \approx\approx_d \max_{t \in \mathfrak{P}(\gamma)} |\, \mathbb{G}_{\varpi,\gamma}(t) \,| \,,$$

and since the quantity on the left can be simulated, this gives a way to simulate the distribution of the quantity of interest in (7.11).

PROOF OF LEMMA (7.14). In view of the equivalent kernel approximation result of § 7, it suffices to prove the lemma for the estimator

$$v^{n\gamma}(t) = \tfrac{1}{n} \sum_{i=1}^{n} s_i \, Y_i \, \mathfrak{R}_{\omega,m,\gamma}(X_i, t) \,, \quad t \in [\,0\,,1\,] \,.$$

Now,

$$\mathbb{V}\text{ar}[\, v^{n\gamma}(t) \,|\, \mathbb{X}_n \,] = n^{-2} \sum_{i=1}^{n} \big\{ \,|\, f_o(X_i) \,|^2 + \sigma^2(X_i) \big\} \,|\, \mathfrak{R}_{\omega,m,\gamma}(X_i, t) \,|^2$$

and

$$\mathbb{V}\text{ar}[\, v^{n\gamma}(t) \,|\, \mathbb{X}_n \,,\, \mathbb{Y}_n] = n^{-2} \sum_{i=1}^{n} |\, Y_i \,|^2 \,|\, \mathfrak{R}_{\omega,m,\gamma}(X_i, t) \,|^2 \,.$$

Consequently, since $|\, Y_i \,|^2 = |\, f_o(X_i) \,|^2 + 2\, f_o(X_i)\, D_i + |\, D_i \,|^2$, it suffices to show that, uniformly in $t \in [\,0\,,1\,]$,

$$\text{(7.16)} \qquad \frac{n^{-2} \sum_{i=1}^{n} D_i \, f_o(X_i) \,|\, \mathfrak{R}_{\omega,m,\gamma}(X_i, t) \,|^2}{\mathbb{V}\text{ar}[\, v^{n\gamma}(t) \,|\, \mathbb{X}_n \,]} = \mathcal{O}_P\big(\sqrt{(n\gamma)^{-1} \log n} \,\big)$$

and, with $\delta_i = |\, D_i \,|^2 - \sigma^2(X_i)$ for all i,

$$\text{(7.17)} \qquad \frac{n^{-2} \sum_{i=1}^{n} \delta_i \,|\, \mathfrak{R}_{\omega,m,\gamma}(X_i, t) \,|^2}{\mathbb{V}\text{ar}[\, v^{n\gamma}(t) \,|\, \mathbb{X}_n \,]} = \mathcal{O}_P\big(\sqrt{(n\gamma)^{-1} \log n} \,\big) \,.$$

Note that the expression above for the denominator looks like a kernel regression estimator for noiseless data, except that the kernel is too large by a factor $n\gamma$. Then, by Theorem (14.6.13),

$$n\gamma \Big| \mathbb{V}\mathrm{ar}[\, v^{n\gamma}(t) \,|\, \mathbb{X}_n \,] - \mathbb{V}\mathrm{ar}[\, v^{n\gamma}(t) \,] \Big| = \mathcal{O}\big(\sqrt{(n\gamma)^{-1} \log n} \,\big)$$

almost surely. Now, since

$$\begin{aligned} & n\gamma \,\mathbb{V}\mathrm{ar}[\, v^{n\gamma}(t) \,] \\ &\quad = \gamma \int_0^1 \{\, |\, f_o(x) \,|^2 + \sigma^2(x) \,\} \,|\, \mathfrak{R}_{\omega,n,\gamma}(x,\, t) \,|^2 \,\omega(x)\, dx \\ &\quad = \{\, |\, f_o(t) \,|^2 + \sigma^2(t) \,\} \,\gamma \,\|\, \mathfrak{R}_{\omega,n,\gamma}(\,\cdot\,,\, t) \,\|^2_{L^2((0,1),\omega)} + \mathcal{O}(\gamma) \\ &\quad \asymp 1 \end{aligned}$$

uniformly in $t \in [\,0\,,\,1\,]$, it follows that

$$\text{(7.18)} \qquad \mathbb{V}\mathrm{ar}[\, v^{n\gamma}(t) \,|\, \mathbb{X}_n \,] \asymp (n\gamma)^{-1} \quad \text{uniformly in } t \in [\,0\,,\,1\,] \,.$$

Now, consider the numerators. Let $q(x,\, t) = f_o(x) \,|\, \mathfrak{R}_{\omega,n,\gamma}(x,\, t) \,|^2$. Omitting a factor n^{-1}, using Lemma (21.3.4) we have

$$\Big| \tfrac{1}{n} \sum_{i=1}^{n} D_i \, q(X_i,\, t) \Big| = \|\, q(\,\cdot\,,\, t) \,\|_{\omega,1,\gamma,1} \cdot \mathcal{O}\big(\sqrt{(n\gamma)^{-1} \log n} \,\big)$$

almost surely. Of course, it is a standard exercise to show that

$$\text{(7.19)} \qquad \|\, q(\,\cdot\,,\, t) \,\|_{\omega,1,\gamma,1} = \mathcal{O}(\gamma^{-1}) \,,$$

and the net result is that, in probability,

$$n^{-2} \sum_{i=1}^{n} D_i \, f_o(X_i) \,|\, \mathfrak{R}_{\omega,m,\gamma}(X_i,\, t) \,|^2 = \mathcal{O}\big((n\gamma)^{-3/2} (\log n)^{1/2} \big) \,.$$

Together with (7.18), this shows the bound $\sqrt{(n\gamma)^{-1} \log n}$ for the quantity on the left of (7.16). The bound (7.17) follows similarly. Q.e.d.

(7.20) EXERCISE. (a) Show (7.19). (b) Show the bound (7.17).

EXERCISES: (7.13), (7.20).

8. Additional notes and comments

Ad § 1: It is interesting to note that initially (1.9) and (4.9) formed the *starting* point of investigations showing that these problems are equivalent to estimation for the white noise process

$$dY(t) = f_o(t)\, dt + \frac{\sigma(t)}{\sqrt{n}}\, dW(t) \,, \quad 0 \leqslant t \leqslant 1 \,,$$

in the sense of Le Cam's theory of experiments. Here, $W(t)$ is the standard Wiener process. See, e.g., BROWN, CAI, and LOW (1996), BROWN, CAI,

Low, and Zhang (2002) and Grama and Nussbaum (1998). The estimation for white noise processes with drift goes back to Pinsker (1980).

Some other methods for constructing confidence bands are discussed by Stadtmüller (1986), Eubank and Speckman (1993), and Claeskens and van Keilegom (2003). For closely related hypothesis tests, see, e.g., Lepski and Tsybakov (2000).

Ad § 2: The proof of Theorem (2.15) largely follows Eubank and Speckman (1993), but the *BV* trick (2.16) simplifies the argument. Diebolt (1993) uses the same equal-in-distribution arguments, but he constructs strong approximations to the *marked empirical process* (without conditioning on the design)

$$F_n(t) \stackrel{\text{def}}{=} \tfrac{1}{n} \sum_{i=1}^{n} D_i \, \mathbb{1}(X_i \leqslant t) \,, \quad t \in [0,1] \,.$$

The connection to our presentation is by way of

$$\tfrac{1}{n} \sum_{i=1}^{n} D_i \, \mathfrak{R}_{\omega m h}(X_i, t) = \int_0^1 \mathfrak{R}_{\omega m h}(x, t) \, dF_n(x)$$

and integration by parts (rather than summation by parts).

Ad §§ 5 and 6: The development is analogous to that of Konakov and Piterbarg (1984), with the additional complication of a location dependent smoothing parameter. For nonuniform designs, our limiting stochastic processes are not stationary, so the asymptotic distribution of the maxima is problematic. Possibly, the limiting processes are locally stationary and satisfy the conditions C1–C6 of Hüsler, Piterbarg, and Seleznjev (2000), so that their Theorem 2 applies. However, the verification of these conditions constitute nontrivial exercises in probability theory themselves. See also Weber (1989) and references therein. We should also mention that "our" Gaussian processes should provide another way to prove limit theorems for the uniform error of nonparametric regression estimators. Regarding asymptotics for the L^2 error in spline smoothing, see also Li (1989) and Horváth (1993).

23

Nonparametric Regression in Action

1. Introduction

After the preparations of the previous chapters, we are ready to confront the practical performance of the various estimators in nonparametric regression problems. In particular, we report on simulation studies to judge the efficacy of the estimators and present more thorough analyses of the data sets introduced in § 12.1.

Of course, the previous chapters suggest a multitude of interesting experiments, but here we focus on smoothing splines and local polynomial estimators. The dominant issue is smoothing parameter selection: the "optimal" choice to see what is possible, and data-driven methods to see how close one can get to the optimal choice, whatever the notion of "optimality". We take it one step further by also considering the selection of the order of the estimators, both "optimal" and data-driven. (For local polynomials, the order refers to the order of the polynomials involved; for smoothing splines, it refers to the order of the derivative in the penalization or the order of the associated piecewise polynomials.) The methods considered are generalized cross-validation (GCV), Akaike's information criterion (AIC), and generalized maximum likelihood (GML). It is interesting to note that Davies, Gather, and Weinert (2008) make complementary choices regarding estimators, smoothing parameter selection procedures, *and* the choice of test examples.

The regression problems considered are mostly of the standard variety with deterministic, uniform designs and iid normal noise, although we also briefly consider nonnormal (including asymmetric) noise and nonuniform designs. The data sets conform to this, although not completely. Interesting questions regarding the utility of *boundary corrections, relaxed boundary splines, total-variation penalized least-squares, least-absolute-deviations splines, monotone estimation* (*or constrained estimation in general*), *and estimation of derivatives* are not considered. We do not pay much attention to random designs, but see § 6.

P.P.B. Eggermont, V.N. LaRiccia, *Maximum Penalized Likelihood Estimation*,
Springer Series in Statistics, DOI 10.1007/b12285_12,

In the simulation studies, the estimators are mainly judged by the discrete L^2 error criterion (the square root of the sum of squared estimation errors at the design points). We compare the "optimal" results with respect to the smoothing parameter when the order of the estimators is fixed and also the optimal results over "all" orders. Results based on data-driven choices of the smoothing parameter for a fixed order of the estimators are included as well. Undeterred by a paucity of theoretical results, we boldly select the order of the estimators in a data-driven way as well, but see STEIN (1993).

The main reason for selecting the discrete L^2 error criterion is convenience, in that it is easy to devise rational methods for selecting the smoothing parameter. Although a case can be made for using the discrete L^∞ error criterion (uniform error bounds), selecting the smoothing parameter in the L^∞ setting is mostly *terra incognita*. However, in the context of confidence bands, it is mandatory to undersmooth (so that the bias is negligible compared with the noise component of the estimators), and the L^2 criterion seems to achieve this; see § 5. Here, the quantities of interest are the width of the confidence bands for a nominal confidence level as well as the actual coverage probabilities. While studying confidence bands, the effect of nonnormal and asymmetric noise is investigated as well.

In the first set of simulation studies, we investigate various smoothing parameter selection procedures for smoothing splines and local polynomial estimators from the discrete L^2 error point of view. The regression problems are of the form

$$(1.1) \qquad y_{in} = f_o(x_{in}) + d_{in} \ , \quad i = 1, 2, \cdots, n \ ,$$

for $n = 100$ and $n = 400$, where the design is deterministic and uniform,

$$(1.2) \qquad x_{in} = \frac{i-1}{n-1} \ , \quad i = 1, 2, \cdots, n \ ,$$

and the noise is iid normal,

$$(1.3) \qquad (\, d_{1,n}, \, d_{2,n}, \, \cdots, d_{n,n} \,)^T \sim \text{Normal}(\, 0 \, , \, \sigma^2 I \,) \ ,$$

with σ^2 unknown but of course known in the simulations. We settled almost exclusively on a noise-to-signal ratio of 10% in the sense that

$$(1.4) \qquad \sigma = \tfrac{1}{10} \left(\max_{1 \leqslant i \leqslant n} f_o(x_{in}) - \min_{1 \leqslant i \leqslant n} f_o(x_{in}) \right) .$$

Since, roughly speaking, the estimation error is a function of σ^2 / n, we refrained from choosing more than one noise level. Both the smoothing spline of order $2m$ and the local polynomial estimator of order m are here denoted by $f^{nh,m}$. The accuracy of $f^{nh,m}$ as an estimator of f_o is measured by

$$(1.5) \qquad \| f^{nh,m} - f_o \|_2 = \left\{ \tfrac{1}{n} \sum_{i=1}^{n} | f^{nh,m}(x_{in}) - f_o(x_{in}) |^2 \right\}^{1/2} .$$

Table 1.1. The nine regression functions to be used in the simulations. The "Cantor", "Gompertz" and "Richards" families are defined in (1.7)–(1.8).

(BR)	$f_o(x) = \text{Richards}(\, x \mid \beta\,) - 5 \exp\Big(-\frac{20}{3}\,(\, x - \frac{2}{5}\,)^2 \Big)$
with	$\beta_0 = 49.\ ,\ \beta_1 = 1.34\ ,\ \beta_2 = 84.\ , \beta_3 = 1.3\ ,$
(CH)	$f_o(x) = \text{Gompertz}(\, x \mid \beta\,)$
with	$\beta_0 = 0.5\ ,\ \beta_1 = 0.9\ ,\ \beta_2 = 4.5\ ,$
(Poly5)	$f_o(x) = \sum_{i=0}^{5} \beta_i\,(x - 0.5)^i$
with	$\beta_0 = 0\,,\ \beta_1 = 0.25\,,\ \beta_2 = -2.61\ ,$ $\beta_3 = 9.34\,,\ \beta_4 = -10.33\,,\ \beta_5 = 3.78\ ,$
(Wa-I)	$f_o(x) = 4.26 * \big(\, \mathrm{e}^{-3x} - 4\,\mathrm{e}^{-6x} + 3\,\mathrm{e}^{-9x}\,\big)\ ,$
(Wa-II)	$f_o(x) = 4.26 * \big(\, \mathrm{e}^{-6x} - 4\,\mathrm{e}^{-12x} + 3\,\mathrm{e}^{-18x}\,\big)\ ,$
(HM)	$f_o(x) = 6x - 3 + 4 \exp(\, -16\,(x - 0.5)^2/\sqrt{2\pi}\,)\ ,$
(sn5px)	$f_o(x) = \sin(5\,\pi\,x\,)\ ,$
(Cantor)	$f_o(x) = \text{Cantor}\big(\, x + \frac{1}{16}\pi\,\big)\ ,$
(Can5x)	$f_o(x) = \text{cantor}\big(\, 5\,x + \frac{1}{16}\pi\,\big)\ .$

Later on, we also consider the uniform discrete error,

$$(1.6) \qquad \big|\, f^{nh,m} - f_o \,\big|_\infty = \max_{1 \leqslant i \leqslant n} \,|\, f^{nh,m}(x_{in}) - f_o(x_{in})\,|\ .$$

At the risk of stating the obvious, note that the quantities $\big|\, f^{nh,m} - f_o \,\big|_2$ and $\big|\, f^{nh,m} - f_o \,\big|_\infty$ are known only in simulation studies. For the aforementioned data sets, these quantities are unknowable.

In the simulation studies, the following test examples are used. Their graphs are shown in Figure 12.1. The first one, labeled "BR" is meant to mimic the Wood Thrush Data Set of Brown and Roth (2004). The second one ("CH") is a mock-up of the Wastewater Data Set of Hetrick and Chirnside (2000). The example "HM" depicts a linear regression function with a small, smooth bump, taken from Müller (1988). The example "Wa-I" is based on Wahba (1990); "Wa-II" is a scaled-up version. The example "sn5px" is a very smooth function with a relatively large scale from Ruppert, Sheather, and Wand (1992). We decided to include a low-order polynomial as well ("Poly5"), inspired by the Wastewater Data Set. At this late stage of the game, we point out that the "usual" results for nonparametric regression assume that the regression function belongs to some Sobolev space $W^{m,p}(a, b)$, or variations thereof, for a fixed (small)

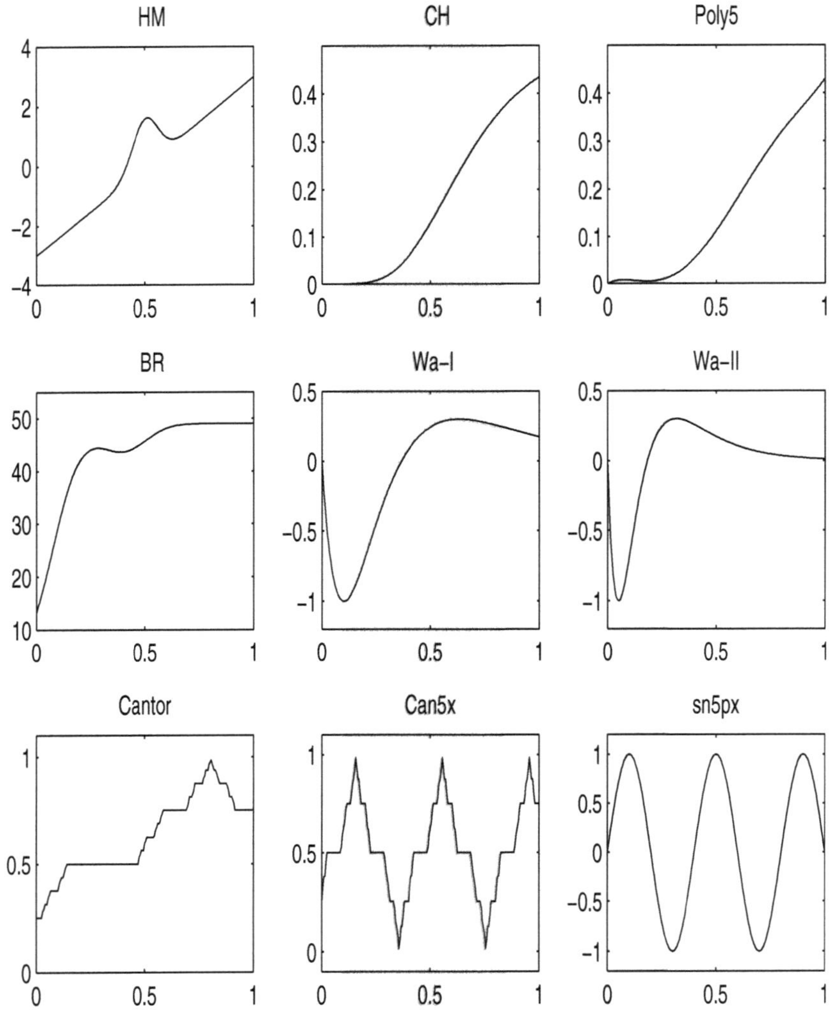

Figure 1.1. The nine regression functions used in the simulations with smoothing splines and local polynomial estimators.

value of m, implicitly saying that it does not belong to $W^{m+1,p}(a,b)$, say. But, as is typical, the examples considered above are in $C^\infty[\,a\,,\,b\,]$ if not in fact real-analytic. In an attempt at subterfuge, we therefore also include a Cantor-like function as well as a scaled-up version ("Cantor" and "Can5x"). Note that "Can5x" and "sn5px" have the same scale.

These nine regression functions are tabulated in Table 1.1. Here, the "Richards" family, introduced by RICHARDS (1959), is defined as

$$\text{Richards}(\,x\,|\,\beta\,) = \beta_0\,\bigl(\,1 + \exp(\,\beta_1 - \beta_2\,t\,)\,\bigr)^{-1/\beta_3}\;; \tag{1.7}$$

the "Gompertz" family, due to GOMPERTZ (1825), is defined as

$$\text{(1.8)} \qquad \text{Gompertz}(\,x\,|\beta\,) = \beta_0 \exp\Bigl(\,-\beta_2^{-1} \exp(\,\beta_2\,(\,x-\beta_1\,)\,)\Bigr) \;,$$

and $\text{Cantor}(x)$ is the standard Cantor function, defined on $[\,0\,,\,1\,]$ as

$$\text{(1.9)} \qquad \text{Cantor}(x) = y \;, \quad \begin{array}{ll} y = (2m-1)\,2^{-n} \;, & m = 1,2,\,\cdots,2^{n-1} \;, \\ |\,x-y\,| \leqslant \tfrac{1}{2}\,3^{-n} \;, & n = 1,2,\,\cdots \;, \end{array}$$

$[\,-1\,,\,0\,]$ by reflection, and on the rest of the line by periodic extension,
(1.10)

$$\text{Cantor}(x) = \begin{cases} \text{Cantor}(\,-x\,) \;, & -1 \leqslant x \leqslant 0 \;, \\ \text{Cantor}(\,x-2k\,) \;, & 2k-1 \leqslant x \leqslant 2k+1 \; (k \in \mathbb{Z}) \;. \end{cases}$$

(1.11) REMARK. The algorithmic construction of $\text{Cantor}(x)$ is more revealing: Divide the interval $[\,0\,,\,1\,]$ into three equal parts (subintervals). On the middle subinterval, define the function to be equal to $\frac{1}{2}$ (which is the coordinate of its midpoint). Now divide each of the remaining intervals in threes and define the function on the middle intervals to be $\frac{1}{4}$ and $\frac{3}{4}$ (again, the coordinates of the midpoints), and proceed recursively.

2. Smoothing splines

We now report on the performance of smoothing splines in the standard regression problem (1.1)–(1.3), with the nine test examples. In all examples, we choose

$$\text{(2.1)} \qquad \sigma = \tfrac{1}{10}\,\Bigl\{\,\max_{1\leqslant i\leqslant n} f_o(x_{in}) - \min_{1\leqslant i\leqslant n} f_o(x_{in})\,\Bigr\} \;.$$

The fact that it is possible to compute smoothing spline estimators of "all" orders makes it possible to investigate whether it would be beneficial to choose the order of the smoothing spline and whether it can be done well in a data-driven way. The answer to the first question is unequivocally yes, but there is room for improvement on the data-driven aspect of it. Of course, the (standard) smoothing parameter h needs choosing as well.

We repeat that interest is in the dependence of the smoothing spline estimator on the smoothing parameter h as well as on the order m of the derivative in the penalization. Thus, the solution of the smoothing spline problem

$$\text{(2.2)} \qquad \begin{array}{ll} \text{minimize} & \frac{1}{n} \sum\limits_{i=1}^{n} |\,f(x_{in}) - y_{in}\,|^2 + h^{2m}\,\|\,f^{(m)}\,\|^2 \\ \text{subject to} & f \in W^{m,2}(0,1) \end{array}$$

is denoted by $f^{nh,m}$. We report on the estimated mean of the quantity ("the error")

$$(2.3)\qquad D(f^{nh,m}, f_o) \stackrel{\text{def}}{=} \frac{\| f^{nh,m} - f_o \|_2}{\max_{1\leqslant i\leqslant n} f_o(x_{in}) - \min_{1\leqslant i\leqslant n} f_o(x_{in})} ,$$

with $\| \cdot \|_2$ as in (1.5), in several scenarios. In connection with (2.1), the scaling in (2.3) of the L^2 error by the range of the regression function makes comparison of results for different examples possible. Also, restricting the order to $m \leqslant 8$ seems reasonable. After all, the pieces of the spline are then polynomials of degree 15. Since the largest sample size considered is $n = 400$, this would seem more than adequate.

In Tables 2.1 and 2.2, for each "order" m $(1 \leqslant m \leqslant 8)$, we report on (the estimated values of)

$$(2.4)\qquad \mathbb{E}\Big[\min_{h>0} D(f^{nh,m}, f_o) \Big] .$$

In the last column, under the heading "ch" (for "chosen"), the optimum over the order m,

$$(2.5)\qquad \mathbb{E}\Big[\min_{1\leqslant m\leqslant 8} \min_{h>0} D(f^{nh,m}, f_o) \Big] ,$$

is reported. Note that when estimating the quantity (2.5), the best order is determined for each replication.

We also considered three data-driven methods for choosing the smoothing parameter h: Generalized maximum likelihood (GML), see (20.2.28), the "improved" Akaike information criterion (AIC), see (18.6.27), and generalized cross validation (GCV), see (18.4.2). Thus, for each m, we report on the estimated values of

$$(2.6)\qquad \mathbb{E}[D(f^{nH,m}, f_o)] ,$$

with the random H chosen by the three methods. Finally, these three methods were also used to select m by choosing, in each replication, that value of m for which the respective functionals are minimized. Thus, the last column under the heading "ch" lists the estimated values of

$$(2.7)\qquad \mathbb{E}[D(f^{nH,M}, f_o)] ,$$

with the random M picked by the three methods. Note that these choices of h and m are *rational.*

In Table 2.1, the results are reported for sample size $n = 100$ based on 1000 replications. Table 2.2 deals with sample size $n = 400$ and is based on 250 replications. Otherwise, it is the same as Table 2.1.

We now discuss some of the simulation results. The surprise is that they are clear-cut. For the "optimal" choices, it is abundantly clear that the order of the smoothing splines should be chosen. In some examples, even though the optimal error for fixed order does not change drastically with

Table 2.1. L2 errors of splines for $n = 100$ (percentage of range).

order	1	2	3	4	5	6	7	8	ch
HM									
opt:	3.18	3.04	3.12	3.21	3.30	3.37	3.44	3.51	2.97
gml:	3.61	3.38	3.77	4.10	4.33	4.32	4.27	4.11	4.11
aic:	3.36	3.26	3.38	3.51	3.65	3.75	3.82	3.84	3.41
gcv:	3.48	3.34	3.46	3.59	3.69	3.75	3.78	3.79	3.59
CH									
opt:	2.51	2.09	2.11	2.20	2.28	2.42	2.60	2.77	1.94
gml:	3.41	2.27	2.28	2.47	2.44	2.50	2.67	2.84	2.41
aic:	2.64	2.33	2.42	2.54	2.55	2.61	2.77	2.92	2.57
gcv:	2.77	2.46	2.53	2.64	2.65	2.69	2.82	2.97	2.74
Poly5									
opt:	2.57	2.03	2.09	2.17	2.26	2.40	2.60	2.77	1.91
gml:	3.39	2.21	2.30	2.41	2.41	2.48	2.66	2.84	2.34
aic:	2.70	2.28	2.42	2.49	2.52	2.59	2.77	2.92	2.53
gcv:	2.84	2.40	2.52	2.59	2.62	2.67	2.81	2.96	2.70
BR									
opt:	3.48	2.61	2.67	2.73	2.73	2.74	2.76	2.84	2.52
gml:	4.06	2.77	2.89	3.20	3.34	3.13	2.95	2.93	2.93
aic:	3.60	2.81	2.93	3.06	3.11	3.08	3.05	3.01	3.05
gcv:	3.76	2.91	3.04	3.15	3.17	3.14	3.09	3.04	3.21
Wa-I									
opt:	4.50	3.57	3.11	2.85	2.72	2.65	2.68	2.78	2.62
gml:	4.97	3.70	3.29	3.05	2.91	2.84	2.81	2.86	2.90
aic:	4.70	3.75	3.33	3.11	3.00	2.96	2.91	2.94	3.02
gcv:	4.90	3.83	3.43	3.21	3.09	3.02	2.96	2.98	3.17
Wa-II									
opt:	5.19	4.38	3.86	3.54	3.33	3.20	3.12	3.07	3.07
gml:	5.56	4.66	4.23	3.95	3.73	3.56	3.41	3.31	3.31
aic:	5.68	4.58	4.06	3.76	3.58	3.46	3.38	3.33	3.39
gcv:	5.80	4.61	4.13	3.84	3.65	3.50	3.40	3.34	3.53
Cantor									
opt:	3.75	3.64	3.69	3.75	3.81	3.85	3.89	3.94	3.55
gml:	4.06	3.90	4.52	4.78	4.85	4.69	4.61	4.51	4.70
aic:	3.89	3.86	3.97	4.09	4.21	4.27	4.29	4.28	4.05
gcv:	4.05	3.93	4.04	4.13	4.21	4.23	4.23	4.22	4.19
Can5x									
opt:	5.70	5.57	5.60	5.59	5.57	6.07	8.15	8.35	5.47
gml:	6.26	6.08	7.93	8.51	8.58	8.68	8.78	8.78	8.04
aic:	6.39	5.89	5.81	5.76	5.74	6.10	8.50	8.57	5.78
gcv:	6.26	5.82	5.81	5.78	5.68	6.07	8.40	8.53	6.11
sn5px									
opt:	4.75	3.46	3.18	3.19	3.08	3.09	3.22	3.24	2.98
gml:	6.73	3.83	3.38	3.29	3.19	3.20	3.38	3.52	3.52
aic:	4.94	3.59	3.33	3.38	3.27	3.31	3.44	3.42	3.39
gcv:	5.21	3.70	3.44	3.50	3.37	3.36	3.46	3.43	3.57

Table 2.2. L2 errors of splines for $n = 400$ (percentage of range).

order	1	2	3	4	5	6	7	8	ch
HM									
opt:	1.88	1.70	1.72	1.76	1.80	1.84	1.87	1.94	1.67
gml:	2.19	1.75	1.83	1.95	2.09	2.20	2.31	2.37	2.37
aic:	1.92	1.78	1.81	1.86	1.91	1.95	1.97	1.99	1.86
gcv:	1.95	1.79	1.82	1.87	1.92	1.95	1.97	1.99	1.88
CH									
opt:	1.44	1.13	1.15	1.17	1.18	1.23	1.32	1.40	1.05
gml:	1.98	1.25	1.22	1.34	1.29	1.28	1.36	1.43	1.35
aic:	1.49	1.26	1.31	1.35	1.36	1.36	1.43	1.50	1.39
gcv:	1.51	1.28	1.32	1.36	1.37	1.36	1.44	1.50	1.40
Poly5									
opt:	1.50	1.11	1.16	1.17	1.16	1.20	1.31	1.40	1.04
gml:	1.98	1.21	1.25	1.35	1.27	1.24	1.34	1.43	1.34
aic:	1.55	1.23	1.32	1.36	1.34	1.33	1.42	1.50	1.39
gcv:	1.57	1.25	1.34	1.37	1.35	1.34	1.42	1.50	1.40
BR									
opt:	2.07	1.50	1.53	1.47	1.44	1.45	1.47	1.49	1.41
gml:	2.39	1.61	1.62	1.72	1.70	1.56	1.58	1.60	1.60
aic:	2.19	1.63	1.74	1.77	1.75	1.73	1.71	1.72	1.72
gcv:	2.16	1.66	1.76	1.75	1.76	1.82	1.77	1.72	1.72
Wa-I									
opt:	2.72	2.05	1.72	1.55	1.45	1.40	1.39	1.42	1.37
gml:	2.95	2.08	1.79	1.63	1.54	1.49	1.49	1.46	1.51
aic:	2.78	2.11	1.83	1.68	1.59	1.57	1.56	1.52	1.59
gcv:	2.77	2.13	1.84	1.69	1.61	1.57	1.56	1.53	1.61
Wa-II									
opt:	3.16	2.51	2.15	1.93	1.79	1.71	1.66	1.63	1.63
gml:	3.22	2.57	2.25	2.07	1.96	1.88	1.81	1.75	1.75
aic:	3.22	2.57	2.24	2.04	1.92	1.83	1.78	1.75	1.78
gcv:	3.22	2.58	2.25	2.05	1.93	1.84	1.79	1.75	1.80
Canto									
opt:	2.41	2.41	2.46	2.50	2.52	2.55	2.58	2.62	2.36
gml:	2.56	2.50	2.60	2.73	2.97	3.14	3.17	3.13	2.76
aic:	2.46	2.50	2.58	2.63	2.65	2.63	2.64	2.66	2.55
gcv:	2.47	2.51	2.60	2.66	2.67	2.64	2.64	2.66	2.56
Can5x									
opt:	3.61	3.59	3.66	3.70	3.96	4.06	6.80	7.45	3.55
gml:	3.79	3.80	4.00	4.04	4.02	4.06	7.75	7.77	3.82
aic:	3.70	3.70	3.84	3.92	4.01	4.06	6.80	7.45	3.78
gcv:	3.68	3.68	3.80	3.87	4.02	4.06	6.80	7.45	3.77
sn5px									
opt:	2.72	1.87	1.71	1.70	1.60	1.62	1.66	1.65	1.55
gml:	3.46	2.16	1.83	1.74	1.68	1.67	1.72	1.70	1.70
aic:	2.73	1.95	1.79	1.79	1.71	1.73	1.76	1.74	1.76
gcv:	2.79	1.96	1.81	1.80	1.73	1.74	1.76	1.74	1.78

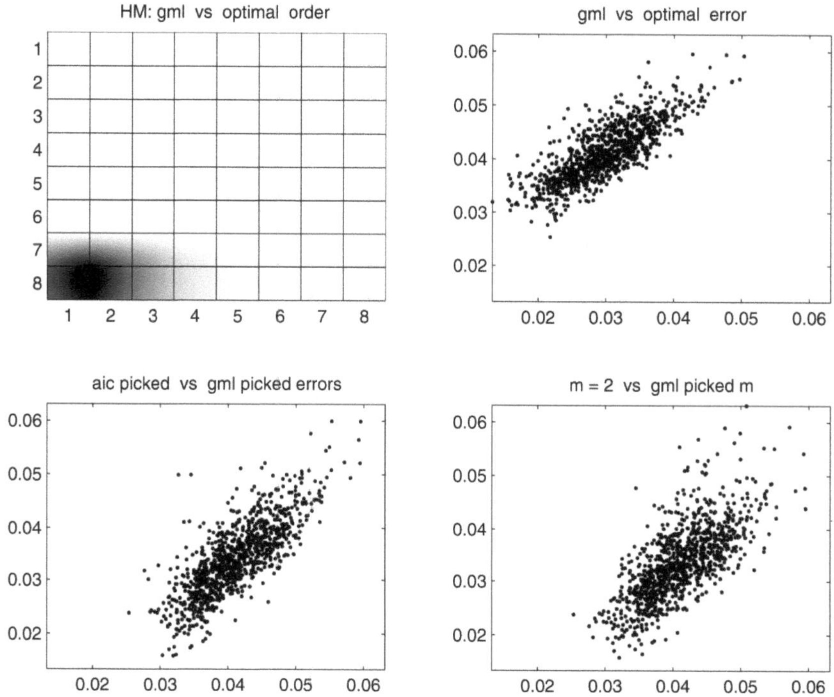

Figure 2.1(a). Spline smoothing of various orders applied to the example "HM" with sample size $n = 100$ and $\sigma = 0.1$. Top left: qualitative 2D-histogram of the optimal order versus the order picked by the GML method in each replication (darker means higher frequency). Indeed, in this example, GML always picked order 8. The other diagrams show scatter plots of the errors $D(f^{nh,m}, f_o)$ in various scenarios. The bottom right diagram deals with errors for the best fixed order according to GML, which changes from case to case, versus the order picked in each replication by GML.

the order of the spline, picking the optimal order for each replication gives a worthwhile improvement. Another surprise is that, for most examples, the cubic smoothing spline is not competitive with higher-order splines. By way of example, if we had to pick an optimal order for the "Wa-I" example, it would be 6 ($\equiv$ the order of the derivative in the penalization). However, for the "BR" example, the best fixed order would be 3 (with cubic splines close to optimal), and for the "Cantor" and "Can5x" examples, it would be 2. (Why it is not 1 the authors cannot explain.) In general, there is no fixed order that works (close to) best for all the examples considered.

The picture changes somewhat when the order of the splines is chosen by the various procedures. Here, the relatively small sample size makes things difficult, but it still seems clear that picking a fixed m regardless of the setting is not a good idea. For the "Wa-I" procedure, GML consistently

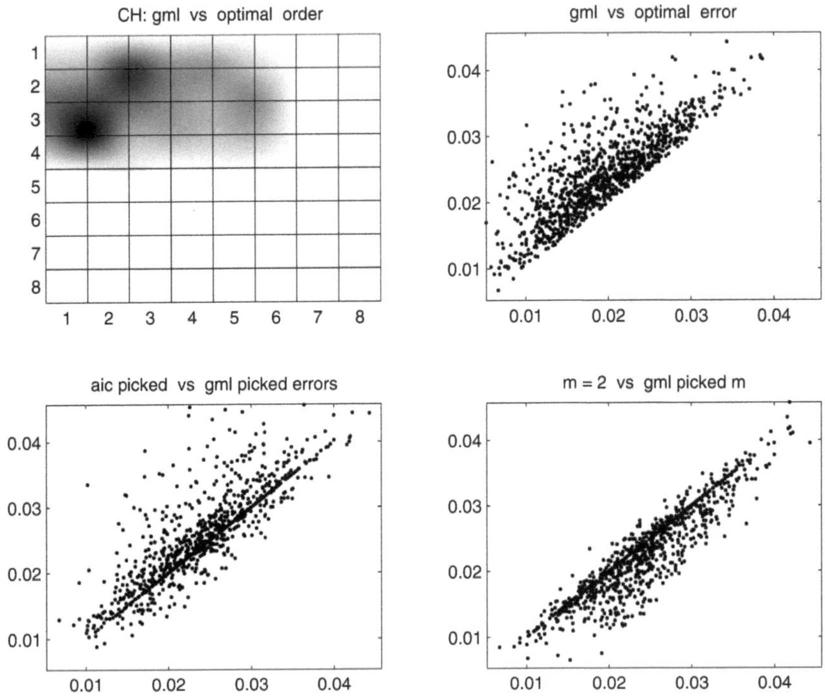

Figure 2.1(b). Spline smoothing of various orders for the example "CH".

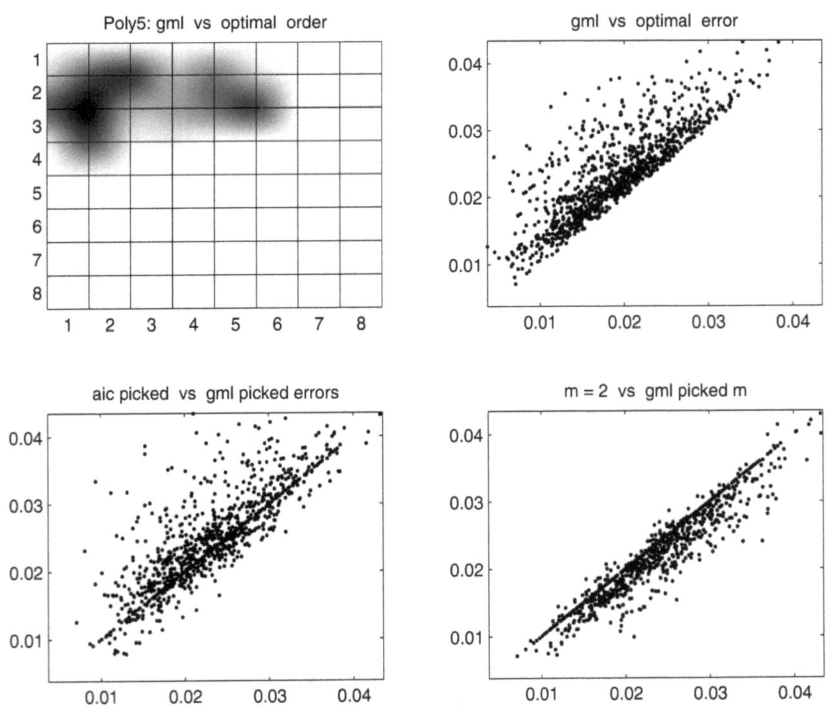

Figure 2.1(c). Spline smoothing of various orders for the example "Poly5".

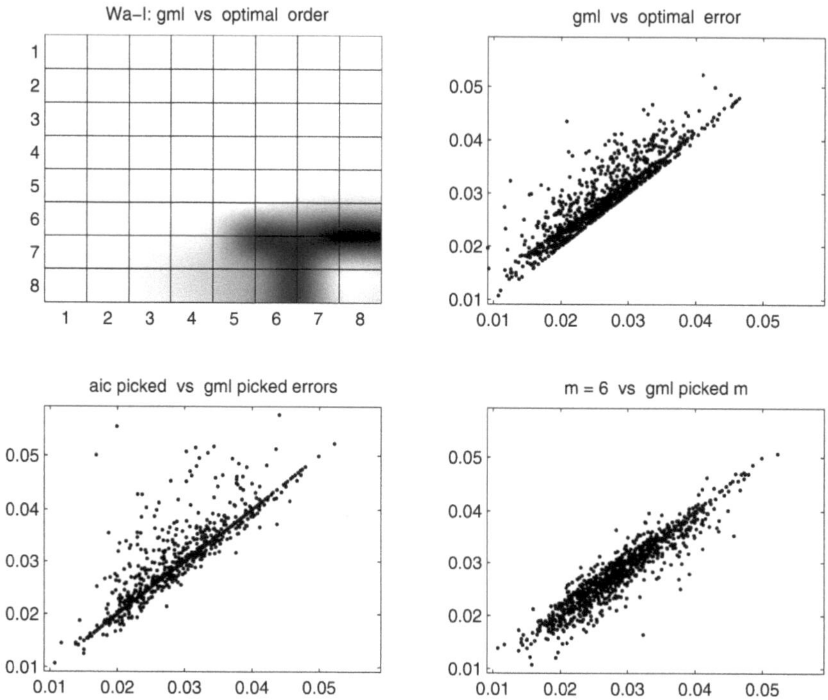

Figure 2.1(d). Spline smoothing of various orders for "Wa-I".

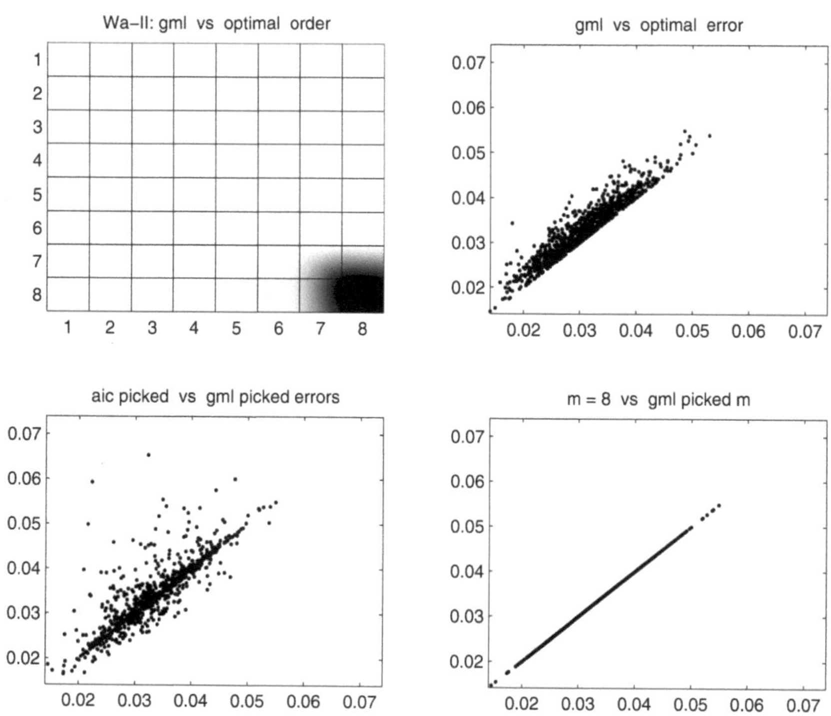

Figure 2.1(e). Spline smoothing of various orders for "Wa-II".

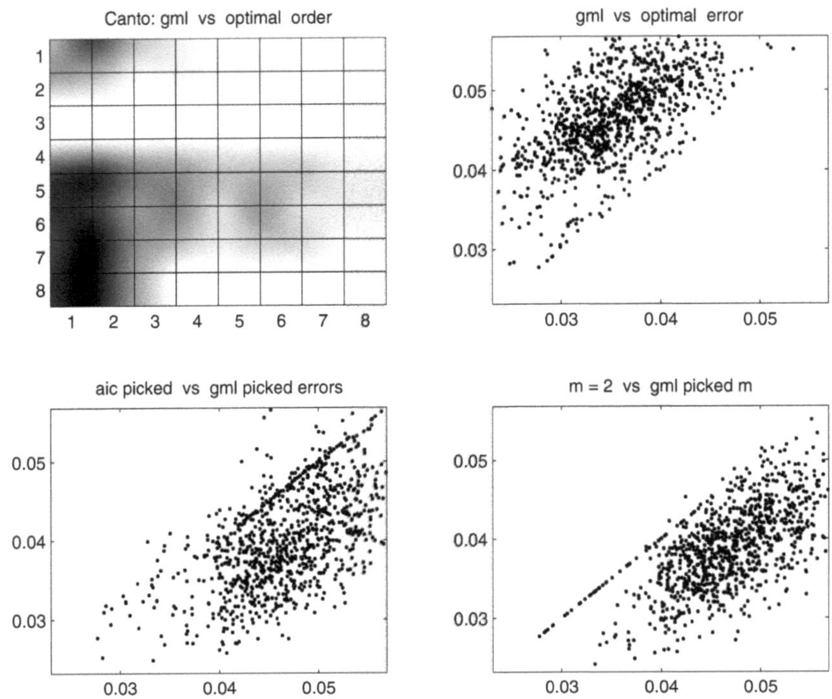

Figure 2.1(f). Spline smoothing of various orders for "Cantor".

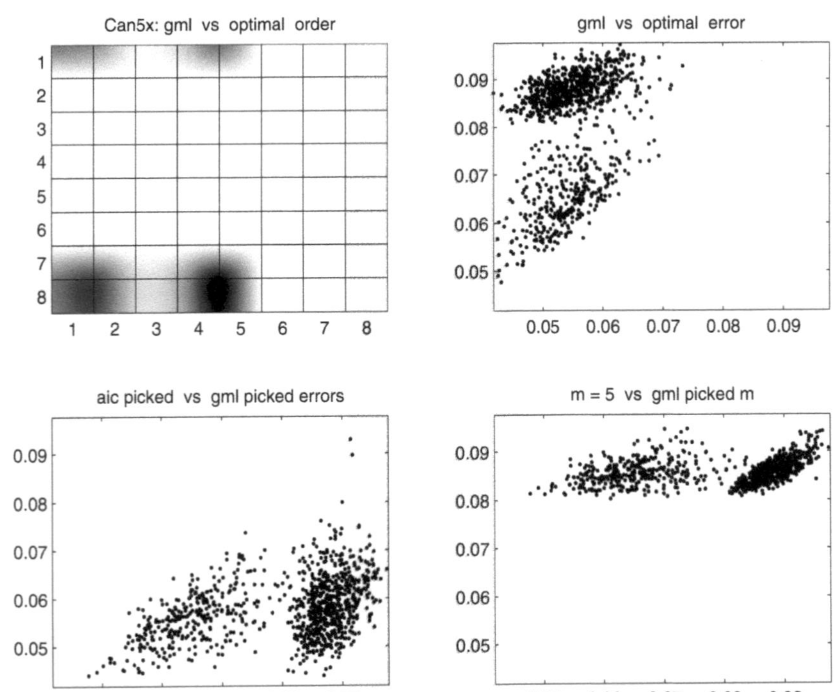

Figure 2.1(g). Spline smoothing of various orders for the example "Can5x".

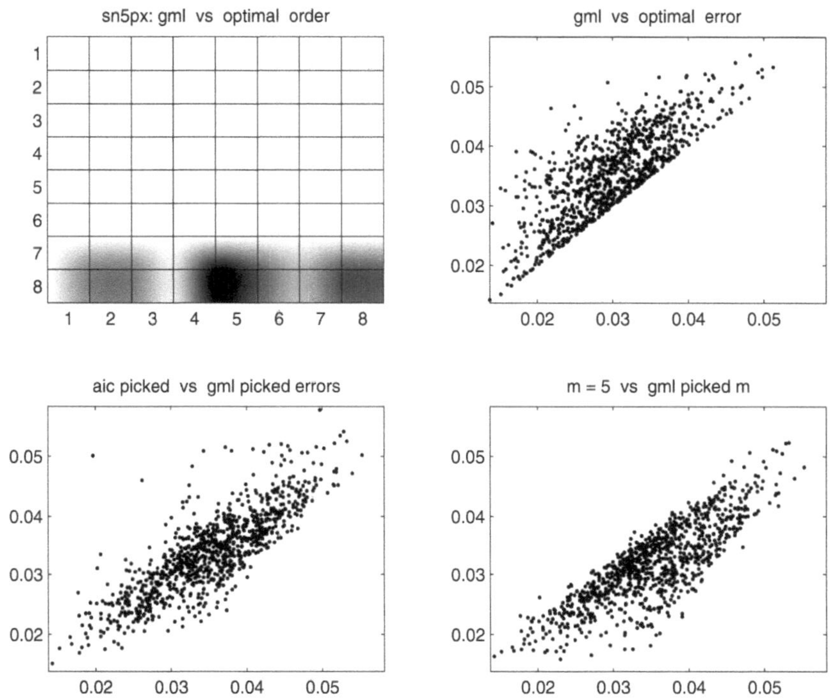

Figure 2.1(h). Spline smoothing of various orders for the example "sn5px".

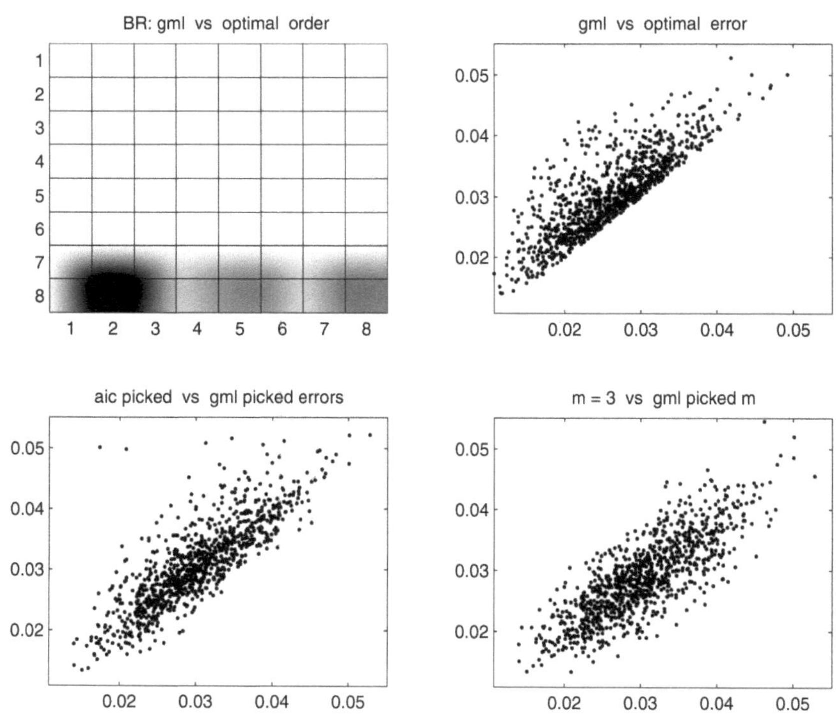

Figure 2.1(i). Spline smoothing of various orders for the example "BR".

Table 2.3. Rankings of the performance from the discrete L^2 error point of view of the various methods for selecting both the smoothing parameter and the order of the estimator for sample sizes $n = 100$ and $n = 400$.

rankings	$n = 100$	$n = 400$
gml aic gcv	CH, Poly5, BR, Wa-I, Wa-II	CH, Poly5, BR, Wa-I, Wa-II, sn5px
aic gml gcv	sn5px	
gml gcv aic		
aic gcv gml	HM, Cantor, Can5x	HM, Cantor, Can5x
gcv aic gml		Can5x
gcv gml aic		

picks in the range 4 to 8. For the "Wa-II" example, it almost always picks $m = 8$, which indeed gives close to the optimal L^2 errors. Some other details of the "optimal" results are given in Figure 2.1(a)–(i). For $n = 400$, things are different: Note that, for all examples except for "Wa-I", "Wa-II", and "sn5px", always picking $m = 2$ (cubic smoothing splines) beats the data-driven choices of the order by a sizable margin. Still, one would assume that better methods for choosing the order must exist. Finally, we note that none of the methods for choosing the smoothing parameter and the order beats itself when the best fixed order is chosen. However, observe that the latter method is not a *rational* one. The conclusion that more work on the selection of the order is needed would seem to be an understatement.

For the smoothing parameter selection procedures, the results are surprisingly clear-cut as well. Ignoring the examples "HM", "Cantor", and "Can5x", the overall ranking is GML, AIC, GCV, with rather large differences between them. For $n = 400$, the differences between AIC and GCV are substantially less, in accord with the observed equivalences in § 18.7. For the excluded examples, the ranking is essentially AIC, GCV, GML. For $n = 400$, without excluding the three examples, a case can be made for considering AIC the best method; see Table 3.2. (That the "Cantor" and "Can5x" examples should be exceptional seems understandable because of their nonsmoothness. However, it is not clear what is special about the "HM" example!) Statistically, the differences between the various selection procedures are significant: For the pairwise comparisons, the t-statistic is typically about 9 (with 999 degrees of freedom). For $n = 400$, the difference between AIC and GCV narrows, as expected by the asymptotic equivalence, as does the difference between GML and AIC, but to a lesser extent. Moreover, the differences are still practically significant. Apparently, this excellent behavior of the GML procedure must be attributed to the smooth prior distribution it assumes of the regression function. Indeed, for the examples based on the Cantor function, this assumption is patently false, and GML behaves poorly. For "sn5px", GML is best for $n = 400$ but not for $n = 100$. One could argue (after the fact) that for $n = 100$ the example "sn5px" is too rough to comply with a smooth prior. The superiority of GML over GCV for very smooth regression functions (but not for regression functions of finite smoothness) was already noted by Wahba (1985) and Kohn, Ansley, and Tharm (1991) and theoretically investigated by Stein (1990)in the guise of estimating the parameters of a stochastic process. This does not contradict the reverse findings of Wahba (1985) for nonsmooth *periodic* examples such as the Beta(5, 10) function, which is smooth only up to order four, when viewed as a *periodic* function.

It is surprising that GCV came in last for *all* the examples, although this agrees with findings in the literature for cubic splines; see Hurvich, Simonoff, and Tsai (1995), Speckman and Sun (2001), and Lee (2003, 2004).

Finally, we highlight some other aspects of the simulation results. In Figures 2.1(a)–(i), for sample size $n = 100$, we show the histograms of the optimal orders and the orders selected by the GML procedure, as well as the scatter plot of the associated errors $D(f^{nh,m}, f_o)$. We also show the scatter plots of the errors for the AIC method versus the GML method and the errors of the best fixed order as determined by GML versus the GML-picked orders. As already mentioned, the GML procedure tends to pick too high an order. The example "Can5x", Figure 2.1(h), is especially noteworthy in that it only picks the smallest, $m = 1$, or the largest, $m = 8$, but nothing in between. This seems to explain the two clouds in the scatter plots involving the GML-picked order. To a somewhat lesser extent, this also happens in the "Cantor" example, Figure 2.1(h).

Although no doubt there are many other interesting facts to be discovered about the simulations, we leave it at this. In § 4, some results on interior L^2 and (interior) L^∞ bounds are discussed and compared with corresponding results for local polynomials.

3. Local polynomials

We now report on the performance with respect to the (discrete) L^2 error of local polynomial estimators in the standard regression problem (1.1)–(1.3) with the nine test examples. Obviously, it is easy to compute local polynomial estimators of "all" orders, so, as for smoothing splines, we investigate whether it would be beneficial to choose the order of the local polynomials and whether it can be done well in a data-driven way. The answer to the first question is unequivocally yes. Again, there is room for improvement on the data-driven aspect of it. Of course, the (standard) smoothing parameter needs choosing as well.

Although, as for smoothing splines, see CUMMINS, FILLOON, and NYCHKA (2001), it is theoretically possible to choose the smoothing parameter and the order in a pointwise fashion, we follow the lead of FAN and GIJBELS (1995a) and do it globally. For some of the examples, it would make sense to do it in a piecewise fashion, as in § 6, but we shall not explore this here. Again, in all examples, we choose

$$\sigma = \tfrac{1}{10} \left\{ \max_{1 \leqslant i \leqslant n} f_o(x_{in}) - \min_{1 \leqslant i \leqslant n} f_o(x_{in}) \right\} . \tag{3.1}$$

Interest is in the dependence of the local polynomial estimator on the smoothing parameter h as well as on the order m of the local polynomials. Thus, denote the solution of the problem

$$\begin{aligned} &\text{minimize} \quad \tfrac{1}{n} \sum_{i=1}^{n} A_h(x_{in} - t) \, | \, p(x_{in}) - y_{in} \, |^2 \\ &\text{subject to} \quad p \in \mathcal{P}_m \end{aligned} \tag{3.2}$$

by $p^{nh,m}(x;t)$ and define the estimator of the regression function by

$$f^{nh,m}(t) = p^{nh,m}(t;t) , \quad t \in [0,1] . \tag{3.3}$$

In the local polynomial computations, only the Epanechnikov kernel is used; i.e., $A_h(x) = h^{-1} A(h^{-1}x)$ for all x, with

$$A(x) = \tfrac{3}{4} (1 - x^2)_+ , \quad -\infty < x < \infty . \tag{3.4}$$

We consider six methods for choosing the smoothing parameter h;

(a) "locgcv": Generalized cross validation for the (global) local error, see (18.8.22), combined with the fudge factor (18.9.17) to get the smoothing parameter for the pointwise error.
(b) "ptwgcv": Generalized cross validation for the (global) pointwise error directly; see (18.8.22).
(c) "aic": The "improved" Akaike's information criterion, see (18.6.27).
(d) "irsc": The IRSC method (irsc) of FAN and GIJBELS (1995a), see (18.8.22). Although one could work out the IRSC procedure for odd-order local polynomials, we restrict attention to even orders 2 through 8.
(e) "irsc + locgcv".
(f) "irsc + ptwgcv".

Apart from the "irsc" method, all of the methods above estimate an estimation error of one kind or another, so all of them can be used for selecting the order as well as the smoothing parameter. In connection with the "irsc" method for choosing h, for each fixed order we also report on the "irsc"

Table 3.1. L2 errors of loco-s for $n = 100$ (percentage of range).

HM	2	4	6	8	ch	CH	4	6	8	ch
opt:	3.11	3.30	3.49	3.65	3.08	2.32	2.22	2.45	2.79	2.09
loc:	3.34	3.62	3.85	4.03	3.75	2.54	2.57	2.64	2.97	2.68
ptw:	3.44	3.67	3.96	4.18	3.60	2.69	2.67	2.78	3.11	2.79
aic:	3.34	3.58	3.81	4.07	3.42	2.60	2.58	2.64	2.97	2.63
irs:	3.36	3.83	4.25	4.36	3.36	2.54	2.46	2.53	2.87	2.54
+ loc:					3.84					2.60
+ ptw:					3.55					2.57

BR						Poly5				
opt:	2.91	2.83	2.87	2.89	2.64	2.24	2.16	2.43	2.79	2.04
loc:	3.09	3.12	3.29	3.18	3.25	2.47	2.49	2.63	2.97	2.62
ptw:	3.21	3.18	3.33	3.25	3.32	2.63	2.59	2.76	3.11	2.72
aic:	3.15	3.10	3.27	3.17	3.21	2.53	2.50	2.63	2.97	2.55
irs:	3.11	3.15	3.37	3.05	3.11	2.46	2.36	2.52	2.87	2.46
+ loc:					3.18					2.53
+ ptw:					3.22					2.48

Wa-I						Wa-II				
opt:	3.66	2.92	2.70	2.79	2.55	4.38	3.55	3.20	3.07	2.97
loc:	3.83	3.18	3.07	2.98	3.04	4.55	3.77	3.50	3.47	3.47
ptw:	3.97	3.31	3.19	3.12	3.30	4.64	3.90	3.65	3.54	3.65
aic:	3.88	3.22	3.11	2.98	3.13	4.62	3.82	3.53	3.48	3.52
irs:	3.87	3.13	3.09	2.88	3.80	4.59	3.75	3.46	3.63	4.48
+ loc:					2.97					3.60
+ ptw:					3.13					3.69

Can5x						sn5px				
opt:	5.70	5.71	5.73	5.78	5.58	4.13	3.53	3.44	3.46	3.33
loc:	6.04	5.90	5.92	5.99	5.98	4.25	3.71	3.63	3.69	3.68
ptw:	5.98	5.98	6.32	7.01	5.96	4.37	3.85	3.82	3.88	3.93
aic:	6.13	5.89	5.93	5.99	5.95	4.30	3.73	3.65	3.70	3.70
irs:	5.94	6.00	6.41	6.98	5.94	4.28	3.67	3.57	3.61	4.28
+ loc:					6.54					3.62
+ ptw:					6.07					3.71

Cantor					
opt:	3.73	3.82	3.91	4.02	3.66
loc:	3.91	4.15	4.34	4.41	4.35
ptw:	4.02	4.19	4.39	4.48	4.18
aic:	3.94	4.12	4.32	4.42	4.08
irs:	3.93	4.39	4.64	4.60	3.93
+ loc:					4.50
+ ptw:					4.20

Table 3.2. L2 errors of loco-s for $n = 400$ (percentage of range).

HM	2	4	6	8	ch	CH	4	6	8	ch
opt:	1.79	1.81	1.89	1.97	1.74	1.31	1.23	1.24	1.41	1.14
loc:	1.86	1.89	1.99	2.08	1.98	1.40	1.41	1.35	1.52	1.38
ptw:	1.91	1.96	2.02	2.10	1.96	1.48	1.45	1.38	1.54	1.46
aic:	1.89	1.94	2.01	2.08	1.94	1.46	1.44	1.35	1.50	1.44
irs:	1.85	1.88	2.06	2.52	1.85	1.39	1.41	1.29	1.44	1.39
+ loc:					2.09					1.37
+ ptw:					1.91					1.40

BR						Poly5				
opt:	1.68	1.54	1.59	1.60	1.47	1.27	1.19	1.21	1.41	1.10
loc:	1.75	1.68	1.74	1.79	1.74	1.37	1.39	1.32	1.52	1.38
ptw:	1.82	1.70	1.78	1.81	1.78	1.44	1.43	1.35	1.53	1.45
aic:	1.80	1.68	1.76	1.79	1.75	1.42	1.41	1.33	1.50	1.43
irs:	1.75	1.64	2.12	1.80	1.75	1.35	1.39	1.27	1.44	1.35
+ loc:					1.79					1.36
+ ptw:					1.76					1.38

Wa-I						Wa-II				
opt:	2.14	1.60	1.44	1.42	1.34	2.55	1.96	1.71	1.62	1.56
loc:	2.20	1.71	1.61	1.54	1.55	2.60	2.05	1.86	1.79	1.80
ptw:	2.24	1.76	1.64	1.55	1.60	2.65	2.10	1.90	1.83	1.86
aic:	2.23	1.73	1.63	1.52	1.56	2.63	2.09	1.89	1.82	1.85
irs:	2.20	1.68	1.57	1.45	2.05	2.60	2.03	1.82	1.82	2.49
+ loc:					1.49					1.84
+ ptw:					1.51					1.85

Can5x						sn5px				
opt:	3.57	3.66	3.73	3.78	3.56	2.35	1.89	1.80	1.80	1.75
loc:	3.64	3.93	3.99	4.05	4.01	2.40	1.96	1.88	1.90	1.89
ptw:	3.66	3.82	3.97	4.06	3.72	2.44	2.02	1.92	1.94	1.97
aic:	3.64	3.82	3.94	4.02	3.70	2.42	2.00	1.92	1.92	1.95
irs:	3.67	4.22	4.07	4.12	3.67	2.39	1.95	1.86	1.89	2.39
+ loc:					4.13					1.90
+ ptw:					3.76					1.92

Cantor					
opt:	2.44	2.53	2.58	2.60	2.43
loc:	2.55	2.67	2.71	2.74	2.71
ptw:	2.54	2.69	2.75	2.77	2.60
aic:	2.53	2.67	2.73	2.73	2.59
irs:	2.59	2.66	2.88	3.30	2.59
+ loc:					2.94
+ ptw:					2.67

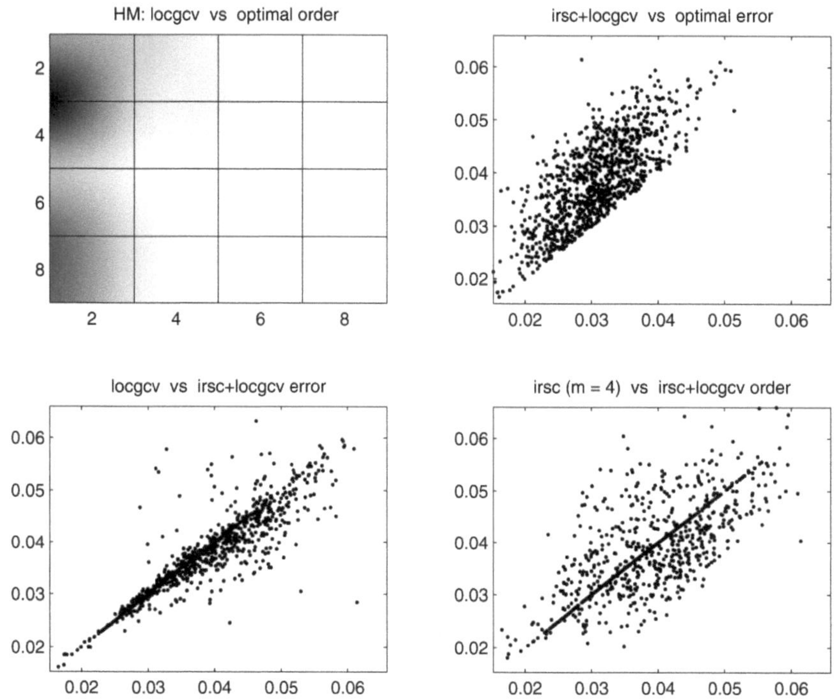

Figure 3.1(a). Local polynomials of various orders applied to the example "HM" with sample size $n = 100$ and $\sigma = 0.1$. Top left: qualitative 2D histogram of the order picked by the "irsc+ptwgcv" method in each replication versus the optimal order for each replication (darker means higher frequency). The other diagrams show scatter plots of the errors $D(f^{nh,m}, f_o)$ in various scenarios. The bottom right diagram deals with errors of the "irsc" method for the best fixed order according to "irsc", which changes from case to case, versus the "irsc" error corresponding to the order picked in each replication by "ptwgcv".

method when the order is chosen by "locgcv" and by "ptwgcv". These are the entries labeled (e) and (f).

We report on the estimated mean of the quantity $D(f^{nh,m} , f_o)$ ("the error") of (2.3) for $m = 2, 4, 6, 8$.

Tables 3.1 and 3.2 present a summary of the simulation results. The first four columns give the optimal error for each order m,

$$(3.5) \qquad \mathbb{E}\Big[\min_{h>0} D(f^{nh,m} , f_o) \Big] , \quad m = 2, 4, 6, 8 .$$

The optimal error over the even orders,

$$(3.6) \qquad \mathbb{E}\Big[\min_{1 \leqslant m \leqslant 8} \min_{h>0} D(f^{nh,m} , f_o) \Big] ,$$

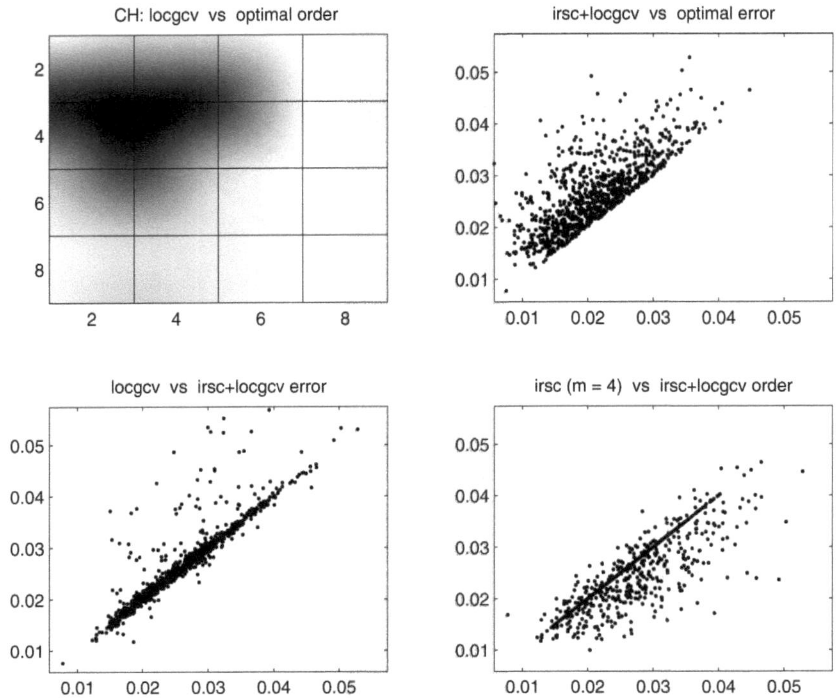

Figure 3.1(b). Local polynomials of various orders for the example "CH".

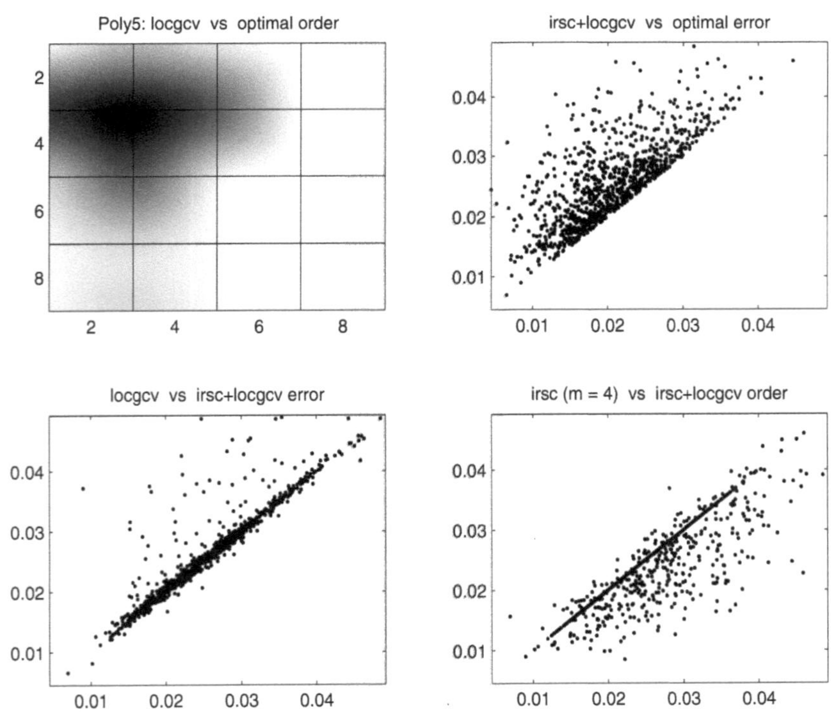

Figure 3.1(c). Local polynomials of various orders for the example "Poly5".

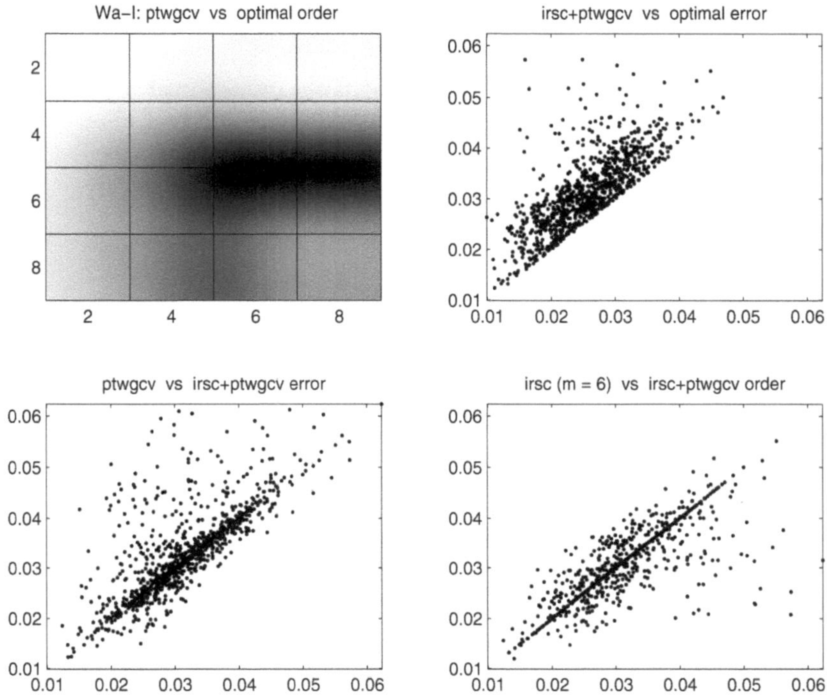

Figure 3.1(d). Local polynomials of various orders for "Wa-I".

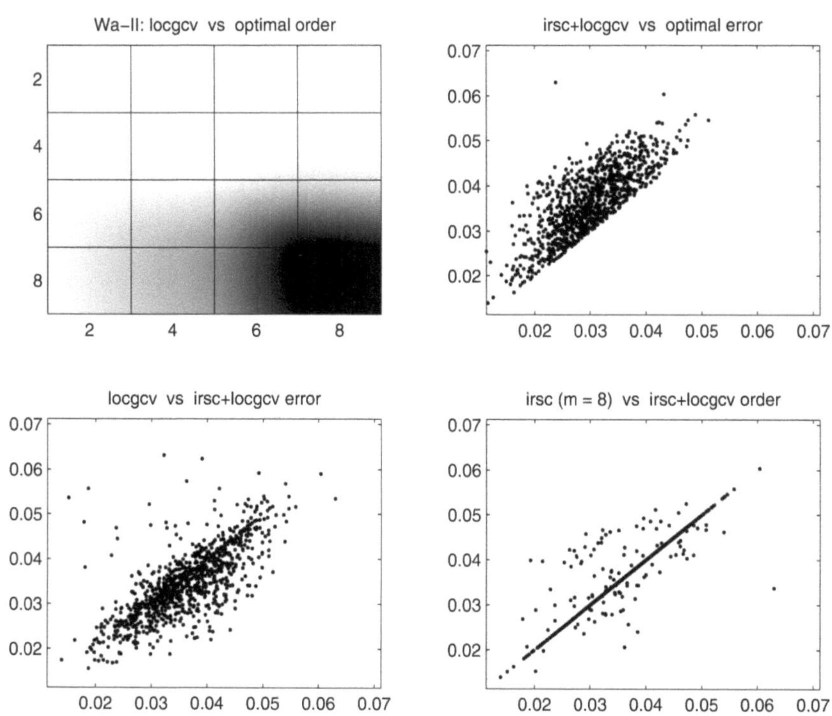

Figure 3.1(e). Local polynomials of various orders for "Wa-II".

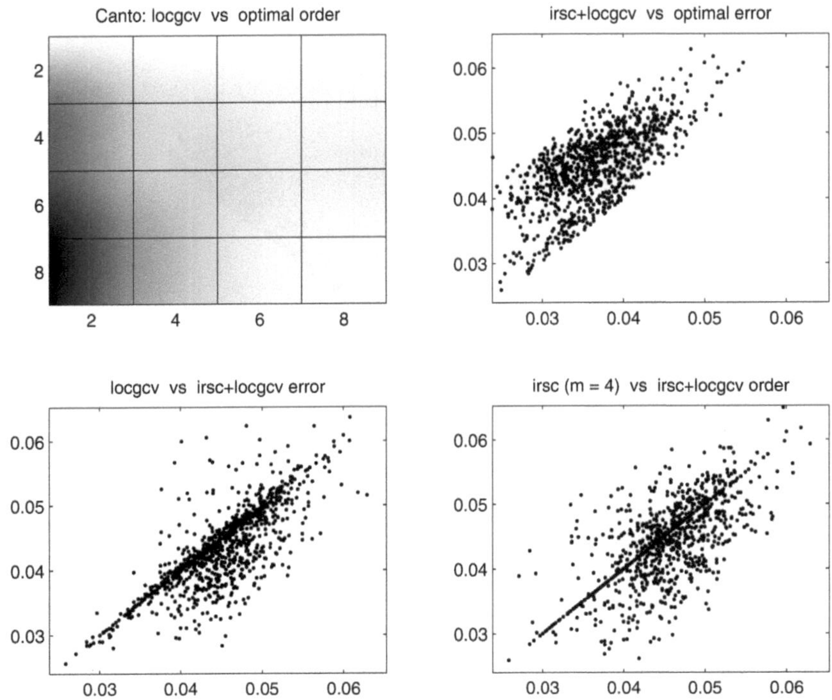

Figure 3.1(f). Local polynomials of various orders for "Cantor".

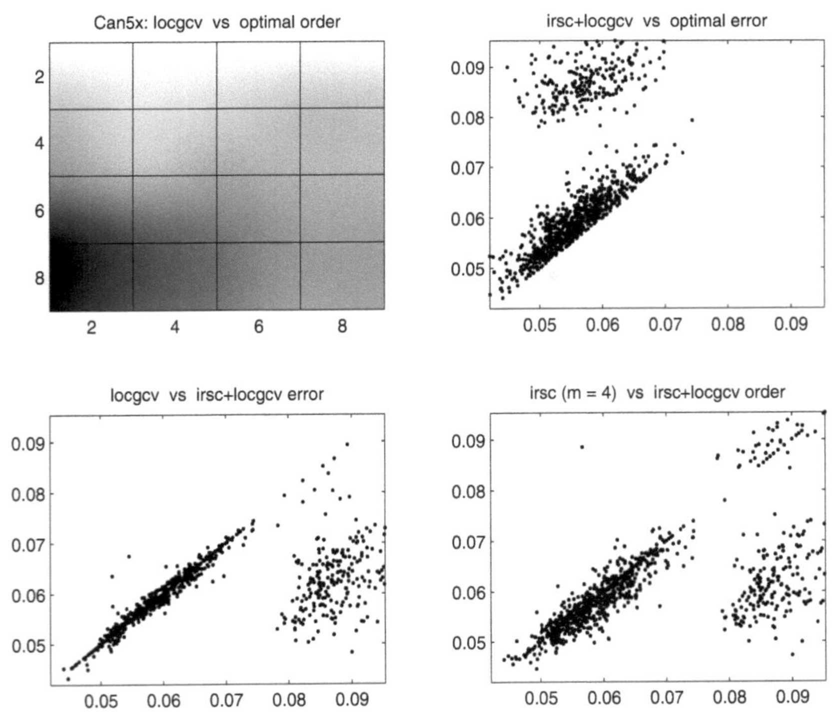

Figure 3.1(g). Local polynomials of various orders for the example "Can5x".

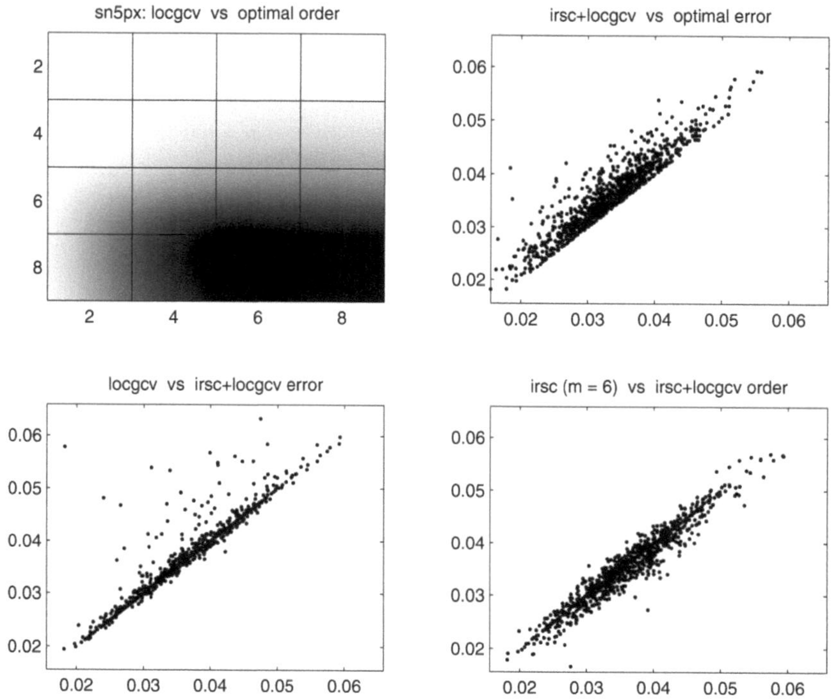

Figure 3.1(h). Local polynomials of various orders for "sn5px".

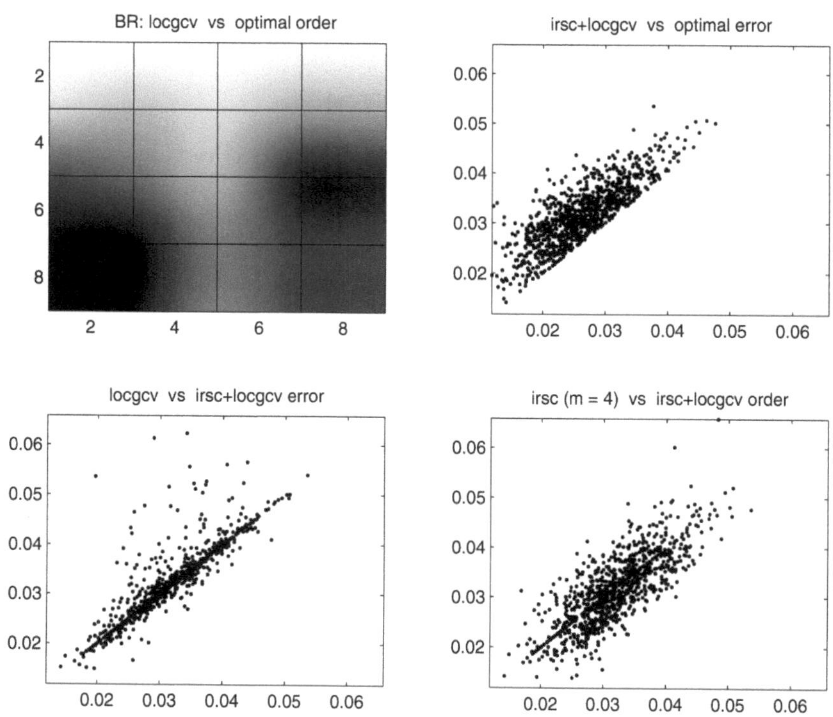

Figure 3.1(i). Local polynomials of various orders for the example "BR".

Table 3.3. Rankings of the performance from the discrete L^2 error point of view of the various methods for sample sizes $n = 100$ and $n = 400$. "loc" and "ptw" refer to "irsc+locgcv" and "irsc+ptwgcv", respectively.

rankings	$n = 100$	$n = 400$
aic ptw loc	HM, Can5x, Cantor	BR, Can5x, Cantor
ptw aic loc		HM
aic loc ptw	Wa-II	
ptw loc aic	CH, Poly5	
loc aic ptw	BR, Wa-I, sn5px	
loc ptw aic		CH, Poly5, Wa-I, Wa-II, sn5px

is given in the last column under the heading "ch" ("chosen" order). Thus, when estimating the quantity (3.6), the best order is determined for each replication.

For each m, Table 3.1 also gives the estimated values of

$$\mathbb{E}[\, D(f^{nH,m} \,, f_o)\,] \,, \tag{3.7}$$

with the random H chosen by "locgcv", "ptwgcv", "aic", and "irsc". The last column reports on

$$\mathbb{E}[\, D(f^{nH,M} \,, f_o)\,] \,, \tag{3.8}$$

with the random M picked by the various methods (the value of m that minimizes the "locgcv", "ptwgcv","aic", and "irsc" functionals in each replication). We also report on the "irsc" error when the order is chosen by the "locgcv" and "ptwgcv" criteria.

In Table 3.1, the results are reported for sample size $n = 100$ based on 1000 replications. Table 3.2 deals with sample size $n = 400$ and is based on 250 replications. Otherwise, it is the same as Table 3.1.

We briefly discuss the simulation results of Tables 3.1 and 3.2. For each regression function, the "optimal" results (in the first row) imply that it is potentially rewarding to choose the order of the local polynomial estimator. In particular, always choosing $m = 2$ is not a good idea. However, the actual method investigated for choosing the order are susceptible to improvement. In particular, no selection methods beats itself when the best fixed order is chosen. (Of course, as for smoothing splines, the latter is not a *rational* method.)

Regarding the various methods used, the striking result is that for each fixed order the perplexing "irsc" method seems to be the best overall, except for the nonsmooth examples based on the Cantor function and the ever-exasperating "HM" example. With these cases included, a case could perhaps be made for the "locgcv" method. Of course, the "irsc" method is no good for selecting the order (it always leads to $m = 2$), but in combination with either "locgcv" or "ptwgcv" it performs very well. In fact, "irsc + locgcv" outperforms "locgcv" by itself, and "irsc + ptwgcv" beats "ptwgcv" by itself, except for the Cantor examples.

The observed rankings from best to worst of the remaining methods ("aic", "irsc + locgcv", and "irsc + ptwgcv") for each test example are tabulated in Table 3.3. The surprise is that the rankings do not appear to be uniformly distributed, even though one would assume that all of these methods are (asymptotically) equivalent. We dare to draw the conclusion that, for sample size $n = 100$, "aic" and "irsc + locgcv" are comparable, but,for $n = 400$, "irsc + locgcv" is best. As always, the exceptions are the nonsmooth examples "Cantor" and "Can5x", as well as the "HM" example, but the otherwise nice example "BR" is contrary as well. (Quite possibly, an additional conclusion is that $n = 100$ is too small a sample size for any data-driven method for selecting the smoothing parameter *and* the order.)

Finally, in Figures 3.1(a)-(i), for sample size $n = 100$, we show the (qualitative) 2D histograms of the optimal orders versus the orders picked by the "ptwgcv" method in each replication (darker means higher value, but the scale is different from those for smoothing splines). Since only four different orders were considered, the 2D histograms are less dramatic than for splines. Note that the best orders are defined in the sense of the errors (3.6) and presumably change with the sample size. In fact, it seems hard to draw any conclusions about the histograms. Some scatter plots of the errors $D(f^{nh,m}, f_o)$ in various scenarios are shown as well. In the scatter plots in the bottom right of each figure, the errors of the "irsc" method for the best fixed overall order for the "irsc" method (not a *rational* procedure) are plotted against the errors for the "irsc+locgcv" method.

4. Smoothing splines versus local polynomials

Finally, we compare smoothing splines with local polynomials. This also seems the right time to add (global) polynomial estimators into the mix.

The comparison of the estimators begins by considering the scaled discrete L^2 error $D(f^{nh,m}, f_o)$; see (2.3). This then constitutes a summary of the results from §§ 2 and 3. That is, we consider the (estimated) means of the optimal errors

$$\mathbb{E}\big[\min_{h\,,\,m} D(f^{nh,m} , f_o) \big] \tag{4.1}$$

for smoothing splines and local polynomials and

$$\mathbb{E}\big[\min_{m} D(p^{n,m} , f_o) \big] \tag{4.2}$$

Table 4.1. Comparing L2 errors of splines, local polynomials and global polynomials for $n = 100$ and $n = 400$ (the L^2 errors expressed as a percentage of the range). For each example, the first row lists the optimal L^2 errors. The second row lists the "chosen" errors: "gml" errors for splines, "irsc+locgcv" errors for local polynomials, and "gcv" errors for global polynomials.

		n = 100			n=400		
		splines	locos	polys	splines	locos	polys
HM	L2	2.97	3.08	3.68	1.67	1.74	2.05
	ch	4.11	3.84	4.22	2.37	2.09	2.23
CH	L2	1.94	2.09	2.30	1.05	1.14	1.23
	ch	2.41	2.60	2.83	1.35	1.37	1.46
Poly5	L2	1.91	2.04	2.24	1.04	1.10	1.19
	ch	2.34	2.53	2.76	1.34	1.36	1.45
BR	L2	2.58	2.64	2.92	1.40	1.47	1.60
	ch	3.01	3.18	3.45	1.64	1.79	1.79
Wa-I	L2	2.62	2.55	2.73	1.37	1.34	1.43
	ch	2.90	2.97	3.15	1.51	1.49	1.63
Wa-II	L2	3.07	2.97	3.14	1.63	1.56	1.65
	ch	3.31	3.60	3.55	1.75	1.84	1.81
Cantor	L2	3.55	3.66	4.08	2.36	2.43	2.65
	ch	4.70	4.50	4.63	2.76	2.94	2.78
Can5x	L2	5.47	5.58	7.83	3.55	3.56	6.74
	ch	8.04	6.54	8.21	3.82	4.13	6.78
sn5px	L2	2.98	3.33	3.31	1.55	1.75	1.72
	ch	3.52	3.62	3.65	1.70	1.90	1.88

for global polynomials, with $1 \leqslant m \leqslant 20$. Here, $p^{n,m}$ is the least-squares polynomial estimator of order m; see § 15.2. Of course, local polynomials encompass global polynomials in the limit as $h \to \infty$, but for local polynomials we only considered orders up to 8.

We also consider data-driven methods for selecting the smoothing parameter(s). For smoothing splines, we use the "gml" method; for local polynomials, "irsc+locgcv"; and for global polynomials the only method considered was the "gcv" method of Exercise (18.2.18).

The results are summarized in Table 4.1. Again, the estimated means are based on 1000 replications for sample size $n = 100$ and on 250 replications for sample size $n = 400$. We emphasize that, for each replication, the various estimators (smoothing splines, local polynomials, and global polynomials) were computed on the same data.

The results are remarkably uniform. For the optimal errors, for six of the nine test examples, the ranking is

$$\text{splines} - \text{local polynomials} - \text{global polynomials}$$

or splines-local-global for short. For "Wa-I" and "Wa-II", it is

$$\text{local} - \text{splines} - \text{global}$$

and for "sn5px" we have

$$\text{splines} - \text{global} - \text{local}\,.$$

For the data-driven methods, the ranking splines-local-global remains overwhelmingly intact, except for "Wa-II", "Cantor", and "sn5px", where it is splines-global-local.

(4.3) REMARK. It is interesting to note that for the examples based on WAHBA (1990), to wit "Wa-I" and "Wa-II", which one would assume would be ideal to put smoothing splines in the best light, local polynomials give better L^2 errors, whereas for the "HM" example, which is featured in many papers on local polynomial estimation, smoothing splines perform better.

(4.4) REMARK. Another surprise here is not that most of the time global polynomials are not competitive with splines or local polynomials but rather that in some examples global polynomials beat local polynomials. This can be avoided, but only at the price of including local polynomial estimators of ridiculously high order.

Thus, for the data-driven methods selected here, "splines+gml" always wins on average. It is not clear whether this entails practical advantages, e.g., when these estimators are used to construct confidence bands, but see § 5. A likely explanation for the dominance of splines may be found in the boundary behavior of the estimators. To check this, for each of the estimators, we also computed the scaled interior L^2 error for the interval $I = [\,\frac{1}{10}\,,\,\frac{9}{10}\,]$,

$$\| f^{nh,m} - f_o \|_{2,I} \Big/ \big(\max_{1 \leqslant i \leqslant n} f_o(x_{in}) - \min_{1 \leqslant i \leqslant n} f_o(x_{in}) \big) \,, \tag{4.5}$$

where, cf. (1.5),

$$\| g \|_{2,I} = \Big(\tfrac{1}{n} \sum \big\{ \, | g(x_{in}) |^2 \; \big| \; x_{in} \in I \, \big\} \Big)^{1/2} . \tag{4.6}$$

Thus, Table 4.2 lists the means of the scaled, optimal, interior L^2 error (4.5) for each estimator (the lines marked "int-L2"). The lines marked

Table 4.2. Interior errors of splines, local, and global polynomials.

		n = 100			n=400		
		splines	locos	polys	splines	locos	polys
HM	L2	2.97	3.08	3.68	1.67	1.74	2.05
	int-L2	2.51	2.50	2.82	1.40	1.40	1.57
	SUP	6.48	7.25	8.90	3.87	4.53	5.81
	int-SUP	5.50	5.46	6.42	3.13	3.17	3.68
CH	L2	1.94	2.09	2.30	1.05	1.14	1.23
	int-L2	1.48	1.50	1.70	0.79	0.81	0.90
	SUP	3.62	4.42	5.21	2.06	2.60	3.01
	int-SUP	2.81	2.95	3.36	1.55	1.67	1.88
Poly5	L2	1.91	2.04	2.24	1.04	1.10	1.19
	int-L2	1.42	1.43	1.63	0.76	0.77	0.87
	SUP	3.63	4.37	5.15	2.13	2.63	3.03
	int-SUP	2.76	2.83	3.21	1.49	1.57	1.75
BR	L2	2.58	2.64	2.92	1.40	1.47	1.60
	int-L2	1.98	1.94	2.15	1.09	1.09	1.20
	SUP	5.49	6.09	7.01	3.19	3.74	4.35
	int-SUP	4.08	4.13	4.69	2.29	2.38	2.66
Wa-I	L2	2.62	2.55	2.73	1.37	1.34	1.43
	int-L2	1.85	1.74	1.98	0.97	0.92	1.03
	SUP	6.50	6.75	7.15	3.87	4.00	4.20
	int-SUP	3.85	3.83	4.27	2.04	2.04	2.22
Wa-II	L2	3.07	2.97	3.14	1.63	1.56	1.65
	int-L2	2.11	1.99	2.29	1.14	1.07	1.21
	SUP	8.51	8.32	8.76	5.26	5.12	5.28
	int-SUP	4.55	4.40	4.97	2.46	2.42	2.65
Cantor	L2	3.55	3.66	4.08	2.36	2.43	2.65
	int-L2	3.00	3.01	3.22	2.04	2.07	2.21
	SUP	8.55	9.13	10.4	6.70	7.20	8.05
	int-SUP	7.74	7.69	8.42	6.37	6.35	7.25
Can5x	L2	5.47	5.58	7.83	3.55	3.56	6.74
	int-L2	4.78	4.83	7.16	3.13	3.13	6.36
	SUP	14.2	14.6	21.1	10.6	10.9	19.4
	int-SUP	13.4	13.6	20.9	10.2	10.0	19.3
sn5px	L2	2.98	3.33	3.31	1.55	1.75	1.72
	int-L2	2.29	2.46	2.48	1.17	1.29	1.28
	SUP	6.68	8.64	9.48	3.75	5.26	5.68
	int-SUP	4.79	5.42	5.57	2.50	2.89	2.92

"L2" contain the optimal L^2 errors, already listed in Table 4.1. Roughly speaking, with respect to the interior L^2 error, smoothing splines and local polynomials are equivalent. Moreover, the differences in the interior L^2 errors are much smaller than for the global versions. Indeed, there are a number of ties for $n = 400$. Thus, the conclusion that the boundary behavior of the estimators is the (main) cause for the dominance of smoothing splines with respect to the global L^2 error seems justified.

Now, to generate more heat if not insight, we also compare scaled, discrete uniform error bounds,

$$\frac{\max\limits_{1\leqslant i\leqslant n} \bigl| f^{nh,m}(x_{in}) - f_o(x_{in}) \bigr|}{\max\limits_{1\leqslant i\leqslant n} f_o(x_{in}) - \min\limits_{1\leqslant i\leqslant n} f_o(x_{in})} , \tag{4.7}$$

as well as the interior uniform error,

$$\frac{\max \bigl\{ \bigl| f^{nh,m}(x_{in}) - f_o(x_{in}) \bigr| \bigm| x_{in} \in I \bigr\}}{\max\limits_{1\leqslant i\leqslant n} f_o(x_{in}) - \min\limits_{1\leqslant i\leqslant n} f_o(x_{in})} \tag{4.8}$$

(the lines marked "SUP" and "int-SUP" in Table 4.2).

The overall rankings for the global uniform errors are even more uniform than for the global L^2 error. The ranking is

splines – local – global ,

the only exception being provided by "Wa-II". Again, boundary behavior seems to play a role: The differences between the interior uniform errors is much smaller than between the global versions. The overall ranking is

splines – local – global

for the examples "HM", "CH", "Poly5", "BR", and "sn5px", and

local – splines – global

for the remaining "Wa-I", "Wa-II", and the two Cantor examples. It should be noted that, for both splines and local polynomials, further analysis of the simulation results suggests that the optimal order for the interior errors tends to be larger than for the global versions. This reinforces the previous assessment that one should consider variable-order estimators and not stick with a preconceived choice of the order of the estimators.

As a concluding remark, we note that from the tables in §§ 2 and 3 one may also gleam a comparison of the two most popular forms of the estimators considered (cubic splines and local linear polynomials). We leave this for the enjoyment of the reader.

5. Confidence bands

In this section, we present some simulation results on confidence bands for the regression function in the standard nonparametric regression problem

(1.1)–(1.2). We consider both iid normal noise as in (1.3) and nonnormal iid noise, to wit $t(6)$, $t(10)$, two-sided exponential, and absolute normal noise centered at its mean. Unfortunately, heteroscedasticity is not considered (but see § 7). The construction of the confidence bands is based on Chapter 22, so that the development is based on smoothing splines of fixed, deterministic order but random (data-driven) smoothing parameters. Recall that there are advantages to using smoothing splines but that the theory predicts boundary trouble.

To demonstrate the boundary trouble, we construct confidence bands for the whole interval $[\,0\,,\,1\,]$ as well as for the interior region $I = [\,\frac{1}{10}\,,\,\frac{9}{10}\,]$. Indeed, boundary problems do emerge with a vengeance, but even in the interior not all is well. All of this can be traced to the fact that the squared bias is not small enough compared with the variance of the estimators. Thus, bias corrections are needed, which in effect amounts to undersmoothing. A remarkably effective way of doing this is introduced, and rather astounding simulation results back this up. The recommendations we put forward should apply to local polynomial estimators as well but were not tried.

The construction of the confidence bands proceeds as follows. First, compute the smoothing spline estimator $f^{nH,m}$ of order m for the problem (1.1)–(1.3), with the random smoothing parameter H chosen by the GCV, AIC, or GML procedure. Next, let $\varphi^{nH,m}$ be the smoothing spline estimator, with the same m and H as before, for the pure-noise problem; i.e., with (1.1) replaced by

$$y_{in} = z_{in}\,, \quad i = 1, 2, \cdots, n\,, \tag{5.1}$$

with

$$(\,z_{1,n}\,,\,z_{2,n}\,,\,\cdots\,,\,z_{n,n}\,)^T \sim \text{Normal}(\,0\,,\,I_{n\times n}\,)\,, \text{ independent of } H\,. \tag{5.2}$$

Then, for $0 < \alpha < 1$ and subsets $A \subset [\,0\,,\,1\,]$, define the critical values $c_{\alpha nm}(H, A)$ by

$$\mathbb{P}\left[\,\max_{x_{in}\in A} \frac{|\,\varphi^{nH,m}(x_{in})\,|}{\sqrt{\mathrm{Var}[\,\varphi^{nH,m}(x_{in})\mid H\,]}} > c_{\alpha nm}(H, A)\,\right] = \alpha\,. \tag{5.3}$$

(We take $\alpha = 0.10$ and $\alpha = 0.05$.) The $100(\,1-\alpha\,)\%$ confidence band for $f_o(x_{in})$, $x_{in} \in A$, is then

$$f^{nH,m}(x_{in}) \pm \widehat{\sigma}\, c_{\alpha nm}(H, A)\,\sqrt{\mathrm{Var}[\,\varphi^{nH,m}(x_{in})\mid H\,]}\,, \quad x_{in} \in A\,. \tag{5.4}$$

In (5.4), we need an estimator $\widehat{\sigma}$ of the standard deviation of the noise. We ignore potential boundary trouble and take $\widehat{\sigma}^2 = \sigma^2_{nH,m}$, with $\sigma^2_{nH,m}$ the estimator of Ansley, Kohn, and Tharm (1990),

$$\sigma^2_{nh} = \frac{\langle\, y_n\,,\, y_n - R_{nh}\, y_n\,\rangle}{\mathrm{trace}(\,I - R_{nh}\,)}\,. \tag{5.5}$$

However, using an estimated variance confuses the issue, so initially we just take the actual variance.

The objective of the simulation study is to estimate the discrepancies between the noncoverage probabilities

$$(5.6) \qquad \mathbb{P}\left[\, T_{nH,m}(A) > c_{\alpha nm}(H,A) \,\right] ,$$

where

$$(5.7) \qquad T_{nH,m}(A) \stackrel{\text{def}}{=} \max_{x_{in} \in A} \frac{\left|\, f^{nH,m}(x_{in}) - f_o(x_{in}) \,\right|}{\sqrt{\widehat{\sigma}^{\,2}_{nH,m} \,\mathbb{V}\mathrm{ar}\left[\, \varphi^{nH,m}(x_{in}) \;\middle|\; H \,\right]}} ,$$

and the nominal confidence level α. Thus, for each of the nine test examples, 1000 replications were run (with the smoothing parameter in each replication chosen by GCV, AIC, or GML), and the number of times the maxima in question exceeded their critical values were counted. For cubic splines ($m = 2$) and sample sizes $n = 100$, 400, and 1600, the simulations are summarized in Table 5.1 (with the exact noise variance).

We briefly discuss the results. Several things stand out. First, for the nonsmooth examples "Cantor" and "Can5x", the confidence bands just do not work. Since the theory of Chapter 22 strongly suggests this, we shall not discuss it further. (But how would one fix it? The authors do not know.) Second, the noncoverage probabilities for the whole interval are bad for all examples. Again, the theory suggests this, but that does not solve the problem. Finally, the interior noncoverage probabilities are remarkably good for the examples "Wa-I", "Wa-II", and "sn5px", and remarkably unimpressive for the other smooth examples. It is hard to tell which one of the two should surprise us more. (To see what one should expect if the theory worked, the number of noncovered cases would be a Binomial$(1000, \alpha)$ random variable with standard deviation approximately 9.5 for $\alpha = 0.10$ and 6.9 for $\alpha = 0.05$. So, for $\alpha = 0.05$, when the mean number of bad cases is 50, observing 61 bad cases is okay, but observing 75 is not.) That the coverage probabilities are more reasonable for sample size $n = 1600$ is good news/bad news: Why do the asymptotics kick in so slowly?

(5.8) Remark. The weirdness of the "HM" example appears to be due to the fact that the bump is barely visible: The size of the bump is about 2 and the standard deviation of the error is 0.6. For $n = 400$ and $n = 1600$, things get better, but they are not good. Making the bump bigger does indeed result in more accurate noncoverage probabilities.

There is an "inverse" way of describing how close the coverage probabilities are to their nominal values, and that is by reporting the appropriate quantiles of the random variable $T_{nH,m}(A)$. Equivalently, we may ask how much the confidence bands need to be inflated in order to get the proper

Table 5.1. The noncoverage probabilities of the standard confidence bands for cubic smoothing splines (in units of 0.001). The sample size is n. For each example, the rows correspond to GML, GCV, and AIC, respectively. For each sample size, the columns correspond to global and interior coverages for $\alpha = 0.10$ and global and interior coverages for $\alpha = 0.05$.

HM		n=100		:		n=400			:		n=1600		
627	650	526	544	:	311	316	231	232	:	175	173	109	108
376	389	289	296	:	282	294	199	201	:	267	276	178	182
461	481	373	382	:	305	315	225	225	:	270	279	180	184
CH													
174	160	93	91	:	135	121	64	60	:	124	109	63	57
219	215	133	128	:	202	178	109	95	:	200	162	109	88
223	218	137	130	:	201	177	109	95	:	202	163	110	89
Poly5													
176	181	103	97	:	147	138	72	63	:	129	113	74	62
250	259	153	155	:	216	198	111	105	:	267	208	161	137
260	270	160	159	:	215	197	116	108	:	267	207	162	138
BR													
195	174	109	106	:	146	129	86	67	:	117	104	66	54
246	230	155	151	:	235	203	131	107	:	252	197	156	115
279	263	192	179	:	237	206	132	108	:	257	201	157	114
Wa-I													
443	112	344	57	:	455	105	339	55	:	348	96	248	58
385	118	297	61	:	508	112	433	54	:	611	100	533	61
525	130	452	72	:	564	109	494	55	:	630	100	546	61
Wa-II													
764	107	687	57	:	732	108	639	57	:	686	91	573	48
477	111	358	58	:	628	116	521	58	:	713	96	636	52
704	111	613	56	:	693	113	594	56	:	745	93	674	51
Cantor													
911	907	861	843	:	1000	1000	998	998	:	999	999	998	999
685	684	556	560	:	881	880	814	812	:	938	942	880	886
799	804	700	692	:	927	929	871	877	:	950	953	900	909
Can5x													
977	974	933	934	:	1000	1000	1000	1000	:	1000	1000	1000	999
824	829	716	723	:	849	844	763	745	:	982	978	958	944
963	959	912	911	:	943	940	884	885	:	994	991	976	968
sn5px													
106	100	52	50	:	107	105	58	54	:	98	92	52	48
170	143	111	84	:	177	157	109	79	:	182	138	102	77
184	153	116	89	:	188	164	119	88	:	187	140	104	78

Table 5.2. The inflation factors by which the critical values must be increased to get the nominal noncoverage probabilities (in percentages). The rows and columns are as in Table 5.1.

HM	n=100			:		n=400			:		n=1600		
196	248	180	226	:	126	147	117	135	:	109	118	102	110
142	170	130	155	:	121	140	112	129	:	117	130	109	120
151	182	139	166	:	125	143	116	132	:	117	130	109	120
CH													
109	120	100	108	:	104	113	95	102	:	103	111	094	102
115	127	105	115	:	111	119	101	109	:	111	119	102	109
116	127	105	115	:	111	120	101	109	:	111	119	102	109
Poly5													
110	120	100	109	:	105	115	96	104	:	104	112	95	102
118	128	107	116	:	113	123	102	112	:	117	126	108	115
118	130	107	118	:	113	125	103	112	:	117	126	108	115
BR													
110	122	101	111	:	106	112	98	103	:	102	110	94	102
117	129	108	118	:	112	118	103	109	:	115	120	106	110
121	136	111	123	:	112	119	103	109	:	115	120	106	110
Wa-I													
144	111	133	103	:	137	109	128	101	:	125	109	117	102
140	112	129	104	:	160	109	149	101	:	157	110	146	103
155	114	143	105	:	162	109	151	101	:	158	110	147	102
Wa-II													
190	109	176	101	:	196	109	182	101	:	187	106	176	100
144	110	133	102	:	179	109	167	101	:	192	107	180	100
170	111	158	103	:	185	109	173	101	:	192	107	181	100
Cantor													
187	211	171	192	:	218	244	202	223	:	257	278	240	259
173	200	159	183	:	195	219	181	201	:	183	207	171	193
185	213	170	194	:	202	224	188	205	:	186	212	174	198
Can5x													
201	225	187	208	:	198	210	185	196	:	199	210	187	198
154	171	143	159	:	165	180	155	167	:	161	170	152	160
181	196	168	182	:	181	194	169	180	:	166	176	156	165
sn5px													
101	108	93	100	:	101	109	94	100	:	100	106	94	100
109	114	101	106	:	108	113	101	105	:	107	112	100	104
109	116	101	107	:	109	115	102	107	:	107	113	100	105

coverage probabilities. Thus, we define the confidence band inflation factors r such that the (estimated) noncoverage probabilities are "correct",

$$(5.9) \qquad \mathbb{P}\left[\,T_{nH,m}(A) \;>\; r\,c_{\alpha nm}(H,A)\,\right] \;=\; \alpha\;.$$

In Table 5.2, we report these inflation factors for each example, for sample sizes $n = 100$, 400, and 1600, for cubic smoothing splines. The conclusion is that we are off by roughly a factor 2 and that this factor decreases as n increases (except for the two nonsmooth examples). In any case, it shows what is going wrong in the nonsmooth cases, where the regression function *never* lies inside the confidence band.

So why do the confidence bands not deliver the advertised coverage probabilities? The only possible and well-known reason is that the bias is not negligible compared with the noise component of the estimators, despite the fact that the theory says it is so asymptotically; see, e.g., EUBANK and SPECKMAN (1993), XIA (1998), CLAESKENS and VAN KEILEGOM (2003), and many more. There are various theoretical remedies. One is to consider (asymptotic) 100% confidence bands, such as the Bonferroni-type bands of EUBANK and SPECKMAN (1993) or the (asymptotic) 100% confidence bands of DEHEUVELS and MASON (2004, 2007), alluded to in § 22.7. Another one is to undersmooth the estimator, but it is not clear how to do this in practice, as bemoaned by CLAESKENS and VAN KEILEGOM (2003). (Asymptotically, it is easy, but asymptotically for smoothing splines, there is no need for it.) Also, compared with GML, the GCV procedure achieves undersmoothing, but apparently not enough to solve the problem. The third possibility is to construct bias corrections, as in EUBANK and SPECKMAN (1993), HALL (1992), NEUMANN (1997), and XIA (1998). EUBANK and SPECKMAN (1993) (for kernel estimators) reckon that the bias behaves as

$$(5.10) \qquad \mathbb{E}[\,f^{nh,m}(x)\,]\big|_{h=H} - f_o(x) = c_m\,H^m\,f_o^{(m)}(x) + \cdots$$

and proceed to estimate $f_o^{(m)}(x)$ with a kernel estimator based on another random smoothing parameter Λ. Thus, for the purpose of exposition, their bias correction is based on a higher-order estimator. Using some poetic license, we denote it here by

$$(5.11) \qquad b^{nH\Lambda}(x) = c_m\,H^m\,(\,f^{n\Lambda,m+1}\,)^{(m)}(x)\;.$$

Then, their bias-corrected estimator is $g^{nH\Lambda,m} = f^{nH,m} - b^{nH\Lambda}$, so that the confidence intervals are based on the distribution of

$$(5.12) \qquad \max_{x_{in}\in A}\;\frac{\left|\,g^{nH\Lambda,m}(x_{in}) - f_o(x_{in})\,\right|}{\sqrt{\mathrm{Var}\left[\,g^{nh\lambda,m}(x_{in})\,\right]\big|_{h=H\,,\,\lambda=\Lambda}}}\;.$$

Note that $g^{nH\Lambda}$ is in effect a higher-order estimator than the original $f^{nH,m}$.

Inspection of the above suggests several other possibilities. Since estimating the m-derivative is a much harder problem than plain nonparametric

Table 5.3. The noncoverage probabilities of the standard confidence bands for quintic smoothing splines (in units of 0.001), with the smoothing parameters from cubic smoothing splines. The sample size is n. For each example, the rows correspond to GML, GCV, and AIC, respectively. For each sample size, the columns correspond to global and interior coverages for $\alpha = 0.10$ and global and interior coverages for $\alpha = 0.05$.

	n=100				:	n=400			
HM									
gml:	321	326	219	235	:	137	130	72	71
gcv:	211	212	133	140	:	146	141	77	74
aic:	230	235	147	155	:	144	142	76	74
CH									
gml:	110	113	49	50	:	116	115	54	54
gcv:	152	150	75	85	:	153	156	79	75
aic:	146	147	73	79	:	148	150	77	73
Poly5									
gml:	118	126	61	60	:	130	121	64	55
gcv:	162	176	86	89	:	177	167	93	83
aic:	153	168	87	86	:	173	159	90	80
BR									
gml:	108	100	56	53	:	110	101	50	57
gcv:	139	130	81	70	:	145	137	76	57
aic:	135	123	77	70	:	142	136	77	59
Wa-I									
gml:	95	88	47	46	:	108	89	55	47
gcv:	112	112	57	56	:	126	109	62	52
aic:	107	100	54	51	:	120	103	62	52
Wa-II									
gml:	109	89	60	51	:	120	103	49	51
gcv:	117	104	65	65	:	130	112	55	60
aic:	114	95	59	53	:	130	106	54	55
Cantor									
gml:	622	626	480	496	:	994	996	964	969
gcv:	453	445	313	325	:	801	809	700	703
aic:	543	547	404	422	:	865	866	769	778
Can5x									
gml:	811	804	703	689	:	991	990	962	961
gcv:	569	587	414	420	:	629	626	507	511
aic:	751	754	604	596	:	803	797	702	711
sn5px									
gml:	93	91	54	49	:	108	106	50	50
gcv:	111	104	54	56	:	117	103	60	52
aic:	96	99	50	52	:	114	102	60	50

regression, one might construct bias corrections by way of more accurate estimators. The first possibility is similar to the double kernel method of DEVROYE (1989) in nonparametric density estimation. Thus, if we may represent the estimator at the design points as the hat matrix operating on the data,

$$(5.13)\qquad (\, f^{nh,m}(x_{1,n}),\, f^{nh,m}(x_{2,n}),\, \cdots ,\, f^{nh,m}(x_{n,n})\,)^T = R_{nh,m}\, y_n \,,$$

where $y_n = (\, y_{1,n}, y_{2,n}, \cdots , y_{n,n}\,)^T$, then the estimator

$$(5.14)\qquad (\, 2\, R_{nh,m} - (\, R_{nh,m}\,)^2\,)\, y_n$$

will have negligible bias compared with $f^{nh,m}$.

The second possibility is already mentioned in the context of kernel estimators by EUBANK (1999), p. 109, and is based on the realization that we already have higher-order estimators, viz. the smoothing spline estimator with the order incremented by one. Thus, our recommendation is to use

$$b^{nH}(x) = f^{nH,m}(x) - f^{n,H,m+1}(x) \;.$$

Then, the bias-corrected estimator is

$$(5.15)\qquad f^{nH,m}(x) - b^{nH}(x) = f^{nH,m+1}(x)$$

and the confidence bands are based on

$$(5.16)\qquad \max_{x_{in}\in A} \frac{|\, f^{nH,m+1}(x_{in}) - f_o(x_{in})\,|}{\sqrt{\widehat{\sigma}^2\, \mathrm{Var}[\, \varphi^{nh,m+1}(x_{in})\,]\,|_{h=H}}} \,,$$

which is just $T_{nH,m+1}(A)$ from (5.7). The crucial difference is that the random H is chosen by the relevant methods for the estimator of order m, not for order $m+1$. In effect, this is undersmoothing!

Early on in the development of confidence bands in nonparametric regression, there was some question whether bias correction or undersmoothing was better, with undersmoothing being the winner; see HALL (1992) and NEUMANN (1997). The derivation above culminating in (5.16) shows that it depends mostly on how the two techniques are implemented.

(5.17) REMARK. The approach above should be applicable in the construction of confidence bands using local polynomials. The only question is whether one should use the pair m and $m+1$ or m and $m+2$. If the analogy with smoothing splines is correct, it would be the latter.

We implemented the above for cubic and quintic splines. The simulation results for sample sizes $n = 100$ and 400 based on 1000 replications are summarized in Table 5.3. Again several things stand out. First, it works remarkably well for the GML method, and only so-so for AIC and GCV. Second, even the boundary trouble is fixed! Moreover, it works even better for $n = 400$ than for $n = 100$. It did not come as a great surprise that

Table 5.4. The loss of efficiency due to the undersmoothing of the quintic smoothing splines. Shown are the discrete L^2 errors as percentages of the range of each regression function for cubic, quintic, and undersmoothed quintic smoothing splines for the GML procedure only. For sample size $n = 100$, 1000 replications are involved; for $n = 400$ only, 250 replications.

	n=100			:	n=400		
	cubic	quintic	under-smoothed	:	cubic	quintic	under-smoothed
HM:	3.38	3.77	3.37	:	1.77	1.83	1.83
CH:	2.27	2.28	2.55	:	1.25	1.22	1.43
Poly5:	2.21	2.30	2.48	:	1.21	1.25	1.36
BR:	2.77	2.89	3.08	:	1.51	1.62	1.70
Wa-I:	3.70	3.29	3.80	:	2.06	1.79	2.13
Wa-II:	4.66	4.23	4.33	:	2.57	2.25	2.51
Cantor:	3.90	4.52	3.86	:	2.50	2.60	2.56
Can5x:	6.08	7.93	5.80	:	3.80	4.00	3.80
sn5px:	3.83	3.38	4.51	:	2.15	1.83	2.45

it is still a no go for the nonsmooth examples. We did not investigate by how much the quantiles differ from the critical values. A final question concerns the loss of efficiency associated with the undersmoothing in the quintic smoothing spline estimators. In Table 5.4, we report the discrete L^2 errors (in percentages of the range of the regression function), for the cubic ($m = 2$ and 1000 replications) and quintic ($m = 3$ and 250 replications) splines using the GML procedure for selecting the smoothing parameter, as well as the undersmoothed quintic smoothing spline estimator (1000 replications). Consequently, in comparison with Table 2.2, there are slight differences in the reported mean L^2 errors for quintic splines.

The relative loss of efficiency seems to increase when the sample size is increased from $n = 100$ to $n = 400$, which one would expect since undersmoothing is a small-sample trick. Also, the undersmoothing results in a *gain* of efficiency for the nonsmooth examples, and even for the "HM" example, since the undersmoothing makes the bump "visible"; cf. Remark (5.8). In general, the authors' are not sure whether this loss of efficiency is acceptable or not. However, *some* loss of efficiency is unavoidable.

It is obvious that the question of optimality of confidence bands still remains. In general, confidence bands are difficult to compare, but in the present setting of undersmoothing for quintic smoothing splines, the shapes of the confidence bands are more or less the same for all h but get narrower with increasing h. (The pointwise variances decrease, as do the critical values.) So the question is, once a good method of undersmoothing has been found, can one do with less undersmoothing? The difficulty is

Table 5.5. Noncoverage probabilities for the "BR" example of the "normal" confidence bands for various distributions of the (iid) noise for cubic ($m = 2$) and quintic ($m = 3$) smoothing splines with smoothing parameter chosen by the GML procedure for cubic splines. Shown are the noncoverage probabilities using the true ("true") and estimated ("esti") variances of the noise. The standard deviations were the same for all noise distributions (10% of the range of the regression function).

	n=100				:	n=400			
normal									
m=2 true	195	174	110	106	:	146	129	86	67
m=2 esti	172	171	93	89	:	139	125	79	64
m=3 true	108	100	56	53	:	110	101	50	51
m=3 esti	138	118	72	72	:	114	103	55	44
Laplace									
m=2 true	195	163	135	116	:	163	152	97	82
m=2 esti	171	139	96	82	:	168	155	88	82
m=3 true	151	120	95	80	:	128	122	80	68
m=3 esti	122	102	79	63	:	1128	125	70	68
Student-t(6)									
m=2 true	212	196	139	125	:	159	132	95	80
m=2 esti	208	190	128	110	:	162	141	99	81
m=3 true	151	129	97	74	:	130	119	75	60
m=3 esti	141	122	85	64	:	130	111	74	60
Student-t(10)									
m=2 true	204	133	123	120	:	155	147	91	81
m=2 esti	194	189	118	101	:	152	140	88	76
m=3 true	121	122	71	64	:	119	117	66	54
m=3 esti	111	110	67	61	:	121	119	66	57
|Z|-E[|Z|]									
m=2 true	178	159	109	89	:	128	110	83	65
m=2 esti	176	155	103	82	:	131	109	78	70
m=3 true	119	94	75	55	:	110	110	62	57
m=3 esti	107	93	63	48	:	112	110	59	54

that this seems to be purely a small-sample problem. We said before that asymptotically the problem goes away.

Nonnormal noise. Next, we briefly consider what happens when the noise is not normal; i.e., the assumption (1.4) is replaced by

$$(5.18) \qquad (d_{1,n} , d_{2,n} , \cdots , d_{n,n})^T \quad \text{are iid with a suitable distribution} .$$

The main concern is whether the normal approximations are accurate enough. If the smoothing parameters are "large", everything is fine, since then the estimators at each point are weighted averages of many iid random variables. So, the question is whether the undersmoothing advocated above gives rise to smoothing parameters that are too "small". We also investigate the effect of estimated variances.

In (5.18), we consider the two-sided exponential, denoted as the Laplace distribution, the Student t distribution with 6 and 10 degrees of freedom, and one-sided normal noise centered at its mean, succinctly described as $|\,Z\,| - \mathbb{E}[\,|\,Z\,|\,]$, with Z the standard normal random variable. In all cases, the noise generated was scaled to have the usual variance given by (2.1). The simulation setup is the same as before, except that here we limit ourselves to one example only (BR). The results are summarized in Table 5.5. The case of normal noise is included also for comparison with the other types of noise, as well as to show the effect of estimated variances on the noncoverage probabilities. The example "Wa-I" was also run, and compared with "BR", the results were noticeably better for $n = 100$ and barely better for $n = 400$.

We make some brief observations regarding the results for nonnormal noise. First, using estimated variances results in slightly improved coverage probabilities for both $m = 2$ and the tweaked $m = 3$ case, irrespective of the sample size or the type of noise. Second, the tails of the noise distribution affect the coverage probabilities; i.e., the coverage probabilities (for $m = 3$ with undersmoothing) deteriorate noticeably for Laplace and Student t noise. While one would expect $t\,(6)$ noise to give poorer results than $t\,(10)$, it is striking that, if anything, $t\,(6)$ noise "beats" Laplace noise. The asymmetry of the noise distribution seems to be mostly harmless, if we may draw such a broad conclusion from the absolute-normal noise case.

We leave confidence bands for nonnormal noise here, but obviously there is more to be done. For example, bootstrapping ideas come to minds; see, e.g., Claeskens and van Keilegom (2003) and references therein. Also, one could explore the use of least-absolute-deviations smoothing splines of Chapter 17, but that carries with it the problem of constructing confidence bands (and smoothing parameter selection) for these splines.

(5.19) Computational Remarks. Some comments on the computations are in order. First, the computation of the variances $\mathbb{V}\mathrm{ar}[\,\varphi^{nH,m}(x_{in}) \mid H\,]$ amounts to, in the language of Chapter 18, the calculation of the diagonal elements of $R_{nH}\,R_{nH}^T$, or $\|\,R_{nH}\,e_i\,\|^2$ for the standard basis $e_1, e_2, \cdots, e_n$ of $\mathbb{R}^n$. Each one of the $R_{nH}\,e_i$ can be computed by the Kalman filter at a cost of $\mathcal{O}(\,n\,)$ totaling $\mathcal{O}(\,n^2\,)$ overall. It is not clear to the authors whether it can be done in $\mathcal{O}(\,n\,)$ operations.

Second, the critical values $c_{\alpha nm}(H, A)$ for the whole and the interior interval are determined by way of simulations. For $\alpha = 0.10$ and $\alpha = 0.05$ and sample size $n = 100$, it was thought that 10,000 replications would be sufficient. Of course, the random H is not known beforehand, so (a fine grid of) all "possible" H must be considered. This is the price one pays for not employing the asymptotic distribution theory.

6. The Wood Thrush Data Set

In this section, we analyze the Wood Thrush Data Set of Brown and Roth (2004), following Brown, Eggermont, LaRiccia, and Roth (2007). The emphasis is on nonparametric confidence bands as well as on the (in)adequacy of some parametric models, especially the misleading accuracy of the estimator as "predicted" by the parametric confidence bands. For smoothing splines, the confidence bands need investigation to sort out the effect that the nonuniform random design might have. The randomness of the design is not a problem since we condition on the design; it is the nonuniformity that requires attention.

As with any real data set, there are plenty of problematic aspects to the data of Brown and Roth (2004). While most of them will be ignored, we do spell out which ones cause problems.

The Wood Thrush Data Set set contains observations on age, weight, and wingspan of the common wood thrush, *Hylocichla mustelina*, over the first 92 days of their life span. The data were collected over a period of 30 years from a forest fragment at the University of Delaware. Early on, birds are collected from the nest, measured and ringed, and returned to the nest. The age of the nestling is guessed (in days). This is pretty accurate since the same nests were inspected until there were nestlings, but inaccurate in that age is rounded off to whole days. The data for older birds are obtained from recapture data. Since the birds were measured before, one has accurate knowledge of their ages. However, because of the difficulty of capturing older birds, there are a lot more data on nestlings (the biologists know where to look for nests) but much less during the critical period when nestlings first leave the nest and stay hidden.

Thus, the collected data consists of the records

$$(6.1) \qquad (t_i , y_i , z_i) , \quad i = 1, 2, \cdots , N ,$$

for $N = 923$, about which we will say more later, and where

$$\begin{aligned} t_i &= \text{age of } i\text{-th bird, in days} , \\ y_i &= \text{weight of } i\text{-th bird, in grams} , \\ z_i &= \text{wingspan of } i\text{-th bird, in centimeters} . \end{aligned}$$

We shall ignore the wingspan data. Since the $n = 923$ observations involve only 76 distinct ages, we present the data in the compressed form

$$(6.2) \qquad (x_j , n_j , \overline{y}_j , s_j^2) , \quad j = 1, 2, \cdots , J ,$$

with $J = 76$, where $x_1 < x_2 < \cdots < x_J$ are the distinct ages, n_j is the number of observations at age x_j,

$$n_j = \sum_{i=1}^{N} \mathbb{1}(t_i = x_j) , \tag{6.3}$$

and $\overline{y}_j$ and s_j^2 are the mean and pure-error variance at age x_j,

$$\begin{aligned} \overline{y}_j &= \frac{1}{n_j} \sum_{i=1}^{N} y_i \, \mathbb{1}(t_i = x_j) , \\ s_j^2 &= \frac{1}{n_j - 1} \sum_{i=1}^{N} | y_i - \overline{y}_j |^2 \, \mathbb{1}(t_i = x_j) , \quad n_j \geqslant 2 ; \end{aligned} \tag{6.4}$$

see Table 6.1.

Of course, we assume that the data (6.1) are independent, conditioned on the t_i. This would not be correct if there were many repeated measurements on a small number of birds since then one would have what is commonly referred to as longitudinal data. The remedy is to take only the measurement for the oldest age and discard the early data. In fact, this has been done. We also assume that the variance of the measured weight is the same throughout the approximately 80 day period of interest. This seems reasonable, but the readers may ascertain this for themselves from the data presented in Table 6.1. Thus, the proposed *nonparametric* model for the data (6.1) is

$$y_i = f_o(t_i) + \varepsilon_i , \quad i = 1, 2, \cdots , N , \tag{6.5}$$

where $f_o(t)$ is the mean weight of the wood thrush at t days of age and ε_i represents the measurement error of the weight, as well as the bird-to-bird variation, with

$$(\varepsilon_1, \varepsilon_2, \cdots , \varepsilon_N)^T \sim \text{Normal}(0 , \sigma^2 I_{N \times N}) \tag{6.6}$$

and σ^2 unknown. The nonparametric assumption is that f_o is a smooth function. However, it is customary in (avian) growth studies to consider *parametric* models, such as the Gompertz, logistic, and Richards families of parametric regression functions. See Table 6.2, where we also list the parameters obtained by least-squares fitting of the Wood Thrush Data Set: For a parametric family $g(\,\cdot \mid \beta)$, depending on the parameter $\beta \in \mathbb{R}^p$ (with $p = 3$ or 4, say), the estimator is obtained by solving

$$\text{minimize} \quad \tfrac{1}{N} \sum_{i=1}^{N} \bigl| y_i - g(x_i \mid \beta) \bigr|^2 . \tag{6.7}$$

We conveniently forget about the possibility of multiple local minima in such problems, as well as the fact that "solutions" may wander off to ∞. Asymptotic confidence bands for parametric regression models are typically based on the linearization of $g(\,\cdot \mid \beta)$ around the estimated β_o; see, e.g., Bates and Watts (1988) or Seber and Wild (2003).

Table 6.1. The Wood Thrush Sata Set of Brown and Roth (2004), compressed for age and weight.

age	replics	mean	PE-var	age	replics	mean	PE-var
7	10	31.52	12.77	46	4	47.38	8.40
8	72	30.56	9.87	48	3	48.83	19.08
9	324	32.29	11.97	49	7	49.21	15.07
10	133	32.70	11.11	50	7	47.29	7.32
11	39	34.43	7.39	51	7	49.71	7.90
12	62	34.05	9.12	52	3	49.33	2.33
13	28	34.63	11.65	53	3	51.67	0.33
14	7	37.36	2.39	54	6	49.92	5.84
15	3	37.17	9.33	56	9	49.56	12.65
16	2	38.25	0.12	57	5	47.40	3.55
17	1	38.50	---	58	7	51.71	16.49
18	3	38.33	14.08	59	5	50.00	14.12
19	5	40.90	10.80	60	5	50.60	14.30
20	5	40.90	52.92	61	2	51.75	0.12
22	3	48.50	22.75	62	6	53.25	5.78
23	3	48.83	5.33	63	5	51.80	5.20
24	4	46.25	18.75	64	3	52.67	0.33
25	6	46.25	3.88	66	3	51.00	4.75
26	7	46.71	12.82	67	1	48.50	---
27	5	49.50	8.75	68	3	50.33	7.58
28	5	48.80	5.07	69	1	55.00	---
29	6	48.08	4.84	70	1	51.50	---
30	4	51.62	7.56	72	3	53.50	16.75
31	1	47.00	---	73	2	52.75	1.12
32	8	49.06	8.60	74	2	49.75	1.12
33	4	45.62	7.23	75	2	47.00	8.00
34	7	48.59	3.03	76	2	53.00	2.00
35	1	48.00	---	77	1	54.00	---
36	8	48.62	25.41	78	1	54.00	---
37	8	48.25	7.86	80	2	51.75	1.12
38	6	45.83	6.97	82	1	55.00	---
39	3	48.17	24.33	83	2	50.25	0.12
40	1	51.00	---	85	1	49.00	---
41	4	46.25	30.75	86	1	54.50	---
42	3	49.67	6.33	89	2	52.50	24.50
43	3	48.17	16.08	90	2	52.50	12.50
44	5	48.40	19.17	91	1	55.50	---
45	2	45.50	24.50	92	1	48.00	---

Table 6.2. The four regression functions to be used in the simulations. The "Gompertz" family was defined in (1.8). The reparametrized "Richards" and "logistic" families are defined in (6.9)-(6.10).

(realBR)	$f_o(x) = \text{Richards}(\, x \mid \beta\,) - 4\exp\Big(-\big(\,x/100 - 0.425\,\big)^2/(0.16)^2\Big)$
with	$\beta_0 = 49.679\ ,\ \beta_1 = 11.404\ ,\ \beta_2 = 43.432\ ,\ \beta_3 = 17.092\ ,$
(Richards)	$f_o(x) = \text{Richards}(\, x \mid \beta\,)$
with	$\beta_0 = 49.679\ ,\ \beta_1 = 8.5598\ ,\ \beta_2 = 43.432\ ,\ \beta_3 = 17.092\ ,$
(Gompertz)	$f_o(x) = \text{Gompertz}(\, x \mid \beta\,)$
with	$\beta_0 = 51.169\ ,\ \beta_1 = -33.176\ ,\ \beta_2 = 0.078256\ ,$
(logistic)	$f_o(x) = \text{logistic}(\, x \mid \beta\,)$
with	$\beta_0 = 50,857\ ,\ \beta_1 = 21.411\ ,\ \beta_2 = 0.093726\ ,$

In Figure 6.1(c), we show the 95% confidence band for the unknown mean function, f_o, based on the undersmoothed quintic smoothing spline with the GML smoothing parameter for cubic splines (denoted as "$m = 3$ w/ $m = 2$".) There, we also show the confidence bands corresponding to the regular quintic smoothing spline, which is indeed narrower and smoother but of course does not have the correct noncoverage probability. (The cubic spline confidence band is still narrower, but only barely.) The rough shape of the band also reflects the necessary loss of efficiency of the undersmoothed quintic spline. As said before, there is a price to be paid for getting the advertised coverage probabilities. The question is whether the price would be less for other methods. In Figure 6.1(d), we again show the confidence band based on the undersmoothed quintic spline, as well as the Richards and Gompertz estimators. The inadequacy of the Gompertz estimator (and the very similar logistic estimator) is apparent. The Richards estimator fails to lie completely inside the spline confidence band. In Figure 6.1(a), we show the confidence band based on the Richards estimator, as well as the Gompertz estimator. In Figure 6.1(b), the situation is reversed. It is interesting that the implied adult weight limit of either one does not lie inside the confidence band of the other.

Note that both the Richards and Gompertz models capture certain aspects of the growth curve very well but completely miss others. For example, the Richards model estimates the mass and growth for ages up to 30 days much better than the Gompertz model. This is true especially in the range 12 to 34 days. However, the Richards model underestimates the mass from day 65 on. On the other hand, in both the Richards and Gompertz models, there is no hint of the dip in mass during the period of 30 to 40 days. As shown in Figure 12.1.2, the cubic or quintic smoothing

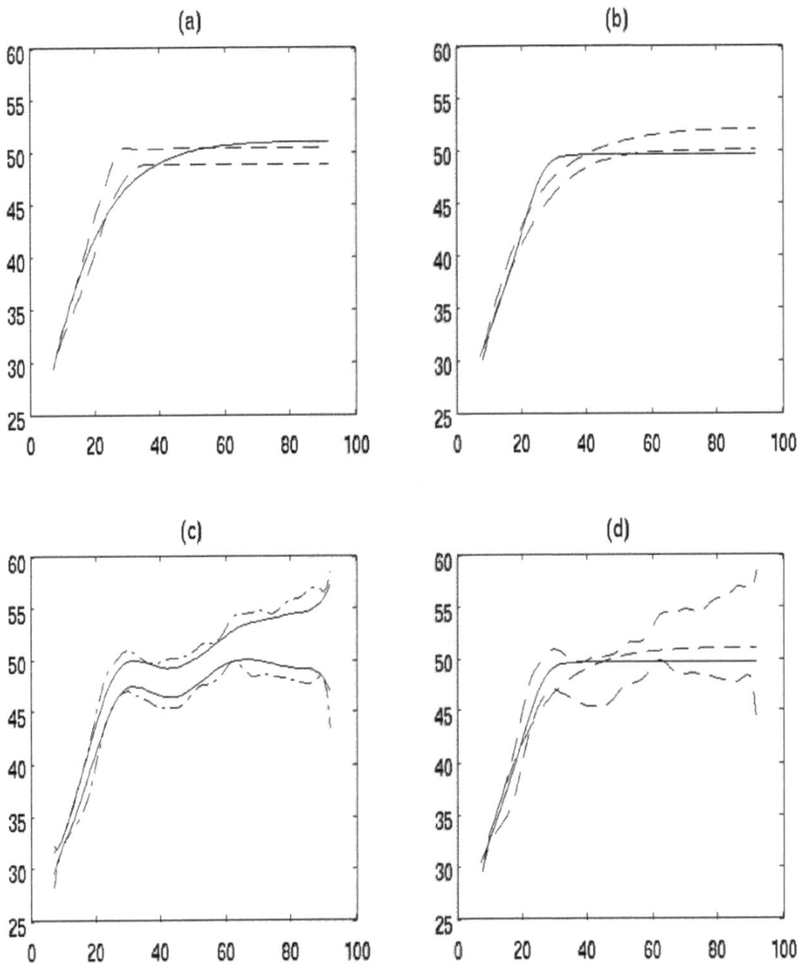

Figure 6.1. 95% confidence bands for the wood thrush growth curve. The figures show the confidence bands for the Richards, Gompertz, and regular and undersmoothed quintic spline ($m = 3$ w/ $m = 2$) estimators. (a) The Richards confidence band. The Gompertz estimator (solid line) does not lie inside the band. (b) The Gompertz confidence band. The Richards estimator does not lie inside the band. (c) The confidence bands based on the undersmoothed quintic spline estimator (dashed) and the one based on the oversmoothed quintic spline (solid). The regular quintic spline confidence band lies completely inside the undersmoothed one. (d) As in (c), with the Gompertz and Richards estimators. The Gompertz estimator clearly does not lie inside the band at approximately 25 days. The Richards curve barely crosses the upper band as well as the lower band (at approximately 35 and 60 days). Not shown is that the Gompertz and logistic confidence bands nearly coincide.

spline estimators capture all of the important aspects over all stages of the growth curve.

Of course, we must gather some evidence as to whether the strange design has an adverse effect on the coverage probabilities. A simulation study should shed some light on this. At the same time, it gives us an opportunity to investigate the parametric models used above. Before proceeding, we mention the pure-error estimator of the variance, $\widehat{\sigma}^2_{\mathrm{PE}}$, of σ^2 in the model (6.5)–(6.6),

$$(6.8) \qquad \widehat{\sigma}^{\,2}_{\mathrm{PE}} = 3.3224 \ .$$

For the "regular" quintic and cubic splines, the residual sums of squares lead to the point estimators 3.3274 and 3.3281.

We employ the simulation setup of § 5, with the parametric regression functions given by the Gompertz, logistic, and Richards fits to the Wood Thrush Data Set. We also incorporate the "true" regression function, a mock-up of the (oversmoothed) quintic smoothing spline estimator; see Table 6.2. The Gompertz function is defined in (1.8), the reparametrized Richards function reads as

$$(6.9) \qquad \mathrm{Richards}(\, x \mid \beta\,) = \beta_0 \left(1 + \beta_3 \exp(\, \beta_1 - \beta_2 \, t \,) \right)^{-1/\beta_3} ,$$

and the logistic function, $\mathrm{logistic}(x \mid \beta)$, is given by

$$(6.10) \qquad \mathrm{logistic}(\, x \mid \beta\,) = \beta_0 \Big/ \left(1 + \left(\frac{\beta_0}{\beta_1} - 1 \right) \exp(\, -\beta_2 \, x \,) \right) .$$

In the simulations, we employed the design from the Wood Thrush Data Set, including the independent replications, as detailed in Table 6.1. Thus, the design is deterministic, whereas in the Wood Thrush Data Set it is of course random. Thus, the sample size was $n = 923$, with $J = 76$ distinct design points. The noise was taken to be independent $\mathrm{Normal}(\,0\,,\, \widehat{\sigma}^2_{\mathrm{PE}}\,)$. For each regression function, the usual 1000 replications were run. The critical values employed in the confidence bands are based on the same design, iid normal noise, and 10,000 replications. (For the computational aspects, see Remark (6.11).)

The estimated noncoverage probabilities are shown in Table 6.3. For loss of efficiency results, see Table 6.4. We briefly comment on the results. First of all, the confidence bands based on the undersmoothed quintic spline ($m = 3$ w/ $m = 2$) seem to work pretty well. Second, the confidence bands for the parametric models work very well when the regression function belongs to the same family but not at all when it is not. This holds even for the Gompertz and logistic families, despite the fact that the estimators are always very close to each other. Also, the L^2 errors of the parametric estimators are always better than those of the spline estimators, provided the parametric model "fits". The Richards estimator always does very well in comparison with Gompertz and logistic because it "contains" the logistic model (for $\beta_3 = 1$) as well as the Gompertz model in the limit

Table 6.3. Noncoverage probabilities (in units of 0.001) for the global Wood Thrush design, using the Gompertz, Richards, logistic, and realBR models, and Gompertz, Richards, logistic, and smoothing spline estimators using GML smoothing parameter selection, with $m = 2$, $m = 3$ with $m = 2$ smoothing and $m = 3$. "true" and "esti" refer to the true and estimated variances used in the confidence bands. The confidence level is 95%.

model	var	Gomp-ertz	Rich-ards	logis-tic	smoothing splines m=2	m=3 w/ m=2	m=3
Gompertz	true:	38	35	161	102	43	102
	esti:	41	37	167	101	49	106
Richards	true:	1000	67	1000	138	44	382
	esti:	1000	67	1000	141	50	378
logistic	true:	108	34	35	78	45	80
	esti:	112	36	39	78	50	77
realBR	true:	1000	1000	1000	166	40	295
	esti:	1000	1000	1000	159	49	296

Table 6.4. The mean discrete L^2 errors in the setup of Table 6.3.

model	Gomp-ertz	Rich-ards	logis-tic	smoothing splines m=2	m=3 w/ m=2	m=3
Gompertz:	.1759	.1903	.2052	.2787	.3772	.2460
Richards:	.9313	.2059	.7962	.3031	.3954	.3313
logistic:	.2034	.1985	.1760	.2733	.3748	.2411
real BR:	.7409	.7000	.6427	.3172	.4084	.3321

as $\beta_3 \to 0$. We do not know why the noncoverage probabilities for the Richards family are significantly different from the nominal 0.050. Perhaps it should be noted that although the $n = 923$ sample size is large, there are only $J = 76$ distinct design points, so the bias in the estimators is likely to be larger than one would think with sample size $n = 923$. Another point is that, for the Richards estimator, the coefficient β_3 tends to be large, which puts the linearization approach alluded to earlier in doubt. However, unfortunately, we shall not pursue this.

(6.11) COMPUTATIONAL REMARKS. The large number of independent replications in the data makes the computations interesting. While the Kalman filter approach described in § 20.9 applies as is, including the determination of the GML functional for choosing the smoothing parameter and the estimation of the variance, it is computationally more efficient to consider the reduced model, see (6.3)–(6.4),

$$\overline{y}_j = f_o(x_j) + r_j\, \varepsilon_j \ , \quad j = 1, 2, \cdots, J \ ,$$

with $r_j = n_j^{-1/2}$. Then, the new ε_j satisfy

$$(\varepsilon_1, \varepsilon_2, \cdots, \varepsilon_J)^T \sim \text{Normal}(0\,,\, \sigma^2 I_{J\times J}) \ ;$$

see Exercises (20.2.16) and (20.2.34).

In Chapter 20, we only considered homoscedastic noise, and it seems that now we are paying the price. However, one may use the homoscedastic version of the Kalman filter on the transformed model

$$r_j^{-1}\, \overline{y}_j = g_o(x_j) + \varepsilon_j \ , \quad j = 1, 2, \cdots, J \ ,$$

with $g_o(x_j) = r_j^{-1}\, f_o(x_j)$ for all j, as follows. The state-space model for the original model was

$$\mathbb{S}_{j+1} = Q_{j+1|j}\, \mathbb{S}_j + \mathbb{U}_j \ , \tag{6.12}$$

with $\mathbb{E}[\, \mathbb{U}_j \mathbb{U}_j^T\,] = \Sigma_j$ determined from the white noise prior. Now, let $\mathbb{S}_j^r = r_j^{-1}\, \mathbb{S}_j$, and $Q_{j+1|j}^r = (\, r_j / r_{j+1}\,)\, Q_{j+1|j}$. Multiplying (6.12) by r_{j+1} then gives

$$\mathbb{S}_{j+1}^r = Q_{j+1|j}^r\, \mathbb{S}_j^r + \mathbb{U}_j^r \ , \tag{6.13}$$

with $\mathbb{E}[\, \mathbb{U}_j^r (\mathbb{U}_j^r)^T\,] = r_j^{-2}\, \Sigma_j$. Thus, the homoscedastic version of the Kalman filter may be applied to obtain the estimator g^{nh} of g_o at the design points, and then

$$f^{nh}(x_j) = r_j\, g^{nh}(x_j) \ , \quad j = 1, 2, \cdots, J \ . \tag{6.14}$$

The leverage values need some attention as well: We need the diagonal entries of $\mathbb{V}\text{ar}[\, \widehat{y}^{nh}\,]$, where $\widehat{y}_j^{nh} = f^{nh}(x_j)$. Now, f^{nh} is the solution to

$$\text{minimize} \quad \sum_{i=1}^{n} r_j^{-2}\, |\, f(x_j) - \overline{y}_j\, |^2 + nh^{2m}\, \|\, f^{(m)}\, \|^2 \quad \text{over} \quad f \in W^{m,2}(0,1) \ .$$

Letting $D = \text{diag}(\, r_1, r_2, \cdots, r_J\,)$, then $\widehat{y}^{nh} = R_{nh}\, \overline{y}$, with the hat matrix

$$R_{nh} = (\, D^{-2} + nh^{2m}\, M\,)^{-2}\, D^{-2} \ .$$

Note that $R_{nh}^T = D^{-2}\, R_{nh}\, D^2$. Then, $\mathbb{V}\text{ar}[\, y^{nh}\,] = \sigma^2\, R_{nh}\, D^2\, R_{nh}^T$ and

$$(\mathbb{V}\text{ar}[\, y^{nh}\,])_{i,i} = \sigma^2\, \|\, D\, R_{nh}^T\, e_i\, \|^2 = \sigma^2\, \|\, D^{-1}\, R_{nh}\, D^2\, e_i\, \|^2 \ , \tag{6.15}$$

where e_i is the i-th unit vector in the standard basis for $\mathbb{R}^n$. So, as before, the leverage values may be computed with the homoscedastic version of the Kalman filter. The computation of the GML functional must be updated as well but will give the same answers as the plain Kalman filter on the original model (6.5)–(6.6).

The computations for the parametric models were done using the MATLAB procedure NLINFIT. The confidence bands for the Gompertz and logistic models were obtained using the MATLAB procedure NLPREDCI. For the Richards family, this does not work very well since the (estimated) Jacobian matrix

$$J \stackrel{\text{def}}{=} \left(\nabla_\beta g(x_1 \,|\, \widehat{\beta}) \,\middle|\, \nabla_\beta g(x_2 \,|\, \widehat{\beta}) \,\middle|\, \cdots \,\middle|\, \nabla_\beta g(x_N \,|\, \widehat{\beta}) \right)^T \in \mathbb{R}^{N \times 4}$$

not infrequently fails to have full rank. (In the extreme case $\beta_3 = \infty$, when the estimator consists of a sloped line segment plus a horizontal line segment, the rank equals 2.) Thus, the variance matrix (projector) $J(J^T J)^{-1} J$ was computed using the singular value decomposition of J. The pointwise variances are then given as the diagonal elements of this matrix. The extra computational cost is negligible.

7. The Wastewater Data Set

We now analyze the Wastewater Data Set of HETRICK and CHIRNSIDE (2000) on the effectiveness of the white rot fungus, *Phanerochaete chrysosporium*, in treating wastewater from soybean hull processing. In particular, we determine confidence bands for the regression function and investigate the effect on the noncoverage probabilities of the apparent heteroscedasticity of the noise in the data. The data show the increase over time of the concentration of magnesium as an indicator of the progress of the treatment. The data were collected from a "tabletop" experiment at 62 intervals of 10 minutes each. Thus, the data consist of the records

$$(7.1) \qquad (x_{in}, y_{in}), \quad i = 1, 2, \cdots, n,$$

with $n = 63$ and $x_{in} = 10\,(i-1)$ minutes. The initial time is 0, the final time is $x_T = 620$. The y_{in} are the measured concentrations. The data are listed in Table 7.1; a plot of the data was shown in Figure 12.1.1(b). It is clear that the noise in the system is heteroscedastic; i.e., there is a dramatic change in the noise level between $x_{35} = 340$ and $x_{36} = 350$. Indeed, this is due to a switch in the way the concentration has to be measured, so that $x_c = 345$ is a (deterministic) change point. To model the heteroscedasticity of the data, we assume that the variance is constant on each of the intervals $[0, x_c)$ and $(x_c, x_T]$. Using second-order differencing, see (7.8) below, one gets the estimators of the variances on the two intervals,

$$(7.2) \qquad \widehat{\sigma}_1^2 = 7.3138_{10}{-04} \quad \text{and} \quad \widehat{\sigma}_2^2 = 1.6322_{10}{-02},$$

Table 7.1. The Wastewater data of HETRICK and CHIRNSIDE (2000). The evolution of the concentration of magnesium in processed wastewater as a function of time. Elapsed time in units of 10 seconds. Concentration in units of 1200 ppm (parts per million).

time	conc.	time	conc.	time	conc.	time	conc.
0	0.000000	16	0.006744	32	0.121932	48	0.318582
1	0.000000	17	0.009656	33	0.133082	49	0.318665
2	0.000003	18	0.013045	34	0.143136	50	0.327380
3	0.000007	19	0.017279	35	0.166818	51	0.363808
4	0.000005	20	0.022261	36	0.206268	52	0.343617
5	0.000005	21	0.027778	37	0.173117	53	0.360273
6	0.000006	22	0.034248	38	0.181616	54	0.350025
7	0.000004	23	0.041758	39	0.197496	55	0.328825
8	0.000033	24	0.050080	40	0.201898	56	0.361943
9	0.000065	25	0.054983	41	0.238238	57	0.402238
10	0.000171	26	0.064672	42	0.238238	58	0.398558
11	0.000450	27	0.074247	43	0.250830	59	0.417468
12	0.000756	28	0.083786	44	0.247585	60	0.388098
13	0.001617	29	0.092050	45	0.268713	61	0.451378
14	0.002755	30	0.105043	46	0.285802	62	0.424770
15	0.004295	31	0.112345	47	0.300963		

with a ratio of about $1 : 500$. In combination with the small sample size, $n = 63$, this large a ratio will prove to be challenging.

The biochemistry of the process suggests that overall the magnesium concentration should increase smoothly with time, but as usual in growth phenomena, early on this does not have to be the case. So, the data may be modeled *nonparametrically* by

$$y_{in} = f_o(x_{in}) + d_{in} \ , \quad i = 1, 2, \cdots, n \ , \tag{7.3}$$

with f_o a smooth, "mostly" increasing function. We assume that the noise $d_n = (d_{1,n}, d_{2,n}, \cdots, d_{n,n})^T$ is multivariate normal with independent components. In view of the discussion of the heteroscedasticity above, set

$$\mathbb{E}[\, d_n \, d_n^T \,] = V_o^2 \ , \tag{7.4}$$

where $V_o = D(\sigma_{o,1}, \sigma_{o,2})$ is an $n \times n$ diagonal matrix. Here, for arbitrary reals s and t, with $k = 35$,

$$D(\, s \, , \, t \,) = \begin{cases} s \ , & 1 \leqslant i \leqslant k \ , \\ t \ , & k+1 \leqslant i \leqslant n \ , \end{cases} \tag{7.5}$$

and all other elements equal to 0. Note that $x_k = 340$ and $x_{k+1} = 350$, and indeed the change point is at time 345. Note that the data are positive, so that the assumption of normality is not quite right, but we shall let it pass. For an attempt to fix this, see Remark (7.21) below.

We now discuss the construction of the smoothing spline estimator of the mean function. Following the maximum penalized likelihood approach of § 12.2, the smoothing spline estimator of order $2m$ is the solution of

$$\text{(7.6)} \qquad \begin{array}{ll} \text{minimize} & \mathfrak{L}(f, \sigma_1, \sigma_2) \\ \text{subject to} & f \in W^{m,2}(0,1), \ \sigma_1 > 0, \ \sigma_2 > 0 \ , \end{array}$$

where

$$\text{(7.7)} \quad \mathfrak{L}(f, \sigma_1, \sigma_2) = \log\det(V) + \frac{1}{n} \sum_{i=1}^{n} \frac{\left| f(x_{in}) - y_{in} \right|^2}{2\, V_{i,i}^2} + \lambda \, \| f^{(m)} \|^2 \ ,$$

in which $V = D(\sigma_1, \sigma_2)$ and λ must be chosen appropriately. As discussed in § 12.2, in (7.6) some restrictions must be imposed on σ_1 and σ_2. However, instead of exactly solving a suitably modified version of (7.7), we take the more pragmatic route of first estimating $\sigma_{o,1}$ and $\sigma_{o,2}$ using second-order differencing as in Exercise (22.4.48). In the present case of a deterministic uniform design, this takes the form

$$\text{(7.8)} \qquad \begin{aligned} \widehat{\sigma}_1^2 &= \frac{1}{6(k-2)} \sum_{i=2}^{k-1} \left| \, y_{i+1} - 2\, y_{in} + y_{i-1} \right|^2 \ , \\ \widehat{\sigma}_2^2 &= \frac{1}{6(n-k-2)} \sum_{i=k+1}^{n-1} \left| \, y_{i+1} - 2\, y_{in} + y_{i-1} \right|^2 \ . \end{aligned}$$

Then, our smoothing spline estimator is the solution of the problem

$$\text{(7.9)} \qquad \text{minimize} \quad \frac{1}{n} \sum_{i=1}^{n} \frac{\left| f(x_{in}) - y_{in} \right|^2}{\widehat{V}_{i,i}^2} + h^{2m} \, \| f^{(m)} \|^2$$

over $f \in W^{m,2}(0,1)$, with $\widehat{V} \equiv D(\widehat{\sigma}_1, \widehat{\sigma}_2)$. However, to have better scaling properties of the smoothing parameter, we changed this to

$$\text{(7.10)} \qquad \widehat{V} \equiv D(\widehat{\varrho}, 1/\widehat{\varrho}) \quad \text{with} \quad \widehat{\varrho} = \sqrt{ \widehat{\sigma}_1 / \widehat{\sigma}_2 } \ .$$

As before, the GML procedure is used to select the smoothing parameter $h = H$. The solution of (7.9) is denoted by $f^{n,m,H}$.

(7.11) REMARK. One wonders whether all of this is worth the trouble. The alternative is to just ignore the heteroscedasticity of the noise; i.e., in (7.9), replace the $\widehat{V}_{i,i}^2$ by 1. We refer to this estimator as the "homoscedastic" smoothing spline. The original one is the "heteroscedastic" estimator. Note that the convergence rates of the usual measures of the error for both estimators are the same, so all one can hope for is a reduction in the asymptotic constant. In Tables 7.2 and 7.3, we judge these two estimators by comparing various measures of the error and their standard deviations. The measures of the error considered are the global discrete L^2 and L^∞

norms, $| f^{n,m,H} - f_o |_p$ for $p = 2$ and $p = \infty$, see (1.5) and (1.6); the interior versions, $| f^{n,m,H} - f_o |_{p,I}$ for $p = 2$ and $p = \infty$, with $I = [\, 62\, ,\, 620 - 62\,]$, cf. (4.6); and weighted versions, where $f^{n,m,H} - f_o$ is replaced by

$$\text{(7.12)} \qquad \frac{f^{n,m,H} - f_o}{\sqrt{\mathbb{V}\mathrm{ar}[\, f^{n,m,h}\,]\big|_{h=H} / (\,\widehat{\sigma}_1\, \widehat{\sigma}_2\,)}} \; .$$

The strange scaling comes from (7.10). Here, the smoothing parameter H is the optimal one for each measure. The results were obtained from 1000 replications for the example

$$\text{(7.13)} \qquad f_o(x) = \beta_o \exp\Bigl(- \Bigl(\frac{x}{\beta_1} \Bigr)^{-\beta_2} \Bigr) \; , \quad 0 \leqslant x \leqslant 620 \; ,$$

where $\beta_0 = 1.0189$, $\beta_1 = 554.39$, $\beta_2 = 1.3361$. This example is like the "CH" model of Table 1.1, except that we sharpened the corner at $x_{15} = 140$.

Inspection of Tables 7.2 and 7.3 reveals that ignoring the heteroscedasticity of the noise is not a good idea. Unreported experiments show that the effect on the shapes and widths of confidence bands is even more impressive.

Next, we consider the construction of the confidence bands. As before, we take them to be of the form

$$\text{(7.14)} \qquad f^{n,m,H}(x_{in}) \pm c_{\alpha,n,m,H} \sqrt{\mathbb{V}\mathrm{ar}[\, f^{n,m,h}(x_{in})\,]\big|_{h=H}} \; .$$

As a slight modification of the development in the Computational Remark (6.11), for $\widehat{y}^{nh} = \bigl(f^{n,m,h}(x_{1,n}), f^{n,m,h}(x_{2,n}), \cdots, f^{n,m,h}(x_{n,n}) \bigr)^T$, we have that

$$\text{(7.15)} \qquad \mathbb{V}\mathrm{ar}[\, \widehat{y}^{nh}\,] = R_{nh} \left(D(\sigma_{o,1}, \sigma_{o,2}) \right)^2 R_{nh}^T \; ,$$

with the hat matrix $R_{nh} = \bigl(\widehat{V}^{-2} + nh^{2m} M \bigr)^{-1} \widehat{V}^{-2}$. Now, one could use the previous estimator of the variances in (7.15), but in hindsight, it is a good idea to estimate them again. The two choices

$$\text{(7.16)} \qquad \begin{aligned} \widehat{\sigma}^2_{o,1,o} &= \tfrac{1}{k} \sum_{i=1}^{k} |\, f^{n,m,H}(x_{in}) - y_{in}\, |^2 \; , \\ \widehat{\sigma}^2_{o,1,1} &= \tfrac{1}{k} \sum_{i=1}^{k} y_{in} \left(y_{in} - f^{n,m,H}(x_{in}) \right) \; , \end{aligned}$$

suggest themselves, and likewise for $\sigma^2_{o,2}$. In view of the small sample size, it is also a good idea to inflate the estimators, so the (rational!) choice

$$\text{(7.17)} \qquad \widehat{\sigma}^2_{o,1} = \max\bigl(\widehat{\sigma}^2_{o,1,o}\, , \, \widehat{\sigma}^2_{o,1,1} \bigr)$$

was made and likewise for $\widehat{\sigma}^2_{o,2}$.

Thus, the confidence bands take the form

$$\text{(7.18)} \qquad f^{n,m,H}(x_{in}) \pm c_{\alpha,n,m,H} \sqrt{ \bigl[\, R_{nH}\, D(\widehat{\sigma}_{o,1}, \widehat{\sigma}_{o,2})\, R_{nH}^T \,\bigr]_{i,i} } \; .$$

Table 7.2. Ignoring the heteroscedasticity of the noise in wastewater-like data. Estimated means and standard deviations of various measures of the optimal estimation error are shown for the "homoscedastic" cubic smoothing spline, see Remark (7.9) and the "heteroscedastic" cubic smoothing spline (7.7) for the example (7.3)–(7.4) and (7.13), with ratios of the two variances 500, 10, and 1. Both global and interior errors are reported. (L2 and SUP: the discrete L^2 and discrete L^∞ errors. L2W and SUPW: the weighted versions. The std lines contain the standard deviations.)

	plain estimator		weighted estimator	
	global error	interior error	global error	interior error
		ratio of sigmas = 500		
L2	3.1085e-03	2.3930e-03	2.2906e-03	1.3320e-03
std	1.1241e-03	9.4126e-04	1.1798e-03	8.9242e-04
L2W	2.2677e-03	2.0725e-03	1.2743e-03	1.0394e-03
std	9.1467e-04	9.3539e-04	2.7134e-04	2.7305e-04
SUP	7.2175e-03	5.1753e-03	5.9624e-03	3.6782e-03
std	3.6037e-03	2.1389e-03	3.1520e-03	2.1182e-03
SUPW	7.2164e-03	7.0135e-03	3.2593e-03	2.8497e-03
std	4.9604e-03	7.0135e-03	1.0232e-03	2.8497e-03
		ratio of sigmas = 10		
L2	3.3012e-03	2.5555e-03	2.8860e-03	2.0349e-03
std	1.1040e-03	9.3338e-04	1.1476e-03	9.1378e-04
L2W	3.0348e-03	2.5158e-03	2.7701e-03	2.2095e-03
std	7.8830e-04	7.6161e-04	7.6195e-04	7.2511e-04
SUP	7.3602e-03	5.3446e-03	6.8292e-03	4.6253e-03
std	3.5305e-03	2.0676e-03	3.4744e-03	2.1018e-03
SUPW	6.6943e-03	5.7240e-03	6.3156e-03	5.1371e-03
std	2.1384e-03	5.7240e-03	2.1361e-03	5.1371e-03
		ratio of sigmas = 1		
L2	1.1687e-03	9.5096e-04	1.1763e-03	9.5757e-04
std	2.8012e-04	2.6834e-04	2.7993e-04	2.6930e-04
L2W	4.0283e-03	3.3636e-03	4.0390e-03	3.3753e-03
std	1.3164e-03	1.2357e-03	1.3152e-03	1.2350e-03
SUP	2.6108e-03	2.0629e-03	2.6319e-03	2.0792e-03
std	8.4246e-04	6.1133e-04	8.3951e-04	6.1225e-04
SUPW	1.0191e-02	8.8100e-03	1.0229e-02	8.8552e-03
std	3.5916e-03	8.8100e-03	3.5950e-03	8.8552e-03

Table 7.3. Ignoring the heteroscedasticity as for Table 7.2, but now for quintic splines.

	plain estimator		weighted estimator	
	global error	interior error	global error	interior error
		ratio of sigmas = 500		
L2	3.2130e-03	2.3910e-03	1.6390e-03	9.4977e-04
std	1.1505e-03	9.9412e-04	1.1395e-03	7.4846e-04
L2W	2.1694e-03	1.9231e-03	1.1808e-03	9.4861e-04
std	6.7026e-04	6.4101e-04	2.7793e-04	2.7558e-04
SUP	7.2490e-03	4.8984e-03	4.5115e-03	2.4783e-03
std	3.1497e-03	1.9951e-03	3.7095e-03	2.0683e-03
SUPW	5.7202e-03	5.3762e-03	2.9627e-03	2.5148e-03
std	1.9829e-03	5.3762e-03	8.9656e-04	2.5148e-03
		ratio of sigmas = 10		
L2	3.4218e-03	2.5729e-03	2.8984e-03	2.0719e-03
std	1.1042e-03	9.6035e-04	1.0520e-03	8.2068e-04
L2W	3.0940e-03	2.4885e-03	2.8109e-03	2.2082e-03
std	7.9989e-04	7.7306e-04	7.6703e-04	7.3126e-04
SUP	7.5373e-03	5.1615e-03	6.5055e-03	4.3120e-03
std	3.0735e-03	1.9150e-03	3.2081e-03	1.8464e-03
SUPW	7.1109e-03	5.5216e-03	6.4066e-03	4.8887e-03
std	2.4165e-03	5.5216e-03	2.3222e-03	4.8887e-03
		ratio of sigmas = 1		
L2	1.1537e-03	9.1891e-04	1.1586e-03	9.2453e-04
std	2.8284e-04	2.6723e-04	2.8319e-04	2.6824e-04
L2W	3.9536e-03	3.2140e-03	3.9655e-03	3.2274e-03
std	1.3244e-03	1.2410e-03	1.3238e-03	1.2401e-03
SUP	2.6013e-03	1.9352e-03	2.6086e-03	1.9469e-03
std	8.8893e-04	5.9121e-04	8.9332e-04	5.9023e-04
SUPW	1.0116e-02	8.1607e-03	1.0139e-02	8.1947e-03
std	4.1860e-03	8.1607e-03	4.1734e-03	8.1947e-03

Table 7.4. Noncoverage probabilities (in units of 1000) of the undersmoothed quintic smoothing spline for the regression function (7.13) using the estimator (7.9)–(7.10). Results are reported for homoscedastic critical values. The columns refer to global and interior noncoverages for $\alpha = 0.10$ and global and interior noncoverages for $\alpha = 0.05$. "Perfect" scores would be 100, 100, 50 and 50.

	n = 63				n = 233			
ratio	glob10	int10	glob05	int05	glob10	int10	glob05	int05
1:1	116	103	69	57	97	88	40	43
1:10	130	115	78	70	107	97	47	51
1:500	149	139	101	85	102	98	53	48

As before, we use $m = 3$, with the smoothing parameter H obtained via the GML procedure for the cubic smoothing spline. The estimators (7.14) are obtained from the quintic smoothing spline with the H obtained via the GML method for the quintic spline itself.

The critical values $c_{\alpha,n,m,h}$ are the ones obtained from simulations for pure, homoscedastic iid normal noise.

In Table 7.4, the noncoverage results are tabulated for simulated data, using the mean function (7.13), based on 1000 replications. For $n = 63$ and the ratio $\sigma_{o,1}^2 : \sigma_{o,1}^2 = 1 : 500$, the results are acceptable but not great; for $n = 233$, they are nice. We also considered the ratios $1 : 1$ and $1 : 10$. Here, the results are acceptable already for $n = 63$. For $n = 233$, the results are perhaps too nice, possibly a result of the overestimation of the variances in (7.17). The conclusion is that this works but that the combination $n = 63$ and the ratio $1 : 500$ of the variances is a bit too much.

We must still exhibit the confidence bands for the mean function of the actual Wastewater Data Set. In view of the simulation results above, to get "honest" confidence bands for the mean function for the ratios of the variances $1 : 500$ and sample size $n = 63$, we determined the inflation factor r from the simulations reported in Table 7.4 to be

$$
\begin{aligned}
r_{\text{glob10}} &= 1.0687 \quad , \quad r_{\text{int10}} = 1.0437 \ , \\
r_{\text{glob05}} &= 1.0874 \quad , \quad r_{\text{int05}} = 1.0748 \ .
\end{aligned} \tag{7.19}
$$

The "honest" confidence band for the mean function in the Wastewater Data Set was then taken to be

$$
f^{n,m,H}(x_{in}) \pm r_{t,\alpha}\, c_{\alpha,n,m,H} \sqrt{\left[R_{nH}\, D(\widehat{\sigma}_{o,1}, \widehat{\sigma}_{o,2})\, R_{nH}^T \right]_{i,i}} \tag{7.20}
$$

for all i in the appropriate range. The global 95% confidence band is shown in Figure 7.1(a). In Figure 7.1(b), one can appreciate the changing

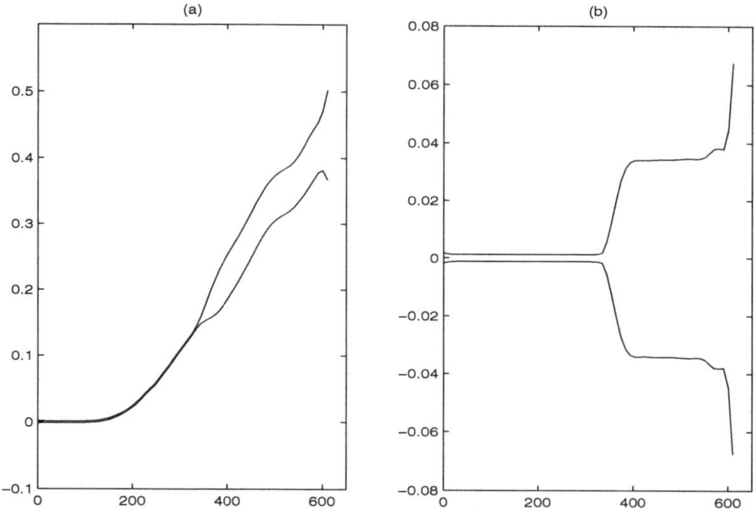

Figure 7.1 (a) The 95% confidence band for the mean function in the Wastewater Data Set, according to the model (7.4)–(7.6). (b) The shape of the confidence band itself.

width of the confidence band more clearly. We note again that we ignored the nonnegativity of the mean function.

(7.21) REMARK. An interesting alternative model for the Wastewater Data Set is given by

$$y_{in} = \mu(x)^{-1}\,\text{Poisson}\big(\,\mu(x)\,f_o(x_{in})\,\big)\;,\quad i = 1, 2, \cdots, n\;,$$

where
$$\mu(x) = \begin{cases} 1000\,\sqrt{500}\;, & 0 \leqslant x \leqslant x_c\;, \\ 1000\;, & x_c < x \leqslant x_T\;. \end{cases}$$

This model guarantees that the data are always positive. It would be interesting to analyze the Wastewater Data Set using maximum penalized likelihood estimation for this model, with both μ and f_o to be estimated; see, e.g., KOHLER and KRZYŻAK (2007) and references therein.

(7.22) REMARK. For the Wastewater Data Set, we decided that one smoothing parameter would be sufficient. The alternative is to have a separate smoothing parameter for each of the two pieces. Thus, in the style of

ABRAMOVICH and STEINBERG (1996), the estimation problem then is

$$\begin{aligned} \text{minimize} \quad & \sigma_1^{-2} \Big\{ \sum_{i=1}^{m} | f(x_{in}) - y_{in} |^2 + h_1^{2m} \, \| f^{(m)} \|_{(0,x_c)}^2 \Big\} + \\ & \sigma_2^{-2} \Big\{ \sum_{i=m+1}^{n} | f(x_{in}) - y_{in} |^2 + h_2^{2m} \, \| f^{(m)} \|_{(x_c,x_T)}^2 \Big\} \\ \text{subject to} \quad & f \in W^{m,2}(0,1) \; ; \end{aligned}$$

see the notation (13.2.9) for $L^2(a,b)$ integrals. So now both h_1 and h_2 must be chosen. A pragmatic approach would be to determine smoothing splines and smoothing parameters for each piece separately. With h_1 and h_2 so determined, one could solve the problem above to smoothly glue the two pieces together. Unfortunately, we shall not pursue this.

Before closing the book on this book, there is one last item on the agenda. All the way back in § 12.1, we briefly discussed the case of logarithmic transformations of the Wastewater Data Set. For the purpose of prediction, it is hard to argue against it, but for estimation things are problematic, as we now briefly discuss.

Ignoring the blemish of having to discard some of the "early" data, taking logarithms in the model (7.3) for some s leads to

$$\widetilde{y}_{in} = g_o(x_{in}) + \delta_{in} \; , \quad i = s, s+1, \cdots, n \; , \tag{7.23}$$

where $\widetilde{y}_{in} = \log y_{in}$, $g_o(x) = \log f_o(x)$, and

$$\delta_{in} = \log\Big(1 + \frac{d_{in}}{f_o(x_{in})} \Big) \; . \tag{7.24}$$

Of course, the δ_{in} are independent, but it is an exercise in elementary convexity to show that

$$\mathbb{E}[\, \delta_{in} \,] < 0 \; , \quad i = 1, 2, \cdots, n \; . \tag{7.25}$$

Consequently, the usual nonparametric estimators of g_o will not be consistent. To emphasize the point, think of the mean-of-the-noise function $m_o(x)$, $x \in [\, 0 \, , \, 1 \,]$, satisfying

$$m_o(x_{in}) = \mathbb{E}[\, d_{in} \,] \; , \quad i = s, s+1, \cdots, n \; . \tag{7.26}$$

Then, the function m_o does not depend on the sample size or the design. Thus, the usual estimators of g_o will be consistent estimators of $g_o + m_o$ but not g_o. While it is undoubtedly possible to construct a consistent estimator of m_o , and thus g_o , this would be more than one bargained for when taking logarithms. So, transformations like (7.25) are not without peril.

(7.27) FINAL REMARK. It is interesting to note that

$$\text{median}(\, d_{in} \,) = 0 \iff \text{median}(\, \delta_{in} \,) = 0 \; .$$

Consequently, the roughness-penalized least-absolute-deviations estimators of §§ 17.4 and 17.5 would be consistent estimators of g_o. Unfortunately, it is hard to think of natural settings where $\mathbb{E}[\, d_{in} \,] = 0$, median$(\, d_{in} \,) = 0$, *and* $f_o(x_{in}) + d_{in} > 0$ all would hold. One can think of d_{in} having a symmetric pdf with support on $[\, -f_o(x_{in}) \,, \, f_o(x_{in}\,) \,]$, but ... .

8. Additional notes and comments

Ad § 1: The collection of test examples considered here seems to be ideally suited to torpedo any strong conclusion one might wish to draw from the simulation experiments. This is good since there is no universally best nonparametric regression procedure.

In the simulations, we avoided showing pictures of how nice (or how nonsmooth) the estimators look. The authors do not think that the purpose of nonparametric regression is to produce pretty pictures.

Ad § 2: Spline smoothing with data-driven choices of the order were apparently first suggested by GAMBER (1979) and WAHBA and WENDELBERGER (1980) (using GCV, of course). See also STEIN (1993).

Ad § 3: FAN and GIJBELS (1995b) investigate choosing the order of local polynomials, but their point of view is somewhat different. In particular, they do not report on L^2 errors, which makes comparisons difficult. Their method is based on Mallows' procedure for the pointwise error (18.8.17), with an estimated pointwise error variance.

Ad § 4: HURVICH, SIMONOFF, and TSAI (1995) show some results regarding cubic smoothing splines and local linear and quadratic polynomials, but their goal was to compare what we call "aic" with "gcv" and other methods for selecting the smoothing parameter, but not the order of the estimators. The same applies to the papers by LEE (2003, 2004).

Ad § 6: It would be interesting to test whether the regression function for the Wood Thrush Data Set is monotone, but that is another story. See, e.g., TANTIYASWASDIKUL and WOODROOFE (1994) and PAL and WOODROOFE (2007).

A4

Bernstein's Inequality

Bernstein's inequality was the primary probabilistic tool for showing uniform error bounds on convolution-like kernel estimators and smoothing splines. Here, we give the precise statement and proof. The simplest approach appears to be via the slightly stronger inequality of BENNETT (1962), with the corrections observed by GINÉ (1974). We follow the proof as given by DUDLEY (1982).

(1.1) BENNETT'S INEQUALITY. *Let $X_1, X_2, \cdots, X_n$ be independent mean zero, bounded random variables. In particular, for $i = 1, 2, \cdots, n$, suppose that*

$$\mathbb{E}[\, X_i \,] = 0 \quad , \quad \mathbb{E}[\, |\, X_i \,|^2 \,] = \sigma_i^2 \quad , \quad |\, X_i \,| \leqslant K_i \; .$$

Let

$$S_n = \sum_{i=1}^{n} X_i \quad , \quad V_n = \sum_{i=1}^{n} \sigma_i^2 \quad , \quad M_n = \max_{1 \leqslant i \leqslant n} K_i \; .$$

Then,

$$\mathbb{P}\Big[\, |\, S_n \,| > t \,\Big] \leqslant 2 \, \exp\Big(-\frac{V_n}{M_n^2} \, \Psi\Big(1 + \frac{M_n \, t}{V_n} \Big) \Big) \; ,$$

where $\Psi(x) = x \log x + 1 - x$.

PROOF. We drop the subscripts n everywhere. It suffices to prove the one-sided version

$$\mathbb{P}[\, S > t \,] \leqslant \exp\Big(-\frac{V}{M} \, \Psi\Big(1 + \frac{t}{V} \Big) \Big) \; .$$

To make the formulas a bit more palatable, let $Y_i = X_i / M$. Then the Y_i are independent and satisfy

$$\mathbb{E}[\, Y_i \,] = 0 \quad , \quad \mathbb{E}[\, |\, Y_i \,|^2 \,] = s_i^2 \quad , \quad |\, Y_i \,| \leqslant 1 \; ,$$

in which $s_i = \sigma_i / M$, for all $i = 1, 2, \cdots, n$. Then, with $\tau = t \,/ M$, and $T = \sum_{i=1}^{n} Y_i$. we have that

$$\mathbb{P}[\, S > t \,] = \mathbb{P}[\, T > \tau \,] \; ,$$

Now, by the Chernoff-Chebyshev inequality, for all $\theta > 0$,

$$\mathbb{P}[\,T > \tau\,] = \mathbb{P}[\,\exp(\theta T) > \exp(\theta\tau)\,] \leqslant \exp(-\theta\tau)\,\mathbb{E}[\,\exp(\theta T)\,]\ .$$

By independence, we get

$$\mathbb{E}[\,\exp(\theta T)\,] = \prod_{i=1}^{n} \mathbb{E}[\,\exp(\theta Y_i)\,]\ , \tag{1.2}$$

and then,

$$\mathbb{E}[\,\exp(\theta Y_i)\,] = \mathbb{E}\Big[\sum_{\ell=0}^{\infty} \theta^\ell\, Y_i^\ell/\ell!\Big] \leqslant 1 + s_i^2 \sum_{\ell=2}^{\infty} \theta^\ell/\ell!\ ,$$

where we used that $\mathbb{E}[\,Y_i\,] = 0$ and, since $|\,Y_i\,| \leqslant 1$, that

$$\mathbb{E}[\,|\,Y_i\,|^\ell\,] \leqslant \mathbb{E}[\,|\,Y_i\,|^2\,] = s_i^2\ , \quad \ell \geqslant 2\ . \tag{1.3}$$

The last infinite series equals $e^\theta - 1 - \theta$, and then using the inequality $1 + x \leqslant \exp(x)$, we get

$$\mathbb{E}[\,\exp(\theta Y_i)\,] \leqslant \exp\big(\,s_i^2\,(e^\theta - 1 - \theta)\,\big)\ , \tag{1.4}$$

and then from (1.2) finally,

$$\mathbb{E}[\,\exp(\theta T)\,] \leqslant \exp\big(\,W\,(e^\theta - 1 - \theta)\,\big)\ , \tag{1.5}$$

where $W = \sum_{i=1}^{n} s_i^2 = V/M^2$.

Summarizing, we have for all $\theta > 0$,

$$\mathbb{P}[\,T > \tau\,] \leqslant \exp\big(-\theta\tau + W\,(e^\theta - 1 - \theta)\,\big) = \exp\big(\,W\,(e^\theta - 1 - \varrho\,\theta)\,\big)\ ,$$

where $\varrho = 1 + \tau/W$. Now, minimize over θ. Differentiation shows that the minimizer is $\theta = \log\varrho$, and then

$$e^\theta - 1 - \varrho\,\theta = \varrho - 1 - \varrho\log\varrho = -\Psi(\varrho)\ .$$

It follows that

$$\mathbb{P}[\,T > \tau\,] \leqslant \exp\big(-W\,\Psi(1 + \tau/W)\,\big)\ .$$

The proof is finished by translating back into terms of the X_i and related quantities. Q.e.d.

(1.6) Bernstein's Inequality. *Under the conditions of* Bennett's Inequality,

$$\mathbb{P}[\,|\,S_n\,| > t\,] \leqslant 2\,\exp\Big(\frac{-\,t^2}{2\,V_n + \frac{2}{3}\,t\,M_n}\Big)\ .$$

Proof. Again, drop the subscripts n everywhere. We have the Kemperman (1969) inequality, familiar to readers of Volume I, Exercise (9.1.22),

$$(\,\tfrac{2}{3}\,x + \tfrac{4}{3}\,)\,(x\log x + 1 - x) \geqslant (x-1)^2\ , \quad x > 0\ ,$$

which implies that

$$\frac{V}{M^2}\,\Psi\big(1+M\,t\,/V\,\big) \geqslant \frac{t^2}{\frac{2}{3}\,M\,t\,+2\,V} \,. \qquad \text{Q.e.d.}$$

(1.7) REMARK. Inspection of the proof of BENNETT'S INEQUALITY reveals that the boundedness of the random variables is used only in (1.3), which in terms of the X_i reads as

$$\mathbb{E}[\,|\,X_i\,|^\ell\,] \leqslant \sigma_i^2\,M^{\ell-2} \quad \text{for all} \quad \ell \geqslant 2\,. \tag{1.8}$$

So, one could replace the boundedness of the random variables by this condition. How one would ever verify this for unbounded random variables is any body's guess, of course. Well, you wouldn't. Note that for any univariate random variable X with distribution P,

$$\Big\{\,\mathbb{E}[\,|\,X\,|^\ell\,]\,\Big\}^{1/\ell} = \Big\{\,\int_{\mathbb{R}} |\,x\,|^\ell\, dP(x)\,\Big\}^{1/\ell} \longrightarrow P\text{-}\sup\,|\,X\,| \quad \text{for} \quad \ell \to \infty\,.$$

If X is unbounded, this says that $\mathbb{E}[\,|\,X\,|^\ell\,]$ grows faster than exponential, and (1.8) cannot hold. Moreover, it shows that in (1.8), the factor M satisfies $M \geqslant P\text{-}\sup\,|\,X\,|$, so that if one wants to get rid of the boundedness assumption, one must *assume* (1.4).

A5

The TVDUAL inplementation

1. TVdual.m

```
% function to compute the solution of the dual problem
% of nonparametric least squares estimation with total
% variation penalization:
%
%  minimize     (1/2) < x , M x >  -  < qs , *x >
%  subject to   -alpha <= x <= alpha
%
%  The matrix M is the standard negative Laplacian
%
   function [ x, grad ] = TVdual( x, alpha, q )
%  Choleski factorization
   n = length( x ) ;
   [ Ldi , Lco ] = choltri( 2*ones(n,1), -ones(n-1,1) ) ;
%  make sure initial guesses are sane
   [ x, active ] = sani_acti( x, alpha ) ;
%
   optimal = 0 ;
   while ( optimal == 0 )
      constr_viol = 1 ;
      while constr_viol > 0
         [ howmany, first, last ] = find_runs( active ) ;
         [ x, active, constr_viol ] = ...
         restricted_trisol( howmany, first,last,...
                                 alpha, x, q, Ldi, Lco ) ;
      end
      [ optimal, active, grad ] = dual_optim( x, q, active ) ;
   end
   return
```

2. restricted_trisol.m

```
%
   function [ x, active, constr_viol ] = ...
   restrcted_trisol( hmb, first, last, alpha, xold, q, Ldi, Lco)
%
   n = length( xold );
   x = xold ; % copy constrained components of  xold
   constr_viol = 0 ; %did not find constarint violations (yet)
%
   for j = 1 : hmb
      blocksize = last(j) - first(j) + 1 ;
%     old xj for this block
      xjold = xold( first(j) : last(j) );
%     construct right hand side
      rhs = q( first(j) : last(j) ) ;
%     first and last may need updating
      if first(j) > 1
        rhs( 1 ) = rhs( 1 ) + xold( first(j) - 1 ) ;
      end
      if last(j) < n
         rhs blocksize) = rhs(blocksize) + xold( last(j) + 1 );
      end
%     back substitution
      xj = tri_backsub( Ldi, Lco, rhs ) ;
%     move from xjold to xj, until you hit constraints
      [ xj, viol ] = move( xjold, xj, alpha );
%     put xj in its place
      x( first(j) : last(j) ) = xj;
      constr_viol = max( constr_viol, viol );
   end
%  sanitize  x  and determine active constraints
   [ x, active ] = sani_acti( x , alpha ) ;
   return
```

3. sani_acti.m

```
% make sure x satisfies the constraints, and
% determine which ones are active
%
  function [ x, active ] = sani_acti( x, alpha )
  x = max( -alpha, min( alpha, x ) );
  active = ( x == alpha ) - ( x == -alpha );
  return
```

4. move.m

```
% for a block of unconstrained x(j),  move from the feasible
% x  towards the proposed  new iterate  z , but stop before
% the boundary of     -alpha <= x <=  +alpha    is crossed.
%
  function [ xnew, viol ] = move( x, z, alpha )
  n = length( x ) ;
  t = 1 ; % largest possible stepsize
  d = z - x ;
  for j = 1 : n
      if ( d(j) ~= 0 )
         t = min( t, ( -x(j) + alpha * sign( d(j)) ) / d(j) ) ;
      end
  end
  if t <= 0
     step_size = t
     error('on boundary, trying to move out')
  end
  viol = ( t < 1 ) ;
  xnew = x + t * d ;
  return
```

5. find_runs.m

```
% find runs of zeros in a linear array
%
  function [ nrb, first, last ] = find_runs( a )
  n = length( a );
  first = nan( 1, n); last = nan( 1, n);
  j = 1 ; nrb = 0 ;
  while j < n
     a( n+1 ) = 0; % so we  w i l l  find a zero
     while a( j ) ~= 0
           j = j + 1 ;
     end
     if j == n + 1
%       we ran out of elements
        break
     end
     nrb = nrb + 1; first( nrb ) = j ;
     a( n + 1 ) = 1 ; % so we  w i l l  find nonzero
     while a( j ) == 0
           j = j + 1 ;
     end
     last( nrb ) = j-1 ;
  end
```

```
% cleanup
  if nrb > 0
     first = first( 1: nrb ); last = last( 1 : nrb ) ;
  else
     first = []; last = [];
  end
  return
```

```
5, dual_optim.m
% check the optimality of the current  feasible iterate
% and remove an active constraint if it  is not. Note that
% 'active'  is supposed to satisfy
%
%  x(j) = +alpha    <==>  active(j) = +1
%  x(j) = -alpha    <==>  active(j) = -1
% -alpha < x(j) < alpha  <==>  active(j) = 0
%
  function [ optimal, active, grad ] = ...
                              dual_optim( x, q, active )
%
  grad = -[ 0; 0; x ] + 2*[ 0; x; 0] - [ x; 0; 0 ] ;
  grad = grad( 2 : length(grad) - 1 ) - q ;
  [ max_viol, ind_viol ] = max( active .* grad );
  if max_viol <= ( eps * max(abs(grad)) )
     optimal = 1;
  else
     optimal = 0;
     active( ind_viol ) = 0 ; % remove an active  constraint
  end
  return
```

```
6. choltri.m
%  Cholesky factorization of a positive definite
%  tridiagonal matrix
%
%  Usage:  [ Ldi, Lco] = choltri(di, co)
%  Input:  the diagonal   di(1:n) and the codiagonal co(1:n-1)
%          of the matrix to be factored
%  Ouput:  the diagonal   Ldi(1:n) and the codiagonal Lco(1:n-
1)
%          of the Choleski factor
%
   function [ Ldi, Lco] = choltri( di, co)
   n = length(di);
   Ldi=zeros(n,1); Lco=zeros(n-1,1);
```

```
%
   for k=1:n-1;
     if di(k) <= 0
        error('matrix not positive definite, computation aborted')
     end
     Ldi(k) = sqrt( di(k) );
     Lco(k) = co(k) / Ldi(k);
     di(k+1) = di(k+1) - Lco(k)^2 ;
   end
   Ldi(n) = sqrt( di(n) ) ; % should check that  di(n) >  0, ...
   return
```

```
7. tri_backsub.m
% Back substitution for solving
%           R'*R*x=y
% when R is an upper trinagular bi-diagonal matrix
% Usage: x = tri_backsol(di, co, y )
% Input: the diagonal   di(1:n) and codiagonal co(1:n-1)
%        of the matrix R, and the right  hand side y
% Ouput: the solution x of the sytem of equations
%
  function [ x ] = tri_backsub( di ,co ,y )
  m=length( di ); n = length( y ) ;
  if length(co) ~= m-1
     error(['the lengths do not match'])
  end
  if n > m
     error('system too small for data')
  end
% backsubstitution
  x = zeros(n,1) ;
  x(1) = y(1) / di(1) ;
  for k = 2 : n
      x(k) = ( y(k) - co(k-1)*x(k-1) ) / di(k) ;
  end
  x(n) = x(n) / di(n) ;
  for k = n-1 : -1 : 1
      x(k) = ( x(k) - co(k)*x(k+1) ) / di(k) ;
  end
  return
```

A6

Solutions to Some Critical Exercises

1. Solutions to Chapter 13: Smoothing Splines

SOLUTION OF EXERCISE (13.4.10). The Arithmetic-Geometric Mean Inequality gives

$$a\,x \leqslant (1/q)\,a^q + (1/p)\,x^p \ .$$

Substitute this into the hypothesized inequality for x and solve for x^p. □

SOLUTION OF EXERCISE (13.4.21). Of course, what we are doing here is studying the variance part of f^{nh}. The first observation is that the Quadratic Behavior Lemma (13.3.1) has an obvious analogue for the problem (13.3.20). Adding this identity to that of the lemma gives, with $\varepsilon \equiv f^{nh} - f_{hn}$,

$$(*) \quad \tfrac{2}{n} \sum_{i=1}^{n} |\,\varepsilon(x_{in})\,|^2 + 2h^{2m} \|\,\varepsilon^{(m)}\,\|^2 =$$

$$\tfrac{1}{n} \sum_{i=1}^{n} |\,f^{nh}(x_{in}) - y_{io}\,|^2 - \tfrac{1}{n} \sum_{i=1}^{n} |\,f_{hn}(x_{in}) - y_{io}\,|^2 +$$

$$\tfrac{1}{n} \sum_{i=1}^{n} |\,f_{hn}(x_{in}) - y_{in}\,|^2 - \tfrac{1}{n} \sum_{i=1}^{n} |\,f^{nh}(x_{in}) - y_{in}\,|^2 \ .$$

Keeping in mind that $y_n = y_o + d_n$, Exercise (13.4.18) and the Random Sum Lemma (13.2.20) show that the right hand side of $(*)$ equals

$$\tfrac{2}{n} \sum_{i=1}^{n} d_{in}\,\varepsilon(x_{in}) = 2\,\langle\,\varepsilon\,,\,\mathfrak{S}^{n,h}\,\rangle_{m,h} \leqslant 2\,\|\,\varepsilon\,\|_{m,h}\,\|\,\mathfrak{S}^{n,h}\,\|_{m,h} \ .$$

Now, by Lemma (13.2.21), the left-hand side of the inequality $(*)$ is bounded from below by $2\,\zeta^{nh}\,\|\,f^{nh} - f_{hn}\,\|_{m,h}^2$, with $\zeta^{nh} \to 1$. Pretending it actually *equals* 1, putting everything together gives

$$\|\,\varepsilon\,\|_{m,h}^2 \leqslant \|\,\mathfrak{S}_{n,h}\,\|_{m,h}\,\|\,\varepsilon\,\|_{m,h} \ .$$

Canceling the common factor and both sides of the inequality results in $\| f^{nh} - f_{hn} \|_{m,h} \leqslant \| S_{n,h} \|_{m,h}$ and completes the proof. □

SOLUTION OF EXERCISE (13.4.22). Adding the analogues of the identity of the Quadratic Behavior Lemma (13.3.1) for the two quadratic minimization problems in question gives, with $\delta \equiv f_{hn} - f_h$,

$$(*)\quad \tfrac{1}{n} \sum_{i=1}^{n} | \delta(x_{in}) |^2 + \| \delta \|^2 + 2\, h^{2m} \| \delta^{(m)} \|^2 \leqslant$$
$$\tfrac{1}{n} \sum_{i=1}^{n} | f_h(x_{in}) - f_o(x_{in}) |^2 - \tfrac{1}{n} \sum_{i=1}^{n} | f_{hn}(x_{in}) - f_o(x_{in}) |^2 +$$
$$\| f_{hn} - f_o \|^2 - \| f_h - f_o \|^2 .$$

Exercise (13.4.20) shows that the right-hand side equals

$$\tfrac{2}{n} \sum_{i=1}^{n} \delta(x_{in}) \left(f_h(x_{in}) - f_o(x_{in}) \right) - 2 \langle \delta , f_h - f_o \rangle - \| \delta \|^2 .$$

Move the term $- \| \delta \|^2$ to the left-hand side of the inequality. By the asymptotic uniformity of the design, what is left may be bounded by

$$c_m\, n^{-1} \| \{ \delta \, (f_h - f_o) \}' \|_1 ,$$

which itself is dominated by

$$C_m\, (nh)^{-1} \| \delta \|_{1,h} \| f_h - f_o \|_{1,h} \leqslant K_m\, (nh)^{-1} \| \delta \|_{m,h} \| f_h - f_o \|_{m,h}$$

for suitable constants C_m and K_m. Using the uniformity of the design once more, we obtain that

$$\zeta^{nh} \| \delta \|_{m,h}^2 \leqslant c_m\, (nh)^{-1} \| \delta \|_{m,h} \| f_h - f_o \|_{m,h}$$

with $\zeta^{nh} \to 1$. The conclusions follow. □

SOLUTION OF EXERCISE (13.4.23). Lemma (13.3.1) applied to the problem (13.4.19) immediately yields the result. □

SOLUTION OF (13.4.24). Use the triangle inequality

$$\| f^{nh} - f_o \|_{m,h} \leqslant \| f^{nh} - f_{hn} \|_{m,h} + \| f_{hn} - f_h \|_{m,h} + \| f_h - f_o \|_{m,h} .$$

2. Solutions to Chapter 14: Kernel Estimators

SOLUTION OF EXERCISE (14.6.28). We again have the representation

$$S^{nh}(x) - S_h(x) = \int_{\mathbb{R}} \Big\{ \Big(h \frac{\partial}{\partial z} + I \Big) \{ A_h(x - z)\, f(z) \} \Big\} \times$$
$$[g_h \circledast (d\Omega_n - d\Omega_o)](z)\, dz .$$

It follows that $\| S^{nh}(x) - S_h(x) \|_\infty \leqslant c \| g_h * (d\Omega_n - d\Omega_o) \|_\infty$, with

$$c = \sup_{x\in\mathbb{R}} \int_{\mathbb{R}} \Bigl| \Bigl\{ \Bigl(-h\frac{\partial}{\partial z} + I \Bigr) \bigl\{ A_h(x-z)\, f(z) \bigr\} \Bigr\} \Bigr| \, dz$$

$$\leqslant \| A_h(x - \cdot\,)\, f(\,\cdot\,) \|_{h,W^{1,1}(0,1)} \; .$$

The required bound on $\| g_h*(dF_n - dF_o) \|_\infty$ follows from Theorem (14.6.13). Q.e.d.

SOLUTION OF EXERCISE (14.7.13). First, define $\mathfrak{B}_m$ by means of

$$\widehat{\mathfrak{B}_m}(h\omega) = \widehat{\mathfrak{B}_{mh}}(\omega) \; , \quad \omega \in \mathbb{R} \; .$$

It suffices to determine the behavior of $\mathfrak{B}_m$. Observe that

$$1 + (2\pi\omega)^{2m} = \prod_{\ell=1}^{2m-1} \text{factor}(\omega,\ell) \; ,$$

where

$$\begin{aligned}
\text{factor}(\omega,\ell) &= 2\pi\omega - \exp\Bigl(\frac{(2\ell+1)\pi i}{2m} \Bigr) \\
&= 2\pi\omega - (-i)\, \exp\Bigl(\frac{(2\ell+m+1)\pi i}{2m} \Bigr) \\
&= (-i)\, \Bigl(2\pi i\omega - \exp\Bigl(\frac{(2\ell+m+1)\pi i}{2m} \Bigr) \Bigr) = (-i)\, \bigl(2\pi i\omega - \varpi_\ell \bigr) \; ,
\end{aligned}$$

with ϖ_ℓ as in Lemma (14.7.11). Then, the partial fraction decomposition of

$$\widehat{\mathfrak{B}_m^{(k)}}(\omega) = (2\pi i\,\omega)^k\, \widehat{\mathfrak{B}_m}(\omega) = \frac{(2\pi i\,\omega)^k}{1 + (2\pi\omega)^{2m}}$$

may be written as

$$\widehat{\mathfrak{B}_{m,h}^{(k)}}(\omega) = \sum_{\ell=0}^{2m-1} \alpha_{\ell,k}\, (\, 2\pi i\,\omega - \varpi_\ell\,)^{-1}$$

for suitable constants $\alpha_{\ell,k}$. Finally, observe that $\{\, 2\pi i\,\omega - \varpi_\ell \,\}^{-1}$ is the Fourier transform of $-\exp(\,x\,\varpi_\ell)\, 1\!\!1(x \leqslant 0)$ or $\exp(\,x\,\varpi_\ell)\, 1\!\!1(x \geqslant 0)$, depending on whether the real part of ϖ_ℓ is positive or negative. Note that $\operatorname{Re}\varpi_\ell \neq 0$ for all ℓ. □

3. Solutions to Chapter 17: Other Estimators

SOLUTION OF (17.4.12). Let $\nu = \nu(d_{1,n}, d_{2,n}, \cdots, d_{n,n})$ denote the numerator of Q_n. Then, ν is a symmetric function of its arguments. Now, following the precepts of the method of bounded differences,

$$(1)\quad \sup_{a,b} \bigl|\, \nu(a, d_{2,n}, \cdots, d_{n,n}) - \nu(b, d_{2,n}, \cdots, d_{n,n}) \,\bigr| =$$

$$\tfrac{1}{n} \sup_{a,b} \bigl|\, |\, \varepsilon(x_{1,n}) - a \,| - |\, a \,| - |\, \varepsilon(x_{1,n}) - b \,| + |\, b \,| \,\bigr| \, .$$

Now, inspection of the graph of $|\,\varepsilon(x_{1,n}) - a\,| - |\,a\,|$ as a function of a shows that it is bounded by $|\,\varepsilon(x_{1,n})\,|$, so that the difference in the last line of (1) is bounded by $2\,|\,\varepsilon(x_{1,n})\,|$. Now apply Theorem (4.4.21). (Beware of the typos.) □

4. Solutions to Chapter 18: Smoothing Parameters

SOLUTION OF EXERCISE (18.6.14). Using the notation of § 19.5, recall that $R_{nh} = (\, I + nh^{2m}\, M\,)^{-1}$ and

$$\|\, (f^{nh})^{(m)} \,\|^2 = \langle\, r_n f^{nh} \,,\, M\, r_n f^{nh} \,\rangle = \langle\, R_{nh} y_n \,,\, M\, R_{nh} y_n \,\rangle \, .$$

Since $nh^{2m}\, M = (\, I + nh^{2m}\, M\,) - I = R_{nh}^{-1} - I$, then

$$nh^{2m}\, \|\, f^{(m)} \,\|^2 = \langle\, R_{nh}\, y_n \,,\, (\, R_{nh}^{-1} - I\,)\, R_{nh}\, y_n \,\rangle = \langle\, R_{nh}\, y_n \,,\, y_n - R_{nh}\, y_n \,\rangle \, .$$

Consequently,

$$\begin{aligned} n\, \sigma_{nh,6}^2 &= \|\, y_n - R_{nh} y_n \,\|^2 + nh^{2m}\, \|\, f^{(m)} \,\|^2 \\ &= \langle\, y_n - R_{nh}\, y_n \,,\, y_n - R_{nh}\, y_n \,\rangle + \langle\, R_{nh}\, y_n \,,\, y_n - R_{nh}\, y_n \,\rangle \\ &= \langle\, y_n \,,\, y_n - R_{nh}\, y_n \,\rangle \, . \end{aligned}$$

Q.e.d.

5. Solutions to Chapter 19: Computing

SOLUTION TO (18.6.23). Since $\varphi'' \in L^2(0,1)$, then

$$(1)\qquad \varphi''(x) = \sum_{k\in\mathbb{N}} \alpha_k \, \sin(\, \pi k x\,) \,, \quad x \in (0,1) \,,$$

with convergence in $L^2(0,1)$. In particular, then

$$(2)\qquad \|\, \varphi'' \,\|^2 = \tfrac{1}{2} \sum_{k\in\mathbb{N}} |\, \alpha_k \,|^2 \, .$$

Integrating (1) twice gives

$$(3)\qquad \varphi(x) = \beta\, x + \gamma - \sum_{k\in\mathbb{N}} (\pi k)^{-2}\, \alpha_k \, \sin(\, \pi k x\,) \,,$$

with the constants β and γ yet to be determined. Since the infinite series in (3) converges absolutely, the conditions $\varphi(0) = \varphi(1) = 0$ imply that

$\gamma = \beta = 0$, and then from (3)

(4) $$\|\varphi\|^2 = \tfrac{1}{2} \sum_{k\in\mathbb{N}} |\pi k|^{-4} |\alpha_k|^2 .$$

Since $\sum_{k\in\mathbb{N}} |\alpha_k|^2 \geqslant \sum_{k\in\mathbb{N}} k^{-4} |\alpha_k|^2$, the desired inequality follows. □

6. Solutions to Chapter 20: Kalman Filtering

SOLUTION OF (20.1.37). (a) Let $X \in \mathbb{R}^n$ and $Y \in \mathbb{R}^m$. If X and Y are jointly normal and have zero means, then the characteristic function may be written as

(1) $$\mathbb{E}[\exp(i u^T X + i w^T Y)] = \exp(-\tfrac{1}{2}(u^T, w^T) V (u^T, w^T)^T)$$

for all $u \in \mathbb{R}^n$, $w \in \mathbb{R}^m$. Here, V is semi-positive-definite. It is useful to partition V as

$$V = \begin{bmatrix} A & B^T \\ B & C \end{bmatrix},$$

with A and C symmetric and semi-positive-definite. In fact, if X is nondegenerate, then A is positive-definite. One verifies that then

$$(u^T, w^T) V (u^T, w^T)^T = u^T A u + 2 w^T B u + w^T C w .$$

Now, in (1), take the gradient with respect to w and set $w = 0$. This gives, for all $u \in \mathbb{R}^n$,

(2) $$\mathbb{E}[i Y \exp(i u^T X)] = -B u \exp(-\tfrac{1}{2} u^T A u) .$$

Taking the gradient in (1) with respect to u and setting $w = 0$ gives likewise

(3) $$\mathbb{E}[i X \exp(i u^T X)] = -A u \exp(-\tfrac{1}{2} u^T A u) ,$$

so that, since A is nonsingular,

(4) $$\mathbb{E}[(Y - B A^{-1} X) \exp(i u^T X)] = 0 \quad \text{for all} \quad u .$$

It follows that $\mathbb{E}[Y \mid X] = B A^{-1} X$. Of course, for zero-mean random variables, we have $\mathbb{C}\mathrm{ov}[Y, X] = \mathbb{E}[Y X^T]$, $\mathbb{V}\mathrm{ar}[X] = \mathbb{E}[X X^T]$. (This solution largely follows the discussion of PARZEN (1962) on conditional expectations. Note that (4) gives, after standard arguments, that

(5) $$\mathbb{E}[(Y - BA^{-1} X)\, \Psi(X)] = 0$$

for all nice scalar functions Ψ. This says that $Y - BA^{-1} X$ is orthogonal to all functions of X if we interpret the expectation as an inner product.)
(b) To solve the maximum likelihood problem, use the Lagrange Multiplier Theorem, see, e.g., Chapter 10 in Volume I or HIRIART-URRUTY and

LEMARÉCHAL (1993). Then, y is the maximum likelihood estimator if there exists a Lagrange multiplier $\lambda \in \mathbb{R}^m$ such that

$$\begin{aligned} V^{-1} y + A^T \lambda &= 0 \\ A\, y \qquad &= b \,, \end{aligned}$$

the solution of which is readily computed as $y = V A^T (A V A^T)^{-1} b$. □

SOLUTION OF (20.3.19). The proof is computational. Since $f \in \mathcal{L}(K)$, we may write it as

$$f(s) = \sum_{j=1}^{k} w_j \, K(\tau_j, s) \,, \quad s \geqslant 0 \,,$$

for suitable k, $w_1, \cdots, w_k$, and $\tau_1, \cdots, \tau_k$, so that

$$\begin{aligned} f(s) &= \sum_{j=1}^{k} w_j \, \langle K(\tau_j, \cdot\,), K(s, \cdot\,) \rangle_K \\ &= \Big\langle \sum_{j=1}^{k} w_j \, K(\tau_j, \cdot\,), K(s, \cdot\,) \Big\rangle_K = \langle f , K(s, \cdot\,) \rangle_K \,. \end{aligned}$$

□

SOLUTION OF (20.3.27). We obviously have, for all $f \in \mathcal{H}(K)$, that

$$\begin{aligned} \Big| f(t) - \sum_{i=1}^{n} w_i \, f(t_i) \Big| &= \Big| \langle f , K(t, \cdot\,) - \sum_{i=1}^{n} w_i \, K(t_i, \cdot\,) \rangle_K \Big| \\ &\leqslant \| f \|_K \, \| K(t, \cdot\,) - \sum_{i=1}^{n} w_i \, K(t_i, \cdot\,) \|_K \,, \end{aligned}$$

with equality for $f = K(t, \cdot\,) - \sum_{i=1}^{n} w_i \, K(t_i, \cdot\,)$. The rest follows. □

SOLUTION OF (20.4.16). Using the constraints to eliminate a in (20.4.15), we get the normal equations $(I + \alpha \, S^T T^{-1} S) \, b = y$. Using the constraint, we may write them as $b + \alpha \, S^T a = y$. Now, multiply by S, and using the constraint again gives $(T + \alpha \, S S^T) \, a = S y$, as required. (One verifies that T is positive-definite under reasonable conditions on the design.) □

SOLUTION OF (20.5.19). Write

$$(\sigma^{-2} \mathbb{B}^T \mathbb{B} + \mathbb{A})^{-1} = \mathbb{A}^{-1} (\sigma^{-2} \mathbb{B}^T \mathbb{B} \mathbb{A}^{-1} + I)^{-1} \,,$$

and use the Sherman-Morrison-Woodbury Lemma (18.3.12) in the form $(I + FG)^{-1} = I - F (I + GF)^{-1} G$, with $F = \mathbb{B}^T$ and $G = \mathbb{B} \mathbb{A}^{-1}$, to obtain

$$(\sigma^{-2} \mathbb{B}^T \mathbb{B} + \mathbb{A})^{-1} = \mathbb{A}^{-1} \big\{ I - \sigma^{-2} \mathbb{B}^T (I + \sigma^{-2} \mathbb{B} \mathbb{A}^{-1} \mathbb{B}^T)^{-1} \mathbb{B} \mathbb{A}^{-1} \big\} \,.$$

Now let this operate on $\mathbb{B}^T$, and use that

$$(I + H)^{-1} H = (I + H)^{-1} (I + H - I) = I - (I + H)^{-1}$$

with $H = \sigma^{-2}\,\mathbb{B}\,\mathbb{A}^{-1}\,\mathbb{B}^T$ to get the desired result. $\square$

SOLUTION OF (20.6.26). (a) Write $\widetilde{y}(t_i)$ as $\widetilde{y}(t_i) = \varepsilon(t_i) + e_1^T\,\widetilde{\mathbb{S}}(t_i)$, where $\widetilde{\mathbb{S}}(t_i) = \mathbb{S}(t_i) - \mathbb{S}_{i|i\text{-}1}$. Now, set $Y_i = \bigl(y(t_1), y(t_2), \cdots, y(t_{i-1})\bigr)^T$, and observe that by Exercise (20.1.35) we may write

$$\mathbb{S}_{i|i\text{-}1} = \mathbb{C}\mathrm{ov}[\,\mathbb{S}(t_i)\,,\,Y_i\,]\,\{\,\mathbb{V}\mathrm{ar}[\,Y_i\,]\,\}^{-1}\,Y_i\;.$$

So then $\mathbb{C}\mathrm{ov}[\,\widetilde{\mathbb{S}}(t_i)\,,\,Y_i\,] = 0$, which implies that for every component $y(t_j)$ of Y_i (i.e., $j \leqslant i-1$),

$$\mathbb{E}[\,\widetilde{y}(t_i)\,y(t_j)\,] = \mathbb{E}[\,\varepsilon(t_i)\,y(t_j)\,] + \mathbb{E}[\,e_1^T\,\widetilde{\mathbb{S}}(t_i)\,y(t_j)\,] = 0\;.$$

This is part (a). Part (b) follows upon inspection from the discussion leading up to (20.6.11). $\square$

SOLUTION OF (20.8.14). (a) Write W_o as $W_o = V_o^{-1/2}\,P\,V_o^{-1/2}$, where

$$P = I - \lambda\,(\,\lambda^T\lambda\,)^{-1}\,\lambda^T\;.$$

Here, $\lambda = V_o^{-1/2}\,\ell$. Obviously, P is an orthogonal projector, and hence it is semi-positive-definite. Then, for any vector $x \in \mathbb{R}^n$, let $y = V_o^{-1/2}\,x$, and we have $x^T W_o\,x = y^T\,P\,y \geqslant 0$.

(b) Part (a) implies that $\mathcal{N}(\,W_o\,) = \mathcal{R}(\,\ell\,)$, and then $\mathcal{R}(\,W_o\,) = \mathcal{N}(\,\ell^T\,)$. Consequently, the matrix W_o has eigenvalue 0 with geometric/algebraic multiplicity m, and the corresponding eigenvectors span $\mathcal{R}(\,\ell\,)$. Moreover, the eigenvectors corresponding to positive eigenvalues are orthogonal to ℓ. Now, it is easy to verify that every eigenvector of W_o is an eigenvector of $\mathbb{W} = W_o + \ell\,(\,\ell^T\ell\,)^{-1}\,\ell^T$, and to determine the corresponding eigenvalue. Then, (b) follows.

(c) The cheap trick is the factorization of a "related" matrix,

$$\begin{bmatrix} I - A\,B^T & A \\ O & I \end{bmatrix}\begin{bmatrix} I & O \\ B^T & I \end{bmatrix} = \begin{bmatrix} I & A \\ B^T & I \end{bmatrix} = \begin{bmatrix} I & O \\ B^T & I - B^T A \end{bmatrix}\begin{bmatrix} I & A \\ O & I \end{bmatrix}.$$

Then just take determinants.

(d) First, write $V_o\,\mathbb{W}$ as

$$V_o\,\mathbb{W} = I + \begin{bmatrix} -\,\ell & V_o\,\ell \end{bmatrix}\begin{bmatrix} (\,\ell^T V_o^{-1}\,\ell\,)^{-1}\,\ell^T V_o^{-1} \\ (\,\ell^T\ell\,)^{-1}\,\ell^T \end{bmatrix}.$$

Then, applying (c) gives $\det(\,V_o\,\mathbb{W}\,) = \det(\,I + C\,)$, where

$$C = \begin{bmatrix} -I & \ell^T\ell\,(\ell^T V_o^{-1}\ell\,)^{-1} \\ -I & (\,\ell^T V_o\,\ell)\,(\,\ell^T\ell)^{-1} \end{bmatrix}.$$

Then, $I + C$ looks like

$$I + C = \begin{bmatrix} O & \ell^T \ell (\ell^T V_o^{-1} \ell)^{-1} \\ -I & \text{stuff} \end{bmatrix} ,$$

and after interchanging rows it is easy to see that

$$\det(I + C) = \det(\ell^T \ell (\ell^T V_o^{-1} \ell)^{-1}) ,$$

and we are in business. □

7. Solutions to Chapter 21: Equivalent Kernels

Solution of Exercise (21.4.9)(b). We use a version of the Dominated Convergence Theorem, but it is easier to just prove it than to locate a precise version.

Let $\varepsilon = f^{nh} - f_o$. The starting point is the inequality (21.4.4). The right-hand side may be bounded by

$$\| \varepsilon \|_{wmh} \{ \| \mathfrak{S}_{\omega nh} \|_{wmh} + h^m \| f_o^{(m)} \| \} .$$

Then, (the proof of) Theorem (21.3.1) yields

$$(1 - \eta_{nh}) \| \varepsilon \|_{wmh}^2 \leqslant \tfrac{1}{n} \sum_{i=1}^{n} | \varepsilon(X_i) |^2 + h^{2m} \| \varepsilon^{(m)} \|^2 ,$$

with $\eta_{nh} \leqslant C \| g_h \circledast (dW_n - dW_o) \|_\infty$ for a suitable constant C. Thus, if $\eta_{nh} < 1$, we get

$$(1) \quad \tfrac{1}{n} \sum_{i=1}^{n} | \varepsilon(X_i) |^2 + h^{2m} \| \varepsilon^{(m)} \|^2 \leqslant \frac{\{ \| \mathfrak{S}_{\omega nh} \|_{wmh} + h^m \| f_o^{(m)} \| \}^2}{1 - \eta_{nh}} .$$

On the other hand, we may split ε into bias and "variance",

$$\tfrac{1}{n} \sum_{i=1}^{n} | \varepsilon(X_i) |^2 \leqslant \tfrac{2}{n} \sum_{i=1}^{n} | f^{nh}(X_i) - f_h(X_i) |^2 + \tfrac{2}{n} \sum_{i=1}^{n} | f_h(X_i) - f_o(X_i) |^2 ,$$

where $f_h = \mathbb{E}[f^{nh} \,|\, \mathbb{X}_n]$. Since $f = f_h$ is the spline smoother of noiseless data and the objective function at the minimizer f_h is smaller than at f_o, this gives

$$\tfrac{1}{n} \sum_{i=1}^{n} | f_h(X_i) - f_o(X_i) |^2 + h^{2m} \| f_h^{(m)} \|^2 \leqslant h^{2m} \| f_o^{(m)} \|^2 ,$$

and we can obviously ignore the second term on the left.

For the first sum, note (again) that $f = f^{nh} - f_h$ is the solution to the spline smoothing problem with pure-noise data, so that, using Exercise (19.5.52),

$$\tfrac{1}{n} \sum_{i=1}^{n} | f^{nh}(X_i) - f_h(X_i) |^2 \leqslant \tfrac{1}{n} \sum_{i=1}^{n} | D_i |^2 .$$

This leads to the nice nonasymptotic bound

$$(2) \qquad \tfrac{1}{n}\sum_{i=1}^{n} |\,\varepsilon(X_i)\,|^2 \leqslant 2\,h^{2m} \|\, f_o^{(m)}\,\|^2 + \tfrac{2}{n}\sum_{i=1}^{n} |\,D_i\,|^2 \; .$$

Now, (1) and (2) allow us to bound the unconditional expectation. First, from (1), with the "arbitrary" cutoff $\frac{1}{2}$ on η_{nh},

$$(3) \quad \mathbb{E}\Bigl[\, 1\!\!1\bigl(\eta_{nh} \leqslant \tfrac{1}{2}\bigr) \cdot \tfrac{1}{n}\sum_{i=1}^{n} |\,\varepsilon(X_i)\,|^2 \Bigr] \leqslant$$

$$2\,\mathbb{E}\Bigl[\Bigl\{ \|\, \mathfrak{S}_{\omega nh}\,\|_{wmh} + h^m \,\|\, f_o^{(m)}\,\| \Bigr\}^2\Bigr] = \mathcal{O}\bigl(\, h^{2m} + (nh)^{-1}\,\bigr) \; .$$

Second, from (2),

$$\mathbb{E}\Bigl[\, 1\!\!1\bigl(\eta_{nh} > \tfrac{1}{2}\bigr) \cdot \tfrac{1}{n}\sum_{i=1}^{n} |\,\varepsilon(X_i)\,|^2 \Bigr]$$

$$\leqslant 2\,h^{2m}\,\|\, f_o^{(m)}\,\|^2 + 2\,\mathbb{E}\Bigl[\, 1\!\!1\bigl(\eta_{nh} > \tfrac{1}{2}\bigr) \cdot \tfrac{1}{n}\sum_{i=1}^{n} |\,D_i\,|^2 \Bigr] \, ,$$

and since η_{nh} depends on the design but not on the noise,

$$\mathbb{E}\Bigl[\, 1\!\!1\bigl(\eta_{nh} > \tfrac{1}{2}\bigr) \cdot \tfrac{1}{n}\sum_{i=1}^{n} |\,D_i\,|^2 \Bigr]$$

$$= \mathbb{E}\Bigl[\, 1\!\!1\bigl(\eta_{nh} > \tfrac{1}{2}\bigr) \cdot \mathbb{E}\Bigl[\, \tfrac{1}{n}\sum_{i=1}^{n} |\,D_i\,|^2 \;\Big|\; \mathbb{X}_n \Bigr]\Bigr]$$

$$\leqslant \sigma^2\,\mathbb{E}[\, \eta_{nh} > \tfrac{1}{2}\,] = \sigma^2\,\mathbb{P}[\, \eta_{nh} > \tfrac{1}{2}\,] \; .$$

The final step is that Bernstein's inequality yields, cf. (14.6.8),

$$\mathbb{P}[\, \eta_{nh} > \tfrac{1}{2}\,] \leqslant 2\,n\,\exp\Bigl(-\frac{\frac{1}{4}\,nh}{w_2 + \frac{1}{3}} \Bigr) \, ,$$

which is negligible compared $(nh)^{-1}$ for $h \gg n^{-1}\log n$. So then

$$(4) \quad \mathbb{E}\Bigl[\, 1\!\!1\bigl(\eta_{nh} > \tfrac{1}{2}\bigr) \cdot \tfrac{1}{n}\sum_{i=1}^{n} |\,\varepsilon(X_i)\,|^2 \Bigr] = o\bigl(\,(nh)^{-1}\,\bigr) \, , \quad nh/\log n \to \infty \; .$$

Now, (3) and (4) combined clinch the argument. Q.e.d.

References

Abramovich, F.; Grinshtein, V. (1999), *Derivation of equivalent kernels for general spline smoothing: A systematic approach*, Bernoulli 5, 359–379.

Abramovich, F.; Steinberg, D.M. (1996), *Improved inference in nonparametric regression using L_k smoothing splines*, J. Statist. Plann. Inference 49, 327–341.

Adams, R.A. (1975), *Sobolev spaces*, Academic Press, New York.

Ahlberg, J.H.; Nilson, E.N.; Walsh, J.L. (1967), *The theory of splines and their applications*, Academic Press, New York.

Akaike, H. (1970), *Statistical predictor identification*, Ann. Inst. Statist. Math. 22, 203–217.

Akaike, H. (1973), *Information theory and an extension of the maximum likelihood principle*, Proceedings of the Second International Symposium on Information Theory (B.N. Petrov, F. Csáki, eds.), Akademiai Kiado, Budapest, pp. 267–281.

Allen, D.M. (1974), *The relationship between variable selection and data augmentation and a method for prediction*, Technometrics 16, 125–127 (1974).

Amstler, C.; Zinterhof, P. (2001), *Uniform distribution, discrepancy, and reproducing kernel Hilbert spaces*, J. Complexity 17, 497–515.

Andrews, D.W.K. (1995), *Nonparametric kernel estimation for semiparametric models*, Econometric Theory 11, 560–596.

Anselone, P.M.; Laurent, P.J. (1968), *A general method for the construction of interpolating and smoothing spline-functions*, Numer. Math. 12, 66–82.

Anselone, P.M.; Sloan, I.H. (1990), *Spectral approximations for Wiener-Hopf operators*, J. Integral Equations Appl. 2, 237–261.

Ansley, C.F.; Kohn, R. (1985), *Estimation, filtering and smoothing in state space models with incompletely specified initial conditions*, Ann. Statist. 11, 1286–1316.

Ansley, C.F.; Kohn, R. (1987), *Efficient generalized cross-validation for state space models*, Biometrika 74, 139–148.

Ansley, C.F.; Kohn, R.; Tharm, D. (1990), *The estimation of residual standard deviation in spline regression*, Working paper 90–021, Australian Graduate School of Management, Sydney.

Antoniadis, A. (2007), *Wavelet methods in statistics: Some recent developments and their applications*, Statist. Surveys 1, 16–55.

Arce, G.R.; Grabowski, N.; Gallagher, N.C. (2000), *Weighted median filters with sigma-delta modulation encoding*, IEEE Trans. Signal Processing 48, 489–498.

Aronszajn, N. (1950), *Theory of reproducing kernels*, Trans. Amer. Math. Soc. 68, 337–404.
Barnes, B.A. (1987), *The spectrum of integral operators on Lebesgue spaces*, J. Operator Theory 18, 115–132.
Barron, A.R.; Birgé, L.; Massart, P. (1999), *Risk bounds for model selection via penalization*, Probab. Theory Relat. Fields 113, 301–413.
Barron, A.; Rissanen, J.; Yu, Bin (1998), *The minimum description length principle in coding and modeling*, IEEE Trans. Inform. Theory 44, 2743–2760.
Barry, D. (1983), *Nonparametric Bayesian regression*, Thesis, Yale University.
Barry, D. (1986), *Nonparametric Bayesian regression*, Ann. Statist. 14, 934–953.
Bates, D.M.; Watts, D.G. (1988), *Nonlinear regression analysis and its applications*, John Wiley and Sons, New York.
Belitser, E.; van de Geer, S. (2000), *On robust recursive nonparametric curve estimation*, High dimensional probability, II (E. Giné, D.M. Mason, J.A. Wellner, eds.), Birkhäuser, Boston, pp. 391–403.
Bennett, G. (1962), *Probability inequalities for the sum of independent random variables*, J. Amer. Statist. Assoc. 57, 33–45.
Beran, R. (1988), *Balanced simultaneous confidence sets*, J. Amer. Statist. Assoc. 83, 679–686.
Berlinet, A.; Thomas-Agnan, C. (2004), *Reproducing kernel Hilbert spaces in probability and statistics*, Kluwer, Dordrecht.
Bianconcini, S. (2007), *A reproducing kernel perspective of smoothing splines*, Department of Statistical Science, University of Bologna.
Bickel, P.J.; Rosenblatt, M. (1973), *On some global measures of the deviations of density function estimates*, Ann. Statist. 1, 1071–1095.
Björck, Å. (1996), *Numerical methods for least squares problems*, SIAM, Philadelphia.
Brown, L.D.; Cai, T.T.; Low, M.G. (1996), *Asymptotic equivalence of nonparametric regression and white noise*, Ann. Statist. 24, 2384–2398.
Brown, L.D.; Cai, T.T.; Low, M.G.; Zhang, C.-H. (2002), *Asymptotic equivalence for nonparametric regression with random design*, Ann. Statist. 30, 688–707.
Brown, L.D.; Levine, M. (2007), *Variance estimation in nonparametric regression via the difference sequence method*, Ann. Statist. 35, 2219–2232.
Brown, W.P.; Roth, R.R. (2004), *Juvenile survival and recruitment of Wood Thrushes Hylocichla mustelina in a forest fragment*, J. Avian Biol. 35, 316–326.
Brown, W.P.; Eggermont, P.; LaRiccia, V.; Roth, R.R. (2007), *Are parametric models suitable for estimating avian growth rates?*, J. Avian Biol. 38, 495–506.
Brown, W.P.; Eggermont, P.; LaRiccia, V.; Roth, R.R. (2008), *Partitioned spline estimators for growth curve estimation in wildlife studies*, Manuscript, University of Delaware.
Buja, A.; Hastie, T.; Tibshirani, R. (1989), *Linear smoothers and additive models*, Ann. Statist. 17, 453–510.
Buckley, M.J.; Eagleson, G.K.; Silverman, B.W. (1988), *The estimation of residual variance in nonparametric regression*, Biometrika 75, 189–200.
Bunea, F.; Wegkamp, M.H. (2004), *Two-stage model selection procedures in partially linear regression*, Canad. J. Statist. 32, 105–118.
Carter, C.K.; Eagleson, G.K. (1992), *A comparison of variance estimators in nonparametric regression*, J. R. Statist. Soc. B 54, 773–780.

Chen, H. (1988), *Convergence rates for parametric components in a partly linear model*, Ann. Statist. 16, 136–146.

Chen, H.; Shiau, J-J. H. (1991), *A two-stage spline smoothing method for partially linear models*, J. Statist. Plann. Inference 27, 187–210.

Chiang, C.-T.; Rice, J.A.; Wu, C.O. (2001), *Smoothing spline estimation for varying coefficient models with repeatedly measured dependent variables*, J. Amer. Statist. Assoc. 96, 605–619.

Claeskens, G.; Van Keilegom, I. (2003), *Bootstrap confidence bands for regression curves and their derivatives*, Ann. Statist. 31, 1852–1884.

Cleveland, W.S. (1979), *Robust locally weighted regression and smoothing scatterplots*, J. Amer. Statist. Assoc. 74, 829–836.

Cleveland, W.S.; Devlin, S.J. (1988), *Locally weighted regression: An approach to regression analysis by local fitting*, J. Amer. Statist. Assoc. 83, 596–610.

Cox, D.D. (1981), *Asymptotics of M-type smoothing splines*, Ann. Statist. 11, 530–551.

Cox, D.D. (1984), *Multivariate smoothing spline functions*, SIAM J. Numer. Anal. 21, 789–813.

Cox, D.D. (1988a), *Approximation of method of regularization estimators*, Ann. Statist. 16, 694–712.

Cox, D.D. (1988b), *Approximation of least-squares regression on nested subspaces*, Ann. Statist. 16, 713–732.

Cox, D.D.; O'Sullivan, F. (1990), *Asymptotic analysis of penalized likelihood and related estimators*, Ann. Statist. 18, 1676–1695.

Craven, P.; Wahba, G. (1979), *Smoothing noisy data with spline functions*, Numer. Math. 31, 377–403.

Csörgő, M.; Révész, P. (1981), *Strong approximations in probability and statistics*, Academic Press, New York.

Cummins, D.J.; Filloon, T.G.; Nychka, D.W. (2001), *Confidence intervals for nonparametric curve estimates: Toward more uniform pointwise coverage*, J. Amer. Statist. Assoc. 96, 233–246.

D'Amico, M.; Ferrigno, G. (1992), *Comparison between the more recent techniques for smoothing and derivative assessment in biomechanics*, Med. Biol. Eng. Comput. 30, 193–204.

Davies, L.; Gather, U.; Weinert, H. (2008), *Nonparametric regression as an example of model choice*, Comm. Statist. Simulation Comput. 37, 274–289.

de Boor, C. (1978), *A practical guide to splines*, Springer-Verlag, New York.

Debón, A.; Montes, F.; Sala, R. (2005), *A comparison of parametric methods for mortality graduation: Application to data from the Valencia region (Spain)*, SORT 29, 269–288.

Debón, A.; Montes, F.; Sala, R. (2006), *A comparison of nonparametric methods in the graduation of mortality: Application to data from the Valencia region (Spain)*, Intern. Statist. Rev. 74, 215–233.

Deheuvels, P. (2000), *Limit laws for kernel density estimators for kernels with unbounded supports*, Asymptotics in statistics and probability (M.L. Puri, ed.), VSP International Science Publishers, Amsterdam, pp. 117–132.

Deheuvels, P.; Derzko, G. (2008), *Asymptotic certainty bands for kernel density estimators based upon a bootstrap resampling scheme*, Statistical models and methods for biomedical and technical systems (F. Vonta; M. Nikulin; C. Huber-Carol, eds.), Birkhäuser, Boston, pp. 171–186.

Deheuvels, P.; Mason, D.M. (2004), *General asymptotic confidence bands based on kernel-type function estimators*, Stat. Inference Stoch. Process. 7, 225–277.

Deheuvels, P.; Mason, D.M. (2007), *Bootstrap confidence bands for kernel-type function estimators*, to appear.

de Jong, P. (1991), *The diffuse Kalman filter*, Ann. Statist. 19, 1073–1083.

DeMicco, F.J.; Lin, Yan; Liu, Ling; Rejtő, L.; Beldona, S.; Bancroft, D. (2006), *The effect of holidays on hotel daily revenue*, J. Hospitality Tourism Res. 30, 117–133.

Devroye, L. (1989), *The double kernel method in density estimation*, Ann. Inst. H. Poincaré Probab. Statist. 25, 533–580.

Devroye, L. (1991), *Exponential inequalities in nonparametric estimation*, Nonparametric functional estimation and related topics (G. Roussas, ed.), Kluwer, Dordrecht, pp. 31–44.

Devroye, L.; Györfi, L. (1985), *Density estimation: The L_1-view*, John Wiley and Sons, New York.

Devroye, L.; Gyorfi, L.; Krzyżak, A.; Lugosi, G. (1994), *On the strong universal consistency of nearest neighbor regression function estimates*, Ann. Statist. 22, 1371–1385.

Devroye, L.; Györfi, L.; Lugosi, G. (1996), *A probabilistic theory of pattern recognition*, Springer-Verlag, New York.

Devroye, L.; Lugosi, G. (2000a), *Variable kernel estimates: On the impossibility of tuning the parameters*, High dimensional probability, II (E. Giné, D.M. Mason, J.A. Wellner, eds.), Birkhäuser, Boston, pp. 405–424.

Devroye, L.; Lugosi, G. (2000b), *Combinatorial methods in density estimation*, Springer-Verlag, New York.

Devroye, L.; Wagner, T.J. (1980), *Distribution-free consistency results in nonparametric discrimination and regression function estimation*, Ann. Statist. 8, 231–239.

Diebolt, J. (1993), *A nonparametric test for the regression function: Asymptotic theory*, J. Statist. Plann. Inference 44, 1–17.

Diewert, W.E.; Wales, T.J. (1998), *A "new" approach to the smoothing problem*, Money, measurement and computation (M.T. Belongin; J.M. Binner, eds.), Palgrave MacMillan, New York, pp. 104–144.

Dolph, C.L.; Woodbury, M.A. (1952), *On the relation between Green's functions and covariances of certain stochastic processes and its application to unbiased linear prediction*, Trans. Amer. Math. Soc. 72, 519–550.

Donoho, D.L.; Johnstone, I.M. (1998), *Minimax estimation via wavelet shrinkage*, Ann. Statist. 26, 879–921.

Dony, J.; Einmahl, U.; Mason, D.M. (2006), *Uniform in bandwidth consistency of local polynomial regression function estimators*, Austrian J. Statist. 35, 105–120.

Dudley, R.M. (1978), *Central limit theorems for empirical measures*, Ann. Probab. 6, 899–929.

Dudley, R.M. (1982), *A course in empirical processes*, École d'Été de Probabilités de Saint-Flour XII – 1982. Lecture Notes in Mathematics 1097 (P.L. Hennequin, ed.), Springer-Verlag, Berlin, pp. 1–148.

Duistermaat, J.J. (1995), *The Sturm-Liouville problem for the operator $(-d^2/dx^2)^m$, with Neumann or Dirichlet boundary conditions*, Technical Report 899, Department of Mathematics, University of Utrecht.

Dym, H.; McKean, H.P. (1972), *Fourier series and integrals*, Academic Press, New York.

Efromovich, S.Yu. (1996), *On nonparametric regression for iid observations in a general setting*, Ann. Statist. 24, 1126–1144.

Efromovich, S.Yu. (1999), *On rate and sharp optimal estimation*, Probab. Theory Relat. Fields 113, 415–419.

Efron, B. (1988), *Computer-intensive methods in statistical regression*, SIAM Rev. 30, 421–449.

Eggermont, P.P.B. (1989), *Uniform error estimates of Galerkin methods for monotone Abel-Volterra integral equations on the half-line*, Math. Comp. 53, 157–189.

Eggermont, P.P.B.; Eubank, R.L.; LaRiccia, V.N. (2005), *Convergence rates for smoothing spline estimators in time varying coefficient models*, Manuscript, University of Delaware.

Eggermont, P.P.B.; LaRiccia, V.N. (1995), *Smoothed maximum likelihood density estimation for inverse problems*, Ann. Statist. 23, 199–220.

Eggermont, P.P.B.; LaRiccia, V.N. (2000a), *Maximum likelihood estimation of smooth monotone and unimodal densities*, Ann. Statist. 29, 922–947.

Eggermont, P.P.B.; LaRiccia V.N. (2000b), *Almost sure asymptotic optimality of cross validation for spline smoothing*, High dimensional probability, II (E. Giné, D.M. Mason, J.A. Wellner, eds.), Birkhäuser, Boston, pp. 425–441.

Eggermont, P.P.B.; LaRiccia V.N. (2003), *Nonparametric logistic regression: Reproducing kernel Hilbert spaces and strong convexity*, Manuscript, University of Delaware.

Eggermont, P.P.B.; LaRiccia V.N. (2006a), *Uniform error bounds for smoothing splines*, High Dimensional probability. IMS Lecture Notes–Monograph Series 51 (V.I. Koltchinskii; Wenbo Li; J. Zinn, eds.), IMS, Hayward, California, 2006, pp. 220–237.

Eggermont, P.P.B.; LaRiccia V.N. (2006b), *Equivalent kernels for smoothing splines*, J. Integral Equations Appl. 18, 197–225.

Eggermont, P.P.B.; Lubich, Ch. (1991), *Uniform error estimates of operational quadrature methods for nonlinear convolution equations on the half-line*, Math. Comp. 56, 149–176.

Einmahl, U.; Mason, D.M. (2000), *An empirical process approach to the uniform consistency of kernel-type function estimators*, J. Theor. Probab. 13, 1–37.

Einmahl, U.; Mason, D.M. (2005), *Uniform in bandwidth consistency of kernel-type function estimators*, Ann. Statist. 33, 1380–1403.

Elfving, T.; Andersson, L.E. (1988), *An algorithm for computing constrained smoothing spline functions*, Numer. Math. 52, 583–595.

Engle, R.F.; Granger, C.W.; Rice, J.A.; Weiss, A. (1986), *Semiparametric estimates of the relation between weather and electricity sales*, J. Amer. Statist. Assoc. 81, 310–320.

Eubank, R.L. (1988), *A note on smoothness priors and nonlinear regression*, J. Amer. Statist. Assoc. 81, 514–517.

Eubank, R.L. (1999), *Spline smoothing and nonparametric regression*, second edition, Marcel Dekker, New York.

Eubank, R.L. (2005), *A Kalman filter primer*, CRC Press, Boca Raton.

Eubank, R.L.; Hart, J.D.; Speckman, P.L. (1990), *Trigonometric series regression estimators with an application to partially linear models*, J. Multivariate Anal. 32, 70–83.

Eubank, R.L.; Huang, C.; Wang S. (2003), *Adaptive order selection for spline smoothing*, J. Comput. Graph. Statist. 12, 382–397.

Eubank, R.; Speckman, P.L. (1990a), *Curve fitting by polynomial-trigonometric regression*, Biometrika 77, 1–9.

Eubank, R.; Speckman, P.L. (1990b), *A bias reduction theorem, with applications in nonparametric regression*, Scand. J. Statist. 18, 211–222.

Eubank, R.L.; Speckman, P.L. (1993), *Confidence bands in nonparametric regression*, J. Amer. Statist. Assoc. 88, 1287–1301.

Eubank, R.; Wang, Suojin (2002), *The equivalence between the Cholesky decomposition and the Kalman filter*, Amer. Statist. 56, 39–43.

Fan, J.; Gijbels, I. (1995a), *Data-driven bandwidth selection in local polynomial fitting: Variable bandwidth and spatial adaptation*, J. R. Statist. Soc. B 57, 371–394.

Fan, J.; Gijbels, I. (1995b), *Adaptive order polynomial fitting: Bandwidth robustification and bias reduction*, J. Comput. Graph. Statist. 4, 213–227.

Fan, J.; Gijbels, I. (1996), *Local polynomial modeling and its applications*, Chapman and Hall, London.

Fan, J.; Gijbels, I.; Hu, T.C.; Huang, L.S. (1996), *A study of variable bandwidth selection for local polynomial regression*, Statist. Sinica 6, 113–127.

Feinerman, R.P.; Newman, D.J. (1974), *Polynomial approximation*, Williams and Wilkins, Baltimore.

Forsythe, G.E. (1957), *Generation and use of orthogonal polynomials for data-fitting with a digital computer*, J. SIAM 5, 74–88.

Fried, R.; Einbeck, J.; Gather, U. (2007), *Weighted repeated median regression*, Manuscript, Department of Statistics, University of Dortmund, Germany.

Gamber, H. (1979), *Choice of optimal shape parameter when smoothing noisy data*, Commun. Statist. A 8, 1425–1435.

Gasser, T.; Müller, H.G. (1979), *Kernel estimation of regression functions*, Smoothing techniques for curve estimation (T, Gasser, M. Rosenblatt, eds.), Lecture Notes in Mathematics 757, Springer-Verlag, New York, pp. 23–68.

Gasser, T.; Müller, H.G.; Kohler, W.; Molinari, L.; Prader, A. (1984), *Nonparametric regression analysis of growth curves*, Ann. Statist. 12, 210–229.

Gebski, V.; McNeil, D. (1984), *A refined method of robust smoothing*, J. Amer. Statist. Assoc. 79, 616–623.

Geman, S.; Hwang, Chii-Ruey (1982), *Nonparametric maximum likelihood estimation by the method of sieves*, Ann. Statist. 10, 401–414.

Giné, E. (1974), *On the central limit theorem for sample continuous processes*, Ann. Probab. 2, 629–641.

Giusti, E. (1984), *Minimal surfaces and functions of bounded variation*, Birkhäuser, Boston.

Golub, G.H.; Heath, M.; Wahba, G. (1979), *Generalized cross-validation as a method for choosing a good ridge parameter*, Technometrics 21, 215–223.

Golubev, G.K.; Nussbaum, M. (1990), *A risk bound in Sobolev class regression*, Ann. Statist. 18, 758–778.

Gompertz, B. (1825), *On the nature of the function expressive of the law of human mortality, and on a new mode of determining the value of life contingencies*, Philos. Trans. R. Soc. B 123, 513–585.

Good, I.J.; Gaskins, R.A. (1971), *Nonparametric roughness penalties for probability densities*, Biometrika 58, 255–277.

Grama, I.; Nussbaum, M. (1998), *Asymptotic equivalence for nonparametric generalized linear models*, Probab. Theory Relat. Fields 13, 984–997.

Green, P.J.; Jennison, C.; Seheult, A. (1985), *Analysis of field experiments by least squares smoothing*, J. R. Statist. Soc. B 47, 299–315.

Green, P.J.; Silverman, B.W. (1990), *Nonparametric regression and generalized linear models. A roughness penalty approach*, Chapman and Hall, London.

Grenander, U. (1981), *Abstract inference*, John Wiley and Sons, New York.

Griffin, J.E.; Steel, M.F.J. (2008), *Bayesian nonparametric modelling with the Dirichlet regression smoother*, Department of Statistics, University of Warwick.

Grünwald, P.D.; Myung, In Jae; Pitt, M.A. (2005), *Advances in Minimum Description Length: Theory and Applications*, MIT Press, Boston, 2005.

Györfi, L.; Kohler, M.; Krzyżak, A.; Walk, A. (2002), *A distribution-free theory of nonparametric regression*, Springer-Verlag, New York.

Hall, P. (1992), *On Edgeworth expansion and bootstrap confidence bands in nonparametric regression*, Ann. Statist. 20, 675–694.

Hansen, M.H.; Yu, Bin (2001), *Model selection and the principle of minimum description length*, J. Amer. Statist. Assoc. 96, 746–774.

Härdle, W. (1990), *Applied nonparametric regression*, Cambridge University Press, Cambridge.

Härdle, W.; Janssen, P.; Serfling, R. (1988), *Strong uniform consistency rates for estimators of conditional functionals*, Ann. Statist. 16, 1428–1449.

Hardy, G.H.; Littlewood, G.E.; Polya, G. (1951), *Inequalities*, Cambridge University Press, Cambridge.

Hart, J.D. (1997), *Nonparametric smoothing and lack-of-fit tests*, Springer, New York.

Hart, J.D.; Wehrly, T.E. (1992), *Kernel regression when the boundary region is large, with an application to testing the adequacy of polynomial models*, J. Amer. Statist. Assoc. 87, 1018–1024.

He, Xuming; Shen, Lixin; Shen, Zuowei (2001), *A data-daptive knot selection scheme for fitting splines*, IEEE Signal Process. Lett. 8, 137–139.

Heckman, N.E. (1988), *Spline smoothing in a partly linear model*, J. R. Statist. Soc. B 48, 244–248.

Hengartner, N.; Wegkamp, M. (2001), *Estimation and selection procedures in regression: An L_1 approach*, Canad. J. Statist. 29, 621–632.

Herriot, J.G.; Reinsch, C.H. (1973), *Procedures for natural spline interpolation*, Commun. ACM 16, 763–768.

Herriot, J.G.; Reinsch, C.H. (1976), *Algorithm 507. Procedures for quintic natural spline interpolation*, ACM Trans. Math. Software 2, 281–289.

Hetrick, J.; Chirnside, A.E.M. (2000), *Feasibility of the use of a fungal bioreactor to treat industrial wastewater*, Department of Bioresource Engineering, University of Delaware.

Hille, E. (1972), *Introduction to general theory of reproducing kernels*, Rocky Mountain J. Math. 2, 321–368.

Hille, E.; Szegö, G.; Tamarkin, J.D. (1937), *On some generalizations of a theorem of A. Markoff*, Duke Math. J. 3, 729–739.

Hiriart-Urruty, J.-B.; Lemaréchal, C. (1993), *Convex analysis and minimization algorithms. Volumes I and II*, Springer-Verlag, New York.

Holladay, J.C. (1957), *A smoothest curve approximation*, Math. Tables Aids Comput. 11, 233–243.

Holmes, R.B. (1975), *Geometric functional analysis and its applications*, Springer-Verlag, New York.

Horváth, L. (1993), *Asymptotics for global measures of accuracy of splines*, J. Approx. Theory 73, 270–287.

Huang, Chunfeng (2001), *Boundary corrected cubic smoothing splines*, J. Statist. Comput. Simulation 70, 107–121.

Huber, P. (1981), *Robust statistics*, John Wiley and Sons, New York.

Hurvich, C.M.; Tsai, C.-L. (1989), *Regression and time series model selection in small samples*, Biometrika 76, 297–307.

Hurvich, C.M.; Simonoff, J.A.; Tsai, C.-L. (1995), *Smoothing parameter selection in nonparametric regression using an improved Akaike information criterion*, J. R. Statist. Soc. B 60, 271–293.

Hüsler, J.; Piterbarg, V.I.; Seleznjev, O. (2000), *On convergence of the uniform norms of Gaussian processes and linear approximation problems*, Ann. Appl. Probab. 13, 1615–1653.

Hutchinson, M.F.; de Hoog, F.R. (1985), *Smoothing noisy data by spline functions*, Numer. Math. 47, 99–106.

Ibragimov, I.A.; Has'minskii, R.Z. (1982), *Bounds for the risk of nonparametric regression estimates*, Theory Probab. Appl. 27, 84–99.

Jandhyala, V.K.; MacNeil, I.B. (1992), *On testing for the constancy of regression coefficients under random walk and change-point alternatives*, Econometric Theory 8, 501–517.

Jones, M.C. (1986), *Expressions for inverse moments of positive quadratic forms in normal variables*, Aust. J. Statist. 28, 242–250.

Jones, M.C. (1987), *On moments of ratios of quadratic forms in normal variables*, Statist. Probab. Lett. 6, 129–136.

Kalman, R.E. (1960), *A new approach to linear filtering and prediction problems*, Trans. ASME–J. Basic Engrg. 82 D, 35–45.

Kemperman, J.H.B. (1969), *On the optimum rate of transmitting information*, Ann. Math. Statist. 40, 2156–2177.

Kimeldorf, G.; Wahba, G. (1970a), *A correspondence between Bayesian estimation on stochastic processes and smoothing by splines*, Ann. Math. Statist. 2, 495–502.

Kimeldorf, G.; Wahba, G. (1970b), *Spline functions and stochastic processes*, Sankhyā A 32, 173–180.

Kincaid, D.; Cheney, W. (1991), *Numerical Analysis*, Brooks/Cole, Pacific Grove.

Klonias, V.K. (1982), *Consistency of two nonparametric maximum penalized likelihood estimators of the probability density function*, Ann. Statist. 10, 811–824.

Koenker, R.; Mizera, I. (2004), *Penalized triograms: Total variation regularization for bivariate smoothing*, J. R. Statist. Soc. B 66, 145–163.

Kohler, M.; Krzyżak, A. (2007), *Asymptotic confidence intervals for Poisson regression*, J. Multivariate Anal. 98, 1072–1094.

Kohn, R.; Ansley, C.F. (1985), *Efficient estimation and prediction in time series regression models*, Biometrika 72, 694–697.

Kohn, R.; Ansley, C.F. (1989), *A fast algorithm for signal extraction, influence and cross-validation in state space models*, Biometrika 76, 65–79.

Kohn, R.; Ansley, C.F.; Tharm, D. (1991), *The performance of cross-validation and maximum likelihood estimators of spline smoothing parameters*, J. Amer. Statist. Assoc. 86, 1042–1050.

Komlós, J.; Major, P.; Tusnády, G. (1976), *An approximation of partial sums of independent* RV*'s and the sample* DF. *II*, Z. Wahrsch. Verw. Gebiete 34, 33–58.

Konakov, V. D.; Piterbarg, V. I. (1984), *On the convergence rate of maximal deviation distribution for kernel regression estimates*, J. Multivariate Anal. 15, 279–294.

Krein, M.G. (1958), *Integral equations on the half-line with kernel depending on the difference of the arguments*, Usp. Mat. Nauk (N.S.) 13, (5) (83): 3–120; Amer. Math. Soc. Transl. 22, 163–288 (1962).

Kroó, A. (2008), *On the exact constant in the L_2 Markov inequality*, J. Approx. Theory 151, 208–211.

Kruskal, J.B. (1964), *Multidimensional scaling by optimizing goodness of fit to a nonmetric hypothesis*, Psychometrika 29, 1–27.

Kufner, A. (1980), *Weighted Sobolev spaces*, B.G. Teubner, Leipzig.

Lawson, C.L.; Hanson, R.J. (1995), *Solving least squares problems*, Prentice-Hall, New York (Reprinted: SIAM, Philadelphia, 1995).

Lee, T.C.M. (2003), *Smoothing parameter selection for smoothing splines: A simulation study*, Comput. Statist. Data Anal. 42, 139–148.

Lee, T.C.M. (2004), *Improved smoothing spline regression by combining estimates of different smoothness*, Statist. Probab. Lett. 67, 133–140.

Lepski, O.V.; Tsybakov, A.B. (2000), *Asymptotically exact nonparametric hypothesis testing in sup-norm and at a fixed point*, Probab. Theory Relat. Fields 117, 17–48.

Li, Ker-Chau (1985), *From Stein's unbiased risk estimates to the method of generalized cross validation*, Ann. Statist. 13, 1352–1377.

Li, Ker-Chau (1986), *Asymptotic optimality of C_L and generalized cross validation in ridge regression with application to spline smoothing*, Ann. Statist. 14, 1101–1112.

Li, Ker-Chau (1989), *Honest confidence regions for nonparametric regression*, Ann. Statist. 17, 1001–1008.

Lin, Xihong; Wang, Naysin; Welsh, A.H.; Carroll, R.J. (2004), *Equivalent kernels of smoothing splines in nonparametric regression for clustered/longitudinal data*, Biometrika 91, 177–193.

Loève, M. (1948), *Fonctions aléatoire du second ordre*, Supplement to: P. Lévy, *Processus stochastiques et mouvement Brownien*, Gauthier-Villars, Paris.

Ma, Yanyuan; Chiou, Jeng-Min; Wang, Naysin (2006), *Efficient semiparametric estimator for heteroscedastic partially linear models*, Biometrika 93, 75–84.

Madsen, K.; Nielsen, H.B. (1993), *A finite smoothing algorithm for linear ℓ_1 estimation*, SIAM J. Optimiz. 3, 223–235.

Mallows, C.L. (1972), *Some comments on C_p*, Technometrics 15, 661–675.

Mammen, E. (1991a), *Estimating a smooth monotone regression function*, Ann. Statist. 19, 724–740.

Mammen, E. (1991b), *Nonparametric regression under qualitative smoothness constraints*, Ann. Statist. 19, 741–759.

Marron, J.S.; Tsybakov, A.B. (1995), *Visual error criteria for qualitative smoothing*, J. Amer. Statist. Assoc. 90, 499–507.

Mason, D.M. (2006), *Private communication.*

Mathews, J.; Walker, R. L. (1979), *Mathematical methods of physics*, Addison-Wesley, New York.

Maz'ja, V.G. (1985), *Sobolev spaces*, Springer-Verlag, Berlin.

McCullagh, P.; Nelder, J. (1989), *Generalized linear models*, Chapman and Hall, London.

McDiarmid, C. (1989), *On the method of bounded differences*, Surveys in combinatorics 1989, Cambridge University Press, Cambridge, pp. 148–188.

Meschkowski, H. (1962), *Hilbertsche Räume mit Kernfunktion*, Springer-Verlag, Berlin.

Messer, K. (1991), *A comparison of a spline estimate to its equivalent kernel estimate*, Ann. Statist. 19, 817–829.

Messer, K.; Goldstein, L. (1993), *A new class of kernels for nonparametric curve estimation*, Ann. Statist. 21, 179–196.

Miyata, S.; Shen, Xiaotong (2005), *Free-knot splines and adaptive knot selection*, J. Japan. Statist. Soc. 35, 303–324.

Mosteller, F.; Wallace, D.L. (1963), *Inference in an authorship problem*, J. Amer. Statist. Assoc. 58, 275–309.

Müller, H.G. (1988), *Nonparametric regression analysis of longitudinal data*, Lecture Notes in Mathematics 46, Springer-Verlag, New York.

Müller, H.G. (1993a), *Smooth optimum kernel estimators near endpoints*, Biometrika 78, 521–530.

Müller, H.G. (1993b), *On the boundary kernel method for nonparametric curve estimation near endpoints*, Scand. J. Statist. 20, 313–328.

Müller, H.G.; Stadtmüller, U. (1999), *Multivariate kernels and a continuous least squares principle*, J. R. Statist. Soc. B 61, 439–458.

Nadaraya, E.A. (1964), *On estimating regression*, Theor. Probab. Appl. 9, 141–142.

Nadaraya, E.A. (1965), *On nonparametric estimation of density functions and regression curves*, Theor. Probab. Appl. 10, 186–190.

Neumann, M. H. (1997), *Pointwise confidence intervals in nonparametric regression with heteroscedastic error structure*, Statistics 29, 1–36.

Neyman, J. (1937), *Smooth test for goodness-of-fit*, Skand. Aktuarietidskr. 20, 149–199.

Nishii, R. (1984), *Asymptotic properties of criteria for selection of variables in multiple regression*, Ann. Statist. 12, 758–765.

Nussbaum, M. (1985), *Spline smoothing in regression models and asymptotic efficiency in L_2*, Ann. Statist. 12, 984–997.

Nychka, D.W. (1988), *Confidence intervals for smoothing splines*, J. Amer. Statist. Assoc. 83, 1134–2243.

Nychka, D.W. (1995), *Splines as local smoothers*, Ann. Statist. 23, 1175–1197.

Oehlert, G.W. (1992), *Relaxed boundary smoothing splines*, Ann. Statist. 20, 146–160.

O'Hagan, A. (1978), *Curve fitting and optimal design for prediction*, J. R. Statist. Soc. B 40, 1–42.

Øksendal, B. (2003), *Stochastic differential equations*, sixth edition, Springer-Verlag, Berlin.

Osborne, M.R.; Prvan, T. (1991), *What is the covariance analog of the Paige and Saunders information filter?*, SIAM J. Sci. Statist. Comput. 12, 1324–1331.

Osborne, M.R.; Söderkvist, I. (2004), *V-invariant methods, generalised least-squares problems, and the Kalman filter*, ANZIAM J. 45 (E), C232–C247.

Oudshoorn, C.G.M. (1998), *Asymptotically minimax estimation of a function with jumps*, Bernoulli 4, 15–33.

Pal, J.K.; Woodroofe, M.B. (2007), *Large sample properties of shape restricted regression estimators with smoothness adjustments*, Statist. Sinica 17, 1601–1616.

Parzen, E. (1961), *An approach to time series analysis*, Ann. Math. Statist. 32, 951–989.

Parzen, E. (1962), *Stochastic processes*, Holden-Day, San Francisco (Reprinted: SIAM, Philadelphia, 1999).

Parzen, E. (1967), *Time series analysis papers*, Holden-Day, San Francisco.

Pickands III, J. (1969), *Asymptotic properties of the maximum in a stationary Gaussian process*, Trans. Amer. Math. Soc. 145, 75–86.

Pinsker, M.S. (1980), *Optimal filtering of square-integrable signals in Gaussian noise (Russian)*, Problems Inform. Transmission 16, (2) 52–68.

Pintore, A.; Speckman, P.L.; Holmes, C.C. (2000), *Spatially adaptive smoothing splines*, Biometrika 93, 113–125.

Pollard, D. (1991), *Asymptotics for least absolute deviation regression estimators*, Econometric Theory 7, 186–199.

Polyak, B.T.; Tsybakov, A.B. (1991), *Asymptotic optimality of the C_p-test in the projection estimation of a regression*, Theory Probab. Appl. 35, 293–306.

Rahman, Q.I.; Schmeisser, G. (2002), *Analytic theory of polynomials*, Clarendon Press, Oxford.

Reinsch, Ch. (1967), *Smoothing by spline functions*, Numer. Math. 10, 177–183.

Ribière, G. (1967), *Regularisation d'operateurs*, Rev. Française Informat. Recherche Opérationelle 1, No. 5, 57–79.

Rice, J.A. (1984a), *Boundary modification for kernel regression*, Comm. Statist. Theor. Meth. 13, 893–900.

Rice, J.A. (1984b), *Bandwidth choice for nonparametric regression*, Ann. Statist. 12, 1215–1230.

Rice, J.A. (1986a), *Convergence rates for partially splined models*, Statist. Probab. Lett. 4, 203–208.

Rice, J.A. (1986b), *Bandwidth choice for differentiation*, J. Multivariate Anal. 19, 251–264.

Rice, J.A.; Rosenblatt, M. (1983), *Smoothing splines regression, derivatives and deconvolution*, Ann. Statist. 11, 141–156.

Richards, F.J. (1959), *A flexible growth function for empirical use*, J. Exp. Botany 10, 290–300.

Riesz, F.; Sz-Nagy, B. (1955), *Functional analysis*, Ungar, New York (Reprinted: Dover, New York, 1996).

Rissanen, J. (1982), *Estimation of structure by minimum description length*, Circuits Systems Signal Process. 1, 395–406.

Rissanen, J. (2000), *MDL denoising*, IEEE Trans. Information Theory 46, 2537–2543.

Ritter, K. (2000), *Average-case analysis of numerical problems*, Lecture Notes in Mathematics 1733, Springer, Berlin.

Rousseeuw, P.J. (1992), *Least median of squares regression*, J. Amer. Statist. Assoc. 79, 871–880.

Ruppert, D.; Sheather, S.J.; Wand, M.P. (1992), *An efficient bandwidth selector for local least-squares regression*, J. Amer. Statist. Assoc. 90, 1257–1270.

Sage, A.P.; Melsa, J.L. (1971), *Estimation theory with applications to communications and control theory*, McGraw-Hill, New York.

Sakhanenko, A.I. (1985), *On unimprovable estimates of the rate of convergence in the invariance principle*, Nonparametric inference, Colloquia Math. Soc. János Bolyai 32, pp. 779–783.

Sánchez, D.A. (1968), *Ordinary differential equations and stability theory: An introduction*, W.H. Freeman and Co., San Francisco (Reprinted: Dover, New York, 1979).

Sansone, G. (1959), *Orthogonal functions*, Wiley Interscience, New York (Reprinted: Dover, New York, 1991).

Schoenberg, I.J. (1973), *Cardinal spline interpolation*, SIAM, Philadelphia.

Schwarz, G. (1978), *Estimating the dimension of a model*, Ann. Statist. 6, 461–464.

Seber, G.A.F. (1977), *Linear regression analysis*, John Wiley and Sons, New York.

Seber, G.A.F.; Wild, C.J. (2003), *Nonlinear regression*, John Wiley and Sons, New York.

Seifert, B.; Gasser, T. (1996), *Finite-sample variance of local polynomials: Analysis and solutions*, J. Amer. Statist. Assoc. 91, 267–275.

Seifert, B.; Gasser, T. (2000), *Variance properties of local polynomials and ensuing modifications*, Statistical theory and computational aspects of smoothing (W. Härdle; M.G. Schimek, eds.), Physica Verlag, Heidelberg, pp. 50–127.

Shao, Q.-M. (1995), *Strong approximation theorems for independent random variables and their applications*, J. Multivariate Anal. 52, 107–130.

Shapiro, H.S. (1969), *Smoothing and approximation of functions*, Van Nostrand Reinhold, New York.

Shibata, R. (1981), *An optimal selection of regression variables*, Biometrika 68, 45–54.

Shorack, G.R. (2000), *Probability for statisticians*, Springer, New York.

Silverman, B.W. (1984), *Spline smoothing: The equivalent variable kernel method*, Ann. Statist. 12, 898–916.

Speckman, P.L. (1981), Manuscript, University of Oregon.

Speckman, P.L. (1982), *Efficient nonparametric regression with cross-validated smoothing splines*, Unpublished manuscript.

Speckman, P.L. (1985), *Spline smoothing and optimal rates of convergence in nonparametric regression models*, Ann. Statist. 13, 970–983.

Speckman, P.L. (1988), *Kernel smoothing in partial linear models*, J. R. Statist. Soc. B 50, 413–436.

Speckman, P.L.; Sun, Dongchu (2001), *Bayesian nonparametric regression and autoregression priors*, Department of Statistics, University of Missouri.

Stadtmüller, U. (1986), *Asymptotic properties of nonparametric curve estimates*, Period. Math. Hungar. 17, 83–108.

Stakgold, I. (1967), *Boundary value problems of mathematical physics. Vol. I.*, Macmillan, New York (Reprinted: SIAM, Philadelphia, 2000).

Stein, M.L. (1990), *A comparison of generalized cross validation and modified maximum likelihood for estimating the parameters of a stochastic process*, Ann. Statist. 18, 1139–1157.

Stein, M.L. (1993), *Spline smoothing with an estimated order parameter*, Ann. Statist. 21, 1522–1544.

Stone, C.J. (1977), *Consistent nonparametric regression (with discussion)*, Ann. Statist. 5, 549–645.

Stone, C.J. (1980), *Optimal rates of convergence for nonparametric estimators*, Ann. Statist. 8, 1348–1360.

Stone, C.J. (1982), *Optimal global rates of convergence for nonparametric regression*, Ann. Statist. 10, 1040–1053.

Stone, M. (1977), *An asymptotic equivalence of choice of model by cross-validation and Akaike's criterion*, J. R. Statist. Soc. B 39, 44–47.

Stone, M. (1979), *Comments on model selection criteria of Akaike and Schwarz*, J. R. Statist. Soc. B 41, 276–278.

Sugiura, N. (1978), *Further analysis of the data by Akaike's information criterion and the finite corrections*, Commun. Statist.-Theor. Meth. A 7, 13–26.

Tantiyaswasdikul, C.; Woodroofe, M.B. (1994), *Isotonic smoothing splines under sequential designs*, J. Statist. Plann. Inference 38, 75–87.

Troutman, J.L. (1983), *Variational calculus with elementary convexity*, Springer-Verlag, New York.

Truong, Y.K. (1989), *Asymptotic properties of kernel estimators based on local medians*, Ann. Statist. 17, 606–617.

Tsybakov, A.B. (1996), *Robust reconstruction of functions by a local approximation method*, Problems Inform. Transmission 22, 75–83.

Utreras, F. (1981), *Optimal smoothing of noisy data using spline functions*, SIAM J. Sci. Statist. Comput. 2, 349–362.

van de Geer, S. (1987), *A new approach to least-squares estimation, with applications*, Ann. Statist. 15, 587–602.

van de Geer, S. (1990), *Estimating a regression function*, Ann. Statist. 18, 907–924.

van de Geer, S. (2000), *Empirical processes in M-estimation*, Cambridge University Press, Cambridge.

van de Geer, S.; Wegkamp, M. (1996), *Consistency for the least squares estimator in nonparametric regression*, Ann. Statist. 24, 2513–2523.

Vorhaus, B. (1948), *The amazing Mr. X.* (*Story by C. Wilbur, produced by Benjamin Stoloff*), Sinister Cinema (Re-released: Alpha Video, 2003).

Wahba, G. (1978), *Improper priors, spline smoothing and the problem of guarding against model errors in regression*, J. R. Statist. Soc. B 40, 364–372.

Wahba, G. (1983), *Bayesian "confidence intervals" for the cross-validated smoothing spline*, J. R. Statist. Soc. B 45, 133–150.

Wahba, G. (1985), *A comparison of GCV and GML for choosing the smoothing parameter in generalized spline smoothing*, Ann. Statist. 13, 1375–1402.

Wahba, G. (1990), *Spline models for observational data*, SIAM, Philadelphia.

Wahba, G.; Wendelberger, J. (1980), *Some new mathematical methods for variational objective analysis using splines and cross validation*, Monthly Weather Rev. 108, 1122–1140.

Walker, J.A. (1998), *Estimating velocities and accelerations of animal locomotion: Simulation experiment comparing numerical differentiation algorithms*, J. Exp. Biol. 201, 981–995.

Wang, Jing; Yang, Lijian (2006), *Polynomial spline confidence bands for regression curves*, Manuscript, Michigan State University.

Wang, Xiao; Li, Feng (2008), *Isotone smoothing spline regression*, J. Comput. Graph. Statist. 17, 1–17.

Watson, G.S. (1964), *Smooth regression analysis*, Sankhyā A 26, 359–372.

Weber, M. (1989), *The supremum of Gaussian processes with a constant variance*, Probab. Theory Relat. Fields 81, 585–591.

Wecker, W.E.; Ansley, C.F. (1983), *The signal extraction approach to nonlinear regression and spline smoothing*, J. Amer. Statist. Assoc. 78, 81–89.

Weinert, H.L.; Kailath, T. (1974), *Stochastic interpretations and recursive algorithms for spline functions*, Ann. Statist. 2, 787–794.

Weinert, H.L.; Byrd, R.H.; Sidhu, G.S. (1980), *A stochastic framework for recursive computation of spline functions: Part II, Smoothing splines*, J. Optimiz. Theory Appl. 30, 255–268.

Weinert, H.L.; Sidhu, G.S. (1978), *A stochastic framework for recursive computation of spline functions: Part I, Interpolating splines*, Trans. IEEE Information Theory 24, 45–50.

Whittaker, E. (1923), *On a new method of graduation*, Proc. Edinburgh Math. Soc. 41, 63–75.

Xia, Yingcun (1998), *Bias-corrected confidence bands in nonparametric regression*, J. R. Statist. Soc. B 60, 797–811.

Yatracos, Y.G. (1985), *Rates of convergence of minimum distance estimators and Kolmogorov's entropy*, Ann. Statist. 13, 768–774.

Yatracos, Y.G. (1988), *A lower bound on the error in nonparametric regression type problems*, Ann. Statist. 16, 1180–1187.

Yosida, K. (1980), *Functional analysis*, Springer-Verlag, New York.

Zaitsev, A.Yu. (2002), *Estimates for the strong approximation in multidimensional central limit theorem*, Proceedings of the International Congress of Mathematicians (Beijing, 2002) III (Tatsien Li, ed.), Higher Education Press, Beijing, pp. 107–116.

Zhang, J.; Fan, J. (2000), *Minimax kernels for nonparametric curve estimation*, J. Nonparam. Statist. 12, 417–445.

Zhou, Shanggang; Shen, Xiaotong (2001), *Spatially adaptive regression splines and accurate knot selection schemes*, J. Amer. Statist. Assoc. 96, 247–259.

Ziegler, K. (2001), *On approximations to the bias of the Nadaraya-Watson regression estimator*, J. Nonparametric Statistics 13, 583–589.

Ziemer, W.P. (1989), *Weakly differentiable functions*, Springer-Verlag, New York.

Zygmund, A. (1968), *Trigonometric series, Volumes I & II*, Cambridge University Press, Cambridge.

Author Index

Subject Index

MIX
Papier aus verantwortungsvollen Quellen
Paper from responsible sources
FSC® C105338

FSC
www.fsc.org

If you have any concerns about our products,
you can contact us on
ProductSafety@springernature.com

In case Publisher is established outside the EU,
the EU authorized representative is:
Springer Nature Customer Service Center GmbH
Europaplatz 3, 69115 Heidelberg, Germany

Printed by Libri Plureos GmbH
in Hamburg, Germany